KT-157-215

MODERN SYSTEMS ANALYSIS AND DESIGN

Jeffrey A. Hoffer • **Joey F. George** • **Joseph S. Valacich**

THE BENJAMIN/CUMMINGS PUBLISHING COMPANY, INC.

READING, MASSACHUSETTS • MENLO PARK, CALIFORNIA
NEW YORK • DON MILLS, ONTARIO • HARLOW, U.K. • AMSTERDAM
BONN • PARIS • MILAN • MADRID • SYDNEY • SINGAPORE • TOKYO
SEOUL • TAIPEI • MEXICO CITY • SAN JUAN, PUERTO RICO

40015608

Executive Editor: Larry Alexander
Senior Acquisitions Editor: Maureen Allaire
Editorial Assistant: Noah Blaustein
Marketing Manager: Melissa Baumwald
Senior Production Editor: Teri Holden
Text Design: Gary Palmatier, Ideas to Images
Cover Design: Yvo Riezebos
Art Supervisor: Karl Miyajima
Copyeditor: Joan Paterson
Proofreader: Cathy Linberg
Manufacturing Coordinator: Janet Weaver
Art and Composition: Thompson Type
Film House: H & S Graphics
Printer: R.R. Donnelley & Sons
Cover Photo: La Belle Jardiniere 1939. Berne. Paul Klee Foundation

Part I photos: form (page 2) courtesy of USA Group, Inc., photo (page 3) © Guy Gillette,
Photo Researchers; Part II photos: (page 74) Seattle Bus Tunnel, Seattle Washington, (page
75) Talmadge Memorial Bridge, Savannah, Georgia, courtesy Atkinson Construction; Part
III photos courtesy of Allison Engine Company; Part IV photos courtesy of McHenry
County; Part V photos: (page 458) PrimePractice CD-ROM; © 1995 IVI Publishing, Inc.
All rights reserved. © 1995 Mayo Foundation for Medical Education and Research.
Reprinted with permission, all rights reserved., (page 459) © 1995 IVI Publishing, Inc.
All rights reserved. Reprinted with permission.; Part VI photos courtesy of Consensys
Group; Part VII photos courtesy of Albertson's Inc.; photo page 821 by Thomas Veneklasen.
Used with permission of the Karl Eller Graduate School of Management/University of
Arizona; photo page 822 © Comshare 1995.

USA Group, Inc., Atkinson Construction, Allison Engine Company, McHenry County, IVI
Publishing, Consensys Group, and Albertson's Inc. are not affiliated with The Benjamin/
Cummings Publishing Company, Inc.

*To Patty, Missi, and Carrie, who are my heroes
and sources of energy and joy in life.
—Jeff*

*To Karen, Evan, and Caitlin, without whose support
and trust this book could never have been completed.
—Joey*

*To Jackie, Jordan, James and the rest of my family,
your love and support are my greatest inspiration.
—Joe*

Preface

DESCRIPTION

This book covers the concepts, skills, methodologies, techniques, tools, and perspectives essential for systems analysts to successfully develop information systems. The primary target audience is upper division undergraduates in a computer information systems curriculum; a secondary target audience is MIS majors in MBA and M.S. programs. Although not explicitly written for the junior college and professional development markets, this book can also be used for these programs.

The book is written assuming that students have taken an introductory course on computer systems and have experience designing programs in several third and/or fourth generation languages. We review basic system principles for those students who have not been exposed to the material on which systems development methods are based. We also assume that students have a solid background in computing literacy and a general understanding of the core elements of a business, including basic terms associated with the production, marketing, finance, and accounting functions.

This book is characterized by the following themes:

1. *Systems development is firmly rooted in an organizational context.* The successful systems analyst requires a broad understanding of organizations, organizational culture, and operation.

2. *Systems development is a practical field.* A coverage of current practices as well as accepted concepts and principles are essential in a textbook.

3. *Systems analysis is a profession.* Standards of practice, a sense of continuing personal development, ethics, and a respect for and collaboration with the work of others are general themes in the textbook.

4. *Systems development has significantly changed with the explosive growth in databases and data-driven architectures for systems.* Systems development and database management can be and possibly should be taught in a highly coordinated fashion; this means that this textbook and the McFadden and Hoffer database text, *Modern Database Management,* fourth edition, also published by Benjamin/Cummings, should be compatible. In fact, the proper linking of these two textbooks is a strategic opportunity to meet the needs of the IS academic field.

5. *Success in systems analysis and design requires not only skills in methodologies and techniques but also in the management of projects: time, resources, and risks.* Thus, learning systems analysis and design requires a thorough understanding of the *process* as well as the techniques and deliverables of the profession.

6. *Systems development is increasingly becoming both automated and more strategic.* Students must understand the capabilities and limitations of such technologies as computer-aided software engineering (CASE) as well as know how systems relate to IS planning, business process re-engineering, and systems integration initiatives.

Given these themes, this textbook emphasizes the following:

- A business rather than a technology perspective.

- The role, responsibilities, and mindset of the systems analyst as well as the systems project manager rather than those of the programmer or business manager.

- The methods and principles of systems development rather than the specific tools or tool-related skills of the field.

OUTSTANDING FEATURES

The following are some of the distinctive features of *Modern Systems Analysis and Design:*

1. This book is organized in parallel to the McFadden and Hoffer *Modern Database Management* text, which will facilitate consistency of frameworks, definitions, methods, examples, and notations to better support SA&D and database courses adopting both texts. Even with the strategic compatibilities between this text and *Modern Database Management,* each of these books is designed to stand alone as a market leader using the standard chapters found in each book.

2. Extensive coverage of oral and written communication skills including systems documentation (one thorough chapter on documentation, training, and support is included), project management, team management, and a variety of systems development and acquisition strategies (e.g., life cycle, prototyping, rapid application development, joint application development, participatory design, and systems re-engineering).

3. A clear linkage of all dimensions of systems description and modeling—process, decision and temporal logic, and data modeling—into a comprehensive and compatible set of systems analysis and design approaches. Such a broad coverage is necessary for students in order to understand the advanced capabilities of many systems development methodologies and tools that are automatically generating a large percentage of code from design specifications.

4. The grounding of systems development in the typical architecture for systems in modern organizations, including database management and distributed and client/server systems.

5. Coverage of rules and principles of systems design, including decoupling, cohesion, modularity, and audits and controls.

6. Consideration of standards for the methodologies of systems analysis and the platforms on which systems are designed.

7. Discussion of systems development and implementation within the context of management of change, conversion strategies, and organizational factors in systems acceptance.

8. Careful attention to human factors in systems design that emphasize usability in both character-based and graphical user interface situations.

9. CASE technology is used throughout the text to illustrate typical systems analysis and design documents and CASE-based systems development is discussed; however, no specific CASE tool is assumed. A wide variety of CASE products are illustrated and the current limitations of CASE technologies are highlighted.

10. In Chapter 3, survey results highlight the talents that good systems analysts have (and previews the main themes of the text) and helps students identify important material throughout the text.

11. The text includes a separate chapter on systems maintenance. Given the type of job many graduates first accept and the large installed base of systems, this chapter covers an important and often neglected topic in SA&D texts.

PEDAGOGY

Several elements in the design and implementation of the text and its supplements make it readable and practical, holding the readers' attention, and assisting the instructor in delivering a better course.

1. Chapter 2 presents a scenario, Pine Valley Furniture (PVF) depicting how a systems analysis and design project is conducted by illustrating the critical success factors and deliverables of such projects. This provides the students with a target or goal for the course and a concrete example in which to better see how each specific topic is applied.

2. Sixteen of the twenty-one chapters include a running case study, Broadway Entertainment Company. This hypothetical, high-technology company provides a rich arena for bringing the concepts, skills, techniques, and tools explained in the chapter to life. Discussion questions for the case are provided in the *Instructor's Manual.*

3. Videotapes are available that will show practicing systems professionals engaged in meetings, interviews, and other tasks during the development of information systems. This series of four videotapes also includes discussions by systems professionals on the critical success factors for systems developers and for the management of systems projects.

4. In addition to PVF in Chapter 2, there are three case situations used throughout the book (several of these match cases used in the McFadden and Hoffer *Modern Database Management* text) to illustrate methods, notations, and design techniques.

5. End-of-chapter review questions and problems and exercises test students' knowledge of the material. An innovative addition is a set of field exercises

that give students an opportunity to explore the practice of SA&D in organizations.

6. Chapter objectives introduce each chapter to help the student identify the main topics within each chapter.

7. A comprehensive *Instructor's Manual* provides answers to all the review questions and problems and exercises from the text, plus teaching suggestions and selected questions and problems which may be used in tests or as supplemental exercises.

8. A case/project companion book is available that contains several projects and extended exercises; these follow the framework for systems development outlined in the text.

9. A comprehensive test bank of over 1,600 objective and short answer questions.

10. Transparency masters for the figures and tables in the book.

USING THIS TEXT

As stated earlier, the book is intended for mainstream SA&D courses. It may be used in a one-semester course on SA&D or over two quarters (first in a systems analysis and then in a systems design course). Because of the consistency with *Modern Database Management,* chapters from this book and from *Modern Database Management* can be used in various sequences suitable for your curriculum. The book will be adopted typically in business schools or departments, not in computer science programs. Applied computer science or computer technology programs may adopt the book.

The typical faculty member who will find this book most interesting is someone

- with a practical, rather than technical or theoretical, orientation

- with an understanding of databases and systems that use databases

- who uses practical projects and exercises in the course.

More specifically, academic programs that are trying to better relate their SA&D and database courses as part of a comprehensive understanding of systems development will be most attracted to this book.

The outline of the book generally follows the systems development life cycle, which allows for a logical progression of topics. However, the book emphasizes that various approaches (e.g., prototyping and iterative development) are also used, so what appears to be a logical progression often is a more cyclic process. Part I of the book provides an overview of systems development, previews the remainder of the book, and shows the student what the process of developing systems is like. Part II covers those skills and concepts that are applied throughout systems development, including systems concepts, project management, and CASE technologies. The remaining five sections provide thorough coverage of the seven phases of a generic systems development life cycle, interspersing coverage of alternatives to the SDLC as appropriate.

Four appendices provide background or extensions to topics covered in the chapters. Appendix A reviews the types of information systems for which systems development projects are conducted. Appendix B overviews rapid application development. Appendix C presents advanced data modeling principles, which extends

Chapter 11. And Appendix D addresses object-oriented systems analysis and design methods, which are of great interest to developers of interactive and real-time systems.

Some chapters may be skipped depending on the orientation of the instructor or the students' background. For example, Chapters 1 (environment of SA&D) and 3 (critical success factors for SA&D) cover topics that are emphasized in some introductory MIS courses. Chapter 6 (project identification and selection) can be skipped if the instructor wants to emphasize systems development once projects are identified or if there are fewer than 15 weeks available for the course. Chapters 11 (conceptual data modeling), 15 (logical data modeling), and 16 (physical database design) can be skipped or quickly scanned (as a refresher) if students have already had a thorough coverage of these topics in a previous database or data structures course. Finally, Chapter 21 (maintenance and re-engineering) can be skipped if these topics are beyond the scope of your course.

Because the material is presented within the flow of a systems development project, it is not recommended that you attempt to use the chapters out of sequence, with a few exceptions: Chapters 9 (process modeling), 10 (logic modeling), and 11 (conceptual data modeling) can be taught in any sequence; and Chapter 15 (logical data modeling) can be taught before Chapters 13 (output design) and 14 (interface design), but Chapters 13 and 14 should be taught in sequence.

SUPPLEMENTS

Instructor's Manual on Disk
by Jeffrey A. Hoffer, Joey F. George, and Joseph S. Valacich
Formatted in Acrobat, the Instructor's Manual accompanying **Modern Systems Analysis and Design** provides answers to all text review questions, problems, and exercises; plus teaching suggestions and transparency masters.

Test Bank on Disk
by Lisa Miller, University of Central Oklahoma
Includes 40–60 multiple choice, 15 matching, and 5 essay questions per chapter. Available for IBM and Macintosh.

Computerized Test Bank
Formatted in Acrobat, the computerized test bank includes the same questions as the printed version in a format that lets you edit questions and generate multiple tests. Available for the IBM PC and the Macintosh.

Electronic Transparencies
Formatted in Acrobat, include art from the text that you can manipulate, print as handouts, make into transparencies, or display directly from a computer in lectures. Available for the IBM PC.

Projects and Cases Workbooks
by George Easton and Annette Easton, San Diego State University
This companion workbook contains 10 cases and extended study questions.

HyperAnalysis Toolkit (HAT)
This CASE (computer-assisted software engineering) tool program gives students the opportunity to view and use the kinds of technologies illustrated in the text. Available bundled with the book.

EDS Video Series
by Electronic Data Systems Corporation (EDS)
This video series, prepared by EDS specifically to accompany **Modern Systems Analysis and Design,** consists of four video segments each approximately 15 minutes in length, that focus on systems analysis and design. Each includes an introduction and prologue from Professors Hoffer, George, and Valacich.

ACKNOWLEDGMENTS

The authors have been blessed by considerable assistance from many people on all aspects of preparation of this text and its supplements. We are, of course, responsible for what eventually appears between the covers, but the insights, corrections, contributions, and proddings of others have greatly improved our manuscript. The people we recognize here all have a strong commitment to students, to the IS field, and to excellence. Their contributions have stimulated us, and frequently rejuvenated us during periods of waning energy for this project.

We would like to recognize the efforts of the many faculty and practicing systems analysts who have been reviewers of the several drafts of our manuscript. We have tried to deal with each reviewer comment, and although we did not always agree with specific points (within the approach we wanted to take in this book), all reviewers made us stop and think carefully about what and how we were writing. The reviewers were: Susan Athey (Colorado State University), Penny Brunner (University of North Carolina, Asheville), Donald Chand (Bentley College), Barry Frew (Naval Post-Graduate School), Jim Gifford (University of Wisconsin), Dale Gust (Central Michigan University), Ellen Hoadley (Loyola College—Baltimore), Robert Jackson (Brigham Young University), Len Jessup (Indiana University), Robert Keim (Arizona State University), Mat Klempa (California State University at Los Angeles), Nancy Martin (USA Group, Indianapolis, Indiana), Mary Prescott (University of South Florida), Terence Ryan (Southern Illinois University), Eugene Stafford (Iona College), Bob Tucker (Antares Alliance, Plano, Texas), Connie Wells (Nicholls State University), Chris Westland (University of Southern California), Charles Winton (University of North Florida), and Terry Zuechow (EDS Corporation, Plano, Texas). All of the reviewers provided honest and helpful comments. We want to especially recognize the in-depth comments from Robert Jackson (Brigham Young University), who always provided constructive comments and challenged us to state our ideas clearly and to say only what needed to be said.

We are very indebted to Len Jessup (Indiana University) for the preparation of the case studies which open each part of the text. Len not only wrote each case study but also identified most of the organizations on which the cases are based. Len and the authors would like to acknowledge the assistance of these organizations and their representatives who helped by being interviewed and editing the cases: Ken Chambless (Atkinson Construction, San Bruno, California), Peg Dawson (EDS Corporation, Indianapolis, Indiana), James Herrett (Atkinson Construction, San Bruno, California), Nancy Martin (USA Group, Indianapolis, Indiana), Craig Olsen (Albertson's, Inc., Boise, Idaho), Carl Pohrte (McHenry County, Illinois), Dori Pelz-Sherman (IVI Publishing, Minneapolis, Minnesota), Pat Steele (Albertson's, Inc., Boise, Idaho), Clay Tyler (Consensys Group, San Diego, California), Hadley Wagner (Albertson's, Inc., Boise, Idaho).

Len Jessup also prepared answers to all of the Problems and Exercises and Field Questions in the text, which appear in the *Instructor's Manual.* Len has been a unique partner in the development of the total product; his contributions are extensive. We especially acknowledge his patience with the authors as we moved material between chapters and appendices and we modified and added to our requests for his assistance. We have all come to appreciate Len's keen sense for quality teaching, the clarity of his thinking, and his friendship.

We are also indebted to our undergraduate and MBA students at Indiana University and Florida State University who have given us many helpful comments as they worked with drafts of this text. Two particular Indiana University doctoral students, Heikki Topi (Helsinki School of Economic, Mikkeli, Finland) and Gary Spurrier (William M. Mercer, Inc., Dallas, Texas) provided considerable feedback on Appendix D and many other parts of the text.

We have also called upon several companies to supply examples of their CASE tools. We thank Antares Alliance, Sterling Software, Texas Instruments, and Visible Systems for their assistance in identifying and supplying various figures for the text.

One unique supplement to this text is a series of four video tapes which illustrate common activities and situations encountered by systems analysts. We are very excited about the pedagogical value of these tapes, and complement EDS Corporation for the sizable commitment of human and financial resources to develop and produce these tapes for exclusive use with our book. Specifically we thank Stu Bailey, Michael Cummings, Vern Olsen, Chris Ryan, and Terry Zuechow of EDS, Bob Tucker of Antares Alliance, and Bill Satterwhite of Whitecap Productions for all of their work on this project.

We also want to thank Lisa Miller from the University of Central Oklahoma who has prepared an extensive test bank for this text. In addition, we thank Ken Griggs for working with the authors and publisher to provide a version of Hyper-Analysis Toolkit for use with our text. We also acknowledge the work of George and Annette Easton (San Diego State University) for producing a case project workbook to accompany this text.

Thanks also go to Fred McFadden (University of Colorado, Colorado Springs) for his assistance in coordinating this text with its companion book—*Modern Database Management.*

Finally, we have been fortunate to work with a large number of creative and insightful people at Benjamin/Cummings who have added much to the development, format, and production of this text. We have been thoroughly impressed with their commitment to this text and to the IS education market. These people include: Michelle Baxter (editor, who signed this book), Larry Alexander (editor, who managed the book through most of the development process), Maureen Allaire (editor, who has supervised the final stages of preparation, marketing, and introduction), Becky Johnson and Shelly Langman (contracted development editors), Teri Holden (senior production editor, whose constant and thorough attention is much appreciated by the authors), Karl Miyajima (art editor), and various editorial assistants (Kathy Galinac, Mark Schmidt, Krista Reid-McLaughlin, Noah Blaustein, and Kathleen Conant) who have filled in all the gaps not handled by others.

The writing of this text has involved thousands of hours of time from the authors and from all of the people listed above. Although our names will be visibly associated with this book, we know that much credit goes to the individuals and organizations listed here for any success this book might achieve. It is important for the reader to recognize all the individuals and organizations who have been committed to the preparation and production of this book.

Jeffrey A. Hoffer (Dayton, Ohio)
Joey F. George (Tallahassee, Florida)
Joseph S. Valacich (Bloomington, Indiana)
September, 1995

Brief Table of Contents

Detailed Table of Contents

Part V   Logical Design   *457*

IVI Publishing
CREATING A CONVERSATION BETWEEN COMPUTER AND USER   458

AN OVERVIEW OF PART V: LOGICAL DESIGN   461

Defining the Context for Systems Development

CHAPTER 1

The Systems Development Environment

CHAPTER 2

A Systems Analysis and Design Project at Pine Valley Furniture

USA Group, Inc.

Application and Promissory Note for Federal Stafford Loans (subsidized and unsubsidized)

Guarantor or Program Identification

WARNING: Any person who knowingly makes a false statement or misrepresentation on this form is subject to penalties which may include fines or imprisonment under the United States Criminal Code and 20 U.S.C. 1097.

Borrower Section

Please print neatly or type. Read the instructions carefully.

1. Last Name / First Name / MI
2. Social Security Number
3. Permanent Street Address (If P.O. Box, see instructions.)
4. Telephone Number
5. Loan Period (Month/Year) From: To:
   City / State / Zip Code
6. Driver's License Number (List state abbreviation first.)
7. Lender Name / City / State / Zip Code
8. Lender Code, if known
9. Date of Birth (Month/Day/Year)

10. **References:** You must provide two separate references with different U.S. addresses. The first reference should be a parent or legal guardian (if living). Both references must be completed fully.
    Name 1. 2.
    Permanent Address
    City, State, Zip Code
    Area Code/Telephone
    Relationship to Borrower

Loan Assistance Requested

11. I request the following loan type(s), to the extent I am eligible (see instructions).
    ☐ a. Subsidized Federal Stafford   ☐ b. Unsubsidized Federal Stafford
12. I request a total amount under these loan types not to exceed (see instructions for loan maximums). My school will certify my eligibility for each loan type for which I am applying. The amount and other details of my loan(s) will be described to me in a disclosure statement.
    $ .00
13. If I check yes, I am requesting postponement (deferment) of repayment for my Stafford and prior SLS loan(s) during the in-school and grace periods. If I check no, I do not want to defer repayment.
    ☐ a. Yes, I want a deferment   ☐ b. No, I do not want a deferment
14. If I check yes, I am requesting that the lender add the interest on my unsubsidized Stafford and prior SLS loan(s) which accrues during the in-school and deferment periods, to my loan principal (capitalization). If I check no, I prefer to pay the interest.
    ☐ a. Yes, I want my interest capitalized   ☐ b. No, I prefer to pay the interest
15. If my school participates in electronic funds transfer (EFT), I authorize the school to transfer the loan proceeds received by EFT to my student account.
    ☐ a. Yes, transfer funds   ☐ b. No, do not transfer funds

Continued on the reverse side.

Promissory Note

Promise to Pay: I promise to pay to the lender, or a subsequent holder of this Promissory Note, all sums disbursed (hereafter "loan" or "loans") under the terms of this Note, plus interest and other fees which may become due as provided in this Note. If I fail to make payments on this Note when due, I will also pay reasonable collection costs, including attorney's fees, court costs, and collection fees. I understand I may cancel or reduce the size of any loan by refusing to accept any disbursement that is issued. I understand that this is a Promissory Note. I will not sign this Note before reading it, including the writing on the reverse side, even if otherwise advised. I am entitled to an exact copy of this Promissory Note and the Borrower's Rights and Responsibilities. My signature certifies I have read, understand, and agree to the terms and conditions of this Application and Promissory Note, including the Borrower Certification and Authorization printed on the reverse side and the accompanying Borrower's Rights and Responsibilities statement.

THIS IS A LOAN(S) THAT MUST BE REPAID.

16. Borrower's Signature   Today's Date (Month/Day/Year)

School Section

To be completed by an authorized school official.

17. School Name
23. School Code/Branch
28. Telephone Number
18. Street Address
24. Cost of Attendance $ .00
29. Recommended Disbursement Date(s) (Month/Day/Year) 1st   2nd
    City / State / Zip Code
25. Federal Expected Family Contribution $ .00
    3rd   4th
19. Loan Period (Month/Day/Year) From: To:
26. Estimated Financial Aid $ .00
30. School Certification (See box on the reverse side.)
20. Grade Level
27. Certified Loan Amounts
    a. Subsidized $ .00
    Signature of Authorized School Official
21. Enrollment Status (Check one.) ☐ Full Time   ☐ At Least Half Time
    b. Unsubsidized $ .00
    Print or Type Name and Title
22. Anticipated Completion (Graduation) Date (Month/Day/Year)
    Date
    Check box if electronically transmitted to guarantor ☐

Lender Section

To be completed by an authorized lending official.

31. Lender Name
32. Lender Code/Branch
33. Telephone Number
34. Lender Use Only
    Street Address
35. Amount(s) Approved a. Subsidized $ .00   b. Unsubsidized $ .00
    City / State / Zip Code
36. Signature of Authorized Lending Official   Print or Type Name, Title, and Date

1/31/94

LENDER COPY

Role of Systems Analysts

USA Group, Inc., in Indianapolis, Indiana, is one of the world's largest student loan management organizations. There are over 300 information systems people at USA Group, including general managers, project managers, systems analysts, programmers, programmer/analysts, technical analysts, and others. Systems analysts at USA Group hold salaried, professional positions, and project team leaders are considered the first line of management.

Analysts at lower grade levels do mostly analysis tasks, such as interviewing, while analysts at higher grade levels often coach junior analysts and serve as the team lead for projects. The team lead is responsible for coordinating the efforts of the team members, ensuring that the team's part of the system is well integrated with other parts of the system, and attending coordination meetings with other team leads, project managers,

and IS managers. Analysts fresh out of school are generally assigned simpler tasks, such as code maintenance on smaller projects, and are supervised by a more senior analyst. Many junior analysts begin as programmer/analysts and are later reclassified as analysts.

The role of the analyst at USA Group changes frequently depending on the project type and the phase of the project in the systems development life cycle. USA Group's largest system project, replacing their Guaranteed Student Loan (GSL) processing system, nicely illustrates the changing role of the analyst. The current system, Eagle 1, is used to process all GSL applications and to manage the loans throughout their lives. USA Group originally implemented Eagle 1 in order to use less paper, reduce errors, and perform loan processing faster. Approximately 150 systems personnel are now working on the new system, Eagle 2. Development cost for Eagle 2 will end up between $30 and $35 million. With Eagle 2, USA Group plans to further improve the loan processing system.

The Eagle 2 project is conducted by small, manageable teams working on subsystems such as (1) Options, Products, and Services, (2) Electronic Funds Transfers, and (3) Loan Approval. Team size changes throughout the life cycle as the need for more or fewer team members dictates. For example, one of the teams currently in the analysis phase includes about a dozen people but will double in size during the construction phase.

Because the project is in the analysis phase, analysts are primarily performing high-level activities, such as business area requirements analysis. Analysts work directly with executives, helping elicit their strategies, goals and objectives, and critical success factors. The analysts will then flesh out more detailed requirements in order to design the conceptual system. During this phase, they will work directly with users and modelers. Together the modelers and analysts draw the conceptual diagrams of how the system will look and function.

The project will then move into the technical design phase, when the systems analysts will work with technical analysts and designers. Next, the project will move into the final specs and code generation phases, when the analysts will work with programmer/analysts and programmers in order to make the system a tangible reality.

Throughout each of the phases of the systems development life cycle, the analysts work closely with project coordinators responsible for managing the CASE tools and related information. The analysts will also work with technical analysts who will be responsible for ensuring that the system and all subsystems have a consistent look and feel.

The systems analysts must know systems development methodologies and have some programming competency. In addition, they must have project management and CASE tool skills and, when acting as the lead on a project, they must have managerial skills, such as supervising team members and organizing work. However, interpersonal skills are perhaps most important. At USA Group, analysts spend a large portion of their time interacting with various people. Although there is some flexibility in assignments to projects and in assignments to tasks within projects, the analysts cannot escape having to interact with people. Systems personnel who prefer working alone are better suited to be programmers or technical analysts.

Analyst Nancy Martin estimates that she spends approximately 20 percent of her time in one-on-one interviews with various business and technical people. Perhaps more importantly, she estimates that she spends an additional, even larger, chunk of time in group meetings with managers, users, and her team members. In these meetings she and the other analysts ask key questions of users to elicit information about their business processes and needs. While the analysts are trying to learn how best to build the system, they are also helping users to better understand their own business processes. These types of meetings are especially prevalent at the front end of the systems projects.

The analysts work together as a team transcribing and translating users' business needs and trying to ask questions without making people feel defensive. The analysts also help the users stop fixating on the current system and the current way of doing things. They help users to form a vision of how they will be able to better accomplish their work.

Martin says that it is amusing to go out for lunch with other analysts from work. She admits that when together off-site they continue to analyze everything. On one lunch outing, they critically analyzed the restaurant layout including the design of the salad bar. They couldn't believe that someone had put the salad dressings, which drip a lot, at the back of the salad bar. Because of this design flaw the "user" was forced to drip salad dressing over everything in the salad bar placed in front of the dressing containers. She laughed and added that "a good system analyst wouldn't have designed it that way."

USA Group likes their systems analysts' penchant for analyzing things, particularly when the analysts are designing and building Eagle 2, the firm's costly, mission-critical loan processing application.

An Overview of Part I:

Defining the Context for Systems Development

Y ou are beginning a journey that will allow you to build on every aspect of your education and experience. Becoming a systems analyst is not a goal, it is a path to a rich and diverse career that will allow you to exercise and continue to develop a wide range of talents. We hope that this introductory part of the text helps open your mind to the opportunities of the systems analysis and design field and to the engaging nature of systems work. In some measure, Part I of the book is a teaser—we give you a taste of the topics and important issues discussed in detail in the rest of the book; at times, we introduce notations and concepts without elaboration. A hint: don't try to read the material in these first two chapters expecting a thorough development of ideas—resist attempting to understand topics in depth at this point. Further, don't expect to learn skills in Part I of the text. Rather, we expect that you will appreciate why you need so many skills as a systems analyst, as well as understand the importance of notations, techniques, and methods, later on in the text. (If you feel a real need to delve into an area, use the Glossary, Index, or Table of Contents to find elaboration on the desired topic.)

Because what you do as a systems analyst occurs within a multifaceted organizational process involving other organizational members and external parties, we begin this text with a view of the "big picture" within which systems are developed. Systems analysis and design does not follow a cookbook, definitive process, and you, as a systems analyst, certainly do not work alone in the analysis and design process. Rather, because of the variety of

- Systems being developed
- People who develop those systems
- Systems analysis and design methods and procedures used by an organization
- Organizational philosophies about systems development
- Skills involved in systems development

you will learn many approaches to systems analysis and design and improve your communication and interpersonal skills as well. Understanding systems development requires not only an understanding of each specific systems analysis and design technique, tool, and method but also requires understanding how these elements cooperate, complement, and support each other within an organizational setting for systems development.

We presume that you have a general understanding of computer-based information systems as typically covered in introductory business information systems texts and courses. (If you do not have this background, you may want to go to a library and look through one or more books on management or business information systems.) If you feel a need for a refresher on this material, Chapter 1 previews systems analysis and design by discussing the environment of information systems and their development. In this chapter, you will read that the general approach to systems development has changed and is changing as we better understand what works, as technology changes, and as users' expectations grow. Today, the modern approach to systems analysis and design integrates consideration of several views of systems with special emphasis on data as the core material.

Systems development is fascinating both to study and to participate in. Variety and change contribute to the fascination. First, contrary to popular belief and possibly inconsistent with your own experiences with computer programming tasks, systems analysis and design is a job with intense and frequent interpersonal contact. You will lead the development of systems working with a diverse team of interested parties: other analysts, IS and business managers, programmers and other IS technical specialists as well as the end users or clients whom you are trying to serve. Understanding how all these people fit into the systems development process will help you understand how to apply the skills, techniques, tools, and methods discussed later in the text.

Second, each system you develop will be different from every other system you have developed before. Systems of different types require emphasizing different aspects of systems analysis and design. Thus, understanding the differences among systems helps you choose the appropriate techniques, tools, and methods to apply in your work.

Finally, the systems analysis and design field is constantly changing and adapting to new situations (a source of job enrichment and a positive challenge for most people in this field). In part, such changes occur because of the strong commitment of the field to constantly improve—an inherent Total Quality Management (TQM) perspective is a tradition in systems analysis and design. Such changes also occur because the cultures of different organizations—with varying degrees of interest in consensus participatory management and leading-edge practices and technologies—favor different approaches to systems development. Thus, we complete our discussion of the environment for systems development with an overview of the various approaches taken in modern organizations. In this chapter and elsewhere, however, it is not our purpose to encourage any one approach over another. Our goal is to provide you with a mosaic of the skills needed to effectively work in whichever environment you find yourself, armed with the knowledge to determine and argue for what you consider to be the best practices in that situation.

Since many readers of this text will have never participated in systems analysis and design efforts, Chapter 2 illustrates a hypothetical systems analysis and design project. Although rather ideal and necessarily brief (at this point of the book), this chapter extends the appreciation for the context of systems analysis and design from Chapter 1, helping you to envision what it might be like to be part of a systems analysis and design project. Chapter 2 shows how the different elements (techniques, practices, and people) of systems analysis and design fit together in a practical setting. Throughout the text, we refer to elements of Chapter 2 as a way to remind you of how individual techniques and methods must be managed and cooperatively used. Chapter 2 also provides a framework for planning a systems development project. A case study or field exercise may be an assignment you need to get started on before you study systems analysis and design techniques in detail.

It is time to begin your journey.

CHAPTER 1

The Systems Development Environment

After studying this chapter, you should be able to:

- Define information systems analysis and design.

- Discuss the modern approach to systems analysis and design that combines both process and data views of systems.

- Describe the organizational roles, including systems analyst, involved in information systems development.

- Describe the different types of information systems.

- Describe the information systems development life cycle (SDLC).

- List alternatives to the systems development life cycle and compare the advantages and deficiencies of the SDLC and its alternatives.

- Explain briefly the role of computer-aided software engineering (CASE) tools in systems development.

INTRODUCTION

Information systems analysis and design is a complex, challenging, and stimulating organizational process that a team of business and systems professionals uses to develop and maintain computer-based information systems. Although advances in information technology continually give us new capabilities, the analysis and design of information systems is driven from an organizational perspective. An organization might consist of a whole enterprise, specific departments, or individual work groups. Organizations can respond to and anticipate problems and opportunities through innovative uses of information technology. Information systems analysis and design is, therefore, an organizational improvement process. Systems are built and rebuilt for organizational benefits. Benefits result from adding value during the process of creating, producing, and supporting the organization's products and services. Thus, the analysis and design of information systems is based on your understanding of the organization's objectives, structure, and processes as well as your knowledge of how to exploit information technology for advantage.

Few business careers present a greater opportunity for significant and visible impact on business as do careers in systems development. Furthermore, analyzing

Information systems analysis and design: The complex organizational process whereby computer-based information systems are developed and maintained.

and designing information systems will give you the chance to understand organizations at a depth and breadth that might take many more years to accomplish in other careers. According to *Money* magazine, the position of systems analyst is the best job in America today (Gilbert, 1994). *Money* predicts that over half a million new systems analyst jobs will be created between 1994 and 2005, more than doubling the total number of positions available today. With such promising career prospects, combined with the challenges and opportunities of dealing with the rapid advances in information technologies, it is difficult to imagine a more exciting career choice than systems analysis and design.

Application software:
Computer software designed to support organizational functions or processes.

An important (but not the only) result of systems analysis and design is **application software;** that is, software designed to support a specific organizational function or process, such as inventory management, payroll, or market analysis. In addition to application software, the total information system includes the hardware and systems software on which the application software runs, documentation and training materials, the specific job roles associated with the overall system, controls, and the people who use the software along with their work methods. Although we will address all these various dimensions of the overall system, we will emphasize application software development—your primary responsibility as a systems analyst.

In the early years of computing, analysis and design was considered to be an art. Now that the need for systems and software has become so great, people in industry and academia have developed work methods that make analysis and design a disciplined process (similar to processes followed in engineering fields). Our goal is to help you develop the knowledge and skills needed to understand and follow such software engineering processes. Central to software engineering processes (and to this book) are various *methodologies, techniques,* and *tools* that have been developed, tested, and widely used over the years to assist people like you during systems analysis and design.

Methodologies are comprehensive, multiple-step approaches to systems development that will guide your work and influence the quality of your final product: the information system. A methodology adopted by an organization will be consistent with its general management style (for example, an organization's orientation toward consensus management will influence its choice of systems development methodology). Most methodologies incorporate several development techniques.

Techniques are particular processes that you, as an analyst, will follow to help ensure that your work is well thought-out, complete, and comprehensible to others on your project team. Techniques provide support for a wide range of tasks including conducting thorough interviews to determine what your system should do, planning and managing the activities in a systems development project, diagramming the system's logic, and designing the reports your system will generate.

Tools are typically computer programs that make it easy to use and benefit from the techniques and to faithfully follow the guidelines of the overall development methodology. To be effective, both techniques and tools must be consistent with an organization's systems development methodology. Techniques and tools must make it easy for systems developers to conduct the steps called for in the methodology. These three elements—methodologies, techniques, and tools—work together to form an organizational approach to systems analysis and design.

Systems analyst: The
organizational role most
responsible for the analysis
and design of information
systems.

Although many people in organizations are responsible for systems analysis and design, in most organizations the **systems analyst** has the primary responsibility. When you begin your career in systems development, you will most likely begin as a systems analyst or as a programmer with some systems analyst responsibilities. The primary role of a systems analyst is to study the problems and needs of an organization in order to determine how people, methods, and information technology can best be combined to bring about improvements in the organization.

A systems analyst helps system users and other business managers define their requirements for new or enhanced information services. As such, a systems analyst is an agent of change and innovation.

In the rest of this chapter, we will examine the systems approach to studying organizations and the systems approach to studying analysis and design. You will learn about the dominant complementary approaches to systems development—the data- and process-oriented approaches. You will also identify the various people who develop systems and the different types of systems they develop. The chapter ends with a discussion of some of the methodologies, techniques, and tools created to support the systems development process.

A MODERN APPROACH TO SYSTEMS ANALYSIS AND DESIGN

The analysis and design of computer-based information systems began in the 1950s. During that period, the focus of the development effort was on the processes the software performed. Since computer power was the critical resource, efficiency of processing became the main goal. Emphasis was placed on automating existing processes such as purchasing or paying, often within single departments. More recently, the focus has shifted to data. As you will see, however, both process and data focuses are necessary for a thorough analysis and design effort. A data focus allows a broader scope in which to address organizational processes (such as materials management or order fulfillment) that cut across departmental boundaries. Such processes may include suppliers, regulators, customers, and competitors as providers or consumers of data and information. The result is not only automation of organizational processes but also empowerment of employees who will have improved access to data and information.

In the rest of this section, we will describe the traditional approach to systems development with its separate treatments of data and the processes that use those data. We will also examine the modern approach whereby data and processes are considered together. You will then read about the Pine Valley Furniture Company, which serves as an example of systems development throughout the book.

Separating Data and Processes That Handle Data

Every information system consists of three key components that must be clearly understood by anyone who analyzes and designs systems: data, data flows, and processing logic (see Figure 1-1). **Data** are raw facts that describe people, objects, and events in an organization, such as a customer's account number, the number of boxes of cereal bought, and whether someone is a Democrat or a Republican. Every information system depends on data in order to produce **information,** which is processed data presented in a form suitable for human interpretation. Systems developers must understand what kind of data a system uses and where the data originate. Data and the relationships among data may be described using various techniques, as we will see later. Figure 1-1 shows the structure of employee data as a simple table of rows (records about different employees) and columns (attributes describing each employee).

Data flows are groups of data that move and flow through a system and include a description of the sources and destinations for each data flow. For example, a customer's account number may be captured when he or she uses a credit card to pay for a purchased item. The account number may then be stored in a file within the system until needed to compile a billing statement or prepare a mailing address

Data: Raw facts about people, objects, and events in an organization.

Information: Data that have been processed and presented in a form suitable for human interpretation, often with the purpose of revealing trends or patterns.

Data flow: Data in motion, moving from one place in a system to another.

Figure 1-1

Differences among data, data flow, and processing logic

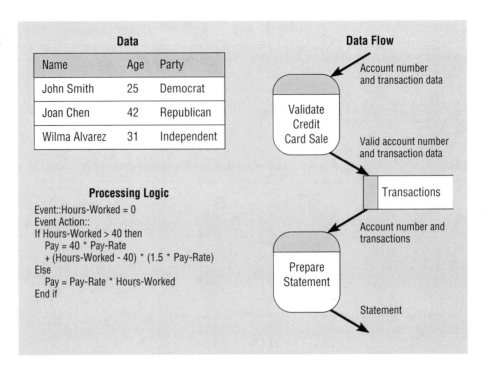

Processing Logic

```
Event::Hours-Worked = 0
Event Action::
If Hours-Worked > 40 then
    Pay = 40 * Pay-Rate
    + (Hours-Worked - 40) * (1.5 * Pay-Rate)
Else
    Pay = Pay-Rate * Hours-Worked
End if
```

for a sales circular. When needed, the account number can be extracted from storage and used to complete a system function. Figure 1-1 illustrates data flows with directional lines that connect rounded rectangles which represent the processing steps that accept input data flows and produce output data flows.

Processing logic, the third component, describes the steps in the transformation of the data and the events that trigger these steps. For example, processing logic in a credit card application will explain how to compute available credit given the current credit balance and the amount of the current transaction. Processing logic will also indicate that the computation of the new credit balance will occur when a clerk presses a key on a credit card scanner to confirm the sales transaction. Figure 1-1, using an English-like language, illustrates the rules for calculating an employee's pay and the event (receipt of a new hours-worked value) that causes this calculation to be made.

Traditionally, an information system's design was based upon what the system was supposed to do, such as billing and inventory control: the focus was on output and processing logic. Although the data the system used as input were important, data were subordinate to the application. The assumption was that we could anticipate all outputs and the proper processing steps with their need for data. Therefore, we could easily derive all data requirements from all known system deliverables. Furthermore, each application contained its own files and data storage capacity. The data had to match the specifications established in each application, and each application was considered separately.

This concentration on the flow, use, and transformation of data in an information system typified the **process-oriented approach** to systems development. The techniques and notations developed from this approach track the movement of data from their sources, through intermediate processing steps, and on to final destinations. Since various parts of an information system work on different schedules and at different speeds, the process-oriented approach also shows where data are temporarily stored until needed for processing. The natural structure of the data is, however, not specified within the traditional process-oriented approach. Until re-

Processing logic: The steps by which data are transformed or moved and a description of the events that trigger these steps.

Process-oriented approach: An overall strategy to information systems development that focuses on how and when data are moved through and changed by an information system.

cently, techniques for the process-oriented approach did not address the timing or triggering of processing steps, only their sequence.

Data processing managers soon realized that there were problems with analyzing and designing systems using only a process-oriented approach. One result was the existence of several specialized files of data, each locked within different applications and programs. Many of the files in these different applications contained the same data elements (see Figure 1-2a). When a single data element changed, it had to be changed in each of these files. If, for example, such a system were in effect at your university and your address changed, it would have to be changed in the files of the library, the registrar's office, the financial aid office, and every other place your address was stored. It also became difficult to combine specialized data files. Even if the files contained the same data elements, each file might use a different name and format for the data. Since it was important to standardize how data elements were represented, data processing managers gradually came to separate the application programs and the data these programs used.

This focus on data typified the **data-oriented approach** to information systems development. The data-oriented approach depicts the ideal organization of data, independent of where and how data are used within a system (see Figure 1-2b). The techniques used for data orientation result in a data model that describes the kinds of data needed in systems and the business relationships among the data. A

Data-oriented approach: An overall strategy of information systems development that focuses on the ideal organization of data rather than where and how data are used.

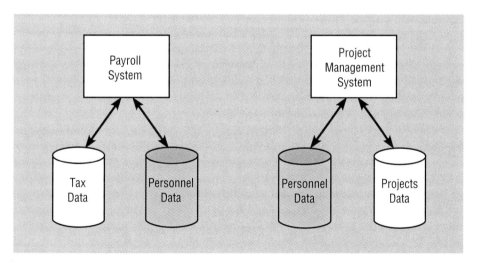

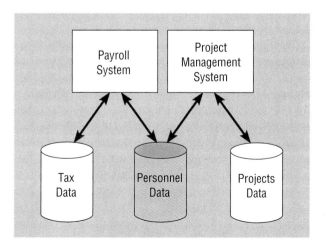

Figure 1-2
Traditional relationship between data and applications, with redundant data, versus the database approach

(a) Traditional approach

(b) Database approach

data model describes the rules and policies of a business. Some people believe a data model to be more permanent than a process model since a data model reflects the inherent nature of a business instead of the way a business operates, which is constantly changing. Some people refer to data-oriented approaches as *information engineering.*

Table 1-1 highlights some of the key distinctions between the process-oriented and the data-oriented approaches to systems development. Although we highlight these two approaches as separate competing orientations, we do so only to emphasize their differences and unique contributions to systems analysis and design and to give you a sense of the historical evolution of systems analysis and design methodologies. As either approach is, by itself, inadequate, this book will cover the techniques and tools you will need to analyze and design both process and data aspects of systems.

Separating Databases and Applications

Database: A shared collection of logically related data designed to meet the information needs of multiple users in an organization.

As data storage management technology advanced, it became possible to represent data not in separate files for each application but in coherent and shared databases. A **database** is a shared collection of logically related data organized in ways that facilitate capture, storage, and retrieval for multiple users in an organization. Databases involve methods of data organization that allow data to be centrally managed, standardized, and consistent. Instead of a proliferation of separate and distinct data files, the database approach allows central databases to be the sole source of data for many varied applications.

Under the data-oriented approach to systems development, databases are designed around *subjects,* such as customers, suppliers, and parts. Designing databases around subjects enables you to use and revise databases for many different independent applications. This focus results in **application independence,** the separation of data and the definition of data from applications.

Application independence: The separation of data and the definition of data from the applications that use these data.

The central point of application independence is that data and applications are separate. For the data-oriented approach to be effective, however, another change in the system design is needed: Organizations that have centrally managed repositories of organizational data must design new applications to work with existing databases. Organizations that do not have centrally managed repositories of organizational data must design databases that will support both current and future applications. You will see an example of the data-oriented approach to designing information systems in the following Pine Valley Furniture Company case.

TABLE 1-1   Key Differences Between the Process-Oriented and Data-Oriented Approaches to Systems Development

Characteristic	Process-Orientation	Data-Orientation
System focus	What the system is supposed to do and when	Data the system needs to operate
Design stability	Limited, as business processes and the applications that support them change constantly	More enduring, as the data needs of an organization do not change rapidly
Data organization	Data files designed for each individual application	Data files designed for the enterprise
State of the data	Much uncontrolled duplication	Limited, controlled duplication

A Modern Approach to Systems Development:
Pine Valley Furniture Company

Pine Valley Furniture Company (PVF) manufactures high-quality, all-wood furniture and distributes it to retail stores within the United States. Their product lines include dinette sets, stereo cabinets, wall units, living room furniture, and bedroom furniture. PVF currently employs about 50 people and is growing rapidly.

When PVF was founded 15 years ago by Donald Knotts, managing data and information systems was relatively simple. Don founded PVF out of his garage where he made custom furniture as a hobby. When business expanded into a rented warehouse, Don hired a part-time bookkeeper. When the company moved into its present location ten years ago, the product line had multiplied, sales volume had doubled, and the size of the staff had increased. Due to the added complexity of the company's operations, Don reorganized the company into functional areas: Manufacturing, which was further sub-divided into three separate functions—Fabrication, Assembling, and Finishing; Sales; Orders; Accounting; and Purchasing. Don and the heads of the functional areas established manual information systems that worked well for a time but, eventually, PVF selected and installed a minicomputer to automate applications.

When the applications were first computerized, each separate application had its own individual data files tailored to the needs of each functional area. As is typical in such situations, the applications closely resembled the manual systems on which they were based. (Three of the computer applications at Pine Valley Furniture are depicted in Figure 1-3.) In the late 1980s, PVF formed a task force to study the possibility of moving to a database approach. After the preliminary study was completed, management decided to convert their information systems to the database approach. They upgraded their minicomputer and implemented a database

Figure 1-3
Three application systems at Pine Valley Furniture (Source: McFadden and Hoffer, 1994)

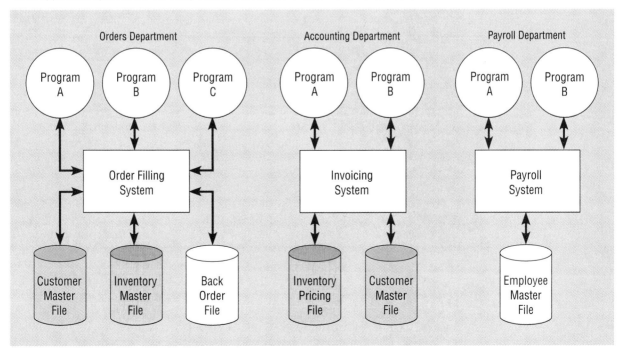

management system. By the time we catch up with Pine Valley Furniture, it has successfully designed and populated a company-wide database and has converted its applications to work with the database. However, PVF is continuing to grow at a rapid rate, putting pressure on its current application systems.

The computer-based applications at PVF parallel the processes that are part of doing business in the company. When customers order furniture, their orders must be processed appropriately so that customers receive what they order and are billed correctly. To fill the orders, furniture has to be built and shipped to customers. Employees have to be paid for their work. Accordingly, most computer-based applications at PVF are in the accounting and financial areas. The applications include order filling, invoicing, accounts receivable, inventory control, accounts payable, payroll, and general ledger. As we stated earlier, each application once had its own data files. For example, there was a customer master file, an inventory master file, a back order file, an inventory pricing file, and an employee master file. The order-filling system used data from three files: customer master, inventory master, and back order. With PVF's new centralized database, data are organized around *entities,* or subjects, such as customers, invoices, and orders (McFadden and Hoffer, 1994).

Pine Valley Furniture, like many firms, decided to develop its application software in-house; that is, it hired staff and bought equipment necessary to build application software suited to its own needs. There are, of course, other ways for firms to obtain application software, as explained more fully in Chapter 12.

YOUR ROLE AND OTHER ORGANIZATIONAL RESPONSIBILITIES IN SYSTEMS DEVELOPMENT

In an organization that develops its own information systems internally, there are several types of jobs involved. In medium to large organizations, there is usually a separate Information Systems (IS) department. Depending on how the organization is set up, the IS department may be a relatively independent unit, reporting to the organization's top manager. Alternatively, the IS department may be part of another functional department, such as Finance, or there may even be an IS department in several major business units. In any of these cases, the manager of an IS department will be involved in systems development. If the department is large enough, there will be a separate division for systems development, which would be homebase for systems analysts, and another division for programming, where programmers would be based (see Figure 1-4). The people for whom the systems are designed are located in the functional departments and are referred to as users or *end users.*

Some organizations use a different structure for their IS departments. Following this model, analysts are assigned and may report to functional departments. In this way, analysts learn more about the business they support. This approach is supposed to result in better systems, since the analyst becomes an expert in both systems development and the business area.

Regardless of how an organization structures its information systems department, systems development is a *team effort.* Systems analysts work together in a team, usually organized on a project basis. Team membership can be expanded to include IS managers, programmers, users, and other specialists who may be involved (throughout or at specific points) in the systems development project. It is rare to find an organizational information system project that involves only one person. Thus, learning how to work with others in teams is an important skill for any IS professional, and we will stress team skills throughout this book.

Figure 1-4

Organization chart for typical IS department

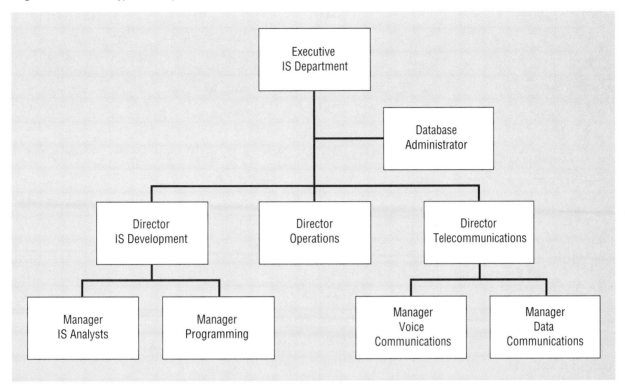

A good team has certain characteristics, some that are a result of how the group is assembled and others that must be acquired through effort on the part of team members (see Table 1-2). A good team is diverse and tolerant of diversity:

- A diverse team has representation from all the different groups interested in a system, and the representation of these groups on the team increases the likelihood of acceptance of the changes a new system will cause.

- Diversity exposes team members to new and different ideas, ideas they might never think of were all team members from the same background, with the same skills and goals.

- New and different ideas can help a team generate better solutions to its problems and defend the course of action it chooses.

- Team members must be able to entertain new ideas without being overly critical, without dismissing new ideas out of hand simply because they are new.

- Team members must be able to deal with ambiguous information as well as with complexity and must learn to play a role on a team (and different roles on different teams) so that the talents of all team members can best be utilized.

In order to work well together, a good team must strive to communicate clearly and completely with its members. Team members will communicate more effectively if they trust each other. Trust, in turn, is built on mutual respect and an ability to place one's own goals and views secondary to the goals and views of the group. To help ensure that a team will work well together, management needs to develop

TABLE 1-2 Characteristics of Successful Teams

- Diversity in backgrounds, skills, and goals
- Tolerance of diversity, uncertainty, ambiguity
- Clear and complete communication
- Trust
- Mutual respect and putting one's own views second to the team
- Reward structure that promotes shared responsibility and accountability

a reward structure that promotes shared responsibility and accountability within the team. In addition to rewards for individual efforts, team members must be rewarded by IS managers for their work as members of an effective work unit.

Team success depends not only on how a team is assembled or the efforts of the group but also on the management of the team. Reward systems are one part of good team management. Effective project management is another key element of successful teams. Project management includes devising a feasible and realistic work plan and schedule, monitoring progress against this schedule, coordinating the project with its sponsors, allocating resources to the project, and sometimes even deciding whether and when a project should be terminated before completing the system.

The characteristics of each systems analysis and design project will dictate which types of individuals should be on the project team. In general, those involved in systems development include IS managers, systems analysts, programmers, end users, and business managers as well as additional IS managers, technicians, and specialists. We will now preview the role of each of these players and stakeholders in systems development.

IS Managers in Systems Development

The manager of an IS department may have a direct role in the systems development process if the organization is small or if that is the manager's style. Typically, IS managers are more involved in allocating resources to and overseeing approved system development projects rather than in the actual development process. Thus, IS managers may attend some project review meetings and certainly will expect written status reports on project progress covering their areas of concern. IS managers may prescribe what methodologies, techniques, and tools are to be used and the procedure for reporting the status of projects. As department leaders, IS managers are also responsible for career planning and development for systems analysts and other employees and for solving problems that arise in the course of development projects.

There are, of course, several IS managers in any medium to large IS department (see Figure 1-4). The manager of an entire IS department may have the title Chief Information Officer and may report to the president or chairman of the firm. Each division of the IS department will also have a manager. Typical titles for these managers are Director of IS Development, IS Operations Manager, and IS Programming Director. The Director of IS Development may be responsible for several development projects at any given time, each of which has a project manager. The responsibilities and focus of any particular IS manager depend on his or her level in the department and on how the organization manages and supports the systems development process.

Systems Analysts in Systems Development

Systems analysts are the key individuals in the systems development process. To succeed as a systems analyst, you will need to develop four skills: analytical, technical, managerial, and interpersonal. *Analytical skills* enable you to understand the organization and its functions, to identify opportunities and problems, and to analyze and solve problems. One of the most important analytical skills you can develop is systems thinking, or the ability to see organizations and information systems as systems. Systems thinking provides a framework from which to see the important relationships among information systems, the organizations they exist in, and the environment in which the organizations themselves exist. *Technical skills* help you understand the potential and the limitations of information technology. As an analyst, you must be able to envision an information system that will help users solve problems and that will guide the system's design and development. You must also be able to work with programming languages, various operating systems, and computer hardware platforms. *Management skills* help you manage projects, resources, risk, and change. *Interpersonal skills* help you work with end users as well as with other analysts and programmers. As a systems analyst, you will play a major role as a liaison among users, programmers, and other systems professionals. Effective written and oral communication, including competence in leading meetings, interviewing, and listening, is a key skill analysts must master. Effective analysts successfully combine these four skills, as Figure 1-5, a typical advertisement for a systems analyst position, illustrates.

As with any profession, becoming a good systems analyst takes years of study and experience. Once hired by an organization, you will generally be trained in the development methodology used by the organization. There is usually a career path for systems analysts that allows them to gain experience and advance into project management and further IS or business management. Many academic IS departments train their undergraduate students to be systems analysts. As your career progresses, you may get the chance to become a manager inside or outside the IS area. In some organizations, you can opt to follow a technical career advancement ladder. As an analyst, you will become aware of a consistent set of professional practices, many of which are governed by a professional code of ethics, similar to other professions.

Programmers in Systems Development

Programmers convert the system specifications given to them by the analysts into instructions the computer can understand. Writing a computer program is sometimes called writing code, or *coding*. Programmers also write program documentation and programs for testing systems. For many years, programming was considered an art. However, computer scientists found that code could be improved if it were structured, so they introduced what is now called *structured programming* (Bohm and Jacopini, 1966). In structured programming, all computing instructions can be represented through the use of three simple structures: sequence, repetition, and selection. Becoming a skilled programmer takes years of training and experience. Many computer information systems undergraduates begin work as programmers or programmer/analysts.

Programming is very labor-intensive, therefore, special-purpose computing tools called *code generators* have been developed to generate reasonably good code from specifications, saving an organization time and money. Code generators do not put programmers out of work; rather, these tools change the nature of programming. Where code generators are in use, programmers take the generated code and fix problems with it, optimize it, and integrate it with other parts of the

Figure 1-5
Typical job ad for a systems analyst

SYSTEMS ANALYST
Distribution Center

We are the world's leading manufacturer of women's intimate apparel products. Our organization in the Far East has an opening for a Systems Analyst.

Requirements:

- Bachelor's degree (or comparable) in Computer Science and/or Business Administrations with 5 (+) years of working experience.

- In-depth understanding of Distribution and Manufacturing concepts (Allocation, Replenishment, Ship Floor Control, Production Scheduling, MRP)

- Working knowledge of project management and all phases of software development life cycle.

- Experience with CASE tools, PC and Bar Code equipment is essential.

- Working knowledge of an AS/400 and/or UNIX environment with the languages C, RPG400 and/or COBOL are desirable

The successful candidate will provide primary interface for all user problems, answer technical questions and requests within the applications development group; work with user areas to establish priorities; and provide recommendations and directions for process improvement through automation; skills in an Asian language is a plus.

We offer an attractive compensation package, relocation assistance and the technical and analytical challenges you would expect in a state-of-the-art environment. The position will report to Senior Management.

Please forward your resume, along with salary requirements to:

---

system. The goal of some computer-aided software engineering (CASE) tools is to provide a variety of code generators that can automatically produce 90 percent or more of code directly from the system specifications normally given a programmer. When this goal is achieved, the role of programmers on systems development teams will be changed further.

End users: Non-information system professionals in an organization who specify the business requirements for and use software applications. End users often request new or modified applications, test and approve applications, and may serve on project teams as business experts.

End Users in Systems Development

In the typical traditional systems development process, systems analysts interview, survey, and observe end users to determine what they need from an information system. **End users** are business professionals who are experts in their fields but who usually do not have the skills, time, desire, or responsibility required to develop information systems. Therefore, as an analyst, you will work with users to convert their knowledge of the business into supportive information systems. End users often are your clients or customers, the people for whom you are building a system. In many cases, end users will also serve on the systems development team, providing their expertise in very active ways. You and other IS professionals will also provide support and assistance for those more sophisticated end users who write, test, and implement their own information or data distribution systems.

Supporting End-User Development   As the number of user requests for new or improved information systems increases, an IS department will have to assign priorities to development projects. Prioritization means that some users will get their systems right away while others must wait. In the 1970s, with the spread of time-sharing systems that enabled several people to use the same computer at the same time, it became technically possible for end users to develop their own systems. As education and experience made end users aware of these technologies and they became skillful in using them, many users, impatient for their requested projects to be scheduled, developed their own applications. Consequently, a significant role for systems analysts and other IS professionals is to help end users develop their own systems, which are often stand-alone systems or data distribution systems designed to enhance an existing system developed by IS professionals.

Support of **end-user development** has several dimensions. First, IS professionals must evaluate and make available suitable tools for end users. Second, IS professionals must train end users in the use of these tools and their proper application. Third, IS professionals must be available to assist end users and to answer questions or perform more complicated systems development work whenever end users have difficulties. Finally, IS professionals must continue to maintain the data capture and transfer applications that manage the databases from which end users extract the data needed in the systems they build.

End-user design and development of information systems has been somewhat controversial in the past (Alavi and Weiss, 1985; Davis, 1982). Some IS managers have worried about the quality of the systems end users produce. However, some end users believe there is no choice, especially if users must wait for the IS department to provide the desired and needed systems. Of course, if an organization is too small to have an IS department and outside consultants are too expensive, end users have little choice but to develop their own systems or settle for off-the-shelf software.

End-user development: An approach to systems development in which users who are not computer experts satisfy their own computing needs through the use of high-level software and languages such as electronic spreadsheets and relational database management systems.

Business Managers in Systems Development

Another group important to systems development efforts is business managers, such as functional department heads and corporate executives. These managers are important to systems development because they have the power to fund development projects and to allocate the resources necessary for the projects' success. Because of their decision-making authority and knowledge of the firm's lines of business, department heads and executives are also able to set general requirements and constraints for development projects. In larger companies where the relative importance of systems projects is determined by a steering committee, these executives have additional power as they are usually members of the steering committees or systems planning groups. Business managers, therefore, have the power to set the direction for systems development, to propose and approve projects, and to determine the relative importance of projects that have already been approved and assigned to other people in the organization.

R.T.C. LIBRARY, LETTERKENNY

005
.12

4 0015608

Other IS Managers/Technicians in Systems Development

In larger organizations where IS roles are more differentiated, there may be several additional IS professionals involved in the systems development effort. A firm with an existing set of databases will most likely have a *database administrator* who is usually involved in any systems project affecting the firm's databases. Network and *telecommunications experts* help develop systems involving data and/or voice communication, either internal or external to the organization. Some organizations have *human factors* departments which are concerned with system interfaces and ease-of-use issues, training users, and writing user documentation and manuals. Overseeing much of the development effort, especially for large or sensitive systems, are an organization's *internal auditors* who ensure that required controls are built into the system. In many organizations, auditors also have responsibility for keeping track of changes in the system's design. The necessary interaction of all these individuals makes systems development very much a team effort.

TYPES OF INFORMATION SYSTEMS AND SYSTEMS DEVELOPMENT

As you can see, several different people in an organization can be involved in developing information systems. Given the broad range of people and interests represented in systems development, you might assume that it could take several different types of information systems to satisfy all an organization's information system needs. Your assumption would be correct.

Up until now we have been talking about information systems in generic terms, but there are actually several different types or classes of information systems. In general, these types are distinguished from each other on the basis of what the system does or by the technology used to construct the system. As a systems analyst, part of your job will be to determine which kind of system will best address the organizational problem or opportunity on which you are focusing. In addition, different classes of systems may require different methodologies, techniques, and tools for development.

From your prior studies and experiences with information systems, you are probably aware of at least four classes of information systems:

- Transaction processing systems

- Management information systems

- Decision support systems (for individuals, groups, and executives)

- Expert systems

In addition, many organizations recognize scientific (or technical) computing and office automation systems. If you need a refresher on what these types of systems mean, read Appendix A at this point. To preview the diversity of systems development approaches, the following sections briefly highlight how systems analysis and design methods differ across the four major types of systems.

Transaction Processing Systems

Transaction processing systems (TPS) automate the handling of data about business activities or transactions, which can be thought of as simple, discrete events in the life of an organization. Data about each transaction is captured, transactions are verified and accepted or rejected, and validated transactions are stored for later aggregation. Reports may be produced immediately to provide standard summarizations of transactions, and transactions may be moved from process to process in order to handle all aspects of the business activity.

The analysis and design of a TPS means focusing on the firm's current procedures for processing transactions, whether those procedures are manual or automated. The focus on current procedures implies a careful tracking of data capture, flow, processing, and output. The goal of TPS development is to improve transaction processing by speeding it up, using fewer people, improving efficiency and accuracy, integrating it with other organizational information systems, or providing information not previously available.

Management Information Systems

A management information system (MIS) takes the relatively raw data available through a TPS and converts them into a meaningful aggregated form that managers need to conduct their responsibilities. Developing an MIS calls for a good understanding of what kind of information managers require and how managers use information in their jobs. Sometimes managers themselves may not know precisely what they need or how they will use information. Thus, the analyst must also develop a good understanding of the business and the transaction processing systems that provide data for an MIS.

Management information systems often require data from several transaction processing systems (for example, customer order processing, raw material purchasing, and employee timekeeping). Development of an MIS can, therefore, benefit from a data-orientation, in which data are considered an organization resource separate from the TPS in which they are captured. Because it is important to be able to draw on data from various subject areas, developing a comprehensive and accurate model of data is essential in building an MIS.

Decision Support Systems

Decision support systems (DSS) are designed to help organizational decision makers make decisions. Instead of providing summaries of data, as with an MIS, a DSS provides an interactive environment in which decision makers can quickly manipulate data and models of business operations. A DSS is composed of a database (which may be extracted from a TPS or MIS), mathematical or graphical models of business processes, and a user interface (or dialogue module) that provides a way for the decision maker, usually a non-technical manager, to communicate with the DSS. A DSS may use both hard historical data as well as judgments (or "what if" scenarios) about alternative histories or possible futures. One form of a DSS, an executive information system (EIS), emphasizes the unstructured capability for senior management to explore data starting at a high level of aggregation and selectively drilling down into specific areas where more detailed understandings of the business are required. In either case, a DSS is characterized by less structured and predictable use; rather, a DSS is a software resource intended to support a certain scope of decision-making activities (from problem finding to choosing a course of action).

The systems analysis and design for a DSS often concentrates on the three main DSS components: database, model base, and user dialogue. As with an MIS, a data-orientation is most often used for understanding user requirements. In addition, the systems analysis and design project will carefully document the mathematical rules that define inter-relationships among different data. These relationships are used to predict future data or to find the best solutions to decision problems. Thus, decision logic must be carefully understood and documented. Also, since a decision maker typically interacts with a DSS, the design of easy-to-use yet thorough user dialogues and screens is important. Because a DSS often deals with situations not encountered every day or situations that can be handled in many different ways, there can be considerable uncertainty on what a DSS should actually do. Thus, systems developers often use methods that prototype the system and iteratively and rapidly redevelop the system based on trial use. The development of a DSS, hence, often does not follow as formal a project plan as is done for a TPS or MIS, since the software deliverable is more uncertain at the beginning of the project.

Expert Systems

Different from any of the other classes of systems we have discussed so far, an expert system (ES) attempts to codify and manipulate knowledge rather than information. If-then-else rules or other knowledge representation forms describe the way an expert would approach situations in a specific domain of problems. Typically, users communicate with an ES through an interactive dialogue. The ES asks questions (that an expert would ask) and the end user supplies the answers. The answers are then used to determine which rules apply and the ES provides a recommendation based on the rules.

The focus on developing an ES is acquiring the knowledge of the expert in the particular problem domain. Knowledge engineers perform knowledge acquisition; they are similar to systems analysts but are trained to use different techniques, as determining knowledge is considered more difficult than determining data.

Summary of Information Systems Types

Many information systems you build or maintain will contain aspects of each of the four major types of information systems. Thus, as a systems analyst, you will likely employ specific methodologies, techniques, and tools associated with each of the four information system types. Table 1-3 summarizes the general characteristics and development methods for each type.

So far, we have concentrated on the context of information systems development, looking at the different organizations where software is developed, the people involved in development efforts, and the different types of information systems that exist in organizations. Now that we have a good idea of context, we can turn to the actual process by which many information systems are developed in organizations, the *systems development life cycle*.

DEVELOPING INFORMATION SYSTEMS AND THE SYSTEMS DEVELOPMENT LIFE CYCLE

Systems development life cycle (SDLC): The traditional methodology used to develop, maintain, and replace information systems.

Like many processes, the development of information systems follows a life cycle. For example, a commercial product follows a life cycle in that it is created, tested, and introduced to the market. Its sales increase, peak, and decline. Finally, the product is removed from the market and replaced by something else. The **systems development life cycle (SDLC)** is a common methodology for systems development in many organizations, featuring several phases that mark the progress of the systems analysis and design effort. Every textbook author and information system development organization uses a slightly different life cycle model, with anywhere from three to almost twenty identifiable phases.

TABLE 1-3 Systems Development for Different IS Types

IS Type	IS Characteristics	Systems Development Methods
Transaction processing system	High-volume, data capture focus; goal is efficiency of data movement and processing and interfacing different TPSs	Process-orientation; concern with capturing, validating, and storing data and with moving data between each required step
Management information system	Draws on diverse yet predictable data resources to aggregate and summarize data; may involve forecasting future data from historical trends and business knowledge	Data-orientation; concern with understanding relationships between data so data can be accessed and summarized in a variety of ways; builds a model of data that supports a variety of uses
Decision support system	Provides guidance in identifying problems, finding and evaluating alternative solutions, and selecting or comparing alternatives; potentially involves groups of decision makers; often involves semi-structured problems and the need to access data at different levels of detail	Data- and decision logic-orientations; design of user dialogue; group communication may also be key and access to unpredictable data may be necessary; nature of systems require iterative development and almost constant updating
Expert system	Provides expert advice by asking users a sequence of questions dependent on prior answers that lead to a conclusion or recommendation	A specialized decision logic-orientation in which knowledge is elicited from experts and described by rules or other forms

Although any life cycle appears at first glance to be a sequentially ordered set of phases, it actually is not. The specific steps and their sequence are meant to be adapted as required for a project, consistent with management approaches. For example, in any given SDLC phase, the project can return to an earlier phase if necessary. Similarly, if a commercial product does not perform well just after its introduction, it may be temporarily removed from the market and improved before being re-introduced. In the systems development life cycle, it is also possible to complete some activities in one phase in parallel with some activities of another phase. Sometimes the life cycle is iterative; that is, phases are repeated as required until an acceptable system is found. Such an iterative approach is especially characteristic of rapid application development methods, such as prototyping, which we introduce later in the chapter. Some people consider the life cycle to be a spiral, in which we constantly cycle through the phases at different levels of detail. The life cycle can also be thought of as a circular process in which the end of the useful life of one system leads to the beginning of another project that will develop a new version or replace an existing system altogether. However conceived, the systems development life cycle used in an organization is an orderly set of activities conducted and planned for each development project. The skills required of a systems analyst apply to all life cycle models. Software is the most obvious end product of the life cycle; other essential outputs include documentation about the system and how it was developed as well as training for users.

Every medium to large corporation and every custom software producer will have its own specific, detailed life cycle or systems development methodology in place. Even if a particular methodology does not look like a cycle, you will probably discover that many of the SDLC steps are performed and SDLC techniques and tools are used. Learning about systems analysis and design from the life cycle approach will serve you well no matter which systems development methodology you use.

When you begin your first job, you are likely to spend several weeks or months learning your organization's SDLC and its associated methodologies, techniques, and tools. In order to make this book as general as possible, we follow a rather generic life cycle model, as described in more detail in Figure 1-6. Notice how our

Figure 1-6
The systems development life cycle

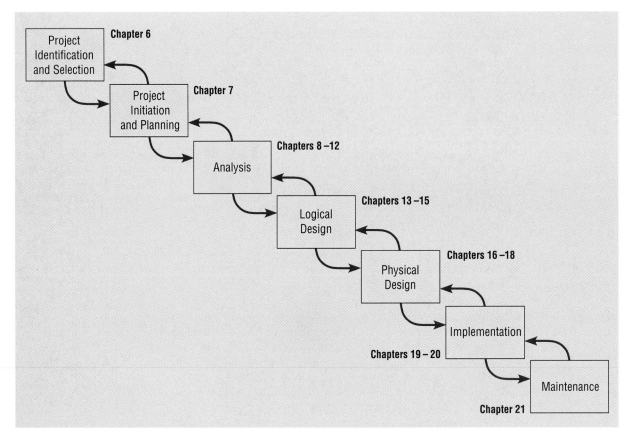

model resembles a staircase with arrows connecting each step to the step before it and to the step after it. This representation of the SDLC is sometimes referred to as the waterfall model. We use this SDLC as one example of a methodology but, more importantly, as a way to arrange the topics of systems analysis and design. Thus, what you learn in this book you can apply to almost any life cycle you might follow. As we describe this SDLC throughout the book, you will see that each phase has specific outcomes and deliverables that feed important information to other phases. At the end of each phase (and sometimes within phases for intermediate steps), a systems development project reaches a *milestone* and, as deliverables are produced, they are often reviewed by parties outside the project team. In the rest of this section we provide a brief overview of each SDLC phase. At the end of the section we summarize this discussion in a table listing the main deliverables or outputs from each SDLC phase.

Project identification and selection: The first phase of the SDLC in which an organization's total information system needs are identified, analyzed, prioritized, and arranged.

The first phase in the SDLC is called **project identification and selection.** In this phase, someone identifies the need for a new or enhanced system. In larger organizations, this recognition may be part of a corporate and systems planning process. Information needs of the organization as a whole are examined, and projects to meet these needs are proactively identified. The organization's information system needs may result from requests to deal with problems in current procedures, from the desire to perform additional tasks, or from the realization that information technology could be used to capitalize on an existing opportunity. These needs can then be prioritized and translated into a plan for the IS department, including a schedule for developing new major systems. In smaller organizations (as well as in large ones), determination of which systems to develop may be affected by ad hoc user requests submitted as the need for new or enhanced systems arises as well as from a formalized information planning process. In either case, during project identification and selection, an organization determines whether or not resources should be devoted to the development or enhancement of each information system under consideration. The outcome of the project identification and selection process is a determination of which systems development projects should be undertaken by the organization, at least in terms of an initial study.

The second phase is **project initiation and planning.** The two major activities in this phase are the formal, yet still preliminary, investigation of the system problem or opportunity at hand and the presentation of reasons why the system should or should not be developed by the organization. A critical step at this point is determining the scope of the proposed system. The project leader and initial team of systems analysts also produce a specific plan for the proposed project which the team will follow using the remaining SDLC steps. This baseline project plan customizes the standardized SDLC and specifies the time and resources needed for its execution. The formal definition of a project is based on the likelihood that the organization's IS department is able to develop a system that will solve the problem or exploit the opportunity and determine whether the costs of developing the system outweigh the benefits it could provide. The final presentation of the business case for proceeding with the subsequent project phases is usually made by the project leader and other team members to someone in management or to a special management committee with the job of deciding which projects the organization will undertake.

> **Project initiation and planning:** The second phase of the SDLC in which a potential information systems project is explained and an argument for continuing or not continuing with the project is presented; a detailed plan is also developed for conducting the remaining phases of the SDLC for the proposed system.

The next phase is **analysis.** During this phase, the analyst thoroughly studies the organization's current procedures and the information systems used to perform organizational tasks. Analysis has several sub-phases. The first is requirements determination. In this sub-phase, you and other analysts work with users to determine what the users want from a proposed system. This sub-phase usually involves a careful study of any current systems, manual and computerized, that might be replaced or enhanced as part of this project. Next, you study the requirements and structure them according to their inter-relationships and eliminate any redundancies. Third, generate alternative initial designs to match the requirements. Then compare these alternatives to determine which best meets the requirements within the cost, labor, and technical levels the organization is willing to commit to the development process. The output of the analysis phase is a description of (but not a detailed design for) the alternative solution recommended by the analysis team. Once the recommendation is accepted by those with funding authority, you can begin to make plans to acquire any hardware and system software necessary to build or operate the system as proposed.

> **Analysis:** The third phase of the SDLC in which the current system is studied and alternative replacement systems are proposed.

The fourth and fifth phases are devoted to designing the new or enhanced system. During design, you and the other analysts convert the description of the recommended alternative solution into logical and then physical system specifications. You must design all aspects of the system from input and output screens to reports, databases, and computer processes. Design occurs in two SDLC phases: logical design and physical design.

Logical design is not tied to any specific hardware and systems software platform. Theoretically, the system could be implemented on any hardware and systems software. The idea is to make sure that the system functions as intended. Logical design concentrates on the business aspects of the system (which is why some life cycles call this phase business design).

> **Logical design:** The fourth phase of the SDLC in which all functional features of the system chosen for development in analysis are described independently of any computer platform.

Physical design: The fifth phase of the SDLC in which the logical specifications of the system from logical design are transformed into technology-specific details from which all programming and system construction can be accomplished.

In **physical design,** you turn the logical design into physical, or technical, specifications. For example, you must convert diagrams that map the origin, flow, and processing of data in a system into a structured systems design that can then be broken down into smaller and smaller units for conversion to instructions written in a programming language. You design the various parts of the system to perform the physical operations necessary to facilitate data capture, processing, and information output. During physical design, the analyst team decides which programming languages the computer instructions will be written in, which database systems and file structures will be used for the data, and which hardware platform, operating system, and network environment the system will run under. These decisions finalize the hardware and software plans initiated at the end of the analysis phase. Now you can proceed with acquiring any new technology not already present in the organization. The final product of the design phase is the physical system specifications in a form ready to be turned over to programmers and other system builders for construction.

Implementation: The sixth phase of the SDLC in which the information system is coded, tested, installed, and supported in the organization.

The physical system specifications are turned over to programmers as the first part of the **implementation** phase. During implementation, you turn system specifications into a working system that is tested and then put into use. Implementation includes coding, testing, and installation. During *coding,* programmers write the programs that make up the system. During *testing,* programmers and analysts test individual programs and the entire system in order to find and correct errors. During *installation,* the new system becomes a part of the daily activities of the organization. Application software is installed, or loaded, on existing or new hardware and users are introduced to the new system and trained. Begin planning for both testing and installation early as the project initiation and planning phase, since both testing and installation require extensive analysis in order to develop exactly the right approach.

Implementation activities also include initial user support such as the finalization of documentation, training programs, and ongoing user assistance. Note that documentation and training programs are finalized during implementation; documentation is produced throughout the life cycle, and training (and education) occur from the inception of a project. Implementation can continue for as long as the system exists since ongoing user support is also part of implementation. Despite the best efforts of analysts, managers, and programmers, however, installation is not always a simple process. Many well-designed systems have failed because the installation process was faulty. Our point is that even a well-designed system can fail if implementation is not well managed. Since the management of implementation is usually done by the project team, we stress implementation issues throughout this book.

Maintenance: The final phase of the SDLC in which an information system is systematically repaired and improved.

The final phase is **maintenance.** When a system (including its training, documentation, and support) is operating in an organization, users sometimes find problems with how it works and often think of better ways to perform its functions. Also, the organization's needs with respect to the system change over time. In maintenance, programmers make the changes that users ask for and modify the system to reflect changing business conditions. These changes are necessary to keep the system running and useful. In a sense, maintenance is not a separate phase but a repetition of the other life cycle phases required to study and implement the needed changes. Thus, you might think of maintenance as an overlay to the life cycle rather than a separate phase. The amount of time and effort devoted to maintenance depends a great deal on the performance of the previous phases of the life cycle. There inevitably comes a time, however, when an information system is no longer performing as desired, when maintenance costs become prohibitive, or when an organization's needs have changed substantially. Such problems indicate that it is time to begin designing the system's replacement, thereby completing the loop and starting the life cycle over again. Often the distinction between major maintenance and new development is not clear, which is another reason maintenance often resembles the life cycle itself.

The SDLC is a highly linked set of phases whose products feed the activities in subsequent phases. Table 1-4 summarizes the outputs or products of each phase based on the above descriptions. The chapters on the SDLC phases will elaborate on the products of each phase as well as on how the products are developed.

Throughout the systems development life cycle, the systems development project itself needs to be carefully planned and managed. The larger the systems project, the greater the need for project management. Several project management techniques have been developed in this century and many have been made more useful through automation. Chapter 4 contains a more detailed treatment of project planning and management techniques. Next, we will discuss some of the criticisms of the systems development life cycle and alternatives developed to address those criticisms.

The Traditional SDLC

There are several criticisms of the traditional life cycle approach to systems development as followed exactly as outlined in Figure 1-6. One criticism relates to the way the life cycle is organized. Although we know that phases of the life cycle can sometimes overlap, traditionally one phase ended and another began once a *milestone* had been reached. The milestone usually took the form of some deliverable or pre-specified output from the phase. For example, the design deliverable is the set of detailed physical design specifications. Once the milestone had been reached and the new phase initiated, it became difficult to go back. Even though business conditions continued to change during the development process and analysts were pressured by users and others to alter the design to match changing conditions, it

TABLE 1-4 Products of SDLC Phases

Phase	Products, Outputs, or Deliverables
Project identification and selection	Priorities for systems and projects; an architecture for data, networks, hardware, and IS management is the result of associated systems planning activities
Project initiation and planning	Detailed steps, or work plan, for project; specification of system scope and high-level system requirements or features; assignment of team members and other resources; system justification or business case
Analysis	Description of current system and where problems or opportunities are with a general recommendation on how to fix, enhance, or replace current system; explanation of alternative systems and justification for chosen alternative
Logical design	Functional, detailed specifications of all system elements (data, processes, inputs, and outputs)
Physical design	Technical, detailed specifications of all system elements (programs, files, network, system software, etc.); acquisition plan for new technology
Implementation	Code, documentation, training procedures, and support capabilities
Maintenance	New versions or releases of software with associated updates to documentation, training, and support

was necessary for the analysts to freeze the design at a particular point and go forward. The enormous amount of effort and time necessary to implement a specific design meant that it would be very expensive to make changes in a system once it was developed. There were no CASE tools, no code generators, and no fourth-generation languages when the SDLC was popularized in the 1960s. If the design was not frozen, the system would never be completed, as programmers would no sooner be done with coding one design than they would receive requests for major changes. The traditional life cycle, then, had the property of locking in users to requirements that had been previously determined, even though those requirements might have changed.

Another criticism of the way the traditional life cycle is often used is that it tends to focus too little time on good analysis and design. The result is a system that does not match users' needs and one that requires extensive maintenance, unnecessarily increasing development costs. According to some estimates, maintenance costs account for as much as 70 percent of the system development costs (Aktas, 1987). Given these problems, people working in systems development began to look for better ways to conduct systems analysis and design.

Structured Analysis and Structured Design

Ed Yourdon and his colleagues developed *structured analysis* and *structured design* in the early 1970s as a way to address some of the problems with the traditional SDLC (Yourdon & Constantine, 1979). By making analysis and design more disciplined, similar to engineering fields, through the use of tools such as data flow diagrams and transform analysis, Yourdon and colleagues sought to emphasize and improve the analysis and design phases of the life cycle. The goal was to reduce maintenance time and effort. Structured analysis and design makes it easier to go back to earlier phases in the life cycle when necessary—for example, when requirements change. Finally, there is also an emphasis on partitioning or dividing a problem into smaller, more manageable units and on making a clear distinction between physical and logical design (DeMarco, 1979; Yourdon & Constantine, 1979). The life cycle used in this book is faithful to these structured principles.

Object-Oriented Analysis and Design

A more recent approach to systems development that is becoming more and more popular is **object-oriented analysis and design** (OOAD). OOAD is often called the third approach to systems development, after the process-oriented and data-oriented approaches. The object-oriented approach combines data and processes (call *methods*) into single entities called **objects.** Objects usually correspond to the real things an information system deals with, such as customers, suppliers, contracts, and rental agreements. Putting data and processes together in one place recognizes the fact that there are a limited number of operations for any given data structure. Putting data and processes together makes sense even though typical systems development keeps data and processes independent of each other. The goal of OOAD is to make system elements more reusable, thus improving system quality and the productivity of systems analysis and design.

Another key idea behind object-orientation is that of **inheritance.** Objects are organized into **object classes,** which are groups of objects sharing structural and behavioral characteristics. Inheritance allows the creation of new classes that share some of the characteristics of existing classes. For example, from a class of objects called "person," you can use inheritance to define another class of objects called

Object-oriented analysis and design (OOAD): Systems development methodologies and techniques based on objects rather than data or processes.

Object: A structure that encapsulates (or packages) attributes and methods that operate on those attributes. An object is an abstraction of a real-world thing in which data and processes are placed together to model the structure and behavior of the real-world object.

Inheritance: The property that occurs when entity types or object classes are arranged in a hierarchy and each entity type or object class assumes the attributes and methods of its ancestors; that is, those higher up in the hierarchy. Inheritance allows new but related classes to be derived from existing classes.

"customer." Objects of the class "customer" would share certain characteristics with objects of the class "person": they would both have names, addresses, phone numbers, and so on. Since "person" is the more general class and "customer" is more specific, every customer is a person but not every person is a customer.

> **Object class:** A logical grouping of objects that have the same (or similar) attributes and behaviors (methods).

As you might expect, you need a computer programming language that can create and manipulate objects and classes of objects in order to create object-oriented information systems. Several object-oriented programming languages have been created (e.g., C++, Eiffel, and ObjectPAL—for Paradox for Windows). In fact, object-oriented languages were developed first and object-oriented analysis and design techniques followed. Because OOAD is still relatively new, there is little consensus or standardization among the many OOAD techniques available. In general, the primary task of object-oriented analysis is identifying objects, defining their structure and behavior, and defining their relationships. The primary tasks of object-oriented design are modeling the details of the objects' behavior and communication with other objects so that system requirements are met and re-examining and redefining objects to better take advantage of inheritance and other benefits of object-orientation.

CONSIDERING DIFFERENT APPROACHES TO DEVELOPMENT

In the continuing effort to improve the systems analysis and design process, several different approaches have been developed. We will describe the more important alternatives in more detail in later chapters. Attempts to make system development less of an art and more of a science are usually referred to as *systems engineering* or *software engineering.* As the names indicate, rigorous engineering techniques are applied to systems development. A very influential practice borrowed from engineering is called prototyping. We will discuss prototyping next, followed by introductions to *Joint Application Design* and *Participatory Design.*

Prototyping

Designing and building a scaled-down but functional version of a desired system is the process known as **prototyping.** You can build a prototype with any computer language or development tool, but special prototyping tools have been developed to simplify the process. A prototype can be developed with some fourth-generation languages (4GLs), with the query and screen and report design tools of a database management system, and with tools called *computer-aided software engineering (CASE) tools.*

> **Prototyping:** An iterative process of systems development in which requirements are converted to a working system that is continually revised through close work between an analyst and users.

Using prototyping as a development technique (see Figure 1-7), the analyst works with users to determine the initial or basic requirements for the system. The analyst then quickly builds a prototype. When the prototype is completed, the users work with it and tell the analyst what they like and do not like about it. The analyst uses this feedback to improve the prototype and takes the new version back to the users. This iterative process continues until the users are relatively satisfied with what they have seen. Two key advantages of the prototyping technique are the large extent to which prototyping involves the user in analysis and design and its ability to capture requirements in concrete, rather than verbal or abstract, form. In addition to being used stand-alone, prototyping may also be used to augment the SDLC. For example, a prototype of the final system may be developed early in analysis to help the analysts identify what users want. Then the final system is

Figure 1-7
The prototyping methodology
(Adapted from Naumann and
Jenkins, 1982)

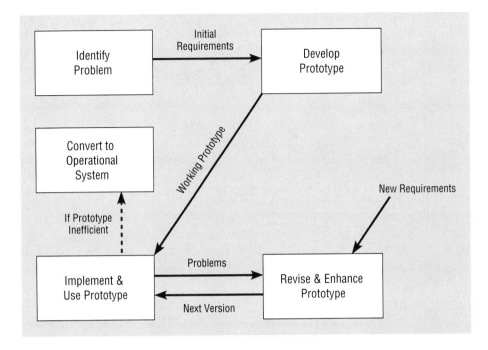

developed based on the specifications of the prototype. We discuss prototyping in greater detail in Chapter 8 and use various prototyping tools in Chapters 13 and 14 to illustrate the design of system outputs.

Prototyping is a form of rapid application development, or RAD. The fundamental principle of any RAD methodology is to delay producing detailed system design documents until after user requirements are clear. The prototype serves as the working description of needs. RAD methodologies emphasize gaining user acceptance of the human-system interface and developing core capabilities as quickly as possible, sacrificing computer efficiency for gains in human efficiency in rapidly building and rebuilding working systems. On the other hand, RAD methodologies can overlook important software engineering principles, the result of which are inconsistencies between system modules, noncompliance with standards, and lack of reusability of system components (Bourne, 1994).

Joint Application Design

Joint Application Design (JAD): A structured process in which users, managers, and analysts work together for several days in a series of intensive meetings to specify or review system requirements.

In the late 1970s, systems development personnel at IBM developed a new process for collecting information system requirements and reviewing system designs. The process is called **Joint Application Design (JAD).** The basic idea behind JAD is to bring structure to the requirements determination phase of analysis and to the reviews that occur as part of design. Users, managers, and systems developers are brought together for a series of intensive structured meetings run by a JAD session leader who maintains the structure and adheres to the agenda. By gathering the people directly affected by an IS in one room at the same time to work together to agree on system requirements and design details, time and organizational resources are better managed. As an added plus, group members are more likely to develop a shared understanding of what the IS is supposed to do. We will discuss JAD in more detail in Chapter 8.

Participatory Design

Developed in Northern Europe, **Participatory Design (PD)** represents a viable alternative approach to the SDLC. PD emphasizes the role of the user much more than traditional North American techniques do. In some instances, PD may involve the entire user community in the development process. Each user has an equal voice in determining system requirements and in approving system design. In other cases, an elected group of users controls the process. These users represent the larger community, much as a legislature represents the needs and wants of the electorate. Typically, under PD, systems analysts work for the users. The organization's management and outside consultants provide advice rather than control (Mumford, 1981). PD is partly a result of the roles of labor and management in the Northern European workplace where labor is more organized, carries more clout, and is more intimately involved with technological changes than is true in North America.

Participatory Design (PD): A systems development approach that originated in Northern Europe in which users and the improvement in their work lives is the central focus.

IMPROVING IS DEVELOPMENT PRODUCTIVITY

Other efforts to improve the system development process have taken advantage of the benefits offered by computing technology itself. The result has been the creation and fairly widespread use of *computer-aided software engineering* or *CASE* tools. CASE tools have been developed for internal use and for sale by several leading firms, including Intersolv (Excelerator™), Arthur Andersen (FOUNDATION™), and Texas Instruments (Information Engineering Facility™, based on James Martin's Information Engineering methodology), to name only a few.

CASE tools are built around a central repository for system descriptions and specifications, including information about data names, format, uses, and locations. The idea of a central repository of information about the project is not new—the manual form of such a repository is called a project dictionary or workbook. The difference is that CASE tools automate the repository for easier updating and for consistency. CASE tools also include diagramming tools for data flow diagrams and other graphical aids, screen and report design tools, and other special-purpose tools. CASE helps programmers and analysts do their jobs more efficiently and more effectively by automating routine tasks. There is more information on CASE in Chapter 5, and we relate many examples of the use of CASE throughout this book.

SUMMARY

This chapter introduced you to information systems analysis and design, the complex organizational process whereby computer-based information systems are developed and maintained. You read about the differences between the process-oriented and data-oriented approaches to systems analysis and design: Process-orientation focuses on what the system is supposed to do while data-orientation focuses on the data the system needs to operate. Process-orientation provides a less stable design than does data-orientation, as business processes change faster than do the data an organization uses. With process-orientation, data files are designed for specific applications whereas data files are designed for the whole enterprise with data-orientation; process-orientation leads to much uncontrolled data redundancy whereas data redundancy is controlled under data-orientation. You studied an example of data-orientation in the Pine Valley Furniture Company. You also learned about application independence, the separation of data from the computer applications that use the data. Data-orientation and application independence frame the way you learn about systems analysis and design in this book.

A major part of this chapter was devoted to examining the context of systems analysis and design. You read about the various people in organizations who develop systems, including systems analysts, programmers, IS managers, business managers, end users, database administrators, human factors experts, telecommunications experts, and auditors. You also learned that there are many different kinds of information systems used in organizations, from transaction processing systems to expert systems to office systems. Development techniques vary with system type.

Finally, you learned about the basic framework that guides systems analysis and design, the systems development life cycle, with its seven major phases: project identification and selection, project initiation and planning, analysis, logical design, physical design, implementation, and maintenance. The life cycle has had its share of criticism, which you read about, and other frameworks have been developed to address the life cycle's problems. These alternative frameworks include prototyping (a Rapid Application Development approach), Joint Application Design, and Participatory Design.

CHAPTER REVIEW

KEY TERMS

Analysis
Application independence
Application software
Data
Database
Data flow
Data-oriented approach
End users
End-user development
Implementation
Information
Information systems
   analysis and design

Inheritance
Joint Application Design
   (JAD)
Logical design
Maintenance
Object
Object class
Object-oriented analysis
   and design (OOAD)
Participatory Design (PD)
Physical design
Process-oriented approach
Processing logic

Project identification and
   selection
Project initiation and
   planning
Prototyping
Systems analyst
Systems development life
   cycle (SDLC)

REVIEW QUESTIONS

1. Define each of the following terms:
   a. systems development life cycle
   b. data
   c. data flows
   d. processing logic
   e. object
   f. object class

   g. inheritance
   h. application software
   i. methodology
   j. technique
   k. tool

2. What is information systems analysis and design?

3. Explain the traditional application-based approach to systems development. How is this different from the data-based approach?

4. What are the organizational roles associated with systems development? Describe the responsibilities of each role.

5. List the different classes of information systems described in this chapter. How do they differ from each other?

6. List and explain the different phases in the systems development life cycle.

7. What are structured analysis and structured design?

8. What is prototyping?

9. What is JAD? What is Participatory Design?

10. What is object-oriented analysis and design?

PROBLEMS AND EXERCISES

1. Match the following terms to the appropriate definitions.

_____ maintenance	a.	phase of the SDLC in which the information system is coded, tested, and installed
_____ analysis	b.	phase of the SDLC in which an organization's total information system needs are analyzed and arranged
_____ design	c.	phase of the SDLC in which the current system is studied and alternative other systems are proposed
_____ project initiation and planning	d.	process of the SDLC in which the system chosen for development is described in detail with a particular physical form in mind
_____ project identification and selection	e.	phase of the SDLC in which an information system is systematically improved
_____ implementation	f.	phase of the SDLC in which a potential project is identified and an argument for the project is presented

2. Why is it important to use systems analysis and design methodologies when building a system? Why not just build the system in whatever way seems to be "quick and easy"? What value is provided by using an "engineering" approach?

3. Choose a business transaction you undertake regularly, such as using an ATM machine, buying groceries at the supermarket, or buying a ticket for a university's basketball game. For this transaction, define the data, draw the data flow diagram, and describe processing logic.

4. How is IBM's Joint Application Design (JAD) approach different from the Participatory Design (PD) approach developed in Northern Europe? You may have to do some digging at the library to answer this question adequately. What are the benefits in using approaches like this in building information systems? What are the barriers to using approaches like this?

5. How would you organize a project team of students to work with a small business client? How would you organize a project team if you were working for a professional consulting organization? How might these two methods of organization differ? Why?

6. How might prototyping be used as part of the SDLC?

7. Describe the difference in the role of a systems analyst in the SDLC versus prototyping.

8. Contrast process-oriented and data-oriented approaches to systems analysis and design. Why does this book make the point that these are complementary, not competing, approaches to systems development?

FIELD EXERCISES

1. Choose an organization with a fairly extensive information systems department. Get a copy of its organization chart (or draw one). At what level in the organization is the highest ranking information systems employee? What is his or her title? To whom does he or she report? What are his or her primary duties? Does the nature of this person's job description, responsibilities, and authority help or hinder this person in doing a good job?

2. Choose an organization and identify personnel who fulfill each of the following roles: IS manager, systems analyst, programmer, end user, business manager, database administrator, network and/or telecommunications manager, information system security manager. Are these roles filled formally or informally? Draw an organization chart linking these people. When these people build, use, and maintain the information systems, do they work together as a team or is their work fairly independent of each other? Why? Are they effective? Why or why not?

3. Choose an organization that you interact with regularly and list as many different "systems" (whether computer-based or not) as you can that are used to process transactions, provide information to managers and executives, help managers and executives make decisions, aid groups to make decisions together, capture knowledge and provide expertise, help design products and/or facilities, and assist people in communicating with each other. Draw a diagram that shows how each of these systems interacts (or should interact) with each other. Are these systems well integrated?

4. Imagine an information system built without using a systems analysis and design methodology and without any thinking about the SDLC. Use your imagination and describe any and all problems that might occur, even if they seem a bit extreme and absurd. Surprisingly, the problems you will describe have probably already happened in one setting or another.

5. Choose a relatively small organization that is just beginning to use information systems. What types of systems are being used? for what purposes? To what extent are these systems integrated with each other? with systems outside the organization? How are these systems developed and controlled? Who is involved in systems development, use, and control?

6. You may want to keep a personal journal of ideas and observations about systems analysis and design while you are studying this book. Use this journal to record comments you hear, summaries of news stories or professional articles you read, original ideas or hypotheses you create, and questions that require further analysis. Keep your eyes and ears open for anything related to systems analysis and design.

Your instructor may ask you to turn in a copy of your journal from time to time in order to provide feedback and reactions. The journal is an unstructured set of personal notes that will supplement your class notes and can stimulate you to think beyond the topics covered within the time limitations of most courses.

REFERENCES

Aktas, A. Z. 1987. *Structured Analysis and Design of Information Systems.* Englewood Cliffs, NJ: Prentice-Hall.

Alavi, M., and I. R. Weiss. 1985. "Managing the Risks Associated with End-User Computing." *Journal of MIS* 2 (Winter): 5–20.

Bohm, C., and I. Jacopini. 1966. "Flow Diagrams, Turing Machines, and Languages with Only Two Formation Rules." *Communications of the ACM* 9 (May): 366–71.

Bourne, K. C. 1994. "Putting Rigor Back in RAD." *Database Programming & Design* 7(8) (Aug.): 25–30.

Davis, G. B. 1982. "CAUTION: User-Developed Systems Can Be Dangerous To Your Organization." *End-User Computing: Concepts, Issues and Applications.* 1989. R. R. Nelson (ed.). New York: Wiley: 209–228.

DeMarco, T. 1979. *Structured Analysis and System Specification.* Englewood Cliffs, NJ: Prentice-Hall.

Gilbert. J. 1994. "The Best Jobs in America." *Money* 23 (March): 70–73.

McFadden, F. R., and J. A. Hoffer. 1994. *Modern Database Management.* 4th edition. Redwood City, CA: Benjamin/Cummings Publishing Company.

Mumford, E. 1981. "Participative Systems Design: A Structure and Method." *Systems, Objectives, Solutions* 1 (1): 5–19.

Naumann, J. D., and A. M. Jenkins. 1982. "Prototyping: The New Paradigm for Systems Development." *MIS Quarterly* 6 (3): 29–44.

Yourdon, E., and L. L. Constantine. 1979. *Structured Design.* Englewood Cliffs, NJ: Prentice-Hall.

A Systems Analysis and Design Project at Pine Valley Furniture

After studying this chapter, you should be able to:

■ Describe the life cycle of a specific systems analysis and design project, including the purpose and deliverables of each step.

■ List the important skills required of a systems analyst in conducting a systems analysis and design project.

■ Explain the purpose of adopting a standard methodology for systems development in an organization.

■ Discuss the roles of individuals who work on a systems project.

■ Explain the differences between the analysis, design, and implementation phases of a systems development project.

■ Describe the purpose of basic documentation produced during a systems development project.

INTRODUCTION

Chapter 1 introduced systems analysis and design, including an overview of the systems development life cycle. Chapter 2 illustrates highlights from a typical successful systems development project that follows this life cycle. Since we must simplify the discussion at this point in the text, our example project is somewhat idealistic. And keep in mind that not all projects go as well as the one described in this chapter. Throughout this book, however, we will raise various technical and managerial issues that can occur during a systems development project. This chapter includes various aspects of project management common in systems development. How the project leaders in our example case, Chris Martin and Juanita Lopez, run the project is their way, not the only way, so do not assume that all projects are managed in just this one way. Due to limited space, the project description is abbreviated (for example, only selected deliverables are illustrated) but you should develop a basic understanding of systems analysis and design from the discussion.

There are several reasons why we present this chapter at this point in the book. First, many students in a systems analysis and design course have not used or even seriously thought about an information system, nor have they participated in a systems development project. The goal of this book is to introduce you to the concepts and the skills you will use to analyze business information requirements and show you how to translate those concepts and skills into systems designs. The goal of this chapter is to briefly illustrate what you will be able to do after you complete a systems analysis and design course using this text. In addition, this chapter introduces the various skills, techniques, and concepts explained in detail in later chapters.

Second, most people learn best when they see specific, concrete examples. All chapters in this book contain numerous examples, illustrations, and documentation written during a systems development project, and each chapter concentrates on a specific aspect of systems development. We have structured this particular chapter to provide you with an overview of a systems project and to explain how the various aspects of systems analysis and design are related.

Finally, many instructors want you to begin the initial steps of a systems development group or individual course project early in the term. Due to the logical progression of topics in this book (project identification, project definition, analysis, design, implementation, and maintenance), you will probably study many pages before you decide on your target for the project. This overview chapter gives you an idea of where you are headed. By studying our example case, you can begin using some basic skills early on, before a more rigorous treatment of topics is presented. For example, this chapter provides a benchmark for your writing of a project proposal or plan, which might be the first step in your course project. Obviously, since this is only Chapter 2 of the book, the examples we use are much simpler than the systems development project in your course or in a real organization.

One note of caution: Do not be concerned with the details of techniques and notations while you study this chapter. You will read about these in depth later in the book. At this point, try to understand the reason for the technique or diagram, and do not be concerned about whether you could use the technique or produce such a diagram.

This chapter illustrates the development process for a modern information system by an in-house development unit. The setting for our illustration is Pine Valley Furniture Company, which was briefly introduced in Chapter 1.

CASE BACKGROUND

Juanita Lopez, head of the manufacturing support unit of the Purchasing Department at Pine Valley Furniture Company (PVF), is concerned about the growing backlog and quality of work in her department as PVF expands. The Purchasing Department manages all interactions between external suppliers and internal PVF departments. These interactions include the following simultaneous processes:

- Identifying and selecting approved vendors for different materials and supplies

- Negotiating prices, order quantities, and other terms with suppliers

- Placing and tracking orders

- Monitoring the quality of materials received

- Working with vendors to identify defects and improve quality to meet PVF standards

- Working with PVF departments in helping them choose a vendor when materials are required

- Working with PVF departments to determine acceptable inventory levels and order lead times to ensure uninterrupted operations

Juanita is concerned about delays in placing vendor orders, inefficiencies in determining the status of vendor orders, and late management summary reports. Furthermore, relationships with vendors are becoming more critical as just-in-time inventory control methods are implemented by PVF manufacturing operations. As Juanita has been charged with solving these issues without adding office staff, she is hoping that new automation can help. A systems analyst for each business unit (like Purchasing) works as a liaison between that unit and the IS development group. Juanita calls Chris Martin, the systems analyst assigned to Purchasing, to discuss her concerns.

Chris has been with PVF for six years. He joined PVF right after graduating from the baccalaureate program in the Computer Information Systems program in the business school of Valley State University. His first job was as a Programmer/ Analyst I, in which he coded and maintained financial application systems written in COBOL.

As Chris became more experienced in his job, he was asked to take on additional program design work and was soon promoted to Programmer/Analyst II. He assumed some supervisory responsibilities on projects and gained more authority in structuring programs, in choosing methods to access databases, and even in recommending functional changes to systems in order to take advantage of technology.

Eventually, the growth in systems work at PVF necessitated the expansion of the information systems development unit, and Chris was promoted to a Junior Systems Analyst position in the group supporting manufacturing operations. In this assignment, Chris participated on a team that developed a five-year plan that would enable information systems to support manufacturing at PVF. This plan considered all aspects of manufacturing and its interactions with other parts of the business, such as marketing, accounting, and purchasing. Chris also worked on one of the initial projects generated from this plan which dealt with product structure (or bill-of-materials) data. As part of this project, Chris was responsible for designing files and databases to maintain this data. Since some product data (primarily physical characteristics) depended on vendor specifications, he had to analyze relevant Purchasing computer files to obtain data descriptions. As a result, at the time of Juanita's call, Chris was already aware of some of the pressures on Purchasing caused by the growth and change in manufacturing operations and had a basic understanding of Purchasing systems at PVF.

IDENTIFYING THE PROJECT

Juanita's career has also progressed rapidly. Her work has been rated very highly and she is well respected. Juanita has never worked on a systems development project, however, and her education at a small liberal arts college did not include any exposure to information systems. Although Juanita knows the operation of her department very well and has a fairly clear idea of some of its problems as well as opportunities for improvement, she does not know how to translate these into precise requests for help from systems development. The IS development group is highly regarded and Juanita believes her open management style and Chris' expertise can complement each other in working together.

At their first meeting, Chris explains that his role at this point is to help Juanita state what her business problem is and to suggest how support from any new or

> **Project Identification and Selection**
>
> *Purpose:* Develop a preliminary understanding of the business situation that has caused the request for a new or an enhanced information system.
> *Deliverable:* A formal request to conduct a project to design and develop an information systems solution to the business problems or opportunities.

improved information systems could solve her problem. In fact, one of their early joint responsibilities is to develop a brief justification for their proposal. Chris points out that, due to the huge demand for new and improved systems at PVF, not all requests can be accommodated. Further, PVF has instituted a charge-out, or transfer pricing, scheme whereby business units requiring systems development have to pay for the work done by the systems development unit. Thus, Juanita has to receive approval from her boss to fund any work performed by Chris' unit.

Chris outlines the process he and Juanita will have to follow. He explains that several years ago PVF had instituted a structured **systems development methodology.** He gives Juanita a copy of a chart summarizing the various steps that have to

Systems development methodology: A standard process followed in an organization to conduct all the steps necessary to analyze, design, implement, and maintain information systems.

Figure 2-1
PVF Systems Development
Methodology (SDM)—
Overview

1. Product Identification and Selection
   - 1.1 System Service Request evaluation
     - 1.1.1 Develop SSR
     - 1.1.2 Analyze SSR by Systems Priority Board
     - 1.1.3 Assign SSR disposition
     - 1.1.4 Allocate resources for Project Initiation and Planning phase
   - 1.2 Establish project management procedures and modifications to SDM

2. Project Initiation and Planning phase
   - 2.1 Develop project plan
     - 2.1.1 Develop overview of user requirements
     - 2.1.2 Analyze stakeholders and risks
     - 2.1.3 Construct work plan
     - 2.1.4 Describe resource requirements for Analysis phase
   - 2.2 Develop business case
   - 2.3 Review Project Initiation and Planning phase*

3. Analysis phase
   - 3.1 Determine requirements
   - 3.2 Structure requirements
   - 3.3 Select system direction
   - 3.4 Refine business case
   - 3.5 Update project plan
   - 3.6 Review Analysis phase*
     - 3.6.1 Review work done to date
     - 3.6.2 Review updated business case
     - 3.6.3 Allocate resources for Logical Design phase

4. Logical Design phase
   - 4.1 Document information flows
   - 4.2 Document data requirements
   - 4.3 Document processing logic
   - 4.4 Document user interfaces
   - 4.5 Refine business case
   - 4.6 Update project plan
   - 4.7 Review Logical Design phase*
     - 4.7.1 Review work done to date

be followed for projects managed by the IS unit (see Figure 2-1). Using this methodology means that all project requests, proposals, and development efforts follow a common and well-defined process. This methodology also specifies how a systems project will be managed and how progress will be measured, the roles of different people involved on the project team, the deliverables due at each step, the tools and techniques used during each step, and the form and content of a project dictionary or repository to document the project. This common development methodology—the PVF Systems Development Methodology (SDM)—allows for

- Consistent management of projects
- Easier estimation of project times and costs

5. Physical Design phase
   5.1 Select technology
   5.2 Design database and files
   5.3 Design programs
   5.4 Develop test plan
   5.5 Plan installation
   5.6 Refine business case
   5.7 Update project plan
   5.8 Review Physical Design phase*
       5.8.1 Review work done to date
       5.8.2 Allocate resources for Implementation phase

6. Implementation phase
   6.1 Programming
   6.2 Testing
   6.3 Education
   6.4 Documentation
   6.5 Installation
   6.6 User acceptance testing
   6.7 Final installation
   6.8 Project termination
       6.8.1 Evaluate personnel
       6.8.2 Review project with management
       6.8.3 Release personnel

7. Maintenance
   7.1 Annual system review
   7.2 System audits
   7.3 Submit System Service Requests for system changes

* These steps may be eliminated by decision of Systems Priority Board.

(revision date: October 1, 1995)

- Higher quality of work since people have been trained in the specifics of the methodology

- A more understandable process for both Chris and Juanita

- Use of automated tools like a CASE product that would yield an easier-to-manage and higher-quality project

The first step is to develop a System Service Request (SSR). An SSR briefly states the business problem or opportunity and presents general ideas on how an information system could deal with the problem or opportunity.

PVF has a standard SSR form, which Juanita and Chris need to complete together. Juanita and Chris will then submit this form to the Systems Priority Board, a diverse group of business and systems managers representing a cross section of PVF. This Board reviews all requests and determines which are worth pursuing. In some cases, the Board suggests that the end users making the request attempt to develop their own systems using user-friendly tools such as an electronic spreadsheet package or simple database management system. The Board also rejects some requests. In most cases, however, the Board assigns a priority to the request and places it in a backlog of such requests for further study. A systems analyst is assigned to these requests in decreasing order of priority. Although this is a very formal process, the systems development group tries to make it as painless as possible for the requesting end user through the help of someone from the liaison staff, like Chris.

Over a two-week period Chris and Juanita meet and talk on the telephone several times as they put together the SSR. The SDM calls for only a cursory analysis of Juanita's situation for the SSR. The idea is to lay out the scope and basic issues involved, not to thoroughly evaluate the situation. Later steps in the methodology refine the analysis. Chris explains to Juanita that one of the basic premises of the methodology is **incremental commitment.** Using incremental commitment, there are many review points during a project, each successively dealing with the information system request in greater depth. The goal is to not sink significant resources into a project until the work is reviewed and the organization is willing to commit further resources to the project. In this way, incremental commitment allows projects to be easily redirected or killed.

Figure 2-2 shows the System Service Request Chris and Juanita develop. As you can see, the request briefly describes the situation in Purchasing. Juanita shows this SSR to her boss, Sal Divario, who signs the form. Sal's signature does not commit him to funding the project, only to supporting the request being sent to the Board. Chris then submits this form to the Systems Priority Board for their decision.

Incremental commitment: A strategy in systems analysis and design in which the project is reviewed after each phase and continuation of the project is rejustified in each of these reviews.

Project Initiation & Planning

Purpose: State business situation and how information systems might help solve a problem or make an opportunity possible.
Deliverable: A written request to study the possible changes to an existing system or the development of a new system.

PROJECT INITIATION AND PLANNING PHASE

The Systems Priority Board finds merit in Juanita's request and decides to assign it the highest priority. One of the advantages of this project is that it fits nicely into the five-year plan for manufacturing systems that had been developed earlier and adopted by the Board. Although meeting needs in the purchasing function had not been identified as an early part of the manufacturing plan, the way in which Chris had helped Juanita write the SSR mentioned some of the same issues that had driven the development of that plan. The Board asks

Figure 2-2
System Service Request for Purchasing Fulfillment System

Pine Valley Furniture
System Service Request

REQUESTED BY _____Juanita Lopez_____   DATE _____November 1, 1994_____

DEPARTMENT _____Purchasing, Manufacturing Support_____

LOCATION _____Headquarters, 1-322_____

CONTACT _____Tel: 4-3267   FAX: 4-3270   e-mail: jlopez_____

TYPE OF REQUEST   URGENCY

[X]   New System   [   ]   Immediate – Operations are impaired or
                                                       opportunity lost
[   ]   System Enhancement   [   ]   Problems exist, but can be worked around
[   ]   System Error Correction   [X]   Business losses can be tolerated until new
                                                       system installed

PROBLEM STATEMENT

Sales growth at PVF has caused greater volume of work for the manufacturing support unit within
Purchasing. Further, more concentration on customer service has reduced manufacturing lead times,
which puts more pressure on purchasing activities. In addition, cost-cutting measures force Purchasing
to be more agressive in negotiating terms with vendors, improving delivery times, and lowering our
investments in inventory. The current modest systems support for manufacturing purchasing is not
responsive to these new business conditions. Data are not available, information cannot be summarized,
supplier orders cannot be adequately tracked, and commodity buying is not well supported. PVF is
spending too much on raw materials and not being responsive to manufacturing needs.

SERVICE REQUEST

I request a thorough analysis of our current operations with the intent to design and build a completely
new information system. This system should handle all purchasing transactions, support display and
reporting of critical purchasing data, and assist purchasing agents in commodity buying.

IS LIAISON _____Chris Martin (Tel: 4-6204   FAX: 4-6200   e-mail: cmartin)_____

SPONSOR _____Sal Divario, Director, Purchasing_____

------------------------ TO BE COMPLETED BY SYSTEMS PRIORITY BOARD ------------------------

[   ]   Request approved   Assigned to _____
                                           Start date _____
[   ]   Recommend revision
[   ]   Suggest user development
[   ]   Reject for reason _____
                           _____
                           _____

that a lead systems analyst be assigned as soon as possible to continue the study of this request.

The Board believes that those closest to the situation outlined in the SSR are well aware of the issues surrounding the request and decides to reduce its direct involvement in the rest of this systems development project. The Board usually requires that the end user and lead systems analyst submitting the request report regularly so that they can assess progress and make decisions on further deployment of resources, ensure that the project fits into systems strategy, and decide whether or not the project should continue. In this case, however, the Board asks for a chance to review this project just one more time, when the complete detailed design for the system is set. In the meantime, the Board suggests that a project steering committee be formed to monitor project progress and to give the project team guidance. This committee consists of the head of the Purchasing Department, Sal Divario; the manager of Manufacturing Systems in the IS development unit, Stacy Morton; and the manager of Database Administration, Greg Sinise. The director of IS Development, a member of the Systems Priority Board, reviews his staffing, determines that Chris is scheduled to complete work on another project soon, and assigns Chris to continue to work on this project as the lead analyst and project manager.

The next major goal for Chris and Juanita is to create a more specific statement of the requirements Juanita and her staff have for information services to deal with the problems in Purchasing. The SDM outlines the general nature of steps to follow during this project initiation and planning phase (see Figure 2-1). Since different projects require different approaches, PVF has chosen to have its staff use the methodology as a guideline rather than a fixed set of steps. Although the details of the SDM documentation outline alternative approaches suitable for different situations (for example, small or large systems, short or longer lead time projects), the designers of the SDM knew they could not anticipate all possible project nuances. Thus, as part of conducting an initial study of the situation in the Purchasing Department, Chris and Juanita must outline a complete project plan of their own to present to the project steering committee.

The Project Plan

Chris takes leadership of the project at this point. To gather information, he conducts several short interviews with purchasing agents and other staff members who work for Juanita, reads recent correspondence about issues in the department, and spends several hours observing a few Purchasing Department staff members as they do their jobs. Chris uses this information to develop a project plan with several key elements:

- A definition of *scope* for the project stating which general functions within the Purchasing Department will be analyzed and which activities outside of Purchasing will be considered.

- A more complete *problem statement* elaborating on the earlier SSR contents.

- An initial *requirements statement* specifying in general terms which types of information and information processing are needed, how urgent the situation is, and what constraints seem to be in place on a systems solution (for example, many of PVF's suppliers are not computerized, so electronic linkages with suppliers are not possible at this time).

- A *request for resources* of people, time, and money to develop the information requirements and system functional specifications. Resources were estimated from looking at similar past projects. The costs for later steps will be presented in a subsequent project status report at the end of the analysis phase, after the nature of the project is better known.

- A *time line* indicating when the project team will perform various steps of the project. This time line is presented in two forms: Gantt charts that show the start date, end date, and duration as well as personnel needed for each project step in each phase; and a PERT (Project Evaluation and Review Technique) chart that shows the precedence relationships between project steps. The time line was also constructed from looking at similar past projects, the SDM, and the above items.

- A *business case* or justification for continuing with the project.

Juanita reviews and edits each of these key elements. She has primary responsibility for preparing the initial **business case,** which has to be formulated to solve a business rather than a technology problem. The business case is based on identifying and, where possible, quantifying the possible benefits and costs of solving the problem. The business case justifies the system project as economically, technically, legally, and politically sound. The time line shows that the project schedule is feasible since each step is reasonable and will result in the project being completed in time to deal with the business problem. That is, the system makes financial sense and can be built and implemented within the organization. Since the actual requirements are not yet known and a precise system design is constructed in a later step, the business case at this point primarily identifies broad categories of costs and benefits such as lower inventory levels, improved service to customers, improved morale in Purchasing, and more favorable pricing from suppliers. The business case is readdressed and refined in each phase of the project.

Their project plan is a ten-page document plus figures. Figure 2-3 contains an excerpt from this plan: a sample of the Gantt and PERT charts specifying the time line of events for the next phase of the project—the analysis phase. The full document shows similar charts for all the remaining phases from analysis through implementation. The total estimated project length is 31 weeks, with the analysis phase taking about the first 10 weeks. The Gantt chart in Figure 2-3a shows that the analysis phase has ten steps and two milestones. Each activity is critical: Some cannot be delayed without the whole project being delayed. Steps 4 and 5 occur during the same period but, as we will see in the PERT chart, they are not independent of each other. Additional Gantt charts not shown here match the requested human resources to each project activity so that everyone can see when each project team member is required. IS and other managers use the human resource Gantt charts to assign their staff to multiple projects and duties throughout the life of the Purchasing Fulfillment System project.

Figure 2-3b is an overview PERT chart for the analysis phase. As we can see from this figure, only steps 7 and 8 occur independently of each other, but neither can start until step 6 is completed, and step 9 cannot begin until both steps 7 and 8 are completed. This chart, along with more detailed PERT charts not included here, shows a clear understanding of the nature of the planned project and gives useful information to anyone trying to understand the project.

Chris and Juanita used several key principles in developing the project plan and schedule: use of frequent review points, scheduling parallel steps where feasible, and thoroughness. First, there are definite review points scheduled (shown by

Business case: The justification for an information system, presented in terms of the tangible and intangible economic benefits and costs, and the technical and organizational feasibility of the proposed system.

Figure 2-3

Project plan excerpt.   (a) Gantt chart

Analysis Phase of Purchasing Fulfillment System Project

ID	Name	December					January					
		11/26	12/3	12/10	12/17	12/24	12/31	1/7	1/14	1/21	1/28	2/4
1	Start Analysis Phase	◆										
2	Overview of Purchasing	▨										
3	Develop Physical DFDs			▨▨▨▨								
4	Develop Logical DFDs					▨▨▨						
5	Develop E-R Diagrams					▨▨▨						
6	Integration & Problem Identification						▨					
7	Interviews							▨▨				
8	JAD Session							▨				
9	Develop Alternative Directions								▨▨			
10	Selection & Justification									▨▨		
11	Prepare for Review Meeting										▨▨	
12	Analysis Phase Review Meeting											◆

Project: Purchasing Fulfillment
Date: 11/15/95 8:00am
Analyst: Chris Martin

Critical ▨▨▨▨	**Progress** ▬▬▬	**Summary** ▼▬▬▼
Noncritical ▬▬▬	**Milestone** ◆	**Rolled Up** ◇

Stakeholder: A person who
has an interest in an existing
or new information system. A
stakeholder is someone who
is involved in the development of a system, in the use
of a system, or someone who
has authority over the parts of
the organization affected by
the system.

the diamonds on the Gantt chart). Each step and phase of the project produces an identifiable deliverable that is reviewed with the project steering committee, the Systems Priority Board, a set of end users, or others interested in the project. Early in the study, Chris identified many **stakeholders** whose needs must be met for the system to be fully implemented. Some of the stakeholders are on either the steering committee or the Priority Board while some have other roles in and around the project. The stakeholders include the following people:

- The *client*, Juanita, who has hired Chris to work on the project and who will play a central role in helping manage the project

- The *funder*, Sal Divario, who as head of Purchasing will allocate funds to pay for the study, systems development, and operating expenses

- The *end users* (including Juanita) who work in Purchasing who will enter data, read reports, and make decisions based on information; these people will be invaluable in determining the information system requirements

- The *technicians* (for example, programmers, systems analysts, database analysts, and data center staff) who will, at different points, work on the project to design, build, and operate the system

- The *end customers* (the manufacturing staff and external suppliers) who will benefit from the new information system and whose interaction with Purchasing may change as a result

Chris knows from past experience that it is essential to involve these stakeholders in the process, to work closely with them in setting realistic expectations about the system changes, to obtain their insights into system requirements and design, and

(b) Overview PERT chart

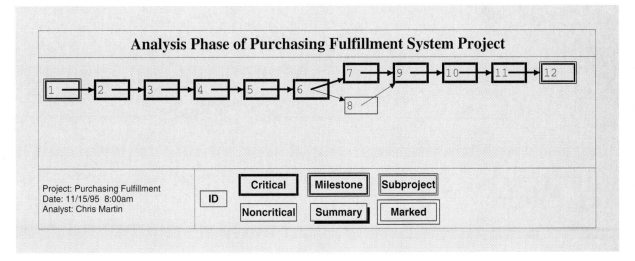

Analysis Phase of Purchasing Fulfillment System Project

Project: Purchasing Fulfillment
Date: 11/15/95 8:00am
Analyst: Chris Martin

ID

| Critical | Milestone | Subproject |
| Noncritical | Summary | Marked |

to gain their commitment to the new system. Input from the stakeholders also helps Chris identify the risks of the project, which might involve potential interfacing problems with other systems, resistance from certain stakeholders, or management policy issues that constrain what a new system might do. Both Chris and Juanita will spend most of their time communicating in writing and orally with these stakeholders at official review points and also informally as needed. The review points serve as a mechanism to check that all stakeholders are reasonably in agreement on the direction of the project.

A second key aspect of the project plan depicted in Figure 2-3 is that the process includes parallel steps, where possible, to decrease overall project time. The project is broken down into identifiable and manageable tasks that can be assigned to different people working in parallel. Although this approach places an extra burden on Chris and Juanita, who must coordinate the activities of these people, compression of the project schedule is necessary so that the problems in Purchasing can be addressed expeditiously.

The final key aspect of the project plan is the thoroughness with which Chris and Juanita have thought out the project. The SDM was a big help in developing the plan because it outlines the standard steps to follow under different circumstances (for example, different system sizes, different types of systems, and constraints on development time) and indicates which techniques and tools to use along the way. Chris knows, however, that even with the slack time built in, he and Juanita cannot anticipate all contingencies. Business needs might change, people involved in the project may not be available when needed, or financial resources might have to be diverted to solve a crisis with another system or elsewhere in Purchasing. Chris and Juanita realize they have to be firm but flexible in managing this schedule.

Chris and Juanita submit their plan and make a brief presentation to the project steering committee. The committee endorses the plan, and Sal Divario authorizes expenditures from his budget to cover the costs of systems analysts and other IS staff to conduct the full analysis phase. Sal also writes a memo to the Purchasing Department staff explaining the project and requesting their full cooperation. Such visible commitment from senior management is often important in gaining the attention of all the personnel needed to conduct a successful project.

Analysis

Purpose: Analyze the business situation thoroughly to determine requirements for a new or an enhanced information system, structure those requirements for clarity and consistency, and select among competing system features those that best meet user requirements within development constraints.
Deliverable: The functional specifications for a system that meets user requirements and is feasible to develop and implement.

ANALYSIS PHASE

Chris and Juanita begin the analysis phase by learning about the Purchasing Department functions that will be addressed in this project. Their goal is to outline the operations at a macro level so that they can structure more detailed work that will be done by a small team of systems analysts assigned to work with them for two weeks. The project plan calls for a top-down approach to systems analysis, typical of the structured philosophy of the SDM. Chris and Juanita develop two charts, a *context diagram* (CD) and a level-0 *data flow diagram* (DFD), to structure the analysts' work (Figure 2-4). The diagrams are built using the CASE tool used by PVF. This CASE tool contains the following features:

- A DFD drawing module

- An entity-relationship (E-R) diagram drawing module (illustrated later in this chapter)

- A data repository that will contain full descriptions of all the elements of a system

- Several notations for describing computer programs

- Prototyping and other code-generation tools for rapidly building working versions of an information system

All documentation of the current and new systems will be created using the CASE tool. By having the analysts use the CASE tool to document their work, all project team members will have access to accurate and consistent documentation on various parts of the system. Thus, combining and comparing their work is simplified.

Context diagram: An overview of an organizational system that shows the system boundary, external entities that interact with the system, and the major information flows between the entities and the system.

The **context diagram** of Figure 2-4a shows the scope of the project. The organizational system the analysts will study is called the Purchasing Fulfillment System, identified by the rectangle with rounded corners in the center of this figure. This open system interacts with three entities external to the Purchasing Fulfillment System: Suppliers—who are outside PVF—and Production Schedulers and Engineering—which are other PVF units. The context diagram shows that the part of Pine Valley Furniture that will be analyzed and redesigned is contained within this one rectangle. As long as the interactions of the Purchasing Fulfillment System with its external entities are preserved and met, the system itself can be redesigned as needed. Since the Purchasing Fulfillment System is an information system, these interactions are data flows between the system and the external entities.

Data flow diagram: A picture of the movement of data between external entities and the processes and data stores within a system.

Figure 2-4a is still at too gross a level to structure the work for the systems analysts. The **data flow diagram** of Figure 2-4b shows the inter-relationships among the major components of the current operations in the Purchasing Department that include the functions of the Purchasing Fulfillment System. Again, major functions called *processes* (for example, process 1.0—Forecast Material Needs) are shown by rectangles with rounded corners; data flows (for example, Material Forecasts) are shown by directional lines. Even at this level of description, data retained by the system, called *data stores*, are not shown (although they could be shown on such diagrams). These will be enumerated in subsequent project steps. Chris and Juanita plan to organize the project so that each systems analyst studies the portion of the Purchasing Department covered by different sections of the diagram in Figure 2-4b.

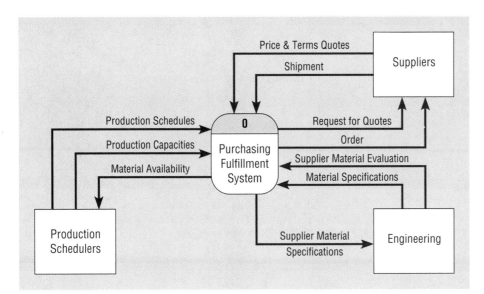

Figure 2-4
Top-down view of Purchasing
Fulfillment System

(a) Context diagram

(b) Data flow diagram

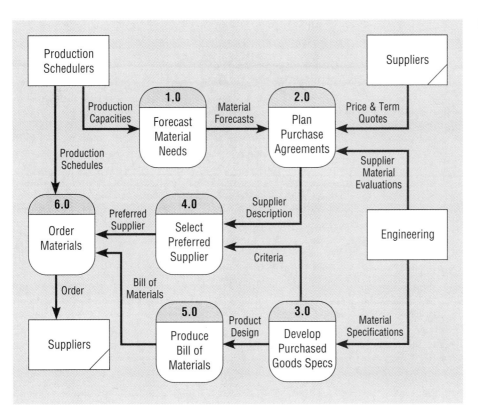

Chris and Juanita meet with the assigned analysts before they begin their work. The team reviews the project and the process they will follow during the analysis phase. Chris portrays the steps as resembling an hour glass in which the analysis will start off very broadly and refine iteratively into more detail until the analysts develop a precise picture of the current purchasing operations. Then they will build the analysis back up to a comprehensive conclusion culminating with the design for a new system. This is consistent with the top-down and bottom-up techniques employed in structured systems development. The analysis steps they follow are listed below:

1. Each systems analyst develops more and more detailed DFDs to show how the functions, data flows, and data stores in his or her portion of the diagram are handled by the current procedures; these additional DFDs describe the current physical system, including the people, business forms, and technology currently used with each process, data flow, and data store. The diagrams identify specific reports, forms, transactions, and other media on which information are stored and transmitted. The systems analysts use interviews, analysis of business documents, and observation to develop these diagrams. Chris emphasizes the importance of not being influenced by how the current system operates. Rather, the purpose of these DFDs is to inventory what the current data processing requirements are so that the team can consider which must be handled in the new system. The data model in the next step will also help the team to not simply duplicate current procedures.

2. Each analyst then rewrites these diagrams to exclude all references to the physical form in which information is processed and develops a logical, or functional, specification of the current system. During this step, each analyst prepares diagrams that explain the natural structure of data independently of how it is processed. This *data model* is described in a notation called an entity-relationship (E-R) diagram which we will illustrate later in this section.

3. Chris then brings the team of analysts together to discuss their independent work and to review the process of combining their logical diagrams (both data flow and entity-relationship) into a consistent and comprehensive description of the current system operations. They then outline the precise problems they discovered with the current system and suggest possible restructuring.

4. Next, the team conducts further data collection by interviewing end users and other stakeholders about problem areas within the current operations and about desired information requirements. They also hold a Joint Application Design (JAD) session in which users, analysts, and systems developers come together to discuss the problems with the current system and ideas for change.

5. Then the team of analysts along with Chris and Juanita restructure the current logical system process structure and data models and identify the major features of two or more alternative new systems. Each alternative takes a different approach to dealing with Purchasing issues: for example, Alternative 1 might assume a preferred supplier for each material item whereas Alternative 2 might assume a supplier is selected from a set of possible vendors for each material item. These alternative structures are described at a general level, sufficient to show major differences but stopping short of a detailed design. This will be done in the next phase of the project—Logical Design. For each alternative, the team identifies its advantages and disadvantages as well as its organizational impacts on Purchasing and other departments.

6. Finally, the team and Chris and Juanita develop a detailed justification to recommend one of the alternatives to the steering committee.

The analysis phase review meeting with the steering committee is well planned. As this review point was in the original project plan, the meeting had already been scheduled with each committee member. During the analysis phase, Chris and Juanita had made it a point to speak with each committee member approximately once a week to keep them abreast of project progress and to give them a chance to provide any insights on issues or questions that had arisen. Due to this two-way communication, the committee already feels some ownership in the team's efforts. And as Chris had sent a copy of the agenda for the meeting to each committee member one week prior to the meeting, everyone knew what to anticipate. The agenda for the analysis phase project status review meeting appears in Figure 2-5.

The purpose of the meeting is to

- Overview what has been done so far

- Highlight findings on problems and opportunities with the current system

- Summarize the information services needed to improve on the current system

- Gain approval for one alternative direction in which to design a new system

- Obtain commitment to the resources needed to continue the project

Two alternative system directions are described at a fairly general level. The objective is to give the committee enough information to make a decision without providing more detail than needed. Also, the team does not want to spend a lot of time developing system details at this point for an alternative that is not acceptable.

The data flow diagrams that the analysts developed help the group focus on areas of concern. Because the diagrams were developed in a top-down matter, following a process called **functional decomposition,** the team has a large number of charts, each concentrating on one area of Purchasing operations. Thus, when a question arises, the team shows one chart with an overview of the relevant part of Purchasing operations and then focuses in on any area that requires more elaboration with a detailed chart.

Functional decomposition: An iterative process of breaking the description of a system down into finer and finer detail which creates a set of charts in which one process on a given chart is explained in greater detail on another chart.

Figure 2-5
Analysis phase review meeting agenda

**Purchasing Fulfillment System
Analysis Phase Project Review**

A. Summary of Project Objectives

B. Progress to Date
1. Steps Performed
2. Project Schedule

C. Findings
1. Problems and Opportunities Identified
2. Information Requirements

D. Alternatives
1. Comparison of Alternatives
2. Details on Recommended Alternative
3. Justification for Recommendation
4. Resource Requirements

E. Plan for Design Phase

F. Questions

G. Conclusions & Action Items

The analysis phase review meeting lasts two hours. The first hour is devoted to presentations by various team members followed by a question and answer period. During the second hour, participants deliberate on the team's recommendation. At the beginning of the meeting, each committee member was given a written project status report including an executive overview page plus a copy of all the presentation overheads. Figure 2-6 shows three key overheads: Figure 2-6a shows an entity-relationship (E-R) diagram, a *conceptual data model,* for part of the data deemed essential in a new Purchasing Fulfillment System while Figures 2-6b and 2-6c contain the justification, or business case, for the recommended alternative new system.

The E-R diagram at this point in the project shows the categories of data needed by the proposed system (for example, supplier, item, product, and shipment data entities are needed). The lines connecting data entities, or rectangles, represent relationships, or direct linkages, between data. For example, a SUPPLIER Supplies one or more ITEMs (to keep the diagram less cluttered, we are leaving off the name of the relationship in the opposite direction, such as an ITEM Supplied by from one to four SUPPLIERs). For simplicity, the chart in Figure 2-6a covers only part of the conceptual data model for the system. The project team will refine this diagram during the logical design phase in the SDM by adding all the individual data items needed to produce computer displays and reports. The E-R diagram shows important rules about the Purchasing function. For example, the marks near the SUPPLIER box on the line (relationship) labeled Supplies indicate that there are at least one and at most four recognized suppliers of an individual purchased item. Also, in PVF terms, a SHIPMENT involves exactly one material item (this is stated by the two marks near ITEM on the line [relationship] labeled Receives). Since the relationship between ITEM and PRODUCT represents which items go into making a product, we call this the Bill of Materials relationship.

Figure 2-6

Analysis phase review meeting excerpts

(a) Entity-relationship (E-R) diagram

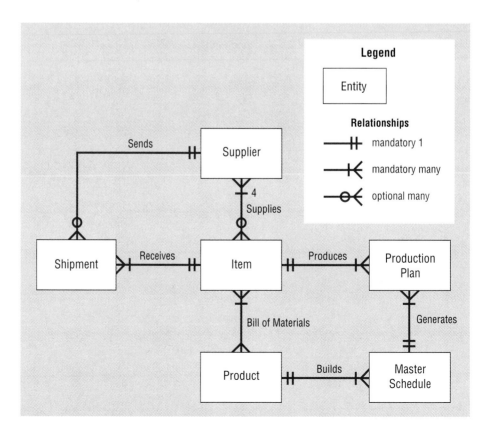

The justification proposes the financial basis for the new system. These benefits and costs are now more specific than in the initial stage of the project since further analysis has been carried out and specific system capabilities are better known. Figure 2-6b lists the identified benefits and costs for the recommended system. Tangible (quantifiable) benefits and costs are broken down into one-time (in year 0) and recurring (annual in each of the following five years) categories. Intangible (non-quantifiable) benefits and costs are also listed for discussion, but these are not part of the financial analysis of the project's net benefits.

Figure 2-6b also includes a list of risks the project team identified for this project and system. An analysis and consideration of risks is important at this and at every review point. Risks show important elements to be managed and also alert stakeholders to issues before they become critical. For example, one risk of the proposed system is that, unless properly handled, suppliers could react negatively to the features of the new system that will cause more careful evaluation of supplier performance. Finally, Figure 2-6c displays the financial analysis Juanita and Chris present to the steering committee.

Also part of the presentation is a request for additional resources in order to continue the project. This request involves the assignment of a database design expert, a LAN specialist, two of the more experienced purchasing agents, and a

(b) System benefits, costs, and risks

TANGIBLE ONE-TIME BENEFITS

Write-off of obsolete inventory:	$ 40,000
Reduction in number of suppliers:	14,000
	$ 54,000

TANGIBLE RECURRING ANNUAL BENEFITS

Lower inventory carrying costs:	$ 23,000
Net materials cost savings:	37,500
Less manufacturing rework:	13,000
Less manufacturing down-time:	25,000
Absorb growth with no additional staff:	32,000
	$130,500

TANGIBLE ONE-TIME COSTS

System development:	$138,000
Equipment:	45,000
Training:	5,000
Conversion and installation:	23,000
	$211,000

TANGIBLE RECURRING ANNUAL COSTS

Data center charges:	$ 39,500

INTANGIBLES
- Foundation for electronic linkage with suppliers in future
- Improved purchasing and manufacturing staff morale
- Improved management reporting and decision making

RISKS
- Possible negative supplier reaction to system changes
- Poor quality data in current systems may necessitate a costly data cleanup project
- Potential delays or problems from possible first use of the Sybase client/server database engine by PVF

Figure 2-6 *(continued)*
Analysis phase review meeting excerpts

(c) Financial justification

	A	B	C	D	E	F	G	H
1	Pine Valley Furniture							
2	Economic Feasibility Analysis							
3	Purchasing Fulfillment System Project							
4								
5				Year of Project				
6		Year 0	Year 1	Year 2	Year 3	Year 4	Year 5	TOTALS
7	Net economic benefit	$54,000	$130,500	$130,500	$130,500	$130,500	$130,500	
8	Discount rate (12%)	1.0000	0.8929	0.7972	0.7118	0.6355	0.5674	
9	PV of benefits	$54,000	$116,518	$104,034	$92,887	$82,935	$74,049	
10								
11	NPV of all BENEFITS	$54,000	$170,518	$274,552	$367,439	$450,374	$524,423	$524,423
12								
13	One-time COSTS	($211,000)						
14								
15	Recurring Costs	$0	($39,500)	($39,500)	($39,500)	($39,500)	($39,500)	
16	Discount rate (12%)	1.0000	0.8929	0.7972	0.7118	0.6355	0.5674	
17	PV of Recurring Costs	$0	($35,268)	($31,489)	($28,115)	($25,103)	($22,416)	
18								
19	NPV of all COSTS	($211,000)	($246,268)	($277,757)	($305,872)	($330,975)	($353,389)	($353,389)
20								
21								
22	Overall NPV							$171,035
23								
24								
25	Overall ROI - (Overall NPV / NPV of all COSTS)							.048
26								
27								
28	Break-even Analysis							
29	Yearly NPV Cash Flow	($157,000)	$81,250	$72,545	$64,772	$57,832	$51,636	
30	Overall NPV Cash Flow	($157,000)	($75,750)	($3,205)	$61,567	$119,399	$171,035	
31								
32	Project break-even occurs between years 2 and 3							
33	Use first year of positive cash flow to calculate break-even fraction - ((64,772 / 61,567) / 64,772) = .05							
34	Actual break-even occurred at 2.05 years (about 2 years and 1 month)							
35								
36	Note: All dollar values have been rounded to the nearest dollar							

programmer to the project team, along with the retention of the analysts throughout the next two phases—logical and physical design. Chris and Juanita explain that the programmer would be useful in the design phases to help facilitate the transition into the implementation phase. They also present an updated project budget, including an estimate of the total development and operating costs.

The estimated return on investment for the new system is large enough that the steering committee has no trouble approving continuation of the project, including support for the new project team members. Sal Divario, Juanita's manager, points out some new purchasing policies under consideration by upper management that will affect the proposed system. As well, Stacy Morton and Greg Sinise suggest some refinements in part of the E-R diagram to accommodate the results of a manufacturing systems plan just developed for one of the production lines.

Logical Design

Purpose: To elicit and structure all information requirements for the new system.
Deliverables: Detailed functional specifications of all data, forms, reports, computer displays, and processing rules for all aspects of the system.

LOGICAL DESIGN PHASE

Chris and Juanita have now managed the Purchasing Fulfillment System project over some significant hurdles. Whereas their roles so far have included not only project leadership but also direct involvement in many project activities, the addition of new team members changes Chris' and Juanita's roles. Now they can concentrate more on coordinating team members, setting work plans and methods, reconciling differences among work done by team members, guiding the project as issues and questions arise, and finalizing project documentation, especially communication with stakeholders. The detailed systems analysis and design work during logical design and subsequent phases will be done by the other team members.

The analysis phase produced a clear statement of the general features of the new Purchasing Fulfillment System, but the detailed steps were purposely left unspecified until steering committee approval of a specific direction for the system. In logical design, a detailed and highly structured set of functional specifications for the system is produced and agreed to by all affected parties (end users, management, outside people and units that have to interact with the new system). Further, these specifications must be unambiguous for the programmers and other IS personnel who will be responsible for physical system design and programming, testing, and installation. Once these requirements are accepted, the subsequent physical design phase develops a technology architecture for the system and programs, computer files, and hardware.

The SDM calls for the development of four deliverables during logical design:

1. Logical DFDs that show all aspects of the movement and processing of data in the new system and with its environment.

2. Precise designs for all business forms, CRT displays, and printed reports—the system inputs and outputs with user-system interfaces. In addition, dialogue scripts are developed that outline the sequence of displays and messages presented to users during their interaction with the system.

3. Specification of the logic necessary to transform inputs to a process into its outputs, including detailing decision rules and data validation standards.

4. Detailed E-R diagrams and other structured data definitions with all data items associated with each identified entity or relationship.

As before, all of these deliverables will be produced with the assistance of the CASE tool, and all descriptions will be stored in the CASE tool's repository.

Chris and Juanita discuss several possible ways to gather the additional information requirements to develop these materials. They decide that the team will use two primary approaches:

1. The project team will develop the detailed system description from information collected during the analysis phase and then conduct meetings with system users, called *structured walkthroughs,* to gain feedback on the logical system design.

2. The project team will use prototyping with a 4th-generation programming language and code-generation capabilities of the CASE tool to iteratively develop and refine the CRT screens and report layouts. The generated code will give them a head start on the implementation phase since much of this can be used in the actual system.

From his experience, Chris knows that logical design is the most crucial stage in systems development. If the project team does not get the requirements right based on the work done in the analysis and logical design phases, any subsequent costs to change the system's capabilities will grow exponentially the further the team gets into the project. On the other hand, Chris also knows from experience that there is no way they will get all the requirements defined, even using prototyping. Due to changing business conditions and the inherent complexity of any system, some elements will be missed entirely or will be inadequate. To deal with the inevitability of not getting all the requirements right in the early stages, Chris encourages the project team to design and build a system that can be easily changed. System maintainability can be accomplished by

- Thoroughly documenting the design and programs

- Using structured programming principles

- Using database technology that will allow for easy evolution of the database

- Creating modular, structured programs

From Chris' experience, it is not uncommon to find that 70 percent of a system's total life cost is spent on maintenance; in order to reduce the lifetime cost of the system, it is important to address system maintainability early in detailed design.

The logical specifications of the system are developed in stages. First, the known outputs and inputs of the system are studied. The outputs and inputs are identified from the DFDs as data flows from and to system users. Reports and displays are generated using CASE tool code generators and the report and computer form designer modules of PVF's database management system. These report and form designs are built with a sample database so that users can see sample output. Also, users can interact with the system by requesting on-demand reports and forms and can use prototypical versions of the buying model. Figure 2-7a shows a mock-up for one system output, the Price Quote report, as produced from the report generator of the database management system PVF uses for prototyping, Paradox for Windows. This report contains information about price quotes from all possible suppliers of selected items. Each item may have multiple vendors and each vendor may have several quotes for the same item, each for different order quantities. The vendor may have certain conditions or restrictions on each quote (for example, seasonal pricing). The report also notes overall experience comments for each vendor. Such report and form designs are reviewed with the relevant users and other interested parties. These parts of the system are iteratively built and rebuilt, using suggestions from users.

As the systems analysts finalize the report and form designs, the database analyst takes these designs and develops a logical data model *for each*. The notation used to describe the data in each system output and input is called the *relational database model*, a standard in the IS field. In this model, data is structured into relations that have desirable data maintenance properties. Figure 2-7b shows the three resulting relations (for now, you can think of a relation as a logical file) for the data in the report of Figure 2-7a. These three relations show all the data attributes as well as the entities and relationships of the required in the database to produce this report. The QUOTE relation links items with vendors. Since the CONDITIONS attribute can appear with each detail line in the report, this is a characteristic of a quote.

Relations developed from *each* Purchasing Fulfillment System input and output design are then combined into a comprehensive set of relations for the whole application. Finally, the conceptual entity-relationship diagram developed in the analysis phase is translated into relational form and compared to the logical data model relations. Differences between the logical and conceptual data models are reconciled and the relations are modified to include all of the anticipated data requirements from known system inputs and outputs as well as more general data about purchasing. It is this reconciled combined set of relations and accompanying entity-relationship data model that are referenced in subsequent system-building work on the Purchasing Fulfillment System. This combined data model is controlled

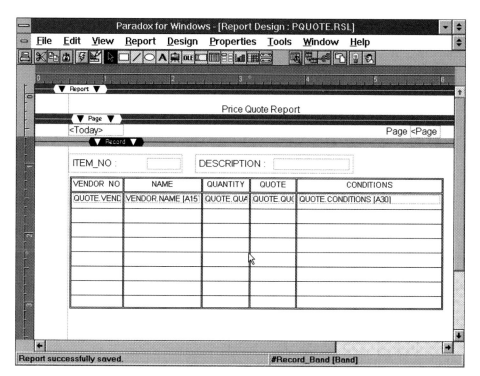

Figure 2-7
Pricing quote report

(a) Report layout

```
ITEM(ITEM_NO, DESCRIPTION)
VENDOR(VENDOR_NO, NAME)
QUOTE(ITEM_NO, VENDOR_NO, QUANTITY, QUOTE, CONDITIONS)
```

(b) Relations

by the data analyst, who also checks to make sure that this data model is consistent with other data models for Pine Valley databases.

As the team consisting of the database analyst, one systems analyst, and several users finalize the logical database design, other analysts and users work on designing the business processes to capture the needed data, including the design of business dialogues for interactions with system users during input and output processes. Dialogues are illustrated using a combination of sample displays and a tree indicating which displays are accessed from which other displays. Display mock-ups are developed from the form generator module of the database management system. Such mock-ups are similar to the sample report layouts illustrated in Figure 2-7a, so we do not illustrate a form for the Purchasing Fulfillment System.

The sample dialogue tree in Figure 2-8 shows the sequence of displays, which displays are accessed from which other displays, and what exit paths exist from each display. This dialogue tree is related to process 4.0 in Figure 2-4b, Select Preferred Supplier. That is, these displays contain data that support a purchasing manager in making the decision about a preferred supplier. This figure is read top-to-bottom with the initial display indicated by the top box. This structure is similar to a menu tree, which you may have seen for an electronic spreadsheet package or other software product. Each display has a numeric label and title. The label numbers are nested so that it is clear which displays are accessed from which other displays. The bottom row of numbers shows the displays the user may exit to from a particular display. Note in Figure 2-8 that the same display layout, Supplier Display, is accessed from two different displays but that the exit options vary depending on how the user gets to this display.

Figure 2-8
Example dialogue tree

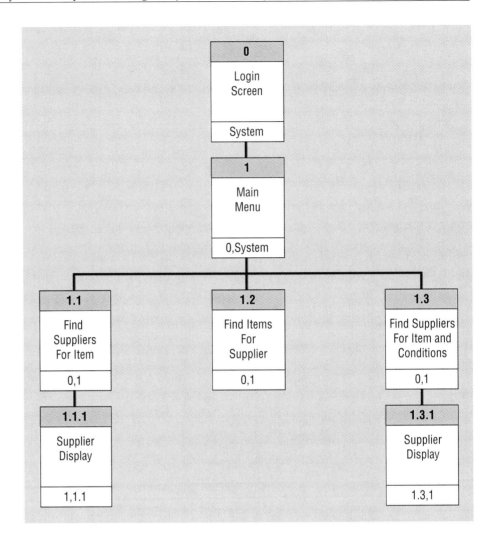

The analysts working on designing the processing steps also develop descriptions of the rules governing how data are processed and transformed. These rules help to guide those involved in physical design. Several forms of such documentation, including *structured English,* are used. Structured English shows which input conditions produce different output. Logic documentation provided via structured English serves two purposes. First, such documentation is reviewed with users to verify that the systems analysts have accurately depicted the rules governing how data are processed. Second, this code-independent description of processing logic is essential for programmers who must write code to perform this logic. Since the particular programming language for the Purchasing Fulfillment System will be selected during the physical design phase of the project, the logic must be specified in a language-independent form for now. Figure 2-9 shows sample structured English used in the Select Preferred Supplier process. This logic is triggered from display 1.1.1 in Figure 2-8. The structured English statements indicate how a supplier can automatically be found for selected values entered by the user.

Although the total logical design phase takes eight weeks, Chris arranges for a second programmer to begin working with the team during the fifth week. He knows that as soon as specific parts of the logical design are finalized, work can begin on designing programs and computer files. Chris directs the database analyst and programmers to begin designing the data entry programs and the database.

```
IF you don't care if item is delivered fast THEN
    Pick vendor with lowest quote for specified quantity of item; if tie, pick most
        recently used vendor
ELSE
    If you want to minimize cost THEN
        Pick vendor with lowest quote for specified quantity of item; if tie, pick
            vendor with best delivery time
    ELSE
        Pick vendor with best delivery time for item; if tie, pick most recently
            used vendor
    ENDIF
ENDIF
```

Figure 2-9
Example of Structured
English

These programs handle the business transactions and populate the Purchasing Fulfillment System database. By running the logical and the physical design phases in parallel, Chris helps to shorten the overall systems development cycle.

Near the end of the logical design phase, Chris assigns one of the systems analysts to be in charge of overall system functional documentation. It is this systems analyst's job to develop training materials and user documentation from the charts, displays, reports, diagrams, and other project documentation. The CASE tool's repository, as well as structured and documented code, will comprise the bulk of the system documentation needed by programmers during maintenance. Chris knows that much of the user documentation cannot be finalized yet since some changes in the functional aspects of the system will occur in later phases as users gain more understanding of the system and tests pinpoint flaws in the design. Thus, there will be some backtracking from implementation to physical and even logical design in order to correct design flaws. However, it is best to start to pull the *user* documentation together now when the reason for certain system elements are fresh in the minds of project team members.

As in the past, Chris and Juanita keep the members of the project steering committee briefed on the project's progress. In addition, Sal Divario sits in on several of the sessions when the system design elements are reviewed with users. Greg Sinise meets with the database analyst and Chris to review the final entity-relationship diagrams and relations. Chris and Juanita meet once with the whole steering committee to discuss a suggestion made by Sal Divario that the team consider the alternative of acquiring a pre-written purchasing system from a leading software vendor instead of building the system. If this alternative is selected, the results of the logical design phase will be used as the requirements listed in a request for proposal (RFP) sent to prospective vendors. After some discussion, the steering committee agrees with Juanita's recommendation that the system should be developed in-house. Her reasons are as follows:

1. The five-year IS plan called for the Purchasing Fulfillment System to be the foundation for electronically linking PVF with its suppliers and as such is to be eventually treated as a system with potential strategic value to PVF; in this case it will be necessary to have a proprietary system.

2. The Purchasing Fulfillment System is not a stand-alone application. In fact, the systems analysis has revealed many interfaces to other existing systems. These interfaces will cause significant reprogramming of any purchased system, minimizing any advantages of acquiring an off-the-shelf package.

3. The Purchasing Department is not in urgent need of a new system. Thus, they can wait for an in-house system to be developed and tested.

Because they have been involved throughout the logical design phase, the steering committee decides to wait for another complete project review until the end of the physical design phase, when the team will make another formal presentation to the Systems Priority Board. In fact, the Board and the steering committee will meet jointly at this project milestone. Stacy Morton assigns two more programmers to the project team for the next phase—physical design. She also sends a memorandum to several hardware and software specialists in the IS unit to request that they be available to assist the project team in making technology selections and system design decisions in the next phase. Using the talents of these programmers during physical design was suggested in the SDM.

PHYSICAL DESIGN PHASE

Physical Design

Purpose: To develop all technology and organizational specifications for the new information system.
Deliverables: Program and database structures, technology purchases, physical site plans, and organization redesigns.

Physical design is the last step before the system is actually built. During physical design, the project team selects all the technology that will be used in the system and produces all the "blueprints" needed by the programmers and other people involved in building the system. In this phase, the team must decide on such physical characteristics for the system as the following:

- What programming language(s) and system software (such as database management, data communication, and network operating systems) will be used to program the system

- Which combination of computer equipment (workstations, central computers, desktop computers, networks, etc.) the system will use, including any new equipment that has to be purchased

- Database structures that work with the chosen database management system

- Structures for each program and flows between programs that coordinate all aspects of the system

Besides these computer system components, work in physical design specifies new facility layouts in the Purchasing Department offices to accommodate new work methods and computer equipment (including lighting, ventilation, and furniture). Testing procedures are established and test data generated. Also, since the system might redistribute work among purchasing agents and other PVF employees, the project team must address organizational and staffing changes as a probable consequence of the new system.

Juanita plays less of a role during physical design than she has played in prior stages, and she starts working on several parallel steps to get the team ready for the implementation phase. She develops a training plan and a system conversion and installation plan. The education plan specifies which people require what kinds of training or awareness education on the new system. Juanita's plan includes a schedule of news bulletins, seminars, and training sessions.

The conversion and installation plan addresses how to make a smooth transition from current operations to those required by the new system. This plan addresses how to switch from the old to the new system and how to deploy the Purchasing Department staff during this transition. It will be necessary not to disrupt department operations during the transition. As coordinating the transfer of data from current PVF files to the new database requires special timing, a database analyst works with Juanita. The final conversion and installation plan Juanita chooses is a type of *single location* or *pilot approach,* illustrated in the Gantt chart of Figure 2-10. This type of plan includes different organizational units in different phases. The first phase involves converting all the vendor data; the second phase

Figure 2-10
Conversion and installation plan

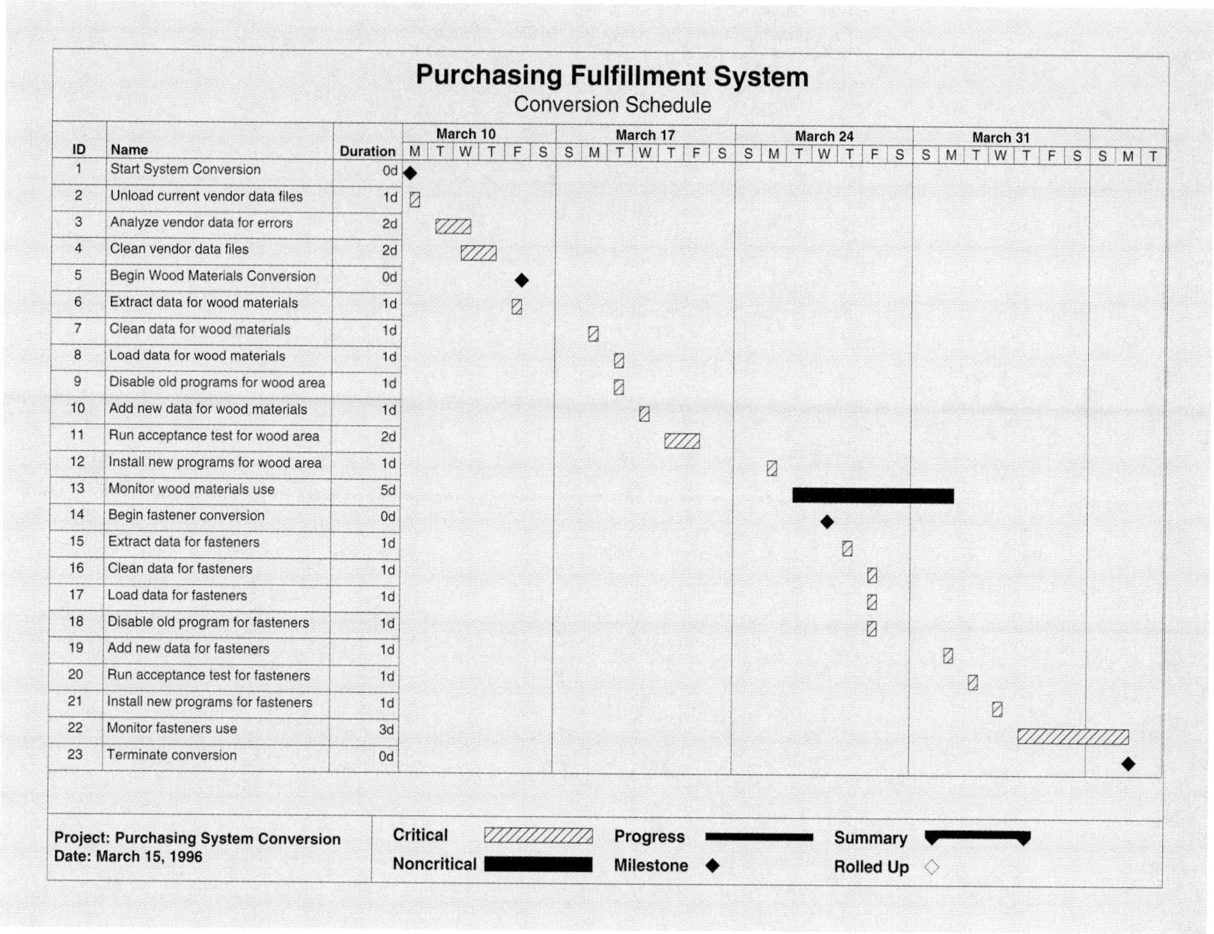
Purchasing Fulfillment System
Conversion Schedule
March 10
March 17
March 24
March 31
ID
Name
Duration
M T W T F S S M T W T F S S M T W T F S S M T W T F S S M T
1 Start System Conversion 0d
2 Unload current vendor data files 1d
3 Analyze vendor data for errors 2d
4 Clean vendor data files 2d
5 Begin Wood Materials Conversion 0d
6 Extract data for wood materials 1d
7 Clean data for wood materials 1d
8 Load data for wood materials 1d
9 Disable old programs for wood area 1d
10 Add new data for wood materials 1d
11 Run acceptance test for wood area 2d
12 Install new programs for wood area 1d
13 Monitor wood materials use 5d
14 Begin fastener conversion 0d
15 Extract data for fasteners 1d
16 Clean data for fasteners 1d
17 Load data for fasteners 1d
18 Disable old program for fasteners 1d
19 Add new data for fasteners 1d
20 Run acceptance test for fasteners 1d
21 Install new programs for fasteners 1d
22 Monitor fasteners use 3d
23 Terminate conversion 0d
Project: Purchasing System Conversion
Date: March 15, 1996
Critical
Noncritical
Progress
Milestone
Summary
Rolled Up

converts data and systems for the wood materials items; and the final phase converts the data and systems for fastener items. This approach allows the project team to concentrate support with a few users at a time. This minimizes errors, isolates system start-up problems to a few users, and allows the team to refine conversion and installation methods as new user groups are incorporated.

The project team makes all the technology choices before designing any of the programs and chooses to use equipment and system software already approved by PVF for two reasons. First, the Purchasing Fulfillment System has no unique requirements that call for additional technology. Second, PVF has standard technology and a special request must be made to deviate from these standards. The chosen platform is Paradox for Windows and Sybase's SQL Server™ database engine on PVF's local area network. Paradox for Windows will be used to build the front-end user interface modules, and the Sybase engine will handle all database management functions. SQL-Link will connect Paradox and SQL Server. This will be the first application at PVF using this combination of technologies, which is somewhat of a risk to the project team. The PVF IS plan calls for movement to the client/server network environment supported by these technologies and, after consulting with various stakeholders, the project team decides that this project is of an appropriate size and complexity to be a good first attempt to build a system within this new technology architecture. Several new PC workstations are needed in Purchasing and additional disk capacity is needed on the network server. An expert in computer capacity planning works with the team during this step to estimate processing volumes for the new system and the load this will place on the network and server.

An important step in physical design is the design of database structures. The database analyst refers to the relations developed in logical design. Although it is not always the most efficient method, the database analyst chooses to define a database file for each relation. The database analyst uses the CASE tool to automatically generate file descriptions in both Paradox and SQL (SQL is the database language understood by SQL Server) from information in the repository. The database analyst then reviews this generated database definition code and modifies some of the field types and lengths, indexes, and other database structures when it is believed more efficient choices can be made. Definitions in Paradox for three of the system files (ITEM, VENDOR, and QUOTE), those needed for the Price Quote Report of Figure 2-7, appear in Figure 2-11. The definition of each file shows the fields and for each, its data type, size (if appropriate), and whether the field has an index. Key fields are also indicated, a requirement of the relational database model. The project team decides that the whole database will be stored on the central file server on PVF's local area network so that all database functions are centralized in this one database managed by the SQL Server software. Since this network is not very busy and the transaction volume for the Purchasing Fulfillment System is not expected to be heavy, the team decides that all data capture programs will be on-line; as a result, no batch data input will be done. Some periodic reports will be built using SQL with SQL Server.

The specifics of all inputs and outputs, as well as a description of all processing steps, have been completely outlined in logical design in such documentation as computer form and report designs, dialogue charts, and structured English. In physical design, all the processing steps are translated into diagrams or various forms of *pseudocode*, which resembles English and is one step short of actually writing in a programming language. Although some CASE tools can generate physical design

Figure 2-11

Sample file descriptions

(a) ITEM file

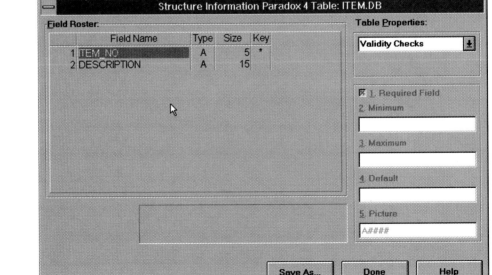

diagrams and pseudocode from repository contents, PVF's tool only helps the team to develop this documentation, which it then stores in the repository. The team of system analysts, programmers, and other specialists involved in physical design use several forms of physical system documentation, including the following:

- *System flow charts*, which show the linkages between different programs and which technologies are used in each part of the system (see Figure 2-12)

- *Structure charts*, which show the hierarchical organization of a program and the flow of data between the different program modules (see Figure 2-13)

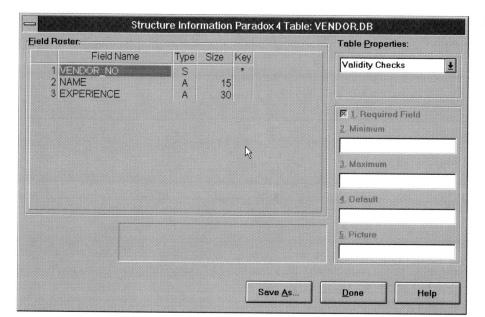

(b) VENDOR file

(c) QUOTE file

Figure 2-12
Example system flow chart

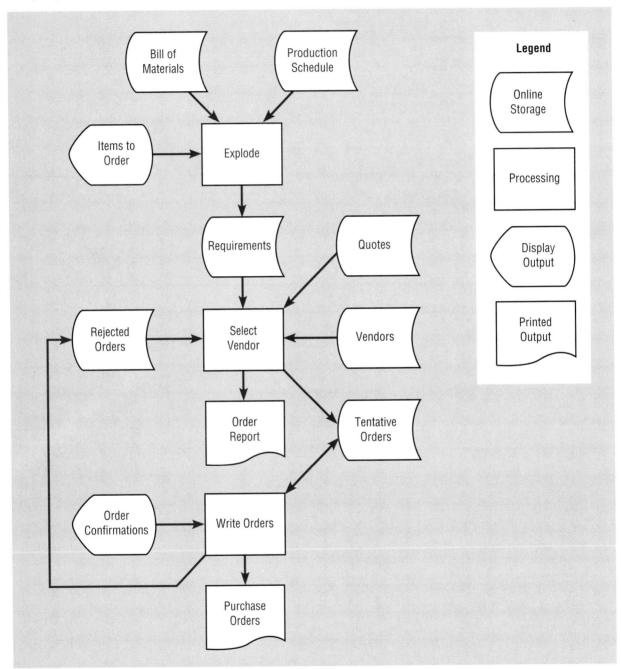

- *Action diagrams*, which show the logic of database accesses required to up-date database files or retrieve data needed in specific screens or reports (see Figure 2-14)

While the team finishes all of the physical system specifications, Chris and Juanita work on their presentation to the steering committee and Systems Priority

Figure 2-13
Example program structure chart

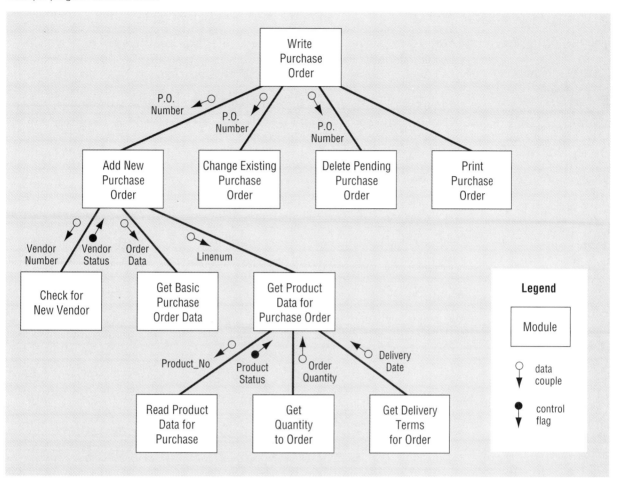

Figure 2-14
Example action diagram

```
  ┌─ READ ITEM RECORD
  │  DISPLAY ITEM_NO, DESCRIPTION
  │  ┌─ READ QUOTE RECORD FOR ITEM
  │  │  READ VENDOR RECORD FOR QUOTE
  │  │  DISPLAY VENDOR_NO, NAME, QUANTITY, QUOTE,
  │  └─    CONDITIONS
  └─
```

Board. Their responsibility is to justify the physical design and to again convince these groups that the project work should continue. Now that more precise costs to build and operate the system can be estimated, the project team again has to justify the system. Chris takes responsibility for the costs to build and operate the system, and Juanita develops improved benefits projections.

PVF maintains data about the time required to develop its systems and Chris uses this history to estimate the programming costs for the Purchasing Fulfillment System. The physical system design shows the proposed system in enough detail that Chris is able to compare the characteristics of the new system to those previously developed at PVF. Chris uses an industry standard called *function points*, which measure units of programming work. The CASE tool is helpful since it is able to calculate the number of function points from the physical design specifications in its repository. Chris also produces estimates for the time needed to write documentation, to develop educational and training materials, to conduct training sessions, to handle the system conversion, and to purchase and maintain new equipment. These numbers go into developing a projected cash flow of one-time and recurring expenses for the new system, similar to those shown in Figure 2-6. Note that the cost of the project so far is not considered. Expenses already incurred are considered sunk costs and, at this point, are not pertinent to the go/no-go decision.

Juanita talks with the various end users in order to update the system benefits estimated earlier. She generates updated estimates for both expected and worst-case savings as well as new income figures. In addition, she refines the previously identified intangible benefits that cannot be factored into the financial analysis. Since most features of the Purchasing Fulfillment System support Purchasing operations, benefits are reasonably easy to quantify from time savings, error reductions, and improved decision-making support. A few system features are aimed more toward improved relationships with vendors and other factors that are more difficult to quantify, so these additional factors remain intangibles.

Juanita is in charge of the review meeting. This is a key decision that Chris and she have agreed to. The reason for Juanita taking the lead is to show that the project has been client-focused and that business needs, not technology, have driven the project. She leads off with a brief review of the project, including a comparison of the project plan with what has occurred. She uses the project Gantt and PERT charts to compare projected versus actual times and costs. She shows that the project to date is on time and within budget, which suggests that it has been well managed and feasible. This message is important in the request for project continuation.

Next, Juanita outlines the business case for continuing the development. The primary element of this business case is the updated financial cost and benefit analysis. In addition, she outlines some organizational changes necessary in order to derive the most benefit from the new system. These include upgrading some Purchasing positions, as higher skills will be required to use the new system, and restructuring the Purchasing staff. This restructuring will organize purchasing agents around supported manufacturing units rather than types of raw materials. This organization is a major intangible selling point for the new system because managers in manufacturing have been frustrated by having to work with so many different purchasing agents to get their materials.

Chris then reviews the essential logical and physical design features. Next, he outlines the timetable for the rest of the project, again using Gantt and PERT charts as in prior steps. He concludes with a request for the allocation of people with certain skills to work on the remaining stages of the project.

Juanita leads the meeting wrap-up. She quickly overviews the education and conversion plans to show that the project team has considered these important remaining elements. She also raises several project risks centering on possible resistance among some purchasing staff members and outlines how she intends to deal with these issues if they arise. Juanita closes the meeting by listing all the people who have contributed to the project, including all the people not on the project team who have participated in JAD sessions, have been interviewed, or have in any way contributed to the project team. This closing reveals the tremendous energy and intelligence behind the recommendations.

Most of the information in this presentation is not totally new to the steering committee and Systems Priority Board members. Both Juanita and Chris have informally met with most of these stakeholders prior to the meeting to review those parts of the presentation of most interest to each member. For example, Chris met with Stacy Morton and Greg Sinise to review the physical design specifications.

Thus, after a forty-minute presentation and twenty minutes of questions, both the committee and the Board approve the allocation of financial and human resources to complete the project.

IMPLEMENTATION AND MAINTENANCE

As you might expect, many important steps remain before the Purchasing Fulfillment System begins operation. In this section, we quickly bring the description of the Purchasing Fulfillment System development project to a conclusion by limiting our presentation to highlights of the role of systems analysts in the remaining project phases.

The Implementation Phase

As outlined in Chapter 1, the implementation phase includes six sub-phases: coding, testing, installation, documentation, training, and support. Coding is performed by the programmers assigned to the project. Often, information systems development staff members with several years of experience perform both systems analyst and programmer functions. In this way the same people who designed the system program the system, which reduces the possibility of any misunderstandings. When different programmers perform each function (whether they are in-house programmers or outside contract programmers), analysts meet periodically with programmers to answer questions about the physical design specifications and to review programs for design compliance.

> **Implementation Phase**
>
> *Purpose:* To program the system, build all data files, test the new system, install system components, convert and cease operation of prior systems, train users, and turn over system to operations.
> *Deliverables:* Programs that work accurately and to specifications, documentation, training materials, and project reviews.

Analysts become more involved during testing. Testing is a multiple-step process crucial to the success of the system. Systems analysts develop the test data necessary to ensure that the programs handle all possible circumstances before testing begins (often as early as the analysis phase when special data maintenance conditions are identified). These test cases cover all valid and invalid input data so that all paths through programs are tested, if possible. The testing process can be aided by computerized testing software which repeatedly presents the same test cases to a program. This ensures that, as programmers create each version of a program (as bugs are fixed and new features added), the program goes through consistent and thorough testing. Analysts also generate test data to stress test the system. A high volume of input is presented to see whether the system breaks down as the database and intermediate files become very large, representing peak loads on the system. Although testing processes vary, systems analysts typically monitor the testing process through the point of user acceptance testing.

During system installation, project team members transfer data from old files to new files and databases, being careful to review old data to validate and correct the data before the data are loaded into the new system. Figure 2-10 shows a typical conversion schedule for a project. As mentioned earlier in this chapter, installation must be planned so that minimal disruption of the business occurs. Installation ends with the new system successfully operating in the organization.

In parallel to these other steps, all documentation—both for systems professionals and users—all training units, and all support procedures are designed and implemented. Many of these design activities were begun much earlier in the project, but they are finalized during implementation. Training for the Purchasing

Fulfillment System includes units on how to enter data, how to interpret reports and displays, and how to override default decisions the system makes. The project team decided that the regular user consultants were sufficient to support this new system but that these consultants needed to be trained on the Purchasing Fulfillment System. They decided to establish a separate telephone number to call for assistance and created a newsletter to announce changes to the system, print user comments, and help the users network with each other.

Another major step in implementation is the close-down of the systems development project. In the case of the Purchasing Fulfillment System, this would involve the following steps:

- Reviewing and evaluating the project, including how well Chris and Juanita managed the project and the quality of the work accomplished by the team of systems analysts, programmers, and other team members

- Turning over all project documentation to an IS department librarian or other staff member to keep for reference as needed in system maintenance

- Releasing project team members as they are no longer needed, including providing personnel appraisals to these individuals and their supervisors

- Submitting a project termination report to Sal Divario and other members of the project steering committee

It will be necessary, however, to monitor the new system closely for a while to ensure that the system actually works in live operation. After this follow-up period, the next phase begins—maintenance.

Maintenance

Purpose: To monitor the operation and usefulness of a system, to repair and enhance the system as required, and to determine when the system is obsolete.
Deliverables: Periodic audits of the system to demonstrate whether the system is accurate and still meets business needs.

The Maintenance Phase

Maintenance is the process of

- Repairing the system as errors, or bugs, are discovered

- Enhancing the system to include new or improved features that were not anticipated or were purposely delayed when the system was first built

- Adapting the software to changes in computer hardware and system software

Recoding to correct minor technical flaws and to make small changes (for example, to create a new report or improve display design) might require only programming skills. Sometimes additional systems analysis and design is necessary to analyze the current system along with the new information needs to determine changes affecting file design or the structure of the system. Any extensive changes require the skills of a systems analyst.

Various individuals can stimulate maintenance. Users might identify flaws and the need for enhancements and then submit a system change request, similar to the System Service Request which instigated the Purchasing Fulfillment System project. Data center staff might find processing errors or inefficiencies and submit a change request. System auditors might identify the need to fix more subtle flaws or to add data protection features like security controls or better system backup. Some organizations employ internal systems auditors who periodically check that systems perform as specified. A system audit might be done as part of an accounting audit by a public accounting firm.

The maintenance phase is typically the longest phase in the life of a system. Whereas the phases through implementation might require 3 to 12 months, maintenance can continue for years thereafter. Thus, most systems analysis and design work falls within the maintenance phase rather than the other life cycle phases. Maintenance is usually done in batches similar to new version releases of purchased software. Each major release is managed like a project, as we have seen in this chapter. Minor or interim releases may be managed more casually, but in some way all the steps (analysis, design, and implementation) must be carried out to ensure that new releases of a system meet specifications.

SUMMARY

In this chapter we used a hypothetical example case to demonstrate some of the depth and texture of a systems analysis and design project. Our goal was to provide you with an overview of the topics within systems analysis and design, to help you anticipate issues you will confront as you work on projects during your systems analysis and design course, and to provide a basis for later chapters.

The PVF case study showed the application of the systems development life cycle and some of the typical principles, such as incremental commitment and functional decomposition, that characterize modern systems development approaches. The systems development life cycle is a common framework that outlines a project from request for new services to the analysis of the current business situation, to identification of information requirements, to the design of the logical, technology-independent system, and ultimately to the design and implementation of the physical system. The life cycle allows for consistent management of projects and greater proficiency through the experience of project team members and proven techniques and tools. Each phase of the project reaches a specific milestone with specific deliverables. These deliverables or products become part of the system documentation and are input to subsequent project phases. We also briefly introduced a variety of typical documentation notations, such as data flow diagrams, entity-relationship diagrams, structured English, and structure charts, all of which can be used to describe the functional and physical specifications of current or new information systems throughout the life cycle.

As part of this discussion we have highlighted the importance of a rigorous approach to project management including the strong need for project planning. We presented various charts, specifically Gantt and PERT, that are used to outline and monitor project progress. We also portrayed the roles of various people—from users to technicians to general managers—during different steps of systems development. In addition, we included glimpses of education, installation, and documentation as important elements of a systems development project.

This chapter also shows the very close working relationship between the lead systems analyst and lead user contact necessary for successful systems projects. The chapter illustrated how Chris and Juanita anticipated issues, addressed needs in business, not technical, terms first, communicated the status of the project regularly and thoroughly to a variety of project stakeholders, planned meetings carefully, and performed many other tasks in a highly professional manner. Several questions at the end of the chapter ask you to discuss why Chris and Juanita were successful project leaders.

For brevity, we presented only a cursory overview of the systems implementation and maintenance phases of the systems development process. In later chapters we will describe in detail the management of programming, different kinds of system testing, alternative ways to switch from an old to a new system, and issues in the long-term maintenance of application software.

CHAPTER REVIEW

KEY TERMS

Business case
Context diagram
Data flow diagram

Functional decomposition
Incremental commitment
Stakeholder

Systems development
    methodology

REVIEW QUESTIONS

1. Define each of the following terms:
    a. Gantt chart
    b. client
    c. funder
    d. end user
    e. maintenance
    f. installation
    g. entity-relationship diagram
    h. system service request
    i. steering committee
    j. test plan
    k. education (training)
    l. system audit

2. What are the advantages of using a standard systems development methodology for all system projects in an organization?

3. Outline the logic behind the incremental commitment philosophy applied in many structured systems analysis and design methodologies.

4. Explain the purpose of the System Service Request as it is applied in Pine Valley Furniture.

5. What was the role of the Systems Priority Board in Pine Valley Furniture? How was this role different from that of the project steering committee created for the Purchasing Fulfillment System?

6. What is the purpose of a context diagram during early stages of systems analysis?

7. What are the general contents of a business case to justify a systems project?

8. Describe the steps necessary to initiate a systems project.

PROBLEMS AND EXERCISES

1. Match the following terms to the appropriate definitions.

    _____ entity-relationship diagram   a. map or diagram that shows a sequence of actions to be performed on a database

_____ data repository

_____ prototyping

_____ structured walkthrough

_____ system flow chart

_____ structure chart

_____ action diagram

b. iteratively building a working version of a system

c. shows the hierarchical structure of a computer program, including messages or data passed between modules

d. database of a CASE tool that contains the definitions for all elements of a system description

e. diagram of the entities and relationships between entities consistent with the rules and policies of an organization

f. step-by-step discussion of documentation about a current or proposed system

g. diagram of the flow of data between computer programs which also shows the computer equipment used within a system

2. Discuss the skills or qualities you recognized in Juanita Lopez that made her a good end-user leader for a systems development project team.

3. Discuss the skills or qualities you recognized in Chris Martin that made him a good systems analyst and project leader for a systems development project team.

4. Discuss why you think Pine Valley Furniture used their systems development methodology as a guideline rather than as a fixed process from which systems development projects could not vary.

5. Explain why it is important to separate logical design and physical design into two identifiable steps.

6. Discuss the importance and role of the informal communication that occurs between systems development project leaders and stakeholders in the system; informal communication is that which occurs outside of formal project review meetings.

7. Discuss the purpose of a written agenda distributed in advance of a project review meeting.

8. Explain why it is important to get as many of the requirements for a new information system right during logical design rather than during later stages of systems development.

9. Discuss why Chris Martin and Juanita Lopez assumed the roles they did during the review meeting at the end of the physical design phase.

10. Discuss why Chris Martin wanted to spend so much time in the logical design phase for the Purchasing Fulfillment System project.

11. Evaluate the System Service Request used at Pine Valley Furniture. What are its strengths and weaknesses? What improvements could be made to the form?

FIELD EXERCISES

1. Ask someone you know in the IS department of your university or other organization to provide you with a copy of the form used in his or her organization to request changes to a system or to create a new information system. How is this form different from the System Service Request format used in Pine Valley Furniture? If

you discover that the organization does not have a form like the SSR, describe how new projects are requested. Design a rough draft of a form that might be used.

2. Many organizations document their systems development life cycle in a manual that outlines each phase and step and who does what, which techniques are used, and what documentation must be produced. IS consulting companies are especially rigorous in outlining standard systems development processes. Through contacts in consulting companies (often a faculty member or student club officer can give you a name and phone number) or literature in your placement office, investigate the definition of the SDLC used by an IS consulting organization. How is this SDLC different from that used by Pine Valley Furniture?

3. If your community has a local chapter of an IS professional organization (for example, DPMA [Data Processing Managers Association] or ASM [Association for Systems Management]), find out when their meetings are and whether students are welcome to attend. If permitted, attend a meeting and, in conversations, ask chapter members to describe positive qualities in lead analysts and project leaders. Compare what you hear to the qualities demonstrated by Chris Martin in this chapter.

4. Almost any new idea that requires funding in an organization, whether it be a new product, an organizational change, the construction of a new facility, or whatever, requires those proposing the idea to make the business case for the innovation. Find an example business case (from someone you know in an organization or possibly from another business course you have taken) and summarize the essential features of this business case. Compared to this business case, was the one made for the Purchasing Fulfillment System insufficient in any way?

5. Interview someone who has been involved in the development of a moderate- to large-sized information system. If you have been involved in such a development project, use your own example for this exercise. In what capacity and in what ways was this person involved in the project? Does his or her involvement resemble that of any of the people in the Pine Valley Furniture project? If so, whom and in what ways? In what ways did the development process for this person's project resemble that used for the Pine Valley Furniture project? How did it differ? Does this person consider the development project he or she was involved in to be a success? Why or why not? Does this person consider the personal development experience that took place to be positive, negative, or neutral? Why?

Preparing and Organizing for Systems Development

Atkinson Construction

IS Project Management and Project Management Software

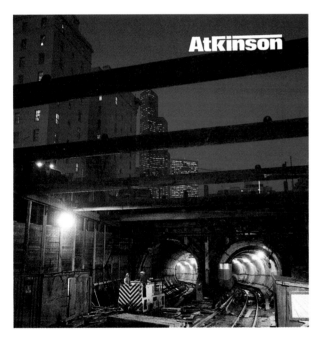

Atkinson Construction, one of several operating units of Guy F. Atkinson Company of California, Inc., is headquartered in San Bruno, California, and generates approximately $200 million in revenue annually. Atkinson is primarily project-based, with projects scattered around the globe.

Atkinson Construction typically has between 15 and 20 construction projects running at a time. These projects exist throughout the country, with many projects currently in the Pacific Northwest. At present, Atkinson has a few projects overseas, with their largest ongoing project in Venezuela. The Venezuela project peaked at

5,000 people. Ken Chambless, controller of Atkinson Construction, says that they typically have at least one large international project at a time, but that the average domestic project has between 150 and 300 people and will last two to three years.

Each project team is organized like a mini-company. The senior management team for a project includes a project manager, a business manager, a project engineer, and an operations manager. They then build a staff around this management team. The size and scope of this staff depends on the size of project team.

To manage the projects, Atkinson Construction has been using a mainframe-based project management system with modules for cost and productivity reporting, general accounting, and other activities. The project teams in the field also use the PC-based Primavera project management software for activities such as scheduling. In addition, the teams can dial up and upload to the mainframe project information such as weekly employee hours for payroll calculations. Electronic-mail is used throughout the company so that people within and between projects can communicate and coordinate.

Over a period of time, Atkinson found that the mainframe hardware and software were too costly to maintain. As a result, three years ago they outsourced the hardware to a firm in Ohio. Controller Chambless says that while this is in some ways easier, it still isn't the optimal solution to providing information systems services to manage the projects. The mainframe system doesn't offer adequate access to data. Chambless characterizes the system as a "batch and wait" process, where users make a request for something and then wait until the next day or the end of an accounting close for a response. Atkinson's goal today is to put information quickly and directly into the hands of users and to make the system more interactive.

As a result of the shortcomings of the current legacy system, Atkinson is in the process of changing to a completely new system that embraces up-to-date technology. Vendor J. D. Edwards will soon implement a new turnkey system that will run project management software on an IBM AS/400 platform. Processing will be distributed throughout the organization. Business Administrator James Herrett temporarily left the field to come in-house and help implement the new system. Herrett explains that the new system will help to manage all aspects of the projects, including project control, job costing, accounts payable, accounts receivable, subcontractor management, and payroll, and will interface with corporate systems. The new system will give users direct, fast access to information and will be less expensive in the long run than the current system. Herrett says that timely information on project costs is critical. Costs can change daily and, if a project manager has to wait several days for data, he or she will not be able to adequately control costs and ensure that the project finishes at or under budget.

The conversion to the new project management system is in itself a massive project, rivaling the size and complexity of Atkinson's construction projects. The system conversion project is divided into seven integrated modules, including payroll, accounts payable, general ledger, and job costing. Each module has its own team of information systems personnel, with some individuals serving on multiple module teams. Each module team has a team leader who reports to the overall project leader.

The conversion project is being managed in much the same way that the construction projects are managed. For example, scheduling of the conversion project is accomplished using the same Primavera software that is used for construction projects. In addition, systems personnel have borrowed an in-house "change management" information system (intended to identify and address problems on construction projects) which they are now using to manage problem resolution on the systems conversion project.

In addition to these module teams, there are separate teams that cut across each of the module teams. For example, there is a large data conversion team that is converting 30 to 40 years' worth of historical project data and current project data for all modules into a consistent format to be used by the new system. The conversion project mirrors the construction projects in that it consists of a complex, interlocking matrix of teams working toward a common goal.

At Atkinson, managing large projects is their business, and information technology plays a significant role in project management. With twenty 150-member project teams spread across the globe, and with some project teams that may comprise thousands of members, Atkinson would not be able to manage mega-projects effectively and efficiently without the most up-to-date information technology available.

An Overview of Part II:

Preparing and Organizing for Systems Development

Every field has some underlying principles that permeate the discipline. In sports, conditioning, flexibility, health, coordination, and dedication are necessary traits. In the law, principles of precision, justice, and protection of rights span all aspects of practice. In systems development, there are certain fundamental principles and concepts that play a role in every phase of the systems development life cycle and that, in a sense, hold the systems development process together. The purpose of Part II is to outline those aspects of systems development that are independent of but apply to each life cycle step. We must cover these principles now since they will be used repeatedly in subsequent chapters, and it can be confusing to introduce them in a piecemeal fashion. A systems analyst and a systems development group must master at least a basic understanding of these principles before useful systems work can occur. We group these fundamental principles, necessary for preparing and organizing for systems development, into three chapters—critical success factors, project management, and automation of systems development.

There are individual factors for a systems analyst that contribute to successful systems development and these critical success factors are the topic of Chapter 3. Various studies of successful systems analysis and design indicate that the systems analyst must possess analytical, technical, management, and interpersonal skills. Although these critical skills are held in common by most successful people in organizations, they take on special forms for systems analysts. Certainly the skill set changes—especially those skills related to technologies—but the changes tend to be more evolutionary than revolutionary. Because many systems analysis and design courses require significant project work, either in groups or in a real organizational setting, you will soon have an opportunity to practice the critical success factors. To help in these endeavors, this chapter sensitizes you to the skills you will soon need to practice. Later in the text, we will elaborate on the general skills outlined in this chapter by presenting specific versions of these general skills as they apply to systems development phases. As Chapter 3 ends, you will develop an appreciation for systems analysis as a profession and understand how these critical skills lead to standards, ethics, and career paths for the field.

Chapter 4 addresses a fundamental characteristic of life as a systems analyst—working within the framework of projects with constrained resources. Although systems development efforts vary considerably in size and complexity, all systems work demands attention to deadlines, working within budgets, and coordinating the work of various people. The very nature of the systems development life cycle implies a systematic approach to managing systems development. This outlook is based on the notion that a project—a group of related activities—leads to a final deliverable. Projects must be started, planned, executed, and completed. The planned work of the project must be represented in such a way that all interested parties can review and understand it. Computer software can help systems analysts and project managers organize and track system project activities.

In your first job as a systems analyst, you may not lead projects and hence you will not actually perform many of the project management steps outlined in Chapter 4. You will, however, have to work within the schedule and other project plans and thus it is important to understand the management process controlling your work. Even though you may have studied project management as part of other courses, this topic is so fundamental to systems development (and so often misunderstood), that we devote Chapter 4 to project management within the systems development setting.

Quality and productivity are essential in any job. For many years, the systems development field has been plagued by meager improvements due to small incremental changes in our tools and techniques. Some believe that the only way to achieve quantum improvements is to provide automated support for systems development, just as we do for the rest of a business in order to achieve organizational benefits. Today, the focus on automated support for systems analysis and design and for the total process of building systems is on computer-aided software engineering, or CASE. The purpose of CASE is to provide an engineering-style discipline with associated tools to enforce standards of practice and to greatly increase the speed by which systems are developed and maintained. Chapter 5 reviews the developments in the rapidly changing and much discussed CASE field. This chapter should help you understand CASE products and their application, the organizational imperatives for CASE, and why CASE may not be a panacea for all organizations and all types of systems. We also preview how CASE is used in each life cycle phase, and we will build on this preview in later chapters that discuss these phases in depth. The goal of Chapter 5 is not to give you skills for using a CASE tool or to promote CASE as a mature and essential tool for systems development; rather, the intent is to make you a knowledgeable consumer of CASE technology, able to know how and when to apply it.

Part II concludes the introductory and background material for the text. Subsequent chapters address the step-by-step processes followed in building systems. You will come across many instances where the fundamental concepts explained in Parts I and II apply as you study and perform systems analysis and design.

Finally, Part II introduces Broadway Entertainment Company, Inc. (BEC). The BEC case, presented in separate sections after selected chapters, serves several purposes. First, whereas the book chapters emphasize the concepts, principles, and methods as being fundamental elements of the field, the BEC case helps demonstrate how these elements might fit into a practical situation. Second, besides illustrating the application of topics, BEC also provides a rich organizational setting for discussing issues about systems analysis and design.

Two BEC case sections are included after Chapter 5; the remaining book chapters through Chapter 20 each have an associated BEC case. The first BEC section introduces the company, summarizes its history, emphasizes its strategic and competitive business setting (which greatly influences the development of systems), and outlines the portfolio of information systems applications in use. The second BEC section provides more detail on existing systems to aid in understanding BEC operations when we look at the requirements and design for new systems in later BEC case sections.

Succeeding as a
Systems Analyst

After studying this chapter, you should be able to:

■ Discuss the analytical skills, including systems thinking, needed for a systems analyst to be successful.

■ Describe the technical skills required of a systems analyst.

■ Discuss the management skills required of a systems analyst.

■ Identify the interpersonal skills required of a systems analyst.

■ Describe the systems analysis profession.

INTRODUCTION

In the first two chapters, you learned about the different types of information systems developed in organizations, the people who develop them, and the project environment in which systems are developed. Before we explore the systems development life cycle in more detail, however, we need to examine the skills needed to succeed as a systems analyst. You will first examine the analytical skills a systems analyst needs, then discuss the technical, management, and interpersonal skills required of a good analyst. One of the key analytical skills you will study is systems thinking, or the ability to see things as *systems.* You probably learned about systems and systems thinking in your introductory information systems class, so we will review here the highlights of systems and systems thinking that directly affect the design of information systems and how a systems analyst develops systems.

As illustrated in Figure 3-1, an analyst works throughout all phases of the systems development life cycle. The life cycle model represents the process of developing information systems, the same process you read about in Chapter 1 and which was followed in the development project in Chapter 2. The skills the analyst needs to be successful are represented by the objects placed in the diagram. The laptop computer represents technical skills; the briefcase represents management skills; the magnifying glass represents analytical skills; and the telephone represents interpersonal skills. As you can see, to follow the guidelines established by any development methodology, an analyst needs to rely on many skills. Although we cannot possibly provide thorough coverage of these skills in this chapter, some will be covered in considerable depth in later chapters while others are discussed more generally. Our goal for these general skills is to sensitize you to abilities that you need to develop from other courses and materials in order to become a successful systems analyst. The chapter ends by stepping back from these specific skills to examine systems analysis as a profession, with its own standards of practice, ethics, and career paths.

ANALYTICAL SKILLS FOR SYSTEMS ANALYSTS

Given the title systems analyst, you might think that analytical skills are the most important. While there is no question that analytical skills are essential, other skills are equally required. First, however, we will focus on the four sets of analytical skills: systems thinking, organizational knowledge, problem identification, and problem analyzing and solving.

Systems Thinking: A Review

If you counted the number of times each key term is used in this book, the key term used most frequently would undoubtedly be *system*. Let's take the time now to exam-

Figure 3-1
The relationship between a systems analyst's skills and the systems development life cycle

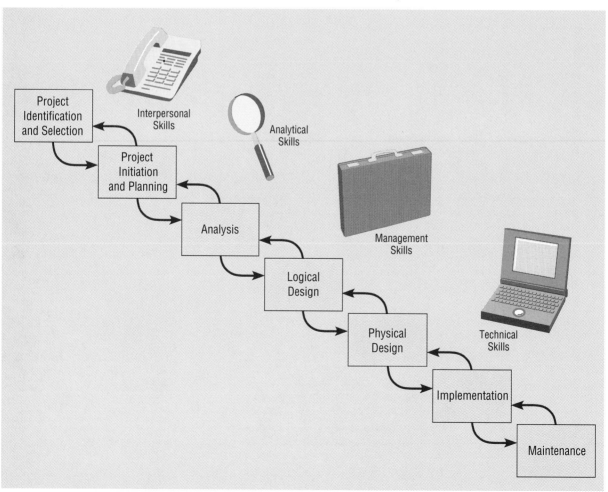

ine systems in general and information systems in particular. (For a more thorough treatment of system concepts, see Martin, et al., 1994). Let's start by examining what we mean by a system and identify the characteristics that define a system.

Definitions of a System and Its Parts   A **system** is an inter-related set of components with an identifiable boundary, working together for some purpose. A system has nine characteristics (see Figure 3-2):

1. Components
2. Inter-related components
3. A boundary
4. A purpose
5. An environment
6. Interfaces
7. Input
8. Output
9. Constraints

A system is made up of components. A **component** is either an irreducible part or an aggregate of parts, also called a *subsystem*. The simple concept of a component

System: An inter-related set of components, with an identifiable boundary, working together for some purpose.

Component: An irreducible part or aggregation of parts that make up a system, also called a subsystem.

Figure 3-2
A general depiction of a system

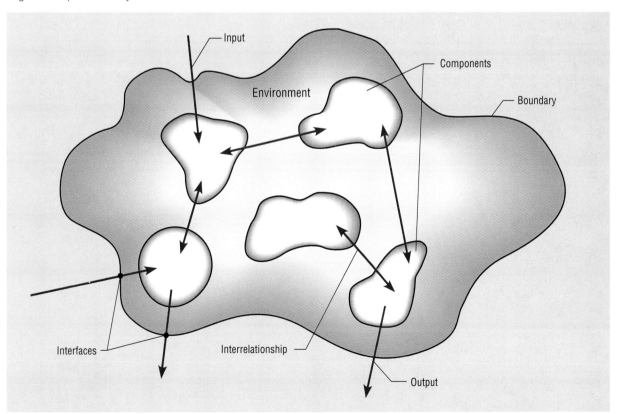

Inter-related components: Dependence of one subsystem on one or more subsystems.

Boundary: The line that marks the inside and outside of a system and which sets off the system from its environment.

Purpose: The overall goal or function of a system.

Environment: Everything external to a system that interacts with the system.

Interface: Point of contact where a system meets its environment or where subsystems meet each other.

Constraint: A limit to what a system can accomplish.

is very powerful. For example, just as with an automobile or a stereo system with proper design, we can repair or upgrade the system by changing individual components without having to make changes throughout the entire system. The components are **inter-related;** that is, the function of one is somehow tied to the functions of the others. For example, the work of one component, such as producing a daily report of customer orders received, may not progress successfully until the work of another component is finished, such as sorting customer orders by date of receipt. A system has a **boundary,** within which all of its components are contained and which establishes the limits of a system, separating the system from other systems. Components within the boundary can be changed whereas things outside the boundary cannot be changed. All of the components work together to achieve some overall **purpose** for the larger system: the system's reason for existing.

A system exists within an **environment**—everything outside the system's boundary. For example, we might consider the environment of a state university to include the legislature, prospective students, foundations and funding agencies, and the news media. Usually the system interacts with its environment, exchanging, in the case of an information system, data and information. The points at which the system meets its environment are called **interfaces,** and there are also interfaces between subsystems (Figure 3-3 provides a list of functions performed by interfaces). An example of a subsystem interface is the clutch subsystem which acts as the point of interaction between the engine and transmission subsystems of a car. As can be seen from Figure 3-3, interfaces may include much functionality. You will spend a considerable portion of time in systems development dealing with interfaces, especially interfaces between an automated system and its users (manual systems) and interfaces between different information systems. It is the design of good interfaces that permits different systems to work together without being too dependent on each other.

A system must face **constraints** in its functioning because there are limits (in terms of capacity, speed, or capabilities) to what it can do and how it can achieve its purpose within its environment. Some of these constraints are imposed inside the

Figure 3-3
Special characteristics of interfaces

INTERFACE FUNCTIONS
Because an interface exists at the point where a system meets its environment, the interface has several special, important functions. An interface provides

- *Security,* protecting the system from undesirable elements that may want to infiltrate it
- *Filtering* unwanted data, both for elements leaving the system and entering it
- *Coding and decoding* incoming and outgoing messages
- *Detecting and correcting errors* in its interaction with the environment
- *Buffering,* providing a layer of slack between the system and its environment, so that the system and its environment can work on different cycles and at different speeds
- *Summarizing* raw data and transforming them into the level of detail and format needed throughout the system (for an input interface) or in the environment (for an output interface)

Because interface functions are critical in communication between system components or a system and its environment, interfaces receive much attention in the design of information systems (see Chapters 13 and 14).

system (for example, a limited number of staff available) and others are imposed by the environment (for example, due dates or regulations). A system takes **input** from its environment in order to function. Mammals, for example, take in food, oxygen, and water from the environment as input. Finally, a system returns **output** to its environment as a result of its functioning and thus achieves its purpose.

Input: Whatever a system takes from its environment in order to fulfill its purpose.

Output: Whatever a system returns to its environment in order to fulfill its purpose.

Now that you know the definition of a system and its nine important characteristics, let's take an example of a system and use it to illustrate the definition and each system characteristic. Consider a system that is familiar to you: a fast food restaurant (see Figure 3-4).

How is a fast food restaurant a system? Let's take a look at the fictional Hoosier Burger Restaurant in Bloomington, Indiana. First, Hoosier Burger has components or subsystems. We can figure out what the subsystems are in many ways but, for the sake of illustration, let's focus on Hoosier Burger's physical subsystems as follows: kitchen, dining room, counter, storage, and office. As you might expect, the subsystems are inter-related and work together to prepare food and deliver it to customers, one purpose for the restaurant's existence. Food is delivered to Hoosier Burger early in the morning, kept in storage, prepared in the kitchen, sold at the counter, and often eaten in the dining room. The boundary of Hoosier Burger is represented by its physical walls and the primary purpose for the restaurant's existence is to make a profit for its owners, Bob and Thelma Mellankamp.

Hoosier Burger's environment consists of those external elements that interact with the restaurant, such as customers (many of whom come from nearby Indiana University), the local labor supply, food distributors (much of the produce is grown locally), banks, and neighborhood fast food competitors. Hoosier Burger has one interface at the counter where customers place orders and another at the back door where food and supplies are delivered. Still another interface is the telephone which managers use regularly to talk with bankers and food distributors. The restaurant faces several constraints. It is designed for the easy and cost-effective preparation

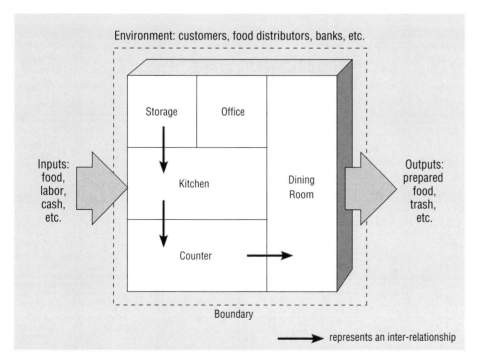

Figure 3-4
A fast food restaurant as a system

of certain popular foods, such as hamburgers and milk shakes, which constrains the restaurant in the foods it may offer for sale. Hoosier Burger's size and its location in the university neighborhood constrain how much money it can make on any given day. The Monroe County Health Department also imposes constraints, such as rules governing food storage. Inputs include, but are not limited to, ingredients for the burgers and other food as well as cash and labor. Outputs include, but are not limited to, prepared food, bank deposits, and trash.

Important System Concepts   Once we have recognized something as a system and identified the system's characteristics, how do we understand the system? Further, what principles or concepts about systems guide the design of information systems? A key aspect of a system for building systems is the system's relationship with its environment. Some systems, called **open systems,** interact freely with their environments, taking in input and returning output. As the environment changes, an open system must adapt to the changes or suffer the consequences. A **closed system** does not interact with the environment; changes in the environment and adaptability are not issues for a closed system. However, all business information systems are open, and in order to understand a system and its relationships to other information systems, to the organization, and to the larger environment, you must always think of information systems as open and constantly interacting with the environment.

There are several other important systems concepts with which systems analysts need to become familiar:

- Decomposition

- Modularity

- Coupling

- Cohesion

In addition, you need to understand the differences between viewing a system at a logical and at a physical level, each with associated descriptions concentrating on different aspects of a system.

Decomposition (or functional decomposition, as defined in Chapter 2) deals with being able to break down a system into its components. These components may themselves be systems (subsystems) and can be broken down into their components as well. (You may want to refer to Figure 2-4 which illustrates a simple example of how an information system can be decomposed.) How does decomposition aid understanding of a system? Decomposition results in smaller and less complex pieces that are easier to understand than larger, complex pieces. Decomposing a system also allows us to focus on one particular part of a system, making it easier to think of how to modify that one part independently of the entire system (Figure 3-5). Figure 3-6 shows the decomposition of a portable compact disc (CD) player. At the highest level of abstraction, this system simply accepts CDs and settings of the volume and tone controls as input and produces music as output. Decomposing the system into subsystems reveals the system's inner workings: there are separate systems for reading the digital signals from the CDs, for amplifying the signals, for turning the signals into sound waves, and for controlling the volume and tone of the sound. Breaking the subsystems down into their components would reveal even more about the inner workings of the system and greatly enhance our understanding of how the overall system works.

Modularity, a direct result of decomposition, refers to dividing a system up into chunks or modules of a relatively *uniform* size. Modules can represent a system simply, making it not only easier to understand but also easier to redesign and rebuild.

Open system: A system that interacts freely with its environment, taking input and returning output.

Closed system: A system that is cut off from its environment and does not interact with it.

Modularity: Dividing a system up into chunks or modules of a relatively uniform size.

Figure 3-5
Purposes of decomposition

DECOMPOSITION FUNCTIONS
Decomposition aids a systems analyst and other systems development
project team members by

- Breaking a system into smaller, more manageable and understandable
  subsystems
- Facilitating the focusing of attention on one area (subsystem) at a time
  without interference from other parts
- Allowing attention to concentrate on the part of the system pertinent to a
  particular audience, without confusing people with details irrelevant to
  their interests
- Permitting different parts of the system to be built at independent times
  and/or by different people

Figure 3-6
An example of system decomposition

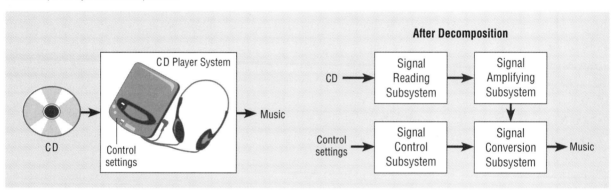

Coupling is the extent to which subsystems are dependent on each other. Subsystems should be as independent as possible. If one subsystem fails and other subsystems are highly dependent on it, the others will either fail themselves or have problems functioning. Looking at Figure 3-6, we would say the components of a portable CD player are tightly coupled. The amplifier and the unit that reads the CD signals are wired together in the same container, and the boundaries between these two subsystems may be difficult to draw clearly. If one subsystem fails, the entire CD player must be sent off for repair. In a home stereo system, the components are loosely coupled since the subsystems, such as the speakers, the amplifier, the receiver, and the CD player, are all physically separate and function independently. For example, if the amplifier in a home stereo system fails, only the amplifier needs to be repaired.

Coupling: The extent to which subsystems depend on each other.

Finally, **cohesion** is the extent to which a subsystem performs a single function. In biological systems, subsystems tend to be well differentiated and thus very cohesive. In man-made systems, subsystems are not always as cohesive as they should be. We will discuss the concepts of coupling and cohesion and how they apply to information systems in detail in Chapter 17.

Cohesion: The extent to which a system or a subsystem performs a single function.

One final key systems concept with which you should be familiar is the difference between logical and physical systems. Any description of a system is abstract since the definition is not the system itself. When we talk about logical and physical systems, we are actually talking about logical and physical system descriptions.

A **logical system description** portrays the purpose and function of the system without tying the description to any specific physical implementation. For example,

Logical system description:
Description of a system that focuses on the system's function and purpose without regard to how the system will be physically implemented.

in developing a logical description of the portable CD player, we describe the basic components of the player (signal reader, amplifier, speakers, controls) and their relations to each other, focusing on the function of playing CDs using a self-contained, portable unit. We do not specify whether the earphone jack contains aluminum or gold, where we could buy the laser that reads the CDs, or how much the jack or the laser cost to produce.

Physical system description: Description of a system that focuses on how the system will be materially constructed.

The **physical system description,** on the other hand, is a material depiction of the system, a central concern of which is building the system. A physical description of the portable CD player would provide details on the construction of each sub-unit, such as the design of the laser, the composition of the earphones, and whether the controls feature digital readouts. A systems analyst should deal with function (logical system description) before form (physical system description), just as an architect does for the analysis and design of buildings.

Benefiting from Systems Thinking   The first step in systems thinking is to be able to identify something as a system. This identification also involves recognizing each of the system's characteristics, for example, identifying where the boundary lies and all of the relevant inputs. But once you have identified a system, what is the value of thinking of something as a system? Visualizing a set of things and their inter-relationships as a system allows you to translate a specific physical situation into more general, abstract terms. From this abstraction, you can think about the essential characteristics of a specific situation. This in turn allows you to gain insights you might never get from focusing too much on the details of the specific situation. Also, you can question assumptions, provide documentation, and manipulate the abstract system without disrupting the real situation.

Let's look again at Hoosier Burger. How can visualizing a fast food restaurant as a system help us gain insights about the restaurant that we might not get otherwise? Let's imagine that Hoosier Burger is facing more demand for its food than it can handle. Some people are convinced that its hamburgers are the best in Bloomington, maybe even in Southern Indiana. Many people, especially IU students and faculty, frequently eat at Hoosier Burger, and the staff is having a difficult time keeping up with the demand. For the owner-managers, Bob and Thelma Mellankamp, the high level of demand is both a problem and an opportunity. The problem is that if the restaurant can't keep up with demand, people will stop coming to eat here, and the owners will lose money. The opportunity is to capitalize on Hoosier Burger's popularity and serve even more customers every day, making larger profits for the owners (which is the purpose of their system).

How does looking at Hoosier Burger as a system help? By decomposing the restaurant into subsystems, we can analyze each subsystem separately and discover if one or more subsystems is at capacity. Capacity is a general problem common to many systems. Let's say, after careful study, we discover that the kitchen, storage, and dining room subsystems have plenty of available capacity. However, the counter is unable to handle the rush of people. Customers have to wait in line for several minutes to place and receive their orders. The counter is the restaurant's bottleneck; thus the capacity of the counter needs to be increased. If we redesign the counter area or the procedures for taking customer orders, then we can increase the counter's capacity and better match it to the kitchen's capacity. Customers will have to wait in line less time to place their orders and they will get their food faster. Fewer customers will turn away because of long lines, which should translate into more food sold and higher profits.

There are other aspects of the system we could have examined, such as outputs, inputs, or environmental conditions, but to make the example more clear and concise, we looked only at subsystems. For this particular problem, decomposing

Hoosier Burger into its subsystems enabled us to determine its problem with demand. Other problems may have required an examination of all aspects of the restaurant system.

Applying Systems Thinking to Information Systems   None of the examples of systems we have examined so far in this chapter have been information systems, even though information systems are the focus of this book. There are two reasons why we have looked at other types of systems first. One is so that you will become accustomed to thinking of some of the many different things you encounter daily as systems and realize how useful systems thinking can be. The second is that thinking of organizations as systems is a useful perspective from which to begin developing information systems. *Information systems can be seen as subsystems in larger organizational systems, taking input from, and returning output to, their organizational environments.*

Let's examine a simplified version of an information system as a special kind of system. In our fast food restaurant example, Hoosier Burger uses an information system to take customer orders, send the orders to the kitchen, monitor goods sold and inventory, and generate reports for management. The information system is depicted as a data flow diagram in Figure 3-7 (data flow diagrams were introduced in Figure 2-4, and you will learn how to draw data flow diagrams in Chapter 9).

As the diagram illustrates, Hoosier Burger's customer order system contains four components or subsystems: Process Customer Food Order, Update Goods Sold File, Update Inventory File, and Produce Management Reports. The arrows in the diagram show how these subsystems are inter-related. For example, the first process produces four outputs: a Kitchen Order, a Receipt, Goods Sold data, and

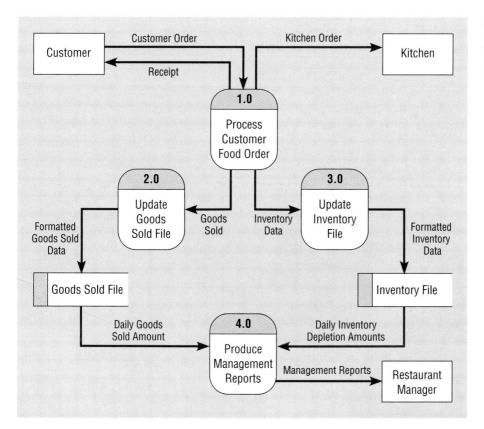

Figure 3-7

A fast food restaurant's customer order information system depicted in a data flow diagram

Inventory Data. The latter two outputs serve as input for other subsystems. The dotted line illustrates the boundary of the system. Notice that the Customer, the Kitchen, and the Restaurant Manager (Bob Mellankamp) are all considered to be outside the customer order system. The specific purpose of the system is to facilitate customer orders, monitor inventory, and generate reports; the system's general purpose is to improve the efficiency of the restaurant's operations.

Since this information system is smaller in scope and purpose than the Hoosier Burger system itself, its environment is also smaller. For our purposes, we can limit the environment to those entities that interact with the system: Customers, the Kitchen, and the Restaurant Manager. Constraints on the system may or may not be apparent from the diagram. For example, the diagram implicitly shows (by omission) that there is no direct data exchange between the customer order system and information systems used by the restaurant's suppliers; this prevents the system from automatically issuing an order for supplies directly to the suppliers when inventory falls below a certain level. We do not know, however, if any other Hoosier Burger system supports such direct data exchange. Another constraint may be the system's inability to provide on-line, real-time information on inventory levels, limiting Bob Mellankamp to receiving nightly batched reports. This is not at all clear from Figure 3-7. In contrast, system input and output are very clear. The only system input is the Customer Order and there are three overall system outputs: a Receipt for the customer, a Kitchen Order, and Management Reports.

On one level of analysis and description, Hoosier Burger's customer order system is a physical system that takes input, processes data, and returns output. The physical system consists of a computerized cash register that a clerk uses to enter a customer order and return a paper receipt to the customer. Another piece of paper, the kitchen order, is generated from a printer in the restaurant's kitchen. The cash register sends data on the order about goods sold and inventory to a computer in Hoosier Burger's office, where computer files on goods sold and inventory are updated by applications software. Other application software uses data in the Goods Sold and Inventory files to generate and print reports on a laser printer in the office.

On another level of analysis and description, Hoosier Burger's customer order system can be explained using a logical description of an information system that focuses on the flow and transformation of data. The physical system is one possible implementation of the more abstract, logical information system description. For the logical information system description, it is irrelevant whether the customer's order shows up in the kitchen as a piece of paper or as lines of text on a monitor screen. What's important is the information that is sent to Hoosier Burger's kitchen. For every logical information system description, there can be several different physical implementations of it.

The way we draw information systems shows how we think of them as systems. Data flow diagrams clearly illustrate inputs, outputs, system boundaries, the environment, subsystems, and inter-relationships. Purpose and constraints are much more difficult to illustrate and must therefore be documented using other notations. In total, all elements of the logical system description must address all nine characteristics of a system.

Organizational Knowledge

As a systems analyst, you will work in organizations. Whether you are an in-house or contract custom software developer, you must understand how organizations work. In addition, you must understand the functions and procedures of the particular organization (or enterprise) you are working for. Furthermore, many of the systems you will build or maintain serve one organizational department and you

TABLE 3-1 Selected Areas of Organizational Knowledge for a Systems Analyst

---

How Work Officially Gets Done in a Particular Organization

Terminology, abbreviations, and acronyms

Policies

Standards and procedures

Standards of practice

Formal organization structure

Job descriptions

Understanding the Organization's Internal Politics

Influence and inclinations of key personnel

Who the experts are in different subject areas

Critical incidents in the organization's history

Informal organization structure

Coalition membership and power structures

Understanding the Organization's Competitive and Regulatory Environment

Government regulations

Competitors, domestic and international

Products, services, and markets

Role of technology

Understanding the Organization's Strategies and Tactics

Short- and long-term strategy and plans

Values and mission

---

must understand how that department operates, its purpose, its relationships with other departments and, if applicable, its relationships with customers and suppliers. Table 3-1 lists various kinds of organizational knowledge that a systems analyst must acquire in order to be successful.

In Chapter 2 you saw how Chris Martin, the lead systems analyst who worked on the Purchasing Fulfillment System for Pine Valley Furniture, was already aware of some of the problems in Purchasing from his past assignments with PVF. Throughout the period covered in Chapter 2, Chris continued to study and learn more about the business operations in Purchasing. A common complaint about systems analysts is that they do not understand the nature of the business. Good analysts take the time to understand the specifics of each business, as Chris did.

Problem Identification

What is a problem? Pounds (1969) defines a problem as the difference between an existing situation and a desired situation. For him, the process of identifying problems is the process of defining differences, so problem solving is the process of finding a way to reduce differences. According to Pounds, a manager defines differences by comparing the current situation to the output of a model that predicts what the output should be. For example, at Hoosier Burger, a certain portion of the food ordered from local produce distributors is expected to go bad before it can be used. Comparing a current food spoilage rate of 10 percent to a desired spoilage rate of 5 percent defines a difference and therefore identifies a problem. In this case, Bob Mellankamp has used a model to determine the desired spoilage rate of 5 percent. The particular model used, showing how fast produce ripens after harvesting, typical delivery times, and how long produce will stay fresh in a refrigerator, has come from research carried out at Purdue University's College of Agriculture. Based on the research, the Mellankamps have set a standard of a 5 percent spoilage rate, with an acceptable variance of 2 percent in either direction. According to this standard, a 5 percent variance between desired and actual is clearly out of line and merits attention. Another model might have indicated that a 10 percent spoilage rate was acceptable. You can see that understanding how managers identify problems is understanding the models they use to define differences.

In order to identify problems that need solving, you must be able to compare the current situation in an organization to the desired situation. You must develop a repertoire of models to define the differences between what is and what ought to be. It is also important that you appreciate the models that information systems users rely on to identify problems. Every functional area of the organization will use different models to find problems; what is helpful in accounting will not necessarily work well in manufacturing. Often you must be able to see problems from a broader perspective. By relying on models from their own particular functional areas, users may not see the real problem from an organizational view.

Problem Analyzing and Solving

Once a problem has been identified, you must analyze the problem and determine how to solve it. Analysis entails finding out more about the problem. Systems analysts learn through experience, with guidance from proven methods, how to get the needed information from people as well as from organizational files and documents. As you seek out additional information, you also begin to formulate alternative solutions to the problem. Devising solutions leads to a search for more information, which in turn leads to improvements in the alternatives. Obviously, such a process could continue indefinitely, but at some point, the alternatives are compared and typically one is chosen as the best solution. Once the analyst, users, and management agree on the general suitability of the solution, they devise a plan for implementing it.

The approach for analyzing and solving problems we describe was formally described by Herbert Simon and colleagues (Simon, 1960) and is the same approach Chris and his project team followed for analyzing the Purchasing problems at PVF in Chapter 2. The approach has four phases: intelligence, design, choice, and implementation. During the intelligence phase, all information relevant to the problem is collected. During the design phase, alternatives are formulated and, during choice the best alternative solution is chosen. The solution is put into practice during the implementation phase.

This problem-analysis and -solving approach should be familiar to you: it is essentially the same general process as that described earlier in the systems development life cycle (see Figure 3-8). Simon's intelligence phase corresponds roughly to the first three phases in the life cycle: project identification and selection, project initiation and planning, and analysis. Simon's design phase corresponds to that part of analysis where alternative solutions are formulated. The detailed solution

Figure 3-8
The systems development life cycle and Simon's problem-solving model

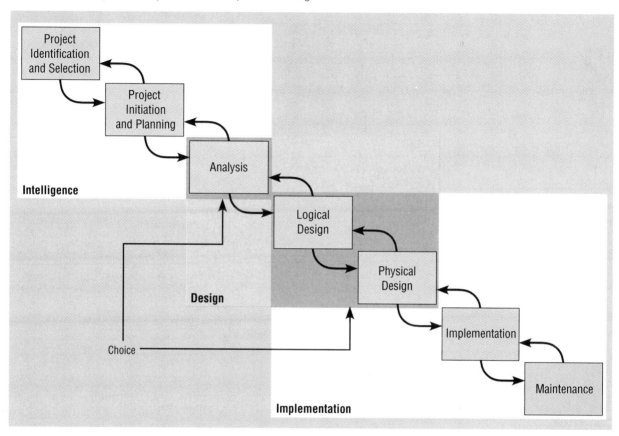

formulation (once the solution is chosen), however, would be performed in the life cycle's logical and physical design phases. Choice of the best solution is made in stages, first at the end of the analysis phase and then during physical design. In our life cycle model, activities that occur in logical design, physical design, implementation, and maintenance correspond to Simon's implementation phase.

Simon's problem-solving model is a useful one which lends insight into how people solve certain kinds of problems, but there are other factors in organizations that influence how problems are solved. Among these are personal interests, political considerations, and limits in time and cognitive ability that affect how much information people can gather and process. We will say more about these factors in later chapters; now we will turn to an examination of the technical skills required of systems analysts.

TECHNICAL SKILLS

Many aspects of your job as a systems analyst are technically oriented. In order to develop computer-based information systems, you must understand how computers, data networks, database management and operating systems, and a host of other technologies work as well as their potential and limitations. Further, you must be technically adept with different notations for representing, or modeling, various aspects of information systems. You need these technical skills not only to perform tasks assigned to you but also to communicate with the other people with whom you work in systems development (see Chapter 1 for a discussion of the roles of various people in systems analysis and design). Rather than develop a single set of technical skills to use throughout your career, you must constantly re-educate yourself about information technology, techniques, and methodologies. These information technology, techniques, and methodologies change quickly, and you must keep up with the changes. You need to understand alternative technologies (like Microsoft Windows and UNIX operating environments) as organizational preferences, since choices vary across companies and over time. Versatility, based on a sound understanding of technical concepts rather than specific tools, gives you the flexibility needed for such a changing skill set.

The following activities will help you stay versatile and up-to-date:

- Read trade publications (for example, *Computerworld, Datamation,* or *PCWeek*) and books.

- Join professional societies (for example, the *Data Processing Managers Association* or the *Association for Computing Machinery*) or other clubs and organizations interested in computing technologies; read their publications and attend their meetings.

- Attend classes or teach at a local college. Teaching is a wonderful way to force yourself to stay current and to learn from others.

- Attend any courses or training sessions offered by your organization.

- Attend professional conferences, seminars, or trade shows.

- Participate in electronic bulletin boards, news groups, or conferences on local, national, or international networks.

Maybe you have seen the cartoon of the person wearing tattered clothes, looking thin, sitting on a park bench feeding the birds. The caption reads, "He was an outstanding systems analyst, but he took a six-month vacation and fell too far behind in his field." Being a systems analyst working in the systems field requires continuous learning.

Because of the rapid changes that occur in technology, we do not dwell on specifics in this section. For example, when this book was being written, object-oriented database technology was considered new and experimental. It is quite possible, however, that this technology may be popular and widespread when you read this book. In general, you should be as familiar as possible with such families of technologies as

- Microcomputers, workstations, minicomputers, and mainframe computers

- Programming languages

- Operating systems, both for single machines and networks

- Database and file management systems

- Data communication standards and software for local and wide area networks

- Systems development tools and environments (such as form and report generators and graphical interface design tools)

- Decision support system generators and data analysis tools as well as modern methods and techniques for describing, modeling, and building systems. How technical you must be will vary by job assignment and where you are in your career. Often, you will be asked to be more technical in the early stages of your career, and then you will assume more managerial responsibilities as you gain experience. We discuss career progression later in this chapter.

MANAGEMENT SKILLS

Systems analysts are almost always members of project teams and are frequently asked to lead teams. Management skills are very useful for anyone in a leadership role. As an analyst, you also need to know how to manage your own work and how to use organizational resources in the most productive ways possible. Self-management, then, is an important skill for an analyst. In this section, we describe four categories of management skills: resource, project, risk, and change management.

Resource Management

Any organizational worker must know how to obtain and work effectively with or-ganizational resources. You saw an example of an analyst practicing sound resource management in Chapter 2, when Chris Martin headed the Purchasing system proj-ect for Pine Valley Furniture. A systems analyst like Chris must know how to get the most out of a wide range of resources: system documentation, information tech-nology, and money. For an analyst leading a team, the most important resource is people. A team leader must learn how to best utilize the particular talents of other team members. He or she must also be able to delegate responsibility, empowering people to do the tasks they have been assigned.

    Resource management includes the following capabilities:

- Predicting resource usage (budgeting)
- Tracking and accounting for resource consumption
- Learning how to use resources effectively
- Evaluating the quality of resources used
- Securing resources from abusive use
- Relinquishing resources when no longer needed and obsoleting resources when they can no longer be useful

Project Management

Effectively managing projects is crucial to a systems analyst's job. Chris Martin was successful in managing the PVF Purchasing system project because he had learned how to manage projects during his six years at PVF. Information systems devel-opment projects range from one-person projects that take very little time and effort to multi-person, multi-year efforts costing millions of dollars. The goal of project management is to prevent projects from coming in late and going over budget. In addition, project management is designed to help managers keep track of the proj-ect's progress.

    Even if you are not a project leader like Chris Martin, you will be given respon-sibilities for parts of a project, or subprojects. In the role of project or subproject manager, you first need to decompose a (sub)project into several independent tasks.

The next step is to determine how the tasks are related to each other and who will be responsible for each task. As we will see in Chapter 4, analysts use established tools and techniques to help manage projects. The most important element, however, is managing the people working on the project. Successful analysts motivate people to work together and instill a sense of trust and interdependence among them. Project management extends beyond the organization to any vendors or contractors working on the project.

Risk Management

Risk management is the ability to anticipate what might go wrong in a project. Once risks to the project have been identified, you must be able to minimize the likelihood that those risks will actually occur. If minimizing risk is not possible, then you try to minimize the damage that might result. Risk management also includes knowing where to place resources (such as people) where they can do the most good and prioritizing activities to achieve the greatest gain. We discuss a key part of risk management that is carried out during project justification—risk assessment—in Chapter 7.

Change Management

Introducing a new or improved information system into an organization is a change process. In general, people do not like change and tend to resist it; therefore, any change in how people perform their work in an organization must be carefully managed. Change management, then, is a very important skill for systems analysts, who are organizational *change agents.* You must know how to get people to make a smooth transition from one information system to another, giving up their old ways of doing things and accepting new ways. Change management also includes the ability to deal with technical issues related to change, such as obsolescence and reusability. You will learn more about managing the change that accompanies a new information system in Chapter 19.

INTERPERSONAL SKILLS

Although, as a systems analyst, you will be working in the technical area of designing and building computer-based information systems, you will also work extensively with all types of people. Perhaps the most important skills you will need to master are interpersonal. In this part of the chapter, we will discuss the various interpersonal skills necessary for successful systems analysis work: communication skills; working alone and with a team; facilitating groups; and managing expectations of users and managers.

Communication Skills

The single most important interpersonal skill for an analyst, as well as for any professional, is the ability to communicate clearly and effectively with others. In Chapter 2, Chris Martin was able to successfully communicate with users, other information systems professionals, and management. Chris established a good, open working relationship with his client Juanita Lopez early in the project and maintained it throughout by communicating effectively. Chris used phone calls and meetings to keep stakeholders informed, involved a variety of people in the project to keep them aware of and committed to the project, and developed and distributed in advance a clear agenda for each meeting.

Communication takes many forms, from written (memos, reports) to verbal (phone calls, face-to-face conversations) to visual (presentation slides, diagrams). The analyst must be able to master as many forms of communication as possible. Oral communication and listening skills are considered by many information system professionals as the most important communication skills analysts need to succeed. Interviewing skills are not far behind. All types of communication, however, have one thing in common: They improve with experience. The more you practice, the better you get. Some of the specific types of communication we will mention are interviewing and listening, the use of questionnaires, and written and oral presentations.

Interviewing, Listening, and Questionnaires   Interviewing is one of the primary ways analysts gather information about an information systems project. Early in a project, you may spend a large amount of time interviewing users about their work and the information they use. There are many ways to effectively interview someone, and becoming a good interviewer takes practice. We will discuss interviewing in more detail in Chapter 8, but it is important to point out now that asking questions is only one part of interviewing. Listening to the answers is just as important, if not more so. Careful listening helps you understand the problem you're investigating and, many times, the answers to your questions lead to additional questions that may be even more revealing and probing than the questions you prepared before your interview.

Although interviews are very effective ways of communicating with people and obtaining important information from them, interviews can also be very expensive and time-consuming. Because questionnaires provide no direct means by which to ask follow-up questions, they are generally less effective than interviews. It is possible, however, though time-consuming, to call respondents and ask them follow-up questions. Questionnaires are less expensive to conduct because the questioner does not have to invest the same amount of time and effort to collect the same information using a questionnaire as he or she does in conducting an interview. For example, using a written questionnaire that respondents complete themselves, you could gather the same information from 100 people in one hour that you could collect from only one person in a one-hour interview. In addition, questionnaires have the advantage of being less biased in how the results are interpreted because the questions and answers are standardized. Creating good questionnaires is a skill that comes only with practice and experience. You will learn more about questionnaire design in Chapter 8.

Written and Oral Presentations   At many points during the systems development process, you must document the progress of the project and communicate that progress to others. This communication takes the following forms:

- Meeting agenda

- Meeting minutes

- Interview summaries

- Project schedules and descriptions

- Memoranda requesting information, an interview, participation in a project activity, or the status of a project

- Requests for proposal from contractors and vendors

and a host of other documents. This documentation is essential to provide a written, not just oral, history for the project, to convey information clearly, to provide details needed by those who will maintain the system after you are off the project team, and to obtain commitments and approvals at key project milestones.

The larger the organization and the more complicated the systems development project, the more writing you will have to do. You and your team members will have to complete and file a report at the end of each stage of the systems development life cycle. The first report will be the business case for getting approval to start the project. The last report may be an audit of the entire development process. And at each phase, the analysis team will have to document the system as it evolves. To be effective, you need to write both clearly and persuasively.

As there are often many different parties involved in the development of a system, there are many opportunities to inform people of the project's status. Periodic written status reports are one way to keep people informed, but there will also be unscheduled calls for ad hoc reports. Many projects will also involve scheduled and unscheduled oral presentations. You saw in Chapter 2 how Chris Martin was involved in scheduled presentations for both making the business case and reviewing the Purchasing system's physical design at the end of the project's physical design phase. Part of oral presentations involves preparing slides, overhead transparencies, or multimedia presentations, including system demonstrations. Another part involves being able to field and answer questions from the audience.

How can you improve your communication skills? We have four simple yet powerful suggestions:

1. Take every opportunity to practice. Speak to a civic organization about trends in computing. Such groups often look for local speakers to present talks on topics of general interest. Conduct a training class on some topic on which you have special expertise. Some people have found participation in Toastmasters, an international organization with local chapters, a very helpful way to improve oral communication skills.

2. Videotape your presentations and do a critical self-appraisal of your skills. You can view videotapes of other speakers and share your assessments with each other.

3. Make use of writing centers located at many colleges as a way to critique your writing.

4. Take classes on business and technical writing from colleges and professional organizations.

Working Alone and with a Team

As a systems analyst, you must often work alone on certain aspects of any systems development project. To this end, you must be able to organize and manage your own schedule, commitments, and deadlines. Many people in the organization will depend on your individual performance, yet you are almost always a member of a team and must work with the team toward achieving project goals. As we saw in Chapter 1, working with a team entails a certain amount of give and take. You need to know when to trust the judgment of other team members as well as when to question it. For example, when team members are speaking or acting from their base of experience and expertise, you are more likely to trust their judgment than when they are talking about something beyond their knowledge. For this reason, the analyst leading the team must understand the strengths and weaknesses of the other team members. To work together effectively and to ensure the quality of the group product, the team must establish standards of cooperation and coordination that guide their work (review Table 1-2 for the characteristics of a successful team). For an example, go back to Chapter 2 and review how Chris worked with his team of information systems professionals during the Purchasing Fulfillment System project.

Facilitating Groups

Sometimes you need to interact with a group in order to communicate and receive information. In Chapter 1, we introduced you to the Joint Application Design (JAD) process in which analysts actively work with groups during systems development. Analysts use JAD sessions to gather systems requirements and to conduct design reviews. The assembled group is the most important resource the analyst has access to during a JAD and you must get the most out of that resource; successful group facilitation is one way to do that. In a typical JAD, there is a trained session leader running the show. He or she has been specially trained to facilitate groups, to help them work together, and to help them achieve their common goals. Facilitation necessarily involves a certain amount of neutrality on the part of the facilitator. The facilitator must guide the group without being part of the group and must work to keep the effort on track by ferreting out disagreements and helping the group resolve differences. Obviously, group facilitation requires training. Many organizations that rely on group facilitation train their own facilitators. Figure 3-9 lists some guidelines for running an effective meeting, a task that is fundamental to facilitating groups.

Managing Expectations

Systems development is a change process, and any organizational change is greeted with anticipation and uncertainty by organization members. Organization members will have certain ideas, perhaps based on their hopes and wishes, about what a new information system will be able to do for them; these expectations about the new system can easily run out of control. Ginzberg (1981) found that successfully managing user expectations is related to successful systems implementation. For

Figure 3-9
Some guidelines for running effective meetings (Adapted from Option Technologies, Inc. [1992])

- Become comfortable with your role as facilitator by gaining confidence in your ability, being clear about your purpose, and finding a style that is right for you.
- At the beginning of the meeting, make sure the group understands what is expected of them and of you.
- Use physical movement to focus on yourself or on the group, depending on which is called for at the time.
- Reward group member participation with thanks and respect.
- Ask questions instead of making statements.
- Be willing to wait patiently for group members to answer the questions you ask them.
- Be a good listener.
- Keep the group focused.
- Encourage group members to feel ownership of the group's goals and of their attempts to reach those goals.

you to successfully manage expectations, you need to understand the technology and what it can do. You must understand the work flows that the technology will support and how the new system will affect them. More important than understanding, however, is your ability to communicate a realistic picture of the new system and what it will do for users and managers. Managing expectations begins with the development of the business case for the system and extends all the way through training people to use the finished system. You need to educate those who have few expectations as well as temper the optimism of those who expect the new system to perform miracles.

SYSTEMS ANALYSIS AS A PROFESSION

Even though systems analysis is a relatively new field, those in the field have established standards for education, training, certification, and practice. Such standards are required for any profession.

Whether or not systems analysis is a profession is open to debate. Some feel systems analysis is not a profession because it simply has not been around long enough to have established the rigorous standards that define a profession. Others feel that at least some standards are already in place. There are guidelines for college curricula and there are standard ways of analyzing, designing, and implementing systems. Professional societies that systems analysts may join include the Society for Information Management, the Data Processing Managers Association (DPMA), and the Association for Computing Machinery (ACM). There is a Certified Data Processing Certificate (CDP) exam, much like the Certified Public Accountant (CPA) exam, that you can take to prove your competency in the field although, unlike the CPA certificate, very few jobs and employers in the IS field require you to have the CDP certificate. Codes of ethics to govern behavior also exist. In this section, we will discuss several aspects of a systems analyst's job: standards of practice, the ACM code of ethics, and career paths for those choosing to become systems analysts.

Standards of Practice

Standard methods or practices of performing systems development are emerging that make systems development less of an art and more of a science. Standards are developed through education and practice and spread as systems analysts move from one organization to another. We will focus here on four standards of practice: an endorsed development methodology, approved development platforms, well-defined roles for people in the development process, and a common language.

There are several different development methodologies now being used in organizations. Although there is no standardization of a single methodology across all organizations, a few prominent methodologies are in common use. An *endorsed development methodology* lays out specific procedures and techniques to be used during the development process. These standards are central to promoting consistency and reliability in methods across all of an organization's development projects. Some methodologies are spread through the work of well-known consultants, such as James Martin's Information Engineering™. Others are spread through major consulting firms, such as Andersen Consulting's Foundation/1™.

Closely associated with endorsed methodologies are approved development platforms. Some methodologies are closely tied to platforms, as is Andersen's methodology and their Foundation® CASE tool. Other methodologies are more adaptable and can work in close accordance with development platforms that exist in the organization, such as database management systems and 4GLs. The point is that organizations, and hence the analysts who work for them, are standardizing around specific platforms, and standards for development emerge from this standardization.

Roles for the various people involved in the development process are also becoming standardized. End users, managers, and analysts are each assigned certain responsibilities for development projects. The training that analysts receive in college, on their first jobs, and during their interactions with other analysts, combine to create a gestalt of the analyst's job. For example, as you study this book and talk about systems development in your class, you are forming certain ideas about what systems analysts do and how systems are developed in organizations. Your ideas are also shaped and reinforced by the other IS courses you take in college. Once you get your first job, you will receive additional training and you will adjust your understanding of systems analysis accordingly. As you gain experience working on projects and interacting with other analysts, who may have been trained at other universities and in other organizations, your ideas will continue to change and grow, but the basic core of what systems analysis means to you will have been established. Many of the experiences you have on the job will reinforce much of what you have already learned about systems analysis. When you leave an organization and go to work elsewhere, you will carry your understanding of systems analysis with you. Over time, as you and other analysts change jobs and move from one organization to another, what it means to be an analyst becomes standardized across organizations, and the standards of practice in the field help define what it means to be an analyst.

Another factor moving the job of the systems analyst toward professionalism is the development of a common language analysts use to talk to each other. Analysts communicate on the job, at meetings of professional societies, and through publications. As analysts develop a special language for communication among themselves, their language becomes standardized. For example, since the late 1970s, systems analysts have begun to rely on data flow diagramming as a means of communication. There are now two primary standards for data flow diagramming: Yourdon, and Gane and Sarson (see Chapter 9). In time, only one may be used. Other examples of communication becoming standardized include the widespread use of common programming languages such as COBOL and C and the spread of SQL as the language of choice for data definition and manipulation for relational databases. As their common language develops, analysts become more cohesive as a group—a characteristic of professions.

Ethics

The ACM is a large professional society made up of information system professionals and academics. It has over 85,000 members. Founded in 1947, the ACM is dedicated to promoting information processing as an academic discipline and to encouraging the responsible use of computers in a wide range of applications. Because of its size and membership, it has much influence in the information systems community. The ACM has developed a code of ethics for its members called the "ACM Code of Ethics and Professional Conduct." The full statement is reproduced in Figure 3-10. The code applies to all ACM members and directly applies to systems analysts.

Note the emphasis in the Code on personal responsibility, on honesty, and on respect for relevant laws. Notice also that compliance with a code of ethics such as this one is voluntary, although article 4.2 calls for, at a minimum, peer pressure for compliance. No one can force an information systems professional to follow these guidelines. However, it is voluntary compliance with the guidelines that makes someone a professional in the first place. Notice that for leaders there is the burden

Figure 3-10
ACM Code of Ethics and Professional Conduct, Revision Draft No. 19 (9/19/91), used with permission

Association for Computing Machinery Professional Code of Ethics

Preamble

These statements of intended conduct are expected of every member (voting members, associate members, and student members) of the Association for Computing Machinery (ACM). Section 1.0 consists of fundamental ethical considerations; section 2.0 includes additional considerations of professional conduct; statements in 3.0 pertain to individuals who have a leadership role; and section 4.0 deals with compliance. ACM shall prepare and maintain an additional document for interpreting and following this Code.

(1.0) General Moral Imperatives

(As an ACM member I will . . .)
(1.1) Contribute to society and human well-being.
(1.2) Avoid harm to others.
(1.3) Be honest and trustworthy.
(1.4) Be fair and take action not to discriminate.
(1.5) Respect property rights (Honor copyrights and patents; give proper credit; not steal, damage, or copy without permission).
(1.6) Respect the privacy of others.
(1.7) Honor confidentiality.

(2.0) Additional Professional Obligations

(As an ACM computing professional I will . . .)
(2.1) Strive to achieve the highest quality in the processes and products of my work.
(2.2) Acquire and maintain professional competence.
(2.3) Know and respect existing law pertaining to my professional work.
(2.4) Encourage review by peers and all affected parties.
(2.5) Give well-grounded evaluations of computer systems, their impacts, and possible risks.
(2.6) Honor contracts, agreements, and acknowledged responsibilities.
(2.7) Improve public understanding of computing and its consequences.

(3.0) Organizational Leadership Imperatives

(As an organizational leader I will . . .)
(3.1) Articulate social responsibilities of members of the organizational unit and encourage full participation in these responsibilities.
(3.2) Shape information systems to enhance the quality of working life.
(3.3) Articulate proper and authorized uses of organizational computer technology and enforce those policies.
(3.4) Ensure participation of users and other affected parties in system design, development and implementation.
(3.5) Support policies that protect the dignity of users and others affected by a computerized system.
(3.6) Support opportunities for learning the principles and limitations of computer systems.

(4.0) Compliance with Code

(4.1) I will uphold and promote the principles of this Code.
(4.2) If I observe an apparent violation of this Code, I will take appropriate action leading to a remedy.
(4.3) I understand that violation of this Code is inconsistent with continued membership in the ACM.

of educating non-IS professionals about computing—about what computing can and cannot do. The Code also expresses concern for the quality of work life and for protecting the dignity and privacy of others when performing professional work, such as developing information systems.

Though not written specifically for systems analysts, the ACM Code of Ethics can easily be adapted to the systems analysis job. Many systems development projects deal directly with many of the issues addressed in the Code: privacy, quality of work life, user participation, and managing expectations. When an analyst must confront one or more of these issues, the Code can be used as a guide for professional conduct.

Career Paths

A typical first job for a recent college graduate who wants to become a systems analyst is as an analyst or programmer/analyst trainee for a corporation or large consulting firm. Other typical entry-level opportunities are as

- End-user support specialist, assisting non-IS professional and clerical staff to better use computer resources

- Decision support analyst, in which you design database queries and data analysis routines to support business analysis and decision making, often for one department, such as market research or investments

- Trainer, in which you prepare and conduct various classes on information systems and technologies

- Computer technology sales and customer support, in which you either sell hardware, software, or services or support the sales staff by installing technology and responding to customer questions

Larger firms usually have their own intensive training programs to instruct trainees in the way the firm develops and maintains systems. Every firm handles systems development a little bit differently. Once trained, the entry-level systems analyst begins a career path, which differs from one firm to another.

For example, in one particular firm where the authors have done research—which we'll call the JKL Corporation—an analyst typically begins as an assistant analyst. The qualifications for this position are a bachelor's degree with little or no experience. Typically after two years of work, the assistant analyst is promoted to an associate analyst. Another two to three years of work experience qualify the associate analyst to become an analyst. The next two steps up the ladder are senior analyst and staff analyst. Of course, not everyone makes it to staff analyst. Many analysts leave after one or two promotions to work for other firms or settle into a position along the way.

At this point in his or her career at JKL Corp., the staff analyst must decide whether he or she wants to pursue a management ladder or a technical ladder. Analysts who are more technically oriented might prefer not to follow a management career path (see Figure 3-11). The management ladder continues upward with the positions of project leader, project manager, and manager. A manager is responsible for planning, developing, and implementing information technology at JKL Corp., for training and staff development, and for establishing and managing budgets. The technical ladder continues upward with the positions of consultant, staff consultant, and senior staff consultant. A senior staff consultant is responsible for major research and development projects. Creativity and vision are necessary for successful performance as a senior staff consultant.

It should be pointed out that not everyone hired by JKL Corp. starts out as an assistant analyst and works his or her way up in the corporation. Some employees are hired directly into higher-level jobs, based on their work experience and education. For example, someone with a master's degree and little or no experience can

Figure 3-11
Information systems career
ladder for the JKL Corp.

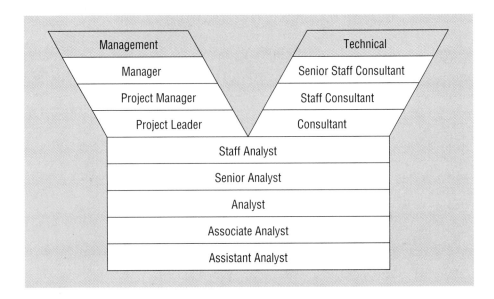

Figure 3-11
Information systems career ladder for the JKL Corp.

be hired directly into the associate analyst position. On the other hand, a person with a bachelor's degree and 14 or more years of proven experience can be hired directly into the manager's position.

Besides progression within the IS organization, some IS professionals choose to make a lateral move into general management positions and then return to senior IS management positions later in their careers. IS is an excellent entry point into any general business career since the process of systems analysis and design gives you a thorough understanding of the business, and your acquired general business knowledge is extremely valuable in general management positions.

For those who enter the consulting world, career progression is typically geared toward becoming a partner or principle of the business. These positions often focus on generating business and maintaining customer relationships. Another interesting option for people with more of an entrepreneurial spirit is to start your own business. You may develop expertise in a certain area or have a unique idea for some software tool that can flourish only in your own company. Many niche products, such as specialized data analysis tools or systems for particular types of functions or businesses, exist in the information systems field. Many of these were created by systems analysts who, after years of experience and continuing development, got a bright idea, networked with contacts, acquired funding, and started their own business. This is one way to become president of the company!

SUMMARY

In this chapter, we have surveyed the skills necessary for success as a systems analyst. The requisite skills are analytical, technical, management, and interpersonal. Analytical skills include the concept of systems thinking, which is one of the most important skills an analyst can learn. Systems thinking provides a disciplined foundation on which all other analyst skills can build. In addition, an analyst needs to understand the nature of business and of the particular enterprise he or she serves and to be able to identify, analyze, and solve problems.

Technical skills change over time as technology changes and analysts need to keep current with changing information technology. This can be accomplished through reading trade journals, joining professional societies, attending or teaching classes, attending conferences, and participating in electronic bulletin boards and news groups. Some technology areas that play a continuing important role are programming languages, operating systems, database management systems, data communications, and systems development techniques and tools.

A useful skill is the ability to manage resources, projects, risk, and change. Interpersonal skills, especially clear communication, are also important. Analysts communicate with team members in interviews, with questionnaires, through written and oral presentations, and through facilitating groups. A key component of communicating about information systems is managing the expectations of both users and managers.

The chapter concluded with an examination of the system analyst's position, the standards of practice, the ACM Code of Ethics, and possible career paths. Systems analysis is becoming more of a science and less of an art as the systems analysis field becomes a profession.

C H A P T E R   R E V I E W

K E Y   T E R M S

Boundary	Environment	Open system
Closed system	Input	Output
Cohesion	Interface	Physical system
Components	Inter-related components	description
Constraints	Logical system description	Purpose
Coupling	Modularity	System

R E V I E W   Q U E S T I O N S

1. Define each of the following terms:
   a. system
   b. interface
   c. boundary
   d. purpose
   e. modularity

2. What is systems thinking? How is it useful for thinking about computer-based information systems?

3. What is decomposition? coupling? cohesion?

4. In what way are organizations systems?

5. What are the differences between problem identification and problem solving?

6. How can a systems analyst determine if his or her technical skills are up-to-date?

7. Explain the management skills needed by systems analysts.

8. Which communication skills are important for analysts? Why?

9. Is systems analysis a profession? Why or why not?

10. What is a code of ethics?

11. What's the difference between a logical system description and a physical system description?

12. Which areas of organizational knowledge are important for a systems analyst to know?

13. What's the difference between an open and a closed system?

14. What kinds of tasks are included in resource management?

PROBLEMS AND EXERCISES

1. Match the following terms to the appropriate definitions.

   _____ closed system

   _____ open system

   _____ components

   _____ constraint

   _____ environment

   a. systems that interact freely with their environments, taking input and returning output

   b. systems that are cut off from their environments and do not interact with them

   c. everything external to a system

   d. limits to what a system can accomplish

   e. the parts or subsystems that make up a system

2. Describe your university or college as a system. What is the input? the output? the boundary? the components? their inter-relationships? the constraints? the pupose? the interfaces? the environment? Draw a diagram of this system.

3. a. A car is a system with several subsystems, including the braking subsystem, the electrical subsystem, the engine, the fuel subsystem, the climate control subsystem, and the passenger subsystem. Draw a diagram of a car as a system and label all of its system characteristics.

   b. Your personal computer is a system. Draw and label a personal computer as a system as you did for a car in part (a).

4. Describe yourself in terms of your abilities at resource, project, risk, and change management. Among these categories, what are your strengths and weaknesses? Why? How can you best capitalize on your strengths and strengthen areas where you are weak? If you do not have managerial or supervisory experience, answer these questions as if you were generalizing from your experiences thus far to your performance later as a manager.

5. Describe yourself in terms of your abilities at each of the following interpersonal skills: working alone versus working with a team, interviewing, listening, writing, presenting, facilitating a group, and managing expectations. Where are your strengths and weaknesses? Why? What can you do to capitalize on your strengths and strengthen areas where you are weak?

6. Use your imagination and hypothesize what a systems analyst would be like if he or she were a person with no personal or professional ethics. What types of systems would that person help to create, and how might they go about building such systems? What would the consequences of these actions be, with what implications for the analyst, for his or her information systems department, for his or her users, for the organization? Specifically, how would a code of ethics and professional conduct help curb the behavior of this person? This may seem like a silly exercise, but even your wildest guesses about the things an unethical analyst might do have probably happened in some setting.

7. You likely receive (and pay) one or more bills each month or semester (for example, tuition, rent, utilities, or telephone). Describe one of these billing systems as an information system. Be sure to list at least one example of each of the nine characteristics of a system for your example billing system.

8. The chapter mentioned that choosing the boundary for a system is a crucial step in analyzing and studying a system. What criteria would you use to determine where to draw a system boundary? What are the ramifications of setting too broad a boundary? Too narrow a boundary?

9. Make a list of the technical skills you have developed at school, as part of any job you've held, and on your own. Using newspaper want ads, trade journals, and other sources, determine if your technical skills are up-to-date. If not, devise a plan to update your technical skills.

FIELD EXERCISES

1. Describe an organization of your choice as an open system. What factors lead you to believe that this system is open? Describe the organization in terms of decomposition, coupling, cohesion, and modularity. What is beneficial about thinking of the organization in this way?

2. Think about a problem you have, perhaps with a grade in a class, with a job you're not satisfied with, or with a co-worker on the job. Describe the problem as a difference between "what is" and "what should be." What must happen to shift your situation from "what is" to "what should be" to bring about a situation that you are satisfied with? What specific actionable steps must you take to make this change happen? What information, if any, will you need to gather about this situation? From where, and/or from whom, must the information come? How can you get this information?

3. Choose a manager you know in any area and describe this person in terms of his or her abilities at resource, project, risk, and change management. Overall, is he or she successful or not? Why or why not?

4. Investigate where on your campus and community you could go to get help and practice with public speaking. Talk with other students, contact your instructors, look in the telephone book and directory of services at your college, and explore avenues to uncover as many sources of public speaking help you can find.

5. Many organizations have an approved technology list from which units are free to purchase hardware and software for application development. Contact the computing services unit at your college (or other organization) and find out what hardware and software are supported on your campus for administrative computing. Given this list of supported technologies, what would you infer are the technical skills required for a systems analyst at your college (or other organization)?

REFERENCES

Ginzberg, M. J. 1981. "Early Diagnosis of MIS Implementation Failure: Promising Results and Unanswered Questions." *Management Science* 27 (April): 459–78.

Martin. E. W., D. W. DeHayes, J. A. Hoffer, and W. C. Perkins. 1994. *Managing Information Technology: What Managers Need to Know.* 2nd ed. New York: Macmillan Publishing Company.

Option Technologies, Inc. 1992. *Just-In-Time Knowledge for Teams.* Mendotta Heights, MN.

Pounds, W. F. 1969. "The Process of Problem Finding." *Industrial Management Review* (Fall): 1–19.

Simon, H. A. 1960. *The New Science of Management Decision.* New York: Harper & Row.

4

Managing the Information Systems Project

After studying this chapter, you should be able to:

■ Explain the process of managing an information systems project.

■ Describe the skills required to be an effective project manager.

■ List and describe the skills and activities of a project manager during project initiation, project planning, project execution, and project close down.

■ Explain what is meant by critical path scheduling and describe the process of creating Gantt and PERT charts.

■ Explain how commercial project management software packages can be used to assist in representing and managing project schedules.

INTRODUCTION

Many aspects of information technology in general and the development of information systems in particular are more glamorous than the management of development projects. This view is underscored by a quote from a book that focuses on the management of information systems projects:

> Project management has rarely received the attention it deserves, and particularly within the computing profession, it has been overshadowed by the battles within the technological arena: Manufacturer versus manufacturer, development language versus development language, mainframe versus micro, operating system versus operating system are the stuff from which great legends are born and great leaders emerge as role models . . . Pity the humble project manager who manages to bring the general ledger system in on time, within budget, and working to the users' satisfaction (Thomsett, 1985).

As the above quotation typifies, some may not view project management to be a glamorous occupation. Yet, project management is an important aspect of the development of information systems and a critical skill for a systems analyst. The focus of project management is to assure that system development projects meet customer expectations and are delivered within budget and time constraints. This chapter describes how you can wear many different hats while managing all or part of a project.

The project manager is responsible for virtually all aspects of a systems development project: what you experience as a project manager is an environment of continual change and problem solving. In some organizations the project manager is a senior systems analyst who "has been around the block" a time or two. In others, both junior and senior analysts are expected to take on this role, managing parts of a project or actively supporting a more senior colleague who is assuming this role. Consequently, it is important that you gain an understanding of the project management process; this will become a critical skill for your future success.

In this chapter we focus on the systems analyst's role in managing information systems projects and will refer to this role as the project manager. The next section will provide you with an understanding of this role and the project management process. The following section examines techniques for reporting project plans using Gantt and PERT charts. The chapter will conclude by discussing the use of commercially available project management software that can be used to assist with a wide variety of project management activities.

MANAGING THE INFORMATION SYSTEMS PROJECT

Project: A planned undertaking of related activities to reach an objective that have a beginning and an end.

Successful projects require managing resources, activities, and tasks needed to complete the project. A **project** is a planned undertaking of a series of related activities to reach an objective that has a beginning and an end. The first question you might ask yourself is "Where do projects come from?" and, after considering all the different things that you could be asked to work on within an organization, "How do I know which projects to work on?" The ways in which each organization answers these questions can vary. For example, Chapter 2 described the process followed by Juanita Lopez and Chris Martin during the development of Pine Valley Furniture's (PVF) Purchasing Fulfillment System. As you might recall, Juanita, having observed problems with the way orders were processed and reported, contacted Chris, a systems analyst within PVF's systems development group. Together they worked on this project.

The first deliverable that Chris and Juanita produced was a System Service Request (SSR), a standard form PVF uses for requesting systems development work. This request was then evaluated by the Systems Priority Board. The Board evaluates development requests in relation to the business problems or opportunities the system will solve or create and also considers how the proposed project fits within the broader scope of the organization's Information Systems architecture and long-range development plans. Both the IS architecture and long-range systems plans are additional sources for projects. Thus, the Board considers all requests with respect to the overall plans and architecture of the organization when choosing projects to approve or reject. Since all organizations have limited time and resources, not all requests can be approved. The review board selects those projects that best meet overall organizational goals. (Chapter 6 provides you with a detailed description of the IS planning process and how projects are identified and selected.) In the case of the Purchasing Fulfillment System request, the Board found merit in the request and approved a more detailed feasibility study and, eventually, the construction of the system. Figure 4-1 provides a graphical view of the steps followed during the project initiation of the Purchasing Fulfillment System.

In summary, systems development projects are undertaken for two *primary* reasons: to take advantage of business *opportunities* or to solve business *problems*. Taking advantage of an opportunity might occur by providing an innovative service to customers through the creation of a new system. Solving a business problem could occur by modifying the way in which an existing system processes data so that more accurate or timely information is provided to users.

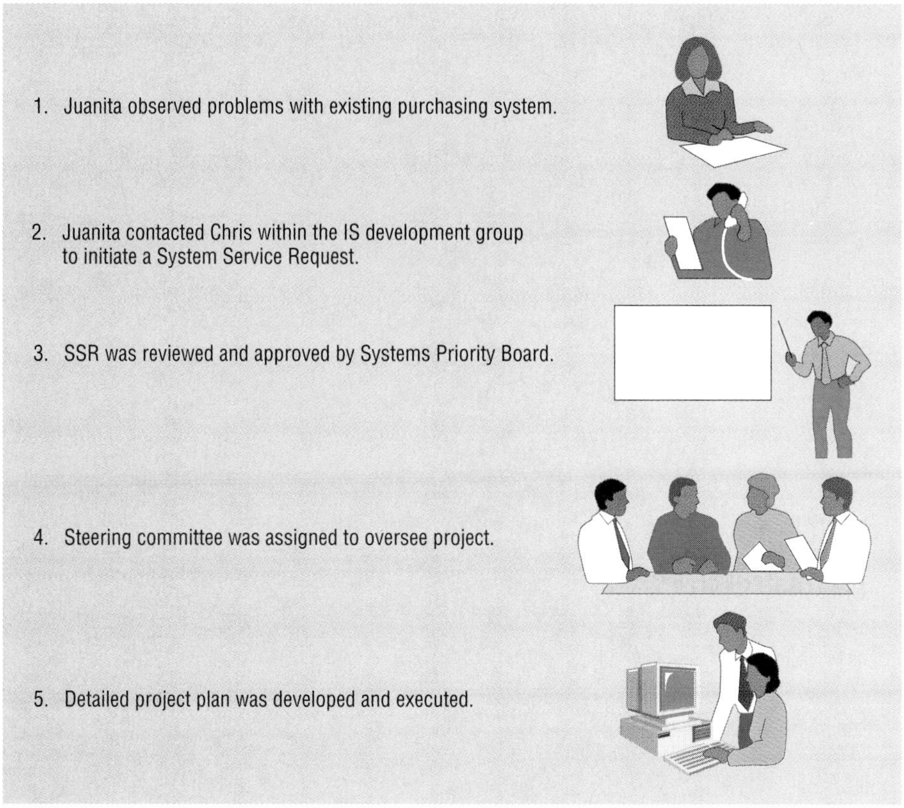

Figure 4-1
Overview of the steps followed during the Purchasing Fulfillment System project

1. Juanita observed problems with existing purchasing system.

2. Juanita contacted Chris within the IS development group to initiate a System Service Request.

3. SSR was reviewed and approved by Systems Priority Board.

4. Steering committee was assigned to oversee project.

5. Detailed project plan was developed and executed.

Projects are not always initiated for the rational reasons (solving business problems or taking advantage of business opportunities) stated above. For example, we have all read or heard of instances in organizations and in government where projects were undertaken to spend resources, to attain or pad budgets, to keep people busy, or to help train people and develop their skills. The goal of this chapter is not to examine issues related to how and why organizations undertake the projects they do but to focus on the project management process in the context of developing information systems. In other words, our focus is not necessarily on *how* and *why* organizations identify projects but on the management of projects *once they have been identified*.

Once a potential project has been identified, an organization must determine the resources required for a project's completion. This understanding is gained by analyzing the scope of the project and determining the probability of successful completion. After gaining this understanding, the organization can then determine whether solving a particular problem or taking advantage of an opportunity is feasible within time and resource constraints. If deemed feasible, a more detailed project analysis is then conducted.

The **project manager** is responsible for creating high-level feasibility plans and detailed project plans as well as staffing the project team. As you will see, determining the size, scope, and resource requirements for a project are just a few of the many skills that a project manager must possess. The skills required to be an effective project manager are varied. A project manager is often referred to as a juggler keeping aloft many balls that reflect the various aspects of a project's development (see Figure 4-2).

In order to be successful in orchestrating the construction of a complex information system, a project manager must have interpersonal skills, leadership skills,

Project manager: An individual with a diverse set of skills—management, leadership, technical, conflict management, and customer relationship—who is responsible for initiating, planning, executing, and closing down a project.

Figure 4-2
A project manager juggles numerous items during a project

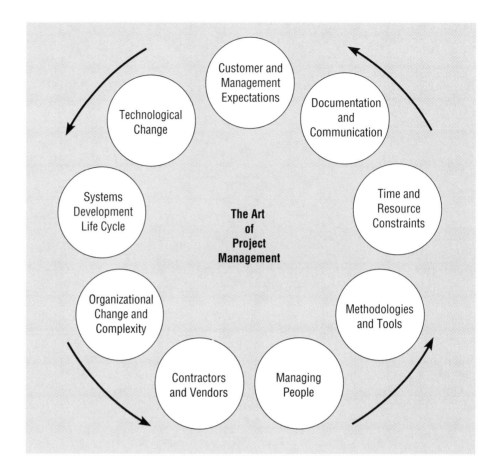

and technical skills. Table 4-1 lists many of the project manager's common skills and activities. Note that many of the skills are personnel- or general management-related, not simply technical skills. The capabilities of an effective project manager augment those of the systems analyst outlined in Chapter 3. Table 4-1 shows that not only must a project manager have varied skills, but is also arguably the most instrumental person to the successful completion of any project (Boehm, 1981). Table 4-1 also shows that a project manager has to deal with a wide variety of issues and tasks that go far beyond simply *constructing* an information system.

Because there are many people with diverse interests involved in a project (as team members, clients, stakeholders), conflict can arise. Although mitigating conflict is typically desirable, conflict can also be used to advantage. For example, Figure 4-3 describes how conflict can be used as a stimulant for creativity. Yet, too much conflict within a team has been found to make people apprehensive of making contributions due to fear of being ridiculed for having a "dumb idea." This means that the project manager must balance the negative and positive influences of conflict.

The focus of the following section is on the **project management** process. The activities involved in managing a project occur in four phases:

Project management: A controlled process of initiating, planning, executing, and closing down a project.

1. Initiating the project

2. Planning the project

3. Executing the project

4. Closing down the project

TABLE 4-1   Common Activities and Skills of a Project Manager

Activity	Description	Skill
Leadership	Influencing the activities of others toward the attainment of a common goal through the use of intelligence, personality, and abilities	Communication; liaison between management, users, and developers; assigning activities; monitoring progress
Management	Getting projects completed through the effective utilization of resources	Defining and sequencing activities; communicating expectations; assigning resources to activities; monitoring outcomes
Customer Relations	Working closely with customers to assure project deliverables meet expectations	Interpreting system requests and specifications; site preparation and user training; contact point for customers
Technical Problem Solving	Designing and sequencing activities to attain project goals	Interpreting system requests and specifications; defining activities and their sequence; making tradeoffs between alternative solutions; designing solutions to problems
Conflict Management	Managing conflict within a project team to assure that conflict is not too high or too low (see Figure 4-3)	Problem solving; smoothing out personality differences; compromising; goal setting
Team Management	Managing the project team for effective team performance	Communication within and between teams; peer evaluations; conflict resolution; team building; self management
Risk and Change Management	Identifying, assessing, and managing the risks and day-to-day changes that occur during a project	Environmental scanning; risk and opportunity identification and assessment; forecasting; resource re-deployment

Researchers and experts in the field of organizational behavior and group dynamics have repeatedly found that some conflict in project teams is not always a bad thing. Conflict can be viewed as a stimulus. For example, many effective teams like to use conflict inducing techniques such as devil's advocacy, a technique where a member of the group tries to improve team performance by surfacing all things that may be wrong with a suggested idea. Thus, at the right levels, conflict has been found to improve overall team performance. An example of conflict having a positive influence on the Microsoft Windows NT software development project team was reported in the *Wall Street Journal*: Innovation requires "dogged system-building and the ability to hold large teams together while allowing—even cultivating—conflict. Conflict lies at the core of innovation . . . When there's no conflict, a lab is no good."

Figure 4-3
Role of conflict in software teams (Adapted from Zachary [1993])

Each phase of the project management process requires that several activities be performed. Research has found that project success is strongly related to the characteristics of the process used to manage the project. In essence, following a formal project management process greatly increases the likelihood of project success. Descriptions of the activities performed during each of the project management processes are presented in the following sections.

Initiating a Project

During **project initiation** the project manager performs several activities that lay the foundation for the rest of the project. Depending upon the size, scope, and complexity of the project, some project initiation activities may be unnecessary and some may be very involved. Your goal should be to initiate the project so that you will be

Project initiation: The first phase of the project management process in which activities are performed to assess the size, scope, and complexity of the project and to establish procedures to support later project activities.

able to effectively perform subsequent project activities. The types of activities you will perform when initiating a project include the following:

1. *Establishing the Project Initiation Team.* This activity focuses on organizing an initial core of project team members who assist in accomplishing the project initiation activities. For example, during the Purchasing Fulfillment System project at PVF, Chris Martin was assigned to support the Purchasing department (see Chapter 2 for a review). It is a PVF policy that all initiation teams consist of at least one user representative, in this case Juanita Lopez, and one member of the IS development group. Therefore, the project initiation team consisted of Chris and Juanita; Chris was the project manager.

2. *Establishing a Relationship with the Customer.* A thorough understanding of your customer builds stronger partnerships and higher levels of trust. At PVF, management has tried to foster strong working relationships between business units (like Purchasing) and the IS development group by assigning a specific individual to work as a liaison between both groups. Since Chris had been assigned to the Purchasing unit for some time, he was already aware of some of the problems with the existing Purchasing systems. PVF's policy of assigning specific individuals to each business unit helped to assure that both Chris and Juanita were comfortable working together prior to the initiation of the project. Many organizations use a similar mechanism for establishing relationships with customers.

3. *Establishing the Project Initiation Plan.* This activity focuses on defining the activities required to organize the initiation team while they are working to define the scope of the project. Chris' role was to help Juanita translate her business requirements into a specification for an improved information system. This required the collection, analysis, organization, and transformation of a lot of information. Since Chris and Juanita were already familiar with each other and their roles within a development project, they next needed to define when and how they would communicate, define deliverables and project steps, and set deadlines. Their initiation plan included agendas for several meetings. These steps eventually led to the creation of their SSR.

4. *Establishing Management Procedures.* Successful projects require the development of effective management procedures. Within PVF, many of these management procedures had been established as standard operating procedures by the Systems Priority Board and the IS development group. For example, all project development work is charged back to the functional unit requesting the work. In other organizations, each project may have unique procedures tailored to its needs. Yet, in general when establishing procedures, you are concerned with developing team communication and reporting procedures, job assignments and roles, project change procedures, and determining how project funding and billing will be handled. It was fortunate for Chris and Juanita that most of these procedures were already established at PVF, allowing them to quickly move on to other project activities.

Project workbook: An online or hardcopy repository for all project correspondence, inputs, outputs, deliverables, procedures, and standards that is used for performing project audits, orientation of new team members, communication with management and customers, scoping future projects, and performing post-project reviews.

5. *Establishing the Project Management Environment and Project Workbook.* The focus of this activity is to collect and organize the tools that you will use while managing the project and to construct the **project workbook.** If the system is being developed with the assistance of CASE, as in the Purchasing Fulfillment System, most diagrams, charts, descriptions, and other repository contents can be printed to produce much of the project workbook contents. Thus, the project workbook serves as a repository for all project correspondence, inputs,

outputs, deliverables, procedures, and standards established by the project team (Rettig, 1990). The project workbook can be stored as an online electronic document or large three-ring binder. The project workbook is used by all team members and is useful for project audits, orientation of new team members, communication with management and customers, scoping future projects, and performing post-project reviews. The establishment and diligent recording of all project information in the workbook is one of the most important activities you will perform as project manager.

Figure 4-4 shows the project workbook for the Purchasing Fulfillment System. It consists of both a large hardcopy binder and electronic diskettes where the system data dictionary and diagrams are stored. For this system, a single binder is currently able to contain all project documents. It is not unusual, however, for project documentation to be spread over several binders. As more information is captured and recorded electronically, however, fewer hardcopy binders may be needed.

Once all these activities have been performed, project initiation is complete (see Table 4-2). Before moving on to the next phase of the project, a review of the work performed during project initiation usually occurs. This review meeting is a formal process that is attended by management, customers, and project team members. An outcome of this meeting is a decision to continue the project, modify it in some way, or to abandon it. In the case of the Purchasing Fulfillment System project, the SSR was accepted by the Board, the Board then selected a project steering committee to monitor project progress and to provide guidance to the team members during subsequent activities. If the scope of the project is modified, it may be necessary to return to project initiation activities and collect additional information. Once a decision is made to continue the project, a much more detailed project plan is developed during the project planning phase.

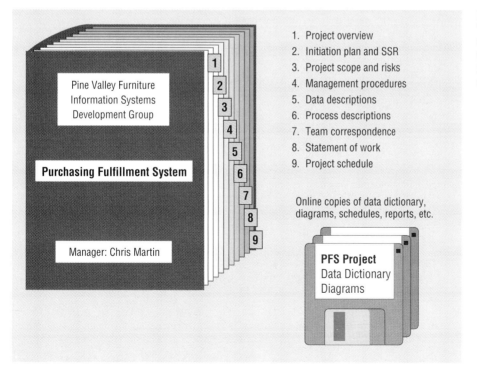

Figure 4-4

The project workbook for the Purchasing Fulfillment System project contains both hardcopy and electronic documents

Pine Valley Furniture
Information Systems
Development Group

Purchasing Fulfillment System

Manager: Chris Martin

1. Project overview
2. Initiation plan and SSR
3. Project scope and risks
4. Management procedures
5. Data descriptions
6. Process descriptions
7. Team correspondence
8. Statement of work
9. Project schedule

Online copies of data dictionary,
diagrams, schedules, reports, etc.

PFS Project
Data Dictionary
Diagrams

TABLE 4-2   Elements of Project Initiation

- Establishing the Project Initiation Team
- Establishing a Relationship with the Customer
- Establishing the Project Initiation Plan
- Establishing Management Procedures
- Establishing the Project Management Environment and Project Workbook

---

Planning the Project

Project planning: The second phase of the project management process which focuses on defining clear, discrete activities and the work needed to complete each activity within a single project.

The next step in the project management process is **project planning.** Distinct from general information systems planning, which focuses on assessing the information systems needs of the *entire* organization (Chapter 6), project planning is the process of defining clear, discrete activities and the work needed to complete each activity within a *single* project. Project planning often requires you to make numerous assumptions about resource availability and potential problems. It is much easier to plan nearer-term activities than those occurring in the future. In actual fact, you often have to construct longer-term plans that are more general in scope and nearer-term plans that are more detailed. The repetitive nature of the project management process requires that plans be constantly monitored throughout the project and periodically updated (usually after each phase) based upon the most recent information.

Figure 4-5 illustrates the principle that nearer-term plans are typically more specific and firm than longer-term plans. For example, it is virtually impossible to rigorously plan activities late in the SDLC without first completing the analysis phase. Also, the outcome of activities performed earlier in the project are likely to have impact on later activities. This means that it is very difficult, and very likely inefficient, to try to plan detailed solutions for activities that will occur far into the future.

The types of activities that must be performed during project planning are varied and numerous (see Table 4-3). As with the project initiation process, the size, scope, and complexity of a project will dictate the comprehensiveness of the project planning process. For example, during the Purchasing Fulfillment System project described in Chapter 2, Chris and Juanita developed a plan that was approximately ten pages in length, not including figures. However, project plans for very large systems may be several hundred pages in length. The types of activities that you can perform during project planning include the following elements:

1. *Describing Project Scope, Alternatives, and Feasibility.* The purpose of this activity is to develop an understanding of the content and complexity of the project. Within PVF's system development methodology, it is required that one of the first meetings focus on defining a project's scope. Although project scope information was not included in the SSR developed by Chris and Juanita, it was important that both shared the same vision for the project before moving too far along. During this activity, you should try to gain answers to and agreement on the following types of questions:

   - What problem or opportunity does the project address?
   - What are the quantifiable results to be achieved?
   - What needs to be done?

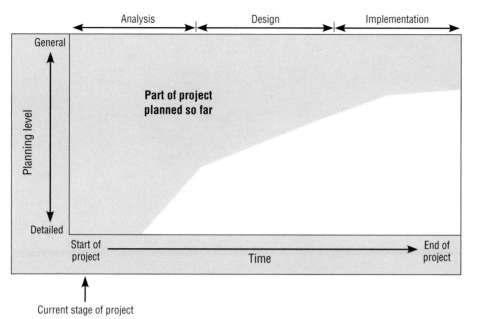

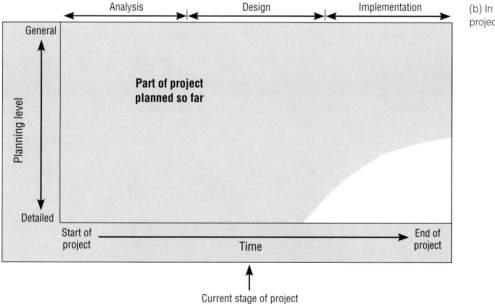

(b) In the middle of the project

- How will success be measured?

- How will we know when we are finished?

After defining the scope of the project, your next objective is to identify and document general alternative solutions for the current business problem or opportunity and assess each for feasibility so that a choice can be made as to which to consider during subsequent SDLC phases. In some instances, an off-the-shelf solution can be found. It is also important that any unique problems, constraints, and assumptions about the project be clearly stated.

TABLE 4-3 Elements of Project Planning

1. Describing Project Scope, Alternatives, and Feasibility

2. Dividing the Project into Manageable Tasks

3. Estimating Resources and Creating a Resource Plan

4. Developing a Preliminary Schedule

5. Developing a Communication Plan

6. Determining Project Standards and Procedures

7. Identifying and Assessing Risk

8. Creating a Preliminary Budget

9. Developing a Statement of Work

10. Setting a Baseline Project Plan

2. *Dividing the Project into Manageable Tasks.* This is a very critical activity during the project planning process. Here, you must divide the entire project into manageable tasks, and then logically order them to ensure a smooth evolution between tasks. The definition of tasks and their sequence is often referred to as the **work breakdown structure.** Some tasks may be performed in parallel while others must follow one another sequentially. Task sequence will depend on which tasks produce deliverables needed in other tasks, when critical resources are available, constraints placed on the project by the client, and the process outlined in the SDLC.

Work breakdown structure:
The process of dividing the project into manageable tasks and logically ordering them to ensure a smooth evolution between tasks.

As you will learn in later chapters, the SDLC consists of several phases. And within each phase, several activities need to be carried out. For example, during the analysis phase, activities must be performed to develop data and process flows. Within each activity, numerous tasks must be completed. For example, when developing a data flow diagram, tasks might include interviewing managers; identifying processes and data inflows, outflows, and transformations; and refining the data flow diagrams. The identification of processes and data flows and their combination into a data flow diagram represent separate tasks. Creating a work breakdown structure requires that you decompose phases into activities and activities into tasks.

It is important that each task be defined at an appropriate level of detail. Defining tasks in too much detail will make the management of the project unnecessarily complex. Discovering the optimal level of detail for representing tasks is a skill that you will develop through experience. For example, it may be very difficult to list tasks that require less than one hour of time to complete in a final work breakdown structure. Alternatively, choosing tasks that are too large in scope (for example, several weeks long) will not provide you with a clear sense of the status of the project or of the interdependencies between tasks. Some guidelines for defining a task are that a task should

- Be done by one person or a well-defined group

- Have a single and identifiable deliverable (The task is, however, the process of creating the deliverable.)

- Have a known method or technique

- Have well accepted predecessor and successor steps

- Be measurable so that percent completed can be determined

3. *Estimating Resources and Creating a Resource Plan.* The focus of this activity is on estimating resource requirements for each project activity and using this information to create a project resource plan. The resource plan is used to help assemble and deploy resources in the most effective manner (for example, you would not want to bring additional programmers onto the project at a rate faster than you could prepare them for useful work). Project estimating and deploying human resources is typically the most important resource planning activity as people are the most expensive element. Commonly used methods for estimating project size and thus the resource requirements for a project are briefly described in Figure 4-6. A detailed review of this topic is beyond the scope of our discussion; the interested reader is encouraged to see Pressman (1992).

Project time estimates for task completion and overall system quality are significantly influenced by the assignment of people to tasks. Although it is important to develop staff by assigning them new tasks in which they can learn new skills, it is equally important to make sure that project members are not in "over their heads" or working on a task that is not well suited to their skills. In essence, the assignment of one person to a task rather than some other person may have significant effects on the completion of an activity. This means that resource estimates may need to be revised based upon the skills of the actual person (or people) assigned to a particular activity. Figure 4-7 shows this relationship by indicating the relative programming speed versus the relative programming quality for three programmers. The figure suggests that Carl should not be assigned tasks in which completion time is critical and that Brenda should be assigned to tasks in which high quality is most vital.

One task assignment approach is to assign a single task type (or only a few task types) to each worker for the duration of the project. For example, you could assign one worker to create all computer displays and another to create all system reports. Such specialization ensures that both workers become efficient at their own particular tasks. Unfortunately, such an assignment may become boring if the task is too specialized or is long in duration. Alternatively, you could assign workers to a wider variety of tasks. This approach

Figure 4-6
Methods for estimating project size

Three commonly used techniques for estimating project size are decomposition, modeling, and function point estimating.

1. *Decomposition* is the process of breaking a project down into small manageable pieces that can be estimated (see Pressman, 1992). Estimates are obtained by "best guesses" and by reviewing historic productivity matrices for similar projects. When assigning human resources, average productivity should reflect the complexity of the project, experience of the person being assigned, or effects of project member turnover (see Abdel-Hamid, 1992). Due to its simplicity, decomposition is the most widely used estimation technique.
2. An example of a *modeling* technique is Boehm's (1981) COnstructive COst MOdel (COCOMO), an estimation model for determining the number of person months required for completing a project. The model uses parameters that were derived from prior projects of differing complexity (that is, the model was empirically developed). COCOMO uses these different parameters to predict human resource requirements for basic, intermediate, and very complex systems.
3. *Function point estimating* obtains a score for a particular project based upon the number and complexity of external inputs, external outputs, internal and external files, and system inquiries (e.g., a help screen) (see Grupe and Clevenger, 1991). Once a function point score is calculated for a given project, it can be compared to the number of function points of prior projects so that a project time estimate can be made.

Figure 4-7

Tradeoffs between the quality of the program code versus the speed of programming (Adapted from Page-Jones, 1985)

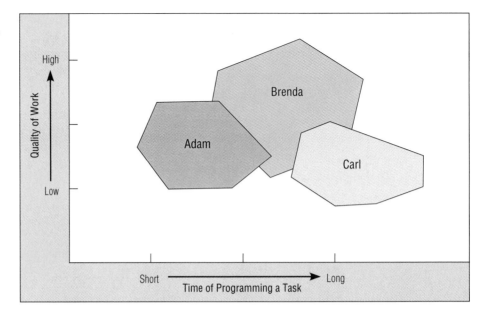

may lead, however, to lowered task efficiency. A middle ground would be to make assignments with a balance of both specialization and task variety. Of course, possible assignments depend upon the size of the development project and the skills of the project team. Regardless of the manner in which you assign tasks, make sure that each team member only works on one task at a time. Exceptions to this rule can occur when a task occupies only a small portion of a team member's time (for example, testing the programs developed by another team member) or during an emergency.

4. *Developing a Preliminary Schedule.* During this activity, you use the information regarding tasks and resource availability to assign time estimates to each activity in the work breakdown structure. This assignment will allow for the creation of target starting and ending dates for the project. Often the development of target dates is an iterative process until a schedule is produced that is acceptable to the customer. Determining an acceptable schedule may require that you find additional or different resources or that the scope of the project be changed. The schedule may be represented as a Gantt or PERT chart, as illustrated in Chapter 2 and discussed later in this chapter.

5. *Developing a Communication Plan.* The focus of this activity is to outline the communication procedures among management, project team members, and the customer. The communication plan includes when and how written and oral reports will be provided by the team, how team members will coordinate work, what messages will be sent to announce the project to interested parties, and what kinds of information will be shared with vendors and external contractors involved with the project. It is important that free and open communication occur among all parties, with respect for proprietary information and confidentiality with the customer (see Kettelhut, 1991).

6. *Determining Project Standards and Procedures.* During this activity, you will specify how various deliverables are produced and tested by you and your project team. For example, the team must decide on which CASE tools to use,

how the standard SDLC might be modified, which SDLC methods will be used, documentation styles (for example, type fonts and margins for user manuals), how team members will report the status of their assigned activities, and terminology. In essence, setting project standards and procedures for work acceptance is a way to assure the development of a high-quality system. Also, you will find it much easier to train new team members when clear standards are in place. Organizational standards for project management and conduct make the determination of individual project standards easier and the interchange or sharing of personnel among different projects feasible.

7. *Identifying and Assessing Risk.* The focus of this activity is to identify sources of project risk and to estimate the consequences of those risks. Risks might arise from the use of new technology, resistance to change, availability of critical resources, competitive or regulatory actions, or team member inexperience with technology or the business area. Thus, you should continually try to identify and assess project risk. We discuss this important topic in greater detail in Chapter 7.

   The identification and assessment of risks are needed when making subsequent planning decisions. Since the identification of project risks is also a requirement of PVF's system development methodology, another early meeting between Chris and Juanita focused on risk assessment. At this meeting, they focused on identifying and describing possible negative outcomes of the project and their corresponding probabilities of occurrence. Although we have listed the identification of risks and the outline of project scope as discrete activities, both are highly related and often concurrently discussed.

8. *Creating a Preliminary Budget.* During this phase, you need to create a preliminary budget that outlines the planned expenses and revenues associated with your project. The project justification, or business case, will demonstrate that the benefits are worth these costs. The project budget is one part of the overall business case for the application and we will fully discuss the development of project budgets in Chapter 7.

9. *Developing a Statement of Work.* A major activity that occurs near the end of the project planning phase is the development of the statement of work. Developed primarily for the customer, this document outlines a description of all work that will be done and makes clear what the project will deliver. The statement of work is useful to make sure that you, the customer, and other project team members have a clear understanding of the intended project size, duration, and outcomes.

10. *Setting a Baseline Project Plan.* Once all of the prior project planning activities have been completed, you will be able to develop a baseline project plan. This baseline plan reflects the best estimate of the project's tasks and resource requirements and is used to guide the next project phase—execution. It is worth noting that the baseline project plan is often changed after the project execution begins. As new information is acquired during project execution, the baseline plan will be updated.

At the end of the project planning phase, a review of the baseline project plan is conducted. The focus of this activity is to double-check all information in the baseline plan. As with the project initiation phase, it may be necessary to modify the plan; this means returning to prior project planning activities before proceeding. As with the Purchasing Fulfillment System project from Chapter 2, you may submit the plan and make a brief presentation to the project steering committee at this time. At this meeting, the committee can endorse the plan, ask for modifications, or determine that it is not wise to continue the project (at least as currently outlined).

Executing the Project

Project execution: The third phase of the project management process in which the plans created in the prior phases (project initiation and planning) are put into action.

Project execution puts the baseline project plan into action. Within the context of the SDLC, project execution occurs primarily during the *analysis, design,* and *implementation* phases. During the development of the Purchasing Fulfillment System, Chris Martin was responsible for several activities during project execution, activities that could be generally described as follows:

1. *Executing the Baseline Project Plan.* Your primary focus is to oversee the execution of the baseline plan. This means that you initiate the execution of project activities, acquire and assign resources, orient and train new team members, keep the project on schedule, and assure the quality of project deliverables. As you might remember from Chris Martin's experience in Chapter 2, this is a formidable task, but a task made much easier through the use of sound project management techniques. During the project, members will come and go from the project team at different times. You are responsible for initiating each team member into the team, including not only providing them with the resources they need but also building and rebuilding the team as a unit. You may want to plan social events, regular team project status meetings, team-level reviews of project deliverables, and other group events to mold the group into an effective team (see the box on "The Project Juke Box").

2. *Monitoring Project Progress Against the Baseline Plan.* One aspect of executing the baseline project plan relates to monitoring your actual progress against the baseline plan. If the project gets ahead of (or behind) schedule, you may have to make adjustments to resources, activities, and budgets. Monitoring project activities not only enables modifications to be made to the current plan but also provides insights for estimating future activities and projects. Additionally, measuring the time and effort expended on each activity will help you improve the accuracy of future estimations. It is possible with project schedule charts, like Gantt charts (see Figure 2-3a), to show progress against a plan; and it is easy with PERT charts (see Figure 2-3b) to understand the ramifications of delays in an activity. Monitoring progress also means that the team leader must evaluate and appraise each team member, occasionally change work assignments or request changes in personnel, and provide feedback to the employee's supervisor.

3. *Managing Changes to the Baseline Project Plan.* You will also encounter pressure to make changes to the baseline plan. At PVF, policies dictate that only approved changes to the project specification can be made and all changes must be reflected in the baseline plan and project workbook, including all

The Project Juke Box

Project leaders have been known to use various means to motivate team members and to create a collective sense of responsibility. One project leader discovered that many of his team members had a common interest in bebop and doowap music and arranged for a juke box filled with such music to be installed in the lunch room of the temporary offices of his team. The juke box became the team's symbol and the envy of other company employees. The leader also planned a racquetball tournament, a trip to a play, and a group reservation at a outdoor challenge camp and at a professional conference. Although the team members came from eight different departments and only two knew each other before the project, team morale remained high, the team was given a productivity award by upper management, and team members formed lasting friendships that had many benefits after the project was over. As a final gift, the leader gave each team member a custom-made pin of a juke box at the project close-down party.

charts (such as the Gantt and PERT charts shown in Figure 2-3). For example, if Juanita suggests a significant change to the existing design of the Purchasing Fulfillment System, a formal change request must be approved by the steering committee. In this request, it should be made clear why changes are desired and include a description of all possible impacts on prior and subsequent activities, project resources, and the overall project schedule. Of course, Chris would have to help Juanita develop such a request. This information will allow the project steering committee to more easily evaluate the costs and benefits of a significant mid-course change.

In addition to changes occurring through formal request, changes may also occur from events far outside your control. In fact, there are numerous events that may motivate making a change to the baseline project plan, including the following possibilities:

- A slipped completion date for an activity

- A bungled activity that must be re-done

- The identification of a new activity that becomes evident later in the project

- An unforeseen change in personnel due to sickness, resignation, or termination

When an event occurs that delays the completion of an activity, you typically have two choices: devise a way to get back on schedule or revise the plan. Devising a way to get back on schedule is the preferred approach because no changes to the plan will have to be made. The ability to head off and smoothly work around problems is a critical skill that you need to master.

As you will see later in the chapter, project schedule charts are very helpful in assessing the impact of change. Using such charts, you can quickly see if the completion time of other activities will be affected by changes in the duration of a given activity or if the whole project completion date will change. Often you will have to find a way to rearrange the activities since the ultimate project completion data may be rather fixed. There may be a penalty to the organization (even legal action) if the expected completion date is not met.

4. *Maintaining the Project Workbook.* As in all project phases, maintaining complete records of all project events is necessary. The workbook provides the documentation new team members require to quickly assimilate project tasks; it provides a history to explain why certain design decisions were made; and it is a primary source of information for producing all project reports.

5. *Communicating the Project Status.* The project manager is responsible for keeping all team members—system developers, managers, and customers—abreast of the project status. Clear communication is a requirement for creating a shared understanding of the activities and goals of the project; such an understanding assures better coordination of activities. This means that the entire project plan should be shared with the entire project team and any revisions to the plan should be communicated to all interested parties so that everyone understands how the plan is evolving. Procedures for communicating project activities can range from formal activities to more informal ones. Keep in mind that different types of procedures are appropriate for different types of communication. Some are useful for *informing* others of project status,

TABLE 4-4   Project Team Communication Methods

Procedure	Formality	Use
Project Workbook	High	Inform Permanent Record
Meetings	Medium to High	Resolve Issues
Seminars and Workshops	Low to Medium	Inform
Project Newsletters	Medium to High	Inform
Status Reports	High	Inform
Specification Documents	High	Inform Permanent Record
Minutes of Meetings	High	Inform Permanent Record
Bulletin Boards	Low	Inform
Memos	Medium to High	Inform
Brown Bag Lunches	Low	Inform
Hallway Discussions	Low	Inform Resolve Issues

others for *resolving* issues, and others for keeping *permanent records* of information and events. Table 4-4 lists numerous communication procedures, their level of formality, and most likely use. Whichever procedure you use, frequent communication helps to assure project success (Kettelhut, 1991).

This section outlined your role as the project manager during the execution of the baseline project plan (see Table 4-5). The ease with which the project can be managed is significantly influenced by the quality of prior project phases. If you develop a high-quality project plan, it is much more likely that the project will be successfully executed. The next section describes your role during project close-down, the final phase of the project management process.

Closing Down the Project

Project close-down: The final phase of the project management process that focuses on bringing a project to an end.

The focus of **project close-down** is to bring the project to an end. Projects can conclude with a *natural* or an *unnatural* termination. A natural termination occurs when the requirements of the project have been met—the project has been completed and is a success. An unnatural termination occurs when the project is stopped before completion. Several events can cause an unnatural termination to a project. For example, it may be learned that the assumption used to guide the project proved to be false, or that the performance of the system or development group was somehow inadequate, or that the requirements are no longer relevant or valid in the customer's business environment. The most likely reasons for the unnatural termination of a project relates to running out of time or money, or both. Regardless of the project termination outcome, there are several activities that must be performed: closing down the project, conducting post-project reviews, and closing the customer contract. Within the context of the SDLC, project close-down occurs after

TABLE 4-5   Elements of Project Execution

1. Executing the Baseline Project Plan
2. Monitoring Project Progress Against the Baseline Plan
3. Managing Changes to the Baseline Project Plan
4. Maintaining the Project Workbook
5. Communicating the Project Status

the *implementation* phase; the system *maintenance* phase typically represents an ongoing series of projects, each needing to be individually managed.

1. *Closing Down the Project.* During close-down, you will perform several diverse activities. If, for example, you have several team members working with you, project completion may signify job and assignment changes for some members. You will likely be required to assess each team member and provide an appraisal for personnel files and for salary determination. You may also want to provide career advice to team members, write letters to superiors praising special accomplishments of team members, and send thank you letters to those who helped but were not team members. As project manager, you must be prepared to handle possible negative personnel issues such as job termination, especially if the project was not successful. When closing down the project, it is also important to notify all interested parties that the project has been completed and to finalize all project documentation and financial records so that a final review of the project can be conducted. You should also celebrate the accomplishments of the team. Some teams will hold a party, and each team member may receive memorabilia (for example, a t-shirt with "I survived the X project"). The goal is to celebrate the team's effort to bring a difficult task to a successful conclusion.

2. *Conducting Post-Project Reviews.* Once you have closed down the project, final reviews of the project should be conducted with management and customers. The objective of these reviews is to determine the strengths and weaknesses of project deliverables, the processes used to create them, and the project management process. It is important that everyone understands what went right and what went wrong in order to improve the process for the next project. Remember, the systems development methodology adopted by an organization is a living guideline that must undergo continual improvement.

3. *Closing the Customer Contract.* The focus of this final activity is to ensure that all contractual terms of the project have been met. A project governed by a contractual agreement is typically not completed until agreed to by both parties, often in writing. Thus, it is paramount that you gain agreement from your customer that all contractual obligations have been met and that further work is either their responsibility or covered under another system service request or contract.

Close-down is a very important activity since a project is not complete until it is closed and it is at close-down that projects are deemed a success or failure. Completion also signifies the chance to begin a new project and apply what you have learned. Now that you have an understanding of the project management process, the next section will describe specific techniques used in systems development for representing and scheduling activities and resources.

REPRESENTING AND SCHEDULING PROJECT PLANS

A project manager has a wide variety of techniques available for depicting and documenting project plans. These planning documents can take the form of graphical or textual reports although graphical reports have become most popular for depicting project plans (see Figure 4-8). A **Gantt chart** is a graphical representation of a project that shows each task activity as a horizontal bar whose length is proportional to its time for completion (Figure 4-8a). Different colors, shades, or shapes can be used to highlight different kinds of tasks. For example, those activities on the critical path (defined below) may be in red and a summary task (a subproject) could have a special bar. Planned versus actual times or progress for an activity can be compared by parallel bars of different colors, shades, or shapes. Gantt charts do not show how tasks must be ordered (precedence), but simply show when an activity should begin and when it should end. Consequently, Gantt charts are often more useful for depicting relatively simple projects or sub-parts of a larger project, the activities of a single worker, and for monitoring the progress of activities compared to scheduled completion dates.

A **PERT** (Program Evaluation Review Technique) **chart** is a graphical depiction of project task activities and their inter-relationships (Figure 4-8b). As with a Gantt chart, different types of tasks can be highlighted by different features on the PERT chart. The distinguishing feature of a PERT is that the ordering of activities is shown by connecting an activity with its predecessor and successor activities. However, the relative size of a node, which represents an activity, or arcs does not imply the activity's duration.

Sometimes a PERT chart is preferable; other times a Gantt chart more easily shows certain aspects of a project:

- Gantt visually shows the duration of activities whereas PERT visually shows the sequence dependencies between activities

- Gantt visually shows the time overlap of activities whereas PERT does not show time overlap but which activities could be done in parallel

- Some forms of Gantt charts can visually show slack time available within an earliest start and latest finish duration whereas PERT shows this by data within activity rectangles

Project managers also use textual reports depicting resource utilization by tasks, project variances, and cost distributions to control activities. For example, Figure 4-9 shows a report from Microsoft Project for Windows® that summarizes all project activities, their durations in weeks, and their scheduled starting and ending dates. Regardless of the project reporting or representation technique, most project managers use computer-based systems to help develop their graphical and textual reports. Later in this chapter, we will discuss these automated systems in more detail.

When managing a project, a manager will periodically review the status of all ongoing project activities. During this review, the manager will assess whether the activities will be completed early, on time, or late. If early or late, the duration of the activity (see column 3 of Figure 4-9) can be updated. Once changed, the scheduled start and finish times of all subsequent activities will also change. Making such a change will also alter a Gantt or PERT chart used to represent the project activities.

Gantt chart: A graphical representation of a project that shows each task activity as a horizontal bar whose length is proportional to its time for completion.

PERT chart: A diagram that depicts project activities and their inter-relationships. PERT stands for Program Evaluation Review Technique.

Figure 4-8

Graphical diagrams for depicting project plans

(a) Gantt chart

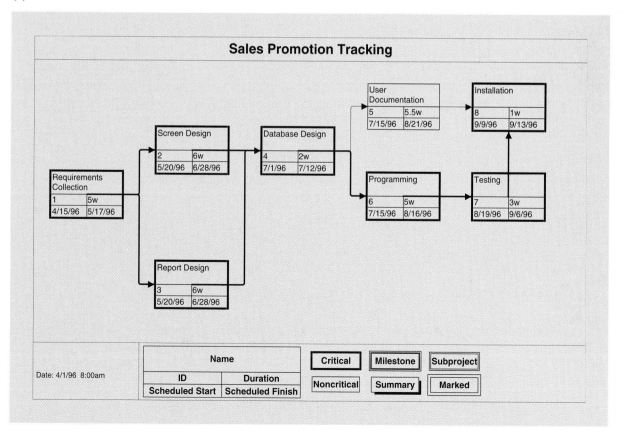

(b) PERT chart

Figure 4-9
Textual report showing project activities, their duration, and scheduled starting and ending dates

Sales Promotion Tracking

ID	Name	Duration	Scheduled Start	Scheduled Finish	Predecessors	Resource Names
1	Requirements Collection	5w	4/15/96 8:00am	5/17/96 5:00pm		
2	Screen Design	6w	5/20/96 8:00am	6/28/96 5:00pm	1	
3	Report Design	6w	5/20/96 8:00am	6/28/96 5:00pm	1	
4	Database Design	2w	7/1/96 8:00am	7/12/96 5:00pm	2,3	
5	User Documentation	5.5w	7/15/96 8:00am	8/21/96 12:00pm	4	
6	Programming	5w	7/15/96 8:00am	8/16/96 5:00pm	4	
7	Testing	3w	8/19/96 8:00am	9/6/96 5:00pm	6	
8	Installation	1w	9/9/96 8:00am	9/13/96 5:00pm	5,7	

Page 1

The ability to easily make changes to a project is a very powerful feature of most project management environments. It allows the project manager to easily gain a clear idea of how activity duration changes impact the overall project completion date. It is also useful for examining "what if" scenarios of adding or reducing resources for an activity and assessing the overall impact on the project.

Representing Project Plans

Resource: Any person, group of people, piece of equipment, or material used in accomplishing an activity.

Critical path scheduling: A scheduling technique where the order and duration of the sequence of activities directly affect the completion date of a project.

Project scheduling and management require that time, costs, and resources be controlled. **Resources** are any person, group of people, piece of equipment, or material used in accomplishing an activity and PERT is a **critical path scheduling** technique used for controlling resources. A critical path refers to a sequence of activities whose order and durations directly affect the completion date of a project. Of the many project scheduling methods, PERT is one of the most widely used and best-known and requires that a project have

- Well-defined activities that have a clear beginning and end point

- Activities that can be worked on independently of other activities

- Activities that are ordered

- Activities that when completed serve the purpose of the project

A major strength of the PERT technique is its ability to represent completion time variability. Because of this, it is more often used than Gantt charts to manage projects such as information systems development where variability in the duration of activities is the norm. PERT charts use a graphical network diagram composed of circles or boxes representing activities and connecting arrows showing required work flows (see Figure 4-10).[1]

---

[1]There are several variations as to how PERT diagrams can be drawn. For example, it is common to use arrows to represent activities and circles or boxes to represent the event of starting or ending an activity. The discussion of these subtle differences is not important for our presentation. We have chosen to use the PERT diagram style most commonly used in project management software.

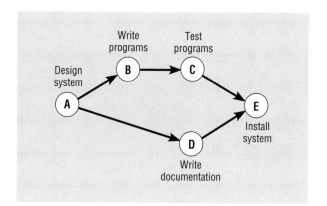

Figure 4-10
PERT chart showing activities and sequence

Constructing a Gantt and PERT Chart at Pine Valley Furniture

Although Pine Valley Furniture has historically been strictly a manufacturing company, they have recently entered the direct sales market for selected target markets. One of the fastest growing of these is economically priced furniture suitable for college students. Management has requested that a new Sales Promotion Tracking System (SPTS) be developed. This project has already successfully moved through project initiation and is currently in the detailed project planning stage, which corresponds to the SDLC phase of *project initiation and planning.* The SPTS will be used to track the sales purchases by college students for the next fall semester. Students typically purchase low-priced beds, bookcases, desks, tables, chairs, and dressers. Since PVF does not normally stock a large quantity of lower-priced items, management feels that a tracking system will help provide information about the college student market that can be used for follow-up sales promotions (for example, a mid-term futon sale).

The project, then, is to design, develop, and implement this information system before the start of the fall term in order to collect sales data at the next major buying period. This deadline gives the project team 24 weeks to develop and implement the system. The Systems Priority Board at PVF wants to make a decision this week based on the feasibility of completing the project within the 24-week deadline. Using PVF's project planning methodology, the project manager, Jim Woo, knows that the next step is to construct Gantt and PERT charts of the project to represent the baseline project plan so that he can use these charts to estimate the likelihood of completing the project within 24 weeks. A major activity of project planning focuses on dividing the project into manageable activities, estimating times for each, and sequencing their order. The steps Jim followed to do this are outlined below.

1. *Identify each activity to be completed in the project.* After discussing the project with PVF's management, sales, and development staff, Jim identified the following major activities for the project:

   • Requirements collection

   • Screen design

   • Report design

   • Database construction

   • Software programming

   • System testing

- User documentation creation

- System installation

2. *Determine time estimates and calculate the expected completion time for each activity.* A standard method for determining the expected completion time for an activity is based on a calculation using three time estimates: the *optimistic time, realistic time,* and *pessimistic time.* The optimistic (o) and pessimistic (p) times reflect the minimum and maximum possible periods of time for an activity to be completed. The realistic time (r) reflects the planner's "best guess" of the amount of time the activity actually will require for completion. Once each of these estimates is made for an activity, an estimated time (ET) can be calculated. Since the estimated completion time should be closest to the realistic time (r), it is weighted four times more than the optimistic (o) and pessimistic (p) times using the following formula:

$$ET = \frac{o + 4r + p}{6}$$

where

ET = estimated time for the completion for an activity
o = optimistic completion time for an activity
r = realistic completion time for an activity
p = pessimistic completion time for an activity

In conclusion, after identifying the major project activities, Jim established optimistic, realistic, and pessimistic time estimates for each activity. These numbers were then used to calculate the expected completion times for all project activities. Figure 4-11 shows the estimated time calculations for each activity of the SPTS project.

3. *Determine the sequence of the activities and precedence relationships among all activities by constructing a Gantt and PERT chart.* This step makes it possible to understand the interrelationships among all activities within the overall project. Jim starts by determining the precedence relationships between activities. Depending upon the complexity and size of the project, several project members may be asked to assist in making activity time and sequencing estimates. The results of this analysis for the SPTS project are shown in Figure 4-12. The first row of this figure shows that no activities precede requirements collection. Row 2 shows that screen design must be preceded by requirements

Figure 4-11
Estimated time calculations for the SPTS project

ACTIVITY	TIME ESTIMATE (in weeks)			EXPECTED TIME (ET) $\frac{o + 4r + p}{6}$
	o	r	p	
1. Requirements Collection	1	5	9	5
2. Screen Design	5	6	7	6
3. Report Design	3	6	9	6
4. Database Construction	1	2	3	2
5. User Documentation	3	6	7	5.5
6. Programming	4	5	6	5
7. Testing	1	3	5	3
8. Installation	1	1	1	1

Figure 4-12
Sequencing of activities
within the SPTS project

ACTIVITY	PRECEDING ACTIVITY
1. Requirements Collection	—
2. Screen Design	1
3. Report Design	1
4. Database Construction	2,3
5. User Documentation	4
6. Programming	4
7. Testing	6
8. Installation	5,7

collection and row 4 shows that both screen and report design must precede database construction. Thus, activities may be preceded by zero, one, or more activities.

Using the estimated time and activity sequencing information (Figures 4-11 and 4-12), Jim can now construct Gantt and PERT charts of the project's activities. To construct the Gantt chart, a horizontal bar is drawn for each activity that reflects its sequence and duration (Figure 4-13). The Gantt chart does not, however, show direct interrelationships between activities. For example, just because the database design activity begins right after the screen design and report design bars finish does not imply that these two activities must finish before database design can begin. To show such precedence relationships, a PERT chart must be used.

PERT charts have two major components: arrows and nodes. Arrows reflect the sequence of activities while nodes reflect activities that consume time and resources. A PERT chart for the SPTS project is shown in Figure 4-14. This diagram has eight nodes labeled 1 through 8.

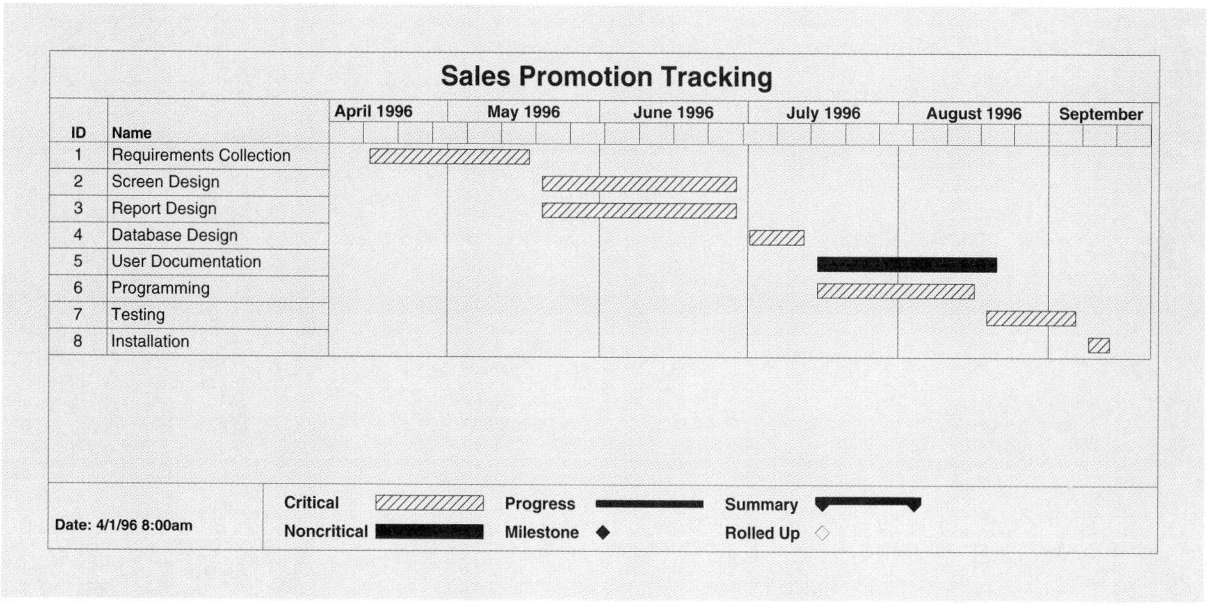

Figure 4-13
Gantt chart for the SPTS project

Sales Promotion Tracking

Figure 4-14
PERT chart for the SPTS project

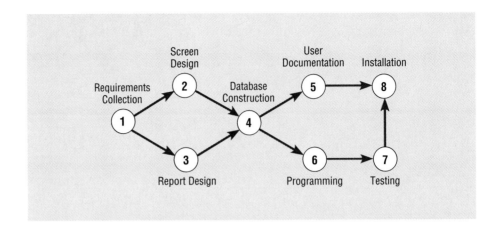

4. *Determine the critical path.* The critical path of a PERT network is represented by the sequence of connected activities that produce the longest overall time period. All nodes and activities within this sequence are referred to as being "on" the critical path. The critical path represents the shortest time in which a project can be completed. In other words, any activity on the critical path that is delayed in completion will result in delaying the entire project. Nodes not on the critical path, however, can be delayed (for some amount of time) without delaying the final completion of the project. These nodes are said to contain **slack time** and allow the project manager some flexibility in scheduling.

Slack time: The amount of time that an activity can be delayed without delaying the project.

Figure 4-15 shows the PERT diagram for the SPTS project in the form Jim used to determine the critical path and expected completion time for the project. To determine the critical path, Jim calculated the earliest and latest expected completion time for each activity. He found each activity's *earliest* expected completion time (T_E) by summing the estimated time (ET) for each activity from left to right (that is, in precedence order), starting at activity 1 and working toward activity 8. In this case T_E for activity 8 is equal to 22 weeks. If two or more activities precede an activity, the largest expected completion time of these activities is used in calculating the new activity's

Figure 4-15
PERT chart for the SPTS project showing estimated times for each activity and their earliest and latest expected completion time

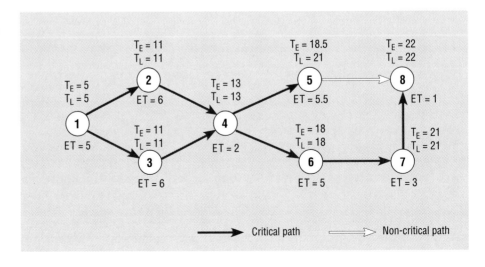

ACTIVITY	T_E	T_L	SLACK $T_L - T_E$	ON CRITICAL PATH
1	5	5	0	✓
2	11	11	0	✓
3	11	11	0	✓
4	13	13	0	✓
5	18.5	21	2.5	
6	18	18	0	✓
7	21	21	0	✓
8	22	22	0	✓

Figure 4-16
Activity slack time calculations for the SPTS project

expected completion time. For example, since activity 8 is preceded by both activities 5 and 7, the largest expected completion time between 5 and 7 is 21, so T_E for activity 8 is 21 + 1, or 22. The earliest expected completion time for the last activity of the project represents the amount of time the project *should* take to complete. Because the time of each activity can vary, however, the projected completion time only represents an estimate. The project may in fact require more or less time for completion.

The *latest* expected completion time (T_L) refers to the time in which an activity can be completed without delaying the project. To find the values for each activity's latest expected completion time (T_L), Jim started at activity 8 and set T_L equal to the final T_E (22 weeks). Next, he worked right to left toward activity 1 and subtracted the expected time for each activity. The slack time for each activity is equal to the difference between its latest and earliest expected completion time ($T_L - T_E$). Figure 4-16 shows the slack time calculations for all activities of the SPTS project. *All activities with a slack time equal to zero are on the critical path.* Thus, all activities except 5 are on the critical path. Part of the diagram shows two critical paths (see Figure 4-15), between activities 1-2-4 and 1-3-4, since both of these *parallel* activities have zero slack.

It is also possible to determine the probability of completing a project by a given date when using PERT. Using the activity completion time estimates, that is, the optimistic, pessimistic, and realistic time estimates for each activity, it is possible to calculate the probability of completing the entire project by a given date. The steps used to conduct this analysis are beyond the scope of this chapter. The interested reader is encouraged to see Miller (1989).

USING PROJECT MANAGEMENT SOFTWARE

A wide variety of automated project management tools are available to help you manage a development project and new versions of these tools are continuously being developed and released by software vendors. Most of the available tools have a common set of features that include the ability to define and order tasks, assign resources to tasks, and easily modify tasks and resources. Project management tools are available to run on IBM-compatible personal computers, the Macintosh, and larger mainframe and workstation-based systems. Key dimensions on which these systems vary include the number of task activities supported, the complexity of relationships, system processing and storage requirements, and, of course, cost. Prices for these systems can range from a few hundred dollars for personal computer-based systems to more than $100,000 for mainframe-based systems. There

Automating Process Management

The most sophisticated project management software do not provide automated support for just project planning but rather for all activities needed to control a project's outcome. These include not only planning (such as with Gantt and PERT charts) but also project tracking, cost estimating, estimating size of software and other project deliverables (which aids in planning), measuring quality and defects, assessing alternative plans, and measuring project productivity. Through a combination of a well-proven systems development methodology (supported by a CASE tool called Process Engineer from LBMS, Inc.) and process management automation (including Microsoft Project) involving support in all of these areas, Holiday Inns Worldwide achieved roughly a 33 percent reduction in the number of tasks needed in systems development projects.

(Adapted from Butler, 1994)

are also several shareware project management programs (Taskman and Project/Event Planner) that can be downloaded from CompuServe and other bulletin boards and whose registration fee is under $100. Since these systems are continuously changing, you should comparison shop before choosing a particular package. Many computer publications such as *PC Magazine, PC Week, MacUser,* and *Datamation* regularly review and contrast available systems (see box "Automating Process Management" for the possible results from using project management software).

To help you get a feel for using project management software, examples of the types of activities you would perform when using a system are illustrated in this section. One particular project management system that has had consistent high marks in computer publication reviews is Microsoft Project for Windows. When using this system to manage a project, you need to perform at least the following activities:

- Establish a project starting or ending date

- Enter activities and assign activity relationships

- Select a scheduling method to review project information

Establishing a Project Starting Date

When using a project management system, the first step is to define the general project information. This information typically includes the name of the project and project manager and the starting or ending date of the project. Starting and ending dates are used to schedule future activities or backdate others (see below) based upon their duration and relationships to other activities. An example from Microsoft Project for Windows of the data entry screen for establishing a project starting or ending date is shown in Figure 4-17. This shows PVF's Purchasing Fulfillment System project from Chapter 2. Here, the starting date for the project is November 27, 1995, at 8:00 in the morning and Chris Martin is named as the project manager.

Entering Tasks and Assigning Task Relationships

The next step in defining a project is to define project tasks and their relationships. For the Purchasing Fulfillment System project, Chris defined twelve tasks to be completed when performing the initial system analysis activities of the project. The task entry screen is shown in Figure 4-18 and is much like a financial spreadsheet program in the way in which information is entered. Specifically, the user moves the cursor to a cell with arrow keys or the mouse and then simply enters a textual *Name* and a numeric *Duration* for each activity. *Scheduled Start* and *Scheduled Finish* are automatically entered based upon the project start date and duration. To set an activity relationship, the *ID* number (or numbers) of the activity that must be completed before the start of the current activity is entered in the *Predecessors* column. Additional codes under this column make the precedence relationships more precise. For example, consider the Predecessor column for ID 6. The entry in this cell

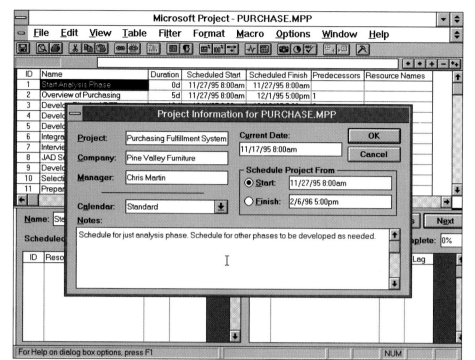

Figure 4-17
Establishing a project starting date in Microsoft Project for Windows

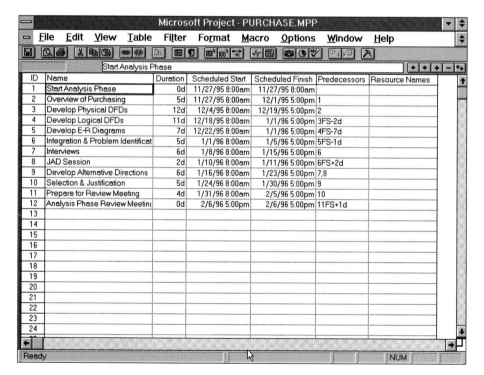

Figure 4-18
Entering tasks and assigning task relationships in Microsoft Project for Windows

says that activity 6 cannot start until one day before the finish of activity 5. Microsoft Project provides many different options for precedence and delays such as in this example, but discussion of these is beyond the scope of our coverage. The project management software uses this information to construct Gantt, PERT, and other project-related reports.

Selecting a Scheduling Method and Reviewing Project Reports

Once information about all the activities for a project has been entered, it is very easy to review the information in a variety of graphical and textual formats using displays or printed reports. For example, Figure 4-19 shows the project information in a Gantt chart screen while Figure 4-20 shows the project information in a condensed PERT chart display. You can easily change how you view the information by making a selection from the *View* menu shown in both figures.

As mentioned in the chapter, interim project reports to management will often compare actual progress to plans. Figure 4-21 illustrates how Microsoft Project shows progress, by a solid line within the activity bar. In this figure, activities 2 and 3 are almost done, but there remains a small percentage of work on each activity, as shown by the incomplete solid lines within the bars for these activities. Assuming that this report represents the status of the project on December 18, 1995, the third activity is approximately on schedule, but the second activity is behind its expected completion date. Tabular reports can summarize the same information.

This brief introduction to project management software has only scratched the surface to show you the power and the features of these systems. Other features that are widely available and especially useful for multi-person projects relate to resource usage and utilization. Resource-related features typically allow you to define

Figure 4-19

Viewing project information as a Gantt chart in Microsoft Project for Windows

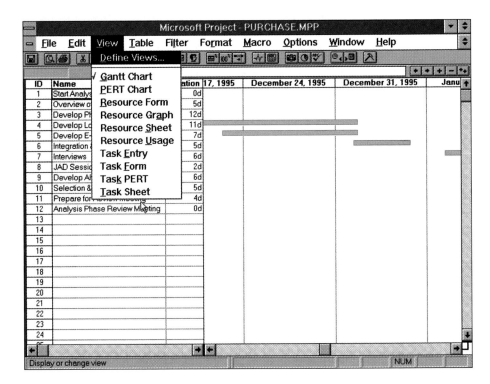

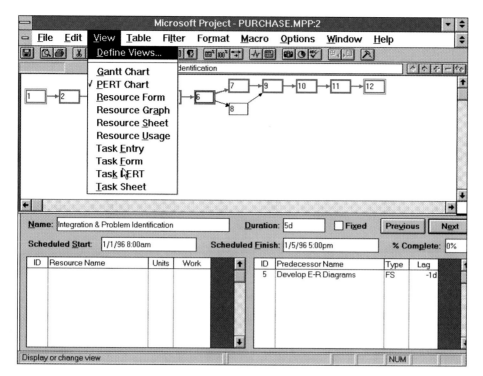

Figure 4-20
Viewing project information as a PERT chart in Microsoft Project for Windows

Figure 4-21
Gantt chart showing progress on activities versus planned durations

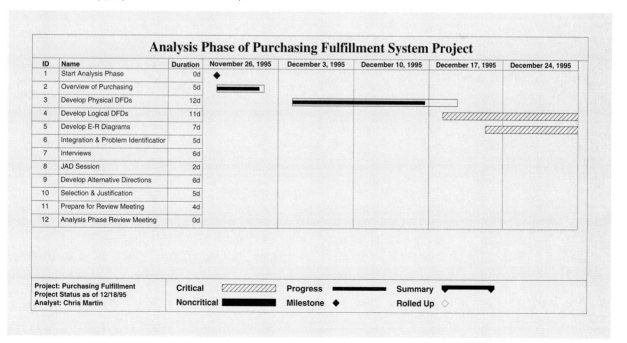

resource characteristics such as standard costing rates and daily availability via a calendar that records holidays, working hours, and vacations—particularly useful for billing and scoping project costs. Often, resources are shared across multiple projects which could significantly affect a project's schedule. Depending upon how projects are billed within an organization, assigning and billing resources to tasks is a very time-consuming activity for most project managers. The features provided in these powerful tools can greatly ease both the planning and managing of projects so that both project and management resources are effectively utilized.

SUMMARY

The focus of this chapter was on managing information system projects and the role of the project manager in this process. A project manager has both technical and managerial skills and is ultimately responsible for determining the size, scope, and resource requirements for a project. Once a project is deemed feasible by an organization, the project manager ensures that the project meets the customer's needs and is delivered within budget and time constraints. To manage the project, the project manager must execute four primary activities: project initiation, project planning, project execution, and project close-down. The focus of project initiation is on assessing the size, scope, and complexity of a project and establishing procedures to support later project activities. The focus of project planning is on defining clear, discrete activities and the work needed to complete each activity. The focus of project execution is on putting the plans developed in project initiation and planning into action. Project close-down focuses on bringing the project to an end.

Gantt and PERT charts are powerful graphical techniques used in planning and controlling projects. Both Gantt and PERT scheduling techniques require that a project have activities that can be defined as having a clear beginning and end, can be worked on independently of other activities, are ordered, and are such that their completion signifies the end of the project. Gantt charts use horizontal bars to represent the beginning, duration, and ending of an activity. PERT is a critical path scheduling method that shows the inter-relationships between activities. Critical path scheduling refers to planning methods whereby the order and duration of the project's activities directly affect the completion date of the project. These charts show when activities can begin and end, which activities cannot be delayed without delaying the whole project, how much slack time each activity has, and progress against planned activities. PERT's ability to use probability estimates in determining critical paths and deadlines makes it a widely used technique for very complex projects.

A wide variety of automated tools for assisting the project manager are available. Most tools have common features including the ability to define and order tasks, assign resources to tasks, and modify tasks and resources. Systems vary regarding the number of activities supported, the complexity of relationships, processing and storage requirements, and cost.

CHAPTER REVIEW

KEY TERMS

Critical path scheduling	Project execution	Project workbook
Gantt chart	Project initiation	Resource
PERT chart	Project management	Slack time
Project	Project manager	Work breakdown
Project close-down	Project planning	structure

REVIEW QUESTIONS

1. Define each of the following terms:
   - a. critical path scheduling
   - b. Gantt chart
   - c. PERT
   - d. project close-down
   - e. project execution
   - f. project initiation
   - g. project planning
   - h. project workbook
   - i. resource
   - j. slack time
   - k. baseline project plan
   - l. earliest expected completion time
   - m. latest expected completion time

2. Contrast the following terms:
   - a. critical path scheduling, Gantt, PERT, slack time
   - b. project, project management, project manager
   - c. project initiation, project planning, project execution, project close-down
   - d. project workbook, resources, work breakdown structure

3. Discuss the reasons why organizations undertake information system projects.

4. List and describe the common skills and activities of a project manager. Which skill do you think is most important? Why?

5. Describe the activities performed by the project manager during project initiation.

6. Describe the activities performed by the project manager during project planning.

7. Describe the activities performed by the project manager during project execution.

8. List various project team communication methods and describe an example of the type of information that might be shared among team members using each method.

9. Describe the activities performed by the project manager during project close-down.

10. What characteristics must a project have in order for critical path scheduling to be applicable?

11. Describe the steps involved in making a Gantt chart.

12. Describe the steps involved in making a PERT chart.

13. In which phase of the systems development life cycle does project planning typically occur? In which phase does project management occur?

14. What are some reasons why one activity may have to precede another activity before the second activity can begin? In other words, what causes precedence relationships between project activities?

PROBLEMS AND EXERCISES

1. Match the following terms to the appropriate definitions.

   _____ project

   _____ project workbook

   _____ Gantt chart
   _____ PERT chart

   _____ baseline project plan

   _____ critical path

   a. those activities whose order and durations directly affect the project completion date
   b. the best estimate of a project's tasks and resource requirements
   c. a repository of all project documentation
   d. a planned undertaking of related activities to reach an objective that has a beginning and an end
   e. a graphical representation of project activities using horizontal bars
   f. program evaluation review technique

2. Which of the four phases of the project management process do you feel is most challenging? Why?

3. A project has been defined to contain the following list of activities along with their required times for completion.

Activity	Time (weeks)	Immediate Predecessors
1 - collect requirements	2	—
2 - analyze processes	3	1
3 - analyze data	3	2
4 - design processes	7	2
5 - design data	6	2
6 - design screens	1	3, 4
7 - design reports	5	4, 5
8 - program	4	6, 7
9 - test and document	8	7
10 - install	2	8, 9

   a. Draw a PERT chart for the activities.
   b. Calculate the earliest expected completion time.
   c. Show the critical path.
   d. What would happen if Activity 6 was revised to take six weeks instead of one week?

4. Construct a Gantt chart for the project defined in problem 3 above.

5. Construct Gantt and PERT charts for a project you are or will be involved in. Choose a project of sufficient depth at either work, home, or school. Identify the activities to be completed, determine the sequence of the activities, and construct a diagram reflecting the starting, ending, duration, and precedence (PERT only) relationships among all activities. For your PERT chart, use the procedure in this chapter to determine time estimates for each activity and calculate the expected time for each activity. Now determine the critical path and the early and late starting and finishing times for each activity. Which activities have slack time?

6. Search computer magazines for recent reviews of project management software. Which packages seem to be most popular? What are the relative strengths and weaknesses of each package software? What advice would you give to someone intending to buy project management software for their PC? Why?

7. What are some sources of risk in a systems analysis and design project and how does a project manager cope with risk during the stages of project management?

8. Look again at the activities outlined in Problem and Exercise 3. Assume that your team is in its first week of the project and has discovered that each of the activity duration estimates is wrong. Activity 2 will take only two weeks to complete. Activities 4 and 7 will each take three times longer than anticipated. All other activities will take twice as long to complete as previously estimated. In addition, a new activity, number 11, has been added. It will take one week to complete and its immediate predecessors are activities 10 and 9. Adjust the PERT chart and recalculate the earliest expected completion times.

9. If given the chance, would you become the manager of an information systems project? If so, why? Prepare a list of the strengths that you would bring to the project as its manager. If not, why not? What would it take for you to feel more comfortable managing an information systems project? Prepare a list and timetable for the necessary training you would need to feel more comfortable about managing an information systems project.

10. For the project you described in Problem and Exercise 5, assume that the worst has happened. A key team member has dropped out of the project and has been assigned to another project in another part of the country. The remaining team members are having personality clashes. Key deliverables for the project are now due much earlier than expected. In addition, you have just determined that a key phase in the early life of the project will now take much longer than you had originally expected. To make matters worse, your boss absolutely will not accept that this project cannot be completed by this new deadline. What will you do to account for these project changes and problems? Begin by reconstructing your Gantt and PERT charts and determine a strategy for dealing with the specific changes and problems described above. If new resources are needed to meet the new deadline, outline the rationale that you will use to convince your boss that these additional resources are critical to the success of the project.

11. How are information system projects similar to other types of projects? How are they different? Are the project management packages you evaluated in Problem and Exercise 6 suited for all types of projects or for particular types of projects? Which package is best suited for information systems projects? Why?

FIELD EXERCISES

1. Identify someone who manages an information systems project in an organization. Describe to him or her each of the skills and activities listed in Table 4-1. Determine which items they are responsible for on the project. Of those they are responsible for, determine which are the more challenging and why. Of those they are not responsible for, determine why not and who is responsible for these activities. What other skills and activities, not listed in Table 4-1, is this person responsible for in managing this project?

2. Identify someone who manages an information systems project in an organization. Describe to him or her each of the project planning elements in Table 4-3. Determine the extent to which each of these elements is part of that person's project planning process. If that person is not able to perform some of these planning activities, or if he or she cannot spend as much time on any of these activities as he or she would like, determine what barriers are prohibitive for proper project planning.

3. Identify someone who manages an information systems project (or other team-based project) in an organization. Describe to him or her each of the project team communication methods listed in Table 4-4. Determine which types of communication

methods are used for team communication and describe which he or she feels are best for communicating various types of information.

4. Identify someone who manages an information systems project in an organization. Describe to them each of the project execution elements in Table 4-5. Determine the extent to which each of these elements is part of that person's project execution process. If that person does not perform some of these activities, or if he or she cannot spend much time on any of these activities, determine what barriers or reasons prevent performing all project execution activities.

5. Interview a sample of project managers. Divide your sample into two small subsamples, one for managers of information systems projects and one for managers of other types of projects. Ask each respondent to identify personal leadership attributes that contribute to successful project management and explain why these are important. Summarize your results. What seem to be the attributes most often cited as leading to successful project management, regardless of the type of project? Are there any consistent differences between the responses in the two subsamples? If so, what are these differences? Do they make sense to you? If there are no apparent differences between the responses of the two subsamples, why not? Are there no differences in the skill sets necessary for managing information system projects versus managing other types of projects?

6. Observe a real information systems project team in action for an extended period of time. Keep a notebook as you watch individual members performing their individual tasks, as you review the project management techniques used by the team's leader, and as you sit in on some of their meetings. What seem to be the team's strengths and weaknesses? What are some areas in which the team can improve?

REFERENCES

Abdel-Hamid, T. K. 1992. "Investigating the Impacts of Managerial Turnover/Succession on Software Project Performance." *Journal of Management Information Systems* 9 (2): 127–144.

Boehm, B. W. 1981. *Software Engineering Economics.* Englewood Cliffs, NJ: Prentice-Hall.

Butler, J. 1994. "Automating Process Trims Software Development Fat." *Software Magazine* 14(8) (Aug): 37–46.

Grupe, F. H., and D. F. Clevenger. 1991. "Using function point analysis as a software development tool." *Journal of Systems Management* 42 (December): 23–26.

Kettelhut, M. C. 1991. "Avoiding group-induced errors in systems development." *Journal of Systems Management* 42 (December): 13–17.

Miller, R. W. 1989. "How to plan and control with PERT." *Managing Projects and Programs,* Harvard Business School Press. Boston, MA: Harvard.

Page-Jones, M. 1985. *Practical Project Management.* New York: Dorset House.

Pressman, R. S. 1992. *Software Engineering.* 3d ed. New York: McGraw-Hill.

Rettig, M. 1990. "Software Teams." *Communications of the ACM* 33 (10): 23–27.

Thomsett, R. 1985. Foreword to *Practical Project Management,* by M. Page-Jones. New York: Dorset House.

Zachary, G. P. 1993. "Agony and Ecstasy Of 200 code writers beget Windows NT." *The Wall Street Journal* (May 26): A1–A6.

5

Automating Development through CASE

After studying this chapter, you should be able to:

■ Identify the trade-offs when using CASE to support systems development activities.

■ Describe organizational forces for and against the adoption of CASE.

■ List and describe the typical components of a comprehensive CASE environment.

■ Describe the general functions of upper CASE, lower CASE, cross life cycle CASE, and the CASE repository.

■ Describe the role of CASE and how it is used to support activities within the SDLC.

INTRODUCTION

In the past, system development was viewed by many as an art that only a few skilled individuals could master. Within many organizations, the techniques employed by each developer could also vary substantially. This lack of consistency in technique and methodology often made it difficult to integrate systems and data or to quickly construct new systems. As a result, many organizations faced a growing backlog of applications to be developed; once developed, many of these systems were error-ridden, over budget, and late. Lack of standards also made maintenance difficult.

To address these problems, information systems professionals concluded that software development needed an engineering-type discipline (Nunamaker, 1992). The goal was to concentrate on developing common techniques, standard methodologies, and automated tools in a manner similar to the traditional engineering field. This chapter covers the evolution and use of automated tools to support the information systems development process. **Computer-aided software engineering (CASE)** refers to automated software tools used by systems analysts to develop information systems. These tools can be used to automate or support activities throughout the systems development process with the objective of increasing productivity and improving the overall quality of systems.

In the following section of this chapter, the objectives of CASE and how these tools are being used within organizations are reviewed. Next, you will learn how

Computer-aided software engineering (CASE): Software tools that provide automated support for some portion of the systems development process.

CASE supports the systems analyst in information systems development by examining the typical components of a comprehensive CASE system. Finally, a discussion of how CASE is applied to support the various phases and activities within the systems development life cycle is presented.

THE USE OF CASE IN ORGANIZATIONS

The purpose of CASE is to make it much easier to enact a single design philosophy within an organization with many projects, systems, and people. CASE can support most of the system development activities; Figure 5-1 highlights selected CASE facilities for each life cycle phase. Although CASE tools run on a variety of mini and mainframe systems, recent advances in microcomputers have made the PC the predominant CASE workstation. CASE helps provide an engineering-type discipline to software development and to the automation of the entire software life cycle process, sometimes with a single family of integrated software tools. In general, CASE assists systems builders in managing the complexities of information system projects and helps assure that high-quality systems are constructed on time and within budget.

Figure 5-1
CASE can provide effective support for most system development activities

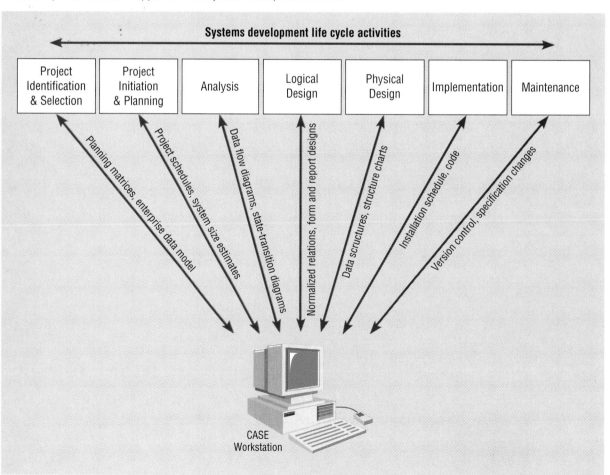

From a systems analyst's perspective, the reason for using CASE may stem from a very straightforward and practical decision such as, "It makes my life easier." However, from an organizational perspective, the reasons for using CASE are much broader (see Table 5-1). Organizations primarily adopt CASE to improve the quality and speed of the systems development process. It is easy to argue that all organizations should openly embrace the adoption and deployment of CASE, since the objectives of CASE are consistent with the objectives of most organizations. Yet, adopting CASE can have numerous and widespread effects on an organization and its information systems development process beyond quality and speed improvements. In this section, we examine the pros and cons of CASE adoption.

The use of CASE is growing as a result of how CASE has evolved to support a wider variety of system development activities. A survey of 408 companies of differing size and industry found that half regularly use CASE, and of those using CASE, most have realized significant productivity improvements (Stickel, 1993). Note, however, that only 50 percent reported to be regular users of CASE. Thus, two important questions for organizations considering CASE are: "Who uses CASE?" and "How are organizations using CASE to improve system development productivity?"

The first question—"Who uses CASE?"—is addressed in Figure 5-2 which shows the results of a survey profiling CASE users (Jones and Arnett, 1992). In Figure 5-2a the number of years of IS experience for the average CASE user is summarized and shows that most CASE users were experienced IS professionals with more than five years' experience. Figure 5-2b summarizes the total number of IS projects the average CASE user has worked on during his or her career. The survey found that more than 50 percent of the users had worked on more than 10 projects. Taken together, these results suggest that the average CASE user is an experienced analyst who has worked on numerous, multi-year projects.

Figure 5-3 addresses the second question—"How are organizations using CASE to improve system development productivity?" This figure shows that the *data dictionary* and the *project management* capabilities of CASE, tools that are used across multiple phases of the SDLC (see the discussion below on *cross life cycle CASE*), are the most widely used features. Features such as documentation support, prototyping, and graphical creation of diagrams were also reported to provide considerable

R.T.C. LIBRARY, LETTERKENNY

TABLE 5-1 Objectives of CASE

Most organizations use CASE to

- Improve the quality of the systems developed
- Increase the speed with which systems are designed and developed
- Ease and improve the testing process through the use of automated checking
- Improve the integration of development activities via common methodologies
- Improve the quality and completeness of documentation
- Help standardize the development process
- Improve the management of the project
- Simplify program maintenance
- Promote reusability of modules and documentation
- Improve software portability across environments

Figure 5-2
A profile of CASE users
(Source: Jones and Arnett,
1992)

(a) Years of IS experience

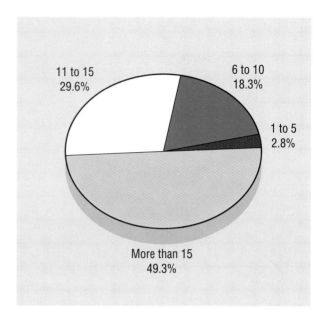

(b) Number of IS projects
(Figures do not add to 100%
due to rounding.)

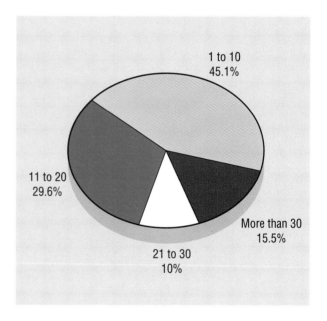

benefits. Based on this data, CASE is most valuable for project requiring these capabilities. In general, CASE is used for medium- to large-sized projects and for which long-term maintenance costs are important to control.

The deployment of CASE within organizations, however, has been slower than expected, with several factors inhibiting widespread application. Most agree that the start-up cost of using CASE is the greatest single factor. Integrated CASE environments range in price from less than $5,000 per analyst to more than $50,000! Lower-end systems are often low in functionality and fail to provide substantial productivity benefits. Furthermore, without adequate CASE training, most people are not able to gain the expertise needed to fully use CASE. For every dollar a company spends on tools, it is projected that from 50 cents to $1 must also be spent on training (Jones, 1993). CASE, an expensive technology, often leaves only large-scale

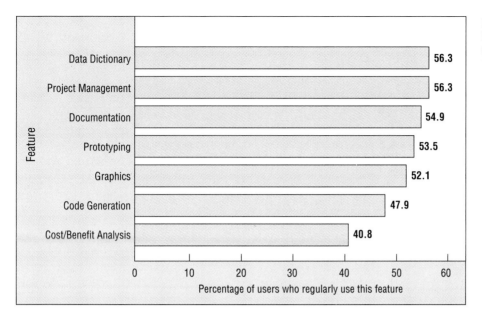

Figure 5-3
Popular uses for CASE
(Source: Jones and Arnett,
1992)

system builders capable of the investment required for organization-wide adoption. For example, it has been estimated that equipping a 150-person IS organization with CASE technology, including software, hardware, training, and maintenance, could exceed $3,000,000 over a five-year period (*I/S Analyzer*, 1993). A practical guideline is that it costs between $5,000 and $15,000 per year to provide CASE to one systems analyst. In spite of this, smaller organizations have effectively deployed CASE by using fewer and less sophisticated tools to automate only a subset of all development activities, thus substantially reducing deployment costs. Nonetheless, adopting CASE is an important decision that must be made at the highest levels of the organization.

Another factor influencing CASE adoption relates to how organizations evaluate their return on investments. The big benefits to using CASE come in the late stages of the SDLC: system construction, testing, implementation, and especially maintenance. Additionally, CASE often lengthens the duration of early stages of the project, sometimes by more than 40 percent (Stone, 1993). This increase in front-end effort is necessitated by the need to completely finish the system design before using automated code generators during system construction. In essence, although CASE provides the potential to significantly shorten the *overall* process, many users and managers are often frustrated by the seemingly long duration of planning, analysis, and design when using CASE. As a result, enthusiasm for CASE can dwindle. Organizations must be patient when making CASE investments—it is likely that the long-term payback from CASE takes more time than some organizations are willing to accept.

Other factors can also influence CASE adoption. One factor, cited to be a significant productivity bottleneck, has been that some CASE tools cannot easily share information between tools. A related issue is the extent to which CASE can support all SDLC activities. Providing a complete set of tools for all aspects of the SDLC has turned out to be more difficult than first imagined. This has left many organizations with distinct tools that "refuse to talk to each other."

The adoption of CASE is also highly related to use of a formal systems development process. Many CASE products force or encourage analysts to follow a specific methodology or philosophy for systems development. For example, Texas Instrument's IEF product is integrally related to the information engineering style of systems development. Thus, an organization without a widely used methodology or one that does not match CASE tools will find it difficult to use CASE. Without a match between a methodology and CASE, CASE is simply another graphical drawing, word processing, and reporting package.

Despite these issues, the long-term prognosis for CASE is very good. The functionality of CASE tools is increasing 5 to 10 percent each year and more than 1,000 stand-alone tools and more than 50 integrated CASE systems currently exist. These estimates imply that CASE environments that *effectively* support all aspects of the SDLC may become a reality by the year 2000. During the next several years, CASE technologies and the market for CASE will begin to mature. This should help improve product offerings and reduce system costs. Additionally, by exposing more systems analysts to CASE technology earlier in their education and career, adoption with less training and better results should result.

Another factor that should stimulate the market for CASE products is the desire of organizations to extend the life of existing systems. Categories of CASE products referred to as *reverse engineering* and *re-engineering* tools are "breathing new life" into existing systems by allowing old programs to be more easily modified to run on new hardware configurations (Pfrenzinger, 1992).

Reverse engineering: Automated tools that read program source code as input and create graphical and textual representations of program design-level information such as program control structures, data structures, logical flow, and data flow.

Reverse engineering refers to the process of creating design specifications for a system or program module from program code and data definitions. For example, CASE tools that support reverse engineering read program source code as input, perform an analysis, and extract information such as program control structures, data structures, logic, and data flow. Once a program is represented at a design level using both graphical and textual representations, the systems analyst can more effectively restructure the code to current business needs or programming practices. An example of a reverse engineering environment is Sterling Software's® VISION:Legacy™, which provides a suite of tools for analyzing, viewing, and restructuring "legacy" systems—systems that have long existed in the organization. Figure 5-4 shows windows about an existing system: the COBOL source code, a control flow graphic, and a type of data flow chart. As with many legacy systems, only the source code may exist, yet additional documentation is necessary to make program maintenance productive.

Re-engineering: Automated tools that read program source code as input, perform an analysis of the program's data and logic, and then automatically, or interactively with a systems analyst, alter an existing system in an effort to improve its quality or performance.

Re-engineering tools are similar to reverse engineering tools but include analysis features that can automatically, or interactively with a systems analyst, alter an existing system in an effort to improve its quality or performance. Although most organizations may have numerous systems that are candidates for reverse engineering or re-engineering, the complexity and effort in using these tools have limited their widespread use. Additionally, most CASE environments do not yet have reverse or re-engineering capabilities. However, as automated development environments evolve to support these features, CASE should evolve to have greater impact beyond what the technology has so far experienced. Figure 5-5 contrasts the growth in the worldwide CASE market between 1992 and 1996 for various types of CASE tools; overall, the market is projected to grow by more than 300 percent.

Besides financial and productivity issues, the culture of an organization can significantly influence the success of CASE adoption. IS personnel with different career orientations have different attitudes toward CASE (Orlikowski, 1989). IS personnel with a managerial orientation welcome CASE because they believe it helps reduce the risk and uncertainty in managing the SDLC. IS personnel with a more technical orientation tend to resist CASE because they feel threatened by the technology's ability to replace some skills they have taken years to master. Table 5-2 lists several possible impacts of CASE on the roles of individuals within organizations. It should be clear to you after reviewing this table that the adoption of CASE should be a well thought-out and highly orchestrated activity.

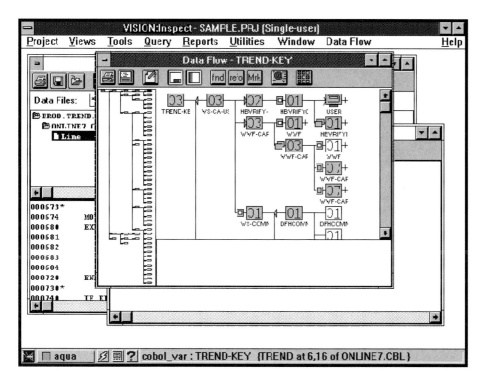

Figure 5-4

Example screen from VISION:Legacy™ by Sterling Software® showing reverse engineering tool

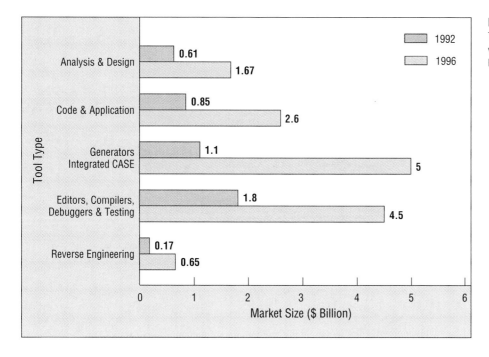

Figure 5-5

The growth of the world-wide CASE market (Source: Pfrenzinger, 1992)

A report prepared by a large accounting firm identified several driving and resisting forces that influence CASE adoption (Chen and Norman, 1992). Forces were identified from both an organization-wide and an individual perspective (see Tables 5-3 and 5-4). Organizational forces came from the pressures to develop and market information-intensive products and services, the increasingly shorter

TABLE 5-2   Common Impacts of CASE on Individuals within Organizations
(Adapted from Chen and Norman, 1992)

Individuals	Common Impact
Systems Analysts	CASE automates many routine tasks of the analyst, making the communication skills (rather than analytical skills) of the analyst most critical.
Programmers	Programmers will piece together objects created by code generators and 4th-generation languages. Their role will include more of maintaining designs rather than source code.
Users	Users will be much more active in the systems development process through the use of upper CASE tools.
Top Managers	Top managers will play a more active role in setting priorities and strategic directions for IS by using CASE-based planning and through user-oriented system development methods.
Functional Managers	Functional managers will play a greater role in leading development projects by using CASE to re-engineer their business processes.
IS Project Managers	IS project managers will have greater control over development projects and resources.

TABLE 5-3   Driving Organizational Forces for the Adoption of CASE

Organizations adopt CASE to

- Provide new systems with shorter development time
- Improve the productivity of the systems development process
- Improve the quality of the systems development process
- Improve worker skills
- Improve the portability of new systems
- Improve the management of the systems development process

TABLE 5-4   Resisting Organizational Forces for the Adoption of CASE

Organizations reject CASE because of

- The high cost of purchasing CASE
- The high cost of training personnel
- Low organizational confidence in the IS department to deliver high-quality systems on time and within budget
- Lack of methodology standards within the organization
- Viewing CASE as a threat to job security
- Lack of confidence in CASE products

time-to-market, the transition to information-based organizations, and the potential for using information systems to support the organization's competitive strategy. Other factors included the increasing pressure to improve developer productivity and system quality. CASE was also viewed as an attractive vehicle for training, for building systems independent of specific implementation platforms, and for more easily managing projects. From the individual perspective, IS professionals who viewed CASE favorably felt that understanding how to use CASE was a valuable and marketable skill, that CASE automated many routine and boring tasks, and that it was an effective method for enforcing a common development methodology.

There were several resisting forces that had acted to preclude many organizations from making the investment in CASE. From an organizational perspective, it was reported that many business managers were losing confidence in the IS department's ability to deliver quality systems on time. Thus, organizations were investing in other options such as end-user computing and outsourcing.

Effective CASE adoption requires using a common design methodology. Thus, if an organization does not *first* have a standard methodology for developing systems, it is unlikely that CASE will be successfully adopted. Additionally, individuals within the IS organization may view CASE as a career threat if no coherent strategy for adoption and deployment is instituted.

This section has described the use of CASE within organizations and discussed several issues surrounding the adoption of CASE. It should now be clear to you that adopting and using CASE is a significant event in any organization. In view of this, the adoption of CASE, or any other technology with so many potential impacts, should be guided with a clear strategy. Table 5-5 outlines many critical issues and implementation strategies. Although all items listed in this table are important, our personal experience is that without top management support, any significant systems project is destined for failure.

TABLE 5-5 CASE Implementation Issues and Strategies
(Adapted from Chen and Norman, 1992)

Top Management Support	CASE requires a substantial long-term investment. Top management support is essential for success.
Contribution	CASE adoption is a business decision that must be justified as bringing business value to the organization by addressing or overcoming some identifiable problem in the development of information systems.
Manage Expectation	History has shown that the benefits of CASE are often long-term (for example, easier maintenance). IS managers must be careful not to oversell CASE as a cure-all for the development of information systems.
Prevent Resistance	The resistance to CASE can come from both inside and outside the IS department. Effectively managing expectations, providing extensive training, and selecting the right pilot projects and personnel will ease adoption and diffusion.
Deploy Carefully	CASE deployment should be done with care. Initial projects and personnel should be selected carefully. Personnel should be respected individuals with high credibility, training, and expertise.
Evaluate Continuously	The effects of CASE (and all other substantial investments) on the organization should be continuously monitored so that timely adjustments in strategy can be made.

COMPONENTS OF CASE

CASE can be used to support a wide variety of SDLC activities. A single CASE product, however, may not have *tools* to provide comprehensive coverage for all aspects of the SDLC. This means that some organizations may use multiple CASE products to gain a broad suite of development tools. CASE tools can be used to help in the project identification and selection, project initiation and planning, analysis, and design phases (**upper CASE**) and/or in the implementation and maintenance phases (**lower CASE**) of the SDLC (see Figure 5-6). A third category of CASE, **cross life cycle CASE,** is tools used to support activities that occur *across* multiple phases of the SDLC. For example, tools used to assist in ongoing activities such as managing the project, developing time estimates for activities, and creating documentation are often considered cross life cycle tools. Over the past several years, vendors of upper, lower, and cross life cycle CASE products have "opened up" their systems through the use of standard databases and data conversion utilities to more easily share information across products and tools. An integrated and standard database called a *repository* is the common method for providing product and tool integration and has been a key factor in the ability of CASE to more easily manage larger, more complex projects and to seamlessly *integrate* data across various tools and products. Integrated CASE or *I-CASE* will be described in more detail later in the chapter. User interface standards such as Microsoft's Windows™ have also greatly eased the integration and deployment of these systems. The general types of CASE tools are listed below:

- *Diagramming tools* that enable system process, data, and control structures to be represented graphically.

- *Computer display and report generators* that help prototype how systems "look and feel" to users. Display (or form) and report generators also make it easier for the systems analyst to identify data requirements and relationships.

Upper CASE: CASE tools designed to support the information planning and the project identification and selection, project initiation and planning, analysis, and design phases of the systems development life cycle.

Lower CASE: CASE tools designed to support the implementation and maintenance phases of the systems development life cycle.

Cross life cycle CASE: CASE tools designed to support activities that occur *across* multiple phases of the systems development life cycle.

Figure 5-6
The relationship between CASE tools and the systems development life cycle

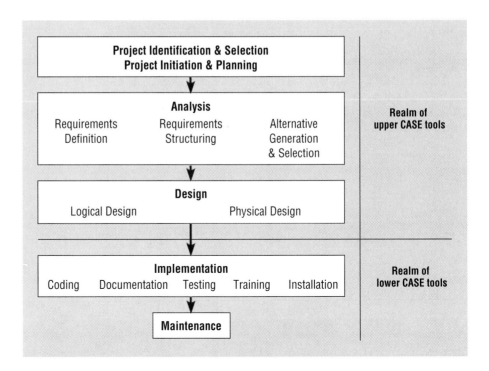

- *Analysis tools* that automatically check for incomplete, inconsistent, or incorrect specifications in diagrams, forms, and reports.

- A central *repository* that enables the integrated storage of specification, diagrams, reports, and project management information.

- *Documentation generators* that help produce both technical and user documentation in standard formats.

- *Code generators* that enable the automatic generation of program and database definition code directly from the design documents, diagrams, forms, and reports.

Besides providing an array of tools, most CASE products also support ad hoc inquiry into and extraction from the repository. Security features, which may be important in some development environments, are also widely available. For example, if you contract with a custom software developer, you may expect them to secure your system specifications so that other project teams may not access your system requirements, design, and code. Some more advanced CASE products also support version control, which allows one repository to contain the description of several versions or releases of the same application system. Also, some CASE products provide import and export facilities to automatically move data between the CASE repository and other software development tools such as word processors, software libraries, and testing environments. Finally, as a shared development database, CASE environments should provide facilities for backup and recovery, user account management, and usage accounting. In the following sections we discuss these different types of CASE tools.

CASE Diagramming Tools

CASE **diagramming tools** allow you to represent a system and its various components visually. Diagrams are very effective for representing process flows, data structures, and program structures. For example, a diagramming technique called data flow diagramming is often used to represent the movement of data between business processes. CASE helps you draw data flow diagrams (DFDs) by providing standard symbols to represent processes, data flows, data stores, and external entities. CASE also assists you in managing the complexity associated with representing higher-level and lower-level processes. For example, Figure 5-7a shows a high-level DFD from the CASE system from Visible Systems Corporation called The Visible Analyst Workbench® (VAW). This view of a business process is the highest level or the most abstract; in DFD terminology this is called the context-level view. This diagram represents an organization's entities—in this instance, those associated with the Department of Motor Vehicles—and the process of providing licenses to potential drivers. Lower-level processes within this system are shown in Figure 5-7b where some business processes are likely to be automated while others may not.

Most CASE systems provide numerous options for representing system information. A common method for representing data is the entity-relationship (E-R) diagram. (E-R diagramming was introduced in Chapter 2.) An example of a CASE-drawn E-R diagram is shown in Figure 5-8 and is from Texas Instruments' Information Engineering Facility® (IEF) CASE environment. In this portion of an E-R diagram, we see four data entities and five relationships including the INCLUDES/INCLUDED IN relationship between the SHIPMENT and PRODUCT entities. This relationship says that a product may be involved in zero to many shipments and that each shipment may be for one customer.

As with DFDs, most CASE environments allow you to display data models at higher or lower levels of detail. In addition, not only can you "link" higher and

Diagramming tools: CASE tools that support the creation of graphical representations of various system elements such as process flow, data relationships, and program structures.

Figure 5-7

Examples of diagramming
tool in Visible Systems
Corporation's VAW

(a) Context-level data flow
diagram

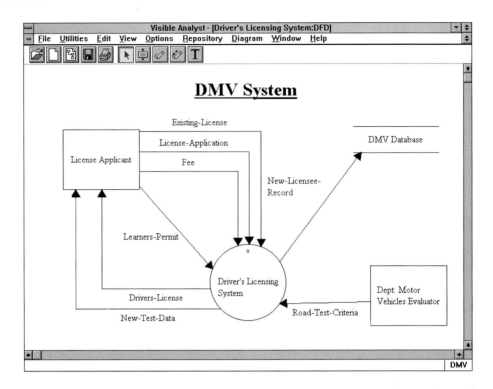

(b) Lower-level data flow
diagram

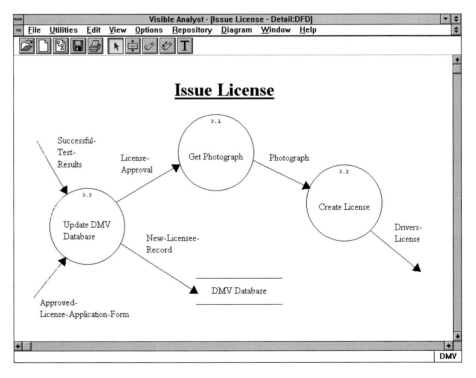

lower level views, you can also associate items in one diagram with items in another. For example, an entity of an E-R diagram can be linked to a data store on a
DFD, and the data elements in a data flow on a DFD can be linked to data elements
of an entity and defined in the repository. The types of diagrams you will use depend upon the methodology standards within your organization and the type of

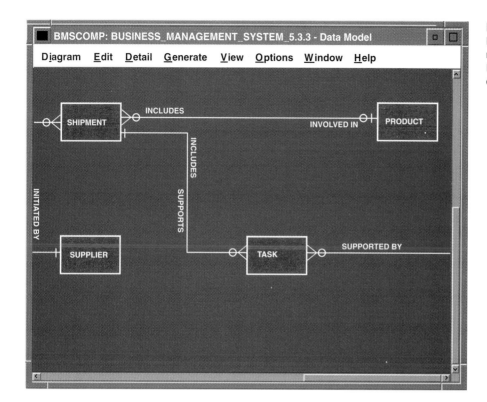

Figure 5-8
Entity-relationship diagramming tool from Texas Instruments' IEF CASE environment

information you are trying to represent. Most IS professionals believe the old proverb that "a picture is worth a thousand words" and have found this to be especially true when representing business processes and complex data relationships. Diagrams are an effective method for developing a common language that users and analysts can use to discuss system requirements. This has resulted in making the diagramming capabilities of most CASE environments a fundamental and indispensable component.

CASE Form and Report Generator Tools

Automated tools for developing computer displays (forms) and reports help the systems analyst design how the user will interact with the new system. These tools, referred to as **form and report generators,** are most commonly used for two purposes: (1) to create, modify, and test prototypes of computer display forms and reports and (2) to identify which data items to display or collect for each form or report. Figure 5-9 shows an example of a form layout design using IEF's window design facility. This facility has numerous features to help you quickly design forms and windows that look and feel consistent to your users. For example, IEF has a template feature that allows you to define common headings, footers, and function key assignments. Once defined, all forms in the system inherit these template definitions. Also, if a common change is desired for all forms, you can simply change the template definition and all system forms will automatically inherit this change. Once you are happy with the design, you can quickly test its usability by converting the design template (Figure 5-9) into a working prototype.

Many CASE environments also allow for the creation of complicated graphical user interfaces or GUIs (pronounced gooie). An example is Sterling Software's VISION:Flashpoint™ that allows the creation of GUIs with multi-windowing capabilities, horizontal and vertical scroll bars, dialogue boxes, embedded graphics, hot

Form and report generators: CASE tools that support the creation of system forms and reports in order to prototype how systems will "look and feel" to users.

Figure 5-9
Window Designer tool from
Texas Instruments' IEF CASE
environment

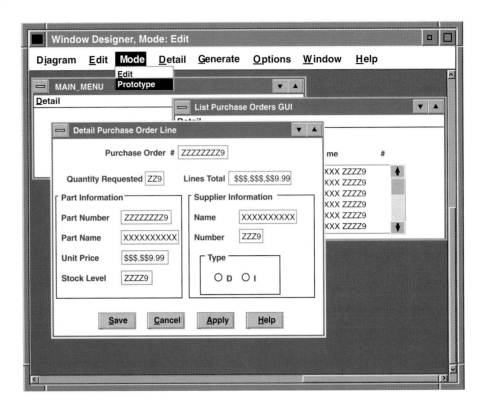

Figure 5-9
Window Designer tool from
Texas Instruments' IEF CASE
environment

spots, check boxes, radio buttons, on-line help, and other common GUI features (see Figure 5-10).

Obviously, many stand-alone tools exist today to automatically generate code for forms and reports in a variety of programming languages. Some support only one language; the form and report generators with Paradox for Windows generate only ObjectPAL code. Other tools, like PowerSoft's PowerBuilder®, can generate code in several languages. The benefit of such tools when they are part of CASE is that these forms and reports are linked with the rest of the systems documentation. For example, each form represents a data flow on some DFD that moves between the system and a user. As a form flow is rerouted, its composition changes or is split apart or combined, and the information necessary to generate the form code is automatically changed by the CASE tool. The code can then quickly be regenerated and shown to users.

You will find that using automated tools for developing forms and reports is useful for both you and the eventual users of your system. For users, interacting with you during the early stages of the SDLC as forms and reports are outlined may help to ease system implementation. Involved users will be more familiar with the system when it is completed. As a result, these users may require less training than uninvolved users. Additionally, they may feel more positive that the system will meet their needs. From your perspective as the analyst, close interaction with users will help you develop a common frame of reference and enable you to better understand their data and processing requirements.

CASE Analysis Tools

One important objective of CASE is to help you handle the complexities of building large systems. We have described how CASE environments automate the creation

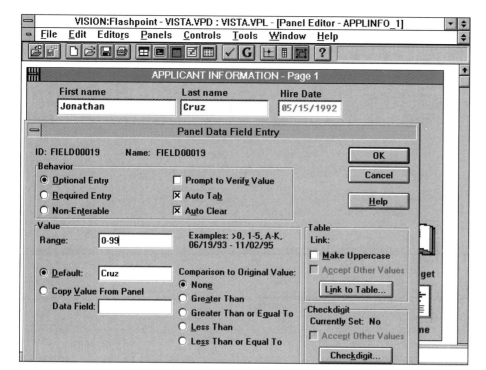

Figure 5-10
Graphical user interface
construction tool from Sterling
Software's CASE environment

of diagrams to represent system process flows, data, and structures in addition to constructing forms and reports. **Analysis tools** generate reports that help you identify possible inconsistencies, redundancies, and omissions in these diagrams, forms, and reports. For example, Table 5-6 provides a list of some of the analysis functions from Intersolv's Excelerator® CASE environment.

Many analysis activities can be performed on the graphical diagrams created by the analyst. Each general diagramming technique has numerous rules that govern how a diagram can be drawn; for example, the "balancing" rule must be followed when creating a lower-level DFD from a higher-level DFD. This rule requires that the number of data flows or arrows flowing into and out of a high-level process must equal those flowing into and out of all this process' lower-level sub-processes. Figure 5-11 shows a violation of the balancing rule: Data flow "C"

Analysis tools: CASE tools that enable automatic checking for incomplete, inconsistent, or incorrect specifications in diagrams, forms, and reports.

TABLE 5-6   A Partial List of the Analysis Functions Performed by Excelerator

Function	Description
Entity Lists	Groups data entities into sets and subsets by identifying common characteristics such as data type, relationships, or some other selection criteria
Extended Analysis	Examines the relationship among the entities defined in diagrams, forms, and reports in a single consolidated analysis
Graph Verification	Examines the diagrams used to represent a system for consistency, completeness, and structural integrity
Graph Summary	Lists the entities that are associated with a particular diagram and all subsequent data and relationships derived in lower-level, more detailed diagrams

Figure 5-11
Out of balance data flow
diagrams

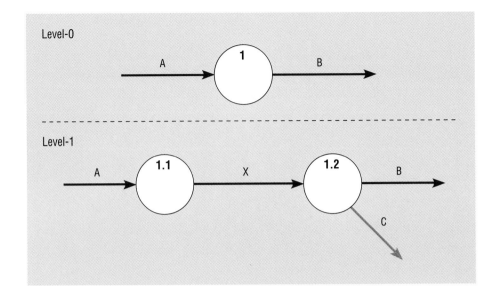

flows out of process 1.2 on the Level-1 diagram, but it does not flow out of process 1 on the Level-0 diagram. Thus, these diagrams are out of balance. An analysis performed on these Level-0 and Level-1 diagrams would alert you to this rule violation.

It is important to note that the types of analyses vary depending upon the organization's development methodology and the features of the CASE environment in use. In addition, many CASE systems support the creation of customized analysis reports. Typically, however, CASE analysis functions focus primarily on data structures and usage and on diagram completeness and consistency. For example, an analysis of all data input forms for a system could be performed. One report generated from this analysis could identify all input forms where the data elements contained on one form were identical to those contained on another (completely redundant forms). A second report could identify those forms where the data elements of one form were completely contained as a subset of the elements on another form (a partially redundant form). Such an analysis may identify that data are being entered into the system from more than one source—a possible data control violation. Such analysis capabilities are especially useful after the work of several systems analysts is combined.

CASE Repository

Substantial benefits when using CASE can only be achieved through the *integration* of various CASE tools and their data. Integrated CASE, or **I-CASE**, tools rely on common terminology, notations, and methods for systems development across all tools. Furthermore, all integrated CASE tools have a common user interface and can share system representations without systems analysts having to convert between different formats used by different tools. Hence, central to I-CASE is the idea of using a common repository for all tools so that this information can be easily shared between tools and SDLC activities. The **repository,** a centralized database, is the nucleus of a comprehensive I-CASE environment and is paramount to the smooth integration of the tools used during the various SDLC phases. This means that the repository holds the complete information needed to create, modify, and evolve a software system from project initiation and planning to code generation and maintenance (see Figure 5-12). With a true I-CASE product, all tools throughout the entire life cycle will use a common repository. In this book, we will interchangeably

I-CASE: An automated systems development environment that provides numerous tools to create diagrams, forms, and reports; provides analysis, reporting, and code generation facilities; and seamlessly shares and integrates data across and between tools.

Repository: A centralized database that contains all diagrams, forms and report definitions, data structure, data definitions, process flows and logic, and definitions of other organizational and system components; it provides a set of mechanisms and structures to achieve seamless data-to-tool and data-to-data integration.

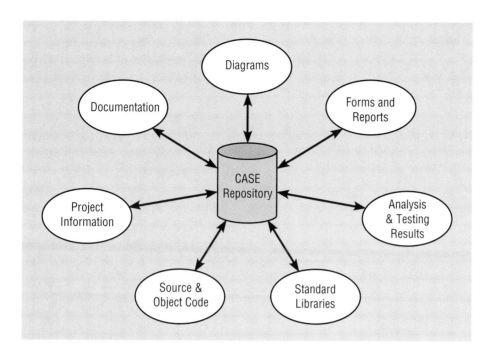

Figure 5-12
System development items
stored in the CASE repository

use both CASE and I-CASE to refer to automated environments used to support the creation of information systems.

For years common development repositories have been used to create information systems independent of CASE. Figure 5-13 reflects the common components of a comprehensive CASE repository. The application development environment is one in which either information specialists or end users use CASE tools, high-level languages, and other tools to develop new applications. The production environment is one in which these same people use applications to build databases, keep the data current, and extract data from databases.

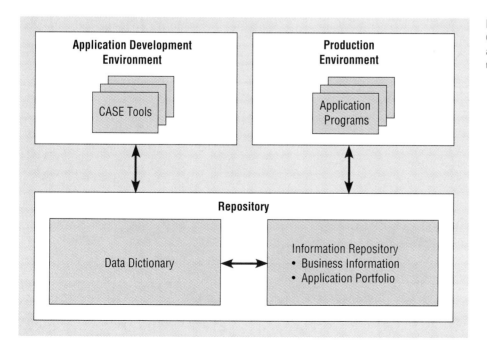

Figure 5-13
Common components of
a comprehensive CASE
repository

Information repository:
Automated tools used to manage and control access to organizational business information and application portfolio as components within a comprehensive repository.

Data dictionary: The repository of all data definitions for all organizational applications.

Cross referencing: A feature performed by a data dictionary that enables one description of a data item to be stored and accessed by all individuals so that a single definition for a data item is established and used.

Within a repository there are two primary segments: the information repository and the data dictionary. The **information repository** combines information about an organization's business information and its application portfolio and provides automated tools to manage and control access to the repository (Bruce, Fuller, and Moriarty, 1989). Business information is the data stored in the corporate databases while the application portfolio consists of the application programs used to manage business information.

The **data dictionary** is a computer software tool used to manage and control access to the information repository. It provides facilities for recording, storing, and processing descriptions of an organization's significant data and data processing resources (Lefkovitz, 1985). Data dictionary features within a CASE repository are especially valuable for the systems analyst when cross referencing data items. **Cross referencing** enables one description of a data item to be stored and accessed by all individuals (systems analysts and end users) so that a single definition for a data item is established and used. Such a description helps to avoid data duplication and makes systems development and maintenance more efficient. For example, if the field length of a data element is changed, the data dictionary can produce a report identifying all programs affected by this change. Within an I-CASE environment, all diagrams, forms, reports, and programs can be automatically updated by the single change to the data dictionary definition. Each entry in a data dictionary has a standard "definition" that can include information such as the following attributes:

1. Element name and any aliases (can include both data items and programs)

2. Textual description of the element

3. List of related elements

4. Element type and format (for example, a calendar date might be of data type "date" and be of the format 12-Jan-96)

5. Range of acceptable values

6. Other information unique to the proper processing of this element

Figure 5-14 shows two computer screens which display information from the repository manager module of the VAW CASE environment where an Entity from an E-R diagram is defined. In this definition, we see that the Entity is named "USER", that it has no alias name, that it has a primary key called "User ID", and an alternate key called "Name". Individual items in the repository can easily be viewed by sequentially moving through each item or by locating them through a query system. Alternatively, exhaustive information on all repository contents can be viewed by using the repository report module (see Figure 5-15).

The term repository is used by many CASE vendors to refer to the combination of both the information repository and the data dictionary. Depending on the nature of the repository and the objectives of the organization, an organization may use a repository in one of three modes. The first, or passive mode, refers to using a repository primarily as a documentation tool by people, and not using it with automated tools such as CASE. However, within organizations that use CASE, the repository will be active in the development of or active in the production of information systems.

Active in Development   Here, the repository is used as a documentation tool and is accessed by various CASE tools during systems development. The automated tools use the repository to generate data structures, documentation, databases, models, and other components of the application environment. As changing busi-

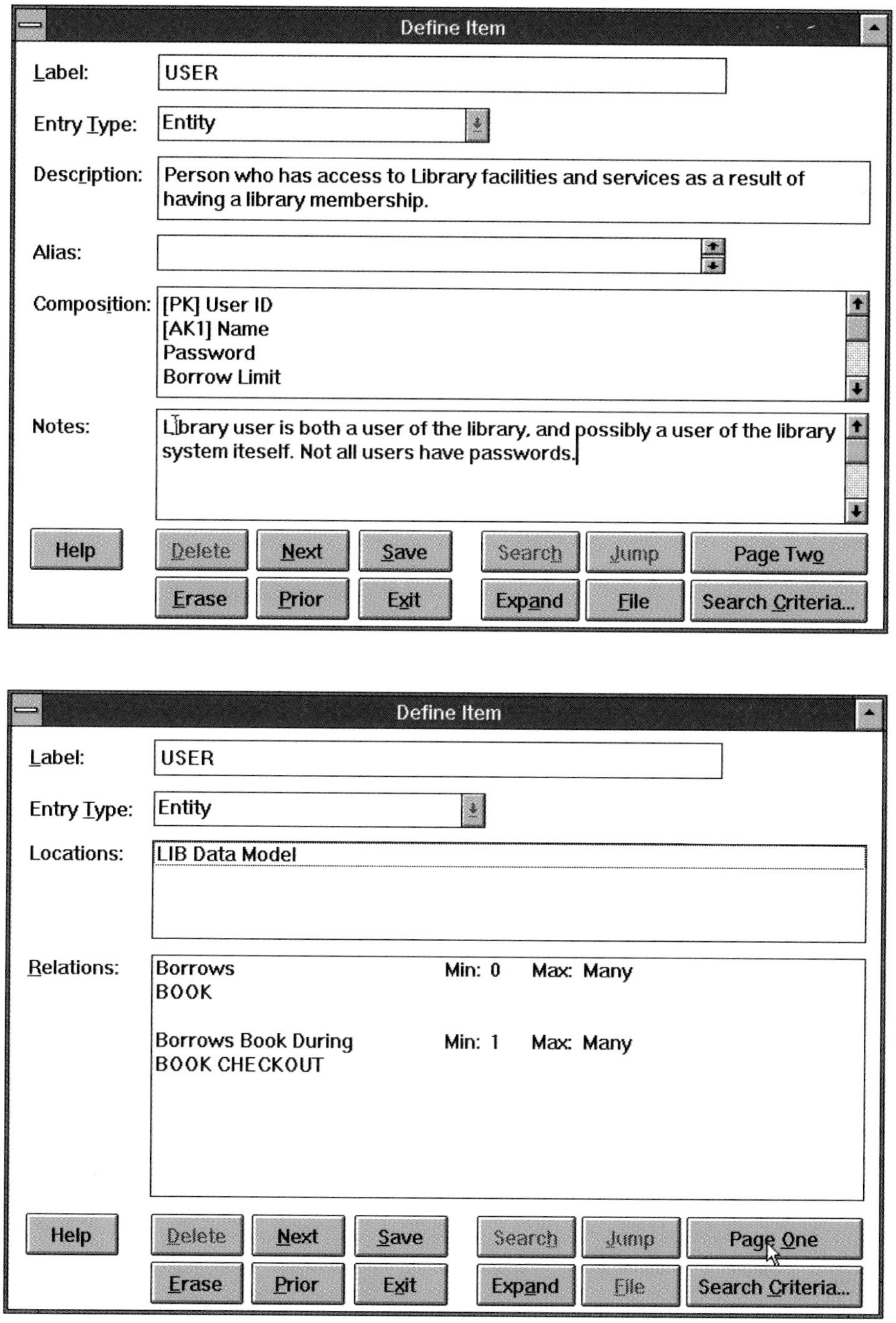

Figure 5-14
Data dictionary definition of a
repository item from Visible
Systems Corporation's VAW
CASE environment

ness conditions necessitate changes to production environment applications, each application is independently modified and these changes are stored in the repository.

Active in Production   In this mode, the repository is more than a documentation and development tool—it is the mechanism through which application programs are constructed, changed, and made consistent with evolving and changing business polices and procedures. This means that during the ongoing use of an information system, all data integrity, data validation, and security access rules are enforced

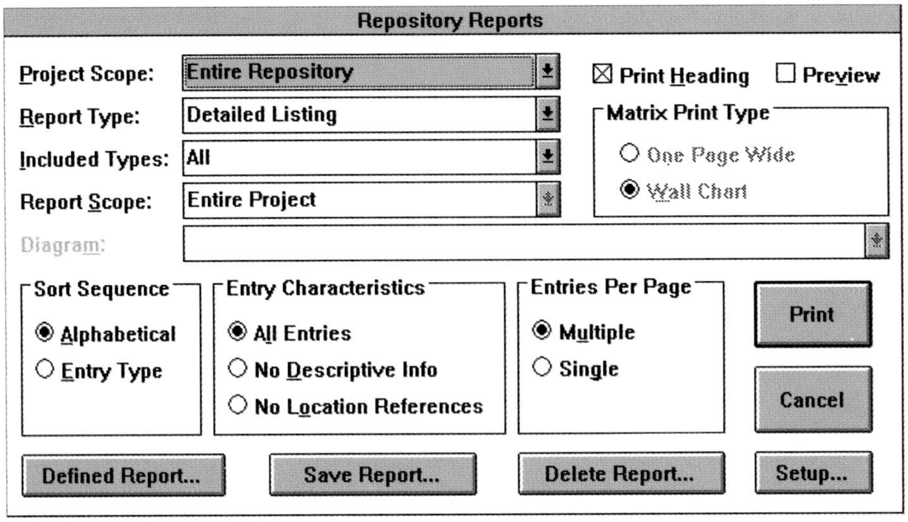

Figure 5-15

Repository report dialogue box from Visible Systems Corporation's VAW CASE environment

Repository Reports		

Project Scope: `Entire Repository`   ☒ Print **H**eading   ☐ Pre**v**iew

Report Type: `Detailed Listing`

Included Types: `All`

Report **Scope:** `Entire Project`

Matrix Print Type
  ○ One Page Wide
  ◉ Wall Chart

Diagra**m**: ` `

Sort Sequence
  ◉ **A**lphabetical
  ○ **E**ntry Type

Entry Characteristics
  ◉ A**l**l Entries
  ○ No **D**escriptive Info
  ○ No L**o**cation References

Entries Per Page
  ◉ M**u**ltiple
  ○ Si**n**gle

[Print]   [Cancel]

[**D**efined Report...] [**S**ave Report...] [**D**elete Report...] [Setup...]

through the information repository. Additionally, any changes made in the repository can be delivered to all relevant systems consistently and simultaneously.

Many organizations use a repository initially in a passive mode, then progress to the active mode in development, and finally go on to the active mode in production. The active mode in production is preferred, since all components of the information systems environment will have a single, centralized source for enacting organization-wide standards and processing procedures.

CASE Repository and the SDLC   During the project initiation and planning phase, the repository is used to store all information, both textual and graphical, related to the problem being solved. Details such as the problem domain, project resources and history, and organizational context are stored in the repository. As the project evolves, the repository becomes the basis for the integration of the various SDLC activities and phases.

During the analysis and design phases of the SDLC, the CASE repository is used to store graphical diagrams and prototype forms and reports. When completeness and consistency analyses are done, the repository allows the data from all diagrams, forms, and reports to be accessed simultaneously for comprehensive analysis. The data stored in the repository is also used as the foundation for the generation of code and documentation. Thus, the CASE repository is the integrating mechanism on all cross life cycle tools and activities.

Additional Advantages of a CASE Repository   Besides specific tool integration, there are two additional advantages to using a comprehensive CASE repository that relate to project management and reusability. The development of most software systems requires that more than one person work on the project; therefore, to most effectively coordinate the activities of multiple developers, project management techniques should be used (Chapter 4). The CASE repository provides a wealth of information to the project manager and allows the manager to exert an appropriate amount of control on the project. For example, on large software development projects, it is customary (and more efficient) to partition the development into distinct subprojects where one or a few people (that is, a team) have primary responsibility for their development. Through the CASE repository, the project manager can restrict members' access to only those aspects of the system for which they are responsible. This reduces the complexity of the system for a given team and provides security such that data are not inadvertently changed or deleted. Partitioning allows multiple teams to work in parallel on different aspects of a single system, potentially reducing total system development time.

Another important use of the CASE repository relates to software **reusability.** In a large organization with many software systems, up to 75 percent of the application programs contain a significant amount of identical functions (Jones, 1986). In addition, as much as 50 percent of systems-level programs and upwards of 70 percent of telecommunications programs contain a significant amount of identical functions. Thus, one easy way for systems developers to enhance their productivity is to stop "reinventing the wheel" (or here, reinventing the function). If all organizational systems were created using CASE technology with a common repository, it would be possible to reuse significant portions of prior systems (or the design of prior systems) in the development of new ones. There are many items that can be reused besides programming code such as design documents (diagrams, specification documents, form and report layouts) and project management modules (schedules, assignments, report formats, project plans). The benefits of reusability are reduced development time and cost and improved software quality by using time-tested modules.

> **Reusability:** The ability to design software modules in a manner so that they can be used again and again in different systems without significant modification.

CASE without a Common Repository   Organizations that do not adopt a single integrated CASE environment must be able to share design and development information among tools. Sharing may be necessary since different, non-integrated tools may be used on the same project or on different projects for coupled systems. Besides using a single CASE repository, you can share data between CASE tools by the following methods:

- Manually entering specifications contained in one repository into another repository (obviously, not a very desirable approach).

- Converting repository contents into some neutral format, like an ASCII file, and then importing these into another repository. This is a better solution in terms of human effort but there still may be considerable loss of information or extra manual effort since the structure of the specifications is lost in such a conversion.

- Converting the specifications in different repositories by using vendor or third-party utilities that translate either directly or through an industry standard exchange format among the repository formats of different CASE tools (IBM External Source Format and the CASE Data Interchange Format provide this type of passive repository sharing, if supported by the CASE tool).

- Allowing one CASE tool to directly read the repository of another CASE tool. This more active type of sharing across CASE tools is possible only if a CASE vendor opens up its database format, as many DBMS vendors have done.

CASE Documentation Generator Tools

Each phase of the SDLC produces documentation. The types of documentation that flow from one phase to the next vary depending upon the organization, methodologies employed, and type of system being built. **Documentation generators** are modules that can create standard reports based upon the contents of the repository. Typically, SDLC documentation includes textual descriptions of needs, solution tradeoffs, diagrams of data and processes, prototype forms and reports, program

> **Documentation generators:** CASE tools that enable the easy production of both technical and user documentation in standard formats.

specifications, and user documentation including application and reference materials. A system that does not have adequate documentation is virtually impossible to use and maintain (Brooks, 1982).

A common problem when developing systems is that ". . . programmers concentrate on getting the application software up and running, rather than producing a document at the end of each development phase" (Hanna, 1992). Thus, documentation is a task that is often left to be dealt with *after* the programs have been completed. This time lag between when development activities occur and when documentation of these activities is produced often results in lower-quality documentation. The value of good documentation in relation to system maintenance is shown in Figure 5-16. This figure shows that the system maintenance effort takes 400 percent longer with poor-quality documentation. High-quality documentation leads to an 80 percent reduction in the system maintenance effort when compared to average-quality documentation. The practical implication of this is that there are some benefits for taking steps to improve the quality of system documentation and severe disadvantages for producing less-than-average-quality documentation.

Documentation generators within a CASE environment provide a method for managing the vast amounts of documentation created during the SDLC. Documentation generators allow the creation of master templates that can be used to verify that the documentation created for each SDLC phase conforms to a standard and that all required documents have been produced. Documentation is an often overlooked aspect of systems development yet, as pointed out in Figure 5-16, is decidedly the most important aspect to building *maintainable* systems. (The interested reader is encouraged to read the classic discussion of systems development and documentation by Brooks [1982] in *The Mythical Man-Month*.)

Code generators: CASE tools that enable the automatic generation of program and database definition code directly from the design documents, diagrams, forms, and reports stored in the repository.

CASE Code Generation Tools

Code generators are automated systems that produce high-level program source code from diagrams and forms used to represent the system. As target environments vary on several dimensions, such as hardware and operating system platforms, many code generators are designed to be special-purpose systems that produce

Figure 5-16
Impact of documentation quality on system maintenance (Source: Hanna, 1992)

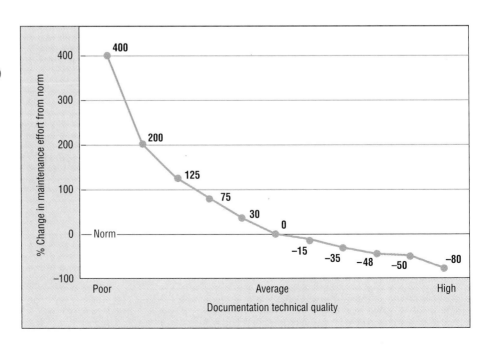

source code for a particular environment in a particular programming language. VAW takes a more flexible approach by producing standard COBOL source code and SQL database definitions (see Figures 5-17 and 5-18). Using standard language conventions, CASE-generated code can typically be compiled and executed on

Notepad - BOOK_MAN.CBL

File Edit Search Help

```
           05  Author-Name                      PIC X(3).
           05  Book-title                       PIC X(60) VALUE spaces.
           05  Birth-Date                       PIC X(6)■.
           05  Death-Date                       PIC X(6)■.

       01  Valid-New-Book.
           05  Call-Number                      PIC X(15).
           05  ISBN                             PIC X(12).
           05  Book-Title                       PIC X(60) VALUE spaces.
           05  Author-Name                      PIC X(3).
           05  Publisher                        PIC X(60).
           05  Date-Published                   PIC X(6)■.
           05  Date-Out                         PIC X(6)■.
           05  Date-Due-Back                    PIC X(6)■.

       01  Valid-User.
           05  User-ID                          PIC X(8).
           05  Name                             PIC X(40).
           05  Password                         PIC X(8).
           05  Borrow-Limit                     PIC 9(1) COMP.
           05  Number-Checked-Out               PIC X(8).
      /
       PROCEDURE DIVISION.


      ******************************************************************
      *                                                                *
      * Paragraph 'Main'.                                              *
      *                                                                *
      *     Derived from VA Module named:                              *
      *         Main                                                   *
      *                                                                *
      * Allow a user to sign on to the Library System                  *
      *                                                                *
      * (Sign_On module)                                               *
      *      Display_Screen(SignOn)                                    *
```

Figure 5-17
View of COBOL code automatically generated by Visible Systems Corporation's VAW CASE environment

Notepad - LIBDEF.SQL

File Edit Search Help

```
CREATE TABLE AUTHOR
 (
    BOOK_TITLE  CHAR(60) NOT NULL WITH DEFAULT,
    BIRTH_DATE  DATE NOT NULL,
    DEATH_DATE  DATE NOT NULL
 );

 CREATE UNIQUE INDEX AK1AUTHOR ON AUTHOR ( BOOK_TITLE );

 CREATE TABLE BOOK
 (
    CALL_NUMBER     CHAR(15) NOT NULL,
    USER_ID         CHAR(8),
    ISBN            CHAR(12) NOT NULL,
    BOOK_TITLE      CHAR(60) NOT NULL WITH DEFAULT,
    PUBLISHER       CHAR(60),
    DATE_PUBLISHED  DATE NOT NULL,
    DATE_OUT        DATE NOT NULL,
    DATE_DUE_BACK   DATE NOT NULL,

    PRIMARY KEY ( CALL_NUMBER )
 );

 CREATE UNIQUE INDEX PKBOOK ON BOOK ( CALL_NUMBER );

 CREATE UNIQUE INDEX AK1BOOK ON BOOK ( ISBN );

 CREATE UNIQUE INDEX AK2BOOK ON BOOK ( BOOK_TITLE );

 CREATE TABLE BOOK_CHECKOUT
 (
    CALL_NUMBER     CHAR(15) NOT NULL,
    CALL_NUMBER     CHAR(15) NOT NULL,
    USER_ID         CHAR(8) NOT NULL,
    CALL_NUMBER     CHAR(15) NOT NULL,
    BOOK_TITLE      CHAR(60) NOT NULL WITH DEFAULT,
```

Figure 5-18
View of SQL database definition code generated from Visible Systems Corporation's VAW CASE environment

numerous hardware and operating system platforms with no, or very minor, changes. Yet standard code and definitions may not take advantage of special hardware or operating system features of specific environments.

CASE AND THE SYSTEMS DEVELOPMENT LIFE CYCLE

Many organizations that use CASE do not use it to support all phases of the SDLC. Some organizations may extensively use the diagramming features but not use code generators. There are a variety of reasons why organizations choose to adopt CASE partially or to not use it at all. These reasons range from a lack of vision for applying CASE to all aspects of the SDLC to the beliefs that CASE technology will fail to meet an organization's unique system development needs. Several key differences between using traditional non-integrated system development processes versus CASE-based development are summarized in Table 5-7.

In traditional systems development, much of the time is spent on coding and testing. When software changes are approved, often the code is first changed and then tested. Once the functionality of the code is assured, the documentation and specification documents are updated to reflect system changes. In effect, this means the systems analyst's job is that of *maintaining code and documentation.* Changes to the system "ripple" through all aspects of the system and supporting documentation and maintaining systems requires that all discrete system documents be updated. For example, if you use a word processing system for logging documentation changes, these files need to be updated. If you use a separate program to graphically reflect process flow and data, changes to these diagrams must occur. Additionally, all other documents must be updated to reflect even the most minuscule changes to the system. The process of keeping all system documentation current can be a very boring and time-consuming activity that is often neglected. This neglect makes future maintenance by the same or (more likely) different programmers difficult at best. Experienced analysts know that complete and consistent documentation is the most important determinant to maintaining large computer systems. Thus, a primary drawback to the traditional development approach is the lack of integration among specification documents, program code, and supporting documentation.

A primary objective of the CASE-based approach to systems development is to overcome the drawbacks of the traditional approach. When using an integrated CASE environment, *your primary role is to maintain the design documents,* as most other aspects of the system can "flow" directly from these diagrams, forms, and report

TABLE 5-7 Traditional Systems Development versus CASE-Based Development (Adapted from McClure, 1989)

Traditional Systems Development	CASE-Based Systems Development
Emphasis on coding and testing	Emphasis on analysis and design
Paper-based specification	Rapid interactive prototyping
Manual coding of programs	Automated code generation
Manual documenting	Automated documentation generation
Intensive software testing	Automated design checking
Maintain code and documentation	Maintain design specifications

templates. The focus of this section is to give you an overview of how CASE can be applied to support each SDLC phase. This overview should enable you to better understand the tradeoffs in using CASE and how CASE changes the role of the systems analyst for each SDLC phase.

Supporting Project Identification and Selection with CASE

The identification and selection of information systems projects is an important activity in most organizations. Projects typically come from two sources: bottom-up and top-down. Bottom-up requests for projects reflect those from managers or business units that desire extensions or changes to an existing system. Top-down requests are those that result from a formal planning process. Regardless of the source of a project request, all must be evaluated. During evaluation, a senior IS manager or, more often in larger organizations, a steering committee assesses all possible projects that an organization could complete and selects those most likely to yield organizational benefits within available resources.

The quality of the project identification and selection process is significantly influenced by the information systems planning process. IS planning provides information that is used not only to identify projects needed to deliver systems to support plans but also to supply the criteria and other information helpful in assessing the potential value of proposed systems. Information systems planning (discussed in Chapter 6) is a top-down process that considers the outside forces that drive the business and the factors critical to the success of the firm. It looks at data and systems in terms of how they help business achieve its objectives. Most planning methodologies have certain techniques in common that include the following elements (Index Technologies, 1988):

1. *A top-down approach.* Planning begins with the business and the data it uses and leads to ideas for specific projects.

2. *Modeling.* Planners define models of business functions, data use, and related factors, often in matrix form.

3. *Involvement and accountability.* A planning team is established and planning involves all levels of management.

4. *Time limits.* Because information technologies (and business conditions) change so rapidly, planning should be completed quickly—in three to six months, if possible.

In general, information systems planning requires collecting, organizing, and analyzing vast amounts of data. Project identification and selection is the process of actually selecting a particular project. Table 5-8 summarizes numerous benefits of using CASE to support this process.

The *purpose* of project identification and selection is to examine the information needs of the organization and to segment and rank these needs so that the most beneficial projects are identified and selected. The desired *output* from this process is a planning document in which the organization decides how it will allocate its IS resources to specific projects. The CASE tools used to support this process include those that help display and structure higher-level organizational information. CASE diagramming capabilities are often extensively used to help communicate ideas between users, management, and the development group. For example, many CASE systems have the ability to display organizational structure charts, function and process decompositions, and matrix tools for gaining insights into the inter-relationships between various items.

TABLE 5-8   Benefits of Using CASE to Support the Project Identification and Selection Process

Benefit	Description
Structure	CASE provides an environment with a common database, approach, vocabulary, and interface.
Access	Various users can work on different parts of the plan simultaneously.
Storage	Information is stored once in a common repository.
Analysis	Complex relationships and "what-if" questions can be easily viewed.
Flexibility	CASE simplifies the process of updating the strategic plan as business conditions change.
Efficiency	Planning information becomes the basis for later development and project management activities.

Figure 5-19 shows a process hierarchy (decomposition diagram) and Figure 5-20 shows an elementary process by entity matrix from the IEF CASE product. In the decomposition diagram, high-level functions are enumerated into successively finer detail until activities with known timing and information needs are identified. In the matrix, relationships indicate which processing capabilities each elementary process has with each entity type in the corporate database. Processing relationships in this example represent whether a process can create new data, delete existing data, change existing data, or have read-only privilege. Numerous high-level views of existing systems or plans for new systems can be viewed in this way to gain insights about current and future system resources. Once this understanding is achieved, specific projects can more easily be identified and selected. For example, if a manager submits a request to enhance customer data, Figure 5-20 shows the potential impact that such an enhancement could extend to business forms, data input forms, and procedures for creating invoices, orders, and customers, processing orders, and updating and deleting customers, since these processes create, update, delete, and read customer data.

Supporting Project Initiation and Planning with CASE

The *purpose* of project initiation and planning is to organize a project team and to create a high-level project plan. During project initiation and planning, an analyst works with customers or users to collect a broad range of information in order to gain an understanding of the project size and its feasibility. During planning a project's scope is defined, a schedule of work is outlined, and the project's feasibility is assessed. Once collected and analyzed, this information is brought together into a summary document called a Baseline Project Plan. Thus the desired *output* from project initiation and planning is the creation of the Baseline Project Plan which is used by the project selection committee to help decide whether the project should be accepted, redirected, or canceled. If selected, the Baseline Project Plan becomes the foundation document for all subsequent SDLC activities.

Project initiation and planning is often referred to as a "mini-analysis" phase. In other words, many of the tools used during the analysis phase of the SDLC may also be used during project initiation and planning (see next section). This, of course, is influenced by the size and complexity of the project. Since the objective of this phase is the creation of a planning document, tools used to create project plans, estimate activity sizes and durations, and create project schedules are paramount. In other words, the planning team will extensively use the cross life cycle capabilities

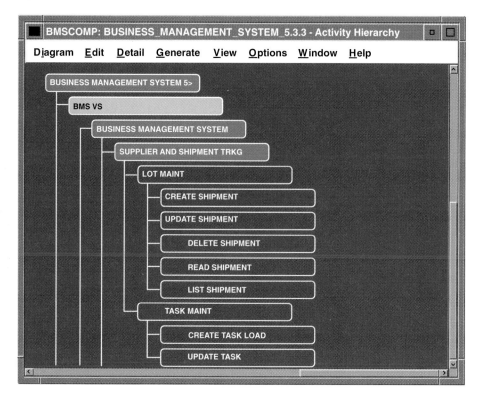

Figure 5-19
Process hierarchy diagram from Texas Instruments' IEF CASE environment

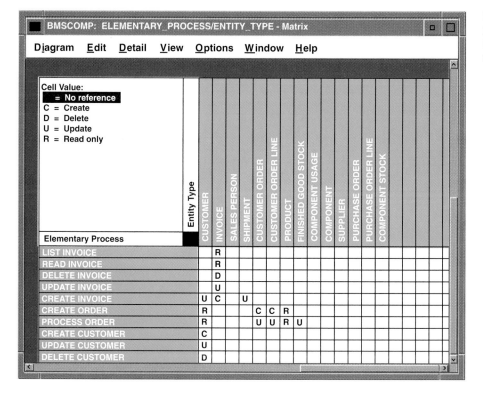

Figure 5-20
Matrix processor from Texas Instruments' IEF CASE environment

of the CASE repository and documentation generator, in addition to project management software, in the development of the Baseline Project Plan.

Supporting Analysis with CASE

The *purpose* of analysis is to determine the requirements for the development project within the cost, labor, and technical constraints of the organization. Analysis has three primary activities: requirements determination, requirements structuring, and alternative generation and selection. During requirements determination, analysts work with users to learn what users want and need from the proposed system. During requirements structuring, requirements are studied and organized according to their inter-relationships. Finally, during alternative generation and selection, the analyst works with users to identify the most effective way to deliver system functions by examining tradeoffs between alternative designs. The desired *output* from these three activities is a set of diagrams, forms, and reports that can be used as the basis for the system design.

Many CASE tools can be used during analysis. Especially important are the diagramming, form and report generators, and, of course, analysis tools. Diagramming tools can be used to model the data and their relationship to other data and processes. For example, after collecting requirements for a new system from users, you can graphically depict your understanding of these requirements. You can use analysis tools to make sure diagrams conform to standard conventions and then you can show the diagrams to users to review and verify that requirements have been accurately captured. Form and report generators can be used in a similar manner.

Supporting Design with CASE

The *purpose* of design is to create detailed plans of the software system. Design is divided into two separate sub-phases: logical design and physical design. Logical design refers to the creation of a system design that is not directly tied to any specific hardware, software, or operating system environments. Physical design refers to a plan for carrying out the logical design using specific hardware, software, and operating system environments.

The desired *output* from logical design is a detailed plan for the system including program algorithms, data structures, and form and report layouts. Alternatively, the desired *output* from physical design is a detailed plan that defines the system specification consistent within the limitations and characteristics of the environment in which the system will be constructed. The CASE tools that can be used to support the design phase include the diagramming, form and report generators, analysis tools, and documentation generators. In addition to these tools, many CASE environments create pseudocode programs called action diagrams that represent detailed processing logic at a much more abstract level than actual programming code, but in much more detail than graphical diagrams. Figure 5-21 shows an action diagram from IEF. This action diagram indicates what data to retrieve or store for the ADD_CUSTOMER_ORDER process.

Supporting Implementation with CASE

The *purpose* of implementation is to translate a system design specification into an information system that becomes part of an organization's business processes. Implementation consists of six sub-phases: coding, testing, installation, documentation, training, and support. Coding refers to the generation of the actual system source

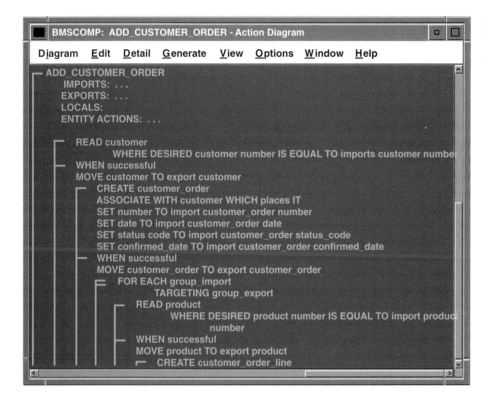

code. Testing is a formal process of validating that programs are error-free and perform their intended functions. Installation is the process of converting to the new
system from the current environment. Documentation is produced for both users
and developers. Training is done for users and data center staff, diffusing the system into the daily activities of the organization. Finally, support means helping
those using or wishing to change the system. In summary, the desired *output* from
implementation is to create, test, and deploy an error-free system that meets the
specifications of the design.

The CASE tools used during implementation include, of course, code generators but also the analysis tools, documentation generators, and form and report
generators commonly used to construct reference materials such as the user, training, and installation guides. The central repository is also a critical component
during the coding and testing processes as the repository is used to store the most
current versions of system modules shared by disparate modules within the overall
system. The repository acts as a library where programmers can "check out" modules and subroutines. It helps eliminate redundancy and assures higher-quality
systems, as only extensively tested and validated modules should be stored for
common use within the repository.

Supporting Maintenance with CASE

The *purpose* of maintenance is to make changes to existing software systems. Changes
may need to be made due to the identification of programming or design errors (or
"bugs"), to satisfy changes in the system requirements, or to adapt to changes in
the implementation environment. Maintenance activities may account for as much
as 70 percent of the development costs for a system over its lifetime. How much
time and effort are devoted to maintenance and its difficulty depend a great deal on

how the system was initially constructed. As was shown in Figure 5-16, the ease or difficulty of system maintenance is directly related to the quality of the system documentation. This means that the desired *output* from the systems development maintenance process is, of course, updated software, but also updated system specifications, diagrams, forms, reports, and documentation. This also means that all CASE tools used to support its initial development are used during maintenance.

SUMMARY

The purpose of this chapter has been to introduce you to the use of automated tools to support the systems development process. First, we examined how CASE is being used within organizations. The objectives of CASE are to aid the systems analyst with automated tools so that higher-quality systems are constructed on time and within budget, maintained economically, and changed rapidly.

Yet, with all the potential benefits of adoption, CASE can affect many individuals within the organization in a variety of ways. Systems development personnel will focus more on their interpersonal and communication skills than solely on their analytical skills. Likewise, users will be much more involved in the development process by actively working with development personnel when defining requirements, finalizing designs, and testing features and capabilities. Additionally, managers will play a more active role in setting project and design priorities. Overall, CASE acts to make the development process more collaborative. It should be clear that implementing CASE is not a small undertaking for most organizations. A long-term perspective must be taken or the potential benefits of CASE may not be realized.

Different categories of CASE tools were introduced, such as reverse and reengineering tools for rebuilding legacy systems. The important database component of CASE tools, the repository, was shown to contain an information repository and a data dictionary. The information repository is used to manage and control access to organizational business information and application systems. The data dictionary is a tool used to manage and control access to all data definitions in the repository. The data dictionary can be a passive tool or can be used actively during development or production (operation) of systems.

The components of a comprehensive CASE system are divided into upper, lower, and cross life cycle CASE tools, covering different segments of the SDLC. Upper CASE tools—diagramming tools, form and report generators, and analysis tools—are used primarily to support project identification and selection, project initiation and planning, analysis, and design. Lower CASE tools—code generators—are used primarily to support system implementation and maintenance. Cross life cycle tools—project management tools—coordinate project activities. The repository and documentation generators are used across multiple life cycle phases to support project management, activity estimation, and documentation creation. We also discussed how each of these tools can be used to support the various activities within the SDLC. In the next chapter, we look in detail at the SDLC by focusing on project identification and selection.

C H A P T E R   R E V I E W

K E Y   T E R M S

Analysis tools
Code generators
Computer-aided software
   engineering (CASE)
Cross life cycle CASE
Cross referencing

Data dictionary
Diagramming tools
Documentation generators
Form and report generators
I-CASE
Information Repository

Lower CASE
Re-engineering
Repository
Reusability
Reverse engineering
Upper CASE

R E V I E W   Q U E S T I O N S

1. Define each of the following terms:

   a. CASE
   b. code generators
   c. cross referencing
   d. data dictionary
   e. I-CASE
   f. information repository
   g. lower CASE
   h. re-engineering
   i. reverse engineering

   j. repository
   k. reusability
   l. upper CASE
   m. version control (within CASE repository)
   n. active in development
   o. active in production
   p. cross life cycle CASE

2. Describe the evolution of CASE and its outlook for the future.

3. What are the objectives for CASE within organizations?

4. How does or can the role of CASE change in relation to the size of an organization? In relation to the type of information systems developed by an organization?

5. Who and how are individuals impacted by the adoption of CASE within an organization?

6. What are the driving forces behind the adoption of CASE? What are the resisting forces?

7. Describe each major component of a comprehensive CASE system. Is any component more important than any other?

8. Contrast the differences between a data dictionary and repository.

9. Contrast the difference between traditional non-automated systems development and CASE-based systems development.

10. Describe how CASE is used to support each phase of the SDLC.

11. Describe the concept of software reusability. Is reusability possible without CASE?

PROBLEMS AND EXERCISES

1. Match the following terms to the appropriate definitions.

   _____ lower CASE

   _____ upper CASE

   _____ I-CASE

   _____ reverse engineering

   _____ re-engineering

   _____ balancing

   _____ data dictionary

   _____ information repository

   a. tools used to manage and control access to a CASE repository

   b. CASE tools designed to support early phases of the SDLC

   c. CASE tools designed to support late phases of the SDLC

   d. maintaining consistency between a more general and a more detailed data flow diagram

   e. the repository of all data definitions

   f. tools which study programs and re-structure them to improve quality or performance

   g. tools which study programs and pro-duce documentation which models the programs and data used in the programs

   h. a CASE environment that allows the seamless sharing of data across CASE tools

2. Review the driving and resisting forces for CASE described in Tables 5-3 and 5-4 and the CASE implementation issues listed in Table 5-5. What do you forecast for CASE evolution and adoption in the future? Why? Would you recommend to col-leagues that they adopt CASE tools? If so, under what circumstances? Why? If you wouldn't recommend using CASE tools now, why not, and what must change to cause you to recommend CASE tool adoption?

3. Review the sample computer forms in Figure 5-9. How is a form generator different from a standard graphics package like, say, Microsoft Windows Paintbrush® or Harvard Graphics®? Why not simply use one of these graphics packages instead of using a CASE tool? What is gained by using a CASE tool rather than a graphics package? To answer these questions adequately, you may need to call a CASE tool vendor directly or find CASE tool product evaluations in the popular press.

4. Using a university or your workplace as the setting, list as many data elements (e.g., student identification number or customer identification number) as you can. Imagine how much information would be gathered and organized in this organiza-tion's database files and data dictionary. Estimate how much computer-based storage space would be needed to store, process, and back up this information. In an environment with computers, but without CASE tools, how would this orga-nization store, organize, and retrieve this information? What are the limitations, weaknesses, and potential problems in trying to do this without the data dictionary component of a CASE tool?

5. What forms of user documentation, either hard copy or on-line, came with your PC, operating system, network operating system, and applications software at work, home, or school? Is this documentation accessible and helpful? At what level would you rate the quality of the documentation—high, average or low? Why? What could be done to improve the documentation? How could CASE help to improve documentation?

6. A goal stated by many vendors of CASE products is to have CASE ultimately be able to automatically generate (and regenerate to any platform and with any changes)

100 percent of the code, error-free, for a new or modified information system. This goal is considered important in order to achieve systems development productivity gains necessary to deal with systems backlog, to improve system quality, and to enhance our ability to maintain systems. Do you think this goal is possible? Why or why not?

7. What problems might occur during systems development if an organization used different CASE tools that did not share a common repository? What parallels can you make between the purpose of shared databases and the purpose of I-CASE?

8. Speculate on the reasons for the results shown in Figure 5-2 about CASE users. What do these reasons imply about the state of CASE tools?

9. Review the data dictionary entries on the Library User data entity in Figure 5-14. What other data about an entity would you suggest be included in such a data dictionary entry? Why?

FIELD EXERCISES

1. Interview a sample of systems analysts, programmers, and IS project managers to elicit their views of CASE. Are they using CASE? If so, what are their evaluations thus far? What effects has using CASE had on their jobs? Do their perceptions fit with those summarized in Table 5-2? If they are not yet using CASE, why not? What must happen before they adopt CASE tools?

2. Find a detailed description of a CASE tool and determine which of the CASE functions discussed in the chapter section "Components of CASE" this tool supports. To do this, you may need to call a CASE tool vendor directly or find CASE tool product evaluations in the popular press.

3. Interview information systems professionals who use CASE tools and find out how they use the tools throughout the SDLC process. Ask them what advantages and disadvantages they see in using a CASE-based system development process rather than other traditional methods.

4. The CASE market is rapidly changing. Search through recent trade publications, like *Datamation, Computerworld, PCWeek, Database Programming & Design, Data Based Advisor,* and *Software Magazine.* Find ads for CASE tools or articles that list CASE tools. Develop a list of these products. What are all the benefits claimed for CASE in these ads and articles? What features of CASE do the vendors promote?

5. Choose an organization that is developing information systems without CASE tools. Talk to the people involved in building the systems and find out more about how they build them. Determine the parts of the systems development process that could be better done using CASE tools. Estimate the time and expense involved in those parts of the process that could be better done using CASE tools. Does it make sense for this organization to adopt CASE tools? Why or why not?

6. Call a CASE tool vendor directly and determine the product's price, functionality, and advantages, and ask about future plans for the product. Are future versions planned? If not, why not? If so, what changes and/or enhancements are planned for future versions? Why?

REFERENCES

Brooks, F. P., Jr. 1982. *The Mythical Man-Month.* Reading, MA: Addison-Wesley.

Bruce, T., J. Fuller, and T. Moriarty. 1989. "So You Want a Repository." *Database Programming & Design* 2(May): 60–69.

Chen M., and R. J. Norman. 1992. "Integrated Computer-aided Software Engineering (CASE): Adoption, Implementation, and Impacts." *Proceedings of the Hawaii International Conference on System Sciences,* edited by J. F. Nunamaker, Jr. Los Alamitos, CA: IEEE Computer Society Press, Vol. 3: 362–73.

Hanna, M. 1992. "Using Documentation as a Life-cycle Tool." *Software Magazine* 12(12) (Dec.): 41–51.

Index Technologies. 1988. *Excelerator Reference Manuals.*

I/S Analyzer. 1993. "The Cost and Benefits of CASE." Rockville, MD: United Communications Group. 31(6) (June).

Jones, C. 1986. *Programming Productivity.* New York, NY: McGraw-Hill.

Jones, M. C., and K. P. Arnett. 1992. "CASE Use is Growing, But in Surprising Ways." *Datamation* 34(May 1): 108–109.

Jones, T. C. 1993. "Equipping the Software Engineer." *Software Magazine* 13(1) (January): 100–109.

Lefkovitz, H. C. 1985. *Proposed American National Standards Information Resource Dictionary System.* Wellesley, MA: QED Information Sciences.

McClure, C. L. 1989. *CASE is Software Automation.* Englewood Cliffs, NJ: Prentice-Hall.

Nunamaker, J. F. 1992. "Build and Learn, Evaluate and Learn." *Informatica* 1(1): 1–6.

Orlikowski, W. J. 1989. "Division Among the Ranks: The Social Implications of CASE Tools for System Developers." *Proceedings of the Tenth International Conference on Information Systems.* 199–210.

Pfrenzinger, S. 1992. "Reengineering Goals Shift Toward Analysis, Transition." *Software Magazine* 12(10) (Oct.): 44–57.

Stickel, E. U. 1993. "An Experience with CASE Tool Support for Financial Product Design." *Data Base* 24(4) (Nov.): 31–35.

Stone, J. 1993. *Inside ADW and IEF: The Promise and Reality of CASE.* New York, NY: McGraw-Hill.

Case Introduction

OUR RUNNING CASE: BROADWAY ENTERTAINMENT COMPANY

Starting with Chapter 5, we will follow all chapters through Chapter 20 with a case example featuring the Broadway Entertainment Company, Inc. Broadway Entertainment is a fictional company in the video rental and recorded music retail industry, but its size, strategies, and business problems (and opportunities) are comparable to those of real businesses in this fast-growing industry.

The first case is designed to introduce you to the company and the people who work for it. The second case follows this one and will present the company's information systems portfolio. The third case will follow Chapter 6 where we discuss the idea for a new information system, which begins the systems development life cycle. A new case will be presented at the end of subsequent chapters. Each case will deal with that phase of the life cycle discussed in the adjoining chapter. The last Broadway case will follow Chapter 20 and covers support procedures for the new system.

Our aim is to provide you with a realistic case example of how the systems development life cycle proceeds through its phases and how analysts, managers, and users work together to develop an information system. Reading about a phase of the life cycle and the techniques and tools that can support that phase is one thing; seeing the techniques and tools in action is another. Our hope is that the book chapters and Broadway Entertainment sections will complement each other as you learn about information systems development following the systems development life cycle.

THE COMPANY

The Broadway Entertainment Company, Inc., owns and operates Broadway Entertainment Company, or BEC, retail outlets. As of January 1996, Broadway owned 2,125 outlets across the United States, Canada, Mexico, and Costa Rica. There is at least one BEC outlet in every state in the U.S. (except Montana) and in each Canadian province. There are thirty-two Canadian stores, two in Mexico, and one in Costa Rica. The company is currently struggling to open a retail outlet in Japan and plans to expand into the European Union (EU) within five years. Broadway is headquartered in Spartanburg, S.C. Canadian operations are headquartered in Vancouver, B.C., and Latin American operations are based in Mexico City.

Each BEC outlet offers for sale two product lines, recorded music and video games, and rents to customers two product lines, recorded videos and video games. Recorded music is sold in two formats: cassette tapes and compact discs (CDs); videos are also rented in two formats: video tapes and laser discs. In calendar year 1995, music sales and video rentals together accounted for over 85 percent of Broadway's U.S. revenues. (See Table 1 for a breakdown of Broadway's domestic revenues for 1995.) Foreign operations added another $17,500,000 to company revenues.

TABLE 1 Calendar 1995 Domestic Revenue by Category

Category	Revenues (in $000s)	Percent
Music Sales	500,000	37.0
Compact Discs	300,000	22.2
Cassettes	200,000	14.8
Video Game Sales	80,000	5.9
Video Game Rentals	120,000	8.9
Video Rentals	650,000	48.2
Video Tapes	585,000	43.4
Laser Discs	65,000	4.8
Total	1,350,000	100.0

The home video and music retail industries are strong and growing, both domestically and internationally. Worldwide retail home video revenue for 1993 exceeded $25 billion. U.S. video rentals are expected to exceed $10.6 billion in 1996, while U.S. video sales are expected to exceed $6 billion for the same year.

For several years, home video has generated more revenue than either theatrical box office or movie pay-per-view. U.S. revenues for the home video industry (rental and sales) are expected to exceed $21 billion by 2000, yet the total domestic revenue for the industry was only $700 million in 1982. The music retail industry, on the other hand, had worldwide revenues of $29 billion in 1992.

Revenues in the U.S. alone are expected to exceed $13 billion in 1997. In 1993, Broadway controlled 5 percent of the domestic market for both home video and music.

To get a good idea of the industry in which Broadway competes, we can use Porter's Five Forces Model of Industry Structure (1980). According to Porter, any industry can be analyzed by focusing on five aspects of its structure: (1) suppliers (2) buyers (3) substitutes (4) barriers to entry and (5) rivalries among the firms competing in the industry (see Figure 1). For suppliers, Broadway deals with all of the major distributors of recorded music (Sony, Matsushita, Time Warner), video games (Nintendo, Sega), and recorded videos (CBS Fox, Viacom). Customers (buyers) consist entirely of individual consumers—Broadway does not provide product for rebroadcast or public showings and does not deal with groups or associations. Substitutes for Broadway's entertainment offerings include television (broadcast, cable, satellite), first-run movies, theater, radio, concerts, and sporting events.

On one level, the barriers to entry into Broadway's business are relatively low. To open a single retail outlet that sells music and video games and rents videos does not take a large amount of capital or experience. On the other hand, to create a company that could compete with Broadway on the same national and international scale would take quite a bit of capital and experience. Broadway's management is not worried about the so-called "Mom and Pop" record and video stores. They are worried about such things as the alliance between USWest and Time Warner to create cable television that lets consumers choose from a large number and variety of videos from a computerized menu system in their homes. Such a system, which can be thought of as a much larger pay-per-view setup, would obviate the need for consumers to ever set foot in a video store. They could see whichever movie or entertainment event they wanted without leaving home and without having to worry about returning any product to the store. Another recent corporate alliance that has Broadway's management worried is the venture by IBM and Blockbuster to provide CD pressing machines in each Blockbuster store. The pressing machines would be linked to large libraries of recorded music. A consumer could search through the library and, once the desired music was found, it would be downloaded to the pressing machine and the CD would be created in the store while the consumer waited. The advantage for Blockbuster would be that they would never have to stock any recorded music, only blank CDs. Inventory carrying costs and stock ordering would practically disappear.

As for rivalries in the industry, Broadway is most concerned about two types of rivals: the large music chains and the large video chains. In the video area, the biggest competitor by far is Blockbuster Entertainment.

Figure 1
Porter's Five Forces Model of Industry Structure

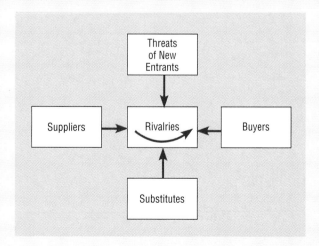

At the end of June 1994, Blockbuster had 3,755 video stores in forty-nine states and nine foreign countries. Of those stores, 2,829 were company-owned, and 926 were franchised. In 1992, Blockbuster entered the music business by acquiring 238 music stores and by entering into a joint agreement with Virgin Retail Group to open 20 "megastores," primarily in Europe. By the end of June 1994, Blockbuster owned 521 music stores in addition to the "megastores," making Blockbuster one of the largest music retailers as well as the largest video rental operator. Other retail music competitors include Musicland, Tower Records, and the Wherehouse. (See Table 2 for the recent revenues of Broadway's chief competitors.)

COMPANY HISTORY

Broadway Entertainment Company, Inc., started as one of the "Mom and Pop" stores that Broadway's management no longer fears. The first BEC outlet opened in the Westgate Mall in Spartanburg, S.C., in 1977 as a music sales store or, as it was popularly known at that time, a record store. The first store exclusively sold recorded music, primarily in vinyl format, but also stocked cassette tapes. Broadway's founder and current chairman of the board, Nigel Broad, had immigrated to South Carolina from his native Great Britain in 1968. In Britain, Nigel had worked in several record stores and as a disc jockey for two different rock radio stations. He moved to South Carolina to be near his last remaining uncle, Albert. After nine years of numerous music-related jobs, Nigel took the money he had been left by his mother, formed Broadway Entertainment Company, Inc., and opened the first BEC outlet in the newly opened Westgate Mall on Spartanburg's west side.

Sales were steady and profits increased. Soon Nigel was able to open a second outlet and then a third.

Predicting that his BEC stores had already met Spartanburg's demand for recorded music, Nigel decided to open his fourth store in nearby Greenville in 1981. At about the same time, he added a new line—Atari video game cartridges. Atari's release of its Space Invaders game cartridge resulted in huge profits for Nigel. The company continued to grow and Broadway expanded beyond South Carolina into neighboring states.

When Nigel first started his business, he had the insight to not stock eight-track tapes. His insight paid off as eight-tracks soon died as a recorded music format. In the early 1980s, Nigel saw the potential in video cassette recorders and video tapes. A few video rental outlets had opened in some of Broadway's markets, but they were all small independent operations. Most video stores had similar business practices: customers had to join a video club and pay some type of membership fee, usually $30 to $40 per family; customers had to return the video within 24 hours of the time they rented it; most stores had limited selections. Nigel saw the opportunity to combine video rentals with music sales in one place. He also decided that he could rent more videos to customers if he changed some of the video store rules such as eliminating the heavy membership fee and allowing customers to keep videos more than one night. Nigel also wanted to offer the best selection of videos anywhere.

Nigel opened his first joint music and video store at the original BEC outlet in Spartanburg in 1985. Customer response was overwhelming. In 1986, Nigel decided to turn all 17 BEC outlets into joint music and video stores. Nigel realized that the expansion would require a large amount of capital to finance the purchase of a large video tape library. Up until this time, Broadway Entertainment had been a closely held company. To move into the video rental business in a big way, Nigel decided to have a public offering later that year. Nigel and his chief financial officer, Bill Patton,

TABLE 2   Revenues for Broadway's Closest Competitors

Competitor	1991 Revenues (in $000s)	1992 Revenues (in $000s)	1993 Revenues (in $000s)
Musicland	932,231	1,020,500	N/A
Blockbuster Entertainment*	961,638	1,315,844	2,227,003
Wherehouse Entertainment	457,400	448,500	471,800

*Blockbuster Entertainment revenue for company-owned stores only.

put together the basic terms of the offer: 1,000,000 shares of stock to sell for $7 each. They were happily surprised when all one million shares sold. The proceeds allowed Broadway to offer video rentals in all 17 BEC outlets and to revive the dying video games line by dropping Atari and adding the newly released Nintendo game cartridges.

Profits from BEC outlets continued to grow throughout the eighties, and Broadway continued to expand. Important events in the expansion are listed in Table 3. One of the most important decisions Nigel made was to acquire existing music and video store chains and convert them to BEC stores. From 1987 through 1993, the number of BEC outlets roughly doubled each year. The decision to go international, made

in 1991, was also important, resulting in 12 Canadian stores that year, with 10 more added in 1992 and 1993. All three Latin American stores were opened in mid-1994. From its beginnings in 1977, with 10 employees and $398,000 in revenues, Broadway Entertainment Company, Inc., grew to 24,225 employees and worldwide revenues of $1,367,500,000 by January 1, 1994.

COMPANY ORGANIZATION

Since 1977 when he founded the company, Nigel Broad has been intimately involved in the growth and development of Broadway Entertainment Company, Inc. In 1992, when the company was about to open its one-thousandth store, Nigel decided that he no longer wanted to be the chief executive officer of the company. Nigel decided to fill only the position of chairman and he promoted his close friend Ira Abramowitz to the offices of president and CEO (see Figure 2).

Most of Broadway's other senior officers have also been promoted from within. Bill Patton, the chief financial officer, had been with Broadway from the beginning, starting out as the fledgling company's bookkeeper and accountant. Karen Gardner had been part of the outside consulting team that had built Broadway's first information system in 1986 and 1987. She became the vice-president in charge of IS for BEC in 1990. Bob Panofsky, the vice-president for human resources, had been with the company since 1981 and had worked his way up to the vice-president level. An exception to the promote-from-within tendency, W. D. Nancy Chen, the vice-president for domestic opera-

TABLE 3   Important Events in Broadway's History

Year	Event
1988	Decision to acquire 30-store MAP Video Store chain
1989	Acquisition of 52-store Buddy's Records chain
1989	Second public offering, 30,000,000 shares offered
1990	Acquisition of 98-store Chang Video Market chain
1991	Decision to expand to Canada, 12 stores opened
1991	Acquisition of 208-store Music World chain
1991	Decision to rent video games
1992	Third public offering, 100,000,000 shares offered
1993	Decision to expand to Latin America

Figure 2
Broadway Entertainment Company, Inc. organization chart

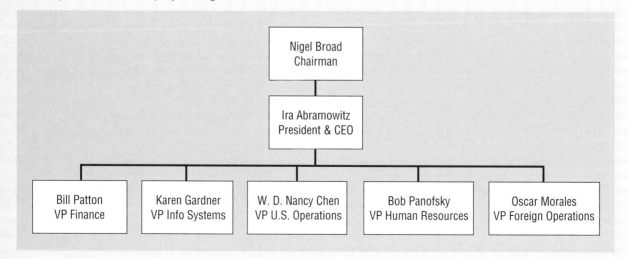

tions, had been recruited from Music World in 1991, shortly before the chain was purchased by Broadway. Oscar Morales had been hired in 1992 from Blockbuster Entertainment, where he had been in charge of Latin American expansion.

INFORMATION SYSTEMS

Broadway Entertainment Company operated from 1977 until 1984 without any information systems support. During its first five years of business, Broadway operated five or fewer music outlets. Like many businesses this size, the owner did not have the expertise or the capital for developing the company's own information systems. Bill Patton, the company's accountant, did all of the company business, including the management of inventory, by hand until he bought an IBM AT in 1984. After he had successfully converted Broadway's accounting and other business functions to work on the AT using pre-packaged software purchased from a local computer store, Bill wondered how he had ever worked without it. Bill told anyone who would listen that computerizing the company had made the expansion to 10 stores in 1984 much easier. In 1985, Nigel expanded the original Spartanburg BEC store to handle video rentals, and by 1986, Broadway had expanded to 17 outlets and was considering its first public offering. The change in the business scope would clearly swamp Bill Patton and his IBM AT. The dramatically increased demands on inventory alone seemed to warrant increased computerization.

BEC had nobody trained in information systems on staff and all the BEC managers were quite busy coping with the business expansion. Nigel and Bill considered hiring a small IS department from outside, but they were reluctant to do so since they did not know how a new group should be managed, how to select quality staff, or what to expect from such employees. From their limited experience with computers, Nigel and Bill realized that computer software could be quite complicated, and building systems for a rapidly changing organization could be quite a challenge. Nigel and Bill also knew that building information systems required discipline and this discipline needed to be a part of systems development from the beginning. Thus, they wanted experienced people, but had no idea how much effort was involved or what expertise would be required. So Nigel, after contacting several other South Carolina businesses, contacted the information consulting firm of Fitzgerald McNally, Inc. about designing and building a custom computer-based information system for Broadway.

Nigel and Bill wanted the new system to perform all of the functions Bill's current system could handle, i.e., accounting, payroll, and inventory. Nigel wanted the system to be readily expandable as he was planning for Broadway's rapid growth. At the operational level, Nigel realized that the video rental business would require new features in their information system, features not needed for pre-recorded music. For one thing, rental customers would not only be taking product from the store, they would also be returning it at the end of the rental period. Further, customers would be required to register with Broadway and attach some kind of deposit to their account in order to help ensure that videos would be returned.

At a managerial level, Nigel wanted the movement of videos in and out of the stores and all customer accounts computerized and easy to track. Being new to the video rental business, Nigel also wanted to be able to search through the data on Broadway's customers describing their rental habits. He wanted to know which videos were the most popular, and he wanted to know who Broadway's most frequent customers were, not only in South Carolina but in every location where Broadway did business.

Fitzgerald McNally, Inc. was happy to get Broadway's account. They assigned Karen Gardner to head up the development team. Karen led a team of her own staff of analysts and programmers, along with participation by several BEC managers, in a thorough analysis and design study. Fitzgerald McNally had recently adopted a CASE tool and associated methodology. This methodology provided the discipline needed for such a major systems development effort. The methodology began with information planning and continued through all phases of the systems development life cycle. Fortunately for the development team, BEC had been doing annual and long-term business planning and was very cooperative at every step of the total process.

After working for almost two years on the project, Karen and her team delivered and installed the system Broadway had requested. The system was centralized, with an IBM 4381 mainframe installed at headquarters in Spartanburg, and three terminals, three light pens, and three dot matrix printers installed in each BEC outlet. The light pens were used to track videotapes, recording when the tapes were rented and when they were returned by reading the bar code on the cassette. The light pens were also used to read the customer's account number which was recorded in a bar code on the customer's BEC account card. The printers generated receipts. The same equipment was used for video game rentals and for video game and music sales. In addition, the system included a small personal computer and printer to handle a few office functions such as the ordering and receiving of goods. The primary software product Fitzgerald McNally built for Broadway, then, monitored and updated inventory levels. Another software product generated and updated the customer database while other parts of the final software package were designed for accounting and payroll. By the time the system was installed and running at the end of 1987, Broadway had expanded to 33 stores, and the system was able to handle Broadway's business demands with few problems.

In 1988, Karen Gardner left Fitzgerald McNally to start her own consulting firm. In 1990, she joined Broadway as the head of its information systems group. Karen led the effort to expand and enhance Broadway's information systems as the company grew to over 2,000 company-owned stores in 1995. Broadway now relies on several IBM 3090 mainframes and in-store point-of-sale computer systems to handle the transaction volume generated by millions of customers at all BEC outlets. Karen and her team have been thinking seriously about decentralizing Broadway's information systems and changing to a client/server network. None of these issues, however, has been resolved.

SUMMARY

Broadway Entertainment Company (BEC) is a $1.35-billion international chain of music, video, and game rental and sales outlets. BEC operates in a highly competitive and rapidly changing industry not only involving some of the largest electronics and entertainment companies but also providing the opportunity for new entrants with innovative delivery systems and service.

BEC started with one store in Spartanburg, S.C., in 1977 and has grown through astute management and capitalization into over 2,000 stores in four countries today. BEC is headed by Chairman of the Board of Directors Nigel Broad who has relinquished all operating duties to a team of executives including Ira Abramowitz (CEO), Bill Patton (CFO), and Karen Gardner (v-p for information systems). Computer systems began with one PC. Through the guidance of an outside consultant, quality systems were developed and eventually an internal IS function was created, headed by Karen Gardner, formerly with the consulting firm. Today, BEC systems run on an IBM mainframe at corporate headquarters and distributed point-of-sale equipment in retail stores. You will learn more about these systems in the next case study.

The Existing Application Portfolio

INTRODUCTION

Broadway Entertainment Company (BEC) has developed computer applications for both local stores as well as corporate functions. BEC has two systems development and support groups, one responsible for in-store applications and the other responsible for corporate applications. The corporate group is further divided by functional area, such as accounting, marketing, and human resources. Each corporate development group also has liaison staff with the in-store group, since data in many corporate systems feed or are fed by in-store applications (for example, market analysis systems depend on transaction data collected by the in-store systems). BEC creates both one-year and three-year IS plans which encompass both store and corporate functions. Regional and country management operations are supported within the corporate IS plan and development group.

The functions of the original in-store systems at BEC have changed very little since they were installed in 1987—customer and inventory tracking is still done by pen-based, bar code scanning of product labels and membership cards, and basic accounting and payroll functions are automated. Daily transactional data (rentals and returns, sales, and other changes in inventory as well as employee time in and out) are all captured at the store in electronic form via a local point-of-sale (POS) computer system and transmitted in batch at night using modems and regular telephone connections to corporate headquarters where all records are centralized on an IBM 3090 mainframe for all stores (Figure 1 depicts BEC's hardware and network architecture). Software updates for the store systems are also distributed electronically via the public data network.

As shown in Figure 1, each BEC store has an AT&T Global Information Solutions (formerly NCR) computer that serves as a host for a number of POS terminals at check-out counters. The computer also serves as the store manager's workstation for generating reports and performing other administrative in-store applications. Some managers have also learned how to use spreadsheet, word processing, and other packages to handle functions not supported by systems provided by BEC for in-store use. Each store computer includes a modem to allow communication with the corporate host computer via a communications front-end processor. The front-end processor at corporate headquarters offloads handling communications traffic from the IBM mainframe so that it can concentrate on data processing applications. BEC's communication protocol is SNA, an IBM standard. Corporate databases are managed by IBM's relational DBMS DB2. BEC uses a variety of programming environments, including C, COBOL, SQL (as part of DB2), and code generators.

All inventory control (e.g., purchasing of new items, retirement of items) is done centrally and, since local stores are corporate owned, employees are paid by the corporation. Today, each store is an isolated outlet; that is, each store has electronic records of only its own activity, including inventory. Profit and loss, balance sheets, and other financial statements are produced for each store by centralized systems. Each store manager is responsible for all local personnel matters (hiring, evaluating, scheduling, and firing employees), local property management (cleaning, day-to-day dealings with property owner if the property is leased, store layout within company guidelines, and so forth), managing local promotions within a budget set by BEC for each store, and customer relations. The local manager also influences stocking decisions, but final decisions are made at headquarters based on store transaction history and market trends. Practically all other functions are managed centrally. The following sections review many of the applications that exist in the stores and at the corporate level.

Figure 1
BEC hardware and network architecture

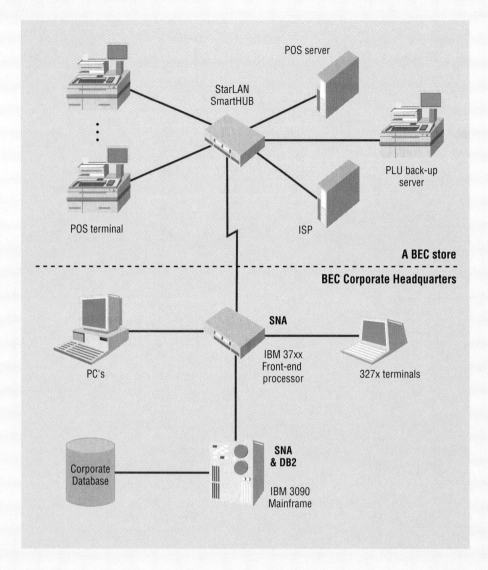

IN-STORE SYSTEMS

Table 1 lists the application systems installed in each store. BEC has developed a turnkey package of hardware and software (called Entertainment Tracker—ET) that is installed in each store worldwide. Besides English, the system also works in Spanish and French; BEC has developed one store system in which the language is set as a parameter when the system is installed in a store. The store system uses AT&T point-of-sale (POS) technology which is Intel microprocessor-based. BEC chose AT&T equipment because AT&T is a worldwide leader in point-of-sale equipment with sales and support offices in all international locations where BEC plans to operate. AT&T equipment and system software is also compatible with the IBM mainframes

used at corporate headquarters, so data transmission between headquarters and stores is not complicated.

As you can see from Table 1, all of these applications are transaction processing systems. In fact, there is a master screen on the POS terminals that leads to all systems, and the type of transaction selected on this master screen determines which ET application is activated. These systems work off a local decentralized database, and there is a similarly structured database for each store. Databases are not distributed. For example, the database of a store in San Diego, California, is not accessible from a store in Ft. Collins, Colorado. Furthermore, store databases are not accessible interactively from corporate headquarters, although various batched data transfers occur between corporate and store systems at night (store

TABLE 1 List of BEC In-Store (Entertainment Tracker) Applications

System Name	Description
Membership	Supports enrollment of new members, issuing membership cards, reinstatement of inactive members, and local data management for transient members
Rental	Supports rentals and returns of all products and outstanding rental reports
Sales	Supports sales and returns of all products (including videos, music, snack food, BEC apparel, and gift certificates)
Inventory Control	Supports all changes in rental and sales inventory that are not sales-based (e.g., receipt of a new tape for rental, rejection of goods damaged in shipment, and transfer of an item from rental to sales categories)
Employee	Supports hiring and terminating hourly employees, as well as all time-reporting activities

transactions, price and membership data updates, etc.). Figure 2 depicts a conceptual view of the database structure in each store. The database contains only current data—the history of customer sales and rentals is retained in a corporate database. Daily sales transactions are transmitted nightly to corporate databases and purged from the local database each night; daily rental data are also transmitted overnight, but a rental transaction is not purged until the items are returned and the completed rental transaction data are transmitted that night. Thus, local stores do not retain any customer sales and rental activity (except for open rentals).

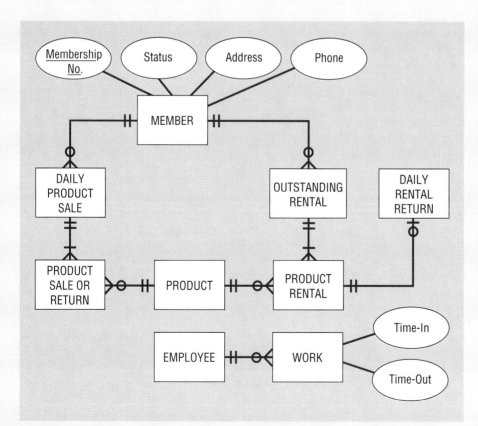

Figure 2
Store database

Data for those members who have had no activity at a local store for more than one year are purged from the local database. Approximately 3 percent of a store's transactions are made by infrequent or transient BEC members. Transient members are those who might use several stores in the same metropolitan area or who rent or purchase items while travelling away from their home. In certain locations, like resorts and college towns, transient customer transactions occur at a much higher percentage. When members use a BEC membership card and no member record exists in the local database, members are asked to provide a local address and phone number where they can be contacted.

All store employees, except the store manager who is on salary, are paid on an hourly basis, so clock-in and clock-out times are entered as a transaction, using employee badges with a bar code strip, on the same point-of-sale terminal used for member transactions. Paychecks are delivered by express mail twice a month. Employee reports (for example, attendance, payroll, and productivity) are produced by corporate systems and sent to store managers. Store managers have been asking for more time and attendance management support functions in the local computer system, but the functionality is currently minimal. Some store managers use stand-alone personal computers to manage personnel scheduling using packaged calendaring and project management software.

All other store record-keeping is manual and local stores do not pay any bills. Most bills, such as rent, utilities, and local advertisements, are sent directly to BEC headquarters; those which do come to a local store are simply forwarded to corporate offices by the store manager. Thus, the local store is not concerned with check writing, accounts payable, banking, or any other financial functions. Accounts receivable is a corporate function. Although the local store manager is responsible for contacting via phone or mail members who are late in returning rented items, the corporate accounts receivable department has final collection responsibilities including cancelling a member's membership. Each night a file of delinquent members is transmitted to each store and, if a member tries to use a delinquent membership, the member is asked to return all outstanding rentals before renting any more items and the current transaction is invalidated. Accounts receivable can also terminate a member for frequent delinquency or for other reasons. When terminated members try to use their cards, a BEC store clerk keeps the membership card and members are given a printed form which explains their rights at that point. Stolen membership cards are handled similarly, except that the store manager deals personally with people using cards that have been reported stolen.

CORPORATE SYSTEMS

Corporate systems run on IBM mainframes using IBM's DB2 relational database management system, although some run on PCs. Application software is written in COBOL, C, SQL (a database processing language), and several 4GLs, and all systems are developed by BEC. Clerks and managers use IBM 327x terminals and PCs for interactive access into corporate systems. There are also many PCs used by managers for stand-alone, end-user applications such as word processing, spreadsheets, specialized databases, and business graphics.

There are more than 20 major corporate systems with over 350 programs and approximately 500,000 lines of code. There are many more specialized systems, often developed for individual managers, projects, or special events. Table 2 lists some of the most active and largest of the major corporate systems. Most of these systems are a combination of transaction processing and MIS or MIS and DSS types. For example, the General Ledger and Financial Accounting application handles both transaction entries as well as management reports which summarize costs and revenues in a variety of ways. Since BEC does not hold any product inventory at the corporate level (all items are shipped directly from suppliers or distributors to the stores), there is no warehousing application.

One interesting aspect of the Banking application is that since stores have no financial responsibilities, BEC uses a local bank only for daily deposits and getting change. BEC's corporate bank, NCNB, arranges correspondent banking relationships for BEC so that local deposits are electronically transferred to BEC's corporate accounts with NCNB.

BEC's applications are still expanding, and they are under constant revision. For example, a current enhancement to the Sales Tracking and Analysis application deals with transient members. In cooperation with several hotel and motel chains that provide VCRs for rental, BEC is undertaking a new marketing campaign aimed at frequent travellers. Another study is underway to assess the costs of redesigning the Sales Tracking and Analysis application to support sales and rentals of business and training products to corporate members; this study will be used to help build the business case for this new market BEC is consider-

TABLE 2 List of BEC Corporate Applications

System Name	Description
Human Resources	Supports all employee functions, including payroll, benefits, employment and evaluation history, training, and career development (including a college scholarship for employees and dependents)
Accounts Receivable	Supports notification of overdue fees and collection of payment from delinquent customers
Banking	Supports interactions with banking institutions, including account management, electronic funds transfers
Accounts Payable, Purchasing, and Shipping	Supports ordering products and all other purchased items used internally and resold/rented, distribution of products to stores, and payment of vendors
General Ledger and Financial Accounting	Supports all financial statement and reporting functions
Property Management	Supports the purchasing, rental, and management of all properties and real estate used by BEC
Member Tracking	Supports record-keeping on all BEC members and transmits and receives member data between corporate and in-store systems
Inventory Management	Supports tracking inventory of items in stores and elsewhere and reordering those items that must be replenished
Sales Tracking and Analysis	Supports a variety of sales analysis activities in support of marketing and product purchasing functions based on sales and rental transaction data transmitted nightly from stores
Store Contact	Supports transmittal of data between corporate headquarters and stores nightly, and the transfer of data to and from corporate and store systems
Fraud	Supports monitoring abuse of membership privileges
Shareholder Services	Supports all shareholder activities, including recording stock purchases and transfers, disbursement of dividends, and reporting
Store and Site Analysis	Supports the activity and profit analysis of stores and the analysis of potential sites for stores

ing entering. At any one time, there are approximately 10 major system changes or new systems under development for corporate applications with over 250 change requests received annually covering requirements from minor bug fixes to reformatting or creating new reports to whole new systems.

STATUS OF SYSTEMS

A rapidly expanding business, BEC has created significant growth for the information systems group Karen Gardner manages. To date, BEC's international operations, headed by Oscar Morales, has not had its own development and support staff, but Karen is considering reorganizing her staff to provide focused attention to this important area. Since BEC wants to invest as much as possible in expansion—especially international—managers have been charged with increasing productivity without adding full-time staff. BEC still uses the services of Fitzgerald McNally, but on a very selective basis, such as when Karen's resources are fully committed. Karen's department includes 33 developers (programmers, analysts, and other specialists in database, networking, etc.) plus data center staff and is now large and technically skilled enough to handle almost all requests. Karen has created as professional an environment in BEC's IS group as exists at Fitzgerald McNally. For example, BEC recently acquired a CASE tool and adopted a standardized systems development methodology consistent with this tool.

Karen's current challenge in managing the IS group is keeping her staff current in the skills they need to successfully support the systems in a rapidly changing and competitive business environment. Besides adapting to new technical skills, Karen's staff needs to be excellent project managers, to completely understand the business, and to exhibit excellent communication with clients and each other. Karen is also concerned about information systems literacy among BEC management and that technology is not being as thoroughly exploited as it could be.

To deal with this situation, Karen is considering several initiatives. First, she has requested a sizeable increase in her training budget and has asked BEC to consider expanding the benefits of its college tuition reimbursement program. Second, Karen is considering instituting a management development program that will better develop junior staff members and will involve various user departments. As part of this program, BEC personnel will rotate in and out of the IS group as part of normal career progression. This program should greatly improve relationships with user departments and increase end-user understanding of technology. The development of this set of technical, managerial, business, and interpersonal skills in and outside IS is a critical success factor for Karen's group in responding to the significant demands and opportunities of the IS area.

SUMMARY

BEC has a hardware and software environment that is similar to that used by many national retail chains. Each store has a computer system with point-of-sale terminals which run mainly sales and rental transaction processing applications, such as product sales and rental, membership, store-level inventory, and employee pay activities. Corporate systems are executed on a mainframe computer at a corporate data center. Corporate systems handle all accounting, banking, property, sales and member tracking, and other applications that involve data from all stores.

BEC is a rapidly growing business with significant demand for information services. To build and maintain systems, BEC has divided its staff into functional area groups for both domestic and international needs. BEC uses modern database management and programming language technologies, including a CASE tool. The BEC IS organization is challenged by keeping current in both business and technology areas. We will see in subsequent case studies after subsequent chapters how BEC responds to a request for a new system within the business and technology environment described in the first two BEC case segments.

Making the Business Case

Allison Engine Company

How Organizational and IS Strategies Are Related

Allison Engine Co., now owned by Rolls Royce, Inc., was formerly Allison Gas Turbine, a division of General Motors (GM), Inc. Allison manufactures gas turbine engines used in helicopters and various aircraft and does work relating to national defense.

While a division of GM, much of Allison's information systems needs were handled "back at corporate." Thus, while Allison employees did some processing locally on personal computers, the bulk of their business information systems was handled by GM. In the mid-80s GM bought Electronic Data Systems, Inc. (EDS), which eventually took over much of GM's information systems.

To process applications from its various divisions, GM uses many mainframe class computers spread throughout the country. For example, some of the processing for Allison is done on a GM corporate-wide computer host in Plano, Texas, while other processing is done at various Information Processing Centers (IPCs) in Indianapolis, Michigan, and other locations. These various hosts are linked through EDSNET, a worldwide network enabling users to access information around the globe.

As with other firms, systems applications involving personnel processes are central to Allison's business processes. Yet nearly all of their personnel applications were handled off site by GM. To the employees at Allison, these personnel applications being run by GM at a distant IPC were to a great extent a "black box." Allison employees sent personnel information (for example, hours worked for each employee) to the GM IPC and GM would later send back output (for example, paychecks) to Allison. The IPC would charge Allison for each use depending on things such as connect time, cycle time, and number of pages in reports. This off-site processing was becoming very expensive to Allison.

Due to increased competitiveness in their industry, Allison management realized that they could not continue to pay for this expensive processing and remain competitive. Their organizational strategy was to become more and more cost-conscious and efficient, as were all other manufacturers in the aircraft and defense areas. In addition, the increased competitiveness necessitated changes in the way Allison reacted to their environment and ran their business. They needed to be more flexible and adaptable, better able to manage

change. Allison needed to have more control over their own business processes and to quickly and easily access their internal business data for decision-making purposes.

Motivated by the changes in their environment and the pressures internally to cut costs, operate more efficiently, and take control of their own processes and data, Allison management decided to evaluate other information systems alternatives for their personnel applications. What they needed was a system that would fit better with their current environment and changing organizational strategy.

To address the concerns for control over processes and data, Allison management decided they needed a system developed and/or managed in-house. Further, there was a great deal of support internally for a client/server approach versus buying their own mainframe with terminals distributed throughout the site. It was thought that a client/server environment would provide more control over processes and better data access than would a mainframe. Given the concerns for costs, the in-house, client/server approach was additionally attractive from a financial perspective.

Allison and EDS ultimately implemented a UNIX-based Sun SparcStation, running People Soft personnel applications, less expensive even than a smaller main-

frame running similar software. These applications are accessible from any PC on the site if it is on the network and if the user has been granted access to the system. This alternative satisfied the criteria of decreased costs, increased control, and increased flexibility better than did other alternatives. Perhaps more importantly, this alternative satisfied the desire to move to a client/server environment.

Access to data is now much faster than before and, in many cases, access is instant. Users can query the system spontaneously and perform ad hoc reporting. Many users have a 486-based PC on their desk connected to the new personnel application through the network. They can retrieve data from the personnel application and paste into their Microsoft Windows applications (for example, into Microsoft Excel) so that they can manipulate and graph the data. As a result, they are more in touch with their own data and in better control over their own business processes.

Allison employees feel that they now have an information system that better fits their organizational strategy for dealing with industry changes. Their new system satisfied objective criteria involving costs, control, and flexibility, and it satisfied the desire to move into new technologies.

Making the Business Case

The demand for new and replacement systems exceeds the ability and resources of most organizations to conduct systems development projects either themselves or by consultants. (An advantage of this is, of course, that there is plenty of work for systems professionals, which is why most surveys of "hot careers" by *Money Magazine* and other sources consistently point to systems analysis as one of the leading current and future job areas.) Thus, organizations must set priorities and a direction for systems development that will yield development projects with the greatest net benefits. As a systems analyst, you must not only analyze user information requirements but also help make the business case, or justify why the system should be built and the development project conducted.

The reason for any new or improved information system is to add value to the organization. As systems analysts, we must choose to use systems development resources to build the mix of systems that adds the greatest value to the organization. How can we determine business value of systems and identify those applications that provide the most critical gains? Part III addresses this topic, which we call making the business case. Business value comes from supporting the most critical business goals and helping the organization deliver on its business strategy. All systems, whether supporting operational or strategic functions, must be linked to business goals. The chapters in this part of the book show how to make this linkage.

The source of systems projects is either initiatives from information systems planning (proactive identification of systems) or requests from users or IS professionals (reactions to problems or opportunities) for new or enhanced systems. In Chapter 6 we outline the linkage between corporate planning, information systems planning, and the identification and selection of projects. We do not include IS planning as part of the SDLC, but the results of IS planning greatly influence the birth and conduct of systems projects. Chapter 6 makes a strong argument that IS planning provides not only insights into which systems an organization needs but also provides information about the objectives and strategies of an organization necessary for evaluating the viability of any potential systems project.

IS planning produces three key deliverables included in CASE repositories and project dictionaries—the information bases referenced throughout the systems development process. These deliverables are listed below.

1. IS plans for specific business areas including general information requirements of projects for new or improved systems

2. Business area analyses that outline business strategy, objectives, critical success factors, and major business processes and data subjects for different business areas and the organization as a whole

3. A framework or architecture for systems that allows systems development and management to work together to build compatible and comprehensive systems

Chapter 6 briefly outlines the process of IS planning and emphasizes these deliverables. Because of its periodic (usually annual) timing and macro orientation, IS planning is not a frequent activity conducted by systems analysts. The results of IS planning, however, directly affect and are used within systems development work, which is why Chapter 6 concentrates on the linkage between IS planning and projects instead of on IS planning per se. The organizations that are most successful in deploying systems are those that link their IS application portfolio to business and IS plans; thus the importance of this topic in our discussion of systems analysis and design.

We return to the Broadway Entertainment Company (BEC) case after Chapter 6. In this case, we show how an idea for a new information system was stimulated from a synergy of corporate strategic planning and the creativity of an individual business manager. We will also show how this idea is initially evaluated and how it leads to the initiation of a systems development project.

The actual SDLC itself starts when a potential development project is identified. As stated above, IS planning can result in an outline for a systems project to build some needed system or to replace a system that is identified as inconsistent with or seriously harming business directions. Another and more frequent source of project identification is system service requests (SSRs) from business managers and IS professionals, usually for very focused systems or incremental improvements in existing systems. (An example of a system service request appeared in Chapter 2 when Juanita Lopez of Pine Valley Furniture requested a new purchasing system.) User managers request a new or replacement system when they believe that improved information services will help them do their jobs. IS professionals may request system updates when technological changes obsolete current system implementations or to improve the performance of an existing system. In either case, the request for service must be understood by management and a justification for the system and associated project must be developed. The SDLC initial phase, project initiation and planning, is the subject of Chapter 7. (Project initiation and planning are also the first two steps of project management as described in Chapter 4 from Part II of the text.)

During project initiation and planning, a brief study is conducted to investigate the proposed system request. This study develops a better understanding of the scope of the potential system change and the nature of the needed system features. From this preliminary understanding of system requirements, a project plan is developed which shows both the detailed steps and resources needed in order to conduct the analysis phase of the life cycle and more general steps for subsequent phases. The feasibility and potential risks of the requested system are also outlined and an economic cost-benefit analysis is conducted to show the potential impact of the system change. In addition to the economic feasibility or justification of the system, technical, organizational, political, legal, schedule, and other feasibilities are assessed. Potential risks—unwanted outcomes—are identified and plans for dealing with these possibilities are identified. Project initiation and planning ends when a formal proposal for the systems development project is completed and submitted for approval to whomever must commit the resources to systems development. If approved, the project moves into the analysis phase.

The reason for the project initiation and planning phase is found, in part, in the notion of incremental commitment, introduced in Chapter 2. The main purpose of project initiation and planning is to assess the worthiness of the identified system and to decide if there is enough potential benefit for the organization to invest in the project further. Thus, project initiation and planning, although critical in the SDLC, must be conducted with minimal time and resources in order to minimize sunk costs. Consequently, the ability to conduct this phase, often done by only one analyst in conjunction with the project requester, is of the utmost importance to successful systems analysis and design.

We illustrate a typical project initiation and planning phase in a BEC case following Chapter 7. In this case, we show how BEC identified one critically important business goal that provided the motivation for a requested system. The case further shows how an analysis of this business goal and the potential system leads to the justification for a system with competitive advantage for BEC and to the associated development project plan.

Identifying and Selecting Systems Development Projects

L E A R N I N G   O B J E C T I V E S

After studying this chapter, you should be able to:

- Describe the project identification and selection process.

- Describe the corporate strategic planning and information systems planning process.

- Explain the relationship between corporate strategic planning and information systems planning.

- Describe how information systems planning can be used to assist in identifying and selecting systems development projects.

- Analyze information systems planning matrices to determine affinity between information systems and IS projects and to forecast the impact of IS projects on business objectives.

INTRODUCTION

The scope of information systems today is the whole enterprise. Managers, knowledge workers, and all other organizational members expect to easily access and retrieve information, regardless of its location. Non-integrated systems used in the past—often referred to as "islands of information"—are being replaced with cooperative, integrated enterprise systems that can easily support information sharing (Morton, 1992). While the goal of building bridges between these "islands" will take some time to achieve, it represents a clear direction for information systems development.

Obtaining integrated enterprise-wide computing presents significant challenges for both corporate and information systems management. For example, given the proliferation of personal and departmental computing wherein disparate systems and databases have been created, how can the organization possibly control and maintain all of these systems and data? In many cases they simply cannot because it is nearly impossible to track who has which systems and what data, where there are overlaps or inconsistencies, and how accurate the information is. The reason that personal and departmental systems and databases abound is that users are either unaware of the information that exists in corporate databases or they cannot easily get at it, so they create and maintain their own information and systems (Reingruber and Spahr, 1992). Intelligent identification and selection of systems development projects, for both new and replacement systems, are critical steps in gaining control of systems and data.

The acquisition, development, and maintenance of information systems consume substantial resources for most organizations. This suggests that organizations can benefit from following a formal process for identifying and selecting projects. The first phase of the systems development life cycle—project identification and selection—deals with this issue. In the next section, you will learn about a general method for identifying and selecting projects and the deliverables and outcomes from this process. This is followed by brief descriptions of corporate strategic planning and information systems planning, two activities that can greatly improve the project identification and selection process.

IDENTIFYING AND SELECTING SYSTEMS DEVELOPMENT PROJECTS

The first phase of the SDLC is project identification and selection. During this activity, a senior manager, a business group, an IS manager, or a steering committee identify and assess all possible systems development projects that an organization unit could undertake. Next, those projects deemed most likely to yield significant organizational benefits, given available resources, are selected for subsequent development activities. Organizations vary in their approach to identifying and selecting projects. In some organizations, project identification and selection is a very formal process in which projects are outcomes of a larger overall planning process. For example, a large organization may follow a formal project identification process whereby a proposed project is rigorously compared to all competing projects. Alternatively, a small organization may use informal project selection processes that allow the highest-ranking IS manager to independently select projects or allow individual business units to decide on projects after agreeing to provide project funding.

There is a variety of sources for information systems development requests. One source is requests by managers and business units for replacing or extending an existing system to gain needed information or to provide a new service to customers. Another source for requests is IS managers who want to make a system more efficient, less costly to operate, or want to move it to a new operating environment. A final source of projects is a formal planning group which identifies projects for improvement that would help the organization meet its corporate objectives (for example, a new system to provide better customer service). Regardless of how a given organization actually executes the project identification and selection process, there is a common sequence of activities that occurs. In the following sections, we describe a general process for identifying and selecting projects and producing the deliverables and outcomes of this process.

The Process of Identifying and Selecting IS Development Projects

As shown in Figure 6-1, project identification and selection consists of three primary activities:

1. Identifying potential development projects

2. Classifying and ranking projects

3. Selecting projects for development

Figure 6-1
Systems development life cycle with project identification and selection highlighted

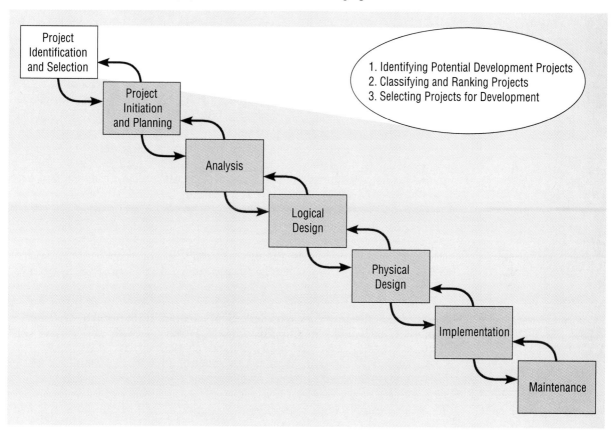

Each of these steps is described below:

1. *Identifying Potential Development Projects.* Organizations vary as to how they identify projects. This process can be performed by

   - A key member of top management, either the CEO of a small- or medium-sized organization or a senior executive in a larger organization

   - A steering committee, composed of a cross section of managers with an interest in systems

   - User departments, in which either the head of the requesting unit or a committee from the requesting department decides which projects to submit (often you, as a systems analyst, will help users prepare such requests)

   - The development group or a senior IS manager

   All methods of identification have been found to have strengths and weaknesses. Research has found, for example, that projects identified by top management more often have a strategic organizational focus. Alternatively, projects identified by steering committees more often reflect the diversity of the committee and therefore have a cross-functional focus. Projects identified by individual departments or business units most often have a narrow, tactical focus. Finally, a dominant characteristic of projects identified by the development group is the ease with which existing hardware and systems will integrate with the proposed project. Other factors, such as project cost, duration, complexity, and risk, are also influenced by the source of a given project. Characteristics of each selection method are briefly summarized in Table 6-1.

Of all the possible project sources, those identified by top management and steering committees most often reflect the broader needs of the organization. This occurs because top management and steering committees are likely to have a broader understanding of overall business objectives and constraints. Projects identified by top management or by a diverse steering committee are therefore referred to as coming from a top-down source.

Projects identified by a functional manager, business unit, or by the information systems development group are often designed for a particular business need within a given business unit. In other words, these projects may not reflect the overall objectives of the organization. This does not mean that projects identified by individual managers, business units, or the IS development group are deficient, only that they may not consider broader organizational issues. Project initiatives stemming from managers, business units, or the development group are generally referred to as coming from a bottom-up source. These are the types of projects in which you, as a systems analyst, will have the earliest role in the life cycle as part of your ongoing support of users. You will help user managers provide the description of information needs and the reasons for doing the project that will be evaluated in selecting, among all submitted projects, which ones will be approved to move into the project initiation and planning phase of the SDLC.

In sum, projects are identified by both top-down and bottom-up initiatives. The formality of the process of identifying and selecting projects can

TABLE 6-1   Characteristics of Alternative Methods for Making Information Systems Identification and Selection Decisions
(Adapted from McKeen, Guimaraes, and Wetherbe, 1994)

Selection Method	Characteristics
Top Management	Greater strategic focus
	Largest project size
	Longest project duration
Steering Committee	Cross-Functional focus
	Greater organizational change
	Formal cost-benefit analysis
	Larger and riskier projects
User Department	Narrow, Non-Strategic focus
	Faster development
	Fewer users, management layers, and business functions
Development Group	Integration with existing systems focus
	Fewer development delays
	Less concern on cost-benefit analysis

vary substantially across organizations. Also, since limited resources preclude the development of all proposed systems, most organizations have some process of classifying and ranking the merit of each project. Those projects deemed to be inconsistent with overall organizational objectives, redundant in functionality to some existing system, or unnecessary will thus be removed from consideration. This topic is discussed next.

2. *Classifying and Ranking IS Development Projects.* The second major activity in the project identification and selection process focuses on assessing the relative merit of potential projects. As with the project identification process, classifying and ranking projects can be performed by top managers, a steering committee, business units, or the IS development group. Additionally, the criteria used when assigning the relative merit of a given project can vary. Commonly used criteria for assessing projects are summarized in Table 6-2. In any given organization, one or several criteria might be used during the classifying and ranking process.

   As with the project identification and selection process, the actual criteria used to assess projects will vary by organization. If, for example, an organization uses a steering committee, it may choose to meet monthly or quarterly to review projects and use a wide variety of evaluation criteria. At these meetings, new project requests will be reviewed relative to projects already identified, and ongoing projects are monitored. The relative ratings of projects are used to guide the final activity of this identification process—project selection.

3. *Selecting IS Development Projects.* The final activity in the project identification and selection process is the actual selection of projects for further development. Project selection is a process of considering both short- and long-term projects and selecting those most likely to achieve business objectives. Additionally, as business conditions change over time, the relative importance of any single project may substantially change. Thus, the identification and selection of projects is a very important and ongoing activity.

   Numerous factors must be considered when making project selection decisions. Figure 6-2 shows that a selection decision requires that the perceived needs of the organization, existing systems and ongoing projects, resource availability, evaluation criteria, current business conditions, and the perspectives of the decision makers will all play a role in project selection decisions.

TABLE 6-2 Possible Evaluation Criteria When Classifying and Ranking Projects

Evaluation Criteria	Description
Strategic Alignment	Extent to which the project is viewed as helping the organization achieve its strategic objectives and long-term goals
Potential Benefits	Extent to which the project is viewed as improving profits, customer service, etc. and the duration of these benefits
Resource Availability	Amount and type of resources the project requires and their availability
Project Size/Duration	Number of individuals and the length of time needed to complete the project
Technical Difficulty/Risks	Level of technical difficulty to successfully complete the project within given time and resource constraints

Figure 6-2

Project selection decisions must consider numerous factors and can have numerous outcomes

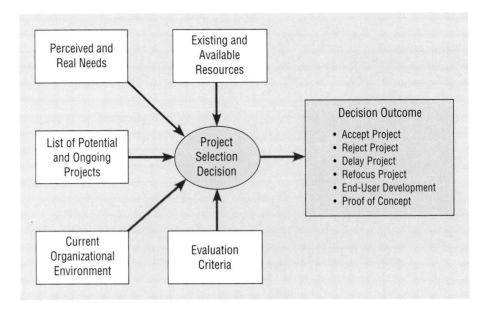

Numerous outcomes can occur from this decision process. Of course, projects can be accepted or rejected. Acceptance of a project usually means that funding to conduct the next phase of the SDLC has been approved. Rejection means that the project will no longer be considered for development. However, projects may also be conditionally accepted; projects may be accepted pending the approval or availability of needed resources or the demonstration that a particularly difficult aspect of the system *can* be developed. Projects may also be returned to the original requesters who are told to develop or purchase the requested system themselves. Finally, the requesters of a project may be asked to modify and resubmit their request after making suggested changes or clarifications.

Deliverables and Outcomes

The primary deliverable from the first SDLC phase is a schedule of specific IS development projects, coming from both top-down and bottom-up sources, to move into the next SDLC phase—project initiation and planning (see Figure 6-3). An outcome of this phase is the assurance that careful consideration was given to project selection, with a clear understanding of how each project can help the organization reach its objectives. Due to the principle of *incremental commitment*, a selected project does not necessarily result in a working system. After each subsequent SDLC phase, you, other members of the project team, and organizational officials will reassess your project to determine whether the business conditions have changed or whether a more detailed understanding of a system's costs, benefits, and risks would suggest that the project is not as worthy as previously thought.

Many organizations have found that in order to make good project selection decisions and to provide sound guidance as issues arise in your work as a systems analyst on a project, a clear understanding of overall organizational business strategy and objectives is required. This means that a clear understanding of the business and the desired role of information systems in achieving organizational goals is a precondition to improving the identification and selection process. In the next section we provide a brief overview of the process many organizations follow, involving corporate strategic planning and information systems planning, when

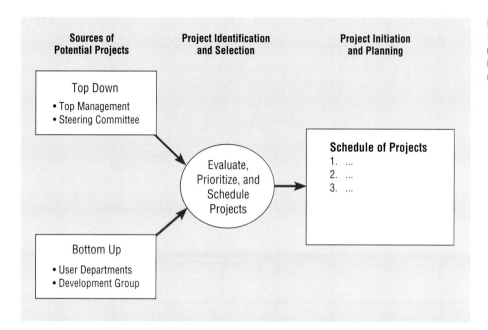

Figure 6-3
Information systems development projects come from both top-down and bottom-up initiatives

setting their business strategy and objectives and when defining the role of information systems in their plans.

CORPORATE AND INFORMATION SYSTEMS PLANNING

Although there are numerous motivations for carefully planning the identification and selection of projects (see Atkinson, 1990; Atkinson and Montgomery, 1990), organizations have not traditionally used a systematic planning process when determining how to allocate IS resources. Instead, projects have often resulted from attempts to solve isolated organizational problems. In effect, organizations have asked the question: "What procedure (application program) is required to solve *this particular problem* as it exists today?" The difficulty with this approach is that the required organizational procedures are likely to change over time as the environment changes. For example, a company may decide to change its method of billing customers or a university may change its procedures for registering students. When such changes occur, it is usually necessary to again modify existing information systems.

In contrast, planning-based approaches essentially ask the question: "What information (or data) requirements will satisfy the decision-making needs or business processes of the enterprise today and well into the future?" A major advantage of this approach is that an organization's informational needs are less likely to change (or will change more slowly) than its business processes. For example, unless an organization fundamentally changes its business, its underlying data structures may remain reasonably stable for more than ten years. However, the procedures used to access and process the data may change many times during that period. Thus, the challenge of most organizations is to design comprehensive information models containing data that are relatively independent from the languages and programs used to access, create, and update them.

To benefit from a planning-based approach for identifying and selecting projects, an organization must analyze its information needs and plan its projects carefully. Without careful planning, organizations may construct databases and systems that support individual processes but do not provide a resource that can be easily shared throughout the organization. Further, as business processes change, lack of data and system integration will hamper the speed at which the organization can effectively make business strategy or process changes.

The need for improved information systems project identification and selection is readily apparent when we consider factors such as the following:

1. The cost of information systems has risen steadily and approaches 40 percent of total expenses in some organizations.

2. Many systems cannot handle applications that cross organizational boundaries.

3. Many systems often do not address the critical problems of the business as a whole nor support strategic applications.

4. Data redundancy is often out of control and users may have little confidence in the quality of data.

5. Systems maintenance costs are out of control as old, poorly planned systems must constantly be revised.

6. Application backlogs often extend three years or more and frustrated end users are forced to create (or purchase) their own systems, often creating redundant databases and incompatible systems in the process.

Careful planning and selection of projects alone will certainly not solve all of these problems. We believe, however, that a disciplined approach, driven by top management commitment, is a prerequisite to most effectively apply information systems in order to reach organizational objectives. The focus of this section is to provide you with a clear understanding of how specific development projects with a broader organizational focus can be identified and selected. Specifically, we describe corporate strategic planning and information systems planning, two processes that can significantly improve the quality of project identification and selection decisions. This section also outlines the types of information about business direction and general systems requirements that can influence the selection decisions and guide the direction of approved projects.

Corporate Strategic Planning

A prerequisite to making effective project selection decisions is to gain a clear idea of where an organization is, its vision of where it wants to be in the future, and how to make the transition to its desired future state. Figure 6-4 represents this as a three-step process. The first step focuses on gaining an understanding of the current enterprise. In other words, if you don't know where you are, it is impossible to tell where you are going. Next, top management must determine where it wants the enterprise to be in the future. Finally, after gaining an understanding of the current and future enterprise, a strategic plan can be developed to guide this transition. The process of developing and refining models of the current and future enterprise as well as a transition strategy is often referred to as **corporate strategic planning.** During corporate strategic planning, executives typically develop a mission statement, statements of future corporate objectives, and strategies designed to help the organization reach its objectives.

All successful organizations have a mission. The **mission statement** of a company typically states in very simple terms what business the company is in. For example, the mission statement for Pine Valley Furniture (PVF) is shown in Figure 6-5. After reviewing the mission statement from PVF, it becomes clear that they are in the business of constructing and selling high-quality wood furniture to the general public, businesses, and institutions such as universities and hospitals. It is

Corporate strategic planning: An ongoing process that defines the mission, objectives, and strategies of an organization.

Mission statement: A statement that makes it clear what business a company is in.

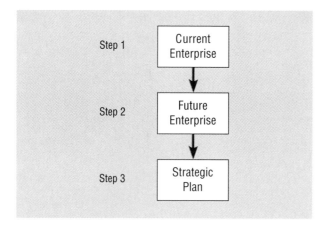

Figure 6-4
Corporate strategic planning is a three-step process

Figure 6-5
Mission statement (Pine Valley Furniture)

Pine Valley Furniture
Corporate Mission Statement

We are in the business of designing, fabricating, and selling to retail stores high-quality wood furniture for household, office, and institutional use. We value quality in our products and in our relationships with customers and suppliers. We consider our employees our most critical resource.

also clear that PVF is not in the business of fabricating steel file cabinets or selling their products through wholesale distributors. Based on this mission statement, you could conclude that PVF does not need a retail sales information system; instead, a high-quality human resource information system would be consistent with their goal.

After defining its mission, an organization can then define its objectives. The **objective statements** refer to "broad and timeless" goals for the organization. These goals can be expressed as a series of statements that are either qualitative or quantitative but that typically do not contain details likely to change substantially over time. Objectives are often referred to as "critical success factors." Here, we will simply use the term "objectives." The objectives for PVF are shown in Figure 6-6, with most relating to some aspect of the organizational mission. For example, objective number two relates to how PVF views its relationships with customers. This goal would suggest that PVF might want to invest in electronic data interchange or on-line order status systems that would contribute to high-quality customer service. Once a company has defined its mission and objectives, a competitive strategy can be formulated.

A **competitive strategy** is the method by which an organization attempts to achieve its mission and objectives. In essence, the strategy is an organization's game plan for playing in the competitive business world. In his classic book on

Objective statements: A series of statements that express an organization's qualitative and quantitative goals for reaching a desired future position.

Competitive strategy: The method by which an organization attempts to achieve its mission and objectives.

Figure 6-6
Statement of corporate objectives (Pine Valley Furniture)

> **Pine Valley Furniture**
> *Statement of Objectives*
>
> 1. PVF will strive to increase market share and profitability (prime objective).
>
> 2. PVF will be considered a market leader in customer service.
>
> 3. PVF will be innovative in the use of technology to help bring new products to market faster than our competition.
>
> 4. PVF will employ the fewest number of the highest-quality people necessary to accomplish our prime objective.
>
> 5. PVF will create an environment that values diversity in gender, race, values, and culture among employees, suppliers, and customers.

competitive strategy, Michael Porter (1980) defined three generic strategies—low-cost producer, product differentiation, and product focus or niche—for reaching corporate objectives (see Table 6-3). These generic strategies allow you to more easily compare two companies in the same industry that may not employ the same competitive strategy. In addition, organizations employing different competitive strategies often have different informational needs to aid decision making. For example, Rolls Royce and GEO are two car lines with different strategies: one is a high-prestige line in the ultra-luxury *niche* while the other is a relatively *low-priced* line for the general automobile market. Rolls Royce may build information systems to collect and analyze information on customer satisfaction to help manage a key company objective. Alternatively, GEO may build systems to track plant and material utilization in order to manage activities related to their low-cost strategy.

To effectively deploy resources such as the creation of a marketing and sales organization, *or to build the most effective information systems,* an organization must clearly understand its mission, objectives, and strategy. A lack of understanding will make it impossible to know which activities are essential to achieving business objectives. From an information systems development perspective, by understanding which activities are most critical for achieving business objectives, an organization has a much greater chance to identify those activities that need to be supported by information systems. In other words, **it is only through the clear understanding of the organizational mission, objectives, and strategies that IS development projects should be identified and selected.** The process of planning how information systems can be employed to assist organizations to reach their objectives is the focus of the next section.

Information systems planning (ISP): An orderly means of assessing the information needs of an organization and defining the systems, databases, and technologies that will best satisfy those needs.

Information Systems Planning

The second planning process that can play a significant role in the quality of project identification and selection decisions is called **information systems planning (ISP).**

TABLE 6-3 Generic Competitive Strategies
(Adapted from Porter, 1980)

Strategy	Description
Low-Cost Producer	This strategy reflects competing in an industry on the basis of product or service cost to the consumer. For example, in the automobile industry, the South Korean-produced Hyundai is a product line that competes on the basis of low cost.
Product Differentiation	This competitive strategy reflects capitalizing on a key product criterion requested by the market (for example, high quality, style, performance, roominess). In the automobile industry, many manufacturers are trying to differentiate their products on the basis of quality (for example, "At Ford, quality is job one.").
Product Focus or Niche	This strategy is similar to both the low-cost and differentiation strategies but with a much narrower market focus. For example, a niche market in the automobile industry is the convertible sports car market. Within this market, some manufacturers may employ a low-cost strategy while others may employ a differentiation strategy on performance or style.

Information systems planning is an orderly means of assessing the information needs of an organization and defining the information systems, databases, and technologies that will best satisfy those needs (Carlson, Gardner, and Ruth, 1989; Parker and Benson, 1989). This means that during ISP you (or, more likely, senior IS managers responsible for the IS plan) must model current and future organization informational needs, and develop strategies and *project plans* to migrate the current information systems and technologies to their desired future state. ISP is a top-down process that takes into account the outside forces—industry, economic, relative size, geographic region, and so on—critical to the success of the firm. This means that ISP must look at information systems and technologies in terms of how they help the business achieve its objectives defined during corporate strategic planning.

The three key activities of this modeling process are represented in Figure 6-7. Like corporate strategic planning, ISP is a three-step process in which the first step is to assess current IS-related assets—human resources, data, processes, and technologies. Next, target blueprints of these resources are developed. These blueprints reflect the desired future state of resources needed by the organization to reach its objectives as defined during strategic planning. Figures 6-8, 6-9, and Table 6-4 describe one popular framework for representing these blueprints—an information systems architecture. Finally, a series of scheduled projects is defined to help move the organization from its current to its future desired state. (Of course, scheduled projects from the ISP process are just one source for projects. Others include bottom-up requests from managers and business units like the System Service Request in Figure 2-2.)

In terms of an information systems architecture as described in Figure 6-9, the purpose of a project is to improve the functionality or operation of information systems in one or more cells of this matrix. For example, a project may focus on reconfiguration of a telecommunications network to speed data communications (row 5 of the Network column), or may restructure work and data flows between business areas (possibly row 3 and below of the process column). Projects can include not only the development of new information systems or the modification of existing ones but also the acquisition and management of new systems, technologies, and platforms. These three activities parallel those of corporate strategic

Figure 6-7

Information systems planning
is a three-step process

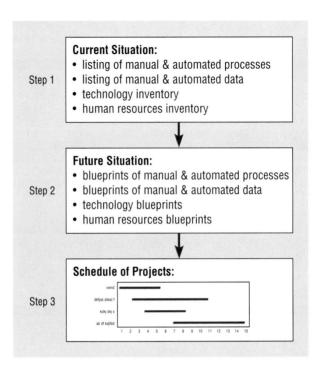

planning and this relationship is shown in Figure 6-10. Numerous methodologies such as Business Systems Planning (BSP) and Information Engineering (IE) have been developed to support the ISP process[1]; most contain the three key activities described below:

1. *Describing the Current Situation.* The most widely used approach for describing the current organizational situation is generically referred to as top-down planning. **Top-down planning** attempts to gain a broad understanding of the informational needs of the entire organization. The approach begins by conducting an extensive analysis of the organization's mission, objectives, and strategy and determining the information requirements needed to meet each objective. This approach to ISP implies by its name a high-level organizational perspective with active involvement of top-level management. The top-down approach to ISP has several advantages over other planning approaches, and these are summarized in Table 6-5 on page 210.

   In contrast to the top-down planning approach, a **bottom-up planning** approach requires the identification of business problems and opportunities which are used to define projects. Using the bottom-up approach for creating IS plans can be faster and less costly to develop than using the top-down approach and also has the advantage of identifying pressing organizational problems. Yet, the bottom-up approach often fails to view the informational needs of the *entire* organization. This can result in the creation of disparate information systems and databases that are redundant or not easily integrated without substantial rework.

   The process of describing the current situation begins by selecting a planning team that includes executives chartered to model the existing situation.

Top-down planning: A generic information systems planning methodology that attempts to gain a broad understanding of the information system needs of the entire organization.

Bottom-up planning: A generic information systems planning methodology that identifies and defines IS development projects based upon solving operational business problems or taking advantage of some business opportunities.

---

[1]For an overview of several information systems planning methodologies, see Koory and Medley (1987) or Martin (1990).

If you were planning to build a new home, you would certainly want to have a blueprint or set of architectural plans before starting construction. In fact, almost any large-scale creative endeavor requires the early development of a vision or architectural plan if that endeavor is to be successfully completed. Designing the blueprints for the future vision of information systems is one of the greatest challenges facing most organizations today. You may be surprised to learn that (unfortunately) many organizations do not have a comprehensive architectural plan for their information systems, although this is changing.

In 1987, John Zachman (then of IBM) developed a comprehensive framework for portraying an information systems architecture (ISA) which is shown in Figure 6-9 (Zachman, 1987). In this framework every information system can be visualized as a combination of three major components: Data, Process, and Network. The Data column represents the "what" in an information system. For example, in a manufacturing system, the "what" might correspond to a bill of materials listing the components required to assemble a finished product. The Process column represents "how." For example, the list of instructions that specify "how" a finished product is assembled from its components. The Network column represents "where," such as the locations of data, processing, and other resources.

The ISA has six rows that represent different views of the data, processes, and network components. Zachman used a construction industry analogy to describe the roles individuals within an organization play when sponsoring, designing, building, and using an information system. Each role corresponds to a row within the framework (see Table 6-4). During its evolution, an information system should move through the cells of the framework from the top of the framework to the bottom. As this occurs, top-level information is transformed into more concrete representations of the system. The lowest level represents an actual information system.

In more recent work, Sowa and Zachman (1992) extend the concepts of the information systems architecture to also include "who, when, and why" components. These three components extend the "what, how, and where" components so that all English language question words are included in the model. Who refers to the concept of determining *who* works with the system. When reflects information related to understanding *when* events occur. Information related to why helps to uncover *why* events occur. The objective of the extended ISA model is to create a systematic taxonomy for relating things in the world to representations in the computer. In other words, the extended ISA framework provides a structured way for viewing information systems from many different perspectives and showing how each is related. Using formal methods such as the ISA framework to represent information systems and plans is becoming more common. If you have not already, it is likely that you will have the opportunity to learn more about this or some similar technique as your career evolves.

To gain this understanding, the team will need to review corporate documents; interview managers, executives, and customers; and conduct detailed reviews of competitors, markets, products, and finances. The type of information that must be collected to represent the current situation includes the identification of all organizational locations, units, functions, processes, data (or data entities), and information systems.

Within Pine Valley Furniture, for example, organizational locations would consist of a list of all geographic areas in which the organization operates (for example, the locations of the home and branch offices). Organizational units represent a list of people or business units that operate within the organization. Thus, organizational units would include v-p manufacturing, sales manager, sales person, and clerk. Functions are cross-organizational collections of activities used to perform day-to-day business operations. Examples of business functions might include research and development, employee development, purchasing, and sales. Processes represent a list of manual or automated procedures designed to support business functions. Examples of

Figure 6-9

Information systems architecture framework (Adapted from Zachman, 1987)

	Data	Process	Network
1 Business Scope	List of Entities Important to the Business	List of Functions the Business Performs	List of Locations in which the Business Operates
2 Business Model	Business Entities and their Inter-relationships	Function and Process Decomposition	Communications Links between Business Locations
3 Information Systems Model	Model of the Business Data and its Inter-relationships	Flows between Application Processes	Distribution Network
4 Technology Model	Database Design	Process Specifications	Configuration Design
5 Technology Definition	Database Schema and Subschema Definition	Program Code and Control Blocks	Configuration Definition
6 Information System	Data and Information	Application Programs	System Configuration

business processes might include payroll processing, customer billing, and product shipping. Data entities represent a list of the information items generated, updated, deleted, or used within business processes. Information systems represent automated and non-automated systems used to transform data into information to support business processes. For example, Figure 6-11 shows portions of the business functions, data entities, and information sys-

TABLE 6-4 Models and Roles in the Information Systems Framework
(Adapted from Moriarty, 1991)

Model Name	Role or Perspective	Role Description
Business Scope	Owner	Provides a strategic overview including business scope, mission, and direction
Business Model	Architect	Develops business models that describe the business scope, mission, and direction
Information Systems Model	Designer	Develops information systems models that support the business
Technology Model	Builder	Converts information systems models into a design that conforms to the features and constraints of the technology
Technology Definition	Contractor	Converts technology models into statements to generate the actual information system
Information System	User	Manages, uses, and operates the completed information system

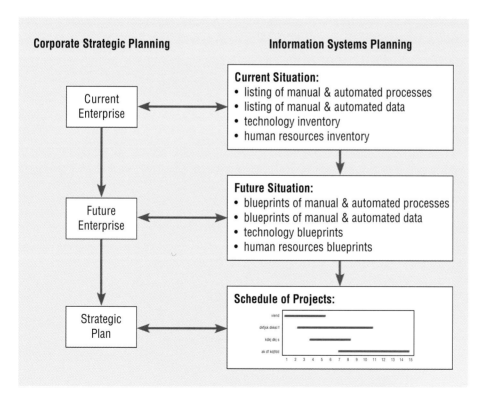

Figure 6-10
Parallel activities of corporate strategic planning and information systems planning

tems of PVF. Once high-level information is collected, each item can typically be decomposed into smaller units as more detailed planning is performed. Figure 6-12 shows the decomposition of several of PVF's high-level business functions into more detailed supporting functions.

TABLE 6-5 Advantages to the Top-Down Planning Approach over Other Planning Approaches (IBM, 1982; pp. 236–37)

Advantage	Description
Broader Perspective	If not viewed from the top, information systems may be implemented without first understanding the business from general management's viewpoint.
Improved Integration	If not viewed from the top, totally new management information systems may be implemented rather than planning how to evolve existing systems.
Improved Management Support	If not viewed from the top, planners may lack sufficient management acceptance of the role of information systems in helping them achieve business objectives.
Better Understanding	If not viewed from the top, planners may lack the understanding necessary to implement information systems across the entire business rather than simply to individual operating units.

Figure 6-11
Information systems planning information (Pine Valley Furniture)

FUNCTIONS:
- business planning
- product development
- marketing and sales
- production operations
- finance and accounting
- human resources

...

DATA ENTITIES:
- customer
- product
- vendor
- raw material
- order
- invoice
- equipment

...

INFORMATION SYSTEMS:
- payroll processing
- accounts payable
- accounts receivable
- time card processing
- inventory management

...

After creating these lists, a series of matrices can be developed to cross reference various elements of the organization. The types of matrices typically developed include the following:

- *Location-to-Function:* This matrix identifies which business functions are being performed at various organizational locations.

- *Location-to-Unit:* This matrix identifies which organizational units are located in or interact with a specific business location.

- *Unit-to-Function:* This matrix identifies the relationships between organizational entities and each business function.

- *Function-to-Objective:* This matrix identifies which functions are essential or desirable in achieving each organizational objective.

- *Function-to-Process:* This matrix identifies which processes are used to support each business function.

- *Function-to-Data Entity:* This matrix identifies which business functions utilize which data entities.

- *Process-to-Data Entity:* This matrix identifies which data are captured, used, updated, or deleted within each process.

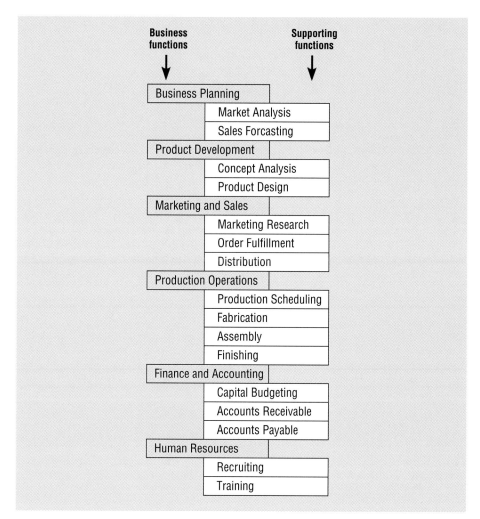

Figure 6-12
Functional decomposition of information systems planning information (Pine Valley Furniture)

- *Process-to-Information System:* This matrix identifies which information systems are used to support each process.

- *Data Entity-to-Information System:* This matrix identifies which data are created, updated, accessed, or deleted in each system.

- *Information System-to-Objective:* This matrix identifies which information systems support each business objective as identified during organizational planning.

For example, Figure 6-13 shows a portion of the Data Entity-to-Function matrix for Pine Valley Furniture. The "X" in various cells of the matrix represents which business functions utilize which data entities. A more detailed picture of data utilization would be shown in the Process-to-Data Entity matrix (not shown here), where the cells would be coded as "C" for the associated process that creates or captures data for the associated data entity, "R" for retrieve (or used), "U" for update, and "D" for delete.

Different matrices will have different relationships depending upon what is being represented. For example, Figure 6-14 shows a portion of the Information

Figure 6-13

Data Entity-to-Function matrix
(Pine Valley Furniture)

Business Functions \ Data Entity Types	Customer	Product	Vendor	Raw Material	Order	Work Center	Equipment	Employees	Invoice	Work Order	...
Marketing and Sales											
Marketing Research	X	X									
Order Fulfillment	X	X			X				X		
Distribution	X	X									
Production Operations											
Production Scheduling						X	X	X		X	
Fabrication						X	X	X		X	
Assembly						X	X	X		X	
Finishing						X	X	X		X	
Finance and Accounting											
Capital Budgeting						X	X				
Accounts Receivable	X	X			X						
Accounts Payable			X	X	X				X		
Human Resources											
Recruiting								X			
Training								X			
...											

X = data entity is used within business function

System-to-Objective matrix. In this matrix, those objectives that are *currently* supported by existing information systems are labeled "C," and objectives intended to be supported by *future* development are represented by "F." One way this matrix would be useful in project selection of a bottom-up project would be as follows. Suppose a request came in to make changes to the Payroll system. You can see from the C in the Payroll row and Profit column that this system currently supports the profit objective well; the requested change should be based on making a contribution to other objectives (for example, personnel) in order to be considered worthwhile. Various impact analyses like this are possible via one matrix or combination of matrices. Through the creation and analysis of these matrices, planners can gain a clear understanding of the organization's current situation (Kerr, 1990). A primer on using matrices for information systems planning is provided in Figure 6-15.

2. *Describing the Target Situation, Trends, and Constraints.* After describing the current situation, the next step in the ISP process is to define the target situation which reflects the desired future state of the organization. This means that the target situation consists of the desired state of the locations, units, functions, processes, data, and information systems (see Figure 6-7). For example, if a desired future state of the organization is to have several new branch offices or a new product line which require several new employee positions, functions, processes, and data, then most lists and matrices will need to be updated to reflect this vision. The target situation must be developed in light of technology and business trends, in addition to organizational constraints. This means

Information System \ Objective	Profit	Service	Innovation	Personnel	Diversity
Transaction Processing					
Order Tracking	F	F		F	
Order Processing	C	C		C	
Plant Scheduling	F	F		F	
Payroll	C				
Accounts Payable	C			C	
Accounts Receivable	C	C		C	
Cash Management	F		F		
. . .					
Management Information Systems					
Sales Management	C	F	F	C	F
Sales Region Analysis	F	F	F	F	F
Inventory Control	C	C	C		
Production Scheduling	F	F	F	F	
. . .					

C = objective currently supported by existing systems
F = objective is planned to be supported by future system

Figure 6-14
Information System-to-Objective matrix (Pine Valley Furniture)

that lists of business trends and constraints should also be constructed in order to help assure that the target situation reflects these issues.

In summary, to create the target situation, planners must first edit their initial lists and record the *desired* locations, units, functions, processes, data, and information systems within the constraints and trends of the organization environment (for example, time, resources, technological evolution, competition, and so on). Next, matrices are updated to relate information in a manner consistent with the desired future state. Planners then focus on the *differences* between the current and future lists and matrices to identify projects and transition strategies.

3. *Developing a Transition Strategy and Plans.* Once the creation of the current and target situations is complete, a detailed transition strategy and plan are developed by the IS planning team. This plan should be very comprehensive, reflecting both broad, long-range issues in addition to providing sufficient detail to guide all levels of management concerning what needs doing, how, when, and by whom in the organization. The components of a typical information systems plan are outlined in Figure 6-16.

The IS plan is typically a very comprehensive document that looks at both short- and long-term organizational development needs. The short- and long-term developmental needs identified in the plan are typically expressed as a series of projects (see Figure 6-17). Projects from the long-term plan tend to build a foundation for later projects (such as transforming databases from old technology into newer technology). Projects from the short-term plan consist of specific steps to fill the gap between current and desired systems or respond

Figure 6-15
Making sense out of planning matrices

During the information systems planning process, before individual projects are identified and selected, a great deal of "behind the scenes" analysis takes place. During this planning period, which can span from six months to a year, IS planning team members develop and analyze numerous matrices like those described in the associated text. Matrices are developed to represent the current and the future views of the organization. Matrices of the "current" situation are called "as is" matrices. In other words, they describe the world "as" it currently "is." Matrices of the target or "future" situation are called "to be" matrices. Contrasting the current and future views provides insights into the relationships existing in important business information, and most importantly, forms the basis for the identification and selection of specific development projects. Many CASE tools provide features that will help you make sense out of these matrices in at least three ways:

1. *Management of Information.* A big part of working with complex matrices is managing the information. Using the dictionary features of the CASE tool repository, terms (such as business functions and process and data entities) can be defined or modified in a single location. All planners will therefore have the most recent information.
2. *Matrix Construction.* The reporting system within the CASE repository allows matrix reports to be easily produced. Since planning information can be changed at any time by many team members, an easy method to record changes and produce the most up-to-date reports is invaluable to the planning process.
3. *Matrix Analysis.* Possibly the most important feature CASE tools provide to planners is the ability to perform complex analyses within and across matrices. This analysis is often referred to as **affinity clustering**. Affinity refers to the extent to which information holds things in common. Thus, affinity clustering is the process of arranging matrix information so that clusters of information with some predetermined level or type of affinity are placed next to each other on a matrix report. For example, an affinity clustering of a Process-to-Data Entry matrix would create roughly a block diagonal matrix with processes that use similar data entities appearing in adjacent rows and data entities used in common by the same processes grouped into adjacent columns. This general form of analysis can be used by planners to identify items that often appear together (or should!). Such information can be used by planners to most effectively group and relate information (e.g., data to processes, functions to locations, and so on). For example, those data entities used by a common set of processes are candidates for a specific database. And those business processes that relate to a strategically important objective will likely receive more attention when managers from those areas request system changes.

Affinity clustering: The process of arranging planning matrix information so the clusters of information with some predetermined level or type of affinity are placed next to each other on a matrix report.

to dynamic business conditions. The top-down (or plan-driven) projects join a set of bottom-up or needs-driven projects submitted as system service requests from managers to form the short-term systems development plan. Collectively, the short- and long-term projects set clear directions for the project selection process. The short-term plan includes not only those projects identified from the planning process but also those selected from among bottom-up requests. The overall IS plan may also influence all development projects. For example, the IS mission and IS constraints may cause projects to choose certain technologies or emphasize certain application features as systems are designed.

In this section, we outlined a general process for developing an IS plan. ISP is a detailed process and an integral part of deciding how to best deploy information systems and technologies to help reach organizational goals. It is beyond the scope of this chapter, however, to extensively discuss ISP, yet it should be clear from our discussion that planning-based project identification and selection will yield substantial benefits to an organization. It is probably also clear to you that, as a systems analyst, you are not usually involved in IS planning, since this process requires senior IS and corporate management participation. On the other hand, the results of IS planning, such as planning matrices like those in Figures 6-13 and 6-14 can be a source of very valuable information as you identify and justify projects.

Figure 6-16

Outline of an information systems plan

I. **Organizational Mission, Objectives, and Strategy**
Briefly describes the mission, objectives, and strategy of the organization. The current and future views of the company are also briefly presented (i.e., where are we, where do we want to be).

II. **Informational Inventory**
This section provides a summary of the various business processes, functions, data entities, and information needs of the enterprise. This inventory will view both current and future needs.

III. **Mission and Objectives of Information Systems**
Description of the primary role IS will play in the organization to transform the enterprise from its current to future state. While it may later be revised, it represents the current best estimate of the overall role for IS within the organization. This role may be as a necessary cost, an investment, or a strategic advantage, for example.

IV. **Constraints on IS Development**
Briefly describes limitations imposed by technology and current level of resources within the company—financial, technological, and personnel.

V. **Overall Systems Needs and Long-Range IS Strategies**
Presents a summary of the overall systems needed within the company and the set of long-range (2–5 years) strategies chosen by the IS department to fill the needs.

VI. **The Short-Term Plan**
Shows a detailed inventory of present projects and systems and a detailed plan of projects to be developed or advanced during the current year. These projects may be the result of the long-range IS strategies or of requests from managers which have already been approved and are in some stage of the life cycle.

VII. **Conclusions**
Contains likely but not-yet-certain events that may affect the plan, an inventroy of business change elements as presently known, and a description of their estimated impact on the plan.

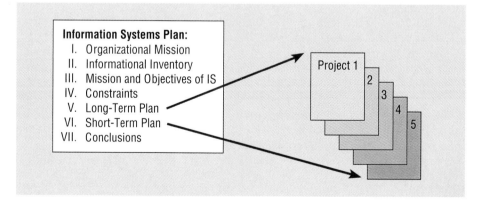

Figure 6-17

Systems development projects flow from the information systems plan

The planning process that we have described requires collecting, organizing, and analyzing vast amounts of data. This process can be eased by using upper CASE tools to more effectively deal with the vast amounts of information and to assist in providing (and enforcing) a common methodology throughout the process. In short, as information is collected during information systems planning, for example, it can be stored in the CASE repository. This information can then be summarized as needed to help identify and select specific development projects. For example, a particular matrix, like Information System-to-Objective, could be generated from the repository to help identify projects that address upgrading or replacing a specific system in order to better achieve prime objectives. In this simple way CASE can be used to improve the project identification and selection process. For more information on the role of CASE during the project identification and selection process, see Chapter 5.

SUMMARY

In this chapter, we described the first phase of the SDLC—project identification and selection. Project identification and selection consists of three primary activities: identifying potential development projects, classifying and ranking projects, and selecting projects for development. A variety of organizational members or units can be assigned to perform this process including top management, a diverse steering committee, business units and functional managers, the development group, or the most senior IS executive. Potential projects can be evaluated and selected using a broad range of criteria such as alignment with business strategy, potential benefits, resource availability and requirements, and risks.

The quality of the project identification and selection process can be improved if decisions are guided by corporate strategic planning and information systems planning. Corporate strategic planning is the process of identifying the mission, objectives, and strategies of an organization. Crucial in this process is selecting a competitive strategy which states how the organization plans to achieve its objectives.

Information systems planning is an orderly means for assessing the information needs of an organization and defining the systems and databases that will best satisfy those needs. ISP is a top-down process that takes into account outside forces that drive the business and the factors critical to the success of the firm. ISP evaluates the current inventory of systems, the desired future state of the organization and its system, and determines which projects are needed to transform systems to meet the future needs of the organization.

Corporate and IS planning are highly inter-related. Conceptually, these relationships can be viewed via various matrices that show how organizational objectives, locations, units, functions, processes, data entities, and systems relate to one another. Selected projects will be those viewed to be most important in supporting the organizational strategy.

The focus of this chapter was to provide you with a clearer understanding of how organizations identify and select projects. The need for improved project identification and selection is apparent for reasons such as the following: the cost of information systems is rising rapidly, systems cannot handle applications that cross organizational boundaries, systems often do not address critical organizational objectives, data redundancy is often out of control, and system maintenance costs continue to rise. Thus, effective project identification and selection is essential if organizations are to realize the greatest benefits from information systems.

C H A P T E R   R E V I E W

K E Y   T E R M S

Affinity clustering

Bottom-up planning

Competitive strategy

Corporate strategic
planning

Information systems
planning (ISP)

Mission statement

Objective statements

Top-down planning

R E V I E W   Q U E S T I O N S

1. Define each of the following terms:

   a. bottom-up planning
   b. competitive strategy
   c. corporate strategic planning
   d. information systems planning

   e. mission statement
   f. objective statements
   g. top-down planning
   h. affinity clustering

2. Contrast the following terms:

   a. mission; objective statements; competitive strategy
   b. corporate strategic planning; information systems planning
   c. top-down planning; bottom-up planning
   d. low-cost producer; product differentiation; product focus or niche
   e. data entity; information system

3. Describe the project identification and selection process.

4. Describe several project evaluation criteria.

5. Discuss several factors that provide evidence for the need for improved information systems planning today.

6. Describe the steps involved in corporate strategic planning.

7. What are three generic competitive strategies?

8. Describe what is meant by information systems planning and the steps involved in the process.

9. List and describe the advantages of top-down planning over other planning approaches.

10. Briefly describe nine planning matrices that are used in information systems planning and project identification and selection.

PROBLEMS AND EXERCISES

1. Match the following terms to the appropriate definitions.

   _____ corporate strategic planning

   _____ mission statement

   _____ objective statement

   _____ competitive strategy

   _____ information systems planning

   _____ top-down planning

   _____ bottom-up planning

   _____ affinity clustering

   a. statement that expresses an organization's qualitative and quantitative goals for reaching a desired future position

   b. method by which an organization attempts to achieve its mission and objectives

   c. statement that makes it clear what business a company is in

   d. ongoing process that defines the mission, objectives, and strategies of an organization

   e. orderly means of assessing the information needs of an organization and defines the systems, databases, and technologies that will best satisfy those needs

   f. process of arranging planning matrix information so that clusters of information with some pre-determined level or type of affinity are placed next to each other on a matrix report

   g. generic information systems planning methodology that attempts to gain a broad understanding of the information system needs of the entire organization

   h. generic information systems planning methodology that identifies and defines the IS development projects based upon solving operational business problems or taking advantage of some business opportunities

2. Write a mission statement for a business that you would like to start. The mission statement should state the area of business you will be in and what aspect of the business you value highly.

3. When you are happy with the mission statement you have developed in response to the prior question, describe the objectives and competitive strategy for achieving that mission.

4. Consider an organization that you believe does not conduct adequate strategic IS planning. List at least six reasons why this type of planning is not done appropriately (or is not done at all). Are these reasons justifiable? What are the implications of this inadequate strategic IS planning? What limits, problems, weaknesses, and barriers might this present?

5. Figure 6-15 introduces the concept of affinity clustering. Suppose that through affinity clustering it was found that three business functions provided the bulk of the use of five data entities. What implications might this have for project identification and subsequent steps in the systems development life cycle?

6. The economic analysis carried out during the project identification and selection phase of the systems development life cycle is rather cursory. Why is this? Conse-

quently, what factors do you think tend to be most important for a potential project to survive this first phase of the life cycle?

7. In those organizations that do an excellent job of IS planning, why might projects identified from a bottom-up process still find their way into the project initiation and planning phase of the life cycle?

8. IS planning, as depicted in this chapter, is highly related to corporate strategic planning. What might those responsible for IS planning have to do if they operate in an organization without a formal corporate planning process?

9. Timberline Technology manufactures membrane circuits in its northern California plant. In addition, all circuit design and R&D work occur at this site. All finance, accounting, and human resource functions are headquartered at the parent company in the upper-Midwest. Sales take place through six sales representatives located in various cities across the country. Information systems for payroll processing, accounts payable, and accounts receivable are located at the parent office while systems for inventory management and computer-integrated manufacturing are at the California plant. As best you can, list the locations, units, functions, processes, data entities, and information systems for this company.

10. For each of the following categories, create the most plausible planning matrices for Timberline Technology, described in Problem and Exercise 9: function-to-data entity, process-to-data entity, process-to-information system, data entity-to-information system. What other information systems not listed is Timberline likely to need?

11. The owners of Timberline Technology (described in Problem and Exercise 9) are considering adding a plant in Idaho and one in Arizona and six more sales representatives at various sites across the country. Update the matrices from Problem and Exercise 10 so that the matrices account for these changes.

FIELD EXERCISES

1. Obtain a copy of an organization's mission statement. (One can typically be found in the organization's Annual Report which are often available in university libraries or in corporate marketing brochures. If you are finding it difficult to locate this material, write or call the organization directly and ask for a copy of the mission statement.) What is this organization's area of business? What does the organization value highly (e.g., high-quality products and services, low cost to consumers, employee growth and development, etc.)? If the mission statement is well written, these concepts should be clear. Do you know anything about the information systems in this company that would demonstrate that the types of systems in place reflect the organization's mission? Explain.

2. Interview the managers of the information systems department of an organization to determine the level and nature of their strategic information systems planning. Does it appear to be adequate? Why or why not? Obtain a copy of that organization's mission statement. To what degree do the strategic IS plan and the organizational strategic plan fit together? What are the areas where the two plans fit and do not fit? If there is not a good fit, what are the implications for the success of the organization? For the usefulness of their information systems?

3. Choose an organization that you have contact with, perhaps your employer or university. Follow the "Outline of an information systems plan" shown in Figure 6-16 and complete a short information systems plan for the organization you chose. Write at least a brief paragraph for each of the seven categories in the outline. If IS personnel and managers are available, interview them to obtain information you

need. Present your mock plan to the organization's IS manager and ask for feedback on whether or not your plan fits the IS reality for that organization.

4. Choose an organization that you have contact with, perhaps your employer or university. List significant examples for each of the items used to create planning matrices. Next, list possible relationships among various items and display these relationships in a series of planning matrices.

5. Write separate mission statements that you believe fit well for Microsoft, IBM, and AT&T. Compare your mission statements with the real mission statements of these companies. Their mission statements can typically be found in their Annual Reports. Were your mission statements comparable to the real mission statements? Why or why not? What differences and similarities are there among these three mission statements? What information systems are necessary to help these companies deliver on their mission statements?

6. Choose an organization that you have contact with, perhaps your employer or university. Determine how information systems projects are identified. Are projects identified adequately? Are they identified as part of the information systems planning or the corporate strategic planning process? Why or Why not?

REFERENCES

Atkinson, R. A. 1990. "The Motivations for Strategic Planning." *Journal of Information Systems Management* 7 (4): 53–56.

Atkinson, R. A., and J. Montgomery. 1990. "Reshaping IS Strategic Planning." *Journal of Information Systems Management* 7 (4): 9–15.

Carlson, C. K., E. P. Gardner, and S. R. Ruth. 1989. "Technology-Driven Long-Range Planning." *Journal of Information Systems Management* 6 (3): 24–29.

IBM. 1982. "Business Systems Planning." In *Advanced System Development/Feasibility Techniques*, edited by J. D. Couger, M. A. Colter, and R. W. Knapp. New York, NY: Wiley: 236–314.

Kerr, J. 1990. "The Power of Information Systems Planning." *Database Programming & Design* 3 (December): 60–66.

Koory, J. L. and D. B. Medley. 1987. *Management Information Systems: Planning and Decision Making,* Cincinnati, OH: South-Western.

Martin, J. (1990). *Information Engineering*. Englewood Cliffs, NJ: Prentice-Hall, Inc.

McKeen, J. D., T. Guimaraes, and J. C. Wetherbe. 1994. "A Comparative Analysis of MIS Project Selection Mechanisms." *Data Base* 25 (February): 43–59.

Moriarty, T. 1991. "Framing Your System." *Database Programming & Design* 4 (June): 57–59.

Morton, C. 1992. "Information Competition: Can OLTP and DSS Peacefully Coexist?" *Data Base Management* 2 (June): 24–28.

Parker, M. M., and R. J. Benson. 1989. "Enterprisewide Information Management: State-of-the-Art Strategic Planning." *Journal of Information Systems Management* 6 (Summer): 14–23.

Porter, M. 1980. *Competitive Strategy: Techniques for Analyzing Industries and Competitors.* New York, NY: Free Press.

Reingruber, M. J. and D. L. Spahr. 1992. "Putting Data Back in Database Design." *Data Base Management* 2 (March): 19–21.

Sowa, J. F. and J. A. Zachman. 1992. "Extending and Formalizing the Framework for Information Systems Architecture." *IBM Systems Journal* 31 (3): 590–616.

Zachman, J. A. 1987. "A Framework for Information Systems Architecture." *IBM Systems Journal* 26 (March): 276–92.

Identifying and Selecting the Customer Activity Tracking System

INTRODUCTION

Over the last two days, all BEC top executives participated in the annual off-site strategic planning session. During the first day, BEC executives reviewed major events and accomplishments of the prior year. They also reviewed and updated BEC corporate objectives and generated ideas on how to continuously improve BEC business practices. On the second day, Nigel Broad, the chair of BEC, worked with the group to develop a forward-looking document that they called BEC's Blueprint for the Decade (Figure 1). In this document, BEC's mission, objectives, and strategy were reviewed, updated, and clearly laid out so that they could be easily expressed to all BEC employees. Nigel believes that all employees need to clearly understand BEC's mission, objectives, and strategy if profit and growth targets are to be realized.

AN IDEA FOR A NEW SYSTEM IS BORN

After completing this intensive two-day session, Nancy Chen, director of U.S. operations, returned home to relax and reflect on the intensive planning session. Upon arrival, she found her mailbox full of new catalogs from companies she had never heard of, and thought to herself, "Where do these catalogs come from? Where do these companies get my name and address? How do these companies know that I like to fly fish in the Rockies? They seem to know exactly what I like to do, I wonder how?" With this, she fixed her dinner and looked forward to browsing through her new catalogs. Later that night while watching a "Next Generation" rerun, she flipped though the catalog titled *Remote B&B's*. While reading, she remembered that she had recently stayed in a beautiful bed and breakfast while on a fishing expedition outside Missoula, Montana. "Ah, maybe that's where they got my name." This also gave her an idea: Could BEC use customer data in a similar way?

While thinking her idea through, she remembered Nigel stressing the importance of "leveraging our existing customer base to get more frequent purchase and rental activity" and the Blueprint's objective of applying technology in innovative ways to provide better customer service. Maybe her idea could serve as a model for the point Nigel was making.

After arriving at work the following day, Nancy searched the Internet for information related to direct marketing and customer profiling. This search yielded several articles and references. In one particular article from the *Wall Street Journal* she read how numerous mail order companies have enhanced promotional efforts and profitability by profiling customer buying habits and then sending their customers specific types of catalogs based upon their profiles. For example, households profiled to have young children are mailed catalogs that include children-related items in addition to adult clothing whereas households profiled as not having children are sent catalogs with no children-related sections. It was reported that companies using these and similar profiling techniques are experiencing substantially improved "hit rates" for promotions. Nancy felt that similar profiling of BEC customers could be performed using customer transaction data. She felt that an analysis of the customer rental and buying patterns could lead to the creation of profiles that might be useful for targeting new product releases and promotions. This would help BEC better utilize its marketing expenditures and provide an innovative customer service through the use of information technology.

To see if her idea were possible, Nancy first contacted Karen Gardner, v-p of Information Systems, to get Karen's views on the project idea. Nancy had been very impressed with Karen at the strategic planning session when Karen had reviewed the strengths and weaknesses of BEC's IS organization. After Nancy gave a brief explanation of her ideas on customer rental and buying patterns, Karen felt that the ideas had merit. She reminded Nancy that, since it appeared that Nancy's ideas would require a costly systems development

Figure 1
Broadway Entertainment Company's mission, objectives, and strategy

---

BLUEPRINT FOR THE DECADE

FORWARD

This blueprint provides guidance to Broadway Entertainment Corporation (BEC), Incorporated for the coming decade. It shows how our mission, objectives, and strategy fit together and provides direction for all individuals and decisions of the Corporation.

OUR MISSION

BEC is a publicly held, for-profit organization focusing on the home entertainment industry that has a global focus for operations. BEC exists to serve customers with a primary goal of enhancing shareholders' investment through the pursuit of excellence in everything we do. BEC will operate under the highest ethical standards, will respect the dignity, rights, and contributions of all employees, and strive to better society.

OUR OBJECTIVES

1. BEC will strive to increase market share and profitability (prime objective).
2. BEC will be a leader in all areas of our business—human resources, technology, operations, and marketing.
3. BEC will be cost-effective in the use of all resources.
4. BEC will rank among industry leaders in both profitability and growth.
5. BEC will be innovative in the use of technology to help bring new products and services to market faster than our competition and to provide better service to our customers.
6. BEC will create an environment that values diversity in gender, race, values, and culture among employees, suppliers, and customers.

OUR STRATEGY

BEC will be a *global* provider of home entertainment products and services by providing the highest quality *customer service*, the *broadest range of products and services,* at the *lowest possible price.*

---

effort and that the new system would need to access corporate databases, a formal request must be developed and approved by the Systems Priority Board (SPB) before proceeding. Nancy remembered Karen's remarks at the strategic planning session about how IS needed to drive its systems development more from corporate strategy and objectives in order to better manage the deluge of requests for new and improved systems. Nancy asked Karen some questions about how to show the linkage between an IS request, corporate strategy and objectives, and the annual IS plan. Karen suggested that Nancy's request could be a good test case for trying out some ways to show this linkage.

Karen assigned Jordan Pippen, a senior systems analyst at BEC, to assist Nancy with the development of a formal Systems Service Request (SSR). As Jordan had led an internal IS study on aligning systems development with corporate plans, she would be especially helpful to Nancy. Over the next two weeks Jordan and Nancy met three times. At the first meeting, Nancy explained her vision of the project to Jordan. This information provided Jordan with adequate information to develop a high-level project statement that she shared with Nancy prior to their second meeting. During their second meeting, they jointly edited this statement as it developed into the Problem Statement within the SSR. At the third meeting, they finalized the wording of the entire SSR before forwarding it to the SPB (see Figure 2).

Figure 2
System service request for the CATS project

**Broadway Entertainment
System Service Request**

REQUESTED BY   Nancy Chen   DATE: April 23, 1995

DEPARTMENT   Operations

LOCATION   Headquarters, 4409

CONTACT   Tel: 6-7755   FAX: 6-7750   e-mail: chen

TYPE OF REQUEST   URGENCY

   [X]   New System   [   ]   Immediate – Operations are impaired or opportunity lost
   [   ]   System Enhancement   [   ]   Problems exist, but can be worked around
   [   ]   System Error Correction   [X]   Business losses can be tolerated until new system installed

PROBLEM STATEMENT

Market, sales, and profit growth at BEC is limited by the extent to which we can leverage each customer to increase the frequency and/or amount of rental or purchase activity. A possible method for BEC to achieve its growth targets has been stated as a desire to "leverage our existing customer base to get more frequent purchase and rental activity." One method to achieve this goal would be to use existing customer information and behavior to develop a purchase and rental profile for each customer. This profile could then be used to target customers who would be more likely to rent a particular movie or purchase a particular music title. Customer profiles would be matched against new product releases. Once a match has been made, customers would then be notified of the pending product release (for example, using a post card) and be informed that the product will be held for the customer at a local store if the customer so desires. To reserve a product, the customer would simply have to call an 800 number, give a code number that is printed on the card to identify the product, store, and customer, and a product would be earmarked for the customer. Inventory management systems would be modified to use this information to assure that specific product is at the right store at the right time. Stores would be given a listing of customers and reserved products so that appropriate measures would be taken to hold back adequate inventory supplies for profiled customers.

SERVICE REQUEST

I request a thorough analysis of this idea be conducted. We currently are underutilizing our information on customer transaction data. This application would provide us with a system that could differentiate BEC from our competitors and help achieve one of our strategic objectives.

IS LIAISON   Jordan Pippen   6-5519   FAX: 6-5550   e-mail: pippen

SPONSOR   Nancy Chen, VP of U.S. Operations

- - - - - - - - - - - - - - - - - - TO BE COMPLETED BY SYSTEMS PRIORITY BOARD -

   [   ]   Request approved   Assigned to _____
                          Start date _____
   [   ]   Recommend revision
   [   ]   Suggest user development
   [   ]   Reject for reason _____

At the last meeting, Jordan and Nancy gave their project a name—the Customer Activity Tracking System (CATS).

About one year ago Karen had received corporate board approval to form the Systems Priority Board. Overall growth at BEC had created extensive demand on the IS organization and Karen wanted general management validation that IS was working on the most important projects. At this time, Karen convenes the SPB, but she is the only IS department representative of the six SPB members. Not all requests must pass before the SPB: requests to fix system errors or to comply with legal/regulatory edicts are simply summarized for SPB awareness. Also, enhancement and new system requests that require less than $2,000 of IS development cost and which do not change existing corporate systems and databases are only summarized for review. The SPB discusses such requests only if an abnormal pattern of these requests arises (for example, more than $20,000 of requests from the same department over 6–12 months).

A primary function of the SPB is to assess an SSR in terms of its alignment with systems development objectives defined in BEC's information systems plan (ISP), updated annually after the corporate strategic planning session. The CATS request was a bottom-up request that was developed outside the normal information systems planning process. The request must now be assessed in light of all projects currently selected for project initiation and planning, the second phase of BEC's systems development life cycle. The SPB must assess how well this request fits in with corporate and IS strategy and existing IS architecture compared to other requests with which it must compete for resources.

If the SPB feels that the request is adequately aligned with their ISP, they will likely approve a more detailed study to outline the development project in detail. If the SPB feels that the request is not adequately aligned with the ISP, they will most likely reject the proposal and provide feedback to Nancy as to why it was rejected. In project initiation and planning, only projects deemed to be adequately aligned with the objectives of BEC's information system plan are given approval for more detailed study.

For this reason, Jordan and Nancy decided to supplement the standard BEC SSR form with information showing the relationship of CATS to the Blueprint for the Decade and to the IS plan (see Figure 3). Although at a very high level, this information nonetheless helped

Figure 3
CATS alignment with ISP

	CATS	Legend
Blueprint Objectives		
1. Increase market share and profitability	C,E	
2. Leader in all areas of business	C,E	C = major contribution
3. Cost-effective use of resources	E	by CATS
4. Industry leader in profitability and growth	C,E	E = major contribution
5. Innovative use of technology	C	by existing systems
6. Value diversity	C	
ISP Objectives		
1. Better align IS development with corporate objectives	P	
2. Deliver global system solutions	S	P = primary goal of CATS
3. Reduce systems development backlog	S	S = secondary goal of CATS
4. Increase skill level of IS staff		
Relationship with Current Systems		
Rental	F,S	
Sales	F,S	F = CATS to be fed data
Inventory Control	S	from this system
Member Tracking	F	S = CATS can provide
Inventory Management	F,S	information to this system
Sales Tracking and Analysis	F,S	

the SPB to see the potential scope of CATS and the potential areas where CATS might have impact. Based on the SSR and this supplemental information, the SPB approved the selection of the CATS project, thereby initiating the development of a Baseline Project Plan to be created during the second phase of the SDLC—project initiation and planning.

The SPB did raise two issues needing to be addressed during the next and subsequent phases. First, Karen noted that many existing systems (see Figure 3) could have some form of interface to CATS, primarily in providing data used by profiling routines in CATS; her concern was that duplication and lack of synchronization of data across these systems might cause problems for CATS. Second, Dwight Ford, BEC legal counsel and member of the SPB, asked that an investigation about the legality and ethics of buying and selling sales activity data be carried out during the project. Dwight had been a strong proponent of the statements on "highest ethical standards" and supporting diversity in the Blueprint for the Decade.

SUMMARY

The objective of this section was to describe how the Customer Activity Tracking System was identified and selected at BEC. CATS, like many systems, was the brainchild of one manager, creatively connecting information she had received from apparently unrelated activities. This idea led to a request to fund further study of a potential new system—CATS. At the end of each subsequent chapter dealing with a phase of the SDLC, the story of CATS will unfold.

7

Initiating and Planning Systems Development Projects

After studying this chapter, you should be able to:

- Describe the steps involved in the project initiation and planning process.

- Explain the need for and the contents of a Statement of Work and Baseline Project Plan.

- List and describe various methods for assessing project feasibility.

- Describe the differences between tangible and intangible benefits and costs and between one-time and recurring benefits and costs.

- Perform cost-benefit analysis and describe what is meant by the time value of money, present value, discount rate, net present value, return on investment, and break-even analysis.

- Describe the general rules for evaluating the technical risks associated with a systems development project.

- Describe the activities and participant roles within a structured walkthrough.

INTRODUCTION

The first phase of the systems development life cycle is project identification and selection, during which the need for a new or enhanced system is recognized. This first life cycle phase does not deal with a specific project but rather identifies the portfolio of projects to be undertaken by the organization. Thus, project identification and selection is a pre-project step in the life cycle. This recognition of potential projects may come as part of a larger planning process, information systems planning, or from requests from managers and business units. Regardless of how a project is identified and selected, the next step is to conduct a more detailed assessment of one particular project selected during the first phase. This assessment does not focus on how the proposed system will operate but rather on understanding the scope of a proposed project and its feasibility of completion given the available resources. It is crucial that organizations understand whether resources should be devoted to a project, otherwise very expensive mistakes can be made. The focus of this chapter is on this process. In other words, project initiation and planning is where projects are either accepted for development, rejected, or redirected. This is also where you, as a systems analyst, begin to play a major role in the systems development process.

In the next section, the project initiation and planning process is briefly reviewed. Next, numerous techniques for assessing project feasibility are described. The information uncovered during feasibility analysis is organized into a document called a Baseline Project Plan. Once this plan is developed, a formal review of the project can be conducted. The process of building this plan is discussed next. Yet, before the project can evolve to the next phase of the systems development life cycle—analysis—the project plan must be reviewed and accepted. In the final major section of the chapter, we provide an overview of the project review process.

INITIATING AND PLANNING SYSTEMS DEVELOPMENT PROJECTS

A key consideration when conducting project initiation and planning (PIP) is deciding when PIP ends and when analysis, the next phase of the SDLC, begins. This is a concern since many activities performed during PIP could also be completed during analysis. Pressman (1992) speaks of three important questions that must be considered when making this decision on the division between PIP and analysis:

1. *How much effort should be expended on the project initiation and planning process?*

2. *Who is responsible for performing the project initiation and planning process?*

3. *Why is project initiation and planning such a challenging activity?*

Finding an answer to the first question, how much effort should be expended on the PIP process, is often difficult. Practical experience has found, however, that the time and effort spent on initiation and planning activities easily pay for themselves later in the project. Proper and insightful project planning, including determining project scope as well as identifying project activities, can easily reduce time in later project phases. For example, a careful feasibility analysis which leads to deciding that a project is not worth pursuing can save a considerable expenditure of resources. The actual amount of time expended will be affected by the size and complexity of the project as well as by the experience of your organization in building similar systems. A rule of thumb is that between 10 and 20 percent of the entire development effort should be expended on the PIP study. Thus, you should not be reluctant to spend considerable time in PIP in order to fully understand the motivation for the requested system.

For the second question, who is responsible for performing the PIP, most organizations assign an experienced systems analyst, or team of analysts for large projects, to perform PIP. The analyst will work with the proposed customers (managers and users) of the system and other technical development staff in preparing the final plan. Experienced analysts working with customers who well understand their information services needs should be able to perform PIP without the detailed analysis typical of the analysis phase of the life cycle. Less experienced analysts with customers who only vaguely understand their needs will likely expend more effort during PIP in order to be certain that the project scope and work plan are feasible.

Third, the project initiation and planning process is viewed as a challenging activity because the objective of the PIP study is to transform a vague system request document into a tangible project description. This is an open-ended process. The analyst must clearly understand the motivation for and objectives of the proposed system. Therefore, effective communication among the systems analyst, users, and management is crucial to the creation of a meaningful project plan. Getting all parties to agree on the direction of a project may be difficult for cross-department projects when different parties have different business objectives. Thus, more complex organizational settings for projects will result in more time required for analysis of the current and proposed systems during PIP.

In the remainder of this chapter, we will describe the necessary activities used to answer these questions. In the next section, we will revisit the project initiation and planning activities originally outlined in Chapter 4 in the section on Managing the Information Systems Project. This is followed by a brief description of the deliverables and outcomes from this process.

The Process of Initiating and Planning Systems Development Projects

As its name implies, two major activities occur during the second phase of the SDLC, project initiation and planning (Figure 7-1). As the steps in the project initiation and planning process were explained in Chapter 4, our primary focus in this chapter is to describe several techniques that are used when performing this process. Therefore, we will only briefly review the PIP process.

Project initiation focuses on activities designed to assist in *organizing* a team to conduct project planning. During initiation, one or more analysts are assigned to work with a customer—that is, a member of the business group that requested or

Figure 7-1
Systems development life cycle with project initiation and planning highlighted

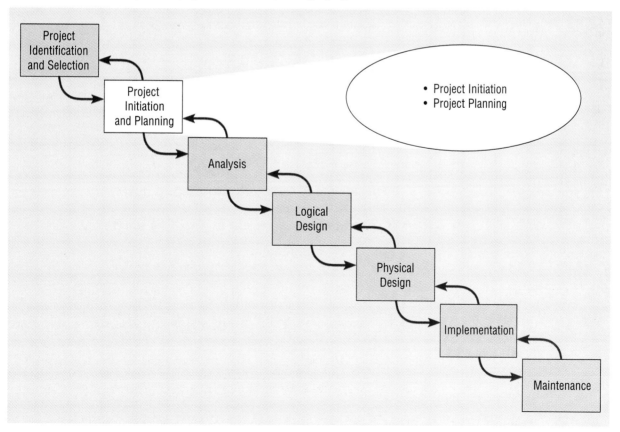

will be impacted by the project—to establish work standards and communication procedures. Examples of the types of activities performed are shown in Table 7-1. Depending upon the size, scope, and complexity of the project, some project initiation activities may be unnecessary or may be very involved. Also, many organizations have established procedures for assisting with common initiation activities.

Project planning, the second activity within PIP, is distinct from general information systems planning which focuses on assessing the information systems needs of the *entire* organization (discussed in Chapter 6). Project planning is the process of defining clear, discrete activities and the work needed to complete each activity within a *single* project. The objective of the project planning process is the development of a *Baseline Project Plan (BPP)* and the *Statement of Work (SOW)*. The BPP becomes the foundation for the remainder of the development project. The SOW produced by the team clearly outlines the objectives and constraints of the project for the customer. As with the project initiation process, the size, scope, and complexity of a project will dictate the comprehensiveness of the project planning process and resulting documents. Further, numerous assumptions about resource availability and potential problems will have to be made. The range of activities performed during project planning are listed in Table 7-2.

Baseline Project Plan (BPP): A major outcome and deliverable from the project initiation and planning phase which contains the best estimate of a project's scope, benefits, costs, risks, and resource requirements.

Deliverables and Outcomes

The major outcomes and deliverables from the project initiation and planning phase are the Baseline Project Plan and the Statement of Work. The **Baseline Project Plan (BPP)** contains all information collected and analyzed during project initiation and

---

TABLE 7-1   Elements of Project Initiation

- Establishing the Project Initiation Team
- Establishing a Relationship with the Customer
- Establishing the Project Initiation Plan
- Establishing Management Procedures
- Establishing the Project Management Environment and Project Workbook

---

TABLE 7-2   Elements of Project Planning

- Describing the Project Scope, Alternatives, and Feasibility
- Dividing the Project into Manageable Tasks
- Estimating Resources and Creating a Resource Plan
- Developing a Preliminary Schedule
- Developing a Communication Plan
- Determining Project Standards and Procedures
- Identifying and Assessing Risk
- Creating a Preliminary Budget
- Developing a Statement of Work
- Setting a Baseline Project Plan

planning. The plan reflects the best estimate of the project's scope, benefits, costs, risks, and resource requirements given the current understanding of the project. The BPP specifies detailed project activities for the next life cycle phase—analysis— and less detail for subsequent project phases (since these depend on the results of the analysis phase). Similarly, benefits, costs, risks, and resource requirements will become more specific and quantifiable as the project progresses. The BPP is used by the project selection committee to help decide whether the project should be accepted, redirected, or canceled. If selected, the BPP becomes the foundation document for all subsequent SDLC activities; however, it is also expected to evolve as the project evolves. That is, as new information is learned during subsequent SDLC phases, the baseline plan will be updated. Later in the chapter we describe how to construct the BPP.

The **Statement of Work (SOW)** is a short document prepared for the customer that describes what the project will deliver and outlines all work required to complete the project. The SOW assures that both you and your customer gain a common understanding of the project and is a very useful communication tool. The SOW is a very easy document to create because it typically consists of a high-level summary of the BPP information (described later). A sample SOW is shown in Figure 7-2. Depending upon your relationship with your customer, the role of the SOW may vary. At one extreme, the SOW can be used as the basis of a formal contractual agreement outlining firm deadlines, costs, and specifications. At the other extreme, the SOW can simply be used as a communication vehicle to outline the current best estimates of what the project will deliver, when it will be completed, and the resources it may consume. A contract programming or consulting firm, for example, may establish a very formal relationship with a customer and use a SOW that is extensive and formal. Alternatively, an internal development group may develop a SOW that is only one-to-two pages in length and is intended to inform customers rather than to set contractual obligations and deadlines.

Statement of Work (SOW): Document prepared for the customer during project initiation and planning that describes what the project will deliver and outlines generally at a high level all work required to complete the project.

ASSESSING PROJECT FEASIBILITY

All projects are feasible given unlimited resources and infinite time (Pressman, 1992). Unfortunately, most projects must be developed within tight budgetary and time constraints. This means that assessing project feasibility is a required activity for all information systems projects and is potentially a large undertaking. It requires that you, as a systems analyst, evaluate a wide range of factors. Typically, some of these factors will be more important than others for some projects and relatively unimportant for other projects. Although the specifics of a given project will dictate which factors are most important, most feasibility factors are represented by the following categories:

- *Economic*
- *Technical*
- *Operational*
- *Schedule*
- *Legal and Contractual*
- *Political*

Together, the culmination of these feasibility analyses form the business case that justifies the expenditure of resources on the project. In the remainder of this section, we will examine various feasibility issues. We begin by examining issues related to

Figure 7-2
Statement of Work for the
Customer Tracking System
(Pine Valley Furniture)

Pine Valley Furniture   **Prepared: 9/20/95**
Statement of Work

Project Name: Customer Tracking Systems
PVF Project Manager: Jim Woo

Customer: Marketing
Project Sponsor: Jackie Judson

Project Start / End (projected):   10/1/95–2/1/96

PVF Development Staff Estimates (man-months):
    Programmers: 2.0
    Jr. Analysts: 1.5
    Sr. Analysts: 0.3
    Supervisors: 0.1
    Consultants: 0.0
    Librarian: 0.1

    TOTAL: **4.0**

Project Description

Goal
    This project will implement a customer tracking system for the
    marketing department. The purpose of this system is to
    automate the … to save employee time, reduce errors, have
    more timely information, …

Objective
    • minimize data entry errors
    • provide more timely information
    • …

Phases of Work
    The following tasks and deliverables reflect the current
    understanding of the project:
    In Analysis, …
    In Design, …
    In Implementation, …

---

economic feasibility and demonstrate techniques for conducting this analysis. This is followed by a discussion of techniques for assessing technical project risk. Finally, issues not directly associated with economic and technical feasibility, but no less important to assuring project success, are discussed.

To help you better understand the feasibility assessment process, we will examine a project at Pine Valley Furniture. For this project, a Systems Service Request (SSR) was submitted by Pine Valley Furniture's (PVF) vice-president of Marketing, Jackie Judson, to develop a Customer Tracking System (Figure 7-3). Jackie feels that this system would allow PVF's marketing group to better track customer purchase

Figure 7-3
System service request for Customer Tracking System (Pine Valley Furniture)

Pine Valley Furniture
System Service Request

REQUESTED BY   Jackie Judson   DATE: August 23, 1995

DEPARTMENT   Marketing

LOCATION   Headquarters, 570c

CONTACT   Tel: 4-3290   FAX: 4-3270   e-mail: jjudson

TYPE OF REQUEST   URGENCY
  [X] New System   [   ] Immediate – Operations are impaired or opportunity lost
  [   ] System Enhancement   [   ] Problems exist, but can be worked around
  [   ] System Error Correction   [X] Business losses can be tolerated until new system installed

PROBLEM STATEMENT

Sales growth at PVF has caused a greater volume of work for the marketing department. This volume of work has greatly increased the volume and complexity of the data we need to deal with and understand. We are currently using manual methods and a complex PC-based electronic spreadsheet to track and forecast customer buying patterns. This method of analysis has many problems: (1) we are slow to catch buying trends as there is often a week or more delay before data can be taken from point of sales system and manually enter it into our spreadsheet; (2) the process of manual data entry is prone to errors (which makes the results of our subsequent analysis suspect); and (3) the volume of data and the complexity of analyses conducted in the system seem to be overwhelming our current system—sometimes the program starts recalculating and never returns while for others it returns information that we know cannot be correct.

SERVICE REQUEST

I request a thorough analysis of our current method of tracking and analysis of customer purchasing activity with the intent to design and build a completely new information system. This system should handle all customer purchasing activity, support display and reporting of critical sales information, and assist marketing personnel in understanding the increasingly complex and competitive business environment. I feel that such a system will improve the competitiveness of PVF, particularly in our ability to better serve our customers.

IS LIAISON   Jim Woo, 4-6207 FAX: 4-6200 e-mail: jwoo

SPONSOR   Jackie Judson, Vice-President, Marketing

---------------------- TO BE COMPLETED BY SYSTEMS PRIORITY BOARD ------------------------
  [   ] Request approved   Assigned to _____
                                 Start date _____
  [   ] Recommend revision
  [   ] Suggest user development
  [   ] Reject for reason _____

activity and sales trends. She also feels that, if constructed, the Customer Tracking System (CTS) would provide many tangible and intangible benefits to PVF. This project was selected by PVF's Systems Priority Board for a project initiation and planning study. During project initiation, senior systems analyst Jim Woo was assigned to work with Jackie to initiate and plan the project. At this point in the project, all project initiation activities have been completed. Jackie and Jim are now focusing on project planning activities in order to complete the BPP.

Assessing Economic Feasibility

Economic feasibility: A process of identifying the financial benefits and costs associated with a development project.

The purpose for assessing **economic feasibility** is to identify the financial benefits and costs associated with the development project; economic feasibility is often referred to as cost-benefit analysis. During project initiation and planning, it will be impossible for you to precisely define all benefits and costs related to a particular project. Yet, it is important that you spend adequate time identifying and quantifying these items or it will be impossible for you to conduct an adequate economic analysis and make meaningful comparisons between rival projects. Here we will describe typical benefits and costs resulting from the development of an information system and provide several useful worksheets for recording costs and benefits. Additionally, several common techniques for making cost-benefit calculations are presented. These worksheets and techniques are used after each SDLC phase as the project is reviewed in order to decide whether to continue, redirect, or kill a project.

Determining Project Benefits   An information system can provide many benefits to an organization. For example, a new or renovated IS can automate monotonous jobs, reduce errors, provide innovative services to customers and suppliers, and improve organizational efficiency, speed, flexibility, and morale. In general, the benefits can be viewed as being both tangible and intangible. **Tangible benefits** refer to items that can be measured in dollars and with certainty. Examples of tangible benefits might include reduced personnel expenses, lower transaction costs, or higher profit margins. It is important to note that not all tangible benefits can be easily quantified. For example, a tangible benefit that allows a company to perform a task in 50 percent of the time may be difficult to quantify in terms of hard dollar savings. Most tangible benefits will fit within the following categories:

Tangible benefit: A benefit derived from the creation of an information system that can be measured in dollars and with certainty.

- Cost reduction and avoidance

- Error reduction

- Increased flexibility

- Increased speed of activity

- Improvement of management planning and control

- Opening new markets and increasing sales opportunities

Within the Customer Tracking System at PVF, Jim and Jackie identified several tangible benefits, summarized on a tangible benefits worksheet shown in Figure 7-4. Jackie and Jim had to establish the values in Figure 7-4 after collecting information from users of the current customer tracking system. They first interviewed the person responsible for collecting, entering, and analyzing the correctness of the current customer tracking data. This person estimated that they spent 10 percent of their time correcting data entry error. Given that this person's salary is $25,000, Jackie and Jim estimated an *error reduction* benefit of $2,500. Jackie and Jim also interviewed

Figure 7-4
Tangible benefits for
Customer Tracking System
(Pine Valley Furniture)

TANGIBLE BENEFITS WORKSHEET *Customer Tracking System Project*	
	Year 1 through 5
A. Cost reduction or avoidance	$ 4,500
B. Error reduction	2,500
C. Increased flexibility	7,500
D. Increased speed of activity	10,500
E. Improvement in management planning or control	25,000
F. Other _____	0
TOTAL tangible benefits	**$50,000**

managers who used the current customer tracking reports. Using this information they were able to estimate other tangible benefits. They learned that *cost reduction or avoidance* benefits could be gained due to better inventory management. Also, *increased flexibility* would likely occur from a reduction in the time normally taken to manually reorganize data for different purposes. Further, *improvements in management planning or control* should result from a broader range of analyses in the new system. Overall, this analysis forecasts that benefits from the system would be approximately $50,000 per year.

Jim and Jackie also identified several intangible benefits of the system. Although they could not quantify these benefits, they will still be described in the final BPP. **Intangible benefits** refer to items that *cannot* be easily measured in dollars or with certainty. Intangible benefits may have direct organizational benefits such as the improvement of employee morale or they may have broader societal implications such as the reduction of waste creation or resource consumption. Potential tangible benefits may have to be considered intangible during project initiation and planning since you may not be able to quantify them in dollars or with certainty at this stage in the life cycle. During later stages, such intangibles can become tangible benefits as you better understand the ramifications of the system you are designing. In this case, the BPP is updated and the business case revised to justify continuation of the project to the next phase. Table 7-3 lists numerous intangible benefits often associated with the development of an information system. Actual benefits will vary from system to system. After determining project benefits, project costs must be identified.

> **Intangible benefit:** A benefit derived from the creation of an information system that cannot be easily measured in dollars or with certainty.

Determining Project Costs   Similar to benefits, an information system can have both tangible and intangible costs. **Tangible costs** refer to items that you can easily measure in dollars and with certainty. From an IS development perspective, tangible costs include items such as hardware costs, labor costs, and operational costs such as employee training and building renovations. Alternatively, intangible costs are those items that you cannot easily measure in terms of dollars or with certainty. **Intangible costs** can include loss of customer goodwill, employee morale, or operational inefficiency. Table 7-4 provides a summary of common costs associated with the development and operation of an information system.

> **Tangible cost:** A cost associated with an information system that can be measured in dollars and with certainty.

> **Intangible cost:** A cost associated with an information system that cannot be easily measured in terms of dollars or with certainty.

TABLE 7-3 Intangible Benefits from the Development of an Information System
(Adapted from Parker and Benson, 1988)

- Competitive necessity
- More timely information
- Improved organizational planning
- Increased organizational flexibility
- Promotion of organizational learning and understanding
- Availability of new, better, or more information
- Ability to investigate more alternatives
- Faster decision making
- Information processing efficiency
- Improved asset utilization
- Improved resource control
- Increased accuracy in clerical operations
- Improved work process that can improve employee morale
- Positive impacts on society

One-time cost: A cost associated with project start-up and development, or system start-up.

Recurring cost: A cost resulting from the ongoing evolution and use of a system.

Besides tangible and intangible costs, you can distinguish IS-related development costs as either one-time or recurring (the same is true for benefits although we do not discuss this difference for benefits). **One-time costs** refer to those associated with project initiation and development and the start-up of the system. These costs typically encompass activities such as system development, new hardware and software purchases, user training, site preparation, and data or system conversion. When conducting an economic cost-benefit analysis, a worksheet should be created for capturing these expenses. For very large projects, one-time costs may be staged over one or more years. In these cases, a separate one-time cost worksheet should be created for each year. This separation will make it easier to perform present value calculations (see below). **Recurring costs** refer to those costs resulting from the ongoing evolution and use of the system. Examples of these costs typically include

- Application software maintenance
- Incremental data storage expense
- Incremental communications
- New software and hardware leases
- Supplies and other expenses (for example, paper, forms, data center personnel)

Both one-time and recurring costs can consist of items that are fixed or variable in nature. Fixed costs refer to costs that are billed or incurred at a regular interval and usually at a fixed rate (a facility lease payment). Variable costs refer to items that vary in relation to usage (long distance phone charges).

During the process of determining project costs, Jim and Jackie identified both one-time and recurring costs for the project. These costs are summarized in Figures 7-5 and 7-6. These figures show that this project will incur a one-time cost of

TABLE 7-4   Possible Information Systems Costs
(Source: King and Schrems, 1978)

Types of Costs	Examples
Procurement	Consulting costs
	Equipment purchase or lease
	Equipment installation costs
	Site preparation and modifications
	Capital costs
	Management and staff time
Start-Up	Operating system software
	Communications equipment installation
	Start-up personnel
	Personnel searches and hiring activities
	Disruption to the rest of the organization
	Management to direct start-up activity
Project-Related	Application software
	Software modifications to fit local systems
	Personnel, overhead, . . . , from in-house development
	Training users in application use
	Collecting and analyzing data
	Preparing documentation
	Managing development
Operating	System maintenance costs (hardware, software, and facilities)
	Rental of space and equipment
	Asset depreciation
	Management, operation, and planning personnel

$42,500 and a recurring cost of $28,500 per year. One-time costs were established by discussing the system with Jim's boss who felt that the system would require approximately four months to develop (at $5,000 per month). To effectively run the new system, the Marketing department would need to upgrade at least five of their current workstations (at $3,000 each). Additionally, software licenses for each workstation (at $1,000 each) and modest user training fees (ten users at $250 each) would be necessary.

As you can see from Figure 7-6, Jim and Jackie believe the proposed system will be highly dynamic and will require, on average, five months of annual maintenance, primarily for enhancements as users expect more from the system. Other ongoing expenses such as increased data storage, communications equipment, and supplies should also be expected. You should now have an understanding of the types of benefit and cost categories associated with an information systems project.

Figure 7-5
One-time costs for Customer Tracking System (Pine Valley Furniture)

ONE-TIME COSTS WORKSHEET *Customer Tracking System Project*	
	Year 0
A. Development costs	$20,000
B. New hardware	15,000

C. New (purchased) software, if any
  1. Packaged applications software
  2. Other _____
	
	5,000
	0

D. User training	2,500
E. Site preparation	0
F. Other _____	0
TOTAL one-time cost	**$42,500**

Figure 7-6
Recurring costs for Customer Tracking System (Pine Valley Furniture)

RECURRING COSTS WORKSHEET *Customer Tracking System Project*	
	Year 1 through 5
A. Application software maintenance	$25,000
B. Incremental data storage required: 20 MB × $50. (estimated cost/MB = $50)	1,000
C. Incremental communications (lines, messages, . . .)	2,000
D. New software or hardware leases	0
E. Supplies	500
F. Other _____	0
TOTAL recurring costs	**$28,500**

It should be clear that there are many potential benefits and costs associated with a given project. Additionally, since the development and useful life of a system may span several years, these benefits and costs must be normalized into present-day values in order to perform meaningful cost-benefit comparisons. In the next section, we address the relationship between time and money.

The Time Value of Money   Most techniques used to determine economic feasibility encompass the concept of the *time value of money* (TVM). TVM refers to the concept of comparing present cash outlays to future expected returns. As previously discussed, the development of an information system has both one-time and recurring costs. Furthermore, benefits from systems development will likely occur sometime in the future. Since many projects may be competing for the same investment dollars and may have different useful life expectancies, all costs and benefits must be viewed in relation to their *present value* when comparing investment options.

A simple example will help in understanding the TVM. Suppose you want to buy a used car from an acquaintance and she asks that you make three payments of $1,500 for three years, beginning next year, for a total of $4,500. If she would agree to a single lump sum payment at the time of sale (and if you had the money!), what amount do you think she would agree to? Should the single payment be $4,500? Should it be more or less? To answer this question, we must consider the time value of money. Most of us would gladly accept $4,500 today rather than three payments of $1,500, because a dollar today (or $4,500 for that matter) is worth more than a dollar tomorrow or next year, because money can be invested. The rate at which money can be borrowed or invested is called the *cost of capital*, and is called the **discount rate** for TVM calculations. Let's suppose that the seller could put the money received for the sale of the car in the bank and receive a 10% return on her investment. A simple formula can be used when figuring out the **present value** of the three $1,500 payments:

Discount rate: The rate of return used to compute the present value of future cash flows.

Present value: The current value of a future cash flow.

$$PV_n = Y \times \left[\frac{1}{(1 + i)^n} \right]$$

where PV_n is the present value of Y dollars n years from now when i is the discount rate.

From our example, the present value of the three payments of $1,500 can be calculated as

$$PV_1 = 1500 \times \left[\frac{1}{(1 + .10)^1} \right] = 1500 \times .9091 = 1363.65$$

$$PV_2 = 1500 \times \left[\frac{1}{(1 + .10)^2} \right] = 1500 \times .8264 = 1239.60$$

$$PV_3 = 1500 \times \left[\frac{1}{(1 + .10)^3} \right] = 1500 \times .7513 = 1126.95$$

where PV_1, PV_2, and PV_3 reflect the present value of each $1,500 payment in year one, two, and three, respectively.

To calculate the *net present value* (NPV) of the three $1,500 payments, simply add the present values calculated above (NPV = PV_1 + PV_2 + PV_3 = 1363.65 + 1239.60 + 1126.95 = $3730.20). In other words, the seller could accept a lump sum payment of $3,730.20 as equivalent to the three payments of $1,500, given a discount rate of 10 percent.

Given that we now know the relationship between time and money, the next step in performing the economic analysis is to create a summary worksheet reflecting the present values of all benefits and costs as well as all pertinent analyses. Due to the fast pace of the business world, PVF's System Priority Board feels that the useful life of many information systems may not exceed five years. Therefore, all cost-benefit analysis calculations will be made using a five-year time horizon as the upper boundary on all time-related analyses. In addition, the management of PVF has set their cost of capital to be 12% (that is, PVF's discount rate). The worksheet constructed by Jim is shown in Figure 7-7.

Cell H11 of the worksheet displayed in Figure 7-7 summarizes the NPV of the total tangible benefits from the project. Cell H19 summarizes the NPV of the total costs from the project. The NPV for the project ($35,003) shows that, overall, benefits from the project exceed costs (see cell H22).

Figure 7-7
Summary spreadsheet reflecting the present value calculations of all benefits and costs for the Customer Tracking System (Pine Valley Furniture)

	A	B	C	D	E	F	G	H
1	*Pine Valley Furniture*							
2	*Economic Feasibility Analysis*							
3	*Customer Tracking System Project*							
4								
5					Year of Project			
6		Year 0	Year 1	Year 2	Year 3	Year 4	Year 5	TOTALS
7	Net economic benefit	$0	$50,000	$50,000	$50,000	$50,000	$50,000	
8	Discount rate (12%)	1.0000	0.8929	0.7972	0.7118	0.6355	0.5674	
9	PV of benefits	$0	$44,643	$39,860	$35,589	$31,776	$28,371	
10								
11	NPV of all BENEFITS	$0	$44,643	$84,503	$120,092	$151,867	$180,239	$180,239
12								
13	*One-time COSTS*	($42,500)						
14								
15	Recurring Costs	$0	($28,500)	($28,500)	($28,500)	($28,500)	($28,500)	
16	Discount rate (12%)	1.0000	0.8929	0.7972	0.7118	0.6355	0.5674	
17	PV of Recurring Costs	$0	($25,446)	($22,720)	($20,286)	($18,112)	($16,172)	
18								
19	NPV of all COSTS	($42,500)	($67,946)	($90,666)	($110,952)	($129,064)	($145,236)	($145,236)
20								
21								
22	*Overall NPV*							$35,003
23								
24								
25	*Overall ROI - (Overall NPV / NPV of all COSTS)*							0.24
26								
27								
28	*Break-even Analysis*							
29	Yearly NPV Cash Flow	($42,500)	$19,196	$17,140	$15,303	$13,664	$12,200	
30	Overall NPV Cash Flow	($42,500)	($23,304)	($6,164)	$9,139	$22,803	$35,003	
31								
32	Project break-even occurs between years 2 and 3							
33	Use first year of positive cash flow to calculate break-even fraction - ((15303 - 9139) / 15303) = .403							
34	*Actual break-even occurred at 2.4 years*							
35								
36	Note: All dollar values have been rounded to the nearest dollar							

The overall return on investment (ROI) for the project is also shown on the worksheet in cell H25. Since alternative projects will likely have different benefit and cost values and, possibly, different life expectancies, the overall ROI value is very useful for making project comparisons on an economic basis. Of course, this example shows ROI for the overall project. An ROI analysis could be calculated for each year of the project.

The last analysis shown in Figure 7-7 is a break-even analysis. The objective of the break-even analysis is to discover at what point (if ever) benefits equal costs (that is, when break-even occurs). To conduct this analysis, the NPV of the yearly cash flows are determined. Here, the yearly cash flows are calculated by subtracting both the one-time cost and the present values of the recurring costs from the present value of the yearly benefits. The overall NPV of the cash flow reflects the total cash flows for all preceding years. Examination of line 30 of the worksheet shows that break-even occurs between years 2 and 3. Since year three is the first in which the general NPV cash flows figure is non-negative, the identification of what point during the year break-even occurs can be derived as follows:

$$\text{Break-Even Ratio} = \frac{\text{Yearly NPV Cash Flow} - \text{general NPV Cash Flow}}{\text{Yearly NPV Cash Flow}}$$

Using data from Figure 7-7,

$$\text{Break-Even Ratio} = \frac{15,303 - 9,139}{15,303} = .404$$

Therefore, project break-even occurs at approximately 2.4 years. A graphical representation of this analysis is shown in Figure 7-8. Using the information from the economic analysis, PVF's Systems Priority Board will be in a much better position to understand the potential economic impact of the Customer Tracking System. It should be clear from this analysis that, without such information, it would be virtually impossible to know the cost-benefits of a proposed system and impossible to make an informed decision regarding approval or rejection of the service request.

There are many techniques that you can use to compute a project's economic feasibility. Because most information systems have a useful life of more than one year and will provide benefits and incur expenses for more than one year, most techniques for analyzing economic feasibility employ the concept of the TVM. Some of these cost-benefit analysis techniques are quite simple while others are more sophisticated. Table 7-5 describes three commonly used techniques for conducting economic feasibility analysis. For a more detailed discussion of TVM or cost-benefit analysis techniques in general, the interested reader is encouraged to review an introductory finance or managerial accounting textbook.

A systems project, to be approved for continuation, may not have to achieve break-even or have an ROI above some organizational threshold as estimated during project initiation and planning. Since you may not be able to quantify many

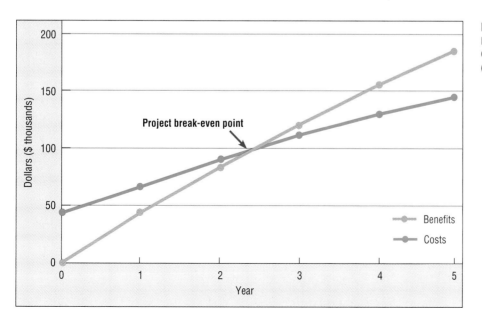

Figure 7-8
Break-even analysis for Customer Tracking System (Pine Valley Furniture)

TABLE 7-5   Commonly Used Economic Cost-Benefit Analysis Techniques

Analysis Technique	Description
Net Present Value (NPV)	NPV uses a discount rate determined from the company's cost of capital to establish the present value of a project. The discount rate is used to determine the present value of both cash receipts and outlays.
Return on Investment (ROI)	ROI is the ratio of the net cash receipts of the project divided by the cash outlays of the project. Tradeoff analysis can be made among projects competing for investment by comparing their representative ROI ratios.
Break-Even Analysis (BEA)	BEA finds the amount of time required for the cumulative cash flow from a project to equal its initial and ongoing investment.

benefits or costs at this point in a project, such financial hurdles for a project may be unattainable. In this case, simply doing as thorough an economic analysis as possible, including producing a long list of intangibles, may be sufficient for the project to progress. One other option is to run the type of economic analysis shown in Figure 7-7 using pessimistic, optimistic, and expected benefit and cost estimates during project initiation and planning. This range of possible outcomes, along with the list of intangible benefits and the support of the requesting business unit, will often be enough to allow the project to continue to the analysis phase. You must, however, be as precise as you can with the economic analysis, especially when investment capital is scarce. In this case, it may be necessary to conduct some typical analysis phase activities during project initiation and planning in order to clearly identify inefficiencies and shortcomings with the existing system and to explain how a new system will overcome these problems. Thus, building the economic case for a systems project is an open-ended activity; how much analysis is needed depends on the particular project, stakeholders, and business conditions.

Assessing Technical Feasibility

Technical feasibility: A process of assessing the development organization's ability to construct a proposed system.

The purpose of assessing **technical feasibility** is to gain an understanding of the organization's ability to construct the proposed system. This analysis should include an assessment of the development group's understanding of the possible target hardware, software, and operating environments to be used as well as system size, complexity, and the group's experience with similar systems. In this section, we will discuss a framework you can use for assessing the technical feasibility of a project in which a level of project risk can be determined after answering a few fundamental questions.

It is important to note that all projects have risk and that risk is not necessarily something to avoid. Yet it is also true that, because organizations typically expect a greater return on their investment for riskier projects, understanding the sources and types of technical risks proves to be a valuable tool when you assess a project. Also, risks need to be managed in order to be minimized; you should, therefore, identify potential risks as early as possible in a project. The potential consequences of not assessing and managing risks can include the following outcomes:

1. Failure to attain expected benefits from the project

2. Inaccurate project cost estimates

3. Inaccurate project duration estimates

4. Failure to achieve adequate system performance levels

5. Failure to adequately integrate the new system with existing hardware, software, or organizational procedures

You can manage risk on a project by changing the project plan to avoid risky factors, assigning project team members to carefully manage the risky aspects, and setting up monitoring methods to determine whether or not potential risk is, in fact, materializing.

The amount of technical risk associated with a given project is contingent on four primary factors: project size, project structure, the development group's experience with the application and technology area, and the user group's experience with development projects and application area. Aspects of each of these risk areas are summarized in Table 7-6. Using these factors for conducting a technical risk assessment, four general rules emerge:

1. *Large projects are riskier than small projects.* Project size, of course, relates to the relative project size that the development group is familiar working with. A "small" project for one development group may be relatively "large" for another. The types of factors that influence project size are listed in Table 7-6.

2. *A system in which the requirements are easily obtained and highly structured will be less risky than one in which requirements are messy, ill-structured, ill-defined, or subject to the judgement of an individual.* For example, the development of a payroll

TABLE 7-6 Project Risk Assessment Factors
(Adapted from Cash, McFarlan, McKenney, and Applegate, 1992)

Risk Factor	Examples
Project Size	Number of members on the project team
	Project duration time
	Number of organizational departments involved in project
	Size of programming effort (e.g., hours, function points)
Project Structure	New system or renovation of existing system(s)
	Organizational, procedural, structural, or personnel changes resulting from system
	User perceptions and willingness to participate in effort
	Management commitment to system
	Amount of user information in system development effort
Development Group	Familiarity with target-hardware, software development environment, tools, and operating system
	Familiarity with proposed application area
	Familiarity with building similar systems of similar size
User Group	Familiarity with information systems development process
	Familiarity with proposed application area
	Familiarity with using similar systems

system has requirements that may be easy to obtain due to legal reporting requirements and standard accounting procedures. On the other hand, the development of an executive support system would need to be customized to the particular executive decision style and critical success factors of the organization, thus making its development more risky (see Table 7-6).

3. *The development of a system employing commonly used or standard technology will be less risky than one employing novel or non-standard technology.* A project has a greater likelihood of experiencing unforeseen technical problems when the development group lacks knowledge related to some aspect of the technology environment. A less risky approach is to use standard development tools and hardware environments. It is not uncommon for experienced system developers to talk of the difficulty of using leading-edge (or in their words, *bleeding edge*) technology (see Table 7-6).

4. *A project is less risky when the user group is familiar with the systems development process and application area than if unfamiliar.* Successful IS projects require active involvement and cooperation between the user and development groups. Users familiar with the application area and the systems development process are more likely to understand the need for their involvement and how this involvement can influence the success of the project (see Table 7-6).

A project with high risk may still be conducted. Many organizations look at risk as a portfolio issue: considering all projects, it is okay to have a reasonable percentage of high-, medium-, and low-risk projects. Given that some high-risk projects will get into trouble, an organization cannot afford to have too many of these. Having too many low-risk projects may not be aggressive enough to make major breakthroughs in innovative uses of systems. Each organization must decide on its acceptable mix of projects of varying risk.

A matrix for assessing the relative risks related to the general rules described above is shown in Figure 7-9. Using the risk factor rules to assess the technical risk level of the Customer Tracking System, Jim and Jackie concluded the following about their project:

1. The project is a relatively *small project* for PVF's development organization. The basic data for the system is readily available so the creation of the system will not be a large undertaking.

2. The requirements for the project are *highly structured* and easily obtainable. In fact, an existing spreadsheet-based system is available for analysts to examine and study.

3. The *development group is familiar* with the technology that will likely be used to construct the system, as the system will simply extend current system capabilities.

4. The *user group is familiar* with the application area since they are already using the PC-based spreadsheet system described in Figure 7-3.

Given this risk assessment, Jim and Jackie mapped their information into the risk framework of Figure 7-9. They concluded that this project should be viewed as having "very low" technical risk (cell 4 of the figure). Although this method is useful for gaining an understanding of technical feasibility, numerous other issues can influence the success of the project. These non-financial and non-technical issues are described in the following section.

		Low Structure	High Structure
High Familiarity with Technology or Application Area	Large Project	(1) Low risk (very susceptible to mismanagement)	(2) Low risk
	Small Project	(3) Very low risk (very susceptible to mismanagement)	(4) Very low risk
Low Familiarity with Technology or Application Area	Large Project	(5) Very high risk	(6) Medium risk
	Small Project	(7) High risk	(8) Medium-low risk

Figure 7-9
Effects of degree of project structure, project size, and familiarity with application area on project implementation risk (Adapted from: Cash et al., 1992)

Assessing Other Feasibility Concerns

In this section, we will briefly conclude our discussion of project feasibility issues by reviewing other forms of feasibility that you may need to consider when formulating the business case for a system during project planning. The first relates to examining the likelihood that the project will attain its desired objectives, called **operational feasibility.** Its purpose is to gain an understanding of the degree to which the proposed system will likely solve the business problems or take advantage of the opportunities outlined in the systems service request or project identification study. For a project motivated from information system planning, operational feasibility includes justifying the project on the basis of being consistent with or necessary for accomplishing the IS plan. In fact, the business case for any project can be enhanced by showing a link to the business or information systems plan. Your assessment of operational feasibility should also include an analysis of how the proposed system will affect organizational structures and procedures. Systems that have substantial and widespread impact on an organization's structure or procedures are typically riskier projects to undertake. Thus, it is important for you to have a clear understanding of how an IS will fit into the current day-to-day operations of the organization.

Another feasibility concern relates to project duration and is referred to as assessing schedule feasibility. The purpose of assessing **schedule feasibility** is for you, as a systems analyst, to gain an understanding of the likelihood that all potential timeframes and completion date schedules can be met and that meeting these dates will be sufficient for dealing with the needs of the organization. For example, a system may have to be operational by a government-imposed deadline, by a particular point in the business cycle (such as the beginning of the season when new products are introduced), or at least by the time a competitor is expected to introduce a similar system. Further, detailed activities may only be feasible if resources are available when called for in the schedule. For example, the schedule should not call for system testing during rushed business periods or for key project meetings during annual vacation or holiday periods. The schedule of activities produced during project initiation and planing will be very precise and detailed for the analysis phase. The estimated activities and associated times for activities after the analysis

Operational feasibility: The process of assessing the degree to which a proposed system solves business problems or takes advantage of business opportunities.

Schedule feasibility: The process of assessing the degree to which the potential time frame and completion dates for all major activities within a project meet organizational deadlines and constraints for affecting change.

phase are typically not as detailed (e.g., it will take two weeks to program the pay-roll report module) but are rather at the life-cycle phase level (e.g., it will take six weeks for physical design, four months for programming, and so on). This means that assessing schedule feasibility during project initiation and planning is more of a "rough-cut" analysis of whether the system can be completed within the constraints of the business opportunity or the desires of the users. While assessing schedule feasibility you should also evaluate scheduling tradeoffs. For example, factors such as project team size, availability of key personnel, subcontracting or outsourcing activities, and changes in development environments may all be considered as having possible impact on the eventual schedule. As with all forms of feasibility, schedule feasibility will be reassessed after each phase, when you can specify with greater certainty the detailed steps and their duration for the next phase.

Legal and contractual feasibility: The process of assessing potential legal and contractual ramifications due to the construction of a system.

A third concern relates to assessing **legal and contractual feasibility** issues. In this area, you need to gain an understanding of any potential legal ramifications due to the construction of the system. Possible considerations might include copyright or nondisclosure infringements, labor laws, antitrust legislation (which might limit the creation of systems to share data with other organizations), foreign trade regulations (for example, some countries limit access to employee data by foreign corporations), and financial reporting standards as well as current or pending contractual obligations. Contractual obligations may involve ownership of software used in joint ventures, license agreements for use of hardware or software, nondisclosure agreements with partners, or elements of a labor agreement (for example, a union agreement may preclude certain compensation or work-monitoring capabilities a user may want in a system). A common situation is that development of a new application system for use on new computers may require new or expanded, and more costly, system software licenses. Typically, legal and contractual feasibility is a greater consideration if your organization has historically used an outside organization for specific systems or services that you now are considering handling yourself. In this case, ownership of program source code by another party may make it difficult to extend an existing system or link a new system with an existing, purchased system.

Political feasibility: The process of evaluating how key stakeholders within the organization view the proposed system.

A final feasibility concern focuses on assessing **political feasibility** in which you attempt to gain an understanding of how key stakeholders within the organization view the proposed system. Since an information system may affect the distribution of information within the organization, and thus the distribution of power, the construction of an IS can have political ramifications. Those stakeholders not supporting the project may take steps to block, disrupt, or change the intended focus of the project.

In summary, depending upon the given situation, numerous feasibility issues must be considered when planning a project. This analysis should consider economic, technical, operational, schedule, legal, contractual, and political issues related to the project. In addition to these considerations, project selection by an organization may be influenced by issues beyond those discussed here. For example, projects may be selected for construction given high project costs and high technical risk if the system is viewed as a strategic necessity; that is, a project viewed by the organization as being critical to its survival. Alternatively, projects may be selected because they are deemed to require few resources and have little risk. Projects may also be selected due to the power or persuasiveness of the manager proposing the system. This means that project selection may be influenced by factors beyond those discussed here and beyond items that can be analyzed. Understanding the reality that projects may be selected based on factors beyond analysis, your role as a systems analyst is to provide a thorough examination of the items that can be assessed. Your analysis will ensure that a project review committee has as much information as possible when making project approval decisions. In the next section, we discuss how project plans are typically reviewed.

BUILDING THE BASELINE PROJECT PLAN

All the information collected during project initiation and planning is collected and organized into a document called the Baseline Project Plan. Once the BPP is completed, a formal review of the project can be conducted with project clients and other interested parties. This presentation is called a *walkthrough* and is discussed later in the chapter. The focus of this review is to verify all information and assumptions in the baseline plan before moving ahead with the project. As mentioned above, the project size and organizational standards will dictate the comprehensiveness of the project initiation and planning process as well as the BPP. Yet, most experienced systems builders have found project planning and a clear project plan to be invaluable to project success. An outline of a Baseline Project Plan is provided in Figure 7-10, which shows that it contains four major sections:

1. Introduction

2. System Description

3. Feasibility Assessment

4. Management Issues

The purpose of the *Introduction* is to provide a brief overview of the entire document and outline a recommended course of action for the project. The entire Introduction section is often limited to only a few pages. Although the Introduction section is sequenced as the first section of the BPP, it is often the final section to be written. It is only after performing most of the project planning activities that a clear overview and recommendation can be created. One activity that should be performed initially is the definition of project scope.

When defining scope for the Customer Tracking System within PVF, Jim Woo first needed to gain a clear understanding of the project's objectives. To do this, Jim briefly interviewed Jackie Judson and several of her colleagues to gain a clear idea of their needs. He also spent a few hours reviewing the existing system's functionality, processes, and data use requirements for performing customer tracking activities. These activities provided him with the information needed to define the project scope and to identify possible alternative solutions. Alternative system solutions can relate to different system scopes, platforms for deployment, or approaches to acquiring the system. We elaborate on the idea of alternative solutions, called design strategies, when we discuss the analysis phase of the life cycle. During project initiation and planning, the most crucial element of the design strategy is the system's scope. In sum, a determination of scope will depend on these factors:

- Which organizational units (business functions and divisions) might be affected by or use the proposed system or system change?

- With which current systems might the proposed system need to interact or be consistent, or which current systems might be changed due to a replacement system?

- Who inside and outside the requesting organization (or the organization as a whole) might care about the proposed system?

- What range of potential system capabilities are to be considered?

Figure 7-10
Outline of a Baseline Project Plan

BASELINE PROJECT PLAN REPORT

1.0 *Introduction*
  A. Project Overview—Provides an executive summary that specifies the project's scope, feasibility, justification, resource requirements, and schedules. Additionally, a brief statement of the problem, the environment in which the system is to be implemented, and constraints that affect the project are provided.
  B. Recommendation—Provides a summary of important findings from the planning process and recommendations for subsequent activities.

2.0 *System Description*
  A. Alternatives—Provides a brief presentation of alternative system configurations.
  B. System Description—Provides a description of the selected configuration and a narrative of input information, tasks performed, and resultant information.

3.0 *Feasibility Assessment*
  A. Economic Analysis—Provides an economic justification for the system using cost-benefit analysis.
  B. Technical Analysis—Provides a discussion of relevant technical risk factors and an overall risk rating of the project.
  C. Operational Analysis—Provides an analysis of how the proposed system solves business problems or takes advantage of business opportunities in addition to an assessment of how current day-to-day activities will be changed by the system.
  D. Legal and Contractual Analysis—Provides a description of any legal or contractual risks related to the project (e.g., copyright or nondisclosure issues, data capture or transferring, and so on).
  E. Political Analysis—Provides a description of how key stakeholders within the organization view the proposed system.
  F. Schedules, Timeline, and Resource Analysis—Provides a description of potential timeframe and completion date scenarios using various resource allocation schemes.

4.0 *Management Issues*
  A. Team Configuration and Management—Provides a description of the team member roles and reporting relationships.
  B. Communication Plan—Provides a description of the communication procedures to be followed by management, team members, and the customer.
  C. Project Standards and Procedures—Provides a description of how deliverables will be evaluated and accepted by the customer.
  D. Other Project-Specific Topics—Provides a description of any other relevant issues related to the project uncovered during planning.

The statement of project scope for the Customer Tracking System project is shown in Figure 7-11.

For the Customer Tracking System (CTS), project scope was defined using only textual information. It is not uncommon, however, to define project scope using di-

Figure 7-11
Statement of project scope (Pine Valley Furniture)

Pine Valley Furniture *Statement of Project Scope*	Prepared by: Jim Woo Date: September 18, 1995

General Project Information
> **Project Name:** Customer Tracking System
> **Sponsor:** Jackie Judson, VP Marketing
> **Project Manager:** Jim Woo

Problem/Opportunity Statement:
> Sales growth has out-paced the marketing department's ability to accurately track and forecast customer buying trends. An improved method for performing this process must be found in order to reach company objectives.

Project Objectives:
> To enable the marketing department to accurately track and forecast customer buying patterns in order to better serve customers with the best mix of products. This will also enable PVF to identify the proper application of production and material resources.

Project Description:
> A new information system will be constructed that will collect all customer purchasing activity, support display and reporting of sales information, aggregate data and show trends in order to assist marketing personnel in understanding dynamic market conditions. The project will follow PVF's systems development life cycle.

Business Benefits:
> Improved understanding of customer buying patterns
> Improved utilizaton of marketing and sales personnel
> Improved utilization of production and materials

Project Deliverables:
> Customer tracking system analysis and design
> Customer tracking system programs
> Customer tracking documentation
> Training procedures

Estimated Project Duration:
> 5 months

---

agrams such as data flow diagrams and entity-relationship models. For example, Figure 7-12 shows a context-level data flow diagram used to define system scope for PVF's Purchasing Fulfillment System described in Chapter 2. The other items in the Introduction section of the BPP are simply executive summaries of the other sections of the document.

The second section of the BPP is the *System Description* where you outline possible alternative solutions in addition to the one deemed most appropriate for the

Figure 7-12

Context-level data flow diagram showing project scope for Purchasing Fulfillment System (Pine Valley Furniture)

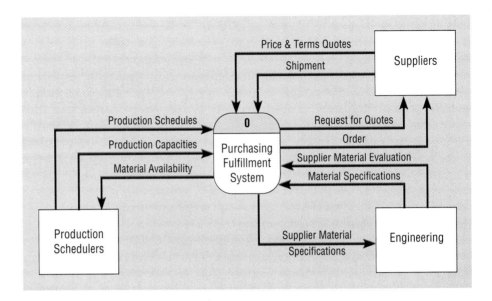

given situation. Note that this description is at a very high level, mostly narrative in form. For example, alternatives may be stated as simply as this:

1. Mainframe with central database

2. Distributed with decentralized databases

3. Batch data input with on-line retrieval

4. Purchasing of a pre-written package

If the project is approved for construction or purchase, you will need to collect and structure information in a more detailed and rigorous manner during the analysis phase and evaluate in greater depth these and other alternative directions for the system. At this point in the project, your objective is only to identify the most obvious alternative solutions.

When Jim and Jackie were considering system alternatives for the CTS, they focused on two primary issues. First, they discussed how the system would be acquired and considered three options: *purchase* the system if one could be found that met PVF's needs, *outsource* the development of the system to an outside organization, or *build* the system within PVF. The second issue focused on defining the comprehensiveness of the system's functionality. To complete this task, Jim asked Jackie to write a series of statements listing the types of tasks that she envisioned marketing personnel would be able to accomplish when using the CTS. This list of statements became the basis of the system description and was instrumental in helping them make their acquisition decision. After considering the unique needs of the marketing group, both decided that the best decision was to build the system within PVF.

In the third section, *Feasibility Assessment*, issues related to project costs and benefits, technical difficulties, and other such concerns are outlined. This is also the section where high-level project schedules are specified using PERT and Gantt charts. Recall from Chapter 4 that this process is referred to as a work breakdown structure. During project initiation and planning, task and activity estimates are not generally detailed. An accurate work breakdown can only be done for the next one or two life cycle activities. After defining the primary tasks for the project, an estimate of the resource requirements can be made. As with defining tasks and activities, this activity is primarily concerned with gaining rough estimates of the human resource requirements, since people are the most expensive resource element. Once you define the major tasks and resource requirements, a preliminary schedule can be developed. Defining an acceptable schedule may require that you find additional or different resources or that the scope of the project be changed. The greatest amount of project planning effort is typically expended on these Feasibility Assessment activities.

The final section, *Management Issues,* outlines a number of managerial concerns related to the project. This will be a very short section if the proposed project is going to be conducted exactly as prescribed by the organization's standard systems development methodology. Most projects, however, have some unique characteristics that require minor to major deviation from the standard methodology. In the Team Configuration and Management portion, you identify the types of people to work on the project, who will be responsible for which tasks, and how work will be supervised and reviewed. In the Communications Plan portion, you explain how the user will be kept informed about project progress (such as periodic review meetings or even a newsletter) and which mechanisms will be used to foster sharing of ideas among team members, such as some form of computer-based conference facility. An example of the type of information contained in the Project Standards and Procedures portion would be procedures for submitting and approving project change requests and any other issues deemed important for the project's success.

You should now have a feel for how a BPP is constructed and the types of information it contains. Its creation is not meant to be a project in and of itself but rather a step in the overall systems development process. Developing the BPP has two primary objectives. First, it helps to assure that the customer and development group share a common understanding of the project. Second, it helps to provide the sponsoring organization with a clear idea of the scope, benefits, and duration of the project.

REVIEWING THE BASELINE PROJECT PLAN

Before the next phase of the SDLC can begin, the users, management, and development group must review the Baseline Project Plan in order to verify that it makes sense. This review takes place before the BPP is submitted or presented to some project approval body, such as an IS steering committee or the person who must fund the project. The objective of this review is to assure that the proposed system conforms to organizational standards and to make sure that all relevant parties understand and agree with the information contained in the Baseline Project Plan. A common method for performing this review (as well as reviews during subsequent life cycle phases) is called a structured **walkthrough.** Walkthroughs are peer group reviews of any product created during the systems development process and are widely used by professional development organizations. Experience has shown that walkthroughs are a very effective way to ensure the quality of an information system and have become a common day-to-day activity for many system analysts.

Walkthrough: A peer group review of any product created during the systems development process. Also called structured walkthrough.

Most walkthroughs are not rigidly formal or exceeding long in duration. It is important, however, that a specific agenda be established for the walkthrough so that all attendees understand what is to be covered and the expected completion time. At walkthrough meetings, there is a need to have individuals play specific roles. These roles are as follows (Yourdon, 1989):

- *Coordinator.* This person plans the meeting and facilitates a smooth meeting process. This person may be the project leader or a lead analyst responsible for the current life cycle step.

- *Presenter.* This person describes the work product to the group. The presenter is usually an analyst who has done all or some of the work being presented.

- *User.* This person (or group) makes sure that the work product meets the needs of the project's customers. This user would usually be someone not on the project team.

- *Secretary.* This person takes notes and records decisions or recommendations made by the group. This may be a clerk assigned to the project team or it may be one of the analysts on the team.

- *Standards bearer.* The role of this person is to ensure that the work product adheres to organizational technical standards. Many larger organizations have staff groups within the unit responsible for establishing standard procedures, methods, and documentation formats. These standards bearers validate the work so that it can be used by others in the development organization.

- *Maintenance oracle.* This person reviews the work product in terms of future maintenance activities. The goal is to make the system and its documentation easy to maintain.

After Jim and Jackie completed their BPP for the Customer Tracking System, Jim approached his boss and requested that a walkthrough meeting be scheduled and that a walkthrough coordinator be assigned to the project. PVF assists the coordinator by providing a Walkthrough Review Form, shown in Figure 7-13. Using this form, the coordinator can more easily make sure that a qualified individual is assigned to each walkthrough role, that each member has been given a copy of the review materials, and that each member knows the agenda, date, time, and location of the meeting. At the meeting, Jim presented the BPP and Jackie added comments from a user perspective. Once the walkthrough presentation was completed, the coordinator polled each representative for his or her recommendation concerning the work product. The results of this voting may result in validation of the work product, validation pending changes suggested during the meeting, or a suggestion that the work product requires major revision before being presented for approval. In this latter case, substantial changes to the work product are usually requested after which another walkthrough must be scheduled before the project can be proposed to the Systems Priority Board (steering committee). In the case of the Customer Tracking System, the BPP was supported by the walkthrough panel pending some minor changes to the duration estimates of the schedule. These suggested changes were recorded by the secretary on a Walkthrough Action List (see Figure 7-14) and given to Jim to incorporate into a final version of the baseline plan presented to the steering committee.

As suggested above, walkthrough meetings are a common occurrence in most systems development groups and can be used for more activities than reviewing the BPP, including the following:

- System specifications

- Logical and physical designs

- Code or program segments

- Test procedures and results

- Manuals and documentation

One of the key advantages to using a structured review process is to ensure that formal review points occur during the project. At each subsequent phase of the

Figure 7-13
Walkthrough review form (Pine Valley Furniture)

Pine Valley Furniture
Walkthrough Review Form

Session Coordinator:

Project/Segment:

Coordinator's checklist:

1. Confirmation with producer(s) that material is ready and stable: _____
2. Issue invitations, assign responsibilities, distribute materials: [ ] Y   [ ] N
3. Set date, time, and location for meeting:

   Date: ___ / ___ / ___   Time: _____ a.m. / p.m. (circle one)

   Location: _____

Responsibilities	*Participants*	*Can Attend*	*Received Materials*
Coordinator	_____	[ ] Y [ ] N	[ ] Y [ ] N
Presenter	_____	[ ] Y [ ] N	[ ] Y [ ] N
User	_____	[ ] Y [ ] N	[ ] Y [ ] N
Secretary	_____	[ ] Y [ ] N	[ ] Y [ ] N
Standards	_____	[ ] Y [ ] N	[ ] Y [ ] N
Maintenance	_____	[ ] Y [ ] N	[ ] Y [ ] N

Agenda:
_____ 1. All participants agree to follow PVF's Rules of a Walkthrough
_____ 2. New material: walkthrough of all material
_____ 3. Old material: item-by-item checkoff of previous action list
_____ 4. Creation of new action list (contribution by each participant)
_____ 5. Group decision (see below)
_____ 6. Deliver copy of this form to the project control manager

Group Decision:
_____ Accept product as-is
_____ Revise (no further walkthrough)
_____ Review and schedule another walkthrough

Signatures		

Figure 7-14
Walkthrough action list (Pine Valley Furniture)

	Pine Valley Furniture *Walkthrough Action List*
Session Coordinator:	
Project/Segment:	
Date and Time of Walkthrough: Date: ___ / ___ / ___ Time: _____ a.m. / p.m. (circle one)	
Fixed (✓)	*Issues raised in review:*

project, a formal review should be conducted (and shown on the project schedule) to make sure that all aspects of the project are satisfactorily accomplished before assigning additional resources to the project. This conservative approach of reviewing each major project activity with continuation contingent on successful completion of the prior phase is called *incremental commitment.* It is much easier to stop or redirect a project at any point when using this approach.

SUMMARY

The project initiation and planning phase is a critical activity in the life of a project. It is at this point that projects are accepted for development, rejected as infeasible, or redirected. The objective of this process is to transform a vague system request into a tangible system description clearly outlining the objectives, feasibility issues, benefits, costs, and time schedules for the project.

Project initiation includes forming the project initiation team, establishing customer relationships, developing a plan to get the project started, setting project management procedures, and creating an overall project management environment. A key activity in project planning is the assessment of numerous feasibility issues associated with the project. Feasibilities include economic, technical, operational, schedule, legal and contractual, and political. These issues are influenced by the project size, the type of system proposed, and the collective experience of the development group and potential customers of the system. High project costs and risks are not necessarily bad; rather it is more important that the organization understands the costs and risks associated with a project and with the portfolio of active projects before proceeding.

After completing all analyses, a Baseline Project Plan can be created. A BPP includes a high-level description of the proposed system or system change, an outline of the various feasibilities, and an overview of management issues specific to the project. Before the development of an information system can begin, the users, management, and development group must review and agree on this specification. The focus of this walkthrough review is to assess the merits of the project and to assure that the project, if accepted for development, conforms to organizational standards and goals. An objective of this process is to also make sure that all relevant parties understand and agree with the information contained in the plan before subsequent development activities begin.

Project initiation and planning is a challenging and time-consuming activity that requires active involvement from many organizational participants. The eventual success of development projects, and the MIS function in general, hinges on the effective use of disciplined, rational approaches such as the techniques outlined in this chapter. In subsequent chapters you will be exposed to numerous other tools that will equip you to become an effective designer and developer of information systems.

C H A P T E R   R E V I E W

KEY TERMS

Baseline Project Plan (BPP)

Discount rate

Economic feasibility

Intangible benefit

Intangible cost

Legal and contractual
  feasibility

One-time cost

Operational feasibility

Present value

Political feasibility

Recurring cost

Schedule feasibility

Statement of Work (SOW)

Tangible benefit

Tangible cost

Technical feasibility

Walkthrough

REVIEW QUESTIONS

1. Define each of the following terms:
   a. Baseline Project Plan
   b. discount rate
   c. one-time cost
   d. recurring cost
   e. walkthrough
   f. Statement of Work

2. Contrast the following terms:
   a. break-even analysis, present value, net present value, return on investment
   b. economic feasibility, legal and contractual feasibility, operational feasibility, political feasibility, schedule feasibility
   c. intangible benefit, tangible benefit
   d. intangible cost, tangible cost

3. List and describe the steps in the project initiation and planning process.

4. What is contained in a Baseline Project Plan? Is the content and format of all baseline plans the same? Why or why not?

5. Describe three commonly used methods for performing economic cost-benefit analysis.

6. List and discuss the different types of project feasibility factors. Is any factor most important? Why or why not?

7. What are the potential consequences of not assessing the technical risks associated with an information systems development project?

8. In what ways could you identify an IS project that was riskier than another?

9. What are the types or categories of benefits from an IS project?

10. What intangible benefits might an organization obtain from the development of an IS?

11. Describe the concept of the time value of money. How does the discount rate affect the value of $1 today versus one year from today?

12. Describe the structured walkthrough process. What roles need to be performed during a walkthrough?

PROBLEMS AND EXERCISES

1. Match the following terms to the appropriate definitions.

   _____ Baseline Project Plan

   _____ economic feasibility

   _____ tangible benefits

   _____ intangible benefits

   _____ one-time cost

   _____ recurring cost

   _____ technical feasibility

   _____ operational feasibility

   _____ schedule feasibility

   _____ legal and contractual feasibility

   _____ political feasibility

   a. benefit derived from the creation of an information system that can generally be easily measured in dollars and with certainty

   b. process of assessing the potential time-frame and completion dates for all major activities within a project

   c. cost associated with project start-up and development, or system start-up

   d. process of assessing the development organization's ability to construct a proposed system

   e. process of identifying the financial benefits and costs associated with a development project

   f. process of assessing the degree to which a proposed system solves business problems or takes advantage of business opportunities

   g. cost resulting from the ongoing evolution and use of system

   h. major outcome and deliverable from the project initiation and planning phase and contains the best estimate of a project's scope, benefits, costs, risks, and resource requirements

   i. process of assessing potential legal and contractual ramifications due to the construction of a system

   j. benefit derived from the creation of an information system that cannot be easily measured in dollars or with certainty

   k. process of evaluating how key stakeholders within the organization view the proposed system

2. Consider the purchase of a PC and laser printer for use at your home and assess risk for this project using the project risk assessment factors in Table 7-6.

3. Consider your use of a PC at either home or work and list tangible benefits from an information system. Based on this list, does your use of a PC seem to be beneficial? Why or why not? Now do the same using Table 7-3, the intangible benefits from an information system. Does this analysis support or contradict your previous analysis? Based on both analyses, does your use of a PC seem to be beneficial?

4. Consider, as an example, buying a network of PCs for a department at your workplace or, alternatively, consider outfitting a laboratory of PCs for students at a university. For your example, estimate the costs outlined in Table 7-4, one-time and recurring costs.

5. Assuming monetary benefits of an information system at $85,000 per year, one-time costs of $75,000, recurring costs of $35,000 per year, a discount rate of 12%, and

a five-year time horizon, calculate the Net Present Value of these costs and benefits of an information system. Also calculate the overall Return on Investment of the project and then present a Break-Even Analysis. At what point does break-even occur?

6. Choose as an example one of the information systems you described in Problem and Exercise 4 above, either buying a network of PCs for a department at your workplace or outfitting a laboratory of PCs for students at a university. Estimate the costs and benefits for your system and calculate the Net Present Value, Return on Investment and present a Break-Even Analysis. Assume a discount rate of 12% and a five-year time horizon.

7. Use the outline for the Baseline Project Plan provided in Figure 7-10 to present the system specifications for the information system you chose for Problems and Exercises 4 and 6 above.

8. For the system you chose for Problems and Exercises 4 and 6, complete section 1.0, A, Project Overview, of the Baseline Project Plan Report. How important is it that this initial section of the Baseline Project Plan Report be done well? What could go wrong if this section is incomplete or incorrect?

9. For the system you chose for Problems and Exercises 4 and 6, complete section 2.0, A, Alternatives, of the Baseline Project Plan Report. Without conducting a full-blown feasibility analysis, what is your gut feeling as to the feasibility of this system?

10. For the system you chose for Problems and Exercises 4 and 6, complete section 3.0, A–F, Feasibility Analysis, of the Baseline Project Plan Report. How does this feasibility analysis compare with your gut feeling from the previous question? What might go wrong if you rely on your gut feeling in determining system feasibility?

11. For the system you chose for Problems and Exercises 4 and 6, complete section 4.0, A–C, Management Issues, of the Baseline Project Plan Report. Why might people sometimes feel that these additional steps in the project plan are a waste of time? What would you say to them to convince them that these steps are important?

FIELD EXERCISES

1. Describe several projects you are involved in or plan to undertake, whether they be related to your education or to your professional or personal life. Some examples are purchasing a new vehicle, learning a new language, renovating a home, and so on. For each, sketch out a Baseline Project Plan like that outlined in Figure 7-10. Focus your efforts on item number 1.0 (Introduction) and 2.0 (System Description).

2. For each project from the previous question, assess the feasibility in terms of economic, operational, technical, schedule, legal and contractual, as well as political aspects.

3. Network with a contact you have in some organization that conducts projects (these might be IS projects, but they could be construction, product development, research and development, or any type of project). Interview a project manager and find out what type of Baseline Project Plan is constructed. For a typical project, in what ways are baseline plans modified during the life of a project? Why are plans modified after the project begins? What does this tell you about project planning?

4. Through a contact you have in some organization that uses packaged software, interview an IS manager responsible for systems in an area that uses packaged application software. What contractual limitations, if any, has the organization encountered with using the package? If possible, review the license agreement for the software and make a list of all the restrictions placed on a user of this software.

5. Choose an organization that you are familiar with and determine what is done to initiate information systems projects. Who is responsible for performing this? Is this process formal or informal? Does this appear to be a top-down or bottom-up process? How could this process be improved?

6. Find an organization that does not use Baseline Project Plans for their IS projects. Why doesn't this organization use this method? What are the advantages and disadvantages of not using this method? What benefits could be gained from implementing the use of Baseline Project Plans? What barriers are there to implementing this method?

REFERENCES

Cash, J. I., F. W. McFarlan, J. L. McKenney, and L. M. Applegate. 1992. *Corporate Information Systems Management.* 3rd ed. Boston, MA: Irwin.

King, J. L. and E. Schrems. 1978. "Cost Benefit Analysis in Information Systems Development and Operation. *ACM Computing Surveys* 10 (1): 19–34.

Parker, M. M. and R. J. Benson. 1988. *Information Economics,* Englewood Cliffs, NJ: Prentice-Hall.

Pressman, R. S. 1992. *Software Engineering.* 3rd ed. New York, NY: McGraw-Hill.

Yourdon, E. 1989. *Structured Walkthroughs,* 4th ed. Englewood Cliffs, NJ: Prentice-Hall.

Initiating and Planning the Customer Activity Tracking System

INTRODUCTION

Having gained approval from the Systems Priority Board to start the Customer Activity Tracking System (CATS) project, Nancy Chen and Jordan Pippen begin the second phase of the systems development life cycle—project initiation and planning. Creation of a Baseline Project Plan (BPP) is the primary outcome and deliverable during this phase. Once the BPP is created, it is then reviewed by the Systems Priority Board in a formal approval session. If the project is approved, it can then move from the project initiation and planning phase to the third SDLC phase—analysis. The objective of this case is to describe the initiation and planning process that occurred at BEC with the Customer Activity Tracking System (CATS).

INITIATING AND PLANNING THE PROJECT

After gaining project selection approval, Jordan and Nancy met numerous times over several weeks to initiate and plan the CATS project. Since Jordan and Nancy had not asked the Systems Priority Board for any additional personnel for the project initiation team, Karen Gardner continued to ensure that Jordan's workload permitted working on initiating and planning CATS. Jordan and/or Nancy held meetings with several BEC managers and IS staff members in order to collect information about data availability and current BEC business processes. Nancy composed and distributed a brief memo to everyone in U.S. operations and to a few key people in other units (for example, international operations) announcing the project and requesting cooperation during the initiation phase.

During initiation and planning, Nancy took the lead in developing a more detailed requirements statement (see Figure 1) and business case that included potential costs and benefits as well as risks (see Figure 2). The system requirements are still described in

words, and details are purposefully vague. These details will be studied during the analysis phase, but Figure 1 will give the analysis team direction for their work. The potential costs and benefits are listed without being quantified, although costs and tangible benefits should be quantified during later systems development phases, probably during the analysis phase. Where possible, benefits are related to BEC's Blueprint for the Decade objectives, listed when the CATS project was requested. The only anticipated element of risk with CATS is that the system requirements are somewhat ill-defined, since customer profiling and target marketing are new concepts at BEC. Problem issues concerning technical, operational, schedule, legal/contractual, or political feasibility are not expected.

In addition to developing a possible schedule for the project (see Figures 3 and 4), Jordan and Nancy jointly worked out a request for resources and an initial project time line. Figure 3 shows an estimate of approximately six months' calendar time for subsequent SDLC phases. The installation is the longest phase for this project in order to handle the rollout of CATS to all stores, including those outside the U.S. operations area. Jordan was instrumental in developing this schedule, in part by studying prior similar project schedules. Figure 4 breaks out the analysis phase activities into more detail. Note that there are activities which respond to the concerns raised by the SPB during project identification and selection about data exchanges among CATS and about existing systems and data privacy. The exchange of data between CATS and existing systems will, however, be addressed in most subsequent phases.

Resource requirements for the thirty-day analysis phase are as follows:

- Two full-time systems analysts and a full-time market researcher from Nancy Chen's staff

- 5 days of assistance from a data analyst from the data administration staff

1. Profile Each Customer
   A profile for each customer will be created based upon prior purchasing and rental activity of both music and videos. Profiles would be updated on a periodic basis, using both past and recent transaction data. Profiles will be ordered based upon rental and purchase volume.

2. Profile New Product Offerings
   Each new product being released by music and video distributors will be profiled prior to its release. For example, a profile for a video release could include information such as rating, movie type (e.g., action, thriller, comedy), actors, and so on.

3. Profile Matching
   An automated system would then match product profiles to customers most likely to rent or purchase a newly released item. Matched profiles are then sorted based on the strength of the association between the customer-product profile.

4. Profile Selection
   Based upon the strength of the association of the profile matches, profiles could be selected that represent those most likely to lead to additional sales. BEC management could use interactive decision support tools to segment profiles by geographic location, customer or product characteristic, in addition to association strength.

5. Produce Product Announcement Card
   An automated system will produce post cards addressed to selected customers announcing the new product releases and the information needed to make a reservation.

6. Track Customer Reservations
   A customer service agent at BEC would be available for customers to contact using an 800 number to make product reservations. The agent would record reservation information into an automated system that would update BEC's inventory control system.

7. Delivery of Reserved Product to Customer
   Products earmarked for profiled customers will be shipped to local BEC stores with a report showing how to notify customers of product arrival.

Figure 1
Statement of system requirement for the CATS project

POSSIBLE COSTS	

One-Time
  System design
  System development
  Changing existing systems
	Ongoing
	  Increased staff
	  System maintenance
	  Operating costs

POSSIBLE BENEFITS	

Tangible
  Increased sales and rentals
    (increased market share and
    profitability)
	Intangible
	  Improved planning
	  New information
	  Better asset utilization
	  Better resource control
	  Better customer service
	    (appeal to diverse customers)
	  Recognition as industry leader

POSSIBLE RISKS	
Lack of experience with target marketing and customer profiling	

Figure 2
Possible costs, benefits, and risks for the CATS project

Figure 3
Gantt chart excerpt for the CATS project

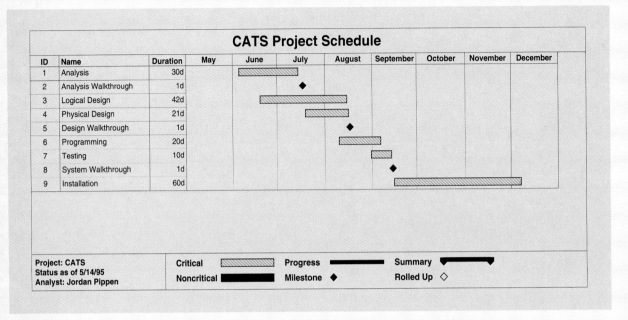

Figure 4
Analysis phase schedule for CATS project

(a) Analysis phase Gantt chart

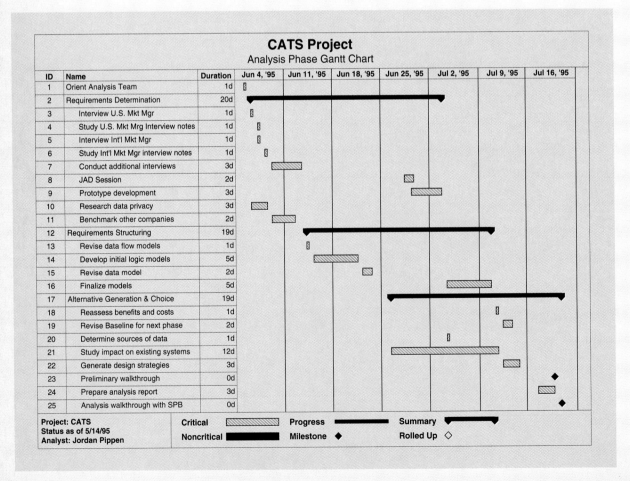

- 10 hours of research time from the office of the legal counsel

- An unspecified amount of time from a variety of managers for gathering system requirements in interviews, during a JAD session, and while reviewing system prototypes

- Unlimited access to documentation on existing systems

- $500 travel budget to visit a nearby mail-order catalog firm that has been experimenting with customer profiling and target marketing

In addition, Nancy will spend about 10 percent of her time on this project and asks that Jordan be assigned full-time as project manager for CATS. As a senior systems analyst, Jordan is quite familiar with standard systems development procedures at BEC, and she recommends that the CATS project follow these procedures

closely. This includes using the Antares Object Star and other CASE tools throughout the project. In total, the resource requirements for the analysis phase represent approximately a $15,000 investment, exclusive of managerial time for interviews, JADs, and other requirements determination activities.

Since the proposed system will tie into existing applications and access several databases, Jordan felt it was necessary, when putting together the project schedule, to develop high-level process and data models for the proposed system that begin to show possible interfaces between CATS and other systems. Figure 5 shows the scope of CATS via a context-level data flow diagram (DFD). The sources/sinks (or entities outside CATS) that will interface with the system are BEC customers and stores. The diagram shows that a customer will receive a notice about a particular product and will then notify BEC if he or she wants to reserve the product for purchase or rental. Local stores will

(b) Analysis phase PERT chart

CATS Project
Analysis Phase PERT Schedule

Project: CATS
Date: 5/14/95 8:00am
Analyst: Jordan Pippen

Figure 5
Context-level data flow diagram for the CATS project

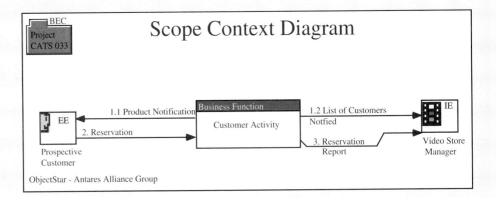

receive lists of customers who have been profiled and contacted. If a customer contacts BEC to reserve a product, this reservation is logged into a reservation database and a report containing all customer reservations is forwarded to each store. Whether or not a customer actually rents or purchases a reserved item is handled by feeds from other existing BEC systems

and is not shown on this diagram, although the analysis phase project schedule (see Figure 4) includes activities to investigate this area.

A level-0 data flow diagram view of CATS is shown in Figure 6. This figure shows that six high-level processes are initially identified within the proposed system: Profiling Customers, Profiling Products, Profile

Figure 6
Level-0 data flow diagram for the CATS project

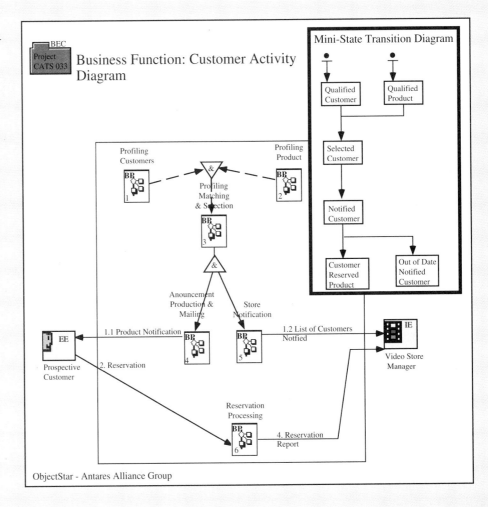

Matching & Selection, Announcement Production & Mailing, Store Notification, and Reservation Processing. Data stores are encapsulated within processes for the data flow notation used by ObjectStar. Once it is known exactly what data are needed from these data stores, the data analyst will identify which databases from the existing systems will map to these data stores. The data analyst will also decide whether changes need to be made in other systems to capture new data that CATS will need; the data and systems analysts will also have to identify the changes that must be made in other systems to accommodate recording sales and rentals of reserved products.

Jordan also developed another view of the scope of CATS via a high-level (or enterprise) data model using entity-relationship (E-R) diagramming (Figure 7). The entities identified included Stores, Customers, Customer Profiles, Customer Transactions, Products, Product Profiles, and Product Reservations. Again, it is not known at this time what data will be kept about each of these entities or which of these data will come from existing databases. The relationship between Stores and Customers, for example, implies that "each Store has one-to-many Customer(s)" and "each Customer is assigned to only one Store" (at least at a particular point in time). The diagram also reflects the fact that Customer(s) have transactions and that a Customer Profile, a combination of customer descriptors,

applies to one or many Customers and a Customer may have several Customer Profiles. A Product Reservation contains information that is stored after a given Customer contacts BEC about reserving a particular Product.

After several weeks of hard work, Nancy and Jordan completed the Baseline Project Plan (BEC's format for a Baseline Project Plan is very similar to the template presented in Figure 10 in Chapter 7). This plan was submitted to BEC's System Priority Board for review. A formal review of the plan was then scheduled for the following Friday morning.

REVIEWING THE BASELINE PROJECT PLAN

At the review meeting of the Baseline Project Plan, both Nancy and Jordan presented information to the SPB. Their presentation was an abridged version of their project plan. An agenda for their presentation is shown in Figure 8.

As with most walkthroughs, Nancy, Jordan, and the SPB had different roles to play in the meeting. Nancy focused on presenting an overview of issues surrounding what CATS was and how CATS would address a key BEC objective. In other words, as the system client and requester, Nancy focused primar-

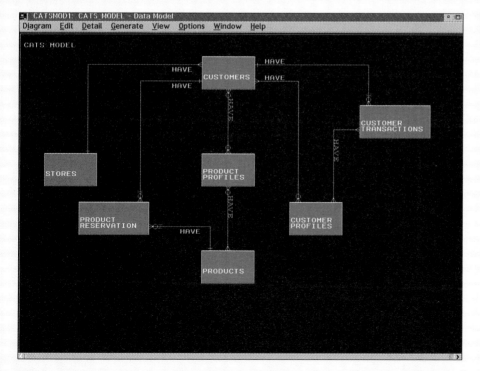

Figure 7

Entity-relationship diagram for the CATS project

Figure 8
Agenda for the Baseline
Project Plan walkthrough

TOPIC	PRESENTER
Introduction to the CATS Project	Nancy (5 minutes)
Functional Description and Scope	Nancy & Jordan (15 minutes)
Schedule	Jordan (5 minutes)
Feasibility Analysis and Risks	Jordan (10 minutes)
Resources Requirements and Costs	Jordan (10 minutes)
Question and Answer Period	Nancy & Jordan (15 minutes)

ily on the business reasons why CATS needed to be constructed. Jordan, on the other hand, focused on explaining the relationship between CATS and the existing information systems portfolio, with particular emphasis on feasibility issues such as system compatibility, complexity, costs, and schedules. The SPB's role was to assess the extent to which CATS was a unique and valuable system that would help BEC reach its strategic objective of providing superior customer service through the use of technology. In other words, the SPB's responsibility is to compare CATS to other projects and, since limited resources preclude all projects from being developed, to select or reject it for continued development.

Both Nancy and Jordan carefully prepared overhead transparencies to assist in making their presentation. They also practiced their presentation several times before the meeting. A common practice at BEC is for presenters to provide copies of their slides as hand-outs to all SPB members. During the question and answer session, Nancy was asked for a rough estimate of the possible benefits. She reported that the *Wall Street Journal* and other articles she had read (which had been the seed for the CATS request) had mentioned a range of 3 to 10 percent increase in sales, which would mean $400,000 to $1,000,000 to BEC at current annual sales and rental levels. Nancy felt that being the first home entertainment company to provide this service could yield returns at the high end of this range—at least until the competition could catch up. The halo effect from better customer service could be longer lasting. Further, the intangible benefit of "better resource control" might be significant, since the reservation system could lead to a significant reduction in inventory levels and would allow more central planning of inventory.

An SPB member then asked Jordan about the potential costs. Jordan responded that, other than the initial $15,000 investment for the analysis phase, she was rather uncertain about subsequent costs. CATS represented a new type of system for BEC (thus the risk statement in Figure 2). Depending on the determination of exactly what a customer profile is and the specifics of interfaces between CATS and existing systems, the cost to design and (re)program CATS and existing systems was uncertain. Jordan saw no need for new hardware or software to support CATS. Although a decision about the development platform for CATS would be made in analysis, Jordan thought that the programming languages, data analysis tools, and database management systems used at BEC would be sufficient to develop CATS. The analysis phase, however, would clarify all of these issues in the recommended design strategy.

After the presentation, Nancy took Jordan to lunch as a small gesture for all the hard work she had done. This lunch turned out to be a celebration, as the CATS project was approved by the SPB to move into the analysis phase. Both Jordan and Nancy knew that this could be the beginning of a long relationship and that much more work was ahead of them before BEC would realize any benefits from the development of this system.

SUMMARY

The project initiation and planning phase of the CATS project demonstrates the need to carefully develop a Baseline Project Plan. CATS has great potential benefit in increased sales and rentals and many potential intangible benefits. However both one-time and ongoing costs as well as the risks of implementing a new system make the success of such projects dubious. Since the motivation for the CATS project was linked to several important BEC objectives, the phased development cycle allows for frequent review of the project, and there is close cooperation between IS and the client organization, the Systems Priority Board approved funding for the analysis phase and is eager to see the results of the analysis phase.

P A R T

IV

Analysis

McHenry County

Using Automated Tools
for Requirements Analysis

McHenry County, northeast of Chicago, has experienced an annual population growth rate of approximately 18 percent. To accommodate this growth, Carl Pohrte, information services administrator for McHenry County, is leading the development of a new data communications infrastructure, part of an ongoing four-year project to completely integrate all county offices, as well as the offices of local cities and municipalities. Carl and others at McHenry County share a vision of an electronically "integrated county."

In addition to the data communications infrastructure, several mission-critical information systems for the county—including a land use system, an integrated finance system, and an integrated justice system, each with several subsystems—are being developed. Currently, Pohrte and his staff are building a data model and a process model for each system, in addition to building the data communication infrastructure which will weave all these systems together across the county.

To develop this data communications infrastructure and the necessary mission-critical systems, Pohrte needed an effective, easy-to-use graphical analysis and modeling tool for client/server application development. He and others at McHenry County decided to use Visible Analyst Workbench (VAW) from Visible Systems, Inc. They chose VAW for its performance, ease of use, low cost, and compatibility with other tools.

The county uses VAW in conjunction with Joint Application Development (JAD) sessions, which are facilitated, structured, dynamic interviews with systems personnel and anywhere from two to sixteen users. The JAD sessions are used to develop overall context diagrams and their subsequent decomposed diagrams and related models. Using VAW, the staff generate data flow diagrams (DFDs) and entity-relationship diagrams (ERDs) to surface all the relevant entities, information, information flows, data stores, events, and their interrelationships.

Interestingly, they have found that it is better to print out a diagram on paper rather than having users focus their attention on a diagram on the computer screen. Pohrte has found that users are more comfortable

critiquing and changing something on paper than on the computer screen. A diagram on the computer monitor seems to be more of a finished product that they do not want to change, even if it is inadequate.

The initial JAD session relies mainly on a facilitator and scribes to develop the overall scope of the business area under study. The "Business Area Analysis" may give rise to one or more systems, but the primary goal is to develop a sense of what information is critical to the operation of the business area. The goal is to identify any external agents (e.g., vendors, governmental agencies, citizens), events, transactions, and data flows within the business area.

After the initial JAD session, VAW is used to capture the diagrams and definitions. The VAW repository is also used to coordinate the big picture, as recorded in the context diagram and the subordinate images in the DFDs that portray the events. All subsequent JAD sessions begin by reviewing the diagrams and text that have been recorded in VAW by the analysts. Users and analysts continue this collaborative, iterative process, moving from event to event. Essentially, they move from the big picture to several smaller pictures.

For logic modeling, they develop an English-like script in AllClear, from Clear Software, Inc. This script is used to generate logic flow charts for the business rules involved in converting input to output flows in the event-level data flow diagrams. The AllClear scripts are actually stored in the VAW repository.

The data base administrator supervises a team of data modelers who build ERDs for each detailed event. VAW has the capability of synthesizing all the event-level ERDs into a single global ERD. VAW can also perform various analyses of the quality of the models, such as key analysis which identifies problems with primary keys. And VAW also ensures proper use of the same entity across event boundaries.

The result of this development work is a set of integrated overall context diagrams with many underlying event diagrams. The repository provides mechanisms to expedite the task of balancing data flows and external agents as they appear in the context diagram and at the event data flow diagram level.

When the infrastructure and systems for the "integrated county" are completed, county residents will be able to dial various offices from their telephone or PC and check on traffic violations or taxes as well as pay fees automatically. The new systems will enable residents of McHenry County to conduct their business quickly and easily, just as McHenry County's use of a CASE tool quickly and easily facilitated the design of the underlying systems.

An Overview of Part IV:

Analysis

Analysis is the first systems development life cycle phase where you begin to understand, in depth, the needs for system changes. Systems analysis involves a substantial amount of effort and cost and is therefore undertaken only after management has decided that the systems development project under consideration has merit and should be pursued through this phase. The project initiation and planning phase provides the basis for a go-ahead decision for analysis and the Baseline Project Plan provides the structure for conducting the analysis phase. The analysis team should not take the analysis process for granted or attempt to speed through it. Most observers today would agree that many of the errors in developed systems are directly traceable to inadequate efforts in the analysis and design phases of the life cycle. As analysis is a large and involved process, we divide it into three main activities to make the overall process easier to understand. As depicted in Figure 1, analysis includes the following steps:

- *Requirements determination.* This is primarily a fact–finding activity.

- *Requirements structuring.* This activity creates a thorough and clear description of current business operations and new information processing services. Requirements structuring has three sub–activities that concentrate on structuring different views or dimensions of the information system.

- *Alternative generation and selection.* This process results in a choice among alternative strategies for subsequent systems design.

The purpose of analysis is to determine what information and information processing services are needed to support selected objectives and functions of the organization; consequently, analysis is fundamentally an intelligence activity in which analysts capture and structure information. Gathering this information about both the current and the replacement systems is called requirements determination, the subject of Chapter 8. The fact-finding techniques in Chapter 8 are used to learn about the current system (if one exists), the organization which the new or replacement system will support, and user requirements or expectations for the replacement system. Appendix B elaborates on one requirements determination method—Rapid Application Development—which can supplement and sometimes replace the other methods discussed in Chapter 8.

During requirements determination, the systems development team attempts to discover important information about how employees perform and will need to

Figure 1

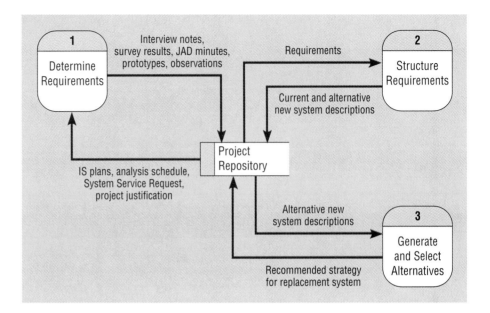

perform their jobs in order to meet future business conditions. Included in this information will be answers to the following questions:

- How does the current system function? Is the system manual or automated?
- What data are necessary for proper functioning of the supported business area?
- What kinds of reports are generated?
- How do people use the system to perform their work?

A study of current operations gives insights on system requirements, grounds analysis in actual information system activity, can be the basis for discovering incremental improvements, and provides necessary information for subsequent steps involved in converting from the current to replacement system. Moreover, analysts must focus on the following additional questions:

- How should a new or replacement system function?
- What data would be needed for it to operate smoothly?
- What kinds of reports would it need to generate?
- How would a new system alter employees' jobs?
- What new or improved information services are needed to support the future organizational goals, objectives, strategies, and functions?

After Chapter 8, the Broadway Entertainment Company (BEC) running case shows how the results of an interview can be used to better understand the requirements for a new system and how different requirements determination techniques can be used in combination to gain a thorough understanding of requirements.

Once the information about current operations and the requirements for a replacement system have been collected, this intelligence must somehow be organized in order to be useful during analysis and, subsequently, to a design team. Organizing, or structuring, system requirements is the second major activity of the analysis phase. The requirements structuring techniques described in Chapters 9–11 are used to unambiguously describe and structure the current system and the alternatives for the replacement system. These techniques produce diagrams and descriptions (models) that can be analyzed to show deficiencies, inefficiencies, missing elements, and illogical components of the current business operation and information systems. Along with user requirements, they are used to determine the strategy for the replacement system.

The results of the requirements determination task during analysis can be structured according to three essential views of the current and replacement information systems:

- *Process*—the sequence of data movement and handling operations within the system

- *Logic and timing*—rules by which data are transformed and manipulated and an indication of what triggers data transformation

- *Data*—the inherent structure of data independent of how or when it is processed

The goal is to capture as complete a specification of the required system as is possible. Ideally, the specification would be so complete that, along with the necessary steps in the design phase, the replacement or new system could be generated without human programming. Today, 100 percent generation from functional specifications is not possible, although reports of 70–90 percent code generation have appeared. Generating computer display and report programs is now fairly common once a data model, as well as display and report layouts, are specified. No matter what percentage of code is automatically generated, the structured specifications provide a clear picture for later design, implementation, and, most importantly, maintenance activities. Maintaining systems from functional specifications is more reliable and significantly more productive than manually modifying code as the inevitable system changes occur.

The *process* view of a system can be represented by data flow diagrams, the subject of Chapter 9. *System logic and timing,* or what goes on inside the black boxes that are identified as processes in data flow diagrams and when these processes occur, can be represented in many ways, including Structured English, decision tables, decision trees, and state-transition diagrams. Modeling system logic is the subject of Chapter 10. Finally, the *data* view of a system shows the rules which govern the structure and integrity of data and concentrates on what data about business entities and relationships among these entities must be accessed within the system. Data modeling pertinent to analysis is the subject of Chapter 11 and advanced concepts appear in Appendix C. In Appendix D we overview the notation used in object-oriented analysis and design for representing system structure.

A Broadway Entertainment Company (BEC) case appears after Chapters 9, 10, and 11. These cases illustrate, for the sample Customer Activity Tracking System, the process, logic, and data models that describe this new system. They also illustrate how diagrams and models for each of these three views of an information system are related to one another to form a consistent and thorough structured description of a proposed system.

Chapter 12 discusses the final step within analysis—how to choose among alternative competing sets of requirements. Chapter 12, in part, returns to tasks similar to those carried out in project initiation and planning activities. That is, once the requirements for the replacement system have been structured and documented, the systems development team realizes that there are many strategies to consider before actually designing and implementing the replacement system. For example, analysts must decide on the proper mix of on-line versus batch processing, what functionality to include when all user requirements cannot be justifiably met, and what scope is necessary for an effective system—the world of computing offers many choices. Before proceeding with design, systems analysts must choose the best design stategy to pursue by balancing a variety of economic, organizational, technical, legal, and other factors.

To find the best strategy, analysts will first generate several competing alternative strategies. One strategy may be high-end, involving state-of-the-art technology and implementing every conceivable function in the new system. Another strategy may be low-end, involving a minimal number of changes and improving functionality while keeping costs and risks low. Still another strategy may be somewhere between these two extremes, and there may be a host of other design strategies that are also contenders.

When alternatives have been identified and documented, the systems development team must present its best choice to management for the purpose of gaining approval for further development. Once the best system design strategy has been identified and recommended, the analysis phase ends. At this point, management has to decide whether to continue the project and proceed to design and, if so, how the design phase of the project should be structured. Management may also decide at this time that packaged software may be a viable option, or that contracting one or more subsequent life cycle steps to a consulting firm might be an appropriate strategy. You will see that the work carried out in analysis is essential to identify and describe the proper system requirements in a unambiguous format for subsequent systems development work.

Chapter 12 shows how project management, incremental commitment, the iterative nature of project management, and the evolving nature of details in project plans apply as we move through the life cycle. Since Chapter 12 brings to a close the section on the analysis phase of the life cycle, a major project milestone is reached. So, Chapter 12 also discusses updating the Baseline Project Plan and the transition from analysis to design.

We conclude Part IV with a Broadway Entertainment Company (BEC) case. This case shows how alternative design strategies for the Customer Activity Tracking System are generated and evaluated and a best strategy selected. The case also explains how the Baseline Project Plan is updated and presented to management to determine if and how the project should continue.

Determining System Requirements

LEARNING OBJECTIVES

After studying this chapter, you should be able to:

- Describe options for designing and conducting interviews and develop a plan for conducting an interview to determine system requirements.

- Design, distribute, and analyze questionnaires to determine system requirements.

- Explain the advantages and pitfalls of observing workers and analyzing business documents to determine system requirements.

- Explain how computing can provide support for requirements determination.

- Participate in and help plan a Joint Application Design session.

- Use prototyping during requirements determination.

- Select the appropriate methods to elicit system requirements.

INTRODUCTION

Systems analysis is the part of the systems development life cycle in which you determine how the current information system functions and assess what users would like to see in a new system. As you learned in Chapter 1, there are three sub-phases in analysis: requirements determination, requirements structuring, and alternative generation and choice.

In this chapter, you will learn about the beginning sub-phase of analysis—determining system requirements. Techniques used in requirements determination have evolved over time to become more structured and, as we will see in this chapter, current methods increasingly rely on the computer for support. We will first study the more traditional requirements determination methods including interviewing, using questionnaires, observing users in their work environment, and collecting procedures and other written documents. We will then discuss modern methods for collecting system requirements. The first of these methods is Joint Application Design (JAD), which you first read about in Chapter 1. Next, you will read about how analysts rely more and more on information systems to help them perform analysis. As you will see, group support systems have been used to support systems analysis, especially as part of the JAD process. CASE tools, discussed in Chapter 5, are also very useful in requirements determination. Finally, you will learn how prototyping can be used as a key tool for some requirements determination efforts.

PERFORMING REQUIREMENTS DETERMINATION

As stated earlier and shown in Figure 8-1, there are three sub-phases to systems analysis: requirements determination, requirements structuring, and generating alternative system design strategies and selecting the best one. We will address these as three separate steps, but you should consider these steps as somewhat parallel and iterative. For example, as you determine some aspects of the current and desired system(s), you begin to structure these requirements or to build prototypes to show users how a system might behave. Inconsistencies and deficiencies discovered through structuring and prototyping lead you to explore further the operation of current system(s) and the future needs of the organization. Eventually your ideas and discoveries converge on a thorough and accurate depiction of current operations and what the requirements are for the new system. As you think about beginning the analysis phase, you probably wonder what exactly is involved in requirements determination. We discuss this process in the next section.

Figure 8-1
Systems development life cycle with analysis phase highlighted

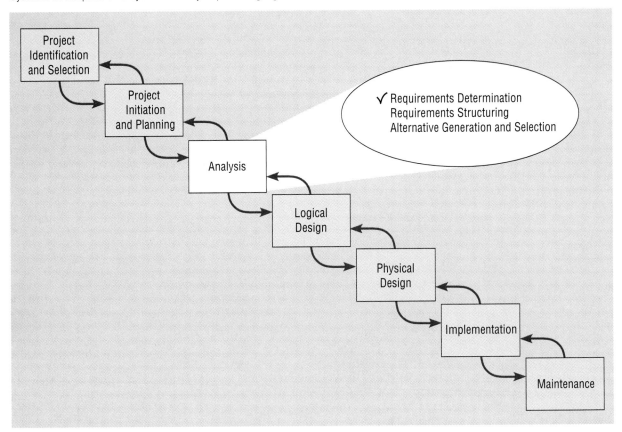

The Process of Determining Requirements

Once management has granted permission to pursue development of a new system (this was done at the end of the project identification and selection phase of the SDLC) and a project is initiated and planned (see Chapter 7), you begin determining what the new system should do. During requirements determination, you and other analysts gather information on what the system should do from as many sources as possible: from users of the current system, from observing users, and from reports, forms, and procedures. All of the system requirements are carefully documented and made ready for structuring, the subject of Chapters 9 through 11.

In many ways, gathering system requirements is like conducting any investigation. Have you read any of the Sherlock Holmes or similar mystery stories? Do you enjoy solving puzzles? From these experiences, we can detect some similar characteristics for a good systems analyst during the requirements determination sub-phase. These characteristics include

- *Impertinence.* You should question everything. You need to ask such questions as: Are all transactions processed the same way? Could anyone be charged something other than the standard price? Might we someday want to allow and encourage employees to work for more than one department?

- *Impartiality:* Your role is to find the best solution to a business problem or opportunity. It is not, for example, to find a way to justify the purchase of new hardware or to insist on incorporating what users think they want into the new system requirements. You must consider issues raised by all parties and try to find the best organizational solution.

- *Relax constraints:* Assume anything is possible and eliminate the infeasible. For example, do not accept this statement: "We've always done it that way, so we have to continue the practice." Traditions are different from rules and policies. Traditions probably started for a good reason but, as the organization and its environment change, traditions may turn into habits rather than sensible procedures.

- *Attention to details:* Every fact must fit with every other fact. One element out of place means that the ultimate system will fail at some time. For example, an imprecise definition of who a customer is may mean that you purge customer data when a customer has no active orders; yet these past customers may be vital contacts for future sales.

- *Reframing:* Analysis is, in part, a creative process. You must challenge yourself to look at the organization in new ways. You must consider how each user views his or her requirements. You must be careful not to jump to this conclusion: "I worked on a system like that once—this new system must work the same way as the one I built before."

Deliverables and Outcomes

The primary deliverables from requirements determination are the various forms of information gathered during the determination process: transcripts of interviews; notes from observation and analysis of documents; analyzed responses from questionnaires; sets of forms, reports, job descriptions, and other documents; and computer-generated output such as system prototypes. In short, anything that the analysis team collects as part of determining system requirements is included in the deliverables resulting from this sub-phase of the systems development life cycle.

Table 8-1 lists examples of some specific information that might be gathered during requirements determination.

These deliverables contain the information you need for systems analysis within the scope of the system you are developing. In addition, you need to understand the following components of an organization:

- The business objectives that drive what and how work is done

- The information people need to do their jobs

- The data (definition, volume, size, etc.) handled within the organization to support the jobs

- When, how, and by whom or what the data are moved, transformed, and stored

- The sequence and other dependencies among different data handling activities

- The rules governing how data are handled and processed

- Policies and guidelines that describe the nature of the business and the market and environment in which it operates

- Key events affecting data values and when these events occur

As should be obvious, such a large amount of information must be organized in order to be useful. This is the purpose of the next sub-phase—requirements structuring.

From just this sub-phase of analysis, you probably already realize that the amount of information to be gathered could be huge, especially if the scope of the system under development is broad. The time required to collect and structure a great deal of information can be extensive and, because it involves so much human effort, quite expensive. Too much analysis is not productive and the term "analysis paralysis" has been coined to describe a systems development project that has bogged down in an abundance of analysis work. Because of the dangers of excessive analysis, today's systems analysts focus more on the system to be developed than on the current system. The techniques you will learn about later in the chapter, JAD and prototyping, were developed to keep the analysis effort at a minimum yet still effective. Other processes have been developed to limit the analysis commitment even more, providing an alternative to the SDLC. One of these processes is called *Rapid Application Development (RAD)* and is the subject of Appendix B. As you will learn from reading the appendix, RAD relies on JAD, prototyping, and integrated CASE tools to be effective. Even RAD, as well as the structured analysis methods discussed in subsequent chapters in the analysis section of this book, rely on a basic understanding of the business area served by an information system. Thus, before you can fully appreciate RAD, you need to learn about traditional fact-gathering techniques. These techniques are the subject of the next section.

TABLE 8-1   Deliverables for Requirements Determination

1. **Information collected from conversations with or observations of users:** interview transcripts, questionnaire responses, notes from observation, meeting minutes

2. **Existing written information:** business mission and strategy statements, sample business forms and reports and computer displays, procedure manuals, job descriptions, training manuals, flow charts and documentation of existing systems, consultant reports

3. **Computer-based information:** results from Joint Application Design sessions, transcripts or files from group support system sessions, CASE repository contents and reports of existing systems, and displays and reports from system prototypes.

TRADITIONAL METHODS FOR DETERMINING REQUIREMENTS

At the core of systems analysis is the collection of information. At the outset, you must collect information about the information systems that are currently being used and how users would like to improve the current systems and organizational operations with new or replacement information systems. One of the best ways to get this information is to talk to the people who are directly or indirectly involved in the different parts of the organizations affected by the possible system changes. In Chapter 2 we identified these people as stakeholders: users, managers, funders, etc. Another way to find out about the current system is to gather copies of documentation relevant to current systems and business processes. In this chapter, you will learn about various ways to get information directly from stakeholders: interviews, questionnaires, group interviews, and direct observation. You will learn about collecting documentation on the current system and organizational operation in the form of written procedures, forms, reports, and other hard copy. These traditional methods of collecting system requirements are listed in Table 8-2.

Interviewing and Listening

Interviewing is one of the primary ways analysts gather information about an information systems project. Early in a project, an analyst may spend a large amount of time interviewing people about their work, the information they use to do it, and the types of information processing that might supplement their work. Other stakeholders are interviewed to understand organizational direction, policies, expectations managers have on the units they supervise, and other non-routine aspects of organizational operations. During interviewing you will gather facts, opinions, and speculation and observe body language, emotions, and other signs of what people want and how they assess current systems.

There are many ways to effectively interview someone and no one method is necessarily better than another. Some guidelines you should keep in mind when you interview, summarized in Table 8-3, are discussed next.

First, you should prepare thoroughly before the interview. Set up an appointment at a time and for a duration convenient for the interviewee. The general nature of the interview should be explained to the interviewee in advance. You may ask the interviewee to think about specific questions or issues or to review certain

TABLE 8-2 Traditional Methods of Collecting System Requirements

- Individually *interview* people informed about the operation and issues of the current system and needs for systems in future organizational activities

- Survey people via *questionnaires* to discover issues and requirements

- *Interview groups* of people with diverse needs to find synergies and contrasts among system requirements

- *Observe workers* at selected times to see how data are handled and what information people need to do their jobs

- *Study business documents* to discover reported issues, policies, rules, and directions as well as concrete examples of the use of data and information in the organization

TABLE 8-3 Guidelines for Effective Interviewing

Plan the Interview

▶ Prepare interviewee: appointment, priming questions

▶ Prepare checklist, agenda, and questions

Listen carefully and take notes (tape record if permitted)

Review notes within 48 hours of interview

Be neutral

Seek diverse views

documentation to prepare for the interview. You should spend some time thinking about what you need to find out and write down your questions. Do not assume that you can anticipate all possible questions. You want the interview to be natural and, to some degree, you want to spontaneously direct the interview as you discover what expertise the interviewee brings to the session.

You should prepare an interview guide or checklist so that you know in which sequence you intend to ask your questions and how much time you want to spend in each area of the interview. The checklist might include some probing questions to ask as follow-up if you receive certain anticipated responses. You can, to some degree, integrate your interview guide with the notes you take during the interview, as depicted in a sample guide in Figure 8-2 on pages 281 and 282. This same guide can serve as an outline for a summary of what you discover during an interview.

The first page of the sample interview guide contains a general outline of the interview. Besides basic information on who is being interviewed and when, you list major objectives for the interview. These objectives typically cover the most important data you need to collect, a list of issues on which you need to seek agreement (for example, content for certain system reports), and which areas you need to explore, not necessarily with specific questions. You also include reminder notes to yourself on key information about the interviewee (for example, job history, known positions taken on issues, and role with current system). This information helps you to be personal, shows that you consider the interviewee important, and may assist you in interpreting some answers. Also included is an agenda for the interview with approximate time limits for different sections of the interview. You may not follow the time limits precisely but the schedule helps you cover all areas during the time the interviewee is available. Space is also allotted for general observations that do not fit under specific questions and for notes taken during the interview about topics skipped or issues raised that could not be resolved.

On subsequent pages you list specific questions; the sample form in Figure 8-2 includes space for taking notes on these questions. Because unanticipated information arises, you will not strictly follow the guide in sequence. You can, however, check off the questions you have asked and write reminders to yourself to return to or skip certain questions as the dynamics of the interview unfold.

Choosing Interview Questions   You need to decide what mix and sequence of open-ended and closed-ended questions you will use. **Open-ended questions** are usually used to probe for information for which you cannot anticipate all possible responses or for which you do not know the precise question to ask. The person being interviewed is encouraged to talk about whatever interests him or her within the general bounds of the question. An example is, "What would you say is the best thing about the information system you currently use to do your job?" or "List the

Open-ended questions:
Questions in interviews and on questionnaires that have no pre-specified answers.

Figure 8-2
Typical interview guide

Interview Outline	
Interviewee: *Name of person being interviewed*	Interviewer: *Name of person leading interview*
Location/Medium: *Office, conference room, or phone number*	Appointment Date: Start Time: End Time:
Objectives: *What data to collect* *On what to gain agreement* *What areas to explore*	Reminders: *Background/experience of interviewee* *Known opinions of interviewee*

Agenda:
   Introduction
   Background on Project
   Overview of Interview
     Topics To Be Covered
     Permission to Tape Record
   Topic 1 Questions
   Topic 2 Questions
   ...
   Summary of Major Points
   Questions from Interviewee
   Closing
	Approximate Time:
	 1 minute
	 2 minutes
	
	 1 minute
	
	 5 minutes
	 7 minutes
	 ...
	 2 minutes
	 5 minutes
	 1 minute

General Observations: Interviewee seemed busy — probably need to call in a few days for follow-up questions since he gave only short answers. PC was turned off —probably not a regular PC user.	
Unresolved Issues, Topics not Covered: He needs to look up sales figures from 1992. He raised the issue of how to handle returned goods, but we did not have time to discuss.	

(continues)

three most frequently used menu options." You must react quickly to answers and determine whether or not any follow-up questions are needed for clarification or elaboration. Sometimes body language will suggest that a user has given an incomplete answer or is reluctant to divulge some information; a follow-up question might yield additional insight. One advantage of open-ended questions in an interview is that previously unknown information can surface. You can then continue exploring along unexpected lines of inquiry to reveal even more new information.

Figure 8-2 *(continued)*
Typical interview guide

Interviewee:	Date:
Questions:	**Notes:**
When to ask question, if conditional *Question number: 1* Have you used the current sales tracking system? If so, how often? *If yes, go to Question 2*	*Answer* Yes, I ask for a report on my product line weekly *Observations* Seemed anxious — may be over-estimating usage frequency
Question: 2 What do you like least about this system?	*Answer* Sales are shown in units, not dollars *Observations* System can show sales in dollars, but user does not know this.

Open-ended questions also often put the interviewees at ease since they are able to respond in their own words using their own structure; open-ended questions give interviewees more of a sense of involvement and control in the interview. A major disadvantage of open-ended questions is the length of time it can take for the questions to be answered. In addition, open-ended questions can be difficult to summarize.

Closed-ended questions provide a range of answers from which the interviewee may choose. Here is an example:

Closed-ended questions:
Questions in interviews and on questionnaires that ask those responding to choose from among a set of specified responses.

Which of the following would you say is the one best thing about the information system you currently use to do your job (pick only one):
a. Having easy access to all of the data you need
b. The system's response time
c. The ability to run the system concurrently with other applications

Closed-ended questions work well when the major answers to questions are well known. Another plus is that interviews based on closed-ended questions do not necessarily require a large time commitment—more topics can be covered. As opposed to collecting such information via questionnaires, you can see body language and hear voice tone which can aid in interpreting the interviewee's responses. Closed-ended questions can also be an easy way to begin an interview and to determine which line of open-ended questions to pursue. You can include an "other" option to encourage the interviewee to add unanticipated responses. A major disadvantage of closed-ended questions is that useful information that does not quite fit into the defined answers may be overlooked as the respondent tries to make a choice instead of providing his or her best answer.

Closed-ended questions, like objective questions on an examination, can follow several forms, including these choices:

- True or false.

- Multiple choice (with only one response or selecting all relevant choices).

- Rating a response or idea on some scale, say from bad to good or strongly agree to strongly disagree. Each point on the scale should have a clear and consistent meaning to each person and there is usually a neutral point in the middle of the scale.

- Ranking items in order of importance.

Interview Guidelines   First, with either open- or closed-ended questions, do not phrase a question in a way that implies a right or wrong answer. The respondent must feel that he or she can state his or her true opinion and perspective and that his or her idea will be considered equally with those of others. Questions such as, "Should the system continue to provide the ability to override the default value, even though most users now do not like the feature?" should be avoided as such wording pre-defines a socially acceptable answer.

The second guideline to remember about interviews is to listen very carefully to what is being said. Take careful notes or, if possible, record the interview on a tape recorder (be sure to ask permission first!). The answers may contain extremely important information for the project. Also, this may be the only chance you have to get information from this particular person. If you run out of time and still need to get information from the person you are talking to, ask to schedule a follow-up interview.

Third, once the interview is over, go back to your office and type up your notes within 48 hours. If you recorded the interview, use the recording to verify the material in your notes. After 48 hours, your memory of the interview will fade quickly. As you type and organize your notes, write down any additional questions that might arise from lapses in your notes or from ambiguous information. Separate facts from your opinions and interpretations. Make a list of unclear points that need clarification. Call the person you interviewed and get answers to these new questions. Use the phone call as an opportunity to verify the accuracy of your notes. You may also want to send a written copy of your notes to the person you interviewed so the person can check your notes for accuracy. Finally, make sure you thank the person for his or her time. You may need to talk to your respondent again. If the interviewee will be a user of your system or is involved in some other way in the system's success, you want to leave a good impression.

Fourth, be careful during the interview not to set expectations about the new or replacement system unless you are sure these features will be part of the delivered system. Let the interviewee know that there are many steps to the project and the perspectives of many people need to be considered, along with what is technically possible. Let respondents know that their ideas will be carefully considered, but that due to the iterative nature of the systems development process, it is premature to say now exactly what the ultimate system will or will not do.

Fifth, seek a variety of perspectives from the interviews. Find out what potential users of the system, users of other systems that might be affected by changes, managers and superiors, information systems staff who have experience with the current system, and others think the current problems and opportunities are and what new information services might better serve the organization. You want to understand all possible perspectives so that in a later approval step you will have information on which to base a recommendation or design decision that all stakeholders can accept.

Administering Questionnaires

Interviews are very effective ways of communicating with people and obtaining important information from them. However, interviews are also very expensive and time-consuming to conduct. Thus, a limited number of questions can be covered and people contacted. In contrast, questionnaires are passive and often yield less depth of understanding than interviews; however, questionnaires are not as expensive to administer per respondent. In addition, questionnaires have the advantage of gathering information from many people in a relatively short time and of being less biased in the interpretation of their results.

Choosing Questionnaire Respondents   Sometimes there are more people to survey than you can handle and you must decide which set of people to send the questionnaire to or which questionnaire to send to which group of people. Whichever group of respondents you choose, it should be representative of all users. In general, you can achieve a representative sample by any one or any combination of these four methods:

1. Those *convenient to* sample: these may be people at a local site, those willing to be surveyed, or those most motivated to respond

2. A *random* group: if you get a list of all users of the current system, simply choose every *n*th person on the list; or, you could select people by skipping names on the list based on numbers from a random number table

3. A *purposeful* sample: here you may specify only people who satisfy certain criteria, such as users of the system for more than two years or users who use the system most often

4. A *stratified* sample: in this case, you have several categories of people whom you definitely want to include—choose a random set from each category (e.g., users, managers, foreign business unit users)

Samples which combine characteristics of several approaches are also common. In any case, once the questionnaires are returned, you should check for non-response bias; that is, a systematic bias in the results since those who responded are different from those who did not respond. You can refer to books on survey research to find out how to determine if your results are confounded by non-response bias.

Designing Questionnaires   Questionnaires are usually administered on paper although they can be administered in person (resembling a structured interview), over the phone (computer-assisted telephone interviewing), or even on diskette.

Questionnaires are less expensive, however, if they do not require a person to administer them directly; that is, if the people answering the questions can complete the questionnaire without help. Also, answers can be provided at the convenience of the respondent, as long as the answers are returned by a specific date.

Questionnaires typically include closed-ended questions, more than can be effectively asked in an interview, and sometimes contain open-ended questions as well. Closed-ended questions are preferable because they are easier to complete and they define the exact coverage required. A few open-ended questions give the person being surveyed an opportunity to add insights not anticipated by the designer of the questionnaire. In general, questionnaires take less time to complete than interviews structured to obtain the same information. In addition, questionnaires are given to many people simultaneously whereas interviews are usually limited to one person at a time.

Questionnaires are generally less rich in information content than interviews, however, because they provide no direct means by which to ask follow-up questions (although it is possible, though time-consuming, to contact respondents after they have returned their completed questionnaires to ask for further information). Also, since questionnaires are written, they do not provide the opportunity to judge the accuracy of the responses. In an interview, you can sometimes determine if people are answering truthfully or fully by the words they use, whether they make direct eye contact, the tone of voice they use, or their body language.

The ability to create good questionnaires is a skill that improves with practice and experience. Because the questions are written, they must be extremely clear in meaning and logical in sequence. When a person is completing a questionnaire, he or she only has the written questions to interpret and answer. You are not there to clarify each question's meaning. For example, what if a closed-ended question were phrased in this way:

> *How often do you back up your computer files?*
> a. *Frequently*
> b. *Sometimes*
> c. *Hardly at all*
> d. *Never*

There are at least two sources of ambiguity in the wording of the question. The first source of ambiguity is the categories offered for the answer: the only nonambiguous answer is "never." "Hardly at all" could mean anything from once per year to once per month, depending on who is answering the question. "Sometimes" could cover the same range of possibilities as "Hardly at all." "Frequently" could be anything from once per hour to once per week. The second source of ambiguity is in the question itself. Does the term "computer files" pertain only to those on my hard disk? Or does it also mean the files I have stored on floppy disk? What if I have more than one PC in my office? And what about the files I have stored on the minicomputer I use for certain applications? I don't back up those files; the system operator does it on a regular basis for all minicomputer files, not just mine. With no questioner present to explain the ambiguities, the respondent is at a loss and must try to answer the question in the best way he or she knows how. Whether the respondent's interpretation is the same as other respondents' is anyone's guess. The respondent cannot be there when the data are analyzed to tell exactly what was meant.

A less ambiguous way to phrase the question and its response categories would be something like this:

> *How often do you back up the computer files stored on the hard disk on the PC you use for over 50% of your work time?*

a. *Frequently (at least once per week)*
b. *Sometimes (from one to three times per month)*
c. *Hardly at all (once per month or less)*
d. *Never*

As you can see from the wording of the question, the phrasing is a bit awkward, but it avoids ambiguity. You may want to break up a single question into multiple questions, or a set of questions and statements, to avoid awkward phrasing. Notice also that the possible responses are much clearer now that they have been specifically defined, and they cover the full range of possibilities, from never to at least once per week with no overlapping time periods.

Obviously, care must be taken in the task of composing closed-ended and open-ended questions. Further, you should be as careful in composing questions for interviews as for questionnaires, since sloppily worded questions cannot be identified every time in an interview unless the interviewee asks for clarification. For both interviews and questionnaires, it is wise to pretest your questions. Pose the questions in a simulated interview and ask the interviewee to rephrase each question as he or she interprets the question. Check responses for reasonableness. You can even ask the same question in what you think are several different ways to see if you receive a materially different response. Use this feedback to adjust the questions to make them less ambiguous.

Questionnaires are most useful in the requirements determination process when used for very specific purposes rather than for more general information gathering. For example, one useful application of questionnaires is to measure levels of user satisfaction with a system or with particular aspects of it. Another useful application is to have several users choose from among a list of system features available in many off-the-shelf software packages. You could ask users to choose the features they most want and quickly tabulate the results to find out which features are most in demand. You could then recommend a system solution based on a particular software package to meet the demands of most of the users.

Choosing Between Interviews and Questionnaires

To summarize the previous sections, you can see that interviews are good tools for collecting rich, detailed information and that interviews allow exploration and follow-up (see Table 8-4). On the other hand, interviews are quite time-intensive and expensive. In comparison, questionnaires are inexpensive and take less time, as specific information can be gathered from many people at once without the personal intervention of an interviewer. The information collected from a questionnaire is less rich, however, and is potentially ambiguous if questions are not phrased precisely. In addition, follow-up to a questionnaire is more difficult as it often involves interviews or phone calls, adding to the expense of the process.

These differences and others are important for you to remember during the analysis phase. Deciding which method to use and what strategy to employ to gather information will vary with the system being studied and its organizational context. For example, if the organization is large and the system being studied is vast and complex, then there will probably be dozens of affected users and stakeholders. If you know little about the system or the organization, a good strategy is to identify key users and stakeholders and interview them. You would then use the information gathered in the interviews to create a questionnaire which would be distributed to a large number of users. You could then schedule follow-up interviews with a few users. At the other extreme, if the system and organization are small and you understand them well, the best strategy may be to interview only one or two key users or stakeholders.

TABLE 8-4   Comparison of Interviews and Questionnaires

Characteristic	Interviews	Questionnaires
Information Richness	High (many channels)	Medium to low (only responses)
Time Required	Can be extensive	Low to moderate
Expense	Can be high	Moderate
Chance for Follow-up and Probing	Good: probing and clarification questions can be asked by either interviewer or interviewee	Limited: probing and follow-up done after original data collection
Confidentiality	Interviewee is known to interviewer	Respondent can be unknown
Involvement of Subject	Interviewee is involved and committed	Respondent is passive, no clear commitment
Potential Audience	Limited numbers, but complete responses from those interviewed	Can be quite large, but lack of response from some can bias results

Interviewing Groups

One drawback to using interviews and questionnaires to collect systems requirements is the need for the analyst to reconcile apparent contradictions in the information collected. A series of interviews may turn up inconsistent information about the current system or its replacement. You must work through all of these inconsistencies to figure out what the most accurate representation of current and future systems might be. Such a process requires several follow-up phone calls and additional interviews. Catching important people in their offices is often difficult and frustrating, and scheduling new interviews may become very time-consuming. In addition, new interviews may reveal new questions that in turn require additional interviews with those interviewed earlier. Clearly, gathering information about an information system through a series of individual interviews and follow-up calls is not an efficient process.

Another option available to you is the group interview. In a group interview, you interview several key people at once. To make sure all of the important information is collected, you may conduct the interview with one or more analysts. In the case of multiple interviewers, one analyst may ask questions while another takes notes, or different analysts might concentrate on different kinds of information. For example, one analyst may listen for data requirements while another notes the timing and triggering of key events. The number of interviewees involved in the process may range from two to however many you believe can be comfortably accommodated.

A group interview has a few advantages. One, it is a much more effective use of your time than is a series of interviews with individuals (although the time commitment of the interviewees may be more of a concern). Two, interviewing several people together allows them to hear the opinions of other key people and gives them the opportunity to agree or disagree with their peers. Synergies also often occur. For example, the comments of one person might cause another person to say, "That reminds me of . . ." or "I didn't know that was a problem." You can benefit from such a discussion as it helps you identify issues on which there is general agreement and areas where views diverge widely.

The primary disadvantage of a group interview is the difficulty in scheduling it. The more people involved, the more difficult it will be finding a convenient time and place for everyone. Modern technology such as video conferences and video phones can minimize the geographical dispersion factors that make scheduling meetings so difficult. Group interviews are at the core of the Joint Application Design process, which we discuss in a later section in this chapter.

Directly Observing Users

All the methods of collecting information that we have been discussing up until now involve getting people to recall and convey information they have about an organizational area and the information systems which support these processes. People, however, are not always very reliable informants, even when they try to be reliable and tell what they think is the truth. As odd as it may sound, people often do not have a completely accurate appreciation of what they do or how they do it. This is especially true concerning infrequent events, issues from the past, or issues for which people have considerable passion. Since people cannot always be trusted to reliably interpret and report their own actions, you can supplement and corroborate what people tell you by watching what they do or by obtaining relatively objective measures of how people behave in work situations. (See the box "Lost Soft Drink Sales" for an example of the importance of systems analysts learning firsthand about the business for which they are designing systems.)

For example, one possible view of how a hypothetical manager does her job is that a manager carefully plans her activities, works long and consistently on solving problems, and controls the pace of her work. A manager might tell you that is how she spends her day. When Mintzberg (1973) observed how managers work, however, he found that a manager's day is actually punctuated by many, many interruptions. Managers work in a fragmented manner, focusing on a problem or on a communication for only a short time before they are interrupted by phone calls or visits from their subordinates and other managers. An information system designed to fit the work environment described by our hypothetical manager would not effectively support the actual work environment in which that manager finds herself.

As another example, consider the difference between what another employee might tell you about how much he uses electronic mail and how much electronic mail use you might discover through more objective means. An employee might tell you he is swamped with e-mail messages and that he spends a significant proportion of his time responding to e-mail messages. However, if you were able to check electronic mail records, you might find that this employee receives only three e-mail messages per day on average, and that the most messages he has ever received during one eight-hour period is ten. In this case, you were able to obtain an

---

Lost Soft Drink Sales

A systems analyst was quite surprised to read that sales of all soft drink products were lower, instead of higher, after a new delivery truck routing system was installed. The software was designed to reduce stock-outs at customer sites by allowing drivers to visit each customer more often using more efficient delivery routes.

Confused by the results, management asked the analyst to delay a scheduled vacation, but he insisted that he could look afresh at the system only after a few overdue days of rest and relaxation.

Instead of taking a vacation, however, the analyst called a delivery dispatcher he had interviewed during the design of the system and asked to be given a route for a few days. The analyst drove a route (for a regular driver actually on vacation), following the schedule developed from the new system. What the analyst discovered was that the route was very efficient, as expected; so at first the analyst could not see any reason for lost sales.

During the third and last day of his "vacation" the analyst stayed overtime at one store to ask the manager if she had any ideas why sales might have dropped off in recent weeks. The manager had no explanation, but did make a seemingly unrelated observation that the regular route driver appeared to have less time to spend in the store. He did not seem to take as much interest in where the products were displayed and did not ask for promotional signs to be displayed, as he had often done in the past.

From this conversation, the analyst concluded that the new delivery truck routing system was, in one sense, too good. It placed the driver on such a tight schedule that a driver had no time left for the "schmoozing" required to get special treatment, that gave the company's products an edge over the competition.

Without first-hand observation of the system in action participating as a system user, the analyst might never have discovered the true problem with the system design. Once time was allotted for not only stocking new products but also for necessary marketing work, product sales returned to and exceeded levels achieved before the new system had been introduced.

accurate behavioral measure of how much e-mail this employee copes with without having to watch him read his e-mail.

The intent behind obtaining system records and direct observation is the same, however, and that is to obtain more firsthand and objective measures of employee interaction with information systems. In some cases, behavioral measures will be a more accurate reflection of reality than what employees themselves believe. In other cases, the behavioral information will substantiate what employees have told you directly. Although observation and obtaining objective measures are desirable ways to collect pertinent information, such methods are not always possible in real organizational settings. Thus, these methods are not totally unbiased, just as no other one data-gathering method is unbiased.

For example, observation can cause people to change their normal operating behavior. Employees who know they are being observed may be nervous and make more mistakes than normal, may be careful to follow exact procedures which they do not typically follow, and may work faster or slower than normal. Moreover, since observation typically cannot be continuous, you receive only a snapshot image of the person or task you observe which may not include important events or activities. Since observation is very time-consuming, you will not only observe for a limited time but also a limited number of people and at a limited number of sites. Again, observation yields only a small segment of data from a possibly vast variety of data sources. Exactly which people or sites to observe is a difficult selection problem. You want to pick both typical and atypical people and sites and observe during normal and abnormal conditions and times to receive the richest possible data from observation.

Analyzing Procedures and Other Documents

As noted above, asking questions of the people who use a system every day or who have an interest in a system is an effective way to gather information about current and future systems. Observing current system users is a more direct way of seeing how an existing system operates, but even this method provides limited exposure to all aspects of current operations. These methods of determining system requirements can be enhanced by examining system and organizational documentation to discover more details about current systems and the organization these systems support.

Although we discuss here several important types of documents that are useful in understanding possible future system requirements, our discussion does not exhaust all possibilities. You should attempt to find all written documents about the organizational areas relevant to the systems under redesign. Besides the few specific documents we discuss, organizational mission statements, business plans, organization charts, business policy manuals, job descriptions, internal and external correspondence, and reports from prior organizational studies can all provide valuable insight.

What can the analysis of documents tell you about the requirements for a new system? In documents you can find information about

- Problems with existing systems (e.g., missing information or redundant steps)

- Opportunities to meet new needs if only certain information or information processing were available (e.g., analysis of sales based on customer type)

- Organizational direction that can influence information system requirements (e.g., trying to link customers and suppliers more closely to the organization)

- Titles and names of key individuals who have an interest in relevant existing systems (e.g., the name of a sales manager who led a study of buying behavior of key customers)

- Values of the organization or individuals who can help determine priorities for different capabilities desired by different users (e.g., maintaining market share even if it means lower short-term profits)

- Special information processing circumstances that occur irregularly that may not be identified by any other requirements determination technique (e.g., special handling needed for a few very large-volume customers and which requires use of customized customer ordering procedures)

- The reason why current systems are designed as they are, which can suggest features left out of current software which may now be feasible and more desirable (e.g., data about a customer's purchase of competitors' products were not available when the current system was designed; these data are now available from several sources)

- Data, rules for processing data, and principles by which the organization operates that must be enforced by the information system (e.g., each customer is assigned exactly one sales department staff member as a primary contact if the customer has any questions)

One type of useful document is a written work procedure for an individual or a work group. The procedure describes how a particular job or task is performed, including data and information that are used and created in the process of performing the job. For example, the procedure shown in Figure 8-3 includes data (list of features and advantages, drawings, inventor name, and witness names) required to prepare an invention disclosure. It also indicates that besides the inventor, the vice-president for research and department head and dean must review the material, and that a witness is required for any filing of an invention disclosure. These insights clearly affect what data must be kept, to whom information must be sent, and the rules that govern valid forms.

Procedures are not trouble-free sources of information, however. Sometimes your analysis of several written procedures will reveal a duplication of effort in two or more jobs. You should call such duplication to the attention of management as an issue to be resolved before system design can proceed. That is, it may be necessary to redesign the organization before the redesign of an information system can achieve its full benefits. Another problem you may encounter with a procedure occurs when the procedure is missing. Again, it is not your job to create a document for a missing procedure—that is up to management. A third and common problem with a written procedure happens when the procedure is out of date. You may realize the procedure is out of date when you interview the person responsible for performing the task described in the procedure. Once again, the decision to rewrite the procedure so that it matches reality is made by management, but you may make suggestions based upon your understanding of the organization. A fourth problem often encountered with written procedures is that the formal procedures may contradict information you collected from interviews, questionnaires, and observation about how the organization operates and what information is required. As in the other cases, resolution rests with management.

All of these problems illustrate the difference between **formal systems** and **informal systems.** Formal systems are systems recognized by the official documentation of the organization; informal systems are the way in which the organization actually works. Informal systems develop because of inadequacies of formal procedures, individual work habits and preferences, resistance to control, and other factors. It is

Formal system: The official way a system works as described in organizational documentation.

Informal system: The way a system actually works.

Figure 8-3
Example of a procedure

GUIDE FOR PREPARATION OF INVENTION DISCLOSURE
(See FACULTY and STAFF MANUALS for detailed Patent Policy and routing procedures.)

(1) DISCLOSE ONLY ONE INVENTION PER FORM.

(2) PREPARE COMPLETE DISCLOSURE.

The disclosure of your invention is adequate for patent purposes ONLY if it enables a person skilled in the art to understand the invention.

(3) CONSIDER THE FOLLOWING IN PREPARING A COMPLETE DISCLOSURE:

(a) All essential elements of the invention, their relationship to one another, and their mode of operation.

(b) Equivalents that can be substituted for any elements.

(c) List of features believed to be new.

(d) Advantages this invention has over the prior art.

(e) Whether the invention has been built and/or tested.

(4) PROVIDE APPROPRIATE ADDITIONAL MATERIAL.

Drawings and descriptive material should be provided as needed to clarify the disclosure. Each page of this material must be signed and dated by each inventor and properly witnessed. A copy of any current and/or planned publication relating to the invention should be included.

(5) INDICATE PRIOR KNOWLEDGE AND INFORMATION.

Pertinent publications, patents or previous devices, and related research or engineering activities should be identified.

(6) HAVE DISCLOSURE WITNESSED.

Persons other than co-inventors should serve as witnesses and should sign each sheet of the disclosure only after reading and understanding the disclosure.

(7) FORWARD ORIGINAL PLUS ONE COPY (two copies if supported by grant/contract) TO VICE PRESIDENT FOR RESEARCH VIA DEPARTMENT HEAD AND DEAN.

important to understand both formal and informal systems since each provides insight into information requirements and what will be required to convert from present to future information services.

A second type of document useful to systems analysts is a business form (see a mock-up of a form in Figure 8-4). Forms are used for all types of business functions, from recording an order to acknowledging the payment of a bill to indicating what goods have been shipped. Forms are important for understanding a system because they explicitly indicate what data flow in or out of a system and which are necessary for the system to function. In the sample invoice form in Figure 8-4, we see data such as the name of the customer, the customer's sold to and ship to addresses, method of payment, data (item number, quantity, etc.) about each line item on the invoice, and calculated data such as tax and totals.

The form gives us crucial information about the nature of the organization. For example, the company can ship and bill to different addresses; item numbers appear to be all-numeric and five digits long; and the freight expense is charged to the customer. A printed form may correspond to a computer display that the system will generate for someone to enter and maintain data or to display data to on-line users. Forms are most useful to you when they contain actual organizational data (as in Figure 8-4), as this allows you to determine the characteristics of data which are actually used by the application. The ways in which people use forms change over time, and data that were needed when a form was designed may no longer be required. You can use the systems analysis techniques presented in Chapters 9 through 11 to help you determine which data are no longer required.

A third type of useful document is a report generated by current systems. As the primary output for some types of systems, a report enables you to work backwards from the information on the report to the data that must have been necessary to generate them. Figure 8-5 presents an example of a typical report prepared for management in a video rental business. This report shows that the system must be able to identify categories for each video and associate customers with the videos they rent. The date near the top of the report is probably the date on which the report was prepared and need not be kept by the system. You would analyze such reports to determine which data need to be captured over what time period and what manipulation of these raw data would be necessary to produce each field on the report.

If the current system is computer-based, a fourth set of useful documents are those that describe the current information systems—how they were designed and how they work. There are a lot of different types of documents that fit this description, everything from flow charts to data dictionaries and CASE tool reports to user manuals. An analyst who has access to such documents is lucky, as many in-house–developed information systems lack complete documentation (unless a CASE tool has been used).

Analysis of organizational documents and observation, along with interviewing and questionnaires, are the methods most used for gathering system requirements. In Table 8-4 we summarized the comparative features of interviews and questionnaires. Table 8-5 summarizes the comparative features of observation and analysis of organizational documents.

MODERN METHODS FOR DETERMINING SYSTEM REQUIREMENTS

Even though we called interviews, questionnaires, observation, and document analysis traditional methods for determining a system's requirements, all of these methods are still very much used by analysts to collect important information. Today, however, there are additional techniques to collect information about the current system, the organizational area requesting the new system, and what the new system should be like. In this section, you will learn about several modern information-gathering techniques for analysis (listed in Table 8-6): Joint Application Design (JAD), group support systems, CASE tools, and prototyping. As we said

Figure 8-4

Example of a business form (Generated from Microsoft Access™)

Company Name | INVOICE

Company Slogan
Enter Address Here
City, State, Zip/Postal Code
Phone: (000) 000-0000

SOLD TO:	Sam Jones
	25 Herky Ave.
	Montreal, H9J1A2 Canada
	(555) 555-7744

INVOICE #	**1**
INVOICE DATE	**3/31/93**
YOUR ORDER #	**84456**
TERMS	**Net 10**
SALESMAN	**George**
F.O.B.	**Hartford, CT**

SHIPPED TO: **Sue Michaels**
60 Krawski Dr.
Mainville, KY 64423 USA
(555) 644-5891

Shipped via
● UPS
○ **US Mail**
○ **Overnight**

Payment type
● Collect
○ **Prepaid**

ITEM #	Quantity	Description	Price	Amount
12223	78	#2 Pine Boards	$12.23	$953.94
14206	5	Widgets	$5.01	$25.05
14456	7	Screws	$12.34	$86.38
17885	1	Nail Gun	$65.00	$65.00
24553	7	Hammers	$37.23	$260.61
26453	3	Gadgets	$10.00	$30.00
41165	5	Nails	$43.23	$216.15
63342	45	Roof Truss	$125.23	$5,635.35
		SUBTOTAL		$7,272.48
		TAX		$42.00
		FREIGHT		$5.50
		PAY THIS AMOUNT ☞		$7,319.98

If you have any questions concerning this invoice, call

Name   Phone Number

MAKE CHECKS PAYABLE TO
Enter Company Name Here

THANK YOU FOR YOUR BUSINESS

Figure 8-5

Example of a report

Who rents what category of videos?

28-Feb-94   1

Video Category	Customer Name
action	
	Eleanor Johnson
	Miles Standish
foreign	
	Alan Alda
	Bob Hope
	Eleanor Johnson
	Jane Smith
	Juan Valdez
sf	
	Alan Alda
	Alexi Kosygin
	Bob Hope
	Eleanor Johnson
	Jane Smith
	John Smith
	Wanda Orlikowski
western	
	Alexi Kosygin
	Bob Hope
	Jane Smith
	Juan Valdez
	Miles Standish
	Wilma Randolph

---

earlier, these techniques can support effective information collection and structuring while reducing the amount of time required for analysis. An alternative to the SDLC, RAD, which combines JAD, CASE tools, and prototyping, is described in more detail in Appendix B.

Joint Application Design

You were introduced to Joint Application Design, or JAD, in Chapter 1. There you learned JAD started in the late 1970s at IBM and that since then the practice of JAD has spread throughout many companies and industries. For example, it is quite popular in the insurance industry in Connecticut where a JAD users group has been formed. In fact, several generic approaches to JAD have been documented and popularized (see Wood and Silver, 1989, for an example). You also learned in Chapter 1 that the main idea behind JAD is to bring together the key users, managers, and systems analysts involved in the analysis of a current system. In that respect, JAD is similar to a group interview; a JAD, however, follows a particular structure of roles and agenda that is quite different from a group interview during

TABLE 8-5 Comparison of Observation and Document Analysis

Characteristic	Observation	Document Analysis
Information Richness	High (many channels)	Low (passive) and old
Time Required	Can be extensive	Low to moderate
Expense	Can be high	Low to moderate
Chance for Follow-up and Probing	Good: probing and clarification questions can be asked during or after observation	Limited: probing possible only if original author is available
Confidentiality	Observee is known to interviewer; observee may change behavior when observed	Depends on nature of document; does not change simply by being read
Involvement of Subject	Interviewees may or may not be involved and committed depending on whether they know if they are being observed	None, no clear commitment
Potential Audience	Limited numbers and limited time (snapshot) of each	Potentially biased by which documents were kept or because document not created for this purpose

TABLE 8-6 Modern Methods for Collecting System Requirements

- Bringing together in a *Joint Application Design (JAD)* session users, sponsors, analysts and others to discuss and review system requirements

- Using *group support systems* to facilitate the sharing of ideas and voicing opinions about system requirements

- Using *CASE tools* to analyze current systems to discover requirements to meet changing business conditions

- Iteratively developing system *prototypes* that refine the understanding of system requirements in concrete by showing working versions of system features

which analysts control the sequence of questions answered by users. The primary purpose of using JAD in the analysis phase is to collect systems requirements simultaneously from the key people involved with the system. The result is an intense and structured, but highly effective, process. As with a group interview, having all the key people together in one place at one time allows analysts to see where there are areas of agreement and where there are conflicts. Meeting with all these important people for over a week of intense sessions allows you the opportunity to resolve conflicts, or at least to understand why a conflict may not be simple to resolve.

JAD sessions are usually conducted in a location other than the place where the people involved normally work. The idea behind such a practice is to keep participants away from as many distractions as possible so that they can concentrate on systems analysis. A JAD may last anywhere from four hours to an entire week and may consist of several sessions. A JAD employs thousands of dollars of corporate resources, the most expensive of which is the time of the people involved. Other expenses include the costs associated with flying people to a remote site and putting them up in hotels and feeding them for several days.

The typical participants in a JAD are listed below:

JAD session leader: The trained individual who plans and leads Joint Application Design sessions.

- **JAD session leader:** The JAD leader organizes and runs the JAD. This person has been trained in group management and facilitation as well as in systems analysis. The JAD leader sets the agenda and sees that it is met. The JAD leader remains neutral on issues and does not contribute ideas or opinions but rather concentrates on keeping the group on the agenda, resolving conflicts and disagreements, and soliciting all ideas.

- Users: The key users of the system under consideration are vital participants in a JAD. They are the only ones who have a clear understanding of what it means to use the system on a daily basis.

- Managers: Managers of the work groups who use the system in question provide insight into new organizational directions, motivations for and organizational impacts of systems, and support for requirements determined during the JAD.

- Sponsor: As a major undertaking due to its expense, a JAD must be sponsored by someone at a relatively high level in the company. If the sponsor attends any sessions, it is usually only at the very beginning or the end.

- Systems Analysts: Members of the systems analysis team attend the JAD although their actual participation may be limited. Analysts are there to learn from users and managers, not to run or dominate the process.

Scribe: The person who makes detailed notes of the happenings at a Joint Application Design session.

- **Scribe:** The scribe takes notes during the JAD sessions. This is usually done on a personal computer or laptop. Notes may be taken using a word processor, or notes and diagrams may be entered directly into a CASE tool.

- IS staff: Besides systems analysts, other IS staff, such as programmers, database analysts, IS planners, and data center personnel, may attend to learn from the discussion and possibly to contribute their ideas on the technical feasibility of proposed ideas or on technical limitations of current systems.

JAD sessions are usually held in special-purpose rooms where participants sit around horseshoe-shaped tables, as in Figure 8-6. These rooms are typically equipped with whiteboards (possibly electronic, with a printer to make copies of what is written on the board). Other audio-visual tools may be used, such as transparencies and overhead projectors, magnetic symbols that can be easily rearranged on a whiteboard, flip charts, and computer-generated displays. Flip chart paper is typically used for keeping track of issues that cannot be resolved during the JAD or for those issues requiring additional information that can be gathered during breaks in the proceedings. Computers may be used to create and display form or report designs or for diagramming existing or replacement systems. In general, however, most JADs do not benefit much from computer support (Carmel, 1991).

When a JAD is completed, the end result is a set of documents that detail the workings of the current system related to the study of a replacement system. Depending on the exact purpose of the JAD, analysts may also walk away from the JAD with some detailed information on what is desired of the replacement system.

Taking Part in a JAD   Imagine that you are a systems analyst taking part in your first JAD. What might participating in a JAD be like? Typically, JADs are held off site, in comfortable conference facilities. On the first morning of the JAD, you and

Figure 8-6

Illustration of the typical room layout for a JAD
(Adapted from Wood and Silver, 1989)

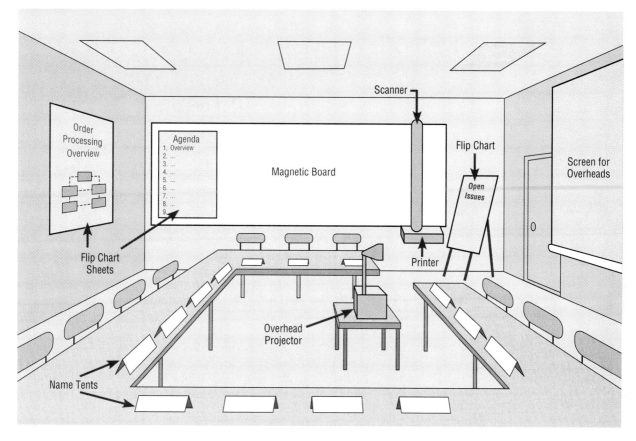

your fellow analysts walk into a room that looks much like the one depicted in Figure 8-6. The JAD facilitator is already there; she is finishing writing the day's agenda on a flip chart. The scribe is seated in a corner at a microcomputer, preparing to take notes on the day's activities. Users and managers begin to enter in groups and seat themselves around the U-shaped table. You and the other analysts review your notes describing what you have learned so far about the information system you are all here to discuss. The session leader opens the meeting with a welcome and a brief run-down of the agenda. The first day will be devoted to a general overview of the current system and major problems associated with it. The next two days will be devoted to an analysis of current system screens. The last two days will be devoted to analysis of reports.

The session leader introduces the corporate sponsor who talks about the organizational unit and current system related to the systems analysis study and the importance of upgrading the current system to meet changing business conditions. He leaves, and the JAD session leader takes over. She yields the floor to the senior analyst who begins a presentation on key problems with the system that have already been identified. After the presentation, the session leader opens the discussion to the users and managers in the room.

After a few minutes of talk, a heated discussion begins between two users from different corporate locations. One user, who represents the office which served as the model for the original systems design, argues that the system's perceived lack of flexibility is really an asset, not a problem. The other user, who represents an office that was part of another company before a merger, argues that the current system is so inflexible as to be virtually unusable. The session leader intervenes and tries to help the users isolate particular aspects of the system that may contribute to the system's perceived lack of flexibility.

Questions arise about the intent of the original developers. The session leader asks the analysis team about their impressions of the original system design. Because these questions cannot be answered during this meeting, as none of the original designers are present and none of the original design documents are readily available, the session leader assigns the question about intent to the "to do" list. This question becomes the first one on a flip chart sheet of "to do" items and the session leader gives you the assignment of finding out about the intent of the original designers. She writes your name next to the "to do" item on the list and continues with the session. Before the end of the JAD, you must get an answer to this question.

The JAD will continue like this for its duration. Analysts will make presentations, help lead discussions of form and report design, answer questions from users and managers, and take notes on what is being said. After each meeting, the analysis team will meet, usually informally, to discuss what has occurred that day and to consolidate what they have learned. Users will continue to contribute during the meetings and the session leader will facilitate, intervening in conflicts, seeing that the group follows the agenda. When the JAD is over, the session leader and her assistants must prepare a report that documents the findings in the JAD and is circulated among users and analysts.

CASE Tools During JAD The CASE tools most useful to analysis during a JAD are those referred to as upper CASE (see Chapter 5), as they apply most directly to activities occurring early in the systems development life cycle. Upper CASE tools usually include planning tools (see Chapter 6), diagramming tools (see Chapter 2 for examples of diagrams used during systems analysis), and prototyping tools, such as computer display and report generators. For requirements determination and structuring, the most useful CASE tools are for diagramming and for display and report generation. The more interaction analysts have with users during this phase, the more useful this set of tools. The analyst can use diagramming and prototyping tools to give graphic form to system requirements, show the tools to users, and make changes based on the users' reactions. The same tools are very valuable for requirements structuring as well. Using common CASE tools during requirements determination and structuring makes the transition between these two sub-phases easier and reduces the time spent. In structuring, CASE tools which analyze requirements information for correctness, completeness, and consistency are also useful. Finally, for alternative generation and selection, diagramming and prototyping tools are key to presenting users with graphic illustrations of what the alternative systems will look like. Such a practice provides users and analysts with better information to select the most desirable alternative system.

Some observers advocate using CASE tools during JADs (Lucas, 1993). Running a CASE tool during a JAD allows analysts to enter system models directly into a CASE tool, providing consistency and reliability in the joint model-building process. The CASE tool captures system requirements in a more flexible and useful way than can a scribe or an analysis team making notes. Further, the CASE tool can be used to project menu, display, and report designs, so users can directly observe old and new designs and evaluate their usefulness for the analysis team. Some CASE tools are too slow for the real-time pace of most JADs, however, so the session leader may want analysts to enter information into the CASE tool after the day's meeting is over. Users and managers can be shown the results of their work the next morning.

Supporting JAD with GSS   The JAD process is typically not well supported by computing, despite suggestions to augment JADs with CASE tools and other computer-based aids. Most traditional JADs rely on one computer for the scribe to use and maybe another for displaying screen and report designs. Since JAD is a structured group process, JAD can benefit from the same computer-based support that can be applied to any group process. Group support systems (GSS) can be used to support group meetings (see Appendix A). Here we will discuss how JAD can benefit from GSS use.

One disadvantage to a JAD session is that it suffers from many of the same problems as any group meeting. For example, the more people in a group, the less time there is for all of them to speak and state their views. Even if you assumed that they all spoke for an equal amount of time, no one would have much time to talk in a one-hour meeting for 12 people (only five minutes each!). The assumption about speaking equally points out a second problem with meetings—one or a few people always dominate the discussion. On the other hand, some people will say absolutely nothing. Whatever outcome the meeting produces tends to be tilted toward those who spoke the most during the meeting and others may not be fully committed to the conclusions reached. A third problem with group meetings is that some people are afraid to speak out for fear they will be criticized. A fourth problem is that most people are not willing to criticize or challenge their bosses in a meeting, even if what the boss is saying is wrong.

JADs suffer from all of these problems with meetings. The result is that important views often are not aired. Such an outcome is unfortunate as the design of the new system could be adversely impacted and the system may have to be reworked at great expense when those important views finally become known.

GSSs have been designed specifically to help alleviate some of the problems with group meetings. In order to provide everyone in the meeting with the same chance to contribute, group members type their comments into computers rather than speak them. The GSS is set up so that all members of the group can see what every other member has been typing. In the one-hour meeting for 12 people mentioned earlier, all 12 can contribute for the full hour, instead of just for five minutes, using a GSS. If everyone in the meeting is typing, not talking, and everyone has the same chance to contribute, then the chances of domination of the meeting by any one individual are greatly reduced. Also, comments typed into a GSS are anonymous. Anonymity helps those who fear criticism because only the comment, and not the person, can be criticized, since no one knows who typed what. Anonymity also provides the ability to criticize your boss.

Supporting a JAD with a GSS has many potential benefits. Using a GSS, a JAD session leader is more likely to obtain contributions from everyone, rather than from just a few. Important ideas are less likely to be missed. Similarly, poor ideas are more likely to be criticized. A study comparing traditional JAD to JAD supported with GSS found that using a GSS did lead to certain enhancements in the JAD process (Carmel, George, and Nunamaker, 1992). Among the findings were that GSS-supported JADs tended to be more time-efficient than traditional JAD and participation was more equal because there was less domination by certain individuals than in traditional JAD. The study also found that introducing a GSS into a JAD session had other, less desirable, effects. GSS-supported JADs tended to be less structured and it was more difficult to identify and resolve conflicts when a GSS was used, due in part to the anonymity of interaction. Supporting a JAD with GSS, then, does seem to provide some benefits through altering how the group works together. Yet a reduction in the JAD leader's ability to resolve conflicts could be a problem, especially since JAD was designed to help uncover and resolve conflicts.

Other Uses of Group Support Systems for Requirements Determination

GSSs have also been used in other requirements determination activities. One interesting case is its use in helping determine the knowledge requirements for an expert system designed to provide help for computer users in an organization (Nunamaker, et al., 1988). The process of obtaining knowledge requirements for an expert system is called **knowledge elicitation,** a difficult and sometimes tedious process. Typically, an expert systems developer, called a knowledge engineer (a job similar to that of a systems analyst), must sit through several sessions of watching an expert solve specific problems before he or she gains a good understanding of what the expert's relevant knowledge is. Problem-solving sessions are augmented by countless interviews. Sometimes a knowledge engineer will have to elicit knowledge from several individual experts, multiplying the complexity and tediousness of the task.

> **Knowledge elicitation:** The process of obtaining knowledge requirements for an expert system.

Jay Nunamaker and colleagues (Nunamaker, et al., 1988) decided to support the process of eliciting knowledge from multiple experts by using a GSS. They conducted multiple GSS sessions, each of which was attended by 10 to 12 experts. The experts were computer consultants who worked in the organization's information center. Their jobs were to provide help to users who called the center asking for assistance in using specific software tools. The purpose of the first set of GSS sessions was to collect background information on the problem domain and to clearly define the scope of the problem. The experts used various software tools in the GSS to generate and consolidate lists of commonly occurring problems and to describe typical consulting sessions with users. Nunamaker and his colleagues had enough information from the first set of sessions to define the expert system's architecture and to build a prototype. The purpose of the second set of GSS sessions was to elicit additional knowledge. The result was a second, improved prototype. The final expert system was tested against the information system consultants to determine how close its suggested solutions were to those suggested by the consultants. The expert system's performance was very favorable, indicating that knowledge elicitation by GSS had worked quite well.

Using Prototyping During Requirements Determination

You were introduced to prototyping in Chapter 1 (see Figure 1-9 for an overview of prototyping). There you learned that prototyping is an iterative process involving analysts and users whereby a rudimentary version of an information system is built and rebuilt according to user feedback. You also learned that prototyping could replace the systems development life cycle or augment it. What we are interested in here is how prototyping can augment the requirements determination process.

In order to gather an initial basic set of requirements, you will still have to interview users and collect documentation. Prototyping, however, will allow you to quickly convert basic requirements into a working, though limited, version of the desired information system. The prototype will then be viewed and tested by the user. Typically, seeing verbal descriptions of requirements converted into a physical system will prompt the user to modify existing requirements and generate new ones. For example, in the initial interviews, a user might have said that he wanted all relevant utility billing information on a single computer display form, such as the client's name and address, the service record, and payment history. Once the same user sees how crowded and confusing such a design would be in the prototype, he might change his mind and instead ask for the information to be organized on several screens, but with easy transitions from one screen to another. He might also be reminded of some important requirements (data, calculations, etc.) that had not surfaced during the initial interviews.

You would then redesign the prototype to incorporate the suggested changes. Once modified, users would again view and test the prototype. And, once again, you would incorporate their suggestions for change. Through such an iterative process, the chances are good that you will be able to better capture a system's requirements. The goal with using prototyping to support requirements determination is to develop concrete specifications for the ultimate system, not to build the ultimate system from prototyping.

Prototyping is possible with several 4GLs and with CASE tools, as pointed out in Chapter 5 and the earlier section on CASE tools and analysis in this chapter. As we saw there, you can use CASE tools as part of a JAD to provide a type of limited prototyping with a group of users.

Prototyping is most useful for requirements determination when

- User requirements are not clear or well understood, which is often the case for totally new systems or systems that support decision making

- One or a few users and other stakeholders are involved with the system

- Possible designs are complex and require concrete form to fully evaluate

- Communication problems have existed in the past between users and analysts and both parties want to be sure that system requirements are as specific as possible

- Tools (such as form and report generators) and data are readily available to rapidly build working systems

Prototyping also has some drawbacks as a tool for requirements determination. These include

- A tendency to avoid creating formal documentation of system requirements which can then make the system more difficult to develop into a fully working system

- Prototypes can become very idiosyncratic to the initial user and difficult to diffuse or adapt to other potential users

- Prototypes are often built as stand-alone systems, thus ignoring issues of sharing data and interactions with other existing systems

- Checks in the SDLC are bypassed so that some more subtle, but still important, system requirements might be forgotten (e.g., security, some data entry controls, or standardization of data across systems)

SUMMARY

As we saw in Chapter 1, there are three sub-phases in the systems analysis phase of the systems development life cycle: requirements determination, requirements structuring, and alternative generation and choice. Chapter 8 has focused on requirements determination, the gathering of information about current systems and the need for replacement systems. Chapters 9 through 11 will address techniques for structuring the requirements elicited during requirements determination. Chapter 12 closes Part IV of the book by explaining how analysts generate alternative design strategies for replacement systems and choose the best one.

For requirements determination, the traditional sources of information about a system include interviews, questionnaires, observation, group interviews, and procedures, forms, and other useful documents. Often many or even all of these sources are used to gather perspectives on the adequacy of current systems and the requirements for replacement systems. Each form of information collection has its advantages and disadvantages, which were summarized in Tables 8-4 and 8-5. Selecting the methods to use depends on the need for rich or thorough information, the time and budget available, the need to probe deeper once initial information is collected, the need for confidentiality for those providing assessments of system requirements, the desire to get people involved and committed to a project, and the potential audience from which requirements should be collected.

Both open- and closed-ended questions can be posed during interviews or in questionnaires. In either case, you must be very precise in formulating a question in order to avoid ambiguity and to insure a proper response. During observation you must try not to intrude or interfere with normal business activities so that the people being observed do not modify their activities from normal processes. The results of all requirements gathering methods should be compared, since there may be differences between the formal or official system and the way people actually work, the informal system.

You also learned about alternative methods to collect requirements information, many of which themselves make use of information systems. Joint Application Design (JAD) begins with the idea of the group interview and adds structure and a JAD session leader to it. Typical JAD participants include the session leader, a scribe, key users, managers, a sponsor, and systems analysts. JAD sessions are usually held off site and may last as long as one week.

One special type of information system, an expert system, uses a knowledge engineer in a knowledge elicitation process to understand how experts process data or make decisions. The knowledge elicitation process is similar to requirements determination for other types of information systems.

Although JAD sessions typically rely little on computer support, systems analysis is increasingly performed with computer assistance, such as group support systems. You also read how information systems can support requirements determination with CASE tools and for prototyping. As part of the prototyping process, users and analysts work closely together to determine requirements which the analyst then builds into a model. The analyst and user then work together on revising the model until it is close to what the user desires.

The result of requirements determination is a thorough set of information, including some charts, that describes the current systems being studied and the need for new and different capabilities to be included in the replacement systems. This information, however, is not in a form that makes analysis of true problems and clear statements of new features possible. Thus, you and other analysts will study this information and structure it into standard formats suitable for identifying problems and unambiguously describing the specifications for new systems. With modern information systems, structuring requires documenting the flow or movement of data throughout the organization and information systems, the logic of transforming data into information (including the timing of events that cause data to be transformed or processed), and the rules that govern the relationships between different data handled by the system. We discuss a variety of popular techniques for structuring requirements in the following three chapters.

CHAPTER REVIEW

KEY TERMS

Closed-ended questions   JAD session leader   Open-ended questions
Formal system   Knowledge elicitation   Scribe
Informal system

REVIEW QUESTIONS

1. Define each of the following terms:

   a. Joint Application Design   e. requirements structuring
   b. prototyping   f. formal system
   c. Group Support System (GSS)   g. informal system
   d. requirements determination

2. Describe systems analysis and the major activities that occur during this phase of the systems development life cycle.

3. Describe four traditional techniques for collecting information during analysis. When might one be better than another?

4. Compare collecting information by interview and by questionnaire. Describe a hypothetical situation in which each of these methods would be an effective way to collect information system requirements.

5. What is JAD? How is it better than traditional information-gathering techniques? What are its weaknesses?

6. How has computing been used to support requirements determination?

7. How has computing been used to support JAD?

8. How can CASE tools be used to support requirements determination? Which type of CASE tools are appropriate for use during requirements determination?

9. Describe how prototyping can be used during requirements determination. How is it better or worse than traditional methods?

10. What unique benefits does a GSS provide for group methods of requirements determination, such as group interviews or JAD?

PROBLEMS AND EXERCISES

1. Match the following terms to the appropriate definitions.

   _____ open-ended questions   a. questions in interviews and on questionnaires that ask those answering the questions to choose from among a set of pre-specified responses

   _____ closed-ended questions   b. trained individual who plans and leads Joint Application Design sessions

_____ JAD session leader

_____ scribe

_____ knowledge elicitation

c. person who makes detailed notes of the happenings at a Joint Application Design session

d. responses to questions in interviews and on questionnaires that have no pre-specified answers

e. process of obtaining knowledge requirements for an expert system

2. Choose either CASE or GSS as a topic and review a related article from the popular press and from the academic research literature. Summarize the two articles and, based on your reading, prepare a list of arguments for why this type of system would be useful in a JAD session. Also address the limits for applying this type of system in a JAD setting.

3. One of the potential problems with gathering information requirements by observing potential system users mentioned in the chapter is that people may change their behavior when observed. What could you do to overcome this potential confounding factor in accurately determining information requirements?

4. Summarize the problems with the reliability and usefulness of analyzing business documents as a method for gathering information requirements. How could you cope with these problems to effectively use business documents as a source of insights on system requirements?

5. Suppose you were asked to lead a JAD session. List 10 guidelines you would follow to assist you in playing the proper role of a JAD session leader.

6. Prepare a plan, similar to Figure 8-2, for an interview with your academic advisor to determine which courses you should take to develop the skills you need to be hired as a programmer/analyst.

7. Write at least three closed-ended questions that you might use on a questionnaire that would be sent to users of a word processing package in order to develop ideas for the next version of the package. Test these questions by asking a friend to answer the questions; then interview your friend to determine why she responded as she did. From this interview, determine if she misunderstood any of your questions and, if so, rewrite the questions to be less ambiguous.

8. An interview lends itself easily to asking probing questions, or asking different questions depending on the answers provided by the interviewee. Although not impossible, probing and alternative questions can be handled in a questionnaire. Discuss how you could include probing or alternative sets of questions in a questionnaire.

9. Figure 8-2 shows part of a guide for an interview. How might an interview guide differ when a group interview is to be conducted?

10. Group interviews and JADs are very powerful ways to collect system requirements but special problems arise during group requirements collection sessions. Summarize the special interviewing and group problems that arise in such group sessions, and suggest ways that you, as a group interviewer or group facilitator, might deal with these problems.

FIELD EXERCISES

1. Effective interviewing is not something that you can learn from just reading about it. You must first do some interviewing, preferably a lot of it, as interviewing skills

only improve with experience. To get an idea of what interviewing is like, try the following: Find three friends or classmates to help you complete this exercise. Organize yourselves into pairs. Write down a series of questions you can use to find out about a job your partner now has or once held. You decide what questions to use, but at a minimum, you must find out the following: (1) the job's title (2) the job's responsibilities (3) whom your partner reported to (4) who reported to your partner, if anyone did, and (5) what information your partner used to do his or her job. At the same time, your partner should be preparing questions to ask you about a job you have had. Now conduct the interview. Take careful notes. Organize what you find into a clear form that another person could understand (you might want to use a systems diagram like Figure 2-4, with boundaries, inputs, and outputs). Now repeat the process, but this time, your partner interviews you.

While the two of you have been interviewing each other, your two other friends should have been doing the same thing. When all four of you are done, switch partners and repeat the entire process. When you are all done, each of you should have interviewed two people, and each of you should have been interviewed by two people. Now, you and the person who interviewed your original partner should compare your findings. Most likely, your findings will not be identical to what the other person found. If your findings differ, discover why. Did you use the same questions? Did the other person do a more thorough job of interviewing your first partner, since it was the second time he or she had conducted an interview? Did you both ask follow-up questions? Did you both spend about the same amount of time on the interview? Prepare a report with this person about why your findings differed. Now find both of the people who interviewed you. Does one set of findings differ from the other? Try and figure out why. Did one of them (or both of them) misrepresent or misunderstand what you told them? Each of you should now write a report on your experience, using it to explain why interviews are sometimes inconsistent and inaccurate and why having two people interview someone on a topic is better than having just one person do the interview. Explain the implications of what you have learned for the requirements determination sub-phase of the systems development life cycle.

2. Choose a work team at your work or university and interview them in a group setting. Ask them about their current system (whether computer-based or not) for performing their work. Ask each of them what information they use and/or need and from where/whom they get it. Was this a useful method for you to learn about their work? Why or why not? What comparative advantages does this method provide as compared to one-on-one interviews with each team member? What comparative disadvantages?

3. For the same work team you used in Field Exercise 2, examine copies of any relevant written documentation (e.g., written procedures, forms, reports, system documentation). Are any of these forms of written documentation missing? Why? With what consequences? To what extent does this written documentation fit with the information you received in the group interview?

4. Interview systems analysts, users, and managers who have been involved in JAD sessions. Determine the location, structure and outcomes of each of their JAD sessions. Elicit their evaluations of their sessions. Were they productive? Why or why not?

5. Survey the literature on JAD in the academic and popular press and determine the "state of the art." How is JAD being used to help determine system requirements? Is using JAD for this process beneficial? Why or why not? Present your analysis to the IS manager at your work or at your university. Does your analysis of JAD fit with his or her perception? Why or why not? Is he or she currently using JAD, or a JAD-like method, for determining system requirements? Why or why not?

REFERENCES

Carmel, E. 1991. *Supporting Joint Application Development with Electronic Meeting Systems: A Field Study.* Unpublished doctoral dissertation, University of Arizona.

Carmel, E., J. F. George, and J. F. Nunamaker, Jr. 1992. "Supporting Joint Application Development (JAD) with Electronic Meeting Systems: A Field Study." *Proceedings of the Thirteenth International Conference on Information Systems.* Dallas, TX, December: 223–32.

Carmel, E., R. Whitaker, and J. F. George. 1993 "Participatory Design and Joint Application Design: A Transatlantic Comparison." *Communications of the ACM* 36 (June): 40–48.

Dennis, A. R., J. F. George, L. Jessup, J. F. Nunamaker, Jr., and D. R. Vogel. 1988. "Information Technology to Support Electronic Meetings." *MIS Quarterly* 12 (December): 591–624.

Lucas, M.A. 1993. "The Way of JAD." *Database Programming & Design* 6 (July): 42–49.

Mintzberg, H. 1973. *The Nature of Managerial Work.* New York: Harper & Row.

Nunamaker, Jr., J. F., B. R. Konsynski, M. Chen, A. S. Vinze, I. I. Chen, and M. M. Heltne. 1988. "Knowledge-based Systems Support for Information Centers." *Journal of Management Information Systems* 5 (Summer): 4–24.

Wood, J. and D. Silver. 1989. *Joint Application Design.* New York, NY: John Wiley & Sons.

Analysis: Requirements Determination for the Customer Activity Tracking System

INTRODUCTION

As seen from Figure 4 in the previous Broadway Entertainment Company case, the Baseline Project Plan for the requirements determination step included several interviews, analysis of relevant business documents, benchmarking customer profiling in other firms, and a (tentative) JAD session, among other steps. Space does not permit us to discuss all of these activities. To illustrate requirements determination, this section focuses on the results of the interview with the U.S. marketing manager conducted by one of the systems analysts assigned to the CATS project.

ANALYZING INTERVIEW RESULTS

Frank Napier walked into Jordan Pippen's office. In his hands, he carried several pages of printed notes. Frank had been with BEC for about nine months since his graduation from the computer information systems program at the University of Western Florida. Frank had worked on his first systems project at BEC with Jordan and Jordan had requested that Frank be assigned as one of the two analysts for the analysis phase of the CATS project. Frank was relatively inexperienced for such a highly placed project as CATS, but he was well trained and knew his limitations. Jordan and Nancy Chen, the CATS project client, had agreed that with Jordan's experience as project manager and Frank's excellent reputation in only nine months at BEC, Frank should work out well on the project.

"Hi Frank. What have you got there? Are those your notes on your interview with Wendy?"

"Yes," Frank said, handing the notes to Jordan. At the top of the first page was typed:

Interview with: Wendy Yoshimuro, Marketing Manager, U.S. Operations

Interviewer: Frank Napier, Junior Systems Analyst
Date: May 31, 1996
Tape: CATS1
Topic: BEC Customer Activity Tracking System (CATS)

Jordan picked up Frank's notes and began leafing through them. Wendy Yoshimuro worked for Nancy Chen in the U.S. Operations office. Wendy was the marketing manager whose job responsibilities were most closely aligned with what the CATS system was about, namely, helping to determine product mix based on customer profiles. Wendy's office was responsible for maintaining and interpreting the customer information that was derived from the raw data on customer purchases and rentals collected at the store level. Jordan began to read Frank's interview notes:

Frank: What do the current customer profiles look like?

Wendy: I can give you a copy of the form we use to collect information from new customers (Figure 1) and I can give you a list of all the attributes we store, but I can tell you, too. We have the basic customer information, name, address, credit card number, that kind of stuff. As for video rentals, you know the consumer privacy law prevents us from recording specific video titles a customer rents, but we can record the category of video that was rented. BEC now has 87 different categories, so we can narrow down preferences quite a bit. For example, we don't have a single science fiction category. Instead, we have one category for Star Trek™ for all of the movies and for tapes of both original and "Next Generation" shows. We have another category for George Lucas movies, another category for science fiction movies from the '50s and '60s, and so on. Instead of a single category for action movies, we have Schwartzenegger movies, Stallone

Figure 1

Customer information form for joining BEC

BEC MEMBERSHIP INFORMATION

Mr. Mrs. Ms. : _____

Street Address: _____

City: _____ State: _____ Zip: _____

Telephone: (____) _____

Credit Card Type: AE   MC   Visa   Discover

Credit Card Number: _____

Expiration Date: _____

Social Security Number (optional): _____

---

movies, James Bond movies, Van Damme movies, etc. We are also allowed to keep a record of how many videos of each type a customer has rented.

Frank: What's the time period? Do you store totals per category for the month? For the year?

Wendy: By the month. We keep data on-line for two years and then we archive them. For the membership data, if someone has not been an active customer over a one-year period, we purge their records from our files. If they want to be a customer again, they have to reapply at their local BEC store.

Frank: What about other rentals, like video games?

Wendy: There are no laws that govern what we keep about the video games people rent, so we keep actual titles instead of categories and counts.

Frank: What about purchases, especially music?

Wendy: We keep a record of the titles customers purchase, whether the media is tape or CD.

Frank: What other kinds of information would you need to implement a system like the Customer Activity Tracking System that is proposed?

Wendy: Correct me if I'm wrong, but from what was in Nancy's memo announcing the project, from what she has told us in staff meetings, and from what I've read, one of the purposes of this system is to match customers with newly released products?

Frank: Yes, that's right.

Wendy: I think it would be helpful to get more customer information, such as income, education, age, those kinds of demographics. You have to understand the difference between what we do with this information now and what we would be doing with it in CATS. Right now, all we do is look at sales and rental trends over time. We want to know what types of products we should stock and in what quantities, at each store. You have to remember tastes vary by region of the country. We will sell a lot of Snoop Doggy Dogg CDs in Los Angeles but very few in Salt Lake City. But I'm digressing. When we use our customer information to determine what products we should be selling, we don't care about individuals. We only look at aggregates. For CATS, we **have** to focus on individuals. We have to know what each person's tastes are, but we also need to know how much product they can afford to buy. One other thing you might consider as part of this system is the frequency with which you send reminders to customers. For some customers, you might not want to send more than one notice per month. Some people have limited incomes and limited tastes in music. I mean, how often does Jimmy Buffett come out with a new album anyway? For others, like teenagers with lots of disposable income and wide musical tastes, you might want to send several notices per week.

Frank: That's a good suggestion. If you have any other ideas about CATS after the interview is over, please let me know. Let's see, we're almost done here. I just have a few questions left.

"This is all very useful information," Jordan said to Frank. "It sounds as if Wendy has a lot of good ideas we can use for CATS."

"Yeah," Frank said, "I think she can help us a lot. I have a follow-up interview with Wendy scheduled for next week; I wanted to talk with her in person to verify what I learned in our first interview and to ask some follow-up questions. I think Wendy will be someone we have to really satisfy to win approval for building the system. I understand that some of Nancy's direct reports are beginning to question the potential benefits of CATS. I think they fear that funding of CATS may lessen the chances for funding other marketing projects they want done. If we can convince Wendy of the system's value, this will give Nancy a strong advocate in her group. On another subject, I still haven't been able to find anything in the trade press about information systems similar to what we have in mind for CATS, but I did find this article on custom tailoring of catalogs."

"Oh, I saw that," said Jordan. "Maybe Jorge will find out more from the calls he's making to Lands' End, L.L. Bean, PowerUp!, and the other firms against which we are benchmarking." (Jorge Lopez is the other analyst assigned to the CATS project. Jorge has been with BEC for five years and has been responsible for developing many of the systems that support international operations.)

"I have a call in to one of the people mentioned in the article," Frank replied. "I want to see if they will send us some information on their system design, at least some charts used in training users, maybe some data flow diagrams. I talked to Jorge about this and he said I could go ahead and make the contact since I was so interested. I'll be sure to share what I get with him."

"What have you found out about product announcements from the record and video companies and about how we do product profiles?"

"Those were both on my list of questions for Wendy, but we ran out of time in the last interview. That's another reason I have the second interview with her scheduled. Jordan, I'm still surprised how talkative most of the BEC managers are. It's so difficult to keep them on the topic. I'll try to be a little more as-

sertive in my next interview so that I get all questions covered to some degree. From the first interview, I think I got a pretty good idea about how customer profiles work, so I'm confident I'll be able to find out a lot of what we need to know about product announcements and product profiles."

"I think we're making good progress," said Jordan. "We already have one interview with Wendy in the can. We have the notes from the formal interview that we decided in our orientation meeting I ought to do with Nancy Chen, plus all of my notes from my many talks with her. We have copies of the customer information form they fill out to get BEC membership cards and you're going to get documents on customer and product profiles. We and Jorge have additional interviews scheduled with people who Nancy, Wendy, and others have suggested we meet, and you may be able to get some information on that catalog system. Did I tell you? Next week, I'm meeting with the managers of the seven BEC stores with the highest sales figures. One of them called Nancy after she distributed the project announcement memo and asked to meet with me. It will be a chance for me to understand their reactions to a system like CATS. You know, if implemented, CATS will put an additional burden on the store manager, since the initial conception makes them responsible for reserving merchandise for targeted customers."

"What about feedback?" asked Frank. "Does the current system conception make provisions for feedback? Don't we want to know if CATS works? If the customers we target actually do come in and either buy the records we have reserved for them or rent the tapes we set aside? Isn't that another burden for the store managers? More record keeping."

"That's a good point," Jordan said. "Look at the initial DFD Nancy and I presented to the Systems Priority Board (Figure 2). There is a reservation report on here, but that goes from the system to the stores. Make a note to remember that we'll probably need to add some flow from the stores back to CATS, or maybe we'll need to modify the Rental and Sales systems to capture this and note reservation sales differently in the transaction files. But I'm not sure exactly what changes would have to be made to our existing systems to accommodate the feedback function. It probably wouldn't entail much, if any, extra record keeping at the store level, but it definitely would affect existing systems. I'll ask the store managers about this when I talk with them. That reminds me. We really need someone else on this team who knows the existing related systems. Jorge knows the international systems, so he can help, but you and Jorge are really busy with

Figure 2
Level-0 data flow diagram for
the CATS project

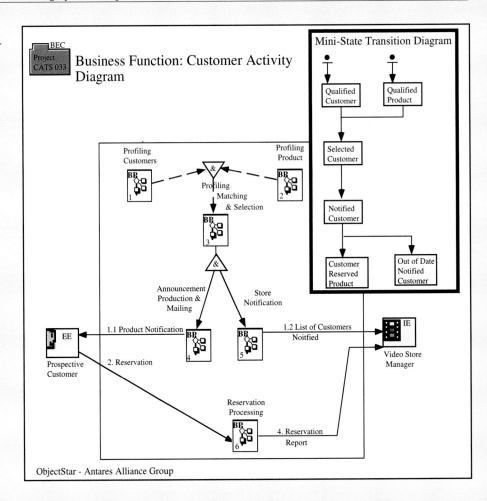

your other assignments. Nancy told me that she can allocate some extra money from her systems budget if the CATS analysis phase budget approved by the SPB comes up short. We can use that to 'borrow' somebody for a few weeks. Let me talk to Karen about it."

"Have you given any more thought to whether we need to have the tentatively scheduled JAD for requirements determination?" Frank asked. "Based on the initial information we have collected so far, I'm beginning to think a one- or two-day JAD really would be a good idea."

"You may be right," Jordan answered. "I'll talk to Karen about that, too. We'll make sure we get a professional JAD facilitator this time. After that rather hostile JAD you and I were part of about six months ago, didn't you tell me there is a professor at Western Florida who might be able to train us on running JADs?"

SUMMARY

Requirements determination is progressing on the CATS project. Interviews, such as the one with Wendy Yoshimuro, yield much insight into possible system requirements and features. Soon Frank, Jorge, and Jordan will need to begin to structure what they have learned into a set of consistent and complete diagrams and descriptions of CATS. In the next BEC segment, we will see how the structuring of data movement and processing requirements can be presented.

9

Structuring System Requirements: Process Modeling

L E A R N I N G   O B J E C T I V E S

After studying this chapter, you should be able to:

- Understand the logical modeling of processes through studying examples of data flow diagrams.

- Draw data flow diagrams following specific rules and guidelines that lead to accurate and well-structured process models.

- Decompose data flow diagrams into lower-level diagrams.

- Balance higher-level and lower-level data flow diagrams.

- Explain the differences among four types of DFDs: current physical, current logical, new physical, and new logical.

- Use data flow diagrams as a tool to support the analysis of information systems.

INTRODUCTION

In the last chapter, you learned about various methods used by systems analysts to collect the information necessary to determine information systems requirements. In this chapter, our focus will be on one tool used to coherently represent the information gathered as part of requirements determination—data flow diagrams. Data flow diagrams allow you to model how data flow through an information system, the relationships among the data flows, and how data come to be stored at specific locations. Data flow diagrams also show the processes that change or transform data. Because data flow diagrams concentrate on the movement of data between processes, these diagrams are called process models.

As the name indicates, a data flow diagram is a graphical tool that allows analysts (and users, for that matter) to depict the flow of data in an information system. The system can be physical or logical, manual or computer-based. In this chapter, you will learn the basic mechanics of drawing and revising data flow diagrams and you will learn the basic symbols and a set of rules for drawing them. You will learn about what to do and what *not* to do when drawing data flow diagrams. You will learn two important concepts related to data flow diagrams: *balancing* and *decomposition*. You will also learn the differences between four different types of data flow diagrams: current physical, current logical, new logical, and new physical. Finally, at the end of the chapter, you will learn how to use data flow diagrams as part of the analysis of an information system.

PROCESS MODELING

Process modeling involves graphically representing the functions, or processes, which capture, manipulate, store, and distribute data between a system and its environment and between components within a system. Over the years, several different tools have been developed for process modeling. In this chapter, we focus solely on data flow diagrams, the traditional process modeling technique of structured analysis and design and the technique most often used today for process modeling.

Data flow diagramming is one of several notations that are called structured analysis techniques. Although not all organizations use each structured analysis technique, collectively techniques like data flow diagrams have had a significant impact on the quality of the systems development process. For example, Raytheon (Gibbs, 1994) has reported a savings from 1988 through 1994 of $17.2 million in software costs by applying structured analysis techniques, due mainly to avoiding rework to fix requirements flaws. This represents a doubling of systems developers' productivity and helped them avoid costly system mistakes.

Modeling a System's Process

As Figure 9-1 shows, there are three sub-phases of the analysis phase of the systems development life cycle: requirements determination, requirements structuring, and generating alternative systems and selecting the best one. The analysis team enters requirements structuring with an abundance of information gathered during requirements determination. During requirements structuring, you and the other team members must organize the information into a meaningful representation of the information system that exists and of the requirements desired in a replacement system. In addition to modeling the processing elements of an information system and how data are transformed in the system, you must also model the processing logic and the timing of events in the system (Chapter 10) and the structure of data within the system (Chapter 11). Thus, a process model is only one of three major complementary views of an information system. Together, process, logic and timing, and data models provide a thorough specification of an information system and, with the proper supporting tools, also provide the basis for the automatic generation of many working information system components.

Deliverables and Outcomes

In structured analysis, the primary deliverables from process modeling are a set of coherent, inter-related data flow diagrams, similar to those presented earlier in Figure 2-4. Table 9-1 provides a more detailed list of the deliverables that result from studying and documenting a system's process. First, a context diagram shows the scope of the system, indicating which elements are inside and which are outside the system. Second, data flow diagrams of the current physical system specify which people and technologies are used in which processes to move and transform data, accepting inputs and producing outputs. These diagrams are developed into sufficient detail to understand the current system and to eventually determine how to convert the current system into its replacement. Third, technology-independent, or

Figure 9-1

Systems development life cycle with the analysis phase highlighted

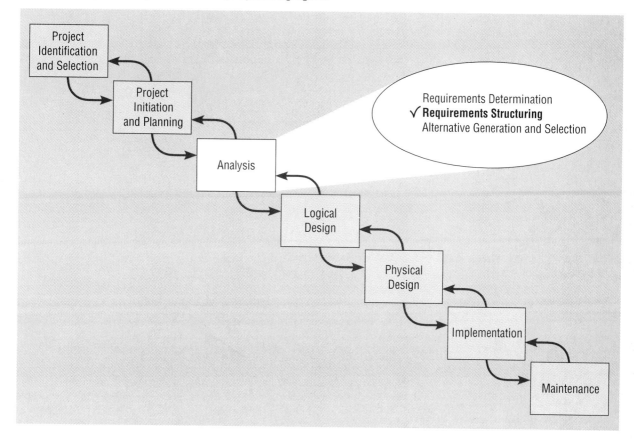

TABLE 9-1 Deliverables for Process Modeling

1. Context data flow diagram (DFD)

2. DFDs of current physical system (adequate detail only)

3. DFDs of current logical system

4. DFDs of new logical system

5. Thorough descriptions of each DFD component

logical, data flow diagrams of the current system show what data processing functions are performed by the current information system. Fourth, the data movement, or flow, structure, and functional requirements of the new system are represented in logical data flow diagrams. Finally, entries for all of the objects included in all diagrams are included in the project dictionary or CASE repository. This logical progression of deliverables allows you to understand the existing system. You can then abstract this system into its essential elements to show the way in which the new system should meet its information processing requirements identified during requirements determination. Remember, the deliverables of process modeling are simply stating *what* you learned during requirements determination; in later steps in the systems development life cycle, you and other project team members will make decisions on exactly *how* the new system will deliver these new requirements in specific manual and automated functions. Since requirements determination and structuring are often parallel steps, data flow diagrams evolve from the more general to the more detailed as current and replacement systems are better understood.

Even though data flow diagrams remain popular tools for process modeling and can significantly increase software development productivity, as reported above for Raytheon, data flow diagrams are not used in all systems development methodologies. Some organizations, like Electronic Data Systems, have developed their own type of diagrams to model processes. Some methodologies, such as Rapid Application Development (RAD), do not model process separately at all. Instead RAD builds process into the prototypes that are created as the core of its development life cycle (see Appendix B). However, even if you never formally use data flow diagrams in your professional career, they remain a part of system development's history. DFDs give you a notation as well as illustrate important concepts about the movement of data between manual and automated steps and a way to depict work flow in an organization. DFDs continue to be beneficial to information systems professionals as tools for both analysis and communication. For that reason, we devote an entire chapter to DFDs.

DATA FLOW DIAGRAMMING MECHANICS

Data flow diagrams are versatile diagramming tools. With only four symbols, you can use data flow diagrams to represent both physical and logical information systems. Data flow diagrams (DFDs) are not as good as flow charts (see Figure 2-12, for example) for depicting the details of physical systems; on the other hand, flow charts are not very useful for depicting purely logical information flows. In fact, flow charting has been criticized by proponents of structured analysis and structured design because it is too physically oriented. Flow charting symbols primarily represent physical computing equipment, such as punch cards, terminals, and tape reels. One continual criticism of system flow charts has been that reliance on them tends to result in premature physical system design. Consistent with the incremental commitment philosophy of the SDLC, you should wait to make technology choices and to decide on physical characteristics of an information system until you are sure all functional requirements are right and accepted by users and other stakeholders.

DFDs do not share this problem of premature physical design because they do not rely on any symbols to represent specific physical computing equipment. They are also easier to use than flow charts as they involve only four different symbols.

Definitions and Symbols

There are two different standard sets of data flow diagram symbols, but each set consists of four symbols that represent the same things: data flows, data stores, processes, and sources/sinks (or external entities). The set of symbols we will use in this book was devised by Gane and Sarson (1979). The other standard set was developed by DeMarco and Yourdon (DeMarco, 1979; Yourdon and Constantine, 1979).

A *data flow* can be best understood as data in motion, moving from one place in a system to another. A data flow could represent data on a customer order form or a payroll check. A data flow could also represent the results of a query to a database, the contents of a printed report, or data on a data entry computer display form. A data flow is data that move together. Thus, a data flow can be composed of many individual pieces of data that are generated at the same time and flow together to common destinations. A **data store** is data at rest. A data store may represent one of many different physical locations for data, for example, a file folder, one or more computer-based file(s), or a notebook. To understand data movement and handling in a system, the physical configuration is not really important. A data store might contain data about customers, students, customer orders, or supplier invoices. A **process** is the work or actions performed on data so that they are transformed, stored, or distributed. When modeling the data processing of a system, it doesn't matter whether a process is performed manually or by a computer. Finally, a **source/sink** is the origin and/or destination of the data. Source/sinks are sometimes referred to as external entities because they are outside the system. Once processed, data or information leave the system and go to some other place. Since sources and sinks are outside the system we are studying, there are many characteristics of sources and sinks that are of no interest to us. In particular, we do not consider the following:

- Interactions that occur between sources and sinks

- What a source or sink does with information or how it operates (that is, a source or sink is a "black box")

- How to control or redesign a source or sink since, from the perspective of the system we are studying, the data a sink receives and often what data a source provides are fixed

- How to provide sources and sinks direct access to stored data since, as external agents, they cannot directly access or manipulate data stored within the system; that is, processes within the system must receive or distribute data between the system and its environment

These principles are consistent with the concepts of system, boundaries, and environment presented in Chapter 3.

The symbols for each set of DFD conventions are presented in Figure 9-2. For both conventions, a data flow is depicted as an arrow. The arrow is labeled with a

Data store: Data at rest, which may take the form of many different physical representations.

Process: The work or actions performed on data so that they are transformed, stored, or distributed.

Source/sink: The origin and/or destination of data, sometimes referred to as external entities.

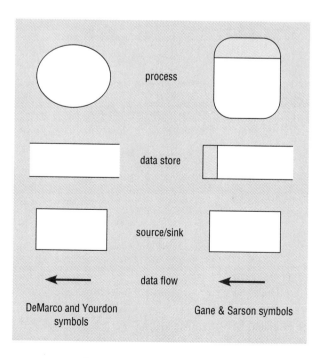

process

data store

source/sink

data flow

DeMarco and Yourdon
symbols

Gane & Sarson symbols

Figure 9-2
Comparison of DeMarco and Yourdan and Gane & Sarson DFD symbol sets

meaningful name for the data in motion; for example, customer order, sales receipt, or paycheck. The name represents the aggregation of all the individual elements of data moving as part of one packet, that is, all the data moving together at the same time. A square is used in both conventions for sources/sinks and has a name which states what the external agent is, such as customer, teller, EPA office, or inventory control system. The Gane & Sarson symbol for a process is a rectangle with rounded corners; it is a circle for DeMarco and Yourdon. The Gane & Sarson rounded rectangle has a line drawn through the top. The upper portion is used to indicate the number of the process. Inside the lower portion is a name for the process, such as generate paycheck, calculate overtime pay, or compute grade point average. The Gane & Sarson symbol for a data store is a rectangle that is missing its right vertical side. At the left end is a small box used to number the data store and inside the main part of the rectangle is a meaningful label for the data store, such as student file, transcripts, or roster of classes. The DeMarco data store symbol consists of two parallel lines, which may be depicted horizontally or vertically.

As stated earlier, sources/sinks are *always* outside the information system and define the boundaries of the system. Data must originate outside a system from one or more sources and the system must produce information to one or more sinks (these are principles of open systems, and almost every information system is an example of an open system). If any data processing takes place inside the source/sink, we are not interested in it, as this processing takes place outside of the system we are diagramming. A source/sink might consist of the following:

- Another organization or organization unit which sends data to or receives information from the system you are analyzing (for example, a supplier or an academic department—in either case, this organization is external to the system you are studying)

- A person inside or outside the business unit supported by the system you are analyzing and who interacts with the system (for example, a customer or loan officer)

- Another information system with which the system you are analyzing exchanges information

Many times students who are just learning how to use DFDs will be confused about whether something is a source/sink or a process within a system. This dilemma occurs most often when the data flows in a system cross office or departmental boundaries so that some processing occurs in one office and the processed data is moved to another office where additional processing occurs. Students are tempted to identify the second office as a source/sink to emphasize the fact that the data have been moved from one physical location to another (Figure 9-3a). However, we are not concerned with where the data are physically located. We are more interested in how they are moving through the system and how they are being processed. If the processing of data in the other office may be automated by your system or the handling of data there may be subject for redesign, then you should represent the second office as one or more processes rather than as a source/sink (Figure 9-3b).

Developing DFDs: An Example

To illustrate how DFDs are used to model the logic of data flows in information systems, we will present and work through an example. Consider Hoosier Burger's

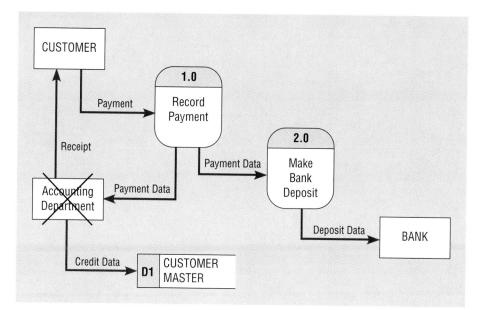

Figure 9-3
Differences between sources/sinks and processes

(a) An improperly drawn DFD showing a process as a source/sink

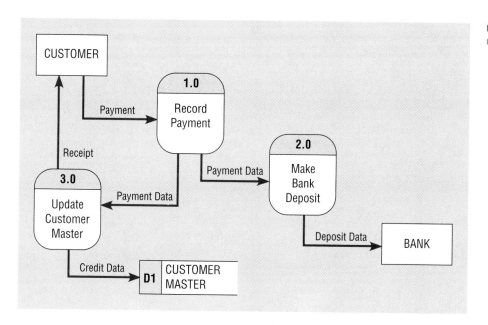

(b) A DFD showing proper use of a process

food ordering system, which you first saw in Chapter 3. The highest-level view of this system, a *context diagram,* is shown in Figure 9-4. You will notice that this context diagram contains only one process, no data stores, four data flows, and three sources/sinks. The single process, labeled "0," represents the entire system; all context diagrams have only one process labeled "0." The sources/sinks represent its environmental boundaries. Since the data stores of the system are conceptually inside the one process, no data stores appear on a context diagram.

The next step for the analyst is to think about which processes are represented by the single process in the context diagram. As you can see in Figure 9-5, we have identified four separate processes, providing more detail of the system we are

Figure 9-4

Context diagram of Hoosier Burger's food ordering system

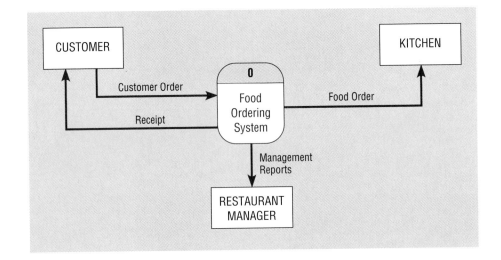

studying. The main processes represent the major functions of the system and these major functions correspond to such actions as the following:

1. Capturing data from different sources (e.g., Process 1.0)

2. Maintaining data stores (e.g. Processes 2.0 and 3.0)

3. Producing and distributing data to different sinks (e.g., Process 4.0)

4. High-level descriptions of data transformation operations (e.g., Process 1.0).

Often, these major functions correspond to the selections of activities on the main system menu.

We see the system begins with an order from a customer, as was the case with the context diagram. In the first process, labeled "1.0," we see that the customer order is processed. The results are four streams or flows of data: (1) the food order is transmitted to the kitchen (2) the customer order is transformed into a list of goods sold (3) the customer order is transformed into inventory data and (4) the process generates a receipt for the customer.

Notice that the sources/sinks are the same in the context diagram and in this diagram: the customer, the kitchen, and the restaurant's manager. This diagram is called a **level-0 diagram** as it represents the primary individual processes in the system at the highest possible level. Each process has a number which ends in .0 (corresponding to the level number of the DFD).

Two of the data flows generated by the first process, "Receive and Transform Customer Food Order," go to external entities so we no longer have to worry about them. We are not concerned about what happens outside of our system. Let's trace the flow of the data represented in the other two data flows. First, the data labeled Goods Sold goes to Process 2.0, Update Goods Sold File. The output for this process is labeled Formatted Goods Sold Data. This output updates a data store labeled Goods Sold File. If the customer order was for two cheeseburgers, one order of fries, and a large soft drink, each of these categories of goods sold in the data store would be incremented appropriately. The daily goods sold amounts are then used as input to Process 4.0, Produce Management Reports. Similarly, the remaining data flow generated by Process 1.0, called Inventory Data, serves as input for Process 3.0, Update Inventory File. This process updates the Inventory File data store, based

Level-0 diagram: A data flow diagram that represents a system's major processes, data flows, and data stores at a high level of detail.

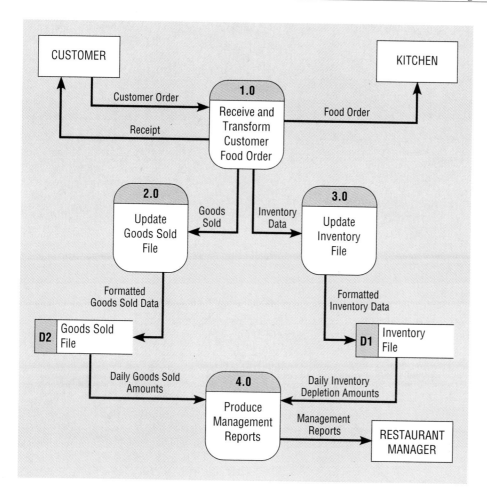

Figure 9-5
Level-0 DFD of Hoosier
Burger's food ordering
system

on the inventory that would have been used to create the customer order. For example, an order of two cheeseburgers would mean that Hoosier Burger now has two fewer hamburger patties, two fewer burger buns, and four fewer slices of American cheese. The Daily Inventory Depletion Amounts are then used as input to Process 4. The data flow leaving Process 4.0, Management Reports, goes to the sink Restaurant Manager.

Figure 9-5 illustrates several important concepts about information movement. Consider the data flow Inventory Data moving from Process 1.0 to Process 3.0. We know from this diagram that Process 1.0 produces this data flow and that Process 3.0 receives it. However, we do not know the timing of when this data flow is produced, how frequently it is produced, or what volume of data is sent. Thus, this DFD hides many physical characteristics of the system it describes. We do know, however, that this data flow is needed by Process 3.0 and that Process 1.0 provides this needed data.

Also implied by the Inventory Data data flow is that whenever Process 1.0 produces this flow, Process 3.0 must be ready to accept it. Thus, Processes 1.0 and 3.0 are coupled to each other. In contrast, consider the link between Process 2.0 and Process 4.0. The output from Process 2.0, Formatted Goods Sold Data, is placed in a data store and, later, when Process 4.0 needs such data, it reads Daily Goods Sold Amounts from this data store. In this case, Processes 2.0 and 4.0 are decoupled by placing a buffer, a data store, between them. Now, each of these processes can work at their own pace and Process 4.0 does not have to be vigilant by being able to accept input at any time. Further, the Goods Sold File becomes a data resource that other processes could potentially draw upon for data.

Data Flow Diagramming Rules

There is a set of rules you must follow when drawing data flow diagrams. Unlike system flow charts, these rules allow you (or a CASE tool) to evaluate DFDs for correctness. The rules for DFDs are listed in Table 9-2. Figure 9-6 illustrates incorrect ways to draw DFDs and the corresponding correct application of the rules. The rules that prescribe naming conventions (rules C, G, I, and P) and those that explain how to interpret data flows in and out of data stores (rules N and O) are not illustrated in Figure 9-6.

Besides the rules of Table 9-2, there are two DFD guidelines that apply most of the time:

- *The inputs to a process are different from the outputs of that process:* The reason is that processes, to have a purpose, typically transform inputs into outputs, rather than simply pass the data through without some manipulation. What may happen is that the same input goes in and out of a process but

---

TABLE 9-2   Rules Governing Data Flow Diagramming

Process:

A. No process can have only outputs. It is making data from nothing (a miracle). If an object has only outputs, then it must be a source.

B. No process can have only inputs (a black hole). If an object has only inputs, then it must be a sink.

C. A process has a verb phrase label.

Data Store:

D. Data cannot move directly from one data store to another data store. Data must be moved by a process.

E. Data cannot move directly from an outside source to a data store. Data must be moved by a process which receives data from the source and places the data into the data store.

F. Data cannot move directly to an outside sink from a data store. Data must be moved by a process.

G. A data store has a noun phrase label.

Source/Sink:

H. Data cannot move directly from a source to a sink. It must be moved by a process if the data are of any concern to our system. Otherwise, the data flow is not shown on the DFD.

I. A source/sink has a noun phrase label.

Data Flow:

J. A data flow has only one direction of flow between symbols. It may flow in both directions between a process and a data store to show a read before an update. The latter is usually indicated, however, by two separate arrows since these happen at different times.

K. A fork in a data flow means that exactly the same data goes from a common location to two or more different processes, data stores, or sources/sinks (this usually indicates different copies of the same data going to different locations).

L. A join in a data flow means that exactly the same data comes from any of two or more different processes, data stores, or sources/sinks to a common location.

M. A data flow cannot go directly back to the same process it leaves. There must be at least one other process which handles the data flow, produces some other data flow, and returns the original data flow to the beginning process.

N. A data flow to a data store means update (delete or change).

O. A data flow from a data store means retrieve or use.

P. A data flow has a noun phrase label. More than one data flow noun phrase can appear on a single arrow as long as all of the flows on the same arrow move together as one package.

Adapted from Celko (1987)

Figure 9-6
Incorrect and correct ways to
draw data flow diagrams

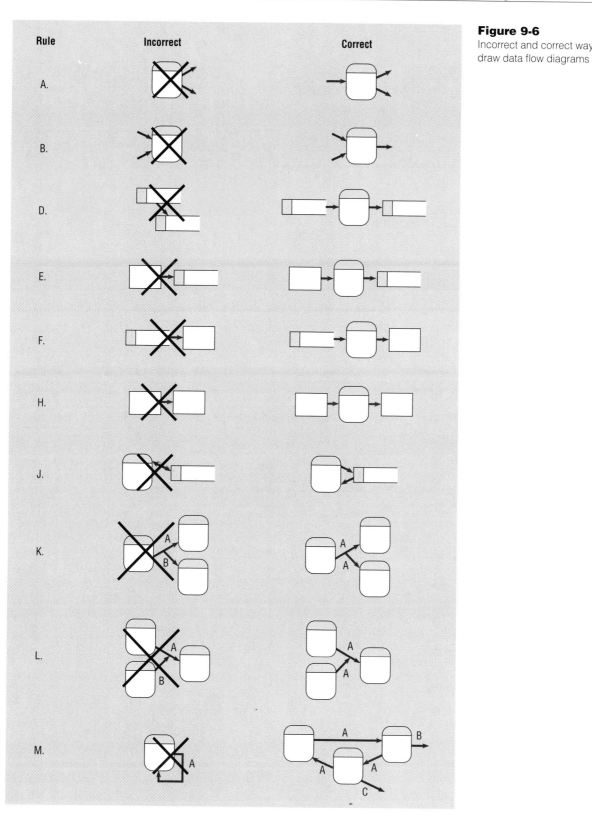

the process also produces other new data flows that are the result of manipulating the inputs.

- *Objects on a DFD have unique names:* Every process has a unique name. There is no reason to have two processes with the same name. To keep a DFD uncluttered, however, you may repeat data stores and sources/sinks. When two arrows have the same data flow name, you must be careful that these flows are exactly the same. It is easy to reuse the same data flow name when two packets of data are almost the same, but not identical. A data flow name represents a specific set of data, and another data flow that has even one more or one less piece of data must be given a different, unique name.

Decomposition of DFDs

In the earlier example of Hoosier Burger's food ordering system, we started with a high-level context diagram. Upon thinking more about the system, we saw that the larger system consisted of four processes. The act of going from a single system to four component processes is called *(functional) decomposition,* which we defined in Chapter 2. For the Hoosier Burger system, we broke down or decomposed the larger system into four processes. Each of those processes (or subsystems) are also candidates for decomposition. Each process may consist of several sub-processes. Each sub-process may also be broken down into smaller units. Decomposition continues until you have reached the point where no sub-process can logically be broken down any further. The lowest level of DFDs is called a *primitive DFD,* which we define later in this chapter.

Let's continue with Hoosier Burger's food ordering system to see how a level-0 DFD can be further decomposed. The first process in Figure 9-5, called Receive and Transform Customer Food Order, transforms a customer's verbal food order (for example, "Give me two cheeseburgers, one small order of fries, and one regular orange soda.") into four different outputs. Process 1.0 is a good candidate process for decomposition. Think about all of the different tasks that Process 1.0 has to perform: (1) receive a customer order (2) transform the entered order into a form meaningful for the kitchen's system (3) transform the order into a printed receipt for the customer (4) transform the order into goods sold data (5) transform the order into inventory data. There are at least these five logically separate functions that occur in Process 1.0. We can represent the decomposition of Process 1.0 as another DFD, as shown in Figure 9-7.

Note that each of the five processes in Figure 9-7 are labeled as sub-processes of Process 1.0: Process 1.1, Process 1.2, and so on. Also note that, just as with the other data flow diagrams we have looked at, each of the processes and data flows are named. You will also notice that there are no sources or sinks represented. Although you may include sources and sinks, the context and level-0 diagrams show the sources and sinks. The data flow diagram in Figure 9-7 is called a level-1 diagram. If we should decide to decompose Processes 2.0, 3.0, or 4.0 in a similar manner, the DFDs we create would also be called level-1 diagrams. In general, a **level-*n* diagram** is a DFD that is generated from *n* nested decompositions from a level-0 diagram.

Level-*n* diagram: A DFD that is the result of *n* nested decompositions of a series of sub-processes from a process on a level-0 diagram.

Processes 2.0 and 3.0 perform similar functions in that they both use data input to update data stores. Since updating a data store is a singular logical function, neither of these processes need to be decomposed further. We can, on the other hand, decompose Process 4.0, Produce Management Reports, into at least three sub-processes: Access Goods Sold and Inventory Data, Aggregate Goods Sold and

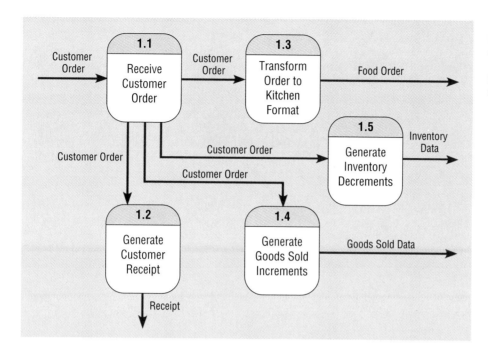

Figure 9-7
Level-1 diagram showing the decomposition of Process 1.0 from the level-0 diagram for Hoosier Burger's food ordering system

Inventory Data, and Prepare Management Reports. The decomposition of Process 4.0 is shown in the level-1 diagram of Figure 9-8.

Each level-1, -2, or -*n* DFD represents one process on a level-*n*-1 DFD; each DFD should be on a separate page. As a rule of thumb, no DFD should have more than about seven processes in it, as too many processes will make the diagram too crowded and more difficult to understand.

To continue with the decomposition of Hoosier Burger's food ordering system, we examine each of the sub-processes identified in the two level-1 diagrams we have produced, one for Process 1.0 and one for Process 4.0. Should we decide

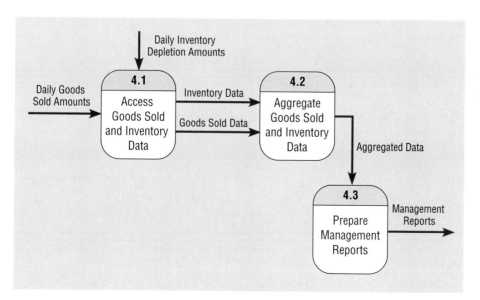

Figure 9-8
Level-1 diagram showing the decomposition of Process 4.0 from the level-0 diagram for Hoosier Burger's food ordering system

that any of these sub-processes should be further decomposed, we would create a level-2 diagram showing that decomposition. For example, if we decided that Process 4.3 in Figure 9-8 should be further decomposed, we would create a diagram that looks like something like Figure 9-9. Again, notice how the sub-processes are labeled.

Just as the labels for processes must follow numbering rules for clear communication, process names should also be clear yet concise. Typically, process names begin with an action verb, such as receive, calculate, transform, generate, or produce. Often process names are the same as the verbs used in many computer programming languages. Examples include merge, sort, read, write, and print. Process names should capture the essential action of the process in just a few words, yet be descriptive enough of the process' action so that anyone reading the name gets a good idea of what the process does. Many times, students just learning DFDs will use the names of people who perform the process or the department in which the process is performed as the process name. This practice is not very useful, as we are more interested in the action the process represents than the person performing it or the place where it occurs.

Balancing DFDs

When you decompose a DFD from one level to the next, there is a conservation principle at work. You must conserve inputs and outputs to a process at the next level of decomposition. In other words, Process 1.0, which appears in a level-0 diagram, must have the same inputs and outputs when decomposed into a level-1 diagram. This conservation of inputs and outputs is called **balancing.**

Balancing: The conservation of inputs and outputs to a data flow diagram process when that process is decomposed to a lower level.

Let's look at an example of balancing a set of DFDs. Look back at Figure 9-4. This is the context diagram for Hoosier Burger's food ordering system. Notice that there is one input to the system, the customer order, which originates with the customer. Notice also that there are three outputs: the customer receipt, the food order intended for the kitchen, and management reports. Now look at Figure 9-5. This is the level-0 diagram for the food ordering system. Remember that all data stores and flows to or from them are internal to the system. Notice that the same single input to the system and the same three outputs represented in the context diagram also appear at level-0. Further, no new inputs to or outputs from the system have been introduced. Therefore, we can say that the context diagram and level-0 DFDs are balanced.

Now look at Figure 9-7, where Process 1.0 from the level-0 DFD has been decomposed. As we have seen before, Process 1.0 has one input and four outputs. The single input and multiple outputs all appear on the level-1 diagram in Figure 9-7. No new inputs or outputs have been added. Compare Process 4.0 in Figure 9-5 to its decomposition in Figure 9-8. You see the same conservation of inputs and outputs.

Figure 9-9

Level-2 diagram showing the decomposition of Process 4.3 from the level-1 diagram for Process 4.0 for Hoosier Burger's food ordering system

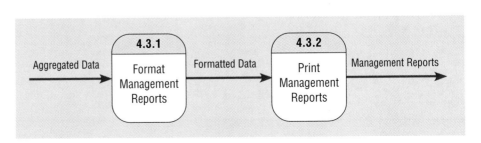

Figure 9-10 shows you one example of what an unbalanced DFD could look like. Here, in the context diagram, there is one input to the system, A, and one output, B. Yet, in the level-0 diagram, there is an additional input, C, and flows A and C come from different sources. These two DFDs are not balanced. If an input appears on a level-0 diagram, it must also appear on the context diagram. What happened in this example? Perhaps, when drawing the level-0 DFD, the analyst realized that the system also needed C in order to compute B. A and C were both drawn in the level-0 DFD, but the analyst forgot to update the context diagram. In making corrections, the analyst should also include "SOURCE ONE" and "SOURCE TWO" on the context diagram. It is very important to keep DFDs balanced, from the context diagram all the way through each level diagram you must create.

A data flow consisting of several sub-flows on a level-n diagram can be split apart on a level-$n+1$ diagram for a process which accepts this composite data flow as input. For example, consider the partial DFDs from Hoosier Burger illustrated in Figure 9-11. In Figure 9-11a we see that a composite, or package, data flow, Payment and Coupon, is input to the process. That is, the payment and coupon always flow together and are input to the process at the same time. In Figure 9-11b the process is decomposed (sometimes called exploded or nested) into two sub-processes, and each sub-process receives one of the components of the composite data flow from the higher-level DFD. These diagrams are still balanced since exactly the same data are included in each diagram.

The principle of balancing and the goal of keeping a DFD as simple as possible lead to four additional, advanced rules for drawing DFDs. These advanced rules are summarized in Table 9-3. Rule Q covers the situation illustrated in Figure 9-11. Rule R covers a conservation principle about process inputs and outputs. Rule S addresses one exception to balancing. Rule T tells you how you can minimize clutter on a DFD.

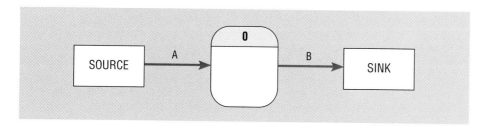

Figure 9-10
An unbalanced set of data flow diagrams

(a) Context diagram

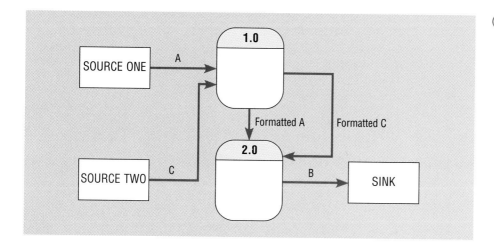

(b) Level-0 diagram

Figure 9-11
Example of data flow splitting

(a) Composite data flow

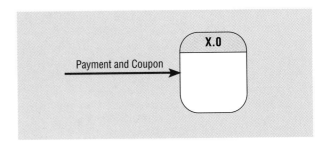

(b) Disaggregated data flows

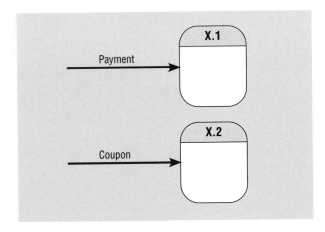

TABLE 9-3   Advanced Rules Governing Data Flow Diagramming

Q. A composite data flow on one level can be split into component data flows at the next level, but no new data can be added and all data in the composite must be accounted for in one or more sub-flows.

R. The inputs to a process must be sufficient to produce the outputs (including data placed in data stores) from the process. Thus, all outputs can be produced, and all data in inputs move somewhere, either to another process or to a data store outside the process or on a more detailed DFD showing a decomposition of that process.

S. At the lowest level of DFDs, new data flows may be added to represent data that are transmitted under exceptional conditions; these data flows typically represent error messages (e.g., "Customer not known; do you want to create a new customer") or confirmation notices (e.g., "Do you want to delete this record").

T. To avoid having data flow lines cross each other, you may repeat data stores or sources/sinks on a DFD. Use an additional symbol, like a double line on the middle vertical line of a data store symbol, or a diagonal line in a corner of a sink/source square, to indicate a repeated symbol.

Adapted from Celko (1987)

FOUR DIFFERENT TYPES OF DFDS

There are actually four different types of data flow diagrams used in the systems development process: (1) current physical (2) current logical (3) new logical (4) new physical. When structured analysis and design was first introduced in the late 1970s, it was argued that analysts should prepare all four types of DFDs, in this particular order.

In a current physical DFD, process labels include the names of people or their positions or the names of computer systems that might provide some of the overall system's processing. That is, the label includes an identification of the "technology" used to process the data. Similarly, data flows and data stores are often labeled with the names of the actual physical media on which data flow or in which data are stored, such as file folders, computer files, business forms, or computer tapes. For the current logical model, the physical aspects of the system are removed as much as possible so that the current system is reduced to its essence, to the data and the processes that transform them, regardless of actual physical form. The new logical model would be exactly like the current logical model if the user were completely happy with the functionality of the current system but had problems with how it was implemented. Typically, though, the new logical model will differ from the current logical model by having additional functions, obsolete functions removed, and inefficient flows reorganized. Finally, the DFDs for the new physical system represent the physical implementation of the new system. The DFDs for the new physical system will reflect the decision of the analysts about which system functions, including those added in the new logical model, will be automated and which will be manual.

To illustrate the differences among the different types of DFDs, we will look at another example from Hoosier Burger. We saw that the food ordering system generates two types of usage data, for goods sold and for inventory. At the end of each day, the manager, Bob Mellankamp, generates the inventory report which tells him how much inventory should have been used for each item associated with sales. The amounts shown on the inventory report are just one input to a largely manual inventory control system Bob uses every day. Figure 9-12 lists the steps involved in Bob's inventory control system.

The data flow diagrams that model the current physical system are shown in Figure 9-13. The context diagram (Figure 9-13a) shows three sources of data outside the system: suppliers, the food ordering system inventory report, and stock-on-hand. Suppliers provide invoices as input, and the system returns payments and orders as outputs to the suppliers. Both the inventory report and the stock-on-hand provide inventory counts as system inputs. The level-0 DFD for Hoosier Burger's inventory system (Figure 9-13b) shows six different processes, most of which also appear in the list of inventory activities in Figure 9-12. You can see from the diagram that when Bob receives invoices from suppliers, he records their receipt on an invoice log sheet and files the actual invoices in his accordion file. Using the invoices, Bob records the amount of stock delivered on the stock logs, which are

---

1. Meet delivery trucks before opening restaurant.
2. Unload and store deliveries.
3. Log invoices and file in accordion file.
4. Manually add amounts received to stock logs.
5. After closing, print inventory report.
6. Count physical inventory amounts.
7. Compare inventory report totals to physical count totals.
8. Compare physical count totals to minimum order quantities; if the amount is less, make order; if not, do nothing.
9. Pay bills that are due and record them as paid.

Figure 9-12

List of activities involved in Bob Mellankamp's inventory control system for Hoosier Burger

paper forms posted near the point of storage for each inventory item. Figure 9-13 also illustrates that a physical DFD shows only data movement, not the movement of materials or other physical items (e.g., energy). Data about physical items such as an Invoice which is data about a shipment to Hoosier Burger, are shown on a physical DFD; the data about a physical item may or may not move with the actual physical item.

Figure 9-13
Hoosier Burger's current physical inventory control system

(a) Context diagram

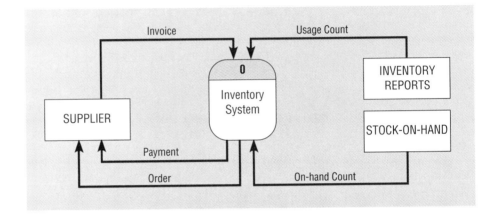

(b) Level-0 data flow diagram

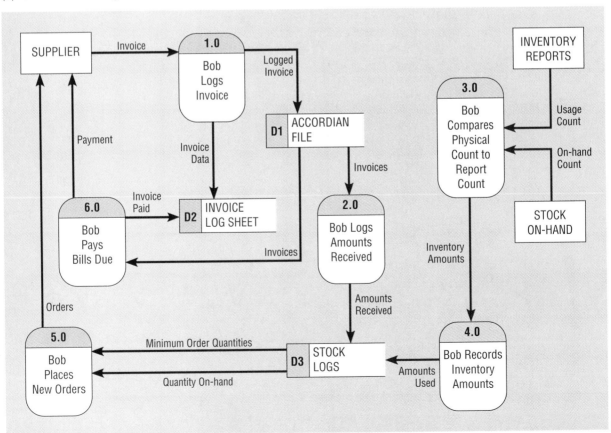

Figure 9-14 gives a partial example of Hoosier Burger's stock log. Notice that the minimum order quantities—the stock level at which orders must be placed in order to avoid running out of an item—appear on the log form. There are also spaces for entering the starting amount, amount delivered, and the amount used for each item. Amounts delivered are entered on the sheet when Bob logs stock deliveries; amounts used are entered after Bob has compared the amounts of stock used according to a physical count and according to the numbers on the inventory report generated by the food ordering system. We should note that Hoosier Burger has standing daily delivery orders for some perishable items that are used every day, like burger buns, meats, and vegetables.

As the DFD in Figure 9-13b shows, Bob uses the minimum order quantities and the amount of stock on hand to determine which orders need to be placed. He uses the invoices to determine which bills need to be paid, and he carefully records each payment.

To continue to the next step, creating the current logical model using DFDs, we need to identify the essence of the inventory system Bob has established. What are the key data necessary to keep track of inventory and to pay bills? What are the key processes involved? We also need to remove all elements of the physical system, such as Bob and the physical file folders he uses. There are at least four key processes that make up the Hoosier Burger's inventory system: (1) account for anything added to inventory (2) account for anything taken from inventory (3) place orders (4) pay bills. Key data used by the system include inventories and stock-on-hand counts, however determined. Major outputs from the system continue to be orders and payments. Focusing on the essential elements of the system results in the DFD for the current logical system shown in Figure 9-15.

The purpose of the new logical DFD is (1) to show any additional functionality necessary in the new system (2) to indicate which, if any, obsolete components have been eliminated, and (3) to describe any changes in the logical flow of data between system components, including different data stores. For Hoosier Burger's inventory system, Bob Mellankamp would like to add three additional functions. First, Bob would like data on new shipments to be entered into an automated system, thus doing away with paper stock log sheets. Bob would like shipment data to be as current as possible by being entered into the system as soon as the new stock arrives at the restaurant. Second, Bob would like the system to determine automatically

Figure 9-14

Hoosier Burger's stock log form

Stock Log					
Date:			Jan 1			Jan 2
	Reorder	Starting	Amount	Amount	Starting
Item	*Quantity*	Amount	Delivered	Used	Amount
Hamburger buns	50 dozen	5	50	43	12
Hot dog buns	25 dozen	0	25	22	3
English muffins	10 dozen	6	10	12	4
Napkins	2 cases	10	0	2	8
Straws	1 case	1	0	1	0

Figure 9-15

Level-0 data flow diagram for Hoosier Burger's current logical inventory control system

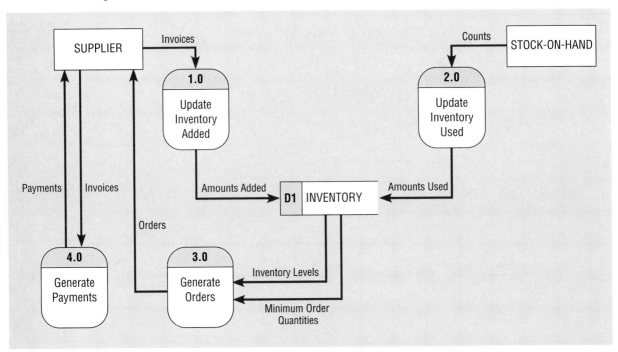

whether a new order should be placed. Automatic ordering would relieve Bob of worrying about whether Hoosier Burger has enough of everything in stock at all times. Finally, Bob would like to be able to know, at any time, the approximate inventory levels for all goods in stock. For some goods, such as hamburger buns, Bob can visually inspect the amount in stock and determine approximately how much is left and how much more is needed before closing time. For other items, however, Bob may need a quick, rough estimate of what is in stock more quickly than it would take him to do a visual inspection.

The new logical data flow diagram for Hoosier Burger's inventory system is shown in Figure 9-16. Notice how the DFD is almost identical to the DFD for current logical, in Figure 9-15. The only difference is a new Process 5.0, which allows for querying the inventory data to get an estimate of how much of an item is in stock. Bob's two other requests for change can both be handled within the existing logical view of the inventory system. Process 1.0, Update Inventory Added, does not indicate whether the updates are in real-time or batched or whether the updates occur on paper or as part of an automated system. Therefore, immediately entering shipment data into an automated system is encompassed by Process 1.0. Similarly, Process 2.0, Generate Orders, does not indicate whether Bob or a computer generates orders or whether the orders are generated on a real-time or batch basis, so Bob's request that orders be generated automatically by the system is already represented by Process 3.0. The next step would be to create a data flow diagram for the new physical system to represent the requests Bob has made. That step will be deferred, however, until several alternative solutions to Bob's inventory problems have been considered in Chapter 12.

You may be asking if it is really necessary for an analyst to construct not just one but four complete sets of data flow diagrams for each system on which he or she works. Many experts today say no, it is not necessary, that analysts should

Figure 9-16

Level-0 data flow diagram for Hoosier Burger's new logical inventory control system

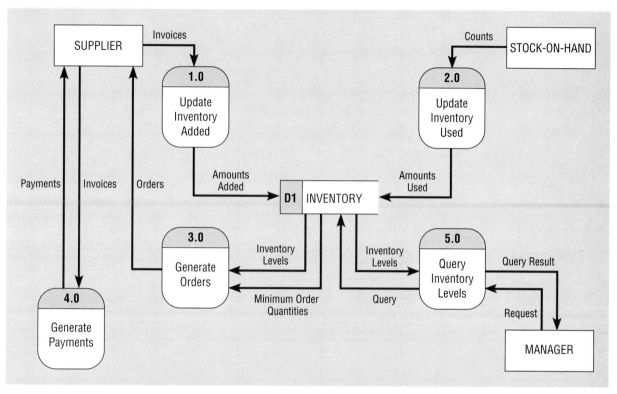

begin as quickly as possible with the new logical DFD. Experts used to recommend that all four levels of DFDs be constructed because of three assumptions:

1. Analysts knew little about the user's business and needed to develop a detailed current physical DFD in order to understand the business.

2. Users were not able to work with a new logical DFD right away.

3. There is not much work in turning current logical DFDs into new logical DFDs.

These assumptions proved to be correct (Yourdon, 1989) but overlooked a greater danger: analysts tended to devote a great deal of time to creating and refining a detailed set of DFDs for the current physical system, most of which was thrown away in the transition to the current logical DFDs. We recommend that you create a set of DFDs for the current physical system but that those DFDs only be detailed enough so that you come away with a good overview of the current system. We agree with the experts that the focus should be on the new logical system.

USING DATA FLOW DIAGRAMMING IN THE ANALYSIS PROCESS

Learning the mechanics of drawing data flow diagrams is important to you, as data flow diagrams have proven to be essential tools for the structured analysis process. Beyond the issues of drawing mechanically correct DFDs, there are other issues related to process modeling with which you as an analyst must be concerned. Such issues, including whether the DFDs are complete and consistent across levels, are dealt with in the next section on guidelines for drawing DFDs. Another issue to consider is how you can use data flow diagrams as a useful tool for analysis, discussed in the final section of the chapter. In these final sections, we also illustrate features of CASE tools that aid in compliance with rules and guidelines, as well as in good systems analysis.

Guidelines for Drawing DFDs

In this section, we will consider additional guidelines for drawing DFDs that extend beyond the simple mechanics of drawing diagrams and making sure that the rules listed in Tables 9-2 and 9-3 are followed. These guidelines include (1) completeness (2) consistency (3) timing considerations (4) the iterative nature of drawing DFDs, and (5) drawing primitive DFDs.

DFD completeness: The extent to which all necessary components of a data flow diagram have been included and fully described.

Completeness   The concept of **DFD completeness** refers to whether you have included in your DFDs all of the components necessary for the system you are modeling. If your DFD contains data flows that do not lead anywhere, or data stores, processes, or external entities that are not connected to anything else, your DFD is not complete. Most CASE tools have built-in facilities that you can run to help you find incompleteness in your DFDs. For example, Figure 9-17a shows a draft of a DFD for the process Bob Mellankamp follows to hire new employees for Hoosier Burger. Figure 9-17b shows the analysis report that a CASE tool would generate indicating the three errors of completeness found on this DFD. Did you identify all the same errors found by the CASE tool? To which of the rules in Tables 9-2 and 9-3 are these errors associated? When you draw many DFDs for a system, it is not uncommon to make errors like those in Figure 9-17; either CASE tool analysis functions or walkthroughs with other analysts can help you identify such problems.

Not only must all necessary elements of a DFD be present, each of the components must be fully described in the project dictionary. With most CASE tools, the project dictionary is linked to the diagram. That is, when you define a process, data flow, source/sink, or data store on a DFD, an entry is automatically created in the repository for that element. You must then enter the repository and complete the element's description. Different descriptive information can be kept about each of the four types of elements on a DFD, and each CASE tool or project dictionary standard an organization adopts has different entry information. Figure 9-18 shows a report of the contents of the VAW repository entry for the Request for Reference data flow on Figure 9-17. In VAW, a data flow repository entry contains

- The label or name (e.g., Request for Reference) for the data flow as entered on DFDs (*Note:* Case and punctuation of the label matter, but if exactly the same label is used on multiple DFDs, whether nested or not, then the same repository entry applies to each reference.)

- A two-line description defining the data flow

- A list of aliases, or other names, by which this same data flow is referred to (e.g., Reference Letter) by different people in the organization

- The composition or list of data elements contained in the data flow

- Notes supplementing the limited space for the description which go beyond defining the data flow to explaining the context and nature of this repository object

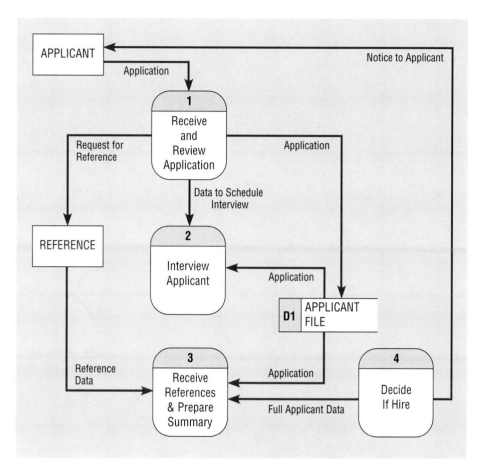

Figure 9-17
Hoosier Burger's hiring
procedures

(a) Data flow diagram

(b) Analysis of completeness report from CASE tool

```
              DFD Analysis Errors [Project 'S330']
Error:  Process labeled 'Interview Applicant' is an input only Process.
Error:  Process labeled 'Receive References & Prepare Summary' is an input only Process.
Error:  Process labeled 'Decide If Hire' is an output only Process.
```

- A list of locations (the names of the DFDs) on which this data flow appears (e.g., the DFD was called "applicant") and the names of the sources and destinations on each of these DFDs for the data flow

By the way, it is this tight linkage between diagrams and the CASE repository which creates much of the value of a CASE tool. Although very sophisticated drawing tools as well as forms and word processing systems exist, these stand-alone tools do not integrate graphical objects with their textual descriptions as CASE tools do.

Consistency   The concept of **DFD consistency** refers to whether or not the depiction of the system shown at one level of a nested set of DFDs is compatible with the depictions of the system shown at other levels. A gross violation of consistency would be a level-1 diagram with no level-0 diagram. Another example of inconsistency would be a data flow that appears on a higher level DFD but not on lower levels (a violation of balancing). Yet another example of inconsistency is a data flow attached to one object on a lower level diagram but attached to another object at a higher level. For example, a data flow named "Payment," which serves as input to

DFD consistency: The extent to which information contained on one level of a set of nested data flow diagrams is also included on other levels.

Figure 9-18
VAW repository entry for a data flow

Date: 5/15/94   *Project:* S330   *Page:* 1
Time: 2:06 PM

Single Entry Listing
Data Flow Diagrams

Request for Reference   Data Flow
    Description:

        A letter sent by Hoosier Burger to individuals or companies listed as
        references on employee applications.

    Alias:

        Reference Letter

    Composition:

        Applicant name
        Date of application
        Position applied for
        Qualifications sought

    Notes:

        This is a personal letter that Bob Mellankamp writes himself. A
        standard part of the letter is a requested date by which the reference is
        to be returned, and this date is two weeks from the date on which Bob's
        letter is sent.

    Location:

        applicant (0)

                Source: Receive and Review Application (Process)
                Dest: REFERENCE (External Entity)

    Date Last Altered: 5/15/94   *Date Created:*   5/15/94

Process 1 on a level-0 DFD, appears as input to Process 2.1 on a level-1 diagram for
Process 2.
    CASE tools also have analysis facilities you can use to detect such inconsisten-
cies across nested data flow diagrams. For example, to help you avoid making DFD

consistency errors when you draw a DFD using a CASE tool, most tools will automatically place the inflows and outflows of a process on the DFD you create when you inform the tool to decompose that process. In manipulating the lower-level diagram, you could accidentally delete or change a data flow which would cause the diagrams to be out of balance; thus, a consistency check facility with a CASE tool is quite helpful.

Timing   You may have noticed in some of the DFD examples we have presented that DFDs do not do a very good job of representing time. On a given DFD, there is no indication of whether a data flow occurs constantly in real-time, once per week, or once per year. There is also no indication of when a system would run. For example, many large transaction-based systems may run several large, computing-intensive jobs in batch mode at night, when demands on the computer system are lighter. A DFD has no way of indicating such overnight batch processing. When you draw DFDs, then, draw them as if the system you are modeling has never started and will never stop. You will learn in Chapter 10 that another type of diagram, a state-transition diagram, is used to show the timing of processes associated with a system.

Iterative Development   The first DFD you draw will rarely capture perfectly the system you are modeling. You should count on drawing the same diagram over and over again, in an iterative fashion. With each attempt, you will come closer to a good approximation of the system or aspect of the system you are modeling. Iterative DFD development recognizes that requirements determination and requirements structuring are interacting, not sequential sub-phases of the analysis phase of the SDLC. One rule of thumb is that it should take you about three revisions for each DFD you draw. Fortunately, CASE tools make revising drawings a lot easier than it would be if you had to draw each revision with pencil and template.

Primitive DFDs   One of the more difficult decisions you need to make when drawing DFDs is when to stop decomposing processes. One rule is to stop drawing when you have reached the lowest logical level; however, it is not always easy to know what the lowest logical level is. Other more concrete rules for when to stop decomposing are

- When you have reduced each process to a single decision or calculation or to a single database operation, such as retrieve, update, create, delete, or read

- When each data store represents data about a single entity, such as a customer, employee, product, or order

- When the system user does not care to see any more detail, or when you and other analysts have documented sufficient detail to do subsequent systems development tasks

- When every data flow does not need to be split further to show that different data are handled in different ways

- When you believe that you have shown each business form or transaction, computer on-line display, and report as a single data flow (This often means, for example, that each system display and report title corresponds to the name of an individual data flow.)

- When you believe there is a separate process for each choice on all lowest-level menu options for the system.

Obviously, the iteration guideline discussed earlier and the various feedback loops in the SDLC (see Figure 9-1) suggest that when you think you have met the above rules for stopping, you may later discover nuances to the system that require you to further decompose a set of DFDs.

By the time you stop decomposing DFDs, a DFD can become quite detailed. Seemingly simple actions, such as generating an invoice, may pull information from several entities and may also return different results depending on the specific situation. For example, the final form of an invoice may be based on the type of customer (which would determine such things as discount rate), where the customer lives (which would determine such things as sales tax), and how the goods are shipped (which would determine such things as the shipping and handling charges).

Primitive DFD: The lowest level of decomposition for a data flow diagram.

At the lowest level DFD, called a **primitive DFD,** all of these conditions would have to be met. Given the amount of detail called for in a primitive DFD, perhaps you can see why many experts believe analysts should not spend their time completely diagramming the current physical information system as much of the detail will be discarded when the current logical DFD is created.

Using these guidelines will help you create DFDs that are more than just mechanically correct. Your data flow diagrams will also be robust and accurate representations of the information system you are modeling. Such primitive DFDs also facilitate consistency checks with the documentation produced from other requirements structuring techniques as well as make it easy for you to transition to system design steps. Having mastered the skills of drawing good DFDs, you can now use them to support the analysis process, the subject of the next section.

Using DFDs as Analysis Tools

We have seen that data flow diagrams are versatile tools for process modeling and that they can be used to model systems that are either physical or logical, current or new. Data flow diagrams can also be used in analysis for a process called *gap analysis.* In gap analysis, the role of the analyst is to discover discrepancies between two or more sets of data flow diagrams, representing two or more states of an information system, or discrepancies within a single DFD.

Once the DFDs are complete, you can examine the details of individual DFDs for such problems as redundant data flows, data that are captured but are not used by the system, and data that are updated identically in more than one location. These problems may not have been evident to members of the analysis team or to other participants in the analysis process when the DFDs were created. For example, redundant data flows may have been labeled with different names when the DFDs were created. Now that the analysis team knows more about the system they are modeling, analysts can detect such redundancies. Such redundancies can be seen most easily from various CASE tool repository reports. For example, many CASE tools can generate a report listing all the processes that accept a given data element as input (remember, a list of data elements is likely part of the description of each data flow). From the label of these processes you can determine whether or not it appears as if the data are captured redundantly or if more than one process is maintaining the same data stores. In such cases, the DFDs may well accurately mirror the activities occurring in the organization. As the business processes being modeled took many years to develop, sometimes with participants in one part of the organization adapting procedures in isolation from other participants, redundancies and overlapping responsibilities may well have resulted. The careful study of the DFDs created as part of analysis can reveal these procedural redundancies and allow them to be corrected as part of system design.

Inefficiencies can also be identified by studying DFDs, and there are a wide variety of inefficiencies that might exist. Some inefficiencies relate to violations of DFD drawing rules. For example, a violation of rule R from Table 9-3 could occur because obsolete data are captured but never used within a system. Other inefficiencies are due to excessive processing steps. For example, consider the correct DFD in item M of Figure 9-6. Although this flow is mechanically correct, such a loop may indicate potential delays in processing data or unnecessary approval operations.

Similarly, a set of DFDs that models the current logical system can be compared to DFDs that model the new logical system to better determine which processes systems developers need to add or revise while building the new system. Processes for which inputs, outputs, and internal steps have not changed can possibly be reused in the construction of the new system. You can compare alternative logical DFDs to identify those few elements which must be discussed in evaluating competing opinions on system requirements. The logical DFDs for the new system can also serve as the basis for developing alternative design strategies for the new physical system. As we saw with the Hoosier Burger example, a process on a new logical DFD can be implemented in several different physical ways.

SUMMARY

Data flow diagrams, or DFDs, are very useful for representing the overall data flows into, through, and out of an information system. Data flow diagrams rely on only four symbols to represent the four conceptual components of a process model: data flows, data stores, processes, and sources/sinks. Data flow diagrams are hierarchical in nature and each level of a DFD can be decomposed into smaller, simpler units on a lower-level diagram. You begin with a context diagram, which shows the entire system as a single process. The next step is to generate a level-0 diagram, which shows the most important high-level processes in the system. You then decompose each process in the level-0 diagram, as warranted, until it makes no logical sense to go any further. When decomposing DFDs from one level to the next, it is important that the diagrams be balanced; that is, inputs and outputs on one level must be conserved on the next level.

There are four sets of data flow diagrams. The first set you create models the current physical system. The next set models the current logical system. The third set models the new logical system, which may be different from the current logical system to the extent that it shows desired functionality, even though the new system may radically re-engineer the flow of data between system components. The fourth set models the new physical information system. Due to the time it takes to create a complete set of DFDs, many experts suggest that you begin work on the new logical DFDs as soon as possible.

Data flow diagrams should be mechanically correct, but they should also accurately reflect the information system being modeled. To that end, you need to check DFDs for completeness and consistency and draw them as if the system being modeled were timeless. You should be willing to revise DFDs several times. Complete sets of DFDs should extend to the primitive level where every component reflects certain irreducible properties; for example, a process represents a single database operation and every data store represents data about a single entity. Following these guidelines, you can produce DFDs to aid the analysis process by analyzing the gaps between existing procedures and desired procedures and between current and new systems.

C H A P T E R   R E V I E W

K E Y   T E R M S

Balancing   Data store   Primitive DFD
DFD completeness   Level-0 diagram   Process
DFD consistency   Level-*n* diagram   Source/sink

R E V I E W   Q U E S T I O N S

1. Define each of the following terms:

   a. new logical data flow diagram   e. level-0 data flow diagram
   b. current logical data flow diagram   f. context diagram
   c. decomposition   g. process modeling
   d. balancing   h. primitive DFD

2. What is a data flow diagram? Why do systems analysts use data flow diagrams?

3. Explain the rules for drawing good data flow diagrams.

4. What is decomposition? What is balancing? How can you determine if DFDs are not balanced?

5. Explain the convention for naming different levels of data flow diagrams.

6. What are the primary differences between current physical and current logical data flow diagrams?

7. Why don't analysts usually draw four complete sets of DFDs?

8. How can data flow diagrams be used as analysis tools?

9. Explain the guidelines for deciding when to stop decomposing DFDs.

10. How do you decide if a system component should be represented as a source/sink or as a process?

11. What unique rules apply to drawing context diagrams?

P R O B L E M S   A N D   E X E R C I S E S

1. Match the following terms to the appropriate definitions.

   _____ source/sink   a. data in motion, moving from one place in a system to another

   _____ data flow   b. data at rest, which may take the form of many different physical representations

   _____ data store   c. work or actions performed on data so that they are transformed or stored or distributed

   _____ process   d. the origin and/or destination of data, sometimes referred to as external entities

_____ DFD completeness

e. extent to which information contained on one level of a set of nested data flow diagrams is also included on other levels

_____ DFD consistency

f. extent to which all necessary components of a data flow diagram have been included

2. Using the example of a retail clothing store in a mall, list relevant data flows, data stores, processes, and sources/sinks. Observe several sales transactions. Draw a context diagram and a level-0 diagram that represent the selling system at the store. Explain why you chose certain elements as processes versus sources/sinks.

3. Choose a transaction that you are likely to encounter, perhaps ordering a cap and gown for graduation, and develop a high-level DFD, or context diagram. Decompose this to a level-0 diagram.

4. Evaluate your level-0 DFD from the previous question using the rules for drawing DFDs in this chapter. Edit your DFD so that it does not break any of these rules.

5. Choose an example like that in the second question and draw a context diagram. Decompose this diagram until it doesn't make sense to continue. Be sure that your diagrams are balanced as discussed in this chapter.

6. Refer to Figure 9-19 which contains drafts of a context and level-0 DFD for a university class registration system. Identify and explain potential violations of rules and guidelines on these diagrams.

7. Why should you develop both logical and physical DFDs for systems? What advantage is there for drawing a logical DFD before a physical DFD for a new information system?

8. What is the relationship between DFDs and entries in the project dictionary or CASE repository?

9. This chapter has shown you how to model, or structure, just one aspect, or view, of an information system, namely the process view. Why do you think analysts have different types of diagrams and other documentation to depict different views (for example, process, logic, and data) of an information system?

10. Consider the DFD in Figure 9-20. List three errors (rule violations) on this DFD.

11. Consider the three DFDs in Figure 9-21. List three errors (rule violations) on these DFDs.

12. Starting with a context diagram, draw as many nested DFDs as you consider necessary to represent all the details of the employee hiring system described in the following narrative. You must draw at least a context and a level-0 diagram. In drawing these diagrams, if you discover that the narrative is incomplete, make up reasonable explanations to complete the story. Supply these extra explanations along with the diagrams. Here is the narrative. Projects, Inc. is an engineering firm with approximately 500 engineers of different types. The company keeps records on all employees, their skills, projects assigned, and departments worked in. New employees are hired by the personnel manager based on data in an application form and evaluations collected from other managers who interview the job candidates. Prospective employees may apply at any time. Engineering managers notify the personnel manager when a job opens and list the characteristics necessary to be eligible for the job. The personnel manager compares the qualifications of the available pool of applicants with the characteristics of an open job, then schedules interviews between the manager in charge of the open position and the three best candidates from the pool. After receiving evaluations on each interview from the manager, the personnel manager makes the hiring decision based upon the evaluations and applications of the candidates and the characteristics of the job, and then notifies the interviewees and the manager about the decision. Applications of rejected

Figure 9-19
Class registration system
(for Problem and Exercise 6)

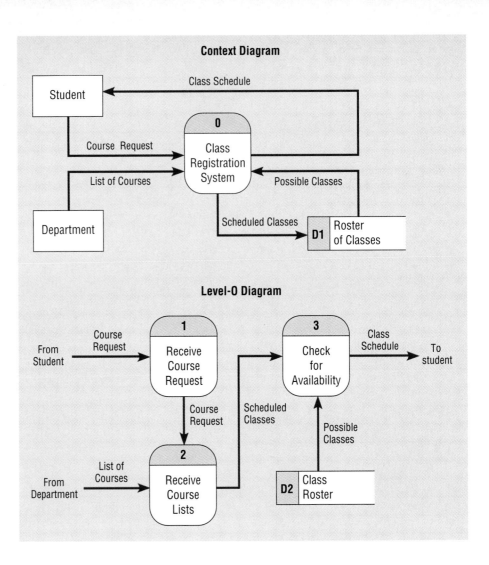

Context Diagram

Student

Class Schedule

0
Class Registration System

Course Request

List of Courses

Possible Classes

Department

Scheduled Classes

D1 Roster of Classes

Level-0 Diagram

From Student

Course Request

1
Receive Course Request

3
Check for Availability

Class Schedule

To student

Course Request

Scheduled Classes

Possible Classes

From Department

List of Courses

2
Receive Course Lists

D2 Class Roster

Figure 9-20
DFD for Problem and
Exercise 10

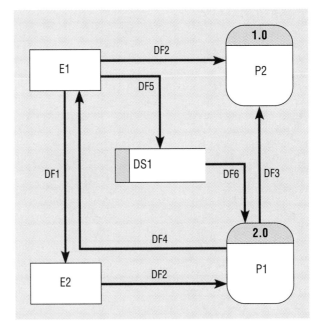

E1

DF2

1.0
P2

DF5

DS1

DF6

DF3

DF1

DF4

2.0
P1

E2

DF2

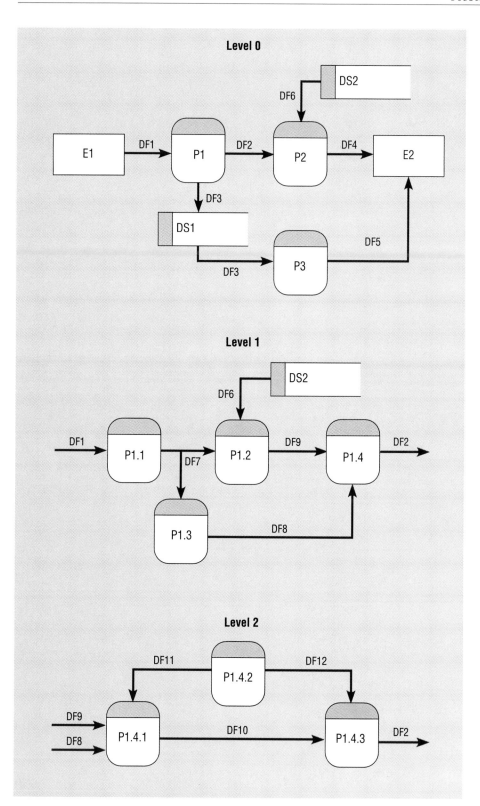

Figure 9-21
DFD for Problem and
Exercise 11

applicants are retained for one year, after which time the application is purged. When hired, a new engineer completes a non-disclosure agreement, which is filed with other information about the employee.

13. a. Starting with a context diagram, draw as many nested DFDs as you consider necessary to represent all the details of the system described in the following narrative. In drawing these diagrams, if you discover that the narrative is incomplete, make up reasonable explanations to make the story complete. Supply these extra explanations along with the diagrams. Here is the narrative. Maximum Software is a developer and supplier of software products to individuals and businesses. As part of their operations, Maximum provides an 800 telephone number help desk for clients who have questions about software purchased from Maximum. When a call comes in, an operator inquires about the nature of the call. For calls that are not truly help desk functions, the operator redirects the call to another unit of the company (such as Order Processing or Billing). Since many customer questions require in-depth knowledge of a product, help desk consultants are organized by product. The operator directs the call to a consultant skilled on the software that the caller needs help with. Since a consultant is not always immediately available, some calls must be put into a queue for the next available consultant. Once a consultant answers the call, he determines if this is the first call from this customer about this problem. If so, he creates a new call report to keep track of all information about the problem. If not, he asks the customer for a call report number, and retrieves the open call report to determine the status of the inquiry. If the caller does not know the call report number, the consultant collects other identifying information such as the caller's name, the software involved, or the name of the consultant who has handled the previous calls on the problem in order to conduct a search for the appropriate call report. If a resolution of the customer's problem has been found, the consultant informs the client what that resolution is, indicates on the report that the customer has been notified, and closes out the report. If resolution has not been discovered, the consultant finds out if the consultant handling this problem is on duty. If so, he transfers the call to the other consultant (or puts the call into the queue of calls waiting to be handled by that consultant). Once the proper consultant receives the call, he records any new details the customer may have. For continuing problems and for new call reports, the consultant tries to discover an answer to the problem by using the relevant software and looking up information in reference manuals. If he can now resolve the problem, he tells the customer how to deal with the problem, and closes the call report. Otherwise, the consultant files the report for continued research and tells the customer that someone at Maximum will get back to him, or if the customer discovers new information about the problem, to call back identifying the problem with a specified call report number.

   b. Analyze the DFDs you created in part (a) of this question. What recommendations for improvements in the help desk system at Maximum can you make based upon this analysis? Draw new logical DFDs that represent the requirements you would suggest for an improved help desk system. Remember, these are to be logical DFDs, so consider improvements independent of technology that can be used to support the help desk.

FIELD EXERCISES

1. Talk to systems analysts who work at an organization. Ask the analyst to show you a complete set of DFDs from a current project. Interview the analyst about his or her views about DFDs and their usefulness for analysis.

2. Interview several people in an organization about a particular system. What is the system like now and what would they like to see changed? Create a complete set of DFDs for the current physical, current logical, and new logical system. Show some of the people you interviewed your DFDs and ask for their reactions. What kinds of comments do they make? What kinds of suggestions?

3. Talk to systems analysts who use a CASE tool. Investigate what capabilities that CASE tool has for automatically checking for rule violations in DFDs. What reports can the CASE tool produce with error and warning messages to help analysts correct and improve DFDs?

4. Find out which, if any, drawing packages, word processors, forms design, and database management systems your university or company supports. Research these packages to determine how they might be used in the production of a project dictionary. For example, do the drawing packages include either set of standard DFD symbols in their graphic symbol palette?

5. At an organization with which you have contact, ask one or more employees to draw a "picture" of the business process they interact with at that organization. Ask them to draw the process using whatever format suits them. Ask them to depict in their diagram each of the components of the process and the flow of information among these components at the highest level of detail possible. What type of diagram have they drawn? Why? In what ways does it resemble (and not resemble) a data flow diagram? Why? When they have finished, help them to convert their diagram to a standard data flow diagram as described in this chapter. In what ways is the data flow diagram stronger and/or weaker than the original diagram?

REFERENCES

Celko, J. 1987. "I. Data Flow Diagrams." *Computer Language* 4(January): 41–43.

DeMarco, T. 1979. *Structured Analysis and System Specification.* Englewood Cliffs, NJ: Prentice-Hall.

Gane, C. and T. Sarson. 1979. *Structured Systems Analysis.* Englewood Cliffs, NJ: Prentice-Hall.

Gibbs, W. W. 1994. "Software's Chronic Crisis." *Scientific American* 271(Sept.): 86–95.

Yourdon, E. 1989. *Managing the Structured Techniques,* 4th ed. Englewood Cliffs, NJ: Prentice-Hall.

Yourdon, E. and L. L. Constantine. 1979. *Structured Design.* Englewood Cliffs, NJ: Prentice-Hall.

Structuring System Requirements: Process Modeling for the Customer Activity Tracking System

INTRODUCTION

Requirements determination had generated a lot of ideas about possible ways to profile customers and products and about various functions to provide in CATS. So far, the only structured descriptions of CATS that the project team had to work with were the original data flow diagrams and entity-relationship diagram developed during the initiation and planning of the CATS project. Although more requirements determination activities would occur, such as a JAD session, it was time for Frank and the other team members to organize what had been discovered so far. Further, the JAD could be a useful opportunity to obtain reactions to the requirements, and the requirements had to be structured and diagrammed to facilitate feedback at the JAD.

DEVELOPING DETAILED DFDS FOR CATS

Frank Napier was studying the level-0 data flow diagram Jordan Pippen had prepared for BEC's Customer Activity Tracking System (CATS) (Figure 1). It was late at night, and Jordan was in Miami getting ready to meet with the managers of the seven stores with the highest sales figures. Frank's job, as part of the "Revise data flow models" activity in the Baseline Project Plan, was to decompose this DFD. Such DFDs were essential so that everyone on the project team could have a common understanding of CATS. Frank sat at his computer, using the ObjectStar CASE tool, drawing level-1 DFDs for Processes 4, 5, and 6. Frank had begun sketching some of these diagrams by hand while studying his notes from several interviews he, Jorge Lopez, and Jordan had conducted. As none of the in-terview notes very clearly described what CATS would have to do, drawing the DFDs would prove challenging and would permit much creativity on Frank's part.

Process 4, Announcement Production & Mailing, seemed straightforward enough, however. Frank understood that the purpose of Process 4 was to take a match that had been made between a given new product and a given customer in Process 3 and print an announcement addressed to that customer. The announcement would indicate that the product could be reserved for that person at his regular BEC store. Looking at the various materials he and the analysis team had gathered during requirements determination, Frank knew that the announcement would contain a product/customer/store (PCS) code, which the customer would call in to an 800 number to ask that the product be held for the customer at the BEC store. Each announcement would have only one PCS, so separate ones would be generated even if a customer was matched with several new products in the same CATS run. Frank decided Process 4 had to contain the logic to generate that code, as he did not see a more appropriate place to create the code anywhere else in the system. Frank drew a sub-process that would create the PCS code and change the customer state from Selected to Notified. Produce Notifications became the sub-process where the announcement would be created. The only process left to do was print and mail the postcard. Frank's decomposition of Process 4 is shown in Figure 2. Recall that the ObjectStar CASE tool encapsulates data inside processes, so data store access is not shown.

The CASE tool automatically created an entry in the repository for each DFD element. Before the analysis phase was complete, the CATS project team would need to enter full descriptions and other information about each process, data flow, and source/sink. For

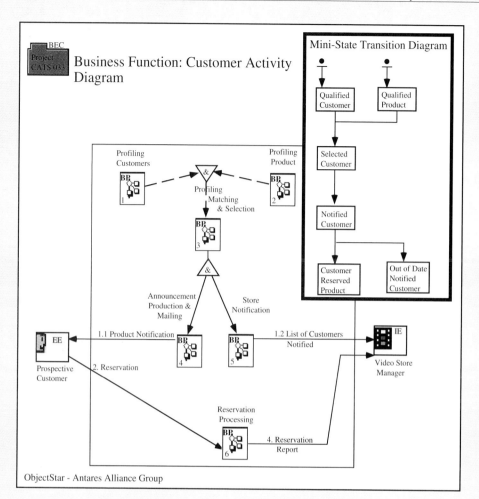

Figure 1
Level-0 data flow diagram for
the CATS project

now, Frank entered a few notes, in outline form, under the description of each element so that other members of the team would have a basic idea of what Frank meant by each term.

Frank next turned to the decomposition of Process 5, Store Notification. Frank reasoned that the customer-product match data would have to be sorted by customer and then by product. The result would be a list of customer-product match instances by customer first and, within customer, by product. As noted above, some customers might have been matched with more than one product each time CATS was run. Once sorted, the customers had to be matched with the store where they were registered. Once all of the customer-product match instances had been grouped by store, the customer profile report could be formatted, generated, and sent to the individual stores. One of the questions Jordan was going to ask the store managers she was meeting with in Miami concerned the utility of the customer profile report. Frank had

pointed out to Jordan before she left that this report showed only potential reservations, but it was the reservation report that told store managers which customers had asked for reservations of which products. Figure 3 shows Frank's decomposition for Process 5, which Frank left still fairly general until the issues mentioned above are resolved.

Process 6, Reservation Processing was more complicated. Within this process, customers had to be able to make reservations, the product reservations data store had to be updated, new product orders had to be initiated via a message sent to the central inventory control system, and the reservation report had to be generated and sent to individual stores. Frank's decomposition for Process 6 appears as Figure 4. This diagram shows two levels of decomposition, including a type of DFD supported by ObjectStar, a task action diagram. As with prior decompositions, as Frank structured the process at a finer detail, he identified flaws in the level-0 diagram which he would have to fix later

Figure 2

Level-1 diagram for
Process 4, Announcement
Production and Mailing

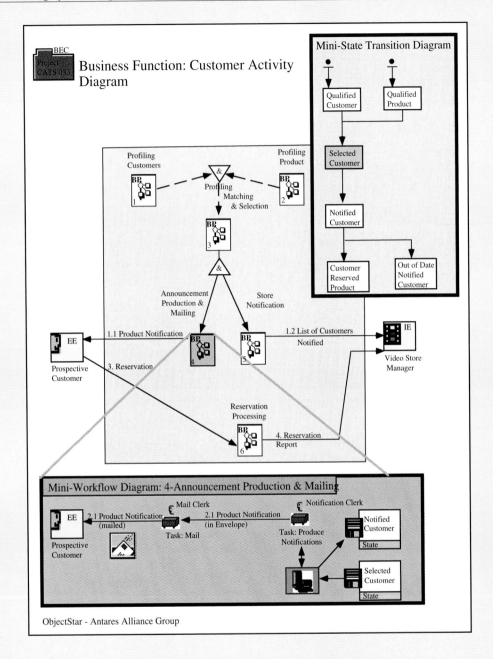

ObjectStar - Antares Alliance Group

in order to eliminate error messages from the CASE tool's DFD validity checker. An example was the need for Process 6 to retrieve data from the Inventory data store and for an interface with the central inventory control system.

Customer reservation confirmations phoned in to the 800 number would be used to update product reservation records. This information would be collected and batched by store; Frank thought this could be suitably done as a batch process, but the decision about whether the Product Reservations data store would be

updated on-line by the 800 number operators or entered on-line for later batch updating would be made during the design phase. Several times per day, the names and amounts of each CATS product each store needed would be generated and compared to current inventory levels for those products. If additional product was needed, that information would be used to order additional product through BEC's centralized inventory control system. Order information and product reservation information would be used to generate a reservation report for the stores.

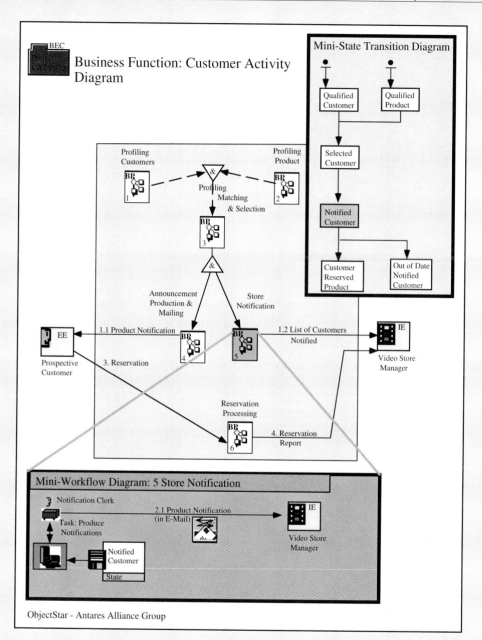

Figure 3
Level-1 diagram for Process 5, Store Notification

The reservation report contained information about the products the store was supposed to reserve and the customers the products were to be reserved for, and information about how much product had been ordered from BEC's warehouses. Frank thought back to a conversation he had had earlier with Jordan Pippen, his boss on the CATS project.

Frank had asked Jordan about feedback and how much of a burden a feedback requirement would place on local stores. Jordan had replied she thought that maybe the reservation report served a feedback func-

tion. Now that he was decomposing the level-0 DFD, Frank realized that the reservation report served only to *tell* the stores which products they had to set aside and for whom and also about any new orders connected with the reservations. Frank also realized that store managers needed to keep track of the reservations and whether customers who made reservations actually came in to the stores and bought the reserved product. Right now, since CATS was a totally new system for BEC, they had no way of really knowing how many of those reservations would turn into sales. Yet

Figure 4

Level-1 diagram for Process 6, Reservation Processing

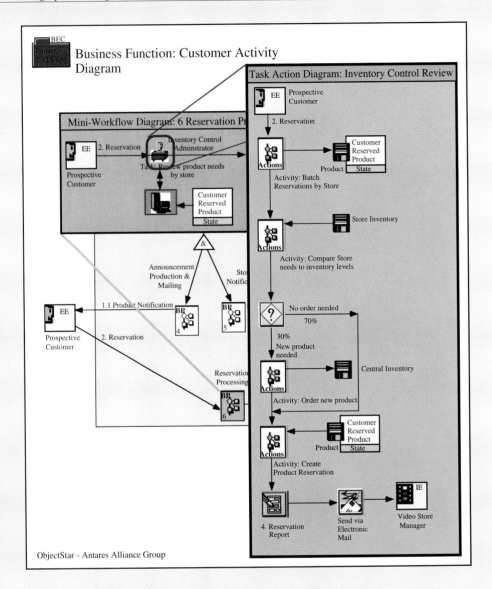

ObjectStar - Antares Alliance Group

the success rate would determine how much product would be automatically ordered for each store, based on reservations. If only 40 percent of the reservations resulted in sales, then less product would be automatically ordered than if 60 or 70 percent of the reservations resulted in sales. Clearly this would be valuable information.

The rental and sales systems could be modified to capture purchases of reserved versus non-reserved items, but modifying these systems would delay the deployment of CATS. An alternative would be to try to capture this in a separate component of CATS at the store level. Store managers would have to assume an additional burden under this alternative design—that is, clerks would enter purchase data through the sale

and rental systems and the manager or a clerk would also have to enter it in the CATS store system—using this approach CATS could be a reality a lot sooner. Frank did not think this would be too much of an extra burden for the manager or clerks in a store. Frank realized his conclusions would be important for Jordan to know before her meetings with the store managers were over since the managers could best judge if such double entry created any problems. Frank created a level-0 DFD for a store-level CATS subsystem, as shown in Figure 5. Frank also made some notes in his personal journal that he kept for each project; he would refer to his notes at the next weekly CATS team meeting and raise the possibility of these alternative design strategies.

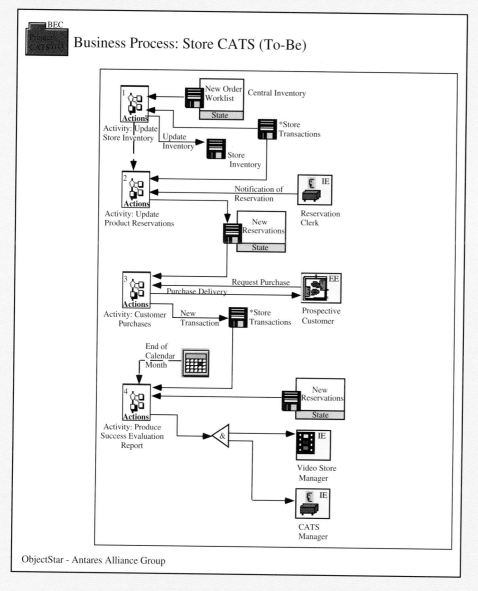

Figure 5

Level-0 data flow diagram
for a store-level CATS
component

The inputs for the store level system would include new order data from BEC's central inventory system, the reservation report from central CATS, and the purchase of reserved products by customers. New order data would be used to update the store inventory records. The reservations report (which would now consist of a transaction file downloaded to the store as well as a printed report) would be used to update product reservations records. Customer purchase data would be used to update customer transaction records for reservation purchases. This current information would serve as input in updating product reservations and store inventory records. The updated product reservations records would then be used to generate a success rate report for the store manager and for the CATS manager. The data in that report would at the very least be used to adjust the ordering thresholds for CATS.

Frank made some notes about the need he had identified for a store-level CATS component and for the relative lightness of the feedback burden for store managers. He printed out his notes and DFDs and faxed them to Jordan at her hotel in Miami. Whether or not she agreed with his conclusions, Jordan at least needed to be made aware of them. After all, her meeting with the store managers was not just for gathering additional information; it was also intended to sell the store managers on CATS and let them know how useful it would be to them and to BEC.

SUMMARY

The diagrams presented in this case are only a few of many DFDs that Frank, Jorge, and Jordan would need to develop to show all of the data movement between CATS components. With the large number of diagrams, the CASE tool and repository are invaluable in ensuring that the diagrams and associated repository entries are consistent and complete. The diagrams show valuable information useful in other analysis and subsequent design steps. Each data flow coming from a source means an automated or manual procedure has to be developed to capture this data, including the possible development of a business form on which to record the data. Each data flow going to a sink implies a file transfer, printed report, or data for computer display. Data flows going into a data store represent database update operations, and data flows coming from a data store represent a query to the database.

The number of processes indicates how much design and programming work must be done in subsequent project phases, which must be outlined in the updated Baseline Project Plan prepared at the end of the analysis phase. No matter how detailed the analysis team makes the DFDs, the actual logic of processing the data (for example, how the Produce Notifications process in Figure 2, works) still needs to be defined in a more algorithmic fashion. Also, since some processes involve multiple data stores, the analysis team will have to develop structures that show how to relate data in associated data stores. The development of the DFDs, however, clarifies many of the elements required in CATS and identifies what general kinds of information users would receive and need to provide. Whether these requirements make sense or not would be validated later in analysis during the JAD and several walkthrough sessions.

10

Structuring System Requirements: Logic Modeling

After studying this chapter, you should be able to:

- Use Structured English as a tool for representing steps in logical processes in data flow diagrams.

- Use decision tables and decision trees to represent the logic of choice in conditional statements.

- Select among Structured English, decision tables, and decision trees for representing processing logic.

- Use state-transition diagrams and state-transition tables to illustrate the event-response relationships between various states of a process or system.

INTRODUCTION

In Chapter 9 you learned how the processes that convert data to information are key parts of information systems. As good as data flow diagrams are for identifying processes, they are not very good at showing the logic inside the processes. Even the processes on the primitive-level data flow diagrams do not show the most fundamental processing steps. Just what occurs within a process? How are the input data converted to the output information? Since data flow diagrams are not really designed to show the detailed logic of processes, you must model process logic using other techniques. This chapter is about the techniques you use for modeling process decision and temporal logic.

First you will be introduced to Structured English, a modified version of the English language that is useful for representing the logic in information system processes. You can use Structured English to represent all three of the fundamental statements necessary for structured programming: choice, repetition, and sequence.

Second, you will learn about decision tables. Decision tables allow you to represent a set of conditions and the actions that follow from them in a tabular format. When there are several conditions and several possible actions that can occur, decision tables can help you keep track of the possibilities in a clear and concise manner.

Third, you will also learn how to model the logic of choice statements using decision trees. Decision trees model the same elements as a decision table but in a more graphical manner.

Fourth, you will have to decide when to use Structured English, decision tables, and decision trees. In this chapter, you will learn about the criteria you can use to make a choice among these three logic modeling techniques.

Finally, since data flow diagrams do not do a very good job of representing temporal aspects of information systems, you will learn about state-transition diagrams. State-transition diagrams are used to show the various states a system component can assume and the events that result in the transition from one state to another. State-transition diagrams are most useful in developing specifications for real-time or on-line information systems. Since many modern systems have multiple on-line components for data input and display, state-transition diagrams are increasingly important models to explain the behavior of information systems.

LOGIC MODELING

In Chapter 8, you learned how the requirements for an information system are collected. Analysts structure the requirements information into data flow diagrams that model the flow of data into and through the information system. Data flow diagrams, though versatile and powerful techniques, are not adequate for modeling all of the complexity of an information system. Although decomposition allows you to represent a data flow diagram's processes at finer and finer levels of detail, the process names themselves cannot adequately represent what a process does and how it does it. For that reason, you must represent the logic contained in the process symbols on DFDs with other modeling techniques.

Logic modeling involves representing the internal structure and functionality of the processes represented on data flow diagrams. These processes appear on DFDs as little more than black boxes, in that we cannot tell from only their names or CASE repository descriptions precisely what they do and how they do it. Yet the structure and functionality of a system's processes are a key element of any information system. Processes must be clearly described before they can be translated into a programming language. In this chapter, we will focus on techniques you can use during the analysis phase to model the logic within processes; that is, data-to-information transformations, decisions, and event-response interactions. In the analysis phase, logic modeling will be complete and reasonably detailed, but it will also be generic in that it will not reflect the structure or syntax of a particular programming language. You will focus on more precise, language-based logic modeling in the design phase of the life cycle.

Modeling a System's Logic

The three sub-phases to systems analysis are requirements determination, requirements structuring, and generating alternative systems and selecting the best one (Figure 10-1). Modeling a system's logic is part of requirements structuring, just as was representing the system with data flow diagrams. Here our focus is on the processes pictured on the data flow diagrams and the logic contained within each. You can also use logic modeling to indicate when processes on a DFD occur (for example, when a process extracts a certain data flow from a given data store). Just as we use logic modeling to represent the logic contained in a data flow diagram's processes, we will use data modeling to represent the contents and structure of a data flow diagram's data flows and data stores.

Figure 10-1

Systems development life cycle with the analysis phase highlighted

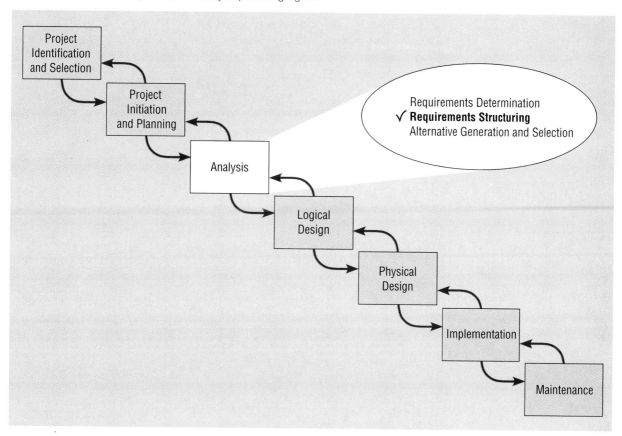

Deliverables and Outcomes

In structured analysis, the primary deliverables from logic modeling are structured descriptions and diagrams that outline the logic contained within each DFD process as well as diagrams that show the temporal dimension of systems—when processes or events occur and how these events change the state of the system. Table 10-1 provides a list of deliverables that result from documenting the logic of a system's processes. Note that the analyst decides if a process requires more than

TABLE 10-1   Deliverables for Logic Modeling

Where appropriate, each process on the lowest- (primitive-) level data flow diagrams will be represented with one or more of the following:

- Structured English representation of process logic
- Decision table representation
- Decision tree representation
- State-transition diagram or table

one representation of its logic. Deliverables can also take the form of new entries into the project dictionary or CASE repository. These entries may update process descriptions or, if possible, store the new diagrams from logic and event-response modeling along with associated repository entries.

Creating diagrams and descriptions of process logic is not an end in itself. Rather, these diagrams and descriptions are created ultimately to serve as part of an unambiguous and thorough explanation of the system's specifications. These specifications are used to explain the system requirements to developers, whether people or automated code generators. Users, analysts, and programmers use logic diagrams and descriptions throughout analysis to incrementally specify a shared understanding of requirements, without regard for programming languages or development environments. Such diagrams may be discussed during JAD sessions or project review meetings. Alternatively, system prototypes generated from such diagrams may be reviewed, and requested changes to a prototype will be implemented by changing logic diagrams and generating a new prototype from a CASE tool or other code generator.

MODELING LOGIC WITH STRUCTURED ENGLISH

You must understand more than just the flow of data into, through, and out of an information system. You must also understand what each identified process does and how it accomplishes its task. Starting with the processes depicted in the various sets of data flow diagrams you and others on the analysis team have produced, you must now begin to study and document the logic of each process. *Structured English* is one method used to illustrate process logic.

Structured English is a modified form of English that is used to specify the contents of process boxes in a DFD. It differs from regular English in that it uses a subset of English vocabulary to express information system process procedures. The same action verbs we listed in Chapter 9 for naming processes are also used in Structured English. These include such verbs as read, write, print, sort, move, merge, add, subtract, multiply, and divide. Structured English also uses noun phrases to describe data structures, such as patron-name and patron-address. Unlike regular English, Structured English does not use adjectives or adverbs. The whole point of using Structured English is to represent processes in a shorthand manner that is relatively easy for users and programmers to read and understand. As there is no standard version, each analyst will have his or her own particular dialect of Structured English.

It is possible to use Structured English to represent all three processes typical to structured programming: sequence, conditional statements, and repetition. Sequence requires no special structure but can be represented with one sequential statement following another. Conditional statements can be represented with a structure like the following:

> BEGIN IF
>   IF Quantity-in-stock is less than Minimum-order-quantity
>   THEN GENERATE new order
>   ELSE DO nothing
> END IF

Another type of conditional statement is a case statement where there are many different actions a program can follow, but only one is chosen. A case statement might be represented as:

Structured English: Modified form of the English language used to specify the logic of information system processes. Although there is no single standard, Structured English typically relies on action verbs and noun phrases and contains no adjectives or adverbs.

```
READ Quantity-in-stock
SELECT CASE
    CASE 1 (Quantity-in-stock greater than Minimum-order-quantity)
        DO nothing
    CASE 2 (Quantity-in-stock equals Minimum-order-quantity)
        DO nothing
    CASE 3 (Quantity-in-stock is less than Minimum-order-quantity)
        GENERATE new order
    CASE 4 (Stock out)
        INITIATE emergency re-order routine
END CASE
```

Repetition can take the form of Do-Until loops or Do-While loops. A Do-Until loop might be represented as follows:

```
DO
    READ Inventory records
    BEGIN IF
        IF Quantity-in-stock is less than Minimum-order-quantity
        THEN GENERATE new order
        ELSE DO nothing
    END IF
UNTIL End-of-file
```

A Do-While loop might be represented as follows:

```
READ Inventory records
WHILE NOT End-of-File DO
    BEGIN IF
        IF Quantity-in-stock is less than Minimum-order-quantity
        THEN GENERATE new order
        ELSE DO nothing
    END IF
END DO
```

Let's look at an example of how Structured English would represent the logic of some of the processes identified in Hoosier Burger's current logical inventory control system. Figure 9-15 is reproduced on the next page as Figure 10-2.

There are four processes depicted in Figure 10-2: Update Inventory Added, Update Inventory Used, Generate Orders, and Generate Payments. Structured English representations of each process are shown in Figure 10-3. Notice that in this version of Structured English the file names are connected with hyphens and file names and variable names are capitalized. Terms that signify logical comparisons, such as greater than and less than, are spelled out rather than represented by their arithmetic symbols. Also notice how short the Structured English specifications are, considering that these specifications all describe level-0 processes. The final specifications would model the logic in the lowest-level DFDs only. From reading the process descriptions in Figure 10-3, it should be obvious to you that much more detail would be required to actually perform the processes described. In fact, creating Structured English representations of processes in higher-level DFDs is one method you can use to help you decide if a particular DFD needs further decomposition.

Notice how the format of the Structured English process description mimics the format usually used in programming languages, especially the practice of indentation. This is the "structured part" of Structured English. Notice also that the language

Figure 10-2

Current logical DFD for Hoosier Burger's inventory control system

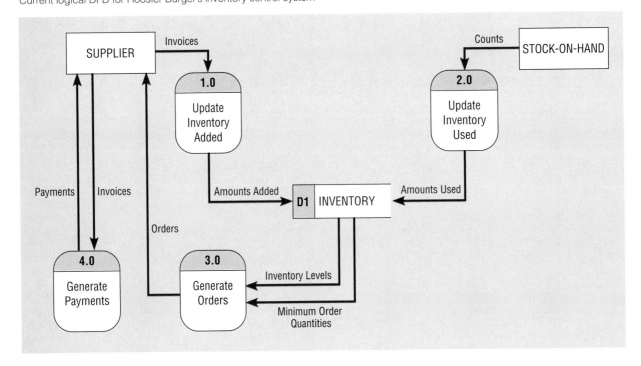

Figure 10-3

Structured English representations of the four processes depicted in Figure 10-2

```
Process 1.0: Update Inventory Added
DO
    READ next Invoice-item-record
    FIND matching Inventory-record
    ADD Quantity-added from Invoice-item-record to Quantity-in-stock on
        Inventory-record
UNTIL End-of-file
```

```
Process 2.0: Update Inventory Used
DO
    READ next Stock-item-record
    FIND matching Inventory-record
    SUBTRACT Quantity-used on Stock-item-record from Quantity-in-stock
        on Inventory-record
UNTIL End-of-file
```

```
Process 3.0: Generate Orders
DO
    READ next Inventory-record
    BEGIN IF
        IF Quantity-in-stock is less than Minimum-order-quantity
        THEN GENERATE Order
    END IF
UNTIL End-of-file
```

```
Process 4.0: Generate Payments
READ Today's-date
DO
    SORT Invoice-records by Date
    READ next Invoice-record
    BEGIN IF
        IF Date is 30 days or greater than Today's-date
        THEN GENERATE Payments
    END IF
UNTIL End-of-file
```

used is similar to spoken English, using verbs and noun phrases. The language is simple enough for a user who knows nothing about computer programming to understand the steps involved in performing the various processes, yet the structure of the descriptions makes it easy to eventually convert to a programming language. Using Structured English also means not having to worry about initializing variables, opening and closing files, or finding related records in separate files. These more technical details are left for later in the design process.

Structured English is intended to be used as a communication technique for analysts and users. Analysts and programmers have their own communication technique, *pseudocode*, which we will study later in this book. Whereas Structured English resembles spoken English, pseudocode resembles a programming language.

MODELING LOGIC WITH DECISION TABLES

Structured English can be used to represent the logic contained in an information system process, but sometimes a process' logic can become quite complex. If several different conditions are involved, and combinations of these conditions dictate which of several actions should be taken, then Structured English may not be adequate for representing the logic behind such a complicated choice. Not that Structured English cannot represent complicated logic; rather, Structured English becomes more difficult to understand and verify as logic becomes more complicated. Research has shown, for example, that people become confused in trying to interpret more than three nested IF statements (Yourdon, 1989a). Where the logic is complicated, then, a diagram may be much clearer than a Structured English statement. A **decision table** is a diagram of process logic where the logic is reasonably complicated. All of the possible choices and the conditions the choices depend on are represented in tabular form, as illustrated in the decision table in Figure 10-4.

The decision table in Figure 10-4 models the logic of a generic payroll system. There are three parts to the table: the **condition stubs**, the **action stubs**, and the **rules.** The condition stubs contain the various conditions that apply in the situation the table is modeling. In Figure 10-4, there are two condition stubs for employee type and hours worked. Employee type has two values: "S," which stands for salaried, and "H," which stands for hourly. Hours worked has three values: less than 40, exactly 40, and more than 40. The action stubs contain all the possible courses of action that result from combining values of the condition stubs. There are four possible courses of action in this table: Pay base salary, Calculate hourly

Decision table: A matrix representation of the logic of a decision, which specifies the possible conditions for the decision and the resulting actions.

Action stubs: That part of a decision table that lists the actions that result for a given set of conditions.

Condition stubs: That part of a decision table that lists the conditions relevant to the decision.

Rules: That part of a decision table that specifies which actions are to be followed for a given set of conditions.

	Conditions/ Courses of Action	Rules					
		1	2	3	4	5	6
Condition Stubs	Employee type	S	H	S	H	S	H
	Hours worked	<40	<40	40	40	>40	>40
Action Stubs	Pay base salary	X		X		X	
	Calculate hourly wage		X		X		X
	Calculate overtime						X
	Produce Absence Report		X				

Figure 10-4

Complete decision table for payroll system example

wage, Calculate overtime, and Produce Absence Report. You can see that not all actions are triggered by all combinations of conditions. Instead, specific combinations trigger specific actions. The part of the table that links conditions to actions is the section that contains the rules.

To read the rules, start by reading the values of the conditions as specified in the first column: Employee type is "S," or salaried, and hours worked are less than 40. When both of these conditions occur, the payroll system is to pay the base salary. In the next column, the values are "H" and "<40," meaning an hourly worker who worked less than 40 hours. In such a situation, the payroll system calculates the hourly wage and makes an entry in the Absence Report. Rule 3 addresses the situation when a salaried employee works exactly 40 hours. The system pays the base salary, as was the case for Rule 1. For an hourly worker who has worked exactly 40 hours, Rule 4 calculates the hourly wage. Rule 5 pays the base salary for salaried employees who work more than 40 hours. Rule 5 has the same action as Rules 1 and 3, and governs behavior with regard to salaried employees. The number of hours worked does not affect the outcome for Rules 1, 3, or 5. For these rules, hours worked is an **indifferent condition** in that its value does not affect the action taken. Rule 6 calculates hourly pay and overtime for an hourly worker who has worked more than 40 hours.

Because of the indifferent condition for Rules 1, 3, and 5, we can reduce the number of rules by condensing Rules 1, 3, and 5 into one rule, as shown in Figure 10-5. The indifferent condition is represented with a dash. Whereas we started with a decision table with six rules, we now have a simpler table that conveys the same information with only four rules.

In constructing these decision tables, we have actually followed a set of basic procedures, as follows:

1. *Name the conditions and the values each condition can assume.* Determine all of the conditions that are relevant to your problem, and then determine all of the values each condition can take. For some conditions, the values will be simply "yes" or "no" (called a limited entry). For others, such as the conditions in Figures 10-4 and 10-5, the conditions may have more values (called an extended entry).

2. *Name all possible actions that can occur.* The purpose of creating decision tables is to determine the proper course of action given a particular set of conditions.

3. *List all possible rules.* When you first create a decision table, you have to create an exhaustive set of rules. Every possible combination of conditions must be

Indifferent condition: In a decision table, a condition whose value does not affect which actions are taken for two or more rules.

Figure 10-5

Reduced decision table for payroll system example

Conditions/ Courses of Action	Rules			
	1	2	3	4
Employee type	S	H	H	H
Hours worked	–	<40	40	>40
Pay base salary	X			
Calculate hourly wage		X	X	X
Calculate overtime				X
Produce Absence Report		X		

represented. It may turn out that some of the resulting rules are redundant or make no sense, but these determinations should be made only *after* you have listed every rule so that no possibility is overlooked. To determine the number of rules, multiply the number of values for each condition by the number of values for every other condition. In Figure 10-4, we have two conditions, one with two values and one with three, so we need 2x3, or 6, rules. If we added a third condition with three values, we would need 2x3x3, or 18, rules.

When creating the table, alternate the values for the first condition, as we did in Figure 10-4 for type of employee. For the second condition, alternate the values but repeat the first value for all values of the first condition, then repeat the second value for all values of the first condition, and so on. You essentially follow this procedure for all subsequent conditions. Notice how we alternated the values of hours worked in Figure 10-4. We repeated "<40" for both values of type of employee, "S" and "H." Then we repeated "40," and then ">40."

4. *Define the actions for each rule.* Now that all possible rules have been identified, provide an action for each rule. In our example, we were able to figure out what each action should be and whether all of the actions made sense. If an action doesn't make sense, you may want to create an "impossible" row in the action stubs in the table to keep track of impossible actions. If you can't tell what the system ought to do in that situation, place question marks in the action stub spaces for that particular rule.

5. *Simplify the decision table.* Make the decision table as simple as possible by removing any rules with impossible actions. Consult users on the rules where system actions aren't clear and either decide on an action or remove the rule. Look for patterns in the rules, especially for indifferent conditions. We were able to reduce the number of rules in the payroll example from six to four, but often greater reductions are possible.

Let's look at an example from Hoosier Burger. The Mellankamps are trying to determine how they reorder food and other items they use in the restaurant. If they are going to automate the inventory control functions at Hoosier Burger, they need to articulate their reordering process. In thinking through the problem, the Mellankamps realize that how they reorder depends on whether the item is perishable. If an item is perishable, such as meat, vegetables, or bread, the Mellankamps have a standing order with a local supplier stating that a pre-specified amount of food is delivered each weekday for that day's use and each Saturday for weekend use. If the item is not perishable, such as straws, cups, and napkins, an order is placed when the stock on hand reaches a certain pre-determined minimum reorder quantity. The Mellankamps also realize the importance of the seasonality of their work. Hoosier Burger's business is not as good during the summer months when the students are off campus as it is during the academic year. They also note that business falls off during Christmas and spring breaks. Their standing orders with all their suppliers are reduced by specific amounts during the summer and holiday breaks. Given this set of conditions and actions, the Mellankamps put together an initial decision table (see Figure 10-6).

Notice three things about Figure 10-6. First, notice how the values for the third condition have been repeated, providing a distinctive pattern for relating the values for all three conditions to each other. Every possible rule is clearly provided in this table. Second, notice we have twelve rules. Two values for the first condition (type of item) times two values for the second condition (time of week) times three values for the third condition (season of year) equals twelve possible rules. Third,

Conditions/ Courses of Action	Rules											
	1	2	3	4	5	6	7	8	9	10	11	12
Type of item	P	N	P	N	P	N	P	N	P	N	P	N
Time of week	D	D	W	W	D	D	W	W	D	D	W	W
Season of year	A	A	A	A	S	S	S	S	H	H	H	H
Standing daily order	X				X				X			
Standing weekend order			X				X				X	
Minimum order quantity		X		X		X		X		X		X
Holiday reduction										X		X
Summer reduction					X		X					

Type of item: Time of week: Season of year:
P = perishable   D = weekday   A = academic year
N = non-perishable   W = weekend   S = summer
                                                        H = holiday

notice how the action for non-perishable items is the same, regardless of day of week or time of year. For non-perishable goods, both time-related conditions are indifferent. Collapsing the decision table accordingly gives us the decision table in Figure 10-7. Now there are only seven rules instead of twelve.

You have now learned how to draw and simplify decision tables. You can also use decision tables to specify additional decision-related information. For example, if the actions that should be taken for a specific rule are more complicated than one or two lines of text can convey, or if some conditions need to be checked only when other conditions are met (nested conditions), you may want to use separate, linked decision tables. In your original decision table, you can specify an action in the action stub that says "Perform Table B." Table B could contain an action stub that returns to the original table and the return would be the action for one or more rules in Table B. Another way to convey more information in a decision table is to use numbers that indicate sequence rather than Xs where rules and ac-

Conditions/ Courses of Action	Rules						
	1	2	3	4	5	6	7
Type of item	P	P	P	P	P	P	N
Time of week	D	W	D	W	D	W	–
Season of year	A	A	S	S	H	H	–
Standing daily order	X		X		X		
Standing weekend order		X		X		X	
Minimum order quantity							X
Holiday reduction					X	X	
Summer reduction			X	X			

tion stubs intersect. For example, for Rules 3 and 4 in Figure 10-7, it would be important for the Mellankamps to account for the summer reduction to modify the existing standing order for supplies. "Summer reduction" would be marked with a "1" for Rules 3 and 4 while "Standing daily order" would be marked with a "2" for Rule 3, and "Standing weekend order" would be marked with a "2" for Rule 4.

You have seen how decision tables can model the relatively complicated logic of a process. Decision tables are more useful than Structured English for complicated logic in that they convey information in a tabular rather than a linear, sequential format. As such, decision tables are compact; you can pack a lot of information into a small table. Decision tables also allow you to check for the extent to which your logic is complete, consistent, and not redundant. Despite the usefulness of Structured English and decision tables, there are still other techniques available for modeling process logic. The next such technique you will learn about is decision trees.

MODELING LOGIC WITH DECISION TREES

A **decision tree** is a graphical technique that depicts a decision or choice situation as a connected series of nodes and branches. Decision trees were first devised as a management science technique to simplify a choice where some of the needed information is not known for certain. By relying on the probabilities of certain events, a management scientist can use a decision tree to choose the best course of action. Although this type of decision tree is beyond the scope of our text, we can use modified decision trees (without the probabilities) to diagram the same sorts of situations for which we used decision tables. Why introduce yet another diagramming technique to do what a decision table does? Both decision tables and decision trees are communication tools designed to make it easier for analysts to communicate with users. Deciding exactly which technique to use depends on various factors which we discuss in detail in the next section, after you have an understanding of decision trees.

As used in requirements structuring, decision trees have two main components: decision points, which are represented by nodes, and actions, which are represented by ovals. Figure 10-8 shows a generic decision tree. To read a decision tree, you begin at the root node on the far left. Each node is numbered, and each number

> **Decision tree:** A graphical representation of a decision situation in which decision points (nodes) are connected together by arcs (one for each alternative on a decision) and terminate in ovals (the action which is the result of all of the decisions made on the path that leads to that oval).

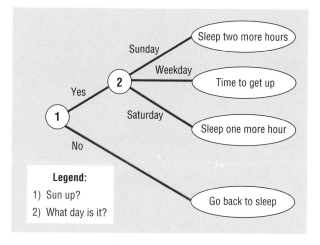

Figure 10-8
Generic decision tree

corresponds to a choice; the choices are spelled out in a legend for the diagram. Each path leaving a node corresponds to one of the options for that choice. From each node, there are at least two paths that lead to the next step, which is either another decision point or an action. Finally, all possible actions are listed on the far right of the diagram in leaf nodes. Each rule is represented by tracing a series of paths from the root node, down a path to the next node, and so on, until an action oval is reached.

Look back at the decision tables we created for the payroll system logic (Figures 10-4 and 10-5). There are at least two ways to represent this same information as a decision tree. The first is shown in Figure 10-9. Here all of the choices are limited to two outcomes, either yes or no. However, looking at how the conditions are phrased in the decision tables, you remember that hours worked has three values, not two. You might argue that forcing a condition with three values into a set of conditions that have only yes or no available as values is somewhat artificial. To preserve the original logic of the decision situation, you can draw your decision tree as depicted in Figure 10-10. Here, there are only two conditions; the first condition has two values, and the second has three values, as is true in the decision table. We leave it as an exercise for the reader to diagram the logic of the situation depicted in Figures 10-6 and 10-7.

We have waited until now to make two important points about decision tables and decision trees. Once you have spent some time creating logic modeling aids such as these, be ready to refine them by drawing the diagrams again and again. As was the case with data flow diagrams, decision tables and decision trees benefit greatly from iteration. The second point is that you should always share your work with other team members and users to get feedback on the mechanical and content

Figure 10-9
Decision tree representation of the decision logic in the decision tables in Figures 10-4 and 10-5, with only two choices per decision point

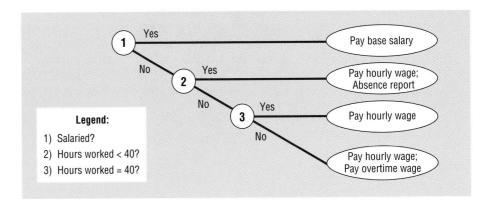

Figure 10-10
Decision tree representation of the decision logic in the decision tables in Figures 10-4 and 10-5, with multiple choices per decision point

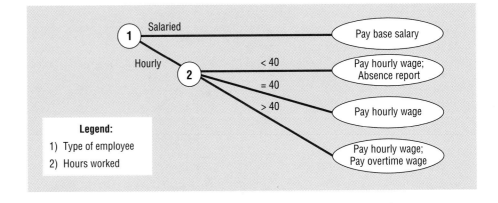

accuracy of your work. Other team members and users will often provide insight into issues you might have overlooked in describing the logic. For that reason, it is not uncommon for the analysis team leader to schedule a walkthrough at some point during the requirements structuring process.

DECIDING AMONG STRUCTURED ENGLISH, DECISION TABLES, AND DECISION TREES

How do you decide whether to use Structured English, decision tables, or decision trees when modeling process logic? On one level, the answer is to use whichever method you prefer and understand best. For example, some analysts and users prefer to see the logic of a complicated decision situation laid out in tabular form, as in a decision table; others will prefer the more graphical structure of a decision tree. Yet the issue actually extends beyond mere preferences. Just because you are very adept at using a hammer doesn't mean a hammer is the best tool for all home repairs—sometimes a screwdriver or a drill is the best tool. The same is true of logic modeling techniques: you have to consider the task you are performing and the purpose of the techniques in order to decide which technique is best. The relative advantages and disadvantages of Structured English, decision tables, and decision trees for different situations are presented in Tables 10-2 and 10-3.

Table 10-2 summarizes the research findings for comparisons of all three techniques. One study summarized in the table compared Structured English to decision tables and decision trees and analyzed the techniques for two different tasks (Vessey and Weber, 1986). The first task was determining the correct conditions and actions from a description of the problem, much the same situation analysts face when defining conditions and actions after an interview with a user. The study found that decision trees were the best technique to support this process as they naturally separate conditions and actions, making the logic of the decision rules more apparent. Even though Structured English does not separate conditions and actions, it was considered the second best technique for this task. Decision tables were the worst technique. The second task was converting conditions and actions to sequential statements, similar to what an analyst does when converting the stated conditions and actions to the sequence of pseudocode or a programming language. Structured English was the best technique for this task, as it is already written sequentially, but researchers found decision trees to be just as good. Decision tables were last again.

Both decision trees and decision tables do have at least one advantage over Structured English, however. Both decision tables and trees can be checked for completeness, consistency, and degree of redundancy. We checked all of our examples of decision tables for completeness when we made sure that each initial table included all possible rules. We knew the tables were complete when we multiplied

TABLE 10-2 Criteria for Deciding among Structured English, Decision Tables, and Decision Trees

Criteria	Structured English	Decision Tables	Decision Trees
Determining conditions & actions	Second Best	Third Best	Best
Transforming conditions & actions into sequence	Best	Third Best	Best
Checking consistency & completeness	Third Best	Best	Best

TABLE 10-3 Criteria for Deciding between Decision Tables & Decision Trees

Criteria	Decision Tables	Decision Trees
Portraying complex logic	Best	Worst
Portraying simple problems	Worst	Best
Making decisions	Worst	Best
More compact	Best	Worst
Easier to manipulate	Best	Worst

the number of values for each condition to get the total number of possible rules. Following the other specific steps outlined earlier in the chapter will also help you check for a decision table's consistency and degree of redundancy. The same procedures can be easily adopted for decision trees. However, there are no such easy means to validate Structured English statements, giving decision tables and trees at least one advantage over Structured English.

Researchers have also compared decision tables to decision trees (Table 10-3). The pioneers of structured analysis and design thought decision tables were best for portraying complex logic while decision trees were better for simpler problems (Gane and Sarson, 1979). Others have found decision trees to be better for guiding decision making in practice (Subramanian, et al., 1992), but decision tables have the advantage of being more compact than decision trees and easier to manipulate (Vanthienen, 1994). If more conditions are added to a situation, a decision table can easily accommodate more conditions, actions, and rules. If the table becomes too large, it can easily be divided into sub-tables, without the inconvenience of using flowchart-like tree connections used with decision trees. Creating and maintaining complex decision tables can be made easier with computer support (e.g., Prologa mentioned in Vanthienen, 1994).

MODELING TEMPORAL LOGIC WITH STATE-TRANSITION DIAGRAMS AND TABLES

State-transition diagram: A diagram that illustrates how processes are related to each other in time. State-transition diagrams illustrate the states a system component can have and the events that cause change from one state to another.

State-transition table: A table that illustrates how processes are related to each other in time. State-transition tables illustrate the states a system component can have and the events that cause change from one state to another. All possible state-event combinations are explored.

While Structured English, decision tables, and decision trees are well-suited to modeling most types of process logic, they are not well-suited for modeling logic that is time-dependent. For applications where time is important, such as on-line and real-time applications, analysts sometimes use **state-transition diagrams** or **state-transition tables** to supplement other logic modeling techniques. A state-transition diagram shows how processes on a data flow diagram, and sometimes different time-dependent states of a single process, are ordered in time. A state-transition table provides much the same information but in table form. These diagrams are an important part of object-oriented systems analysis and design (see Appendix D), but they can also be useful for process-oriented systems development (Flaatten, et al., 1992).

State-Transition Diagrams

A state can be thought of as a mode or condition of existence for a process or other system component, as determined by current circumstances. In object-oriented thought, a state consists of all an object's properties, which are static, and the values of those properties, which are dynamic (Booch, 1994). Transitions from one state to another are caused by stimuli we refer to as events. A wide range of phenomena could qualify as events, from the occurrence of a transaction to an inventory level reaching a reorder point, to a user clicking a mouse button on a field on a data entry screen, to the change in time from one day (or hour or month or year) to the next. Once a new state is entered, an action associated with the state takes place (Shlaer and Mellor, 1992). An event, then, triggers a transition to a new state and causes an action to occur.

Like data flow diagrams, there are few symbols used in creating state-transition diagrams. Each state is represented with a rectangle; each transition is represented with an arrow (Figure 10-11). The event that triggers the transition from one state to another appears as a label next to the transition arrow. The action that occurs when the new state is entered appears as a list of instructions written under the corresponding state rectangle. Actions can be described in the pseudocode or Structured English of the analyst's choice. Each state is ordered in time (Conger, 1994). The second state follows the first, a third follows the second, and so on, according to the diagram. There can be many paths through a state-transition diagram, but the transition to each new state is triggered by one and only one event.

Figure 10-11 is a state-transition diagram for an automatic coffee maker, a coffee-making system. Notice that each state rectangle is labeled with a name and a number: "1. Idle" and "2. Making coffee." Notice that each event also has a label (C1 and C2, where the "C" refers to the coffee maker) and a short command for what the event entails. Actions are listed by each state rectangle. When each state is entered, these actions occur. Finally, notice that the only event that triggers a transition to State 2 is Event C1—the only way to make coffee is to turn the switch to "on." The only event that triggers a transition to State 1 is Event C2, turning the switch to "off." The coffee-making system is in one or the other of these two states at any given time. The state the system is in right now is called its current state. Our coffee maker can transition back and forth between being idle and making coffee indefinitely—there is always an exit from each state. In some systems, there may be one or more states from which there are no exits. These are called final states. You should think of events as control signals that can carry data. These data are then supplied to the action on arrival in the state, much in the way parameters are passed between parts of a computer program.

To better illustrate these points, let's look at an example from Hoosier Burger. We saw in previous chapters that Hoosier Burger utilizes a real-time food ordering system; each transaction with Hoosier Burger customers is an order for food. Orders are taken at special cash registers that have push buttons for each menu item.

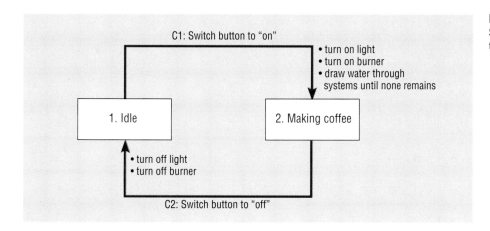

Figure 10-11

State-transition diagram for a two-state coffee maker

Clerks push a menu item button once each time the item is ordered, e.g., if the customer orders two hamburgers, the clerk pushes the hamburger button twice. As we saw before, once the order is completed, the kitchen receives a copy of it. The clerk then totals the dollar amount of the order, asks for payment, and returns the customer's change. We can represent the food ordering process at Hoosier Burger with the state-transition diagram shown in Figure 10-12.

In the diagram, there are seven different states and seven different events. States are named and numbered in the same fashion that you saw in Figure 10-11. Actions associated with the states appear below state rectangles. As with the coffee maker example, some actions are made up of many individual steps. Each transition arrow is labeled with the associated event, and all event labels begin with "R" for "register." There is no final state in this system, as all states have exits. Notice that only one event can lead to a transition to a particular state, no matter where that event originated. For example, only event R2: Clear Button Pushed, results in a transition to State 1, Idle. Event R2 can be initiated by State 2, Opening Order; State 3, Error State; State 4, Voiding Order; and State 7, Recording Order. Only event R5, Unexpected Button Pushed, can lead to a transition to State 3, Error State.

There are two ways to proceed when constructing state-transition diagrams. One way is to think of all possible states the system may assume and then deter-

Figure 10-12

State-transition diagram for Hoosier Burger's food-ordering system

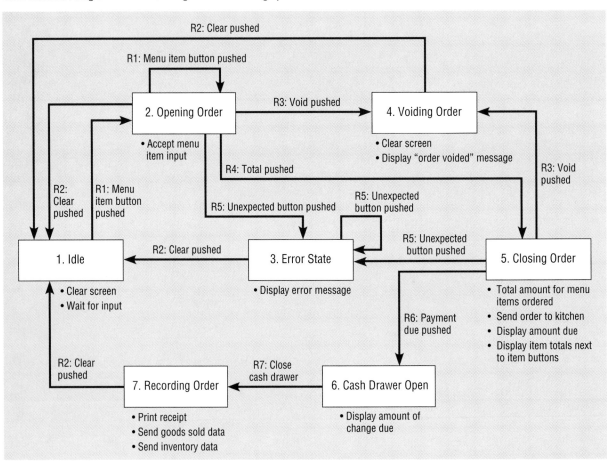

mine all possible connections among the states. The second way is to start with the first state, determine which states are possible to enter from the first state, and then continue methodically to determine which states are possible to enter from each of those states, until the entire network has been mapped (Yourdon, 1989b). Using either method, it is also important to consider how the system reacts if inappropriate actions are taken or if other errors occur.

To draw state-transition diagrams, follow this set of rules:

1. Identify the possible states or the initial state.

2. Draw rectangles representing each state.

3. Connect the states with arrows to show transitions.

4. Each state should lead to one or more other states.

5. Label the transition arrows with event names.

6. List the appropriate actions under each state rectangle.

7. Consider system reactions to unexpected events.

8. Study the diagram to determine if it should be decomposed.

9. Discuss the diagram with team members to ensure accuracy and consistency.
   a. Ask if all states have been defined.
   b. Ask if you can reach all of the states.
   c. Ask if you can exit from all of the states (except the final state[s]).
   d. Ask if the system responds properly to all possible events.

Looking at the way we used state-transition diagrams to model the order-taking system in the Hoosier Burger example, you can see how the same techniques could also be used to model menus and navigational paths for dialogues in on-line systems. For a menu model, the initial state would be the main menu display, and arrows would lead to each of the other states that could be assumed; that is, to the choices contained in the menu and/or buttons on this display. Thus, each state is a menu or submenu with its associated display; each event coming off a state is a choice for leaving or changing that display. The complete menu and option selection structure can be shown on one state-transition diagram; for many on-line systems, however, this would create an unwieldy diagram. As an alternative, you can create a separate state-transition diagram for each menu and display. An even more complex state-transition diagram could show what happens for more detailed events on a display, such as by moving the cursor to a particular field or by entering data into a field. Other extensions include listing actions that occur not only when the state is entered but also when the state is exited.

This section has presented only the basics of state-transition diagrams. The interested reader is referred to Harel (1987) for a thorough explanation.

State-Transition Tables

It is possible to represent the information in a state-transition diagram in a state-transition table. In a state-transition table, each row represents one of the possible states; each column represents one of the possible events. Cells are filled to specify what happens when an instance of an entity or object in a given state receives a particular event. Cells are filled with one of three choices: (1) the number of the new state (2) "event ignored" or (3) "can't happen." You may need to add footnotes to more fully explain the "can't happen" occurrences. An event can cause an instance

Figure 10-13

State-transition table for Hoosier Burger's food-ordering system

States	R1: Menu item pushed	R2: Clear pushed	R3: Void pushed	R4: Total pushed	R5: Odd button pushed	R6: Payment due pushed	R7: Cash drawer closed
1. Idle	2	event ignored	event ignored	event ignored	event ignored	event ignored	can't happen
2. Opening Order	2	1	4	5	3	event ignored	can't happen
3. Error State	event ignored	1	event ignored	event ignored	3	event ignored	can't happen
4. Voiding Order	event ignored	1	event ignored	event ignored	event ignored	event ignored	can't happen
5. Closing Order	event ignored	event ignored	4	event ignored	3	6	can't happen
6. Cash Drawer Open	event ignored	event ignored	event ignored	event ignored	event ignored	event ignored	7
7. Recording Order	event ignored	1	event ignored	event ignored	event ignored	event ignored	can't happen

Note: In the R7 column, the event Close Cash Drawer can't happen because the cash drawer is already closed for all states except State 6, Cash Drawer Open.

to transition to the state it is currently in, as you can see with State 2, Opening Order, in Figure 10-12. In such a case, the transition causes the action associated with entering the state to re-execute. Some analysts (Shlaer and Mellor, 1992) believe that both state-transition diagrams and tables should be used, even though they convey much of the same information. State-transition tables are better for checking completeness and inconsistency than diagrams, as you must consider every possible state-event combination to successfully complete a table. Figure 10-13 shows the state-transition information about Hoosier Burger's food ordering system in table format.

SUMMARY

Logic modeling is one of the three key activities in the requirements structuring phase of systems analysis. There are various techniques available for modeling the decision logic in information system processes. One method, Structured English, is a special form of spoken English analysts use to illustrate the logic of processes depicted in data flow diagrams. Structured English is primarily a communication technique for analysts and users. Structured English dialect varies with analysts, but it must represent sequence, conditional statements, and repetition in information system processes.

Decision tables and decision trees are graphical methods for representing process logic. In decision tables, conditions are listed in the condition stubs, possible actions are listed in the action stubs, and rules link combinations of conditions with the actions that should result. A first pass at a decision table involves listing all possible rules that result from all values of the conditions listed. Second and subsequent passes reduce the complexity by eliminating rules that don't make sense and by combining rules with indifferent conditions. The same logic portrayed in a decision table can be portrayed in a decision tree but, in a decision tree, the conditions are represented by decision points and the values are represented by paths between decision points and ovals that contain actions.

Several research studies have compared Structured English, decision tables, and decision trees as techniques for representing decision logic. Most studies show that decision trees are the best for many criteria while decision tables are the worst on some criteria and the best on other criteria. Because there is no best technique to use for structuring requirements, an analyst must be proficient at all three techniques, as required by different situations.

To depict temporal logic, we use both state-transition diagrams and state-transition tables. State-transition diagrams and tables allow you to consider the various states a system component can assume at specific points in time, triggered by specific events. State-transition diagrams and tables are most useful in modeling information systems where time is important, such as on-line and real-time systems. State-transition tables are more useful than state-transition diagrams for checking consistency and completeness as every possible state-event combination must be considered in constructing the table.

C H A P T E R   R E V I E W

K E Y   T E R M S

Action stubs	Decision tree	State-transition diagram
Condition stubs	Indifferent condition	State-transition table
Decision table	Rules	Structured English

R E V I E W   Q U E S T I O N S

1. Define each of the following terms:

   a. Structured English   d. state-transition diagram
   b. decision table   e. event
   c. decision tree   f. state

2. What is the purpose of logic modeling? What techniques are used to model decision logic and what techniques are used to model temporal logic?

3. What is Structured English? How can Structured English be used to represent sequence, conditional statements, and repetition in an information systems process?

4. What is the difference between Structured English and pseudocode?

5. What are the steps in creating a decision table? How do you reduce the size and complexity of a decision table?

6. Explain the structure of a decision tree.

7. What is the purpose of a state-transition diagram? Where did state-transition diagrams originate? How are they useful in structured systems analysis and design?

8. How do you know when to use Structured English, decision tables, or decision trees? Which is best for what situation?

9. What verbs are used in Structured English? What type of words are not used in Structured English?

10. What does the term "limited entry" mean in a decision table?

11. What is the formula that is used to calculate the number of rules a decision table must cover?

12. Explain the differences between state-transition diagrams and state-transition tables.

PROBLEMS AND EXERCISES

1. Match the following terms to the appropriate definitions.

_____ rules	a. that part of a decision table listing the actions that result for a given set of conditions
_____ action stubs	b. that part of a decision table listing the conditions relevant to the decision
_____ condition stubs	c. that part of a decision table specifying which actions are to be followed for a given set of conditions
_____ indifferent condition	d. in a decision table, a condition whose value does not affect the decision

2. Represent the decision logic in the decision table of Figure 10-5 in Structured English.

3. Look at the set of data flow diagrams created in Chapter 9 for Hoosier Burger's food ordering system.
   a. Write Structured English to represent the logic in each process, at each level data flow diagram.
   b. Represent the decision logic of one or more of the processes as decision tables.

4. Diagram the decision logic illustrated in the decision tables in Figures 10-6 and 10-7 as decision trees.

5. Represent Hoosier Burger's inventory payment process with a state-transition diagram. (See Process 4.0 in Figure 10-2.)

6. What types of questions need to be asked during requirements determination in order to gather the information needed for logic modeling? Give examples.

7. In one company the rules for buying personal computers are such that if the purchase is over $15,000.00, it has to go out for bid and the Request for Proposals must be approved by the Purchasing Department. If the purchase is under $15,000.00, the personal computers can simply be bought from any approved vendor; however, the Purchase Order must still be approved by the Purchasing Department. If the purchase goes out for bid, there must be at least three proposals received for the bid. If not, the RFP must go out again. If there still are not enough proposals, then the process can continue with the one or two vendors that have submitted proposals. The winner of the bid must be on an approved list of vendors for the company and, in addition, must not have any violations against them for affirmative action or

environmental matters. At this point, if the proposal is complete, the Purchasing Department can issue a Purchase Order. Use Structured English to represent the logic in this process. Notice the similarities between the text in this question and format of your answer.

8. If you were going to develop a computer-based tool to help an analyst interview users and quickly and easily create and edit Structured English logic models, what type of tool would you build? What features would it have, and how would it work? Why?

9. Present the logic of the business process described in Problem and Exercise 7 in a decision table and then in a decision tree. For this example, how do these two techniques compare with one another? Why? How do they compare with the Structured English technique? Why? Do either of these techniques seem better suited to helping us diagnose problems and/or inefficiencies in the business process? Why or why not?

10. If you were going to develop a computer-based tool to help an analyst interview users and quickly and easily create and edit decision table and decision tree logic models, what type of tool would you build? What features would it have, and how would it work? Why? Would this type of tool be more or less useful than a tool built for this purpose which was based on Structured English? Why?

11. In a relatively small company that sells thin, electronic keypads and switches, the rules for selling products are such that sales representatives are assigned to unique regions of the country. Sales come either from cold calling, referrals, or current customers with new orders. A sizable portion of their business comes from referrals from larger competitors, who send their excess and/or "difficult" projects to this company. The company tracks these and, similarly, returns the favors to these competitors by sending business their way. The sales reps receive a 10% commission on purchases, not on orders, in their region. They can collaborate on a sale with reps in other regions and share the commissions, with 8% going to the "home" rep, and 2% going to the "visiting" rep. For any sales beyond the rep's previously stated and approved individual annual sales goals, he or she receives an additional 5% commission, an additional end-of-the-year bonus determined by management, and a special vacation for his or her family. Customers receive a 10% discount for any purchases over $100,000.00 per year, which are factored into the rep's commissions. In addition, the company focuses on customer satisfaction with the product and service, so there is an annual survey of customers in which they rate the sales rep. These ratings are factored into the bonuses such that a high rating increases the bonus amount, a moderate rating does nothing, and a low rating can lower the bonus amount. The company also wants to ensure that the reps close all sales. Any differences between the amount of orders and actual purchases is also factored into the rep's bonus amount. As best you can, present the logic of this business process first using Structured English, then using a decision table, and then using a decision tree. Write down any assumptions you have to make. Which of these techniques is most helpful for this problem? Why?

12. The following is an example which demonstrates the rules of the tenure process for faculty at many universities. Present the logic of this business process first using Structured English, then using a decision table, and then using a decision tree. Write down any assumptions you have to make. Which of these techniques is most helpful for this problem? Why?

A faculty member applies for tenure in his or her sixth year by submitting a portfolio summarizing his or her work. In rare circumstances he or she can come up for tenure earlier than the sixth year, but only if the faculty member has permission of the department chair and college dean. New professors, who have worked at other universities before taking their current jobs, rarely, if ever, come in with tenure.

They are usually asked to undergo one "probationary" year during which they are evaluated and only can then be granted tenure. Top administrators coming in to a new university job, however, can often negotiate for retreat rights that enable them to become a tenured faculty member should their administrative post end. These retreat arrangements generally have to be approved by faculty. The tenure review process begins with an evaluation of the candidate's portfolio by a committee of faculty within the candidate's department. The committee then writes a recommendation on tenure and sends it to the department's chairperson who then makes a recommendation, and passes the portfolio and recommendation on to the next level, a college-wide faculty committee. This committee does the same as the department committee and passes its recommendation, the department's recommendation, and the portfolio on to the next level, a university-wide faculty committee. This committee does the same as the other two committees and passes everything on to the provost (or sometimes the academic vice-president). The provost then writes his or her own recommendation and passes everything to the president, the final decision maker. This process, from the time the candidate creates his or her portfolio until the time the president makes a decision, can take an entire academic year. The focus of the evaluation is on research, which could be grants, presentations, and publications, though preference is given for empirical research which has been published in top-ranked, refereed journals and where the publication makes a contribution to the field. The candidate must also do well in teaching and service (i.e., to the university, the community, or to the discipline) but the primary emphasis is on research.

13. An organization is in the process of upgrading micro computer hardware and software for all employees. Hardware will be allocated to each employee in one of three packages. The first hardware package includes a standard micro computer with a color monitor of moderate resolution and moderate storage capabilities. The second package includes a high-end micro computer with high-resolution color monitor and a great deal of RAM and ROM. The third package is a high-end notebook-sized micro computer. Each computer comes with a network interface card so that it can be connected to the network for printing and e-mail. The notebook computers come with a modem for the same purpose. All new and existing employees will be evaluated in terms of their computing needs (e.g., the types of tasks they perform, how much and in what ways they can use the computer). Light users receive the first hardware package. Heavy users receive the second package. Some moderate users will receive the first package and some will receive the second package, depending on their needs. Any employee who is deemed to be primarily mobile (e.g., most of the sales force) will receive the third package. Each employee will also be considered for additional hardware. For example, those who need scanners will receive them and those needing their own printers will receive them. A determination will be made regarding whether or not the user receives a color or black and white scanner, and whether or not they receive a slow or fast, or color or black and white printer. In addition, each employee will receive a suite of software, including a word processor, spreadsheet, and presentation maker. All employees will be evaluated for their additional software needs. Depending on their needs, some will receive a desktop publishing package, some will receive a database management system (and some will also receive a developer's kit for the DBMS), and some will receive a programming language. Every eighteen months those employees with the high-end systems will receive new hardware and then their old systems will be passed on to those who previously had the standard systems. All those employees with the portable systems will receive new notebook computers. Present the logic of this business process first using Structured English, then using a decision table, and then using a decision tree. Write down any assumptions you have to make. Which of these techniques is most helpful for this problem? Why?

14. Draw a state-transition diagram for the following situation at Hoosier Burger. One class of objects Bob Mellankamp handles daily is purchase orders for food and other supplies. Bob creates a purchase order when he assigns a purchase order number and begins to fill in details of the order (such as Hoosier Burger purchase order number and item numbers and quantities for the materials being ordered). After filling in all the other details on a created order, Bob selects a vendor. Bob can void a created order at any time before it is sent to the vendor. No record is kept of voided purchase orders. A purchase order is considered finalized when it is sent to the vendor. Bob cancels an order if the vendor does not confirm the order within a waiting period of one business day for perishable items and three business days for all other items. Bob voids a canceled order and then creates another identical purchase order with a different vendor. Within the waiting period, the vendor will either confirm a finalized order or will indicate that the order cannot be filled, in which case Bob cancels the order. An order remains confirmed until any item on the order is received, at which time it becomes an open order. Items on an open order may arrive in several shipments. When an item arrives, Bob checks the shipment against the order and marks on the order the balance remaining for receipt. An order remains open until Hoosier Burger receives all ordered items. Once all items are received, Bob considers the order filled. Bob does not pay for a filled order until a matching invoice is received. Once all items and an invoice are received, the order is considered pending for payment. If the invoice and items received do not agree, Bob places the order into suspense until he is able to resolve the discrepancies with the vendor. Once he resolves any discrepancies, Bob indicates that the order is ready for payment. When Bob sends the payment check to the vendor, the purchase order is considered closed and Bob records comments about the vendor in records he keeps on all vendors.

FIELD EXERCISES

1. Choose a transaction that you are likely to encounter, perhaps ordering a cap and gown for graduation or ordering clothing from a mail order catalogue. Present the processes in this transaction using Structured English. Based on these examples, what appear to you to be the relative strengths and weaknesses of this technique versus decision tables or decision trees?

2. Why can't process logic simply be represented with regular written English? Research this issue in the library and write a position paper that explains the problems with representing procedure and action with the English language.

3. Explain the history of structured programming. How did it start? What was the basis for it? Why is it considered so important? How does structured programming fit within fourth-generation programming languages? To what extent is structured programming used in organizations with which you have come in contact?

4. Take the same transaction you used in Field Question 1 and draw it as a state-transition diagram.

5. Visit a shopping mall near you and observe the many transactions taking place in the various stores. What techniques presented in this chapter for modeling the logic of processes are most helpful for each of these various processes? Why? Are any of the modeling techniques robust enough to model all of these processes well? Why or why not?

REFERENCES

Booch, G. 1994. *Object-Oriented Analysis and Design, with Applications.* 2d ed. Redwood City, CA: The Benjamin/Cummings Publishing, Co., Inc.

Conger, S. 1994. *The New Software Engineering.* Belmont, CA: Wadsworth Publishing Company.

Davis, W. S. 1983. *Systems Analysis and Design.* Reading, MA: Addison-Wesley Publishing Company.

Flaatten, P. O., D. J. McCubbrey, P. D. O'Riordan, and K. Burgess. 1992. *Foundations of Business Systems,* 2d ed. Fort Worth, TX: Dryden.

Gane, C. and T. Sarson. 1979. *Structured Systems Analysis.* Englewood Cliffs, NJ: Prentice-Hall.

Harel, D. 1987. "Statecharts: A Visual Formalism for Complex Systems." *Science of Computer Programming* 8: 231–74.

Shlaer, S. and S. J. Mellor. 1992. *Object Lifecycles: Modeling the World in States.* Englewood Cliffs, NJ: Prentice-Hall.

Subramanian, G. H., J. Nosek, S. P. Raghunathan, and S. S. Kanitkar. 1992. "A Comparison of the Decision Table and Tree." *Communications of the ACM* 35 (January): 89–94.

Vanthienen, J. 1994. "Technical Correspondence." *Communications of the ACM* 37 (February): 109–111.

Vessey, I. and R. Weber. 1986. "Structured Tools and Conditional Logic." *Communications of the ACM* 29 (January): 48–57.

Yourdon, E. 1989a. *Managing the Structured Techniques.* Englewood Cliffs, NJ: Prentice-Hall.

Yourdon, E. 1989b. *Modern Structured Analysis.* Englewood Cliffs, NJ: Prentice-Hall.

Structuring System Requirements: Logic Modeling for the Customer Activity Tracking System

INTRODUCTION

The CATS project team continues to structure the requirements for the system by documenting the logic of processing rules and decisions that must be made within the system. These logic models will be validated during JAD sessions and walkthroughs. Probably the key decision in CATS is which product announcements should be sent to which customers. This case addresses how the project team represented this decision.

MATCHING PRODUCTS WITH CUSTOMERS

Frank Napier walked into Jorge Lopez's office. "Jorge," Frank said, "I'd like you to take a look at this."

"What is it, Frank?" Jorge asked. Jorge was reviewing some Structured English statements of process logic for BEC's Customer Activity Tracking System (CATS). "I've spent the morning looking at the Structured English statements you asked me to review. Most of this makes sense to me, but I think we need to schedule a walkthrough with Wendy Yoshimuro soon, before we go any further. So, what have you got there?"

"Yesterday I got the marketing information I had asked Wendy for. I've been trying to map out the logic in the process that matches customer profiles and music product profiles. Even though Wendy decided that we would send a customer just one reservation notice that covered all new releases in a given music category rather than a notice about particular artists, there's still a lot of information there, and the overall logic is pretty complicated. I started by looking at just three customer profile characteristics—income, taste in movies, and age—and how the three of them in combination can indicate a person's tastes in music. I started with only four music categories. There are so

many different kinds of music, some kinds filling very narrow niches. If I tried to put all of this information and all of the relationships..."

"Well, let me see what you have now. We can build on your past work and try to simplify the overall logic."

"Here's my first cut at a decision table," Frank said. He showed Jorge his notes. "I assigned values to each of the three customer characteristics I started out with. For income, I started with two categories: affluent and middle class. For movie tastes, I have four categories: adventure/action, romance, horror, and comedy. For age, there are three groupings: teenagers, young adults, and mature."

"How did you define each category?" Jorge asked.

"All of my information came from Wendy. Here's how I broke it down, based on the marketing categories Wendy uses:

Income:
    A = Affluent (> $75,000 per year)
    M = Middle Class ($30,000 to $75,000)
Movie tastes:
    A = Adventure/Action (e.g., "Die Hard" movies)
    R = Romance (e.g., "Sleepless in Seattle")
    H = Horror (e.g., "Friday the Thirteenth" movies)
    C = Comedy (e.g., "Naked Gun" movies)
Age:
    T = Teenager (10 to 20)
    Y = Young adult (21 to 29)
    M = Mature (30 or older)."

"So how does this translate into a decision table?" Jorge asked.

"Since I have 2×4×3 categories in this first cut, I need 24 different decision rules to cover all the possible combinations. That's what I have in my first decision table. The decision rules are based on information Wendy gave me in a memo and also from a procedure description of the music purchasing agent's job that

Wendy sent. The job description described what mix of music to purchase given the demographics of our members, and Wendy thought this might be relevant for matching customers and products, too.

"The four basic music categories I'm using are rock, AOR or adult-oriented radio, urban, and country. Rock is a broad category, ranging from Pearl Jam to the Rolling Stones, from Janis Joplin to the Breeders. AOR includes such performers as Michael Bolton, Mariah Carey, Jimmy Buffett, and Frank Sinatra. The urban music category is represented by such performers as Snoop Doggy Dogg, Ice-T, and N.W.A. Country is also a broad category which includes performers ranging from Garth Brooks to Patsy Cline, from Hank Williams Sr. to Reba McIntyre. Clearly, the final system will have to include narrower categories for both music products and personal tastes in movies. The narrower the categories, the more precise the mailings can be. Anyway, here's the decision table (Figure 1). We can also include categories of videos a member has rented and many other factors, but Wendy wanted to see what resulted from just these factors for now."

"Hmmmm, interesting," replied Jorge. "It looks as if the only group that would get all four new release music reservation notices would be affluent teenagers. They also look like the group that would be the most interested in urban music. Hardly any young adults or mature customers would be interested in urban music. How did you figure this out?"

"All of these decision rules are based on the studies of demographic data that Wendy collects and uses. Teenagers have the most disposable income, so affluent teenagers have even more. They also have eclectic musical tastes, so sending reservation notices for all new musical selections would have the biggest payoff with this group. This is kind of what the music clubs do. Maybe we could modify the 'BEC Membership In-

formation' form Wendy gave me during the original interview I did with her to include music preferences, but the added cost of sending all notices to this group with the most eclectic tastes does not seem that much. Mature customers, on the other hand, don't buy much pre-recorded music and their tastes are narrower so, for this group, more selective notices mean better payoffs. Again, actual movie renting and buying behavior might be a better predictor of music interests than a one-time and possibly rather old membership form."

"What's the connection between movies and music?"

"Lots of movies these days rely on popular music from many different artists in their soundtracks. In fact, some soundtracks become big sellers at the record store. Remember that movie 'The Bodyguard'? The soundtrack was a big seller. Of course it didn't hurt that lots of the songs won Grammies."

"Interesting," said Jorge. "The more I look at this decision table, the more patterns I see."

"Right," said Frank. "In the second version of the table, I have only 12 rules (Figure 2)."

"And to make the entire decision process easier to understand on a more graphical level, I've prepared a decision tree that represents the decision logic in my second decision table (Figure 3)."

"Good work, Frank," Jorge replied. "But it looks as if there is a lot more work to do. Like you said, your musical categories are very broad."

"Yeah, you're right. I found out Broadway actually uses 25 different music categories."

"There has to be at least one more step," volunteered Jorge. "The announcements of new music releases also have to be categorized so that the customer receives a notice of a particular artist's new release based on the customer's profile. Remember, although we legally can't keep records about what

Figure 1

Complete decision table for deciding which music release announcements to send to which customers

Income	A	M	A	M	A	M	A	M	A	M	A	M	A	M	A	M	A	M	A	M	A	M	A	M
Movie tastes	A	A	R	R	H	H	C	C	A	A	R	R	H	H	C	C	A	A	R	R	H	H	C	C
Age	T	T	T	T	T	T	T	T	Y	Y	Y	Y	Y	Y	Y	Y	M	M	M	M	M	M	M	M
Rock notice	X	X	X		X	X	X	X	X	X	X		X	X	X	X						X		X
AOR notice	X		X	X	X		X	X	X	X	X	X			X	X	X	X	X	X	X		X	X
Urban notice	X	X	X		X	X	X						X	X							X			
Country notice	X		X	X	X		X		X		X	X		X		X		X	X	X		X		

Figure 2
Reduced version of decision table from Figure 1

Income	A	A	M	M	M	M	M	M	A	A	A	A
Movie tastes	–	–	R	H	C	A	A	A	A	R	H	C
Age	T	M	–	–	–	T	Y	M	Y	Y	Y	Y
Rock notice	X			X	X	X	X		X	X	X	X
AOR notice	X	X	X		X		X	X	X	X		X
Urban notice	X			X		X					X	
Country notice	X	X	X						X	X	X	X

Figure 3
Decision tree representation of logic in Figure 2

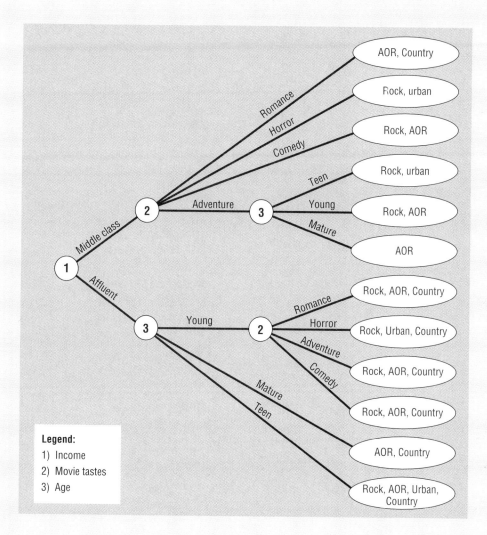

Legend:
1) Income
2) Movie tastes
3) Age

videos customers have bought, we can do this for music. And all of this has to be repeated for new video releases too."

"That's right," Frank replied. "I'm working on it. I'll add a few more factors, as you suggest, and then show the logic models to Jordan to see if she thinks I'm on the right track. Thanks for your help. By the way, as soon as you and the data analyst have a preliminary diagram for the CATS database, let me know and I'll be glad to give you feedback on that."

SUMMARY

Frank, Jorge, and Jordan, the core members of the CATS project team, are now well into some of the most challenging work of systems analysis: transforming what are generally imprecise and vague statements of what people want a system to do into clear and unambiguous models of system requirements. These models allow all interested parties at BEC to know exactly *what* CATS will do and what rules it will follow. Such clarity is necessary for systems design as well as for approval by the Systems Priority Board at the end of the analysis phase. The next BEC case will review some of the work in the last element of structuring requirements—data modeling.

CHAPTER

11

Structuring System Requirements: Conceptual Data Modeling

L E A R N I N G   O B J E C T I V E S

After studying this chapter, you should be able to:

- Concisely define each of the following key data modeling terms: entity, attribute, relationship, degree, cardinality, gerund, business rule, and trigger.

- Draw an entity-relationship (E-R) diagram to represent common business situations.

- Explain the role of conceptual data modeling in the overall analysis and design of an information system.

- Distinguish between unary, binary, and ternary relationships, and give an example of each.

- Define four basic types of business rules in an E-R diagram.

- Explain the role of CASE technology in the analysis and documentation of data required in an information system.

- Relate data modeling to process and logic modeling as different views of describing an information system.

INTRODUCTION

In Chapters 9 and 10 you learned how to model and analyze two important views of an information system: (1) the flow of data between manual or automated steps and (2) the decision and temporal logic of processing data. None of the techniques discussed so far, however, has concentrated on the data that must be retained in order to support the data flows and processing described. For example, you learned how to show data stores, or *data at rest,* in a data flow diagram. The natural *structure* of data, however, was not shown. Data flow diagrams and various processing logic techniques show *how, where,* and *when* data are used or changed in an information system, but these techniques do not show the *definition, structure,* and *relationships* within the data. Data modeling develops this missing, and crucial, piece of the description of an information system.

In fact, some systems developers believe that a data model is the most important part of the statement of information system requirements. This belief is based on the following reasons. First, the characteristics of data captured during data modeling are crucial in the design of databases, programs, computer screens, and printed reports. For example, facts such as these—a data element is numeric, a product can be in only one product line at a time, a line item on a customer order can never be moved to another customer order, customer region name is limited to a specified set of values—are all essential pieces of information in ensuring data integrity in an information system.

Second, data rather than processes are the most complex aspects of many modern information systems and hence require a central role in structuring system requirements. Transaction processing systems can have considerable process complexity in validating data, reconciling errors, and coordinating the movement of data to various databases. These types of systems have been in place for years in most organizations; current systems development focuses more on management information systems (such as sales tracking), decision support systems (such as short-term cash investment), and executive support systems (such as product planning). MIS, DSS, and ESS are more data-intensive and require extracting data from various data sources. The exact nature of processing is also more ad hoc than with transaction processing systems, so the details of processing steps cannot be anticipated. Thus, the goal is to provide a rich data resource that might support any type of information inquiry, analysis, and summarization.

Third, the characteristics about data (such as length, format, and relationships with other data) are reasonably permanent. In contrast, the paths of data flow are quite dynamic. Who receives which data, the format of reports, and what reports are used change considerably and constantly over time. A data model explains the inherent nature of the organization, not its transient form. So, an information system design based on a data orientation, rather than a process or logic orientation, should have a longer useful life. Finally, structural information about data is essential for automatic generation of programs. For example, the fact that a customer order has many line items on it instead of just one line item affects the automatic design of a computer screen for entry of customer orders. Thus, although a data model specifically documents the file and database requirements for an information system, the business meaning, or semantics, of data included in the data model have broader impact on the design and construction of a system.

The most common format used for data modeling is entity-relationship (E-R) diagramming. A similar format used with object-oriented analysis and design methods is briefly reviewed in Appendix D. Data modeling using the E-R notation explains the characteristics and structure of data independent of how the data may be stored in computer memories. A data model using E-R notation is usually developed iteratively. Often IS planners use E-R notation to develop an enterprise-wide data model with very broad categories of data and little detail. Next, during the definition of a project, a specific E-R model is built to help explain the scope of a particular systems analysis and design effort. During requirements structuring, an E-R model represents conceptual data requirements for a particular system. Then, after system inputs and outputs are fully described during logical design, the conceptual E-R data model is refined before it is translated into a logical format (typically a relational data model) from which database definition and physical database design are done. Thus, you will use E-R diagramming in many systems development project steps, and most IS project members need to develop and read E-R diagrams. Therefore, mastery of the requirements structuring methods and techniques addressed in this chapter is critical to your success on a systems development project team.

CONCEPTUAL DATA MODELING

A **conceptual data model** is a representation of organizational data. The purpose of a conceptual data model is to show as many rules about the meaning and inter-relationships among data as are possible.

You typically do conceptual data modeling in parallel with other requirements analysis and structuring steps during systems analysis (see Figure 11-1). You collect the explanations of the business necessary for conceptual data modeling from information-gathering methods like interviewing, questionnaires, and JAD sessions. On larger systems development teams, a subset of the project team concentrates on data modeling while other team members focus attention on process or logic modeling. You develop (or use from prior systems development) a conceptual data model for the current system and build a conceptual data model that supports the scope and requirements for the proposed or enhanced system.

The work of all team members is coordinated and shared through the project dictionary or repository. As discussed in Chapter 5, this repository is often maintained by a common CASE tool, but some organizations still use manual documentation. Whether automated or manual, it is essential that the process, logic, and data model descriptions of a system be consistent and complete since each describes different but complementary views of the same information system. For example, the names of data stores on the primitive-level DFDs often correspond to the names of data

> **Conceptual data model:**
> A detailed model that captures the overall structure of organizational data while being independent of any database management system or other implementation considerations.

Figure 11-1
Systems development life cycle with analysis phase highlighted

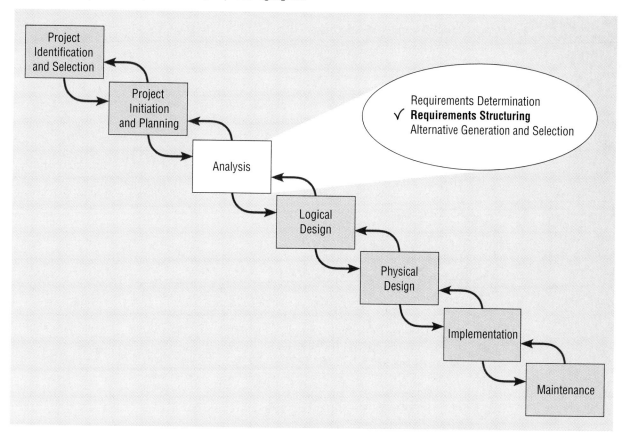

entities in entity-relationship diagrams, and the data elements associated with data flows on DFDs must be accounted for by attributes of entities and relationships in entity-relationship diagrams.

The Process of Conceptual Data Modeling

You typically begin the process of conceptual data modeling by developing a conceptual data model for the system being replaced, if a system exists. This is essential for planning the conversion of the current files or database into the database of the new system. Further, this is a good, but not a perfect, starting point for your understanding of the data requirements of the new system. Then, you build a new conceptual data model which includes all of the data requirements for the new system. You discovered these requirements from the fact-finding methods employed during requirements determination. Today, given the popularity of prototyping, these requirements often evolve through various iterations of a prototype, so the data model is constantly changing.

Conceptual data modeling is one kind of data modeling and database design carried out throughout the systems development process. Figure 11-2 shows the different kinds of data modeling and database design that go on during the whole systems development life cycle. The conceptual data modeling methods we discuss in this chapter are suitable for project identification and selection (including information systems planning which may lead to project identification), project initiation and planning, and analysis phases. These phases of the SDLC address issues of system scope, general requirements, and content—all independent of technical implementation. E-R diagramming is suited for this since E-R diagrams can be translated into a wide variety of technical architectures for data, such as relational, network, and hierarchical. An E-R data model evolves from project identification and selection through analysis as it becomes more specific and is validated by more detailed analysis of system needs.

In the logical design phase, the final E-R model developed in analysis is matched with designs for systems inputs and outputs and is translated into a format from which physical data storage decisions can be made in the physical design phase. During physical design, specific data storage architectures are selected and then, in implementation, files and databases are defined as the system is coded. Through the use of the project repository, a field in a physical data record can, for example, be traced back to the conceptual data attribute which represents it on an E-R diagram. Thus, the data modeling and design steps in each of the SDLC phases are linked through the project repository.

Deliverables and Outcomes

Most organizations today do conceptual data modeling using entity-relationship modeling which uses a special notation to represent as much meaning about data as possible. Thus, the primary deliverable from the conceptual data modeling step within the analysis phase is an E-R diagram, similar to Figure 11-3. This figure shows the major categories of data (rectangles on the diagram) and the business relationships between them (lines connecting rectangles). For example, Figure 11-3 describes that, for the business represented by this diagram, a SUPPLIER *sometimes* Supplies ITEMs to the company, and an ITEM is *always* Supplied by one to four SUPPLIERS. The fact that a supplier only sometimes supplies items implies that the business wants to keep track of some suppliers without designating what they can

Figure 11-2

Relationship between data modeling and the systems development life cycle

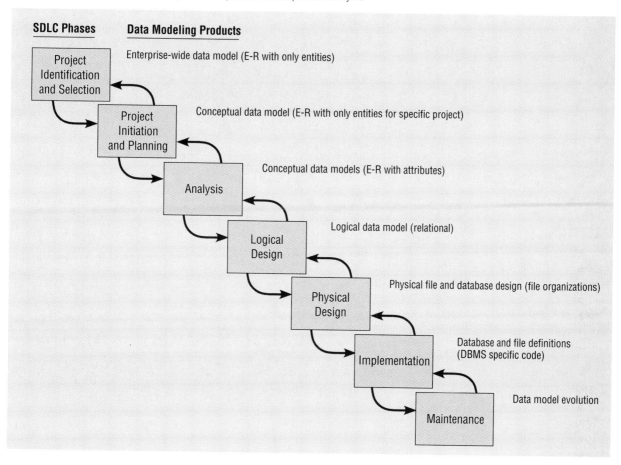

supply. This diagram includes two names on each line giving you explicit language to read a relationship in each direction. For simplicity, we will not typically include two names on lines in E-R diagrams in this book; however, this is a standard used in many organizations.

There may be as many as four E-R diagrams produced and analyzed during conceptual data modeling:

1. An E-R diagram that covers just the data needed in the project's application. (This allows you to concentrate on the data requirements of the project's application without being constrained or confused by unnecessary details.)

2. An E-R diagram for the application system being replaced. (Differences between this diagram and the first show what changes you have to make to convert databases to the new application.) This is, of course, not produced if the proposed system supports a completely new business function.

3. An E-R diagram that documents the whole database from which the new application's data is extracted. (Since many applications possibly share the same database or even several databases, this and the first diagram show how the new application shares the contents of more widely used databases.)

Figure 11-3

Sample conceptual data
model diagram

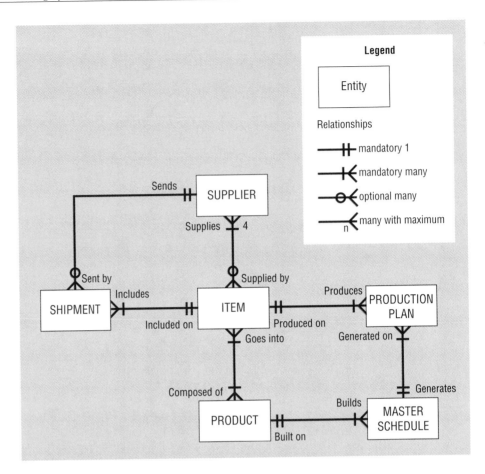

4. An E-R diagram for the whole database from which data for the application system being replaced is drawn. (Again, differences between this diagram and the third show what global database changes you have to make to implement the new application.) Even if there is no system being replaced, an understanding of the existing data systems is necessary to see where the new data will fit in or if existing data structures must change to accommodate new data.

The other deliverable from conceptual data modeling is a full set of entries about data objects to be stored in the project dictionary or repository. The repository is the mechanism to link data, process, and logic models of an information system. For example, there are explicit links between a data model and a data flow diagram. Some important links are briefly explained here.

- Data elements included in data flows also appear in the data model and vice versa. You must include in the data model any raw data captured and retained in a data store and a data model can include only data that has been captured or is computed from captured data. Since a data model is a general business picture of data, both manual and automated data stores will be included.

- Each data store in a process model must relate to business objects (what we will call data entities) represented in the data model. For example, in Figure 9-5, the Inventory file data store must correspond to one or several data objects on a data model.

Similar to what was shown in Figure 9-17, you can use an automated repository to verify these linkages.

GATHERING INFORMATION FOR CONCEPTUAL DATA MODELING

Requirements determination methods must include questions and investigations that take a data, not only a process and logic, focus. For example, during interviews with potential system users—during JAD sessions or within requirements questionnaires—you must ask specific questions in order to gain the perspective on data needed to develop a data model. In later sections of this chapter, we will introduce some specific terminology and constructs used in data modeling. Even without this specific data modeling language, you can begin to understand the kinds of questions that must be answered during requirements determination. These questions relate to understanding the rules and policies by which the area supported by the new information system operates. That is, a data model explains what the organization does and what rules govern how work is performed in the organization. You do not, however, need to know how or when data is processed or used to do data modeling.

You typically do data modeling from a combination of perspectives. The first perspective is generally called the *top-down approach.* This perspective derives the business rules for a data model from an intimate understanding of the nature of the business, rather than from any specific information requirements in computer displays, reports, or business forms. There are several very useful sources of typical questions that elicit the business rules needed for data modeling (see Aranow, 1989; Sandifer and von Halle, 1991a and 1991b). Table 11-1 summarizes a few key questions you should ask system users and business managers so that you can develop an accurate and complete data model. The questions in this table are purposely posed in business terms. In this chapter you will learn the more technical terms included in bold at the end of each set of questions. Of course, these technical terms do not mean much to a business manager, so you must learn how to frame your questions in business terms for your investigation.

You can also gather the information you need for data modeling by reviewing specific business documents—computer displays, reports, and business forms—handled within the system. This process of gaining an understanding of data is often called a *bottom-up approach.* These items will appear as data flows on DFDs and will show the data processed by the system and, hence, probably the data which must be maintained in the system's database. Consider, for example, Figure 11-4, which shows a customer order form used at Pine Valley Furniture. From this form, we determine that the following data must be kept in the database:

ORDER NO	CUSTOMER NO
ORDER DATE	NAME
PROMISED DATE	ADDRESS
PRODUCT NO	CITY-STATE-ZIP
DESCRIPTION	
QUANTITY ORDERED	
UNIT PRICE	

We also see that each order is from one customer, and an order can have multiple line items, each for one product. We will use this kind of understanding of an organization's operation to develop data models.

TABLE 11-1 Requirements Determination Questions for Data Modeling

1. *What are the subjects/objects of the business?* What types of people, places, things, materials, etc. are used or interact in this business, about which data must be maintained? How many instances of each object might exist?—**data entities and their descriptions**

2. *What unique characteristic (or characteristics) distinguishes each object from other objects of the same type?* Might this distinguishing feature change over time or is it permanent? Might this characteristic of an object be missing even though we know the object exists?—**primary key**

3. *What characteristics describe each object?* On what basis are objects referenced, selected, qualified, sorted, and categorized? What must we know about each object in order to run the business?—**attributes and secondary keys**

4. *How do you use this data?* That is, are you the source of the data for the organization, do you refer to the data, do you modify it, and do you destroy it? Who is not permitted to use this data? Who is responsible for establishing legitimate values for this data?—**security controls and understanding who really knows the meaning of data**

5. *Over what period of time are you interested in this data?* Do you need historical trends, current "snapshot" values, and/or estimates or projections? If a characteristic of an object changes over time, must you know the obsolete values?—**cardinality and time dimensions of data**

6. *Are all instances of each object the same?* That is, are there special kinds of each object that are described or handled differently by the organization? Are some objects summaries or combinations of more detailed objects?—**supertypes, subtypes, and aggregations**

7. *What events occur that imply associations between various objects?* What natural activities or transactions of the business involve handling data about several objects of the same or different type?—**relationships, and their cardinality and degree**

8. *Is each activity or event always handled the same way or are there special circumstances?* Can an event occur with only some of the associated objects, or must all objects be involved? Can the associations between objects change over time (for example, employees change departments)? Are values for data characteristics limited in any way?—**integrity rules, minimum and maximum cardinality, time dimensions of data**

Figure 11-4

Sample customer order

```
                          PVF CUSTOMER ORDER
      ORDER NO: 61384                        CUSTOMER NO:  1273

              NAME:            Contemporary Designs
              ADDRESS:         123 Oak St.
              CITY-STATE-ZIP:  Austin, TX 28384

      ORDER DATE: 11/04/93      PROMISED DATE:  11/21/93


       PRODUCT                      QUANTITY      UNIT
       NO         DESCRIPTION       ORDERED       PRICE

       M128       Bookcase          4             200.00
       B381       Cabinet           2             150.00
       R210       Table             1             500.00
```

INTRODUCTION TO E-R MODELING

The basic entity-relationship modeling notation (Chen, 1976) uses three main constructs: data entities, relationships, and their associated attributes. The E-R model notation has subsequently been extended to include additional constructs by Chen and others; for example, see Teorey et al. (1986) and Storey (1991). Several different E-R notations exist, and many CASE tools support multiple notations. For simplicity, we have adopted one common notation for this book, the so-called crow's foot notation. If you use another notation in courses or work, you should be able to easily translate between notations.

An **entity-relationship data model** (or E-R model) is a detailed, logical representation of the data for an organization or for a business area. The E-R model is expressed in terms of entities in the business environment, the relationships or associations among those entities, and the attributes or properties of both the entities and their relationships. An E-R model is normally expressed as an **entity-relationship diagram** (or E-R diagram), which is a graphical representation of an E-R model. The notation we will use for E-R diagrams appears in Figure 11-5, and subsequent sections explain this notation.

> **Entity-relationship data model (E-R model):** A detailed, logical representation of the entities, associations, and data elements for an organization or business area.

> **Entity-relationship diagram (E-R diagram):** A graphical representation of an E-R model.

Entities

An entity (see the first question in Table 11-1) is a person, place, object, event, or concept in the user environment about which the organization wishes to maintain data. An entity has its own identity which distinguishes it from each other entity. Some examples of entities follow:

- Person: EMPLOYEE, STUDENT, PATIENT
- Place: STATE, REGION, COUNTRY, BRANCH
- Object: MACHINE, BUILDING, AUTOMOBILE, PRODUCT
- Event: SALE, REGISTRATION, RENEWAL
- Concept: ACCT, COURSE, WORK CENTER

There is an important distinction between entity *types* and entity *instances*. An **entity type** (sometimes called an *entity class*) is a collection of entities that share common properties or characteristics. Each entity type in an E-R model is given a name. Since the name represents a class or set, it is singular. Also, since an entity is an object, we use a simple noun to name an entity type. We use capital letters in naming an entity type and, in an E-R diagram, the name is placed inside a rectangle representing the entity:

> **Entity type:** A collection of entities that share common properties or characteristics.

An **entity instance** (or **instance**) is a single occurrence of an entity type. An entity type is described just once in a data model while many instances of that entity type may be represented by data stored in the database. For example, there is one EMPLOYEE entity type in most organizations, but there may be hundreds (or even thousands) of instances of this entity type stored in the database.

> **Entity instance (instance):** A single occurrence of an entity type.

A common mistake made when you are just learning to draw E-R diagrams, especially if you already know how to do data flow diagramming, is to confuse data entities with sources/sinks or system outputs and relationships with data flows. A

Figure 11-5

Entity-relationship notation

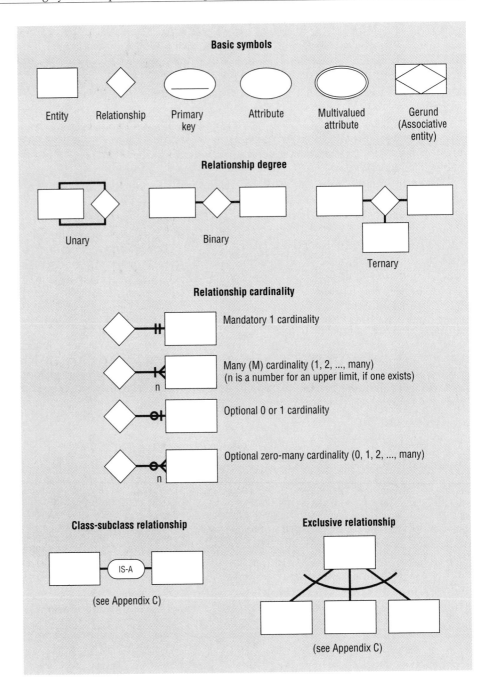

simple rule to avoid such confusion is that a true data entity will have many possible instances, each with a distinguishing characteristic, as well as one or more other descriptive pieces of data. Consider the entity types below that might be associated with a sorority expense system.

In this situation, the sorority treasurer manages accounts, and records expense transactions against each account. However, do we need to keep track of data about the Treasurer and her supervision of accounts as part of this accounting system? The Treasurer is the person entering data about accounts and expenses and making inquiries about account balances and expense transactions by category. Since there is only one Treasurer, TREASURER data does not need to be kept. On the other hand, if each account has an account manager (for example, a sorority officer) who is responsible for assigned accounts, then we may wish to have an ACCOUNT MANAGER entity type, with pertinent attributes as well as relationships to other entity types.

In this same situation, is an expense report an entity type? Since an expense report is computed from expense transactions and account balances, it is a data flow, not an entity type. Even though there will be multiple instances of expense reports over time, the report contents are already represented by the ACCOUNT and EXPENSE entity types.

Often when we refer to entity types in subsequent sections we will simply say entity. This is common among data modelers. We will clarify that we mean an entity by using the term entity instance.

Attributes

Each entity type has a set of attributes (see the third question in Table 11-1) associated with it. An **attribute** is a property or characteristic of an entity that is of interest to the organization (relationships may also have attributes, as we will see in the section on Relationships). Following are some typical entity types and associated attributes:

> **Attribute:** A named property or characteristic of an entity that is of interest to the organization.

    STUDENT: STUDENT NO., NAME, ADDRESS, PHONE NO.

    AUTOMOBILE: VEHICLE ID, COLOR, WEIGHT, HORSEPOWER

    EMPLOYEE: EMPLOYEE NO., NAME, ADDRESS, SKILL

We use capital letters and nouns in naming an attribute. In E-R diagrams, we can visually represent an attribute by placing its name in an ellipse with a line connecting it to the associated entity. Many CASE tools, to avoid placing a large number of symbols on a diagram, do not include attributes on an E-R diagram. Rather, the attributes of an entity (sometimes called its composition) are listed in the repository entry for the entity; then, each attribute may be separately defined as another object in the repository. This is similar to how the composition of a data flow is handled in Visible Analyst Workbench for Windows (see Figure 9-18).

Candidate Keys and Primary Keys

Every entity type must have an attribute or set of attributes that distinguishes one instance from other instances of the same type (see the second question in Table 11-1). A **candidate key** is an attribute (or combination of attributes) that uniquely identifies each instance of an entity type. A candidate key for a STUDENT entity type might be STUDENT NO.

> **Candidate key:** An attribute (or combination of attributes) that uniquely identifies each instance of an entity type.

Sometimes more than one attribute is required to identify a unique entity. For example, consider the entity type GAME for a basketball league. The attribute TEAM NAME is clearly not a candidate key, since each team plays several games. If each team plays exactly one home game against each other team, then the combination of the attributes HOME TEAM and VISITING TEAM is a candidate key for GAME.

Some entities may have more than one candidate key. One candidate key for EMPLOYEE is EMPLOYEE NO.; a second is the combination of NAME and ADDRESS (assuming that no two employees with the same name live at the same address). If there is more than one candidate key, the designer must choose one of the candidate keys as a primary key. A **primary key** is a candidate key that has been selected to be used as the identifier for an entity type. Bruce (1992) suggests the following criteria for selecting primary keys:

Primary key: A candidate key that has been selected as the identifier for an entity type. Primary key values may not be null. Also called an *identifier*.

1. Choose a candidate key that will not change its value over the life of each instance of the entity type. For example, the combination of NAME and ADDRESS would probably be a poor choice as a primary key for EMPLOYEE because the values of ADDRESS and NAME could easily change during an employee's term of employment.

2. Choose a candidate key such that, for each instance of the entity, the attribute is guaranteed to have valid values and not be null. To insure valid values, you may have to include special controls in data entry and maintenance routines to eliminate the possibility of errors. If the candidate key is a combination of two or more attributes, make sure that all parts of the key will have valid values.

3. Avoid the use of so-called intelligent keys, whose structure indicates classifications, locations, and so on. For example, the first two digits of a key for a PART entity may indicate the warehouse location. Such codes are often modified as conditions change, which renders the primary key values invalid.

4. Consider substituting single-attribute surrogate keys for large composite keys. For example, an attribute called GAME NO. could be used for the entity GAME instead of the combination of HOME TEAM and VISITING TEAM.

For each entity, the name of the primary key is underlined on an E-R diagram. The following diagram shows the representation for a STUDENT entity type using E-R notation:

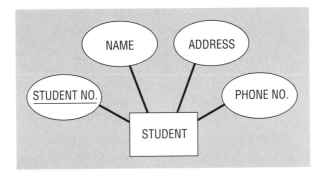

Multivalued Attributes

Multivalued attribute: An attribute that can have more than one value for each entity instance.

A **multivalued attribute** can have more than one value *for each entity instance*. Suppose that SKILL is one of the attributes of EMPLOYEE. If each employee can have more than one skill, SKILL is a multivalued attribute. During conceptual design, it is common to use a special symbol or notation to highlight multivalued attributes. Two ways of showing multivalued attributes are common. The first is to use a double-lined ellipse, so that the EMPLOYEE entity with its attributes is diagrammed as follows:

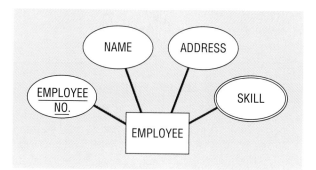

The second approach is to separate the repeating data into another entity, called a *weak* (or *attributive*) entity, and then using a relationship (relationships are discussed in the next section), link the weak entity to its associated regular entity. The approach also easily handles several attributes that repeat together, called a **repeating group.** For example, consider again an employee entity with multivalued attributes for data about each employee's dependents. In this situation, data such as dependent name, age, and relation to employee (spouse, child, parent, etc.) are multivalued attributes about an employee, and these attributes repeat together. We could show these multivalued attributes using the ellipse notation as

Repeating group: A set of two or more multivalued attributes that are logically related.

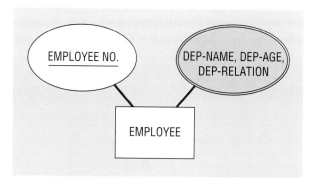

We can show this using an attributive entity, DEPENDENT, and a relationship, shown here simply by a line between DEPENDENT and EMPLOYEE. The crow's foot next to DEPENDENT means that there may be many DEPENDENTs for the same EMPLOYEE.

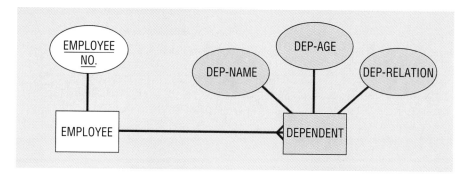

Relationships

Relationships are the glue that hold together the various components of an E-R model (see the fifth, seventh, and eighth questions in Table 11-1). A **relationship** is

Relationship: An association between the instances of one or more entity types that is of interest to the organization.

an association between the instances of one or more entity types that is of interest to the organization. An association usually means that event has occurred or that there exists some natural linkage between entity instances. For this reason, relationships are labeled with verb phrases. For example, a training department in a company is interested in tracking which training courses each of its employees has completed. This leads to a relationship (called Completes) between the EMPLOYEE and COURSE entity types that we diagram as follows:

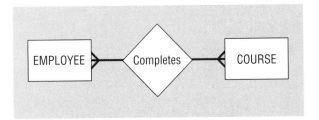

As indicated by the arrows, this is a many-to-many relationship: Each employee may complete more than one course, and each course may be completed by more than one employee. More significantly, we can use the Completes relationship to determine the specific courses that a given employee has completed. Conversely, we can determine the identity of each employee who has completed a particular course.

Many CASE tools, to avoid cluttering an E-R diagram with excess symbols, will not include the relationship diamond and simply place the verb phrase for the relationship name near the line (see Figure 11-3 for an example). We use both representations interchangeably in this book and, as noted earlier, we sometimes use two verb phrases so that there is an explicit name for the relationship in each direction. The standards you will follow will be determined by your organization.

CONCEPTUAL DATA MODELING AND THE E-R MODEL

The last section introduced the fundamentals of the E-R data modeling notation—entities, attributes, and relationships. The goal of conceptual data modeling is to capture as much of the meaning of data as is possible. The more details (business rules) about data that we can model, the better the system we can design and build. Further, if we can include all these details in a CASE repository, and if a CASE tool can generate code for data definitions and programs, then the more we know about data, the more code can be generated automatically. This will make system building more accurate and faster. More importantly, if we can keep a thorough repository of data descriptions, we can regenerate the system as needed as the business rules change. Since maintenance is the largest expense with any information system, the efficiencies gained by maintaining systems at the rule, rather than code, level drastically reduce the cost.

In this section, we explore more advanced concepts needed to more thoroughly model data and learn how the E-R notation represents these concepts.

Degree of a Relationship

Degree: The number of entity types that participate in a relationship.

The **degree** of a relationship (see question 7 in Table 11-1) is the number of entity types that participate in that relationship. Thus, the relationship Completes illustrated above is of degree two, since there are two entity types: EMPLOYEE and COURSE. The three most common relationships in E-R models are *unary* (degree one), *binary* (degree two), and *ternary* (degree three). Higher-degree relationships are possible, but they are rarely encountered in practice, so we restrict our discussion to these three cases. Examples of unary, binary, and ternary relationships appear in Figure 11-6.

Unary Relationship   Also called a *recursive relationship,* a **unary relationship** is a relationship between the instances of one entity type. Two examples are shown in Figure 11-6. In the first example, Is Married to is shown as a one-to-one relationship between instances of the PERSON entity type. That is, each person may be currently married to one other person. In the second example, Manages is shown as a one-to-many relationship between instances of the EMPLOYEE entity type. Using this relationship, we could identify (for example) the employees who report to a particular manager, or reading the Manages relationship in the opposite direction, who the manager is for a given employee.

Figure 11-7 shows an example of another common unary relationship, called a *bill-of-materials structure.* Many manufactured products are made of subassemblies, which in turn are composed of other subassemblies and parts, and so on. As shown in Figure 11-7a, we can represent this structure as a many-to-many unary relationship. In this figure, we use Has Components for the relationship name. The attribute QUANTITY, which is a property of the relationship, indicates the number of each component that is contained in a given assembly.

Unary relationship (recursive relationship): A relationship between the instances of one entity type.

Figure 11-6
Example relationships of different degrees

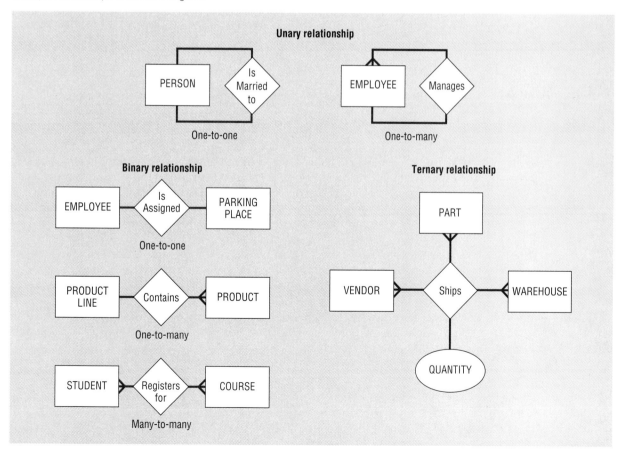

Figure 11-7
Bill-of-materials unary
relationship

(a) Many-to-many relationship

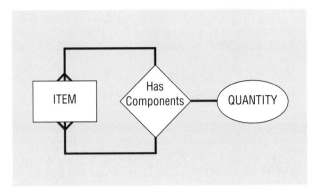

(b) Two instances

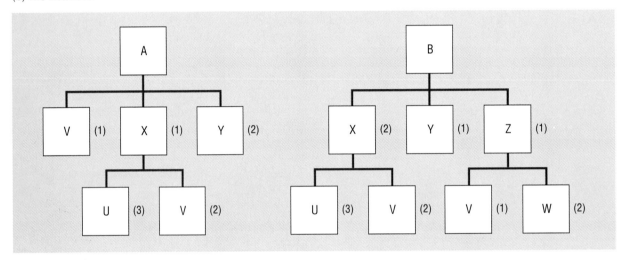

Two occurrences of this structure are shown in Figure 11-7b. Each diagram shows the immediate components of each item as well as the quantities of that component. For example, item X consists of item U (quantity 3) and item V (quantity 2). You can easily verify that the associations are in fact many-to-many. Several of the items have more than one component type (for example, item A has three immediate component types: V, X, and Y). Also, some of the components are used in several higher-level assemblies. For example, item X is used in both item A and item B. The many-to-many relationship guarantees, for example, that the same subassembly structure of X is used each time item X goes into making some other item.

Binary relationship: A relationship between instances of two entity types. This is the most common type of relationship encountered in data modeling.

Binary Relationship   A **binary relationship** is a relationship between instances of two entity types and is the most common type of relationship encountered in data modeling. Figure 11-6 shows three examples. The first (one-to-one) indicates that an employee is assigned one parking place, and each parking place is assigned to one employee. The second (one-to-many) indicates that a product line may contain several products, and each product belongs to only one product line. The third (many-to-many) shows that a student may register for more than one course, and that each course may have many student registrants.

Ternary Relationship   A **ternary relationship** is a *simultaneous* relationship among instances of three entity types. In the example shown in Figure 11-6, the relationship Ships tracks the quantity of a given part that is shipped by a particular vendor to a selected warehouse. Each entity may be a one or a many participant in a ternary relationship (in Figure 11-6, all three entities are many participants).

Ternary relationship: A simultaneous relationship among instances of three entity types.

Note that a ternary relationship is not the same as three binary relationships. For example, QUANTITY is an attribute of the Ships relationship in Figure 11-6. QUANTITY cannot be properly associated with any of the three possible binary relationships among the three entity types (such as that between PART and VENDOR) because QUANTITY is the amount of a particular PART shipped from a particular VENDOR to a particular WAREHOUSE.

Cardinalities in Relationships

Suppose that there are two entity types, A and B, that are connected by a relationship. The **cardinality** of a relationship (see the fifth, seventh, and eighth questions in Table 11-1) is the number of instances of entity B that can (or must) be associated with each instance of entity A. For example, consider the following relationship for video movies:

Cardinality: The number of instances of entity B that can (or must) be associated with each instance of entity A.

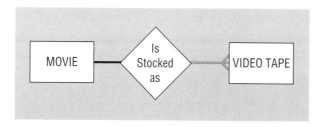

Clearly, a video store may stock more than one video tape of a given movie. In the terminology we have used so far, this example is intuitively a "many" relationship. Yet it is also true that the store may not have a single tape of a particular movie in stock. We need a more precise notation to indicate the *range* of cardinalities for a relationship. This notation was introduced in Figure 11-5, which you may want to review at this point.

Minimum and Maximum Cardinalities   The *minimum* cardinality of a relationship is the minimum number of instances of entity B that may be associated with each instance of entity A. In the preceding example, the minimum number of video tapes available for a movie is zero, in which case we say that VIDEO TAPE is an *optional participant* in the Is Stocked as relationship. When the minimum cardinality of a relationship is one, then we say entity B is a *mandatory participant* in the relationship. The *maximum* cardinality is the maximum number of instances. For our example, this maximum is "many" (an unspecified number greater than one). Using the notation from Figure 11-5, we diagram this relationship as follows:

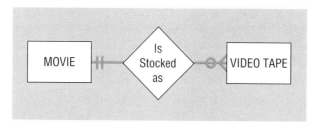

The zero through the line near the VIDEO TAPE entity means a minimum cardinality of zero, while the crow's foot notation means a "many" maximum cardinality.

Examples of three relationships that show all possible combinations of minimum and maximum cardinalities appear in Figure 11-8. A brief description of each relationship follows.

1. PATIENT Has PATIENT HISTORY (Figure 11-8a). Each patient has one or more patient histories (we assume that the initial patient visit is always recorded as an instance of PATIENT HISTORY). Each instance of PATIENT HISTORY "belongs to" exactly one PATIENT.

2. EMPLOYEE Is Assigned to PROJECT (Figure 11-8b). Each PROJECT has at least one assigned EMPLOYEE (some projects have more than one). Each EMPLOYEE may or (optionally) may not be assigned to any existing PROJECT, or may be assigned to several PROJECTs.

3. PERSON Is Married to PERSON (Figure 11-8c). This is an optional 0 or 1 cardinality in both directions, since a person may or may not be married.

It is possible for the maximum cardinality to be a fixed number, not an arbitrary "many" value. For example, suppose corporate policy states that an employee may work on at most five projects at the same time. We could show this business rule by placing a 5 above or below the crow's foot next to the PROJECT entity in Figure 11-8b.

Figure 11-8
Examples of cardinalities in relationships

(a) Mandatory cardinalities

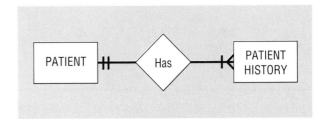

(b) One optional, one mandatory cardinality

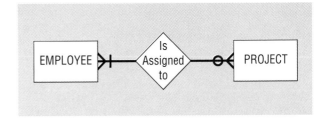

(c) Optional cardinalities

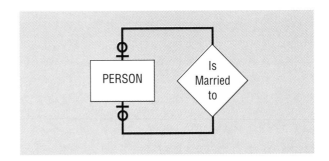

Gerunds

As seen in the examples of the Ships relationship in Figure 11-6 and the Has Components relationship of Figure 11-7, attributes may be associated with a many-to-many relationship as well as with an entity. For example, suppose that the organization wishes to record the date (month and year) when an employee completes each course. Some sample data follows:

EMPLOYEE NO.	COURSE NAME	DATE COMPLETED
549-23-1948	Basic Algebra	March 1994
629-16-8407	Software Quality	June 1994
816-30-0458	Software Quality	Feb 1994
549-23-1948	C Programming	May 1994

From this limited data you can conclude that the attribute DATE COMPLETED is not a property of the entity EMPLOYEE (since a given employee such as 549-23-1948 has completed courses on different dates). Nor is DATE COMPLETED a property of COURSE, since a particular course (such as Software Quality) may be completed on different dates. Instead, DATE COMPLETED is a property of the relationship between EMPLOYEE and COURSE. The attribute is associated with the relationship and diagrammed as follows:

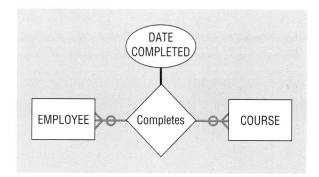

Since many-to-many and one-to-one relationships may have associated attributes, the E-R data model poses an interesting dilemma: is a many-to-many relationship actually an entity in disguise? Often the distinction between entity and relationship is simply a matter of how you view the data. A **gerund** (sometimes called a *composite* or *associative entity*) is a relationship that the data modeler chooses to model as an entity type. Figure 11-9 shows the E-R notation for representing the Completes relationship as a gerund. The diamond symbol is included within the entity rectangle as a reminder that the entity was derived from a relationship. The lines from COMPLETES to the two entities are not two separate binary relationships but rather the

Gerund: A many-to-many (or one-to-one) relationship that the data modeler chooses to model as an entity type with several associated one-to-many relationships with other entity types.

Figure 11-9
Example associative entity

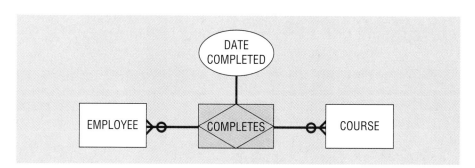

two ends of one binary relationship. The implicit primary key of COMPLETES is the combination of the primary keys of EMPLOYEE and COURSE, EMPLOYEE NO. and COURSE NAME, respectively.

An example of the use of a gerund for a ternary relationship appears in Figure 11-10. This figure shows an alternative (and equally correct) representation of the ternary Ships relationship shown in Figure 11-6. In Figure 11-10, the entity type (gerund) SHIPMENT replaces the Ships relationship from Figure 11-6. Each instance of SHIPMENT represents a real-world shipment by a given vendor of a particular part to a selected warehouse. The QUANTITY of that shipment is an attribute of SHIPMENT. A shipment number is assigned to each shipment and is the primary key of SHIPMENT, as shown in Figure 11-10.

Note that we do not use a diamond along the lines from the gerund to the entities and there are no marks on the lines near the gerund. This is because these lines *do not* represent binary relationships. To keep the same meaning as the ternary relationship of Figure 11-10, we cannot break the Ships relationship from Figure 11-6 into three binary relationships between SHIPMENT and VENDOR, PART, and WAREHOUSE, respectively.

One situation in which a relationship *must* be turned into a gerund is when the gerund has other relationships with entities besides the relationship which caused the creation of the gerund. For example, consider the following E-R model which represents price quotes from different vendors for purchased parts stocked by Pine Valley Furniture:

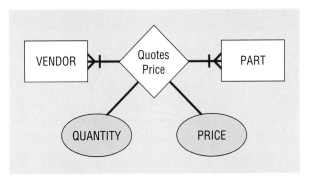

Figure 11-10
SHIPMENT entity type
(a gerund)

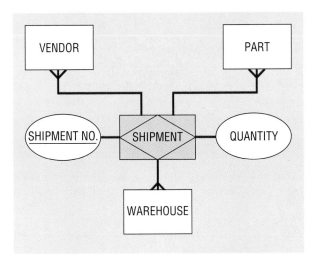

Now, suppose that we also need to know which price quote is in effect for each part shipment received. This additional data requirement *necessitates* that the Quotes Price relationship be transformed into a gerund, and this new situation is represented as follows:

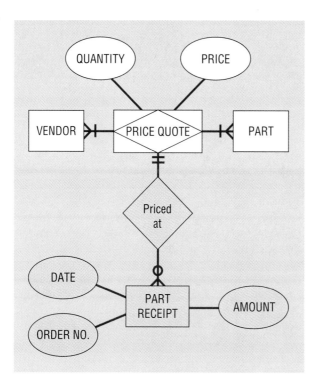

In this case, PRICE QUOTE is not a ternary relationship. Rather, PRICE QUOTE is a binary many-to-many relationship (gerund) between VENDOR and PART. In addition, each PART RECEIPT, based on AMOUNT, has an applicable, negotiated PRICE. Each PART RECEIPT is for a given PART from a specific VENDOR, and the AMOUNT of the receipt dictates the purchase price in effect by matching with the QUANTITY attribute. Since the PRICE QUOTE pertains to a given PART and given VENDOR, PART RECEIPT does not need direct relationships with these entities.

Summary of Conceptual Data Modeling with E-R Diagrams

The purpose of E-R diagramming is to capture the richest possible understanding of the meaning of data necessary for an information system or organization. Besides the aspects shown in this chapter, there are many other semantics about data that E-R diagramming can represent. Some of these more advanced capabilities are explained in Appendix C. Appendix C addresses modeling time, representing multiple relationships between the same entities, and the data abstractions of generalization and aggregation. The interested reader is also referred to Batini, Ceri, and Navathe (1992) and McFadden and Hoffer (1994) for further coverage of E-R diagramming. The following section presents one final aspect of conceptual data modeling: capturing the rules by which the organization operates.

BUSINESS RULES

Business rules: Specifications that preserve the integrity of a conceptual or logical data model.

Conceptual data modeling is a step-by-step process for documenting information requirements that is concerned with both the structure of data and with rules about the integrity of that data (see the eighth question in Table 11-1). **Business rules** are specifications that preserve the integrity of the logical data model. There are four basic types of business rules:

1. *Entity integrity.* Each instance of an entity type must have a unique identifier (or primary key value) that is not null.

2. *Referential integrity constraints.* Rules concerning the relationships between entity types.

3. *Domains.* Constraints on valid values for attributes.

4. *Triggering operations.* Other business rules that protect the validity of attribute values.

The entity-relationship model that we have described in this chapter is concerned primarily with the structure of data rather than with expressing business rules (although some elementary rules are implied in the E-R model). Generally the business rules are captured during requirements determination and stored in the CASE repository (described in Chapter 5) as they are documented. Entity integrity was described earlier in this chapter and referential integrity is described in Chapter 15 since it applies most to logical data modeling. In this section, we briefly describe two types of rules: domains and triggering operations. These rules are illustrated with a simple example from a banking environment, shown in Figure 11-11a. In this example, an ACCOUNT entity has a relationship (Is for) with a WITHDRAWAL entity.

Domains

Domain: The set of all data types and values that an attribute can assume.

A **domain** is the set of all data types and ranges of values that attributes may assume (Fleming and von Halle, 1990). Domain definitions typically specify some (or all) of the following characteristics of attributes: data type, length, format, range, allowable values, meaning, uniqueness, and null support (whether an attribute value may or may not be null).

Figure 11-11b shows two domain definitions for the banking example. The first definition is for ACCT NO. Since ACCT NO. is a primary key attribute, the definition specifies that ACCT NO. must be unique and also must not be null (these specifications are true of all primary keys). The definition specifies that the attribute data type is character and that the format is nnn-nnnn. Thus any attempt to enter a value for this attribute that does not conform to its character type or format will be rejected, and an error message will be displayed.

The domain definition for the AMOUNT attribute (dollar amount of the requested withdrawal) also may not be null, but is not unique. The format allows for two decimal places to accommodate a currency field. The range of values has a lower limit of zero (to prevent negative values) and an upper limit of 10,000. The latter is an arbitrary upper limit for a single withdrawal transaction.

The use of domains offers several advantages:

1. Domains verify that the values for an attribute (stored by insert or update operations) are valid.

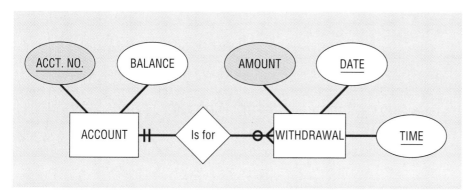

Figure 11-11
Examples of business rules

(a) Simple banking
relationship

```
Name: ACCT NO.                        Name: AMOUNT
Meaning: Customer account number in bank   Meaning: Dollar amount of transaction
Data type: Character                  Data type: Numeric
Format: nnn-nnnn                      Format: 2 decimal places
Uniqueness: Must be unique            Range: 0-10,000
Null support: Non-null                Uniqueness: Nonunique
                                      Null support: Non-null
```

(b) Typical domain definitions

```
User rule: WITHDRAWAL AMOUNT may not exceed ACCOUNT BALANCE
Event: Insert
Entity Name: WITHDRAWAL
Condition: WITHDRAWAL AMOUNT> ACCOUNT BALANCE
Action: Reject the insert transaction
```

(c) Typical triggering
operation

2. Domains ensure that various data manipulation operations (such as joins or unions in a relational database system) are logical.

3. Domains help conserve effort in describing attribute characteristics.

Domains can conserve effort because we can define domains and then associate each attribute in the data model with an appropriate domain. To illustrate, suppose that a bank has three types of accounts, with the following primary keys:

Account Type	Primary key
CHECKING	CHECKING ACCT NO.
SAVINGS	SAVINGS ACCT NO.
LOAN	LOAN ACCT NO.

If domains are not used, the characteristics for each of the three primary key attributes must be described separately. Suppose, however, that the characteristics for all three of the attributes are identical. Having defined the domain ACCT NO. once (as shown in Figure 11-11b), we simply associate CHECKING ACCT NO., SAVINGS ACCT NO., and LOAN ACCT NO. with ACCT NO. Other common domains such as DATE, SOCIAL SECURITY NUMBER, and TELEPHONE NUMBER also need to be defined just once in the model.

Triggering Operations

A **triggering operation** (or **trigger**) is an assertion or rule that governs the validity of data manipulation operations such as insert, update, and delete. The scope of triggering operations may be limited to attributes within one entity, or it may

Triggering operation (trigger): An assertion or rule that governs the validity of data manipulation operations such as insert, update, and delete.

extend to attributes in two or more entities. Complex business rules may often be stated as triggering operations.

A triggering operation normally includes the following components:

1. *User rule:* a concise statement of the business rule to be enforced by the triggering operation

2. *Event:* the data manipulation operation (insert, delete, or update) that initiates the operation

3. *Entity name:* the name of the entity being accessed and/or modified

4. *Condition:* condition that causes the operation to be triggered

5. *Action:* action taken when the operation is triggered

Figure 11-11c shows an example of a triggering operation for the banking situation. The business rule is a simple (and familiar) one: the amount of an attempted withdrawal may not exceed the current account balance. The event of interest is an attempted insert of an instance of the WITHDRAWAL entity type (perhaps from an automated teller machine). The condition is

AMOUNT (of the withdrawal) > ACCOUNT BALANCE

When this condition is triggered, the action taken is to reject the transaction. You should note two things about this triggering operation: first, it spans two entity types; second, the business rule could not be enforced through the use of domains.

The use of triggering operations is an increasingly important component of database strategy. With triggering operations, the responsibility for data integrity lies within the scope of the database management system rather than with application programs or human operators. In the banking example, tellers could conceivably check the account balance before processing each withdrawal. Human operators would be subject to human error and, in any event, manual processing would not work with automated teller machines. Alternatively, the logic of integrity checks could be built into the appropriate application programs, but integrity checks would require duplicating the logic in each program. There is no assurance that the logic would be consistent (since the application programs may have been developed at different times by different people) or that the application programs will be kept up to date as conditions change.

As stated earlier, business rules should be documented in the CASE repository. Ideally, these rules will then be checked automatically by database software. Removing business rules from application programs and incorporating them in the repository (in the form of domains, referential integrity constraints, and triggering operations) has several important advantages:

1. Provides faster application development with fewer errors, since these rules can be generated into programs or enforced by the DBMS

2. Reduces maintenance effort and expenditures

3. Provides faster response to business changes

4. Facilitates end-user involvement in developing new systems and manipulating data

5. Provides for consistent application of integrity constraints

6. Reduces time and effort required to train application programmers

7. Promotes ease of use of a database

THE ROLE OF CASE IN CONCEPTUAL DATA MODELING

CASE tools provide two important functions in conceptual data modeling: (1) maintaining E-R diagrams as a visual depiction of structured data requirements and (2) linking objects on E-R diagrams to corresponding descriptions in a repository. Most CASE tools support one or more of several standard E-R diagramming notations, such as the crow's foot notation used in this text. Many tools do not support drawing ternary or higher relationships, so you may have to model these higher degree relationships as several binary relationships, even though this is not semantically correct.

Figure 11-12 lists the typical data model elements that are placed in the project dictionary or CASE repository during conceptual data modeling. A CASE tool will typically allow you to move directly to the repository entry for an object once you select it on an E-R diagram. The precise list of object characteristics will vary by CASE tool or the standard you use for a project dictionary. Figure 11-12, however, provides a basic set of repository contents used in almost any circumstance. Later data modeling and database design steps develop additional data model elements for the repository, as we will show in subsequent chapters.

Figure 11-12
Typical conceptual data model elements in a project dictionary

Entity (major category of data)	
Name	A short and a long name that uniquely label the entity
Description	Explanation so that it is clear what objects are covered by this entity
Alias	Alternative names used for this entity (that is, synonyms)
Primary key	Name(s) of attribute(s) that form the unique identifier for each instance of this entity
Attributes and repetition	List of attributes associated with this entity and the number of instances of each attribute for each entity instance
Abstraction	Indication of any superclasses or subclasses or composition of entity types involving this entity
Attribute (entity characteristic)	
Name	A short and a long name that uniquely label the attribute
Description	Explanation of the attribute so that its meaning is clearly different from all other attributes
Alias	Alternative names used for this attribute (that is, synonyms)
Domain	The permitted values that this attribute may assume
Computation	If this is not raw data, the formula or method to calculate the attribute's value
Aggregation	Indication of any groupings of attributes involving this attribute (e.g., a month attribute as part of a date attribute)
Relationship (association between entity instances)	
Name	A short and a long name that uniquely label the relationship
Description	Explanation of the relationship so that its meaning is clearly different from all other relationships
Degree	Names of entities involved in the relationship
Cardinality	The potential number of instances of each entity involved in the relationship
Insertion rules	Business rules that control the inclusion of entity instances in this relationship
Deletion rules	Business rules that control the elimination of entity instances from this relationship

AN EXAMPLE OF CONCEPTUAL DATA MODELING AT HOOSIER BURGER

Chapters 9 and 10 structured the process and logic requirements for a new inventory control system for Hoosier Burger. The data flow diagram and decision table (repeated here as Figures 11-13 and 11-14) describe requirements for this new system. The purpose of this system is to monitor and report changes in raw material inventory levels, and to issue material orders and payments to suppliers. Thus, the central data entity for this system will be an INVENTORY ITEM, corresponding to data store D1 in Figure 11-13.

Changes in inventory levels are due to two types of transactions: receipt of new items from suppliers and consumption of items from sales of products. Inventory is added upon receipt of new raw materials, for which Hoosier Burger receives a supplier INVOICE (see Process 1.0 in Figure 11-13). Each INVOICE indicates that the supplier has sent a specific quantity of one or more INVOICE ITEMs, which correspond to Hoosier's INVENTORY ITEMs. Inventory is used when customers order and pay for PRODUCTs. That is, Hoosier makes a SALE for one or more SALE ITEMs, each of which corresponds to a food PRODUCT. Since the real-time customer order processing system is separate from the inventory control system, a source, STOCK ON-HAND on Figure 11-13, represents how data flows from the order processing to the inventory control system. Finally, since food PRODUCTs are made up of various INVENTORY ITEMs (and visa versa), Hoosier maintains a RECIPE to indicate how much of each INVENTORY ITEM goes into making one

Figure 11-13

Level-0 data flow diagram for Hoosier Burger's new logical inventory control system (same as Figure 9-16)

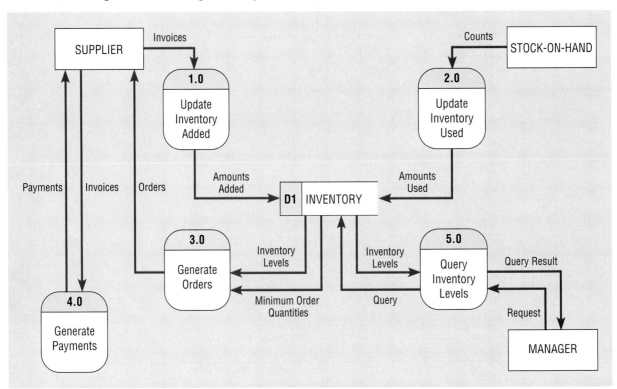

Conditions/	Rules						
Courses of Action	1	2	3	4	5	6	7
Type of item	P	P	P	P	P	P	N
Time of week	D	W	D	W	D	W	–
Season of year	A	A	S	S	H	H	–
Standing daily order	X		X		X		
Standing weekend order		X		X		X	
Minimum order quantity							X
Holiday reduction					X	X	
Summer reduction			X	X			

Figure 11-14

Reduced decision table for Hoosier Burger's inventory reordering (Same as Figure 10-7)

PRODUCT. From this discussion, we have identified the data entities required in a data model for the new Hoosier Burger inventory control system: INVENTORY ITEM, INVOICE, INVOICE ITEM, PRODUCT, SALE, SALE ITEM, and RECIPE. To complete the data model, we must determine necessary relationship between these entities as well as attributes for each entity.

The wording in the previous description tells us much of what we need to know to determine relationships:

- An INVOICE includes one or more INVOICE ITEMs, each of which corresponds to an INVENTORY ITEM. Obviously, an INVOICE ITEM cannot exist without an associated INVOICE, and over time there will be zero to many receipts, or INVOICE ITEMs, for an INVENTORY ITEM.

- Each PRODUCT has a RECIPE of INVENTORY ITEMs. Thus, RECIPE is an associative entity supporting a bill-of-materials type relationship between PRODUCT and INVENTORY ITEM.

- A SALE indicates that Hoosier sells one or more ITEM SALES, each of which corresponds to a PRODUCT. An ITEM SALE cannot exist without an associated SALE, and over time there will be zero to many ITEM SALES for a PRODUCT.

Figure 11-15 shows an E-R diagram with the entities and relationships described above. We include on this diagram two labels for each relationship, one to be read in either relationship direction (e.g., an INVOICE Includes one to many INVOICE ITEMS, and an INVOICE ITEM Is Included on exactly one INVOICE). RECIPE, since it is a gerund, also serves as the label for the many-to-many relationship between PRODUCT and INVENTORY ITEM. Now that we understand the entities and relationships, we must decide which data elements are associated with the entities and gerunds in this diagram.

You may wonder at this point why only the INVENTORY data store is shown in Figure 11-13 when there are seven entities and gerunds on the E-R diagram. The INVENTORY data store corresponds to the INVENTORY ITEM entity in Figure 11-15. The other entities are hidden inside other processes for which we have not shown lower-level diagrams. In actual requirements structuring steps, you would have to match all entities with data stores: each data store represents some

Figure 11-15

Preliminary E-R diagram for Hoosier Burger's inventory control system

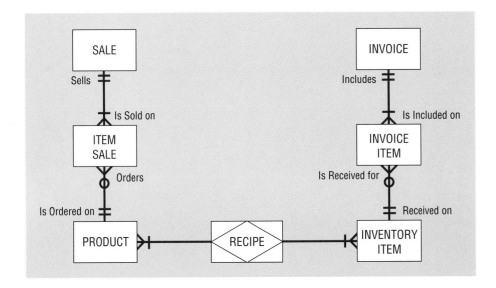

subset of an E-R diagram, and each entity is included in one or more data stores. Ideally, each data store on a primitive DFD will be an individual entity.

To determine data elements for an entity, we investigate data flows in and out of data stores that correspond to the data entity, and supplement this with a study of decision logic and temporal logic that use or change data about the entity. There are six data flows associated with the INVENTORY data store in Figure 11-13. The description of each data flow in the project dictionary or CASE repository would include the data flow's composition, which then tells us what data are flowing in or out of the data store. For example, the Amounts Used data flow coming from Process 2.0 indicates how much to decrement an attribute Quantity in Stock due to use of the INVENTORY ITEM to fulfill a customer sale. Thus, the Amounts Used data flow implies that Process 2.0 will first read the relevant INVENTORY ITEM record, then update its Quantity in Stock attribute, and finally store the updated value in the record. Structured English for Process 2.0 would depict this logic. Each data flow would be analyzed similarly (space does not permit us to show the analysis for each data flow).

The analysis of data flows for data elements is supplemented by a study of decision logic. For example, consider the decision table of Figure 11-14. One condition used to determine the process of reordering an INVENTORY ITEM involves the Type of Item. Thus, Process 3.0 in Figure 11-13 (to which this decision table relates) needs to know this characteristic of each INVENTORY ITEM, so this identifies another attribute of this entity.

Although we do not illustrate a state-transition diagram for this system, the analysis of such a chart could also reveal additional data requirements. For example, a state-transition diagram on an inventory item might show states of below reorder point, above reorder point, and projected above reorder point. This last state occurs when an invoice is received for a new shipment but before the shipment's quantity is verified. Such a state could imply the need for a new attribute on an INVENTORY ITEM to specify whether the Quantity in Stock is an actual value or an estimate.

After having considered all data flows in and out of data stores related to data entities, plus all decision and temporal logic related to inventory control, we derive the full E-R data model, with attributes, shown in Figure 11-16.

Figure 11-16

Final E-R diagram for Hoosier Burger's inventory control system

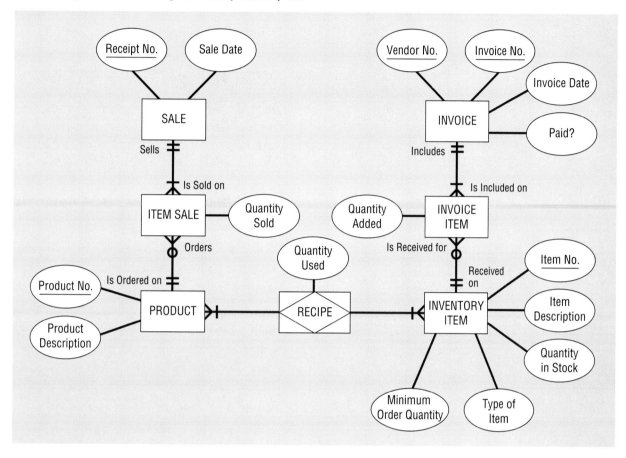

SUMMARY

We have presented the process and notation used to model the data requirements of an information system. We outlined the structuring of conceptual data models using the entity-relationship notation and also discussed how the components of a conceptual data model relate to data flows and data stores as well as states on a state-transition diagram.

Conceptual data modeling is based on certain constructs about the structure, not use, of data. These constructs include: entity, relationship, degree, and cardinality. A data model shows the relatively permanent business rules that define the nature of an organization. Rules define such characteristics of data as the legitimate domain of values for data attributes, the unique characteristics (primary key) of entities, the relationships between different entities, and the triggering operations that protect the validity of attributes during data maintenance.

A data model shows major categories of data, called entities, the associations or relationships between entities, and the attributes of both entities and relationships. A special type of entity called a gerund (or associative entity) is often necessary to represent a many-to-many relationship between entities. Entity types are distinct from entity instances. Each entity instance is distinguished from other instances of the same type by a primary key attribute (or attributes).

Relationships are the glue that hold a data model together. Three common relationship types are unary, binary, and ternary. The minimum and maximum number of entity instances that participate in a relationship represent important rules about the nature of the organization, as captured during requirements determination.

This chapter completes our coverage of the techniques used to structure information system requirements. The next chapter, the last in the section on the analysis phase of the life cycle, addresses the selection among alternative directions for designing the new or replacement system.

CHAPTER   REVIEW

KEY TERMS

Attribute

Binary relationship

Business rules

Candidate key

Cardinality

Conceptual data model

Degree

Domain

Entity-relationship data
    model (E-R model)

Entity-relationship
    diagram (E-R diagram)

Entity instance (instance)

Entity type

Gerund

Multivalued attribute

Primary key

Relationship

Repeating group

Ternary relationship

Triggering operation
    (trigger)

Unary relationship
    (recursive relationship)

REVIEW QUESTIONS

1. Define each of the following terms:

   a. entity
   b. business rule
   c. instance
   d. primary key

   e. relationship
   f. minimum cardinality
   g. attributive entity

2. Discuss why some systems developers believe that a data model is one of the most important parts of the statement of information system requirements.

3. Distinguish between the data modeling done during information systems planning, project initiation and planning, and analysis phases of the systems development life cycle.

4. What elements of a data flow diagram should be analyzed as part of data modeling?

5. Explain why a ternary relationship is not the same as three binary relationships.

6. When must a many-to-many relationship be modeled as a gerund?

7. What is the significance of triggering operations business rules in the analysis and design of an information system?

8. Which of the following types of relationships can have attributes associated with them: one-to-one, one-to-many, many-to-many?

9. What are the linkages between data flow diagrams, decision tables, state-transition diagrams, and entity-relationship diagrams?

PROBLEMS AND EXERCISES

1. Match the following terms to the appropriate definitions.

   _____ entity type

   _____ alias

   _____ many-to-many relationship

   _____ domain

   _____ composite entity

   a. set of all data types and range of values that an attribute can assume
   b. alternative name for an element of a data model
   c. binary relationship in which the maximum number of entity instances related to one instance of the other type of entity is greater than one
   d. another name for a gerund
   e. collection of entities that share common properties

2. Obtain a copy of an invoice, order form, or bill used in one of your recent business transactions. Create an E-R diagram to describe your sample document.

3. Using Table 11-1 as a guide, develop the complete script (questions and possible answers) of an interview between analysts and users within the order entry function at Pine Valley Furniture.

4. An airline reservation is an association between a passenger, a flight, and a seat. Select a few pertinent attributes for each of these entity types and represent a reservation in an E-R diagram.

5. Choose from your own experiences with organizations and draw an E-R diagram for a situation that has a ternary relationship.

6. Assume that at Pine Valley Furniture each product (described by Product No., Description, and Cost) is comprised of at least three components (described by Component No., Description, and Unit of Measure) and components are used to make one or many products (that is, must be used in at least one product). In addition, assume that components are used to make other components and that raw materials are also considered to be components. In both cases of components being used to make products and components being used to make other components, we need to keep track of how many components go into making something else. Draw an E-R diagram for this situation and place minimum and maximum cardinalities on the diagram.

7. Much like Pine Valley Furniture's sale of products, stock brokerages sell stocks and the prices are continually changing. Draw an E-R diagram which takes into account the changing nature of stock price.

8. Study the E-R diagram of Figure 11-17. Based on this E-R diagram, answer the following questions:
   a. How many PROJECTs can an employee work on?
   b. What is the degree of the Includes relationship?
   c. Are there any associative entities on this diagram? If so, name them.
   d. How else could the attribute SKILL be modeled?
   e. Is it possible to attach any attributes to the Includes relationship?
   f. Could TASK be modeled as a gerund?

9. In the Purchasing Department at one company, each purchase request is assigned to a "case worker" within the Purchasing Department. This case worker follows the purchase request through the entire purchasing process and acts as the sole contact person with the person or unit buying the goods or services. The Purchasing Department refers to its fellow employees buying goods and services as "customers."

Figure 11-17
E-R diagram for Problem and Exercise 8

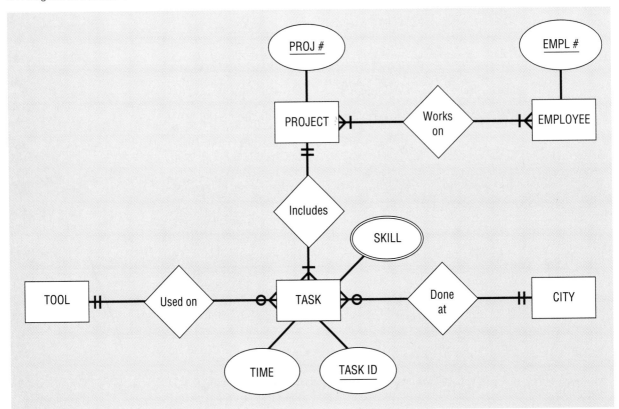

The purchasing process is such that purchase requests over $1,500.00 must go out for bid to vendors, and the associated Request for Bids for these large requests must be approved by the Purchasing Department. If the purchase is under $1,500.00, the product or service can simply be bought from any approved vendor, but the purchase request must still be approved by the Purchasing Department and they must issue a Purchase Order. For large purchases, once the winning bid is accepted, the Purchasing Department can issue a Purchase Order. List the relevant entities and attributes and draw an entity-relationship diagram for this business process. List whatever assumptions you must make to define primary keys, assess cardinality, and so on.

10. If you were going to develop a computer-based tool to help an analyst interview users and quickly and easily create and edit entity-relationship diagrams, what type of tool would you build? What features would it have, and how would it work? Why?

11. For the entity-relationship diagram provided in Figure 11-18, draw in the relationship cardinalities and describe them. Describe any assumptions you must make about relevant business rules. Are there any changes or additions you would make to this diagram to make it better? Why or why not?

12. For the entity-relationship diagram provided in Figure 11-18, assume that this company decided to assign each sales representative to a small, unique set of customers; some customers can now become "members" and receive unique benefits; small manufacturing teams will be formed and each will be assigned to the production of a small, unique set of products; and each purchasing agent will be assigned to a

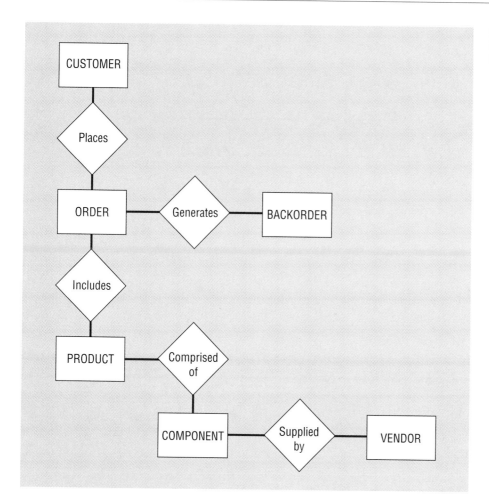

Figure 11-18
E-R diagram for Problem and
Exercise 11

small, unique set of vendors. Make the necessary changes to the entity-relationship diagram and draw and describe the new relationship cardinalities.

13. As best you can, describe the process logic and business rules underlying the entity-relationship diagrams from Problems and Exercises 11 and 12. How are entity-relationship diagrams similar to and different from the logic modeling techniques from the previous chapter (e.g., Structured English, decision tables, decision trees)? In what ways are these data and logic modeling techniques complimentary? What problems might be encountered if either data or logic modeling techniques were not performed well or not performed at all as part of the systems development process?

14. A software training program is divided into training modules and each module is described by module name and the approximate practice time. Each module sometimes has prerequisite modules. Model this situation of training programs and modules with an E-R diagram.

15. Each semester, each student must be assigned an advisor who counsels students about degree requirements and helps students register for classes. Students must register for classes with the help of an advisor, but if their assigned advisor is not available, they may register with any advisor. We must keep track of students, their assigned advisor, and with whom the student registered for the current term. Represent this situation of students and advisors with an E-R diagram.

Figure 11-19
E-R diagram for Problem and Exercise 16

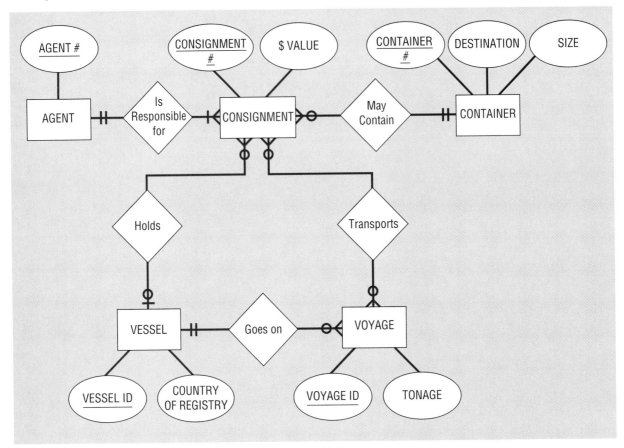

16. Consider the E-R diagram in Figure 11-19. Are all three relationships—Holds, Goes on, and Transports—necessary (i.e., can one of these be deduced from the other two)? Are there reasonable assumptions which make all three relationships necessary? If so, what are these assumptions?

FIELD EXERCISES

1. Interview a friend or family member to elicit from them each of the entities, attributes, relationships, and relevant business rules they come into contact with at work. Use this information to construct and present to this person an entity-relationship diagram. Revise the diagram until it seems appropriate to you and to your friend or family member.

2. Visit an organization that provides primarily a service, such as a dry cleaners, and a company that manufactures a more tangible product. Interview employees from these organizations to elicit from them each of the entities, attributes, relationships, and relevant business rules that are commonly encountered by these companies. Use this information to construct entity-relationship diagrams. What differences

and similarities are there between the diagrams for the service- and the product-oriented companies? Does the entity-relationship diagramming technique handle both situations equally well? Why or why not? What differences, if any, might there be in the use of this technique for a public agency?

3. Discuss with a systems analyst the role of conceptual data modeling in the overall systems analysis and design of information systems at his or her company. How, and by whom, is conceptual data modeling performed? What training in this technique is given? At what point(s) is this done in the development process? Why?

4. Ask a systems analyst to give you examples of unary, binary, and ternary relationships that they have heard of or dealt with personally at their company. Ask them which is the most common. Why?

5. Talk to MIS professionals at a variety of organizations and determine the extent to which CASE tools are used in the creation and editing of entity-relationship diagrams. Try to determine whether or not they use CASE tools for this purpose, what CASE tools are used, and why, when and how they use CASE tools for this. In companies that do not use CASE tools for this purpose, determine why not and what would have to change to have them use CASE tools.

REFERENCES

Aranow, E. B. 1989. "Developing Good Data Definitions." *Database Programming & Design* 2 (8) (August), 36–39.

Batini, C., S. Ceri, and S. B. Navathe. 1992. *Conceptual Database Design.* Redwood City, CA: Benjamin/Cummings Publishing.

Bruce, T. A. 1992. *Designing Quality Databases with IDEF1X Information Models.* New York, NY: Dorset House Publications.

Chen, P. P-S. 1976. "The Entity-Relationship Model—Toward a Unified View of Data." *ACM Transactions on Database Systems* 1 (March), 9–36.

Fleming, C. C., and B. von Halle. 1990. "An Overview of Logical Data Modeling." *Data Resource Management* 1 (1) (Winter), 5–15.

McFadden. F. R., and J. A. Hoffer. 1994. *Modern Database Management.* Redwood City, CA: Benjamin/Cummings Publishing.

Sandifer, A., and B. von Halle. 1991a. "A Rule by Any Other Name." *Database Programming & Design* 4 (2) (February), 11–13.

Sandifer, A., and B. von Halle. 1991b. "Linking Rules to Models." *Database Programming & Design* 4 (3) (March), 13–16.

Storey, V. C. 1991. "Relational Database Design Based on the Entity-Relationship Model." *Data and Knowledge Engineering* 7 (1991), 47–83.

Teorey, T. J., D. Yang, and J. P. Fry. 1986. "A Logical Design Methodology for Relational Databases Using the Extended Entity-Relationship Model." *Computing Surveys* 18 (2) (June), 197–221.

Structuring System Requirements: Conceptual Data Modeling for the Customer Activity Tracking System

INTRODUCTION

The third parallel requirements structuring activity in the analysis phase of the CATS project addresses developing a conceptual data model for the system. Jorge Lopez, one of the systems analysts on the project team, and Buffy Jarvis, a data analyst on loan to the team, are jointly responsible for developing the conceptual data model. Buffy specializes in conceptual data modeling in the data administration staff at BEC. She has reviewed the conceptual data models for the parts of the BEC databases used in the current systems with which CATS is likely to interact. Buffy and Jorge have developed a preliminary E-R data model for CATS and are ready for some feedback from other CATS project team members. We start this case near the end of a meeting Jorge and Buffy have had to prepare a revised E-R diagram consistent with the process and logic models.

CONCEPTUAL DATA MODELING FOR CATS

"Buffy, this has been a really good session. It looks like the estimate of five days of your time to work on this project was just about right. Having all the E-R models for various company databases available via the CASE tool has allowed you to answer most of my questions very quickly. After I get some feedback from Frank and Jordan, we ought to be able to finalize an E-R model of the central CATS system for discussion at the JAD later this month. I've really appreciated how you've been able to accommodate some different names for entities and attributes from the standard ones in the repository. I know our client also will appreciate this. Now, if we could only . . ." A knock on Jorge's cubicle interrupts his last thought.

"Jorge . . . oh, I'm sorry, I didn't know you had someone with you," interjected Frank Napier. "I'll come back later."

"Frank, no problem. Buffy and I were about done," Jorge replied. "You remember Buffy Jarvis from our CATS orientation meeting? Buffy's the data analyst that's working with me on the E-R model for CATS."

"It's good to see you again, Buffy," replied Frank. "Have the two of you come up with a model you think works?"

Buffy took the lead. "Jorge thinks we are ready to get you and Jordan Pippen involved to see if our data model for central CATS is consistent with the process and logic modeling you've been doing. From what we see in the repository, we think the E-R model is fine, but we really need your feedback. You got a couple of minutes to take a quick look?"

"Sure. But, first tell me a little about how you developed your E-R model, which might help me to think about angles you may have missed," replied Frank.

"Buffy and I have tried to consider all the input we could. We started by reviewing the original CATS data model Jordan and Nancy Chen developed for the Baseline Project Plan (see Figure 1). We quickly discovered that it did not conform to the standards set by data administration for E-R models. You may recall from the company data modeling training you went through during your first few months at BEC that we are pretty particular about our standards. So we redrew the original E-R diagram with unique relationship names; two names for each relationship so it can be easily read in either direction; and singular nouns for each entity label. We also reversed the cardinality notation for associative entities, to conform to the notation used by our CASE tool. We talked through all of the relationships to see if their cardinalities made sense. Then, since I've learned a lot about CATS from the various activities in the requirements determination step, I just talked to Buffy about what CATS was and she was able to identify some additional entities and relationships. We didn't have specific forms or reports to look at, but simply in the process of my describing CATS, Buffy heard me mention business entities and rules, like there are three types of products—music, games, and video—and we can't legally keep track of video purchases and rentals. This led us to subtypes of the Product entity."

Buffy added, "Then we tried to match each of the entities with an entity from the consolidated data model we keep in data administration that cuts across all BEC systems. I was glad that Jordan and Nancy Chen had thought to identify some possible interacting systems in their request to the Systems Priority Board since this gave us some clues on where to look for matching entities. Also, the keyword search capability in the CASE tool was a big help. We found a match for many entities, but a few entities, like the profile- and reservation-related entities, were new. We then created CASE repository entries for each new CATS entity and, for entities in the consolidated model, we made sure that the CATS entity name was an alias. We still don't know if the CATS database will physically have separate files for some of these entities, but at least we now know conceptually where the commonality is with existing database entities.

"Fortunately, since Jordan got me involved from the get-go on the project, you'll recall that I was able to give you, Jorge, and Jordan some ideas for questions to ask during the interviews and ways to analyze documents that emphasized data aspects of CATS. You all did a nice job of writing your interview notes to highlight data mentioned or shown to you in some document. I'm glad you decided to keep all your notes in a shared directory open to all CATS project team members on the LAN. Jorge and I read these notes, made a few phone calls to a couple of the interviewees to ask follow-up questions, and looked over several documents again; we were able to get a pretty good idea of the data requirements for CATS."

Jorge continued, "Then as the process and logic models were developed, we studied them to find data store and data flow contents and data elements used in Structured English. We studied decision tables, trees, and state-transition diagrams to see if additional data elements were mentioned. We then assigned all of these data elements to a data entity on our E-R diagram. We wanted to make sure, for example, that every data

Figure 1

Entity-relationship diagram for the CATS project

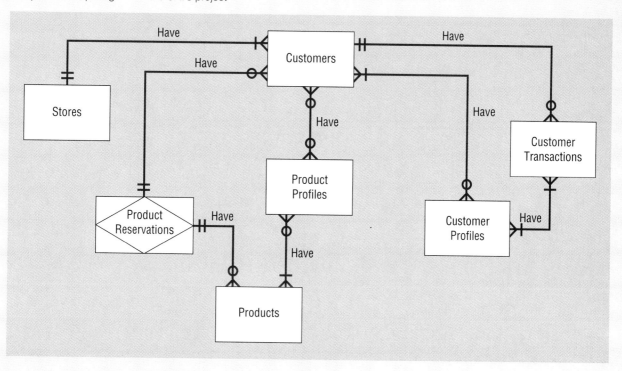

store was a subset of our data model, that every data element brought into CATS from a source was stored either in raw or aggregate form in the database, and that every data element sent out of CATS could be computed from data in the data model or data just passing through CATS. At this point, we were not concerned about how to access data as long as there were relationships in the data model that linked data needed in a process or logic step. We made a few adjustments this morning and were just about ready to send Jordan a note asking for some time at our next project status meeting to share our model with you and her. But, while you are here, we would appreciate a quick look at what we've developed." (Figure 2 contains this E-R diagram and Figure 3 a list of data elements for each entity.)

"You can see a few significant changes right away," added Buffy. "First, we changed the name of the associative entity to Product Announcement to reflect the term used in most of the other CATS models; an attribute of the entity, Date Reserved, indicates whether the announcement turns into a reservation. Some DFDs refer to a Product Reservation data store, and that data store relates to this associative entity [see DFDs in the BEC case after Chapter 9].

"Second, we are able to tell what music products a customer buys or rents via the 'Is Purchased on' relationship between the Customer Transaction and the Music or Games Product entities. A customer may buy

or rent a product that was announced even if the customer does not reserve the item. We thought some type of report about the number of these transactions would be interesting. You might want to add that to a DFD somewhere. Unfortunately, we can't legally track what video products customers rent or buy, even if reserved, so we can associate a video item on a Customer Transaction only with a Product Profile. We made a rather critical decision here. A Customer Transaction could be associated with either a customer or a product, but not both, for a video product. We chose to make the transaction-product association vague through the Product Profile, rather than the customer-transaction association.

"Third, we thought the relationship between Customer Profile and Customer Transaction in the original E-R diagram was redundant since the same information could be found via the Describes and Buys relationships. Similarly, the relationship between Customer and Product Profile in the original E-R diagram is redundant. However, we added the 'Products for Customer' relationship, which could represent the decision rules on how to match customers to products.

"Finally, we included an Inventory associative entity to keep track of quantity on hand in each store. We saw that you had included an Inventory data store on one of your DFDs. This type of data is needed for the product ordering part of CATS. Given the way you show this on your DFD [Figure 4 from the BEC case

Figure 2
Revised CATS E-R model (Note: some relationship names are truncated due to automatic layout by CASE tool)

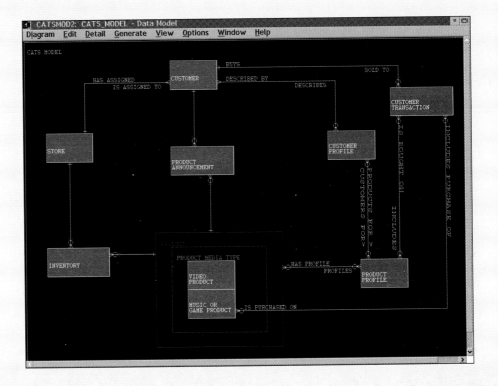

Figure 3

Data elements for entities in CATS E-R model

CATS_MODEL - Data Model Element List

Type	Name
Subject Area	⌐CATS_MODEL
Entity Type	⌐CUSTOMER
Attribute	I CUSTOMER_NUMBER
Attribute	CUSTOMER_NAME
Attribute	CUSTOMER_STREET
Attribute	CUSTOMER_STATE
Attribute	CUSTOMER_ZIP
Attribute	TELEPHONE_NUMBER
Attribute	CREDIT_CARD_TYPE
Attribute	CREDIT_CARD_NUMBER
Attribute	CREDIT_CARD_EXPIRATION_DATE
Attribute	SOCIAL_SECURITY_NUMBER
Relationship	Sometimes DESCRIBED_BY One or More CUSTOMER_PROFILE
Relationship	Sometimes HAVE One or More PRODUCT_ANNOUNCEMENT
Relationship	Sometimes BUYS One or More CUSTOMER_TRANSACTION
Relationship	Always IS_ASSIGNED_TO One STORE
Entity Type	⌐CUSTOMER_PROFILE
Attribute	I CUSTOMER_PROFILE_CATEGORY
Attribute	I CUSTOMER_CATEGORY_LEVEL
Relationship	Sometimes PRODUCTS_FOR_CUSTOMER One or More PRODUCT_PROFILE
Relationship	Always DESCRIBES One or More CUSTOMER
Entity Type	⌐CUSTOMER_TRANSACTION
Attribute	I TRANSACTION_DATE
Relationship	Sometimes INCLUDES_PURCHASE_OF One or More MUSIC_OR_GAME_PRODUCT
Relationship	Sometimes IS_BOUGHT_ON One or More PRODUCT_PROFILE
Relationship	I Always SOLD_TO One CUSTOMER
Entity Type	⌐INVENTORY
Attribute	QUANTITY_IN_STOCK
Relationship	I Always HAVE One STORE
Relationship	I Always HAVE One PRODUCT
Entity Type	⌐PRODUCT
Attribute	I PRODUCT_NUMBER
Attribute	PRODUCT_TITLE
Attribute	PRODUCT_MEDIA_TYPE
Attribute	DATE_PRODUCT_IS_AVAILABLE
Relationship	Sometimes HAS_PROFILE One or More PRODUCT_PROFILE
Relationship	Sometimes HAVE One or More INVENTORY
Relationship	Sometimes HAVE One or More PRODUCT_ANNOUNCEMENT
Partitioning	⌐PRODUCT_MEDIA_TYPE
Subtype	⌐MUSIC_OR_GAME_PRODUCT
Relationship	Sometimes IS_PURCHASED_ON One or More CUSTOMER_TRANSACTION
Subtype	VIDEO_PRODUCT
Entity Type	⌐PRODUCT_ANNOUNCEMENT
Attribute	I DATE_ANNOUNCEMENT_SENT
Attribute	CATS_RUN_NUMBER_FOR_ANNOUNCEMENT
Attribute	DATE_CUSTOMER_RESERVED_PRODUCT
Attribute	PCS_CODE
Relationship	I Always HAVE One PRODUCT
Relationship	I Always HAVE One CUSTOMER
Entity Type	⌐PRODUCT_PROFILE
Attribute	I PRODUCT_CATEGORY_NAME
Attribute	SUPERIOR_PRODUCT_CATEGORY_NAME
Relationship	Sometimes PROFILES One or More PRODUCT
Relationship	Sometimes CUSTOMERS_FOR_PRODUCT One or More CUSTOMER_PROFILE
Relationship	Sometimes INCLUDES One or More CUSTOMER_TRANSACTION
Entity Type	⌐STORE
Attribute	I STORE_NUMBER
Attribute	STORE_MANAGER_NAME
Relationship	Sometimes HAVE One or More INVENTORY
Relationship	Always HAS_ASSIGNED One or More CUSTOMER

after Chapter 9], I think that it is likely that this entity will, in fact, be a view into the database of the inventory control system, which indicates another interface between CATS and an existing system. Exactly how we'll implement this entity will be decided in a later project phase. One of the things I'd like to do for Jorge before we meet next is to take the entities we've identified and create a CRUD matrix showing which entities are Created in, Retrieved from, Updated by, and Deleted by which BEC system. This is the same type of matrix used in our information systems planning method. This will help to better understand the relationships between CATS and existing systems."

"All your changes make sense to me," replied Frank. "I was not sure, however, when I saw the original E-R diagram that the cardinalities next to the profile entities coming from Customer and Product, respectively, were right. I see you've kept these the same. What did you think about those cardinalities?"

"That gave us pause, too," interjected Jorge. "What we have tentatively decided, for example, is that a Customer Profile entity instance is a kind of atomic profile, that is, a single characteristic, like affluent income level. Thus, a customer has a set of these which describes that customer. Similarly, a product might be categorized as both jazz and R&B, which are individual Product Profile entity instances. So, the Describes relationship says that a customer has a set of profiles (a logical AND between the associated Customer Profile instances), and the 'Described by' relationship says which customers have that particular profile instance as one of their associated profiles. We think this gives us the greatest freedom in creating and tracking profiles of customers and products."

"You are right about flexibility, but I don't know what will happen when we have to process database queries and need to retrieve so many records to compose the total profile for a customer," said Frank.

Buffy responded, "Remember, an E-R model is only a conceptual picture of the database. When we normalize the data in logical database design and then pick data structures during physical database design, this conceptual picture of a customer's total profile as potentially several Customer Profile instances might be implemented all in one physical record. But, let's not deal with that until we better understand form and report design, processing volumes, and the other factors that affect physical database design."

"Okay, but why don't you make a note in the repository entry for the profile entities that this issue should be addressed during physical design. Who knows, none of us may be involved in this project then and we want to make sure the physical system design-

ers know about this issue," Frank said. "I'll go back and look at the DFDs and associated Structured English to make sure that the process flows treat profile data in data stores the same way you've represented them in the E-R diagram."

"Good idea about the repository note, Frank," replied Jorge. "I'll enter a note like that right away."

"Jorge and Frank, I've got a question for you," said Buffy. "Something you said before, Jorge, makes me wonder if we don't need a unary relationship on the Product Profile entity. The idea is that maybe there are hierarchical categories of products. For example, some people may have an interest in Jazz in general, so they'd like to know about any new jazz release. Others may be into fusion jazz, New Orleans jazz, big band jazz, or another specialty jazz category. The decision rules on which customer profile to match with what product profile could be adjusted to match aggregate or subordinate categories. This is exactly the kind of association a unary relationship can show. What do you think?"

"Buffy, that's the kind of creative idea someone with a data focus can provide," replied Jorge. "I'll call Wendy and Nancy and see what they think."

"What about the data requirements for the store-level CATS?" asked Frank. "Did you leave that off this data model because you think that will be a client database at the stores?"

"That is one possibility," responded Jorge. "We'll deal with that when we discuss alternative design strategies. Candidly, we've not thought much about the store CATS, yet. Could you give some thought to that before our next CATS staff meeting, when we hope to present this E-R model in a walkthrough with the whole team?"

"I'll be glad to. Besides the questions I've already asked, I don't see any other problems based on this quick review," responded Frank. "If you need any help working on business rules, besides those on the E-R diagram, just ask. I know we'll have to come back to this E-R diagram during logical design, once we design forms and reports and discover that users want some data we did not anticipate, but this level of work seems great for the analysis phase.

"Jorge, I had stopped by to confirm when we were going to meet about the study we are doing on CATS' impact on existing systems. I need to fit in a phone interview with one of the store managers Jordan talked to in Miami since he has some concerns about the store CATS system. Can we meet at 3:00 instead of 2:00 this afternoon?"

"That's fine, and thanks for your reactions to our E-R model. See you at 3:00, your office, right?"

"Yep, my office. See you then, Jorge. Good to see you again, Buffy."

"Thanks for your help," replied Buffy. "See you at the staff meeting."

SUMMARY

You have seen in this and the prior two BEC cases how the CATS team members have structured the requirements for the system. Requirements structuring is based on the information in interview notes and other fact-gathering deliverables from requirements determination as well as on other models of the system, such as decision tables. Structuring requirements, like most systems development work, is quite iterative. As one analysis is done, this raises new issues not seen in other analyses, so the team must re-analyze deliverables they thought were complete. Eventually, all the pieces will come together at the end of the analysis phase into a design strategy for CATS, which is the subject of the next BEC case after Chapter 12.

12

Selecting the Best Alternative Design Strategy

After studying this chapter, you should be able to:

- Describe the different sources of software.

- Assemble the various pieces of an alternative design strategy.

- Generate at least three alternative design strategies for an information system.

- Select the best design strategy using both qualitative and quantitative methods.

- Update a Baseline Project Plan based on the results of the analysis phase.

INTRODUCTION

You have now reached the point in the analysis phase where you are ready to transform all of the information you have gathered and structured into some concrete ideas about the nature of the design for the new or replacement information system. This is called the *design strategy*. From requirements determination, you know what the current system does and you know what the users would like the replacement system to do. From requirements structuring, you know what forms the replacement system's process flow, process logic, and data should take, at a logical level independent of any physical implementation. For example, every data flow from a system process to the external environment represents an on-line display, printed report, or business form or document that the system can produce, often for use by a human. Every data flow from a data store to a process implies a retrieval capability of the system's files and database. Most processes represent a capability of the system to transform inputs into outputs. And data flows from the environment to system processes indicate capabilities to capture data on on-line displays, or some batch method, to validate the data and protect access to the system from the environment and to route raw data to the appropriate processing points.

Thus, at this point in the systems development process you have a preliminary specification of what the new information system should do and you understand why a replacement system is necessary to fix problems in the current system and to respond to new needs and opportunities to use information. Actually, there still may be some uncertainty about the capabilities of a new system. This uncertainty is due to competing ideas from different users and stakeholders on what they would like the system to do and to existing alternatives for an implementation environment for the new system. To bring analysis to a conclusion, your job is to take these structured requirements and transform them into several competing design strategies, one of which will be pursued in the design phase of the life cycle.

Part of generating a design strategy is determining how you want to acquire the replacement system using a combination of sources inside and outside the organization. If you decide to proceed with development in-house, you will have to answer general questions about software, such as whether all of the software should be built in-house or whether some software components should be bought off-the-shelf or contracted to software development companies. You will have to answer general questions about hardware and system software, such as whether the new system will run on a mainframe platform, stand-alone personal computers, or on a client/server platform, and whether the system can run on existing hardware. It is also not too early to begin thinking about data conversion issues, which must be addressed as you move from your current system to the new one. You even have to start thinking about how much training will be required for users, and how easy or difficult the system will be to implement. You have to determine whether you can build and implement the system you desire given the funding and management support you can count on. And you have to address these concerns for each alternative you generate. These issues need to be addressed so that you can update the Baseline Project Plan with detailed activities and resource requirements for the next life cycle phase—logical design—and probably for the physical design phase as well. That is, in this step of the analysis phase you bring the current phase to a close, prepare a report and presentation to management concerning continuation of the project, and get ready to move the project into the design phases.

In this chapter, you will learn why you need to come up with alternative design strategies and about guidelines for generating alternatives. You will then learn about the different issues that must be addressed for each alternative. Once you have generated your alternatives, you will have to choose the best design strategy to pursue. We include a discussion of one technique that analysts and users often use to decide among system alternatives and to help them agree on the best approach for the new information system.

Throughout this chapter we emphasize the need for sound project management. Now that you have seen the various techniques and steps of the analysis phase, we outline what a typical analysis phase project schedule might look like and discuss the execution of the analysis phase and the transition from analysis to design.

SELECTING THE BEST ALTERNATIVE DESIGN STRATEGY

Design strategy: A high-level statement about the approach to developing an information system. It includes statements on the system's functionality, hardware and system software platform, and method for acquisition.

Selecting the best alternative system involves at least two basic steps: (1) generating a comprehensive set of alternative design strategies and (2) selecting the one that is most likely to result in the desired information system, given all of the organizational, economic, and technical constraints that limit what can be done. In a sense then, the most likely strategy is the best one. A system **design strategy** is an approach to developing the system. The strategy includes the system's functionality, hardware and system software platform, and method for acquisition. We use the term design strategy in this chapter rather than the term alternative system because, at the end of analysis, we are still quite a long way from specifying an actual system. This delay is purposeful since we do not want to invest in design efforts until there is agreement on which direction to take the project and the new system. The best we can do at this point is to outline rather broadly the approach we can take in moving from logical system specifications to a working physical system. The overall process of selecting the best system strategy and the deliverables from this step in the analysis process are discussed next.

The Process of Selecting the Best Alternative Design Strategy

As Figure 12-1 shows, there are three sub-phases to systems analysis: requirements determination, requirements structuring, and generating alternative system design strategies and selecting the best one. After the system requirements have been structured in terms of process flow, process logic (decision and temporal), and data, analysts again work with users to package the requirements into different system configurations. Shaping alternative system design strategies involves the following processes:

- Dividing requirements into different sets of capabilities, ranging from the bare minimum that users would accept (the required features) to the most elaborate and advanced system the company can afford to develop (which includes all the features desired across all users). Alternatively, different sets of capabilities may represent the position of different organizational units with conflicting notions about what the system should do.

- Enumerating different potential implementation environments (hardware, system software, and network platforms) that could be used to deliver the different sets of capabilities. (Choices on the implementation environment may place technical limitations on the subsequent design phase activities.)

- Proposing different ways to source or acquire the various sets of capabilities for the different implementation environments.

In theory, if there are three sets of requirements, two implementation environments, and four sources of application software, there would be twenty-four possible design strategies. In practice, some combinations are usually infeasible or uninteresting. Further, usually only a small number—typically three—can be easily considered. Selecting the best alternative is usually done with the help of a quantitative procedure. Analysts will recommend what they believe to be the best alternative but management (a combination of the steering committee and those who will fund the rest of the project) will make the ultimate decision about which system design strategy to follow. At this point in the life cycle, it is also certainly possible for management to end a project before the more expensive phases of design and implementation are begun if the costs or risks seem to outweigh the benefits, if the needs of the organization have changed since the project began, or if other competing projects appear to be of greater worth and development resources are limited.

Deliverables and Outcomes

The primary deliverables from generating alternative design strategies and selecting the best one are outlined in Table 12-1. The primary deliverable that is carried forward into design is an updated Baseline Project Plan detailing the work necessary to turn the selected design strategy into the desired replacement information system. Of course, that plan cannot be assembled until a strategy has been selected, and no strategy can be selected until alternative strategies have been generated and compared. Therefore, all three objects—the alternatives, the selected alternative, and the plan—are listed as deliverables in Table 12-1. Further, these three deliverables plus the supporting deliverables from requirements determination and structuring steps are necessary information to conduct systems design, so all of this

Figure 12-1
Systems development life cycle with the analysis phase highlighted

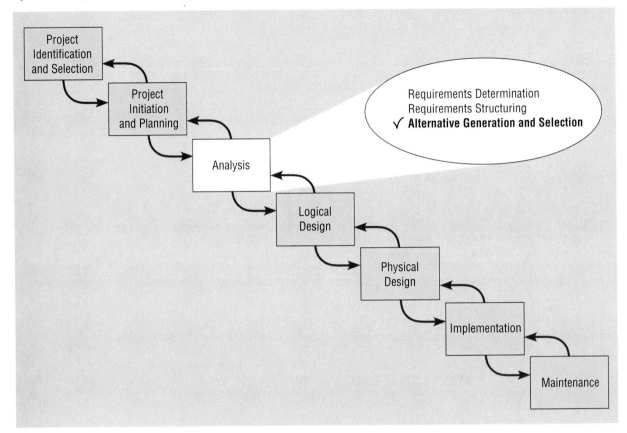

TABLE 12-1 Deliverables for Generating Alternatives and Selecting the Best One

1. At least three substantively different system design strategies for building the replacement information system

2. A design strategy judged most likely to lead to the most desirable information system

3. An Baseline Project Plan for turning the most likely design strategy into a working information system

information is carried in the project dictionary and CASE repository for reference in subsequent phases.

GENERATING ALTERNATIVE DESIGN STRATEGIES

In many cases, it may seem to an analyst that the solution to an organizational problem is obvious. Typically, the analyst is very familiar with the problem, having conducted an extensive analysis of it and how it has been solved in the past, or the analyst is very familiar with a particular solution that he or she attempts to apply to all organizational problems encountered. For example, if an analyst is an expert at using advanced database technology to solve problems, then there is a tendency for the analyst to recommend advanced database technology as a solution to every possible problem. Or if the analyst designed a similar system for another customer or business unit, the "natural" design strategy would be the one used before. Given the role of experience in the solutions analysts suggest, analysis teams typically generate at least two alternative solutions for every problem they work on.

A good number of alternatives for analysts to generate is three. Why three? Three alternatives can neatly represent both ends and the middle of a continuum or spectrum of potential solutions. One alternative represents the low end of the spectrum. Low-end solutions are the most conservative in terms of the effort, cost, and technology involved in developing a new system. In fact, low-end solutions may not involve computer technology at all, focusing instead on making paper flows more efficient or reducing the redundancies in current processes. A low-end strategy provides all the required functionality users demand with a system that is minimally different from the current system.

Another alternative represents the high end of the spectrum. High-end alternatives go beyond simply solving the problem in question and focus instead on systems that contain many extra features users may desire. Functionality, not cost, is the primary focus of high-end alternatives. A high-end alternative will provide all desired features using advanced technologies which often allow the system to expand to meet future requirements. Finally, the third alternative lies between the extremes of the low-end and high-end systems. Such alternatives combine the frugality of low-end alternatives with the focus on functionality of high-end alternatives. Mid-range alternatives represent compromise solutions. There are certainly other possible solutions that exist outside of these three alternatives. Defining the bottom, middle, and top possibilities allows the analyst to draw bounds around what can be reasonably done.

How do you know where to draw bounds around the potential solution space? The analysis team has already gathered the information it needs to identify the solution space, but first that information must be systematically organized. There are two major considerations. The first is determining what the minimum requirements are for the new system. These are the mandatory features, any of which, if missing, make the design strategy useless. Mandatory features are those that everyone agrees are necessary to solve the problem or meet the opportunity. Which features are mandatory can be determined from a survey of users and other stakeholders who have been involved in requirements determination. You would conduct this survey near the end of the analysis phase after all requirements have been structured and analyzed. In this survey, stakeholders rate features discovered during requirements determination or categorize features on some desirable-mandatory scale, and an arbitrary breakpoint is used to divide mandatory from desired features. Some organizations will break the features into three categories: mandatory, essential, and desired. Whereas mandatory features screen out possible solutions, essential features are the important capabilities of a system which will serve as the primary basis for comparison of different design strategies. Desired features are those that users could live without but which are used to select between design strategies that are of almost equal value in terms of essential features. Features can take many different forms. Features might include

- Data kept in system files (for example, multiple customer addresses so that bills can be sent to addresses different from where we ship goods)

- System outputs (printed reports, on-line displays, transaction documents—for example, a paycheck or sales summary graph)

- Analyses to generate the information in system outputs (for example, a sales forecasting module or an installment billing routine)

- Expectations on accessibility, response time, or turnaround time for system functions (for example, on-line, real-time updating of inventory files)

The second consideration in drawing bounds around alternative design strategies is determining the constraints on system development. Constraints may exist on such factors as

- A date when the replacement system is needed

- Available financial and human resources

- Elements of the current system that cannot change

- Legal and contractual restrictions

- The importance or dynamics of the problem which may limit how the system can be acquired (for example, a strategically important system that uses highly proprietary data probably cannot be outsourced or purchased)

Remember, be impertinent and question whether stated constraints are firm; you may want to consider some design alternatives that violate constraints you consider to be flexible.

Both requirements and constraints must be identified and ranked in order of importance. The reason behind such a ranking should be clear. Whereas you can design a high-end alternative to fulfill every wish users have for a new system, you design low-end alternatives to fulfill only the most important wishes. The same is true of constraints. Low-end alternatives will meet every constraint; high-end alternatives will ignore all but the most daunting constraints.

ISSUES TO CONSIDER IN GENERATING ALTERNATIVES

The required functionality of the replacement system and the constraints that limit that functionality form the basis for the many issues that must be considered in putting together all of the pieces that comprise alternative design strategies. That is, most of the substantive debate about alternative design strategies hinges on the relative importance of system features. Issues of functionality lead, however, to other associated issues such as whether the system should be developed and run in-house, software and hardware selection, implementation, and organizational limitations such as available funding levels. This list is not comprehensive, but it does remind you that an information system is more than just software. Each issue must be considered when framing alternatives. We will discuss each consideration in turn, beginning with the outsourcing decision.

Outsourcing

Outsourcing: The practice of turning over responsibility of some to all of an organization's information systems applications and operations to an outside firm.

If another organization develops or runs a computer application for your organization, that practice is called **outsourcing.** Outsourcing includes a spectrum of working arrangements (*Business Week,* June 19, 1989; Moad, 1993). At one extreme is having a firm develop and run your application on their computers—all you do is supply input and take output. A common example of such an arrangement is a company that runs payroll applications for clients so that clients don't have to develop an independent in-house payroll system. Instead they simply provide employee payroll information to the company and, for a fee, the company returns completed paychecks, payroll accounting reports, and tax and other statements for employees. For many organizations, payroll is a very cost-effective operation when outsourced in this way. In another example of outsourcing arrangements, you hire a company to run your applications at your site on your computers. In some cases, an organization employing such an arrangement will dissolve some or all of its information systems unit and fire all of its information systems employees. Most of the time, though, the company brought in to run the organization's computing will hire many of the information systems unit employees.

Why would an organization outsource its information systems operations? As we saw in the payroll example, outsourcing may be cost-effective. If a company specializes in running payroll for other companies, it can leverage the economies of scale it achieves from running one very stable computer application for many organizations into very low prices. But why would an organization dissolve its entire information processing unit and bring in an outside firm to manage its computer applications? One reason may be to overcome operating problems the organization faces in its information systems unit. For example, the City of Grand Rapids, Michigan, hired an outside firm to run its computing 20 years ago in order to overcome personnel problems made difficult by union contracts and civil service constraints. Another reason for total outsourcing is that an organization's management may feel its core mission does not involve managing an information systems unit and that it might achieve more effective computing by turning over all of its operations to a more experienced, computer-oriented company. Kodak decided in the late 1980s that it was not in the computer applications business and turned over management of its mainframes to IBM and management of its personal computers to Businessland (Applegate and Montealegre, 1991).

Outsourcing is an alternative analysts need to be aware of. When generating alternative system development strategies for a system, you as an analyst should consult organizations in your area that provide outsourcing services. It may well be that at least one such organization has already developed and is running an application very close to what your users are asking for. Perhaps outsourcing the replacement system should be one of your alternatives. Knowing what your system requirements are before you consider outsourcing means that you can carefully assess how well the suppliers of outsourcing services can respond to your needs. However, should you decide not to consider outsourcing, you need to consider whether some software components of your replacement system should be purchased and not built.

Sources of Software

We can group organizations that produce software into four major categories: hardware manufacturers, packaged software producers, custom software producers, and in-house developers.

Hardware Manufacturers   At first it may seem counter-intuitive that hardware manufacturers would develop information systems or software. Yet hardware manufacturers are among the largest producers of software. For example, IBM is a leader in software development, controlling 36 percent of the software market represented by the top U.S. computer companies (Table 12-2). However, IBM actually develops relatively little application software (roughly 15 percent of their software revenue is from application software). Rather, IBM's leadership comes from its development of operating systems and utilities (like sort routines or database management systems) for the hardware it manufactures.

TABLE 12-2 The Top 10 U.S. Software Companies in the 1993 *Datamation* 100

Rank among Software Companies	Company	Revenues from Software, 1993	Software Market Share
1	IBM	$10,953.0	36
2	Microsoft	3,740.0	12
3	Computer Associates	2,054.8	7
4	Oracle	1,337.5	4
5	Novell	1,033.0	3
6	Digital	955.0	3
7	Lotus	883.0	3
8	Unisys	779.9	3
9	WordPerfect	707.0	2
10	Hewlett-Packard	499.2	2

Notes: Revenue in millions. Market share is percent of Datamation 100 software revenues.
Adapted from: Datamation, June 15, 1994.

Packaged Software Producers The growth of the software industry has been phenomenal since its beginnings in the mid-1960s. Now, some of the largest computer companies in the world, as measured by the *Datamation* 100, are companies that produce software exclusively (see Table 12-2). Consulting firms, such as American Management Systems and Andersen Consulting, also rank in the top 40 packaged software producers.

Software companies develop what are sometimes called *pre-packaged* or *off-the-shelf* systems. Microsoft's Project and Intuit's Quicken™, QuickPay™, and QuickBooks™ are popular examples of such software. The packaged software development industry serves many market segments. Their software offerings range from general, broad-based packages, such as general ledger, to very narrow, niche packages, such as software to help manage a day care center. Software companies develop software to run on many different computer platforms, from microcomputers to large mainframes. The companies range in size from just a few people to thousands of employees. Software companies consult with system users after the initial software design has been completed and an early version of the system has been built. The systems are then beta-tested in actual organizations to determine whether there are any problems with the software or if any improvements can be made. Until testing is completed, the system is not offered for sale to the public. Unfortunately, the software is sometimes put on the market before it is ready. Figure 12-2 describes the problems Ashton-Tate encountered when it released dBASE IV too soon.

Some off-the-shelf software systems cannot be modified to meet the specific, individual needs of a particular organization. Such application systems are sometimes called *turnkey systems.* The producer of a turnkey system will only make changes to the software when a substantial number of users ask for a specific change. Other off-the-shelf application software can be modified or extended, however, by the producer or by the user, to more closely fit the needs of the organization. Even though many organizations perform similar functions, no two organizations do the same thing in quite the same way. A turnkey system may be good enough for a

Figure 12-2
Ashton-Tate's dBASE IV
disaster

In October 1988, Ashton-Tate Corporation released its long-awaited version of dBASE IV (TM), the successor to its popular dBASE III (TM) database management system. dBASE IV was four times larger than dBASE III, with roughly 500,000 lines of code. Getting the program to operate on a standard IBM-compatible personal computer required ingenuity on the part of programmers. They had to figure out how to chop the program into pieces that could be swapped in and out of computer memory, a difficult task. The program was so complex, it outpaced Ashton-Tate's testing procedures at the time. The result: the company shipped a product containing thousands of errors. Customers complained dBASE IV was too slow and too likely to crash. Further, it lacked features customers were looking forward to having.

Ashton-Tate had once been considered one of the Big Three microcomputer software companies, along with Lotus Development Corp. and Microsoft Corp. In 1985, Ashton-Tate could boast of having 62.5% of the database market for microcomputers. In 1988, before the release of dBASE IV, Ashton-Tate controlled 43% of that market, but market share slipped to 35% in 1989 and to 30% in 1990. The company suffered through three quarterly losses in 1990, ending the year with a slight profit in the fourth quarter.

It took two years for Ashton-Tate to fix the problems with dBASE IV. They finally released a new version in the fall of 1990, two years after the fiasco. By that time, however, it was too late. They had lost many of their customers to fast-growing companies like Oracle Corp. To add to their problems, in December 1990, Ashton-Tate lost their copyright protection for their dBASE products. Ashton-Tate had sued rival Fox Software Inc. for copyright infringement. The judge in the case ruled that Ashton-Tate's products were initially based on a database system in the public domain, therefore invalidating the copyright protection they had for the products. Without such protection, Ashton-Tate had no right to sue Fox.

Ashton-Tate rewon their copyright protection for dBASE in 1991. That same year, however, the company was purchased by another rival, Borland International Inc.

Sources: Zachary, G. P. "How Ashton-Tate Lost Its Leadership in PC Software Arena."
Wall Street Journal, April 11, 1990, A1–A2.
Zachary, G. P. and Bulkeley, W. M. "Ashton-Tate Loses Flagship Software's Copyright Shield."
Wall Street Journal, December 14, 1990, B1, B4.

certain level of performance but it will never perfectly match the way a given organization does business. A reasonable estimate is that off-the-shelf software can at best meet 70 percent of an organization's needs. Thus, even in the best case, 30 percent of the software system doesn't match the organization's specifications.

Custom Software Producers   If a company has a need for an information system but does not have the expertise or the personnel available to develop the system in-house and a suitable off-the-shelf system is not available, the company will likely consult a custom software company. Consulting firms, such as Andersen Consulting or EDS, will help a firm develop custom information systems for internal use. These firms employ people with expertise in the development of information systems. Their consultants may also have expertise in a given business area. For example, consultants who work with banks understand financial institutions as well as information systems. Consultants use many of the same methodologies, techniques, and tools that companies use to develop systems in-house. The 10 largest world-wide computer-services firms (based on revenues not only of custom software development but also outsourcing and other services) are listed in Table 13-3. Other large U.S. firms are ADP, TRW, First Data, and AT&T.

In-House Development   We have talked about three different types of external organizations that serve as sources of software, but in-house development remains an option. Of course, in-house development need not entail development of all of the

TABLE 12-3 Top 10 World-Wide Computer-Services Firms in 1992

Rank	Company	Revenues ($ millions)	Market Share
1	EDS	5,800	12.5
2	IBM	5,175	11.1
3	Andersen Consulting	2,233	4.8
4	Digital Equipment	2,056	4.4
5	Cap Gemini Sogeti	1,729	3.7
6	Nippon Telegraph	1,500	3.2
7	Unisys	1,368	2.9
8	Computer Sciences	1,234	2.7
9	Martin Marietta	850	1.8
10	Hewlett-Packard	700	1.5

Adapted from: Dataquest, Inc., as reported in King and Rigdon (1994)

software that will comprise the total system. Hybrid solutions involving some purchased and in-house software components are common. Table 12-4 compares the different software sources.

If you choose to acquire software from outside sources, this choice is made at the end of the analysis phase. Choosing between a package or external supplier will be driven by your needs, not by what the external party has to sell. As we will discuss, the results of your analysis study will define the type of product you want to buy and will make working with an external supplier much easier, productive, and worthwhile.

Choosing Off-the-Shelf Software

Once you have decided to purchase off-the-shelf software rather than write some or all of the software for your new system, how do you decide what to buy? There are several criteria to consider, and special criteria may arise with each potential software purchase. For each criterion, an explicit comparison should be made between the software package and the process of developing the same application in-house. The most common criteria are as follows:

- Cost

- Functionality

- Vendor support

- Viability of vendor

- Flexibility

- Documentation

- Response time

- Ease of installation

TABLE 12-4   Comparison of Four Different Sources of Software Components

Producers	Source of Application Software?	When to Go to This Type Organization for Software	Internal Staffing Requirements
Hardware Manufacturers	Generally not	For system software and utilities	Varies
Packaged Software Producers	Yes	When supported task is generic	Some IS and user staff to define requirements and evaluate packages
Custom Software Producers	Yes	When task requires custom support and system can't be built internally	Internal staff may be needed, depending on application
In-House Developers	Yes	When resources and staff are available and system must be built from scratch	Internal staff necessary though staff size may vary

These criteria are presented in no particular order. The relative importance of the criteria will vary from project to project and from organization to organization. If you had to choose two criteria that would always be among the most important, those two would probably be vendor viability and vendor support. You don't want to get involved with a vendor that might not be in business tomorrow. Similarly, you don't want to license software from a vendor with a reputation for poor support. How you rank the importance of the remaining criteria will very much depend on the specific situation in which you find yourself.

Cost involves comparing the cost of developing the same system in-house to the cost of purchasing or licensing the software package. You should include a comparison of the cost of purchasing vendor upgrades or annual license fees with the costs you would incur to maintain your own software. Costs for purchasing and developing in-house can be compared based on the economic feasibility measures outlined in Chapter 7 (for example, a present value can be calculated for the cash flow associated with each alternative). Functionality refers to the tasks the software can perform and the mandatory, essential, and desired system features. Can the software package perform all or just some of the tasks your users need? If some, can it perform the necessary core tasks? Note that meeting user requirements occurs at the end of the analysis phase because you cannot evaluate packaged software until user requirements have been gathered and structured. Purchasing application software is not a substitute for conducting the systems analysis phase; rather, purchasing software is part of one design strategy for acquiring the system identified during analysis.

As we said earlier, vendor support refers to whether and how much support the vendor can provide. Support occurs in the form of assistance to install the software, to train user and systems staff on the software, and to provide help as problems arise after installation. Recently, many software companies have significantly reduced the amount of free support they will provide customers, so the cost to use telephone, on-site, fax, or computer bulletin board support facilities should be considered. Related to support is the vendor's viability. You don't want to get stuck with software developed by a vendor that might go out of business soon. This latter point should not be minimized. The software industry is quite dynamic, and innovative application software is created by entrepreneurs working from home offices—the classic cottage industry. Such organizations, even with outstanding software, often do not have the resources or business management ability to stay in business very long. Further, competitive moves by major software firms can render the products of smaller firms outdated or incompatible with operating systems. One software firm we talked to while developing this book was struggling to survive just trying to make their software work on any supposedly IBM-compatible PC (given the infinite combination of video boards, monitors, BIOS chips, and other components). Keeping up with hardware and system software change may be more than a small firm can handle, and good off-the-shelf application software is lost.

Flexibility refers to how easy it is for you, or the vendor, to customize the software. If the software is not very flexible, your users may have to adapt the way they work to fit the software. Are they likely to adapt in this manner? Purchased software can be modified in several ways. Sometimes, the vendor will be willing to make custom changes for you, if you are willing to pay for the redesign and programming. Some vendors design the software for customization. For example, the software may include several different ways of processing data and, at installation time, the customer chooses which to initiate. Also, displays and reports may be easily redesigned if these modules are written in a fourth-generation language. Reports, forms, and displays may be easily customized using a process whereby your company name and chosen titles for reports, displays, forms, column headings, etc. are selected from a table of parameters you provide. You may want to employ some of these same customization techniques for in-house developed systems so that the software can be easily adapted for different business units, product lines, or departments.

Documentation includes the user's manual as well as technical documentation. How understandable and up-to-date is the documentation? What is the cost for multiple copies, if required? Response time refers to how long it takes the software package to respond to the user's requests in an interactive session. Another measure of time would be how long it takes the software to complete running a job. Finally, ease of installation is a measure of the difficulty of loading the software and making it operational.

Validating Purchased Software Information   One way to get all of the information you want about a software package is to collect it from the vendor. Some of this information may be contained in the software documentation and technical marketing literature. Other information can be provided upon request. For example, you can send prospective vendors a questionnaire asking specific questions about their packages. This may be part of a request for proposal (RFP) or request for quote (RFQ) process your organization requires when major purchases are made. Space does not permit us to discuss the topic of RFPs and RFQs here; you may wish to refer to purchasing and marketing texts if you are unfamiliar with such processes (additional references about RFPs and RFQs are found at the end of the chapter).

There is, of course, no replacement for actually using the software yourself and running it through a series of tests based on the criteria for selecting software. Remember to test not only the software but also the documentation, training materials, and even the technical support facilities. One requirement you can place on prospective software vendors as part of the bidding process is that they install (free or at an agreed-upon cost) their software for a limited amount of time on your computers. This way you can determine how their software works in your environment, not in some optimized environment they have.

One of the most reliable and insightful sources is other users of the software. Vendors will usually provide a list of customers (remember, they will naturally tell you about satisfied customers, so you may have to probe for a cross section of customers) and people who are willing to be contacted by prospective customers. And here is where your personal network of contacts, developed through professional groups, college friends, trade associations, or local business clubs, can be a resource; do not hesitate to find some contacts on your own. Such current or former customers can provide a depth of insight on use of a package at their organizations.

To gain a range of opinion about possible packages, you can use independent software testing and abstracting services that periodically evaluate software and collect user opinions. Such surveys are available for a fee either as subscription services or on demand (two popular services are Auerbach Publishers and DataPro); occasionally, unbiased surveys appear in trade publications. Often, however, articles in trade publications, even software reviews, are actually seeded by the software manufacturer and are not unbiased.

If you are comparing several software packages, you can assign scores for each package on each criterion and compare the scores using the quantitative method we demonstrate at the end of the chapter for comparing alternative system design strategies.

Hardware and Systems Software Issues

The first question you need to ask yourself about hardware and system software is whether the new system that follows a particular design strategy can be run on your firm's existing hardware and software platform. System software refers to such key components as operating systems, database management systems, programming languages, code generators, and network software. To determine if current hardware and system software is sufficient, you should consider such factors as the age and capacity of the current hardware and system software, the fit between the hardware and software and your new application's goals and proposed functionality and, if some of your system components are off-the-shelf software, whether the software can run on the existing hardware and system software. The advantages to running your new system on the existing platform are persuasive:

1. Lower costs as little, if any, new hardware and system software has to be purchased and installed

2. Your information systems staff is quite familiar with the existing platform and how to operate and maintain it

3. The odds of integrating your new application system with existing applications are enhanced

4. No added costs of converting old systems to a new platform, if necessary, or of translating existing data between current technology and the new hardware and system software you have to acquire for your system

On the other hand, there are also very persuasive reasons for acquiring new hardware or system software:

1. Some software components of your new system will only run on particular platforms with particular operating systems

2. Developing your system for a new platform gives your organization the opportunity to upgrade or expand its current technology holdings

3. New platform requirements may allow your organization to radically change its computing operations, as in moving from mainframe-centered processing to a database machine or a client-server architecture

As the determination of whether or not to acquire new hardware and system software is so context-dependent, providing platform options as part of your design strategy alternatives is an essential practice.

At this point, if you decide that new hardware or system software is a strong possibility, you may want to issue a request for proposal (RFP) to vendors. The RFP will ask the vendors to propose hardware and system software that will meet the requirements of your new system. Issuing an RFP gives you the opportunity to have vendors carry out the research you need in order to decide among various options. You can request that each bid submitted by a vendor contain certain information essential for you to decide on what best fits your needs. For example, you can ask for performance information related to speed and number of operations per second. You can ask about machine reliability and service availability and whether there is an installation nearby which you can visit for more information. You can ask to take part in a demonstration of the hardware. And of course the bid will include information on cost. You can then use the information you have collected in generating your alternative design strategies.

Implementation Issues

As you will see in Chapter 19, implementing a new information system is just as much an organizational change process as it is a technical process. Implementation involves more than installing a piece of software, turning it on, and moving on to the next software project. New systems often entail new ways of performing the same work, new working relationships, and new skills. Users have to be trained. Disruptions in work procedures have to be found and addressed. In addition, system implementation may be phased in over many weeks or even months. You must address the technical and social aspects of implementation as part of any alternative design strategy. Management and users will want to know how long the implementation will take, how much training will be required, and how disruptive the process will be.

Organizational Issues

One reason management is so interested in the outlook for implementation is that all of the concerns we listed previously cost management money. The longer the implementation process, the more training required; the more disruption expected, the more it will cost to implement the system. Implementation costs are just one cost management has to consider for the new system. Management must also consider the costs of the design process that precedes implementation and the cost of maintaining the system once implementation is over. Overall cost and the availability of funding is just one of the organizational issues to consider in developing alternative design strategies.

A second organizational issue is determining what management will support. Even if adequate funding is available, your organization's management may not be willing or able to support one or another of your alternatives. For example, if your new system differs dramatically from what the corporate office has determined is adequate or from the corporate standard computing environment, your local management may not be willing to support a system that will irritate the corporate office. Most organizations like to have a manageably small number of basic technologies since only a few choices on hardware and software platforms can be supported well. If your new system calls for high levels of cooperation across departments when current operations involve very little of such cooperation, your management may not be willing to support such a system. Your management may also have politically inspired reasons not to support your new system. For example, if your information systems unit reports to the organization's chief financial officer and your new system strengthens a rival department such as manufacturing, management may not support such a system because they prefer to continue the status quo whereby finance is stronger than manufacturing.

A third organizational issue is the extent to which users will accept the new system and use it as designed. A high-end, high-technology solution that represents a radical break in what users are familiar with may have less of a chance of acceptance than a system that is closer to what users know and use. On the other hand, acceptance of a high-end alternative will depend on the users. Some users may demand nothing less than what leading-edge technology can deliver.

These three organizational issues are not the only organizational issues you could, or should, consider. Your assessment of operational and political feasibility (see Chapter 7) which is updated during analysis may identify other organizational issues. Most likely, such organizational issues will affect all of the alternative design strategies you will develop, but it may be possible to offer a range of options in your alternatives that will allow management to make trade-offs between different approaches to various organizational issues. For example, management may be willing to support a system that requires making a little more funding available if it means a little less of a change in the status quo than you might believe is necessary.

DEVELOPING DESIGN STRATEGIES FOR HOOSIER BURGER'S NEW INVENTORY CONTROL SYSTEM

As an example of alternative generation and selection, let's look again at Hoosier Burger's inventory control system. Figure 12-3 lists ranked requirements and constraints for the enhanced information system being considered by Hoosier Burger. The requirements represent a sample of those developed from the requirements determination and structuring carried out in prior analysis steps. The system in question is an upgrade to the company's existing inventory system which was used as an example in Chapters 9, 10, and 11. As you remember, there were several steps to Bob Mellankamp's largely manual inventory control system (Figure 12-4).

Remember that when Bob receives invoices from suppliers, he records their receipt on an invoice log sheet, and he puts the actual invoices in his accordion file. Using the invoices, Bob records the amount of stock delivered on the stock logs, paper forms posted near the point of storage for each inventory item. The stock logs include minimum order quantities as well as spaces for posting the starting

SYSTEM REQUIREMENTS (in descending priority)	SYSTEM CONSTRAINTS (in descending order)
1. Must be able to easily enter shipments into system as soon as they are received.	1. System development can cost no more than $50,000.
2. System must automatically determine whether and when a new order should be placed.	2. New hardware can cost no more than $50,000.
3. Management should be able to determine at any time approximately what inventory levels are for any given item in stock.	3. The new system must be operational in no more than six months from the start of the contract.
	4. Training needs must be minimal, i.e., the new system must be very easy to use.

Figure 12-3

Ranked system requirements and constraints for Hoosier Burger's inventory system

Figure 12-4

The steps in Hoosier Burger's inventory control system

1. Meet delivery trucks before opening restaurant
2. Unload and store deliveries
3. Log invoices and file in accordion file
4. Manually add amounts received to stock logs
5. After closing, print inventory report
6. Count physical inventory amounts
7. Compare inventory reports totals to physical count totals
8. Compare physical count totals to minimum order quantities; if the amount is less, make order; if not, do nothing
9. Pay bills that are due and record them as paid

amount, amount delivered, and the amount used for each item. Amounts delivered are entered on the sheet when Bob logs stock deliveries; amounts used are entered after Bob has compared the amounts of stock used, according to a physical count, and according to the numbers on the inventory report generated by the food ordering system. You remember too that some Hoosier Burger items, especially perishable goods, have standing orders for daily (or Saturday) delivery.

The Mellankamps want to improve their inventory system so that new orders are immediately accounted for, so that the system can determine when new orders should be placed, and so that management can obtain accurate inventory levels at any time of the day, as we saw in Chapter 9. All three of these system requirements have been ranked in order of descending priority in Figure 12-3.

The constraints on developing an enhanced inventory system at Hoosier Burger are also listed in Figure 12-3, again in order of descending priority. The first two constraints cover costs for systems development and for new computer hardware. Development can cost no more than $50,000. New hardware can cost no more than $50,000. The third constraint involves time for development—Hoosier Burger wants the system to be installed and in operation in no more than six months from the beginning of the development project. Finally, Hoosier Burger would prefer that training for the system be simple; the new system must be designed so that it is easy to use. However, as this is the fourth most important constraint, the demands it makes are more flexible than those contained in the other three.

Any set of alternative solutions to Hoosier Burger's inventory system problems must be developed with the company's prioritized requirements and constraints in mind. Figure 12-5 illustrates how each of three possible alternatives meets (or exceeds) the criteria implied in Hoosier Burger's requirements and constraints. Alternative A is a low-end solution. It only meets the first requirement completely and partially satisfies the second requirement, but it does not meet the final one. However, Alternative A is relatively inexpensive to develop and requires hardware that is much less expensive than the largest amount Hoosier Burger is willing to pay. Alternative A also meets the requirements for the other two constraints: it will only take three months to become operational and users will require only one week of training. Alternative C is the high-end solution. Alternative C meets all of the requirements criteria. On the other hand, Alternative C violates two of the four constraints: development costs are high at $65,000 and time to operation is nine months. If Hoosier Burger really wants to satisfy all three of its requirements for its new inventory system, the company will have to pay more than it wants and wait longer for development. Once operational, however, Alternative C will take just as much time to train people to use as Alternative A. Alternative B is in the middle. This alternative solution meets the first two requirements, partially satisfies the third, and does not violate any of the constraints.

Figure 12-5
Description of three alternative systems that could be developed for Hoosier Burger's inventory system

CRITERIA	ALTERNATIVE A	ALTERNATIVE B	ALTERNATIVE C
Requirements			
1. Easy real-time entry of new shipment data	Yes	Yes	Yes
2. Automatic re-order decisions	For some items	For all items	For all items
3. Real-time data on inventory levels	Not available	Available for some items only	Fully available
Constraints			
1. Cost to develop	$25,000	$50,000	$65,000
2. Cost of hardware	$25,000	$50,000	$50,000
3. Time to operation	Three months	Six months	Nine months
4. Ease of training	One week of training	Two weeks of training	One week of training

Now that three plausible alternative solutions have been generated for Hoosier Burger, the analyst hired to study the problem has to decide which one to recommend to management for development. Management will then decide whether to continue with the development project (incremental commitment) and whether the system recommended by the analyst is the system that should be developed.

SELECTING THE MOST LIKELY ALTERNATIVE

One method we can use to decide among the alternative solutions to Hoosier Burger's inventory system problem is illustrated in Figure 12-6. On the left, you see that we have listed all three system requirements and all four constraints from Figure 12-3. These are our decision criteria. We have weighted requirements and constraints equally; that is, we believe that requirements are just as important as

Figure 12-6
Weighted approach for comparing the three alternative systems for Hoosier Burger's inventory system

Criteria	Weight	Alternative A		Alternative B		Alternative C	
		Rating	Score	Rating	Score	Rating	Score
Requirements							
Real-time data entry	18	5	90	5	90	5	90
Auto re-order	18	3	54	5	90	5	90
Real-time data query	14	1	14	3	42	5	70
	50		158		222		250
Constraints							
Development costs	20	5	100	4	80	3	60
Hardware costs	15	5	75	4	60	4	60
Time to operation	10	5	50	4	40	3	30
Ease of training	5	5	25	3	15	5	25
	50		250		195		175
Total	100		408		417		425

constraints. We do not have to weigh requirements and constraints equally; it is certainly possible to make requirements more or less important than constraints. Weights are arrived at in discussions among the analysis team, users, and sometimes managers. Weights tend to be fairly subjective and, for that reason, should be determined through a process of open discussion to reveal underlying assumptions, followed by an attempt to reach consensus among stakeholders. We have also assigned weights to each individual requirement and constraint. Notice that the total of the weights for both requirements and constraints is 50. Our weights correspond with our prioritization of the requirements and constraints.

The next step we have taken is to rate each requirement and constraint for each alternative, on a scale of 1 to 5. A rating of one indicates that the alternative does not meet the requirement very well or that the alternative violates the constraint. A rating of five indicates that the alternative meets or exceeds the requirement or clearly abides by the constraint. Ratings are even more subjective than weights and should also be determined through open discussion among users, analysts, and managers. The next step we have taken is to multiply the rating for each requirement and each constraint by its weight, and we have followed this procedure for each alternative. The final step is to add up the weighted scores for each alternative. Notice that we have included three sets of totals: for requirements, for constraints, and overall totals. If you look at the totals for requirements, Alternative C is the best choice, as it meets or exceeds all requirements. However, if you look only at constraints, Alternative A is the best choice, as it does not violate any constraints. When we combine the totals for requirements and constraints, we see that the best choice is Alternative C, even though it had the lowest score for constraints, as it has the highest overall score.

Alternative C, then, appears to be the best choice for Hoosier Burger. Whether Alternative C is actually chosen for development is another issue. The Mellankamps may be concerned that Alternative C violates two constraints, including the most important one, development costs. On the other hand, the owners (and chief users) at Hoosier Burger may so desire the full functionality Alternative C offers that they are willing to ignore the constraints violations. Or Hoosier Burger's management may be so interested in cutting costs they prefer Alternative A, even though its functionality is severely limited. What may appear to be the best choice for a systems development project may not always be the one that ends up being developed.

UPDATING THE BASELINE PROJECT PLAN

You will recall that the Baseline Project Plan was developed during project initiation and planning (see Chapter 7) to explain the nature of the requested system and the project to develop it. The plan includes (we presented this originally in Figure 7-10 and reproduce it here as Figure 12-7) a preliminary description of the system as requested, an assessment of the feasibility or justification for the system (the business case), and an overview of management issues for the system and project. It was this plan that was presented to a steering committee or other body who approved the commitment of funds to conduct the analysis phase just completed. Thus, it is time to report back (in written and oral form) to this group on the project's progress and to update the group on the findings from analysis. This group will make the final decision on the design strategy to be followed and approve the commitment of resources outlined from the logical (and possibly physical) design steps. Of course, this group could determine that the business case has not developed as originally thought and either stop or drastically redirect the project.

The outline of the Baseline Project Plan can still be used for the analysis phase status report. The updated plan will typically be longer as more is known on each

Figure 12-7
Outline of Baseline Project Plan

BASELINE PROJECT PLAN REPORT

1.0 *Introduction*
   A. Project Overview—Provides an executive summary that specifies the project's scope, feasibility, justification, resource requirements, and schedules. Additionally, a brief statement of the problem, the environment in which the system is to be implemented, and constraints that affect the project are provided.
   B. Recommendation—Provides a summary of important findings from the planning process and recommendations for subsequent activities.

2.0 *System Description*
   A. Alternatives—Provides a brief presentation of alternative system configurations.
   B. System Description—Provides a description of the selected configuration and a narrative of input information, tasks performed, and resultant information.

3.0 *Feasibility Assessment*
   A. Economic Analysis—Provides an economic justification for the system using cost-benefit analysis.
   B. Technical Analysis—Provides a discussion of relevant technical risk factors and an overall risk rating of the project.
   C. Operational Analysis—Provides an analysis of how the proposed system solves business problems or takes advantage of business opportunities in addition to an assessment of how current day-to-day activities will be changed by the system.
   D. Legal and Contractual Analysis—Provides a description of any legal or contractual risks related to the project (e.g., copyright or nondisclosure issues, data capture or transferring, and so on).
   E. Political Analysis—Provides a description of how key stakeholders within the organization view the proposed system.
   F. Schedules, Timeline, and Resource Analysis—Provides a description of potential timeframe and completion date scenarios using various resource allocation schemes.

4.0 *Management Issues*
   A. Team Configuration and Management—Provides a description of the team member roles and reporting relationships.
   B. Communication Plan—Provides a description of the communication procedures to be followed by management, team members, and the customer.
   C. Project Standards and Procedures—Provides a description of how deliverables will be evaluated and accepted by the customer.
   D. Other Project-Specific Topics—Provides a description of any other relevant issues related to the project uncovered during planning.

topic. Further, the various process, logic, and data models are often included to make the system description more specific. Usually only high-level versions of the diagrams are included within section 2.0, and more detailed versions are provided as appendices.

Every section of the Baseline Project Plan Report is updated at this point. For example, section 1.0.B will now contain the recommendation for the design strategy chosen by the analysis team. Section 2.0.A provides the descriptions of the competing strategies studied during alternative generation and selection, often including the types of comparison charts shown earlier in this chapter. Section 3.0 is typically significantly changed since you now know much better than you did during project initiation and planning what the needs of the organization are. For example, economic benefits which were intangible before may now be tangible. Risks, especially operational ones, are likely better understood.

Section 3.0.F will now show the actual activities and their durations during the analysis phase, as well as include a detailed schedule for the activities in the design phases and whatever additional details can be anticipated for later phases. Many Gantt charting packages can show actual progress versus planned activities. It is important to show in this section how well the actual conduct of the analysis phase matched the planned activities. This helps you and management understand how well the project is understood and how likely it is that the stated future schedule will occur. Those activities whose actual durations differed significantly from planned durations may be very useful to you in estimating future activity durations. For example, a longer than expected task to analyze a certain process on a DFD may suggest that the design of system features to support this process may take longer than originally anticipated.

Often the design phase activities will be driven by the capabilities chosen for the recommended design strategy. For example, you will place specific design activities on the schedule for such design deliverables as the following:

- Layout of each report and data input and display screens (the DFDs include data flows for each of these)

- Structuring of data into logical tables or files (the E-R diagrams identify what data entities are involved in this)

- Programs and program modules that need to be described (processes on DFD and process and temporal logic models explain how complicated these tasks will be)

- Training on new technologies to be used in implementing the system

Many design phase activities result in developing design specifications for one or more examples of the types of design deliverables listed previously.

Section 4.0 is also updated. It is likely that the project team needs to change as new skills are needed in the next and subsequent project phases. Also, since project team members are often evaluated after each phase, the project leader may request the reassignment of a team member who has not performed as required. The communication plan needs to be reassessed to see if other communication methods need to be employed (see Table 4-4 for a list of common communication methods). New standards and procedures will be necessary as the team discovers that some current procedures are inadequate for the new tasks. Section 4.0.D is often used to outline issues for management that have been discovered during analysis. Recall, for example, that we discussed in Chapter 8 how you might find redundancies and inconsistencies in job descriptions and the way people actually do their jobs. Since these issues must be resolved by management, and must be addressed before you can progress into detailed system design, now is the last time to call these issues to the attention of management.

As the project leader, you and other analysts also must ensure that the project workbook and CASE repository are completely up-to-date as you finalize the analysis phase. Since the project team composition will likely change and, as time passes, you forget facts learned in earlier stages, the workbook and repository are necessary to transfer information between phases. This is also a good time for the project leader to do a final check that all elements of project execution have been properly handled (see Table 4-5).

Besides the written Baseline Project Plan Report update, an oral presentation is typically made, and it may be at this meeting that a decision to approve your recommendations, redirect your recommendations, or kill the project is made. We discussed such an analysis phase review meeting in Chapter 2 for a hypothetical project in Pine Valley Furniture (see Figure 2-5 for an agenda for such a meeting). It is not uncommon for the analysis team to follow this project review meeting with a suitable celebration for reaching an important project milestone.

Before and After Baseline Project Plans for Hoosier Burger

Even though their Inventory Control System was relatively small, Hoosier Burger developed a Baseline Project Plan for the project. The plan included information for each area listed in the outline in Figure 12-7, parts of which are reproduced below. Now that the analysis phase of the life cycle has ended, the plan must be updated, and those updated sections are also reproduced below. The first item we will consider is the cost-benefit analysis prepared as part of Section 3.0.A, on economic analysis. Hoosier Burger's initial cost-benefit analysis for the inventory project is shown in Figure 12-8. The format is the same as you saw in Chapter 7.

Figure 12-8
Hoosier Burger's initial cost-benefit analysis for its Inventory Control System project

Hoosier Burger							
Economic Feasibility Analysis							
Inventory Control System							
				Year of Product			
	Year 0	Year 1	Year 2	Year 3	Year 4	Year 5	TOTALS
Net economic benefit	$0	$30,000	$30,000	$30,000	$30,000	$30,000	
Discount rate (12%)	1	0.8928571	0.7971939	0.7117802	0.6355181	0.5674269	
PV of benefits	$0	$26,786	$23,916	$21,353	$19,066	$17,023	
NPV of all BENEFITS	$0	$26,786	$50,702	$72,055	$91,120	$108,143	$108,143
One-time COSTS	($100,000)						
Recurring Costs	$0	($2,000)	($2,000)	($2,000)	($2,000)	($2,000)	
Discount rate (12%)	1	0.8928571	0.7971939	0.7117802	0.6355181	0.5674269	
PV of Recurring Costs	$0	($1,786)	($1,594)	($1,424)	($1,271)	($1,135)	
NPV of all COSTS	($100,000)	($101,786)	($103,380)	($104,804)	($106,075)	($107,210)	($107,210)
Overall NPV							$934
Overall ROI - (Overall NPV / NPV of all COSTS)							0.01

The numbers in the spreadsheet are based in part on the constraints listed in Figure 12-3. Part of the worksheet the Mellankamps used to determine the values in the spreadsheet in Figure 12-8 is shown in Table 12-5. The Mellankamps estimated that the benefits of the new system could be quantified in two ways: first, instant entry of shipment data would lead to more accurate inventory data; second, Hoosier Burger would be less likely to run out of stock with an automatic order determination as part of the system. Savings from more accurate data amount to $1,500 per month or $18,000 per year; savings from fewer stockouts amount to $1,000 per month or $12,000 per year. As you can see from Figure 12-8, although a new inventory control system for Hoosier Burger would break even, it is not a very good investment, with only a 1% return, given a 12% discount rate.

Figure 12-9 shows the cost-benefit analysis after the analysis phase has ended, which appears in the updated Baseline Project Plan. Notice that developing the new system, represented by Alternative C, is now a better investment, with a 15% return. Yet the overall costs of Alternative C exceed the costs of the original estimation of the system. What happened?

Much of Figure 12-9 is the same as Figure 12-8. Recurring costs are the same; the discount rate is the same. What has changed, in addition to larger one-time costs, is that net benefits are now estimated to be larger than in Figure 12-8. Details on these new estimates are shown in Table 12-6. The Mellankamps re-estimated the savings from the new system, using more accurate data, and found they had been too optimistic. The expected savings from instantly logging new shipments of supplies were reduced from $18,000 per year to $15,000 per year. But the Mellankamps also found a new benefit: they realized that better management information, and its ready availability through a new query capability, could be quantified at about $1,000 per month. The Mellankamps did not just dream up the $1,000 per month savings estimate. They developed it from thinking about how timely, accurate management information would affect their ability to prepare reports as well as their ability to improve their operations through better inventory control. So even though Alternative C was more expensive to develop than the other alternatives, it actually resulted in a higher level of tangible benefits.

Figure 12-10 shows the part of the project schedule from the initial version of the Baseline Project Plan that applies to subsequent steps. Notice that the schedule covers only the design and implementation phases of the life cycle and that the schedule is very general. The physical design sub-phase is not broken down into its constituent parts. The task times in the schedule are also driven by two of the constraints listed in Figure 12-3. The entire schedule spans exactly six months of activity

TABLE 12-5   Hoosier Burger's Initial Economic Analysis Worksheet

One-time costs: Development	$50,000
One-time costs: Hardware	$50,000
Recurring costs: Maintenance	$ 2,000 per year
Savings: Fewer stockouts due to automatic reordering	$12,000 per year
Savings: More accurate data from shipment logging	$18,000 per year
Intangible benefit: Better management information	

Figure 12-9

Hoosier Burger's revised cost-benefit analysis for its Inventory Control System project

Hoosier Burger							
Economic Feasibility Analysis							
Inventory Control System							
				Year of Product			
	Year 0	Year 1	Year 2	Year 3	Year 4	Year 5	TOTALS
Net economic benefit	$0	$39,000	$39,000	$39,000	$39,000	$39,000	
Discount rate (12%)	1	0.8928571	0.7971939	0.7117802	0.6355181	0.5674269	
PV of benefits	$0	$34,821	$31,091	$27,759	$24,785	$22,130	
NPV of all BENEFITS	$0	$34,821	$65,912	$93,671	$118,457	$140,586	$140,586
One-time COSTS	($115,000)						
Recurring Costs	$0	($2,000)	($2,000)	($2,000)	($2,000)	($2,000)	
Discount rate (12%)	1	0.8928571	0.7971939	0.7117802	0.6355181	0.5674269	
PV of Recurring Costs	$0	($1,786)	($1,594)	($1,424)	($1,271)	($1,135)	
NPV of all COSTS	($115,000)	($116,786)	($118,380)	($119,804)	($121,075)	($122,210)	($122,210)
Overall NPV							$18,377
Overall ROI - (Overall NPV / NPV of all COSTS)							0.15

TABLE 12-6   Hoosier Burger's Updated Economic Analysis Worksheet

One-time costs: Development	$65,000
One-time costs: Hardware	$50,000
Recurring costs: Maintenance	$ 2,000 per year
Savings: Fewer stockouts due to automatic reordering	$12,000 per year
Savings: More accurate data from shipment logging	$15,000 per year
Savings: Better management information & availability	$12,000 per year

and training takes only one week. The estimates of how long each task should take to complete are all very rough.

Compare Figure 12-10 to Figure 12-11, the revised schedule that goes in the updated Baseline Project Plan. The schedule is more detailed and it more closely reflects the development time necessary for Alternative C. Training still takes only

Figure 12-10
Hoosier Burger's initial schedule for its Inventory Control System project

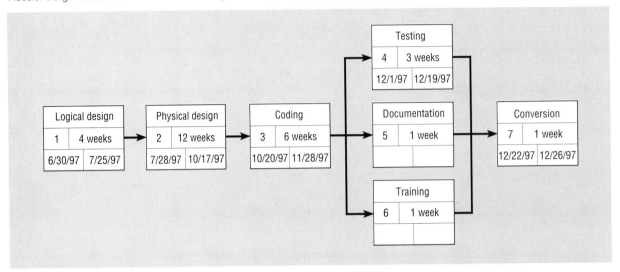

Figure 12-11
Hoosier Burger's revised schedule for its Inventory Control System project

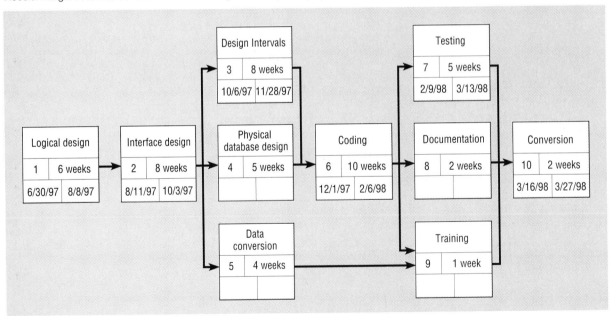

one week, but now that estimate is based on a clear understanding of the requirements of a particular system rather than based primarily on positive thinking. Also, the entire schedule now spans nine months, the time necessary to fully develop and implement Alternative C. Some design tasks in Figure 12-11 have been decomposed into four different sub-tasks, many of which can be worked on concurrently: interface design, designing the internals, physical database design, and data conversion. Note that even this schedule presents the project at a very high level. It would be typical in actual projects to show not only the major steps but also the individual activities needed to complete each step. For example, Interface design might be broken down into many steps for each display or report and for different activities, such as meetings with users to walkthrough the tentative designs.

We have shown you only two parts to the Hoosier Burger Baseline Project Plan for their Inventory Control System. Even for a project this small, a complete Baseline Project Plan would be too much to include in this book. From these examples, though, you should get a good general idea both of what an initial Baseline Project Plan contains and how it changes when a major life cycle phase, like analysis, ends.

SUMMARY

In the alternative generation and choice phase of systems analysis, you develop alternative solutions, or design strategies, to the organization's information system problem. A design strategy is a combination of system features, hardware and system software platform, and acquisition method that characterize the nature of the system and how it will be developed. A good number of alternative design strategies to develop is three, as three alternatives can represent the high-end, middle, and low-end of the spectrum of possible systems that can be built.

When developing design strategy alternatives, you must become aware of all of the possible options available. You must be aware of where you can obtain software that meets some or all of an organization's needs. You can obtain application (and system) software from hardware vendors, packaged software vendors, and custom software developers, as well as from internal systems development resources. You can even hire an organization to handle all of your systems development needs, called outsourcing. You must also know which criteria to use to be able to choose among off-the-shelf software products. These criteria include cost, functionality, vendor support, vendor viability, flexibility, documentation, response time, and ease of installation. You must also determine whether new hardware and system software are needed. Requests for proposals are one way you can collect more information about hardware and system software, their performance, and costs. In addition to software and hardware issues, you must consider implementation issues and broader organizational concerns, such as the availability of funding and management support.

Alternative design strategies are developed after a system's requirements and constraints have been identified and prioritized. Once developed, alternatives can be compared to each other through quantitative methods, but the actual decision may depend on other criteria, such as organizational politics.

Since generating and selecting alternative design strategies completes the analysis phase of the SDLC, the systems development project has reached a major milestone. Once an analysis of alternative design strategies is completed, you and other members of the analysis team will present your findings to a management steering committee and/or the client requesting the system change. In this presentation (both written and oral), you will summarize the requirements discovered, evaluate alternative design strategies and justify the recommended alternative as well as present an updated Baseline Project Plan for the project to follow if the committee decides to fund the next life cycle phase.

C H A P T E R   R E V I E W

K E Y   T E R M S

Design strategy
Outsourcing

R E V I E W   Q U E S T I O N S

1. Define each of the following terms:
   a. outsourcing
   b. request for proposal (RFP)
   c. design strategy
   d. turnkey system
   e. custom software producer
   f. vendor support
   g. Baseline Project Plan

2. What are the deliverables from generating alternatives and selecting the best one?

3. Why generate at least three alternatives?

4. Describe the four sources of software.

5. How do you decide among various off-the-shelf software options? What criteria do you use?

6. What issues are considered when analysts try to determine whether new hardware or system software is necessary? What is an RFP and how do analysts use one to gather information on hardware and system software?

7. What issues other than hardware and software must analysts consider in preparing alternative system design strategies?

8. How do analysts generate alternative solutions to information systems problems?

9. How do managers decide which alternative design strategy to develop?

10. Which elements of a Baseline Project Plan might be updated during the alternative generation and selection step of the analysis phase of the SDLC?

11. What methods can a systems analyst employ to verify vendor claims about a software package?

P R O B L E M S   A N D   E X E R C I S E S

1. Match the following terms to the appropriate definitions.

   _____ alternative generation and selection

   _____ requirements structuring

   a. statement of system features, hardware and software platform, and acquisition approach for a new system

   b. process of collecting information about what the current system and what capabilities users and others want in a new information system

_____ requirements determination

c. process of organizing information about an information system using standard process, logic, and data modeling notations

_____ design strategy

d. process of developing choices for how to proceed with the design and implementation of an information system

2. Find the most current issue of _Datamation_ that includes the most recent _Datamation_ 100 listings. How much has the rank order of the top software companies changed compared to the list printed here? Read the issue and determine why your list is different from the list in this chapter. What changes are occurring in the computer industry that might affect this list?

3. Research how to prepare a Request for Proposal. Prepare an outline of an RFP for Hoosier Burger to use in collecting information on its new inventory system hardware.

4. Recreate the spreadsheet in Figure 12-6 in your spreadsheet package. Change the weights and compare the outcome to Figure 12-6. Change the rankings. Add criteria. What additional information does this "what if" analysis provide for you as a decision maker? What insight do you gain into the decision-making process involved in choosing the best alternative system design?

5. Prepare a list for evaluating computer hardware and system software that is comparable to the list of criteria for selecting off-the-shelf application software presented earlier.

6. The method for evaluating alternatives used in Figure 12-6 is called weighting and scoring. This method implies that the total utility of an alternative is the product of the weights of each criterion times the weight of the criterion for the alternative. What assumptions are characteristic of this method for evaluating alternatives? That is, what conditions must be true for this to be a valid method of evaluation alternatives?

7. Weighting and scoring (see Problem and Exercise 6) is only one method for comparing alternative solutions to a problem. Go to the library and find a book or articles on qualitative and quantitative decision making and voting methods and outline two other methods for evaluating alternative solutions to a problem. What are the pros and cons of these methods compared to the weighting and scoring method? Under weighting and scoring and the other alternatives you find, how would you incorporate the opinions of multiple decision makers?

8. Prepare an agenda for a meeting at which you would present the findings of the analysis phase of the SDLC to Bob Mellankamp concerning his request for a new inventory control system. Use information provided in Chapters 9–12 as background in preparing this agenda. Concentrate on which topics to cover, not the content of each topic.

9. Review the criteria for selecting off-the-shelf software presented in this chapter. Use your experience and imagination and describe other criteria that are or might be used to select off-the-shelf software in the "real world." For each new criterion, explain how use of this criterion might be functional (i.e., it is useful to use this criterion), dysfunctional, or both.

10. The owner of two pizza parlors located in adjacent towns wants to computerize and integrate sales transactions and inventory management within and between both stores. The point of sale component must be very easy to use and flexible enough to accommodate a variety of pricing strategies and coupons. The inventory management, which will be linked to the point of sale component, must also be easy to use and fast. The systems at each store must be linked so that sales and inventory levels

can be determined instantly for each store and for both stores combined. The owner can allocate $40,000 for hardware and $20,000 for software and must have the new system operational in three months. Training must be very short and easy. Briefly describe three alternative systems for this situation and explain how each would meet the requirements and constraints. Are the requirements and constraints realistic? Why or why not?

11. Compare the alternative systems from Problem and Exercise 10 using the weighted approach demonstrated in Figure 12-6. Which system would you recommend? Why? Was the approach taken in this and Problem and Exercise 10 useful even for this relatively small system? Why or why not?

12. Suppose that an analysis team did not generate alternative design strategies for consideration by a project steering committee or client. What might the consequences be of having only one design strategy? What might happen during the oral presentation of project progress if only one design strategy is offered?

13. In the section on Choosing Off-the-Shelf Software there are eight criteria proposed for evaluating alternative packages. Suppose the choice was between alternative custom software developers rather than pre-written packages. What criteria would be appropriate to select and compare among competing bidders for custom development of an application? Define each of these criteria.

FIELD EXERCISES

1. Consider the purchase of a new PC to be used by you at your work (or by you at a job that you would like to have). Describe in detail three alternatives for this new PC that represent the low, mid-, and high points of a continuum of potential solutions. Be sure that the low-end PC meets at least your minimum requirements and the high-end PC is at least within a reasonable budget. At this point, without quantitative analysis, which alternative would you choose?

2. For the new PC described above, developed ranked lists of your requirements and constraints as displayed in 12-6. Display the requirements and constraints, along with the three alternatives, in a diagram like the one displayed in Figure 12-6, and note how each alternative is rated on each requirement and constraint. Calculate scores for each alternative on each criterion and compute total scores. Which alternative has the highest score? Why? Does this choice fit with your selection in the previous question? Why or why not?

3. One of the most competitive software markets today is electronic spreadsheets. Pick three packages (for example, Microsoft Excel, Lotus 1-2-3, and Novell's Quattro Pro—but any three spreadsheet packages would do). Study how you use spreadsheet packages for school, work, and personal financial management. Develop a list of criteria important to you on which to compare alternative packages. Then contact each vendor and ask for the information you need to evaluate their package and company. Request a demonstration copy or trial use of their software. If they cannot provide a sample copy, then try to find a computer software dealer or club where you can test the software and documentation. Based on the information you receive and the software you use, rate each package using your chosen criteria. Which package is best for you? Why? Talk to other students and find out which package they rated as best. Why are there differences between what different students determined as best?

4. Interview businesspeople who participate in the purchase of off-the-shelf software in their organizations. Review with them the criteria for selecting off-the-shelf soft-

ware presented in this chapter. Have them prioritize the list of criteria as they are used in their organization and provide an explanation of the rationale for the ranking of each criterion. Ask them to list and describe any other criteria that are used in their organization.

5. Obtain copies of actual Requests for Proposals used for information systems developments and/or purchases. If possible, obtain RFPs from public and private organizations. Find out how they are used. What are the major components of these proposals? Do these proposals seem to be useful? Why or why not? How and why are the RFPs different for the public versus the private organizations?

REFERENCES

Applegate, L. M. and R. Montealegre. 1991. "Eastman Kodak Company: Managing Information Systems Through Strategic Alliances." Harvard Business School case 9-192-030. Cambridge, MA: President and Fellows of Harvard College.

Harper, D. 1994. "Seek a Partner, Not a Vendor." *Industrial Distribution* 83 (April): 97.

Hartmann, C. R. 1993. "How to Write a Proposal." *D & B Reports* 42 (March-April): 62.

King, R. T. Jr. and J. E. Rigdon. 1994. "Hewlett Prints Computer Services in Big Capital Letters." *Wall Street Journal* (June 3): B8.

"Microcomputer Procurement Guidelines." 1994. *Public Works* 125 (April 15): G23 +.

Mikulski, F. A. 1993. *Managing Your Vendors: The Business of Buying Technology.* Englewood Cliffs, NJ: Prentice-Hall.

Moad, J. 1993. "Inside an Outsourcing Deal." *Datamation* 39 (February 15): 20–27.

"More Companies Are Chucking Their Computers." 1989. *Business Week* (June 19): 72–74.

Semich, J. W. "Is It Bye-Bye Borland?" *Datamation* 40 (June 15): 52–53.

Stein, M. 1993. "Don't Bomb Out When Preparing RFPs." *Computerworld* 27 (February 15): 102.

Zachary, G. P. 1990. "How Ashton-Tate Lost Its Leadership in PC Software Arena." *Wall Street Journal* (April 1): A1–A2.

Zachary, G. P. and W. M. Bulkeley. 1990. "Ashton-Tate Loses Flagship Software's Copyright Shield." *Wall Street Journal* (December 14): B1, B4.

Analysis: Selecting the Best Alternative Design Strategy for the Customer Activity Tracking System

INTRODUCTION

Frank Napier, Jordan Pippen, and Wendy Yoshimuro were meeting in Jordan's office to prepare for a preliminary analysis phase walkthrough meeting with Nancy Chen and Karen Gardner in preparation for the meeting with the Systems Priority Board (SPB). The walkthrough with the SPB would end the analysis phase of their Customer Activity Tracking System (CATS) project. Frank, Jordan, and Jorge had completed all of their work and had developed some alternative design strategies for CATS. Wendy, the marketing manager for U.S. Operations, had become so interested in CATS that she volunteered to be a part of the presentation team at the preliminary walkthrough, to show the strong user commitment to CATS.

PREPARING FOR THE PRELIMINARY CATS ANALYSIS WALKTHROUGH

"Frank," Jordan began, "You will start the meeting by presenting an overview of our three alternative system design strategies for CATS. Then Jorge, Wendy, and I will explain each of the alternative strategies, pointing out the good and bad points of each one."

"How are we going to organize that?" Wendy asked.

"I think the best way," Jordan answered, "is for Jorge and me to explain the information systems issues and for you to explain the functionality issues. After our overview of the alternatives, I will talk about the alternative we've selected and argue for its adoption."

"That's where I show the preliminary budget and schedule for the design phases," Frank said. "I'll also include a very preliminary budget and schedule for implementation and maintenance. Jorge will be pre-pared to go over any of the requirements models we've developed, in case we get any questions about system details. Buffy Jarvis, from data administration, will be present to answer any questions about consistency with other databases. We'll have transparency masters made as backup, but Jorge will be able to show most of the charts on-line through a projection system tied to a PC running the CASE tool. Jorge is also coordinating a full-color computer-based slide show, for all of our presentation materials. We'll test all of this in our meeting with Nancy and Karen."

"You know this is the first time I've been involved in a systems development project at Broadway," Wendy stated. "Tell me again why both my boss, Nancy, and your boss, Karen, are involved in this meeting."

"Nancy has to decide if she is willing to lend the managerial and financial support it will take to complete the life cycle for CATS," Jordan responded. "Nancy is the one who has to pay for the development process. If she thinks what we are proposing is too expensive given the level of benefit we expect, she may cancel the whole project now. Design and implementation are more expensive than analysis, and Nancy may not want her area to take on that kind of expense right now. Karen, on the other hand, will have to decide where and if CATS as contemplated will fit into the overall information systems model for Broadway. For example, she will have to affirm to the SPB that CATS is compatible with the IS architecture and chosen platforms. She also has to decide if she is willing to let Frank, Jorge, and me continue on this project. It's possible one or more of us might get reassigned. Karen also has to think about the additional staffing demands for the design phase. We need to get both Karen and Nancy to buy into our ideas before we take them to the SPB for approval."

"Let's go over the material we've prepared once again, OK?" Wendy asked.

"No problem," replied Jordan. "That's why we're here. After all of our work in analysis, we've come up with three alternative design strategies for CATS. The first alternative is our high-end system—completely automated, all the bells and whistles, completely designed and implemented in-house so it meets our every need."

"And very expensive," Frank added.

"Our third alternative is our low-end system. Some of the work would be done manually, and functionality would be limited . . ."

"But it's very cheap," Frank said.

"Yes," replied Jordan, "It's cheap. Anyway, the middle alternative is as middle-of-the-road as we could get on basic functionality, extra features . . ."

"And cost," Frank said.

"Show me again the requirements and constraints lists," Wendy said.

"Here are the requirements [Figure 1], and here are the constraints [Figure 2]," said Frank.

"I recognize these requirements," Wendy said. "I helped Nancy put them together. This pretty much details the expected system functionality, right?"

"Yes, that's right," Jordan replied, "at least at a high level. You may not have seen the constraints before, however. We developed this list working with Karen, although Nancy had a lot of input, too, especially about the costs involved."

Figure 1
CATS requirements

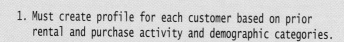

1. Must create profile for each customer based on prior rental and purchase activity and demographic categories.

2. Must create profile for new releases of videos, games, and music.

3. Must match customer profiles to product profiles.

4. Must select customer/product matches most likely to lead to additional sales.

5. Must produce postcards about new products, addressed to customers matched with products.

6. Must be able to record reservation information that can be used to update BEC's inventory control system.

7. Must initiate shipment of reserved products to individual stores.

Figure 2
CATS constraints

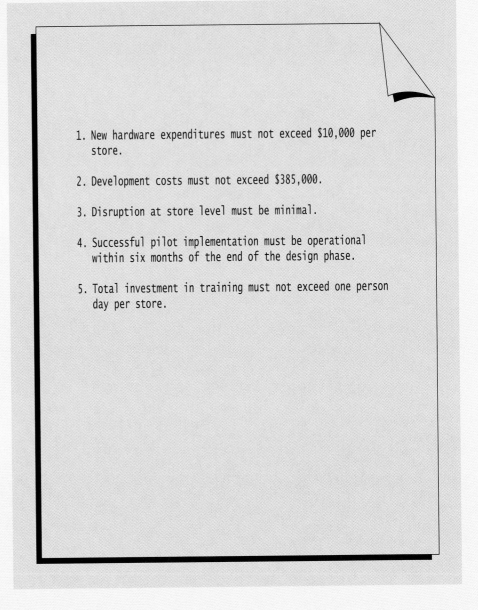

1. New hardware expenditures must not exceed $10,000 per store.

2. Development costs must not exceed $385,000.

3. Disruption at store level must be minimal.

4. Successful pilot implementation must be operational within six months of the end of the design phase.

5. Total investment in training must not exceed one person day per store.

"As you can see from the list, most of the constraints are monetary. But there is also an emphasis on how CATS will affect the stores," Jordan said.

"Where did that emphasis come from?" asked Wendy.

"Some of it is from Nancy, but it became clear to me in my meetings with store managers that they will not tolerate much disruption," Jordan replied. "Some of them are very skeptical about CATS. Oh, by the way, these lists are already prioritized from most to least important."

"Now here is how we estimate if each of the alternatives either meets or does not meet the requirements

and stays faithful to or violates the constraints [Figure 3]. On this chart, our high-end system is Alternative 1, the middle-of-the-road system is Alternative 2, and Alternative 3 is the low-end system."

"Can you go over these alternatives for me one more time?" asked Wendy. "I want to think about them this time from Nancy's perspective."

"Sure," said Frank. "Alternative 1 is our own customized CATS, where we design and build the CATS so that it does everything Nancy and others have told us they want it to do, using state-of-the-art technology, from headquarters on down to the store level. We are going to base the system at the store level on imaging,

Figure 3
Alternative design strategies
for CATS

Criteria	Alternative 1	Alternative 2	Alternative 3
Requirements			
Customer profiles	Yes	Yes	Yes
Product profiles	Yes	Yes	Yes
Profile matching	Yes	Yes	Yes
Match selection	Yes	Yes	Yes
Generate cards	Yes	Yes	No
Track reservations	Yes	No	No
Initiate delivery	Yes	No	No
Constraints			
Hardware costs	$15K/store	$10K/store	$8K/store
Development costs	$500K	$400K	$300K
Store disruption	Medium	Low	Medium
Time to operation	12 months	6 months	6 months
Store training	Two days	One day	Two days

with CD-ROM storage. Eventually we'd like to have Broadway set up a node on the Internet so we can use e-mail instead of postcards and have samples of the video or music available on a bulletin board so customers could download it. Then they could skip the 800 number to make a reservation and just order the product directly over e-mail. They wouldn't have to send their credit card numbers over e-mail because we have them on file."

"So that's why this system violates every constraint," Wendy observed. "What about the middle system?"

"Alternative 2 is very interesting," Frank began. "I did some research and found a company doing a lot of what we want to do. They have a system that creates customer and product profiles, matches and selects profiles, and generates postcards. And the system is for sale. We'd have to do some customization, of course, and there is no tie-in to inventory control for tracking or initiating deliveries."

"What company is this?" Wendy asked.

"It's the TwoCat Petfood Company."

"What?" exclaimed Wendy.

"Yep, that's right. They have a really diverse line of cat foods and related products, and they are always developing new products. They do the same thing we want to do. They use data on loyal customers, create profiles, match those profiles with product profiles, all of the basic things we want. I heard about them from a friend at a local Association for Computing Machinery chapter meeting. We were talking about what projects we were working on and she mentioned she received

similar types of notices from this company for her cat, Memories. I think my friend is a candidate for a Broadway show soundtrack profile category, don't you? Anyhow, I called the company and got all the information; they even sent me a copy of the software to try."

"And the third alternative?"

"It will do the profile generation and matching, but it won't generate postcards. After the matching and selection, a report will be generated and sent to each store. The store will have to send postcards or call customers, keep track of reservations, and order product to fill their reservation needs," said Frank. "That's how we keep development costs down, by making some of the work manual and transferring it to the stores."

"That's why the cheapest alternative results in store disruption and more training time, too," said Jordan.

"OK," said Wendy. "What's our best alternative?"

"Here's how we determined it," said Frank, showing Wendy a spreadsheet output (Figure 4).

"We decided to make the requirements worth 60 percent of the choice, with constraints at 40 percent. We also decided to give weights of 10 to the four major requirements and the first two constraints. As both requirements and constraints are prioritized, the weights decrease in order after that. We determined these weights based on a survey we conducted during the JAD session. It was a small, but representative, sample. You can see that the high-end alternative, Alternative 1, did a great job on the requirements, but since it violated all of the constraints, Alternative 1

Figure 4
CATS alternatives evaluated and compared

Criteria	Weight	Alternative 1		Alternative 2		Alternative 3	
		Rating	Score	Rating	Score	Rating	Score
Requirements							
Customer Profiles	10	10	100	10	100	10	100
Product Profiles	10	10	100	10	100	10	100
Profile Matching	10	10	100	10	100	10	100
Profile Selection	10	10	100	10	100	10	100
Generate Cards	8	8	64	8	64	1	8
Track Reservations	6	6	36	1	6	1	6
Initiate Delivery	6	6	36	1	6	1	6
	60		536		476		420
Constraints							
Hardware Costs	10	5	50	10	100	10	100
Development	10	5	50	8	80	10	100
Store Disruption	8	4	32	8	64	4	32
Time to Operation	7	3	21	7	49	7	49
Store Training	5	2	10	5	25	2	10
	40		163		318		291
Total	100		699		794		711

came in last among the three alternatives in terms of scores."

"Right," said Jordan. "Alternative 3 was last in terms of requirements. It fared OK for the constraints, and it would have done better, but there were problems with store disruption and training."

"So the winner is Alternative 2, the cat food system?" asked Wendy.

"Yep," said Frank, "TwoCat is our winner. The difference in scores is impressive in that Alternative 2 outscored the other two alternatives pretty well. We'll have a good bit of work to do to redesign some on-line displays, add some reports they don't have, and link this system into our other existing application portfolio. Next we just have to convince Nancy and Karen. Now, here is my budget and schedule for the design phase. Here is where you come in. . . ."

"Well, I'd like to stay and talk about the next step, but it's time for my appointment with one of our major suppliers who is probably waiting in my office. I need to go. By the way, would it help if I told him a little about our system? I've read a little about EDI, electronic data interchange, and maybe some of the ordering functions within CATS could be handled through EDI. I'll send you an e-mail if he sees a possible connection. See you later."

SUMMARY

The analysis phase of the CATS project is about to come to a close. The CATS project team, along with several key stakeholders, have identified and evaluated alternative design strategies, deciding on the purchase of a software package from another company, TwoCat Petfood, as part of the recommended design strategy. Although TwoCat is not a software company, this does not matter since BEC is not looking for a software partner. Rather, Jordan Pippen and the other CATS team members expect to adapt the purchased software. In fact, TwoCat will sell BEC the source code for the software, which will facilitate BEC's customization. Thus, the CAT project must continue with subsequent design and implementation steps, although the purchased software will save considerable development time. The purchased software will provide most of the functionality BEC wants, and it is compatible with BEC's hardware and software platform.

The BEC Systems Priority Board (SPB) is impressed with the analysis done by the BEC team and the strong commitment from the client organization, evidenced by the integral role both Nancy Chen and Wendy Yoshimuro played in the analysis phase and the presentation of the design strategy and updated Baseline Project Plan. After gaining SPB approval, the CATS team is now ready to proceed to the design phases.

Logical Design

IVI Publishing

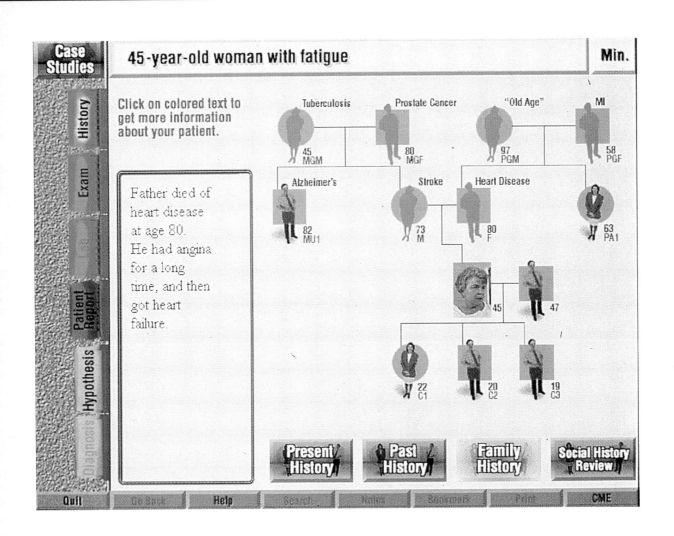

Case Studies

45-year-old woman with fatigue

Min.

History

Exam

Patient Report

Hypothesis

Diagnosis

Click on colored text to get more information about your patient.

Father died of heart disease at age 80. He had angina for a long time, and then got heart failure.

Tuberculosis
45
MGM

Prostate Cancer
80
MGF

"Old Age"
97
PGM

MI
58
PGF

Alzheimer's
82
MU1

Stroke
73
M

Heart Disease
80
F

63
PA1

45

47

22
C1

20
C2

19
C3

Present History

Past History

Family History

Social History Review

Quit   Go Back   Help   Search   Notes   Bookmark   Print   CME

Creating a Conversation Between Computer and User

IVI Publishing was formed in 1991 to publish interactive multimedia software titles, initially on CD-ROM, with a focus on health and medical information. IVI now has dozens of award-winning products and is the leading publisher of health and medical information for professionals (for example, doctors) and consumers. IVI has approximately 100 employees, a relatively flat organizational structure, and a team-based approach.

For professionals, IVI produces *PrimePractice,* a quarterly interactive magazine which provides continuing education to primary care physicians. IVI developed this software with Mayo Clinic doctors and the appropriate accrediting agencies so that doctors can receive continuing medical education credit for using the software. For adult consumers, IVI and the Mayo Clinic produce a line of interactive CD-ROMs, including *Family Health Book, The Total Heart, Sports Health and Fitness,* and *Family Pharmacist.* For children, IVI produces *Anna Tommy and Safety Monkey* (also with the Mayo Clinic) as well as *What Is a Belly Button?, Welcome to Body Land,* and *The Virtual Body* (with Time-Life).

IVI is as innovative in their business processes as they are with their products. They are continually trying to improve their internal business processes. Recently they completed a business process re-engineering project which involved a team of 17 people who focused on completely redesigning the software design process for their products. As a result of this re-engineering process, they produced a "Clean Plate" document which describes their new streamlined software design process. IVI previously took nine months to complete a new product. The length of the new software development process—including software design, building, and testing—is shorter than the previous process by one third.

Whereas other software publishers have one person design a new product alone, literally in a vacuum with no input from other designers or users, IVI uses a team approach. An executive producer sponsors a new product project and a senior designer is assigned as the project leader. The senior designer consults with system engineers, artists, graphic designers, a product manager from marketing, content experts (for example, doctors), and users to design the new software. The core team then works with a professional facilitator to conduct extensive user testing in a special laboratory environment with computers and audio/visual recording equipment. IVI will often conduct such product testing in separate cities for a demographic sampling of various users' reactions to new products, features, and functionality.

To support this collaborative design approach, each project is assigned to a large dedicated collaborative workroom equipped with personal computers and peripherals, whiteboards, noteboards, conference tables and chairs, and phones. IVI also supports individual workspaces in addition to these team rooms. IVI has many of these collaborative workrooms, with seven such rooms in one wing of the building alone. Project team members commonly spend several hours a day in the workroom collaborating on their project.

The innovation of IVI's business processes is matched by the innovation of their design philosophy. Dori Pelz-Sherman, senior designer, explains that while pure graphic designers provide useful input into the interactive software design process, graphic design skills do not necessarily transfer well to the interactive software design process. Designers like Pelz-Sherman are described as "conceptual interactive designers." They have to be concerned not only about the aesthetics of the software but also about whether or not the software is functional, easily understood, and useful. Pelz-Sherman believes that, for interactive software, the designer must get into the head of the user and create interfaces that lighten the "cognitive load." Content of the software must drive the visuals. She constantly asks herself, "What is the point of this software? What will the user want and need to do with it, and how?"

While print conventions have taken over 500 years to evolve, interactive software is very new. Pelz-Sherman argues that in the most closely related discipline, graphic design, designers have been trained almost exclusively in the fine arts and have had little experience or training with the analysis and presentation of data, cognitive science, or user-machine interactivity. She cites the work of Wurman who has argued that the graphic design profession bestows awards for appearance rather than for understandability. She borrows a line from Wurman that there are no Oscars, Emmys, or Tonys for making graphics comprehensible.

Pelz-Sherman represents a new breed of designer at IVI. She has an undergraduate double major in Visual Arts and English, with a teaching degree for both areas and a master's degree in Design and Instructional Technology at San Diego State University. Because programs in interactive software design are so rare, she customized an advanced degree by combining coursework in instructional design, computing, graphic design, and psychology.

IVI Chief Executive Officer Ron Buck has said that he would like people to think of IVI for the way they work rather than for what they produce. Given the success of IVI's products and its innovative management practices, IVI is likely to be remembered equally for their software products, business processes, and management practices.

An Overview of Part V:

Logical Design

L ogical design is often the first phase of the systems development life cycle in which you and the user develop a concrete understanding of how the system will operate. Although previous phases may have resulted in prototypes for selected system functions, it is during logical design that you define how the system will appear to users; you describe the "look and feel" of all system inputs, outputs, and interfaces and dialogues. Also during logical design you supplement the conceptual data model from the analysis phase with new data requirements. These are identified as you design system inputs and outputs and then transform all data requirements into a new type of data model, the relational database model, which has certain desirable properties.

As depicted in Figure 1, logical design includes the following steps:

- *Designing forms (hard copy and computer displays) and reports,* which describe how data will appear to users in system inputs and outputs

- *Designing interfaces and dialogues,* which describe the pattern of interaction between system users and software

- *Designing logical databases,* which describe a standard structure for the database of a system that is easy to implement in a variety of database technologies

Logical design is tightly linked to previous systems development phases, especially analysis. For example, every system input and output you develop during logical design will appear as a data flow between a manual and automated process or between a sink/source and an automated process in data flow diagrams. In fact, you must develop a form or report design for every data flow between a user and the system. Also, the E-R diagram from conceptual data modeling is the typical starting point for logical data modeling. Because you often discover additional data requirements during logical design as the users review prototypes of the system's inputs and outputs, logical data modeling is not simply a transformation process from the E-R data model. In addition, the events on a state-transition diagram which correspond to user actions (for example, cancel an order) prescribe menu selections, buttons, or commands on computer displays.

As with most life cycle phases and subphases, the three subphases of logical design are not necessarily sequential. The logical design of system inputs and outputs, interfaces, and databases interact to identify flaws and missing elements. Thus,

Figure 1

Steps in Logical Design

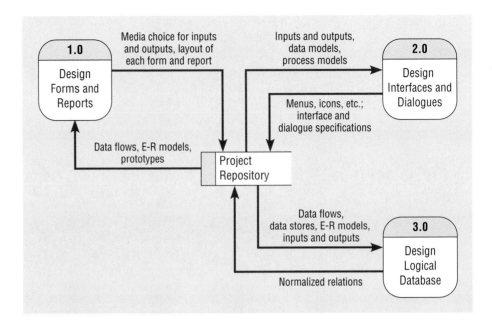

the project dictionary or CASE repository becomes an active and evolving component of system development management during logical design. It is only when each logical design element (forms and reports, interface and dialogue, and logical database) is consistent with others and satisfactory to the end user that you know logical design is complete.

In Chapter 13 you will learn the principles and guidelines that make for usable system inputs and outputs. Your overall goal in formatting the presentation of data to users should be usability: helping users of all types to use the system efficiently, accurately, and with satisfaction. The achievement of these goals can be greatly improved if you follow certain guidelines on the way to present data on business forms, visual display screens, printed documents, and other kinds of media. At a general level, Chapter 13 addresses the relative advantages of batch and on-line media and whether data are intended for users within or outside your organization.

Fortunately, there has been considerable research on how to present data to users, and Chapter 13 summarizes and illustrates the most useful of these guidelines. You will see how to place titles on system inputs and outputs, how and when to highlight data for users, how to effectively use color, how to display textual data, how to organize lists and tables for readability and use, and how to design graphical output formats. You will also understand how poor data input and output designs can bias users' understanding of data. The material in Chapter 13 applies to both graphical user interface environments, like Microsoft Windows or Apple Macintosh, as well as character-based systems.

Chapter 14 addresses principles you should follow in tying all the system inputs and outputs together into an overall pattern of interaction between users and the system. System interfaces and dialogues form a conversation that provides users access to and navigation between each system function. Chapter 14 discusses the principles of building on-line systems in which users interact directly with the application system. One choice you must make is the style of interaction: command language, menus, forms, objects or icons, and natural language. The most useful application of each interaction style is discussed. You must also choose among alternative hardware, such as a keyboard, mouse, or graphics tablet, and you will briefly review when to apply each.

As part of system interface design, there are additional choices you must make about how a user can enter, edit, and display data during data entry. The choices include providing for cursor control, editing characters, exiting system tasks, and providing help to users. Central to all interfaces is providing user feedback. You must choose the type of status information, prompting messages, and error and warning indications. Data entry must be controlled for accuracy, and we review various controls that you can include in the interface design to improve data entry accuracy. All of these interface elements must be designed during logical design.

One of the specifications that you must provide to those who will program an on-line information system is the sequence of on-line forms and when each is displayed. Guidelines for consistency, shortcuts, reversal of actions, and control will help you decide on usable dialogues. Dialogue diagramming is a graphical tool that specifies which on-line form is accessible from a given form.

Principles of good interface and dialogue design apply in almost every computer environment. Graphical user interface (GUI) environments, however, present special issues since they permit so many variations and options. Although GUI forms, reports, and dialogues are illustrated throughout Chapters 13 and 14, Chapter 14 closes with a review of special interface design topics and design guidelines for a GUI environment.

One of the core principles of the systems development process is that analysis and design is broken down into many steps, each permitting identifiable progress. Every aspect of an information system specification passes through such an incremental process, although different methodologies like the structured life cycle, prototyping, and RAD arrange these steps in different ways. Data are a core system element studied in all systems development methodologies. For data, you have seen how data flow diagrams and E-R diagrams are used to depict the data requirements of a system. Both of these diagrams are very flexible and give you considerable latitude in how you represent data. For example, you can use one or many data stores with a process in a DFD. E-R diagrams provide more structure, but an entity can still be either very detailed or rather aggregate. During the data modeling subphase of logical design, discussed in Chapter 15, you define data in its most fundamental form, called normalized data. Normalization is a well-defined method of identifying relationships between each data attribute and representing all the data so that they cannot logically be broken into more detail. The goal is to rid the data design of unwanted anomalies which would make a database susceptible to errors and inefficiencies.

Normalization is driven by the concept of functional dependencies, which are atomic relationships among data. For example, if a system knows a student's ID and it can unambiguously find the student's name but not a unique course, then the student ID and name are more fundamentally related than the student ID and course. In Chapter 15 you will learn about different normal forms, the principles behind each, and how to transform unnormalized data into well-structured data called relations. Also, you will learn how to identify functional dependencies in system inputs and outputs and how to translate these and E-R models into relations. For example, there are well-defined rules for how to transform a one-to-many relationship on an E-R diagram into a relational database model. In addition, you will learn how to merge the relations that represent several system inputs and outputs and an E-R diagram into one consolidated set of relations, which will then be the basis for physical database design.

The deliverables of logical design include detailed, functional specifications for system inputs, outputs, interfaces, dialogues, and databases. Often these elements are represented in prototypes, or working versions. Users must be active during logical design to verify that the system is usable and meets the requirements defined in analysis. The project dictionary or CASE repository is updated to include each form, report, interface, dialogue, and relation design. Due to considerable user involvement in reviewing prototypes and specifications during logical design and due to the fact that logical and physical design phases can be scheduled with considerable overlap in the project baseline plan, there often is not a formal review milestone or walkthrough after the logical design phase. If prototyping is not done, however, then you should conduct a logical design walkthrough similar to the walkthrough at the end of the analysis phase.

Each chapter in the logical design section of the book concludes with a case from Broadway Entertainment Company (BEC). These cases illustrate the development of the Customer Activity Tracking System (CATS) as specific inputs, outputs, dialogues, and database are specified.

Designing Forms and Reports

After studying this chapter, you should be able to:

■ Explain the process of designing forms and reports and the deliverables for their creation.

■ Describe and contrast the differences between internal information, external information, and turnaround documents.

■ Apply the general guidelines for formatting forms and reports.

■ Use color and know when color improves the usability of information.

■ Format text, tables, and lists effectively.

■ Describe how the formatting of information can bias users' understanding.

■ Explain how to assess usability and describe how variations in users, tasks, technology, and environmental characteristics influence the usability of forms and reports.

INTRODUCTION

In this chapter, you will learn guidelines to follow when designing forms and reports. In general, forms are used to present or collect information on a single item such as a customer, product, or event. Forms can be used for both input and output. Reports, on the other hand, are used to convey information on a collection of items. Form and report design is a key ingredient for successful systems. As users often equate the quality of a system to the quality of its input and output methods, you can see that the design process for forms and reports is an especially important activity. And since information can be collected and formatted in many ways, gaining an understanding of the dos and don'ts and the trade-offs between various formatting options is a useful skill for all systems analysts.

In the next section, the process of designing forms and reports is briefly described and we also provide guidance on the deliverables produced during this process. Guidelines for formatting information are then provided which serve as the building blocks for designing all forms and reports. This is followed by a short discussion on how information can be presented in order to bias understanding.

The final major section describes methods for assessing the usability of form and report designs.

DESIGNING FORMS AND REPORTS

This is the first chapter focusing on logical design within the systems development life cycle (see Figure 13-1). In this chapter, we describe issues related to the design of system inputs and outputs—forms and reports. In Chapter 14, we focus on the logical design of dialogues and interfaces, which are how users interact with systems. Due to the highly related topics and guidelines in these two chapters, they form one conceptual body of guidelines and illustrations that jointly guide the design of all aspects of system inputs and outputs. Chapter 15 focuses on logical database design. In each of these chapters, your objective is to gain an understanding of how you can transform information gathered during analysis into a coherent design. Although all logical design issues are related, topics discussed in this chapter on designing forms and reports are especially relative to those in the following chapter—the design of dialogues and interfaces.

  System inputs and outputs—forms and reports—were identified during requirements structuring. The kinds of forms and reports the system will handle were established as part of the design strategy formed at the end of the analysis

Figure 13-1
Systems development life cycle with logical design phase highlighted

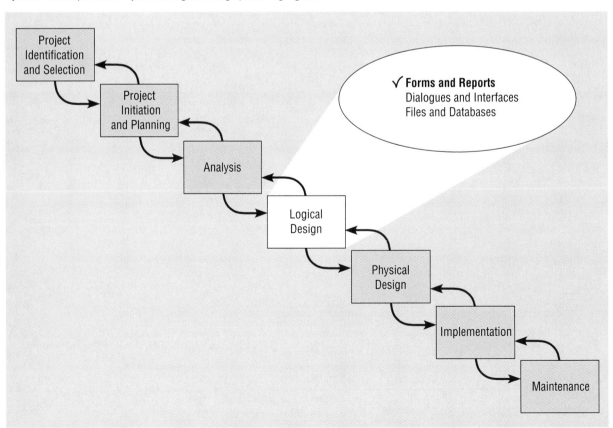

phase of the systems development process. During analysis, however, you may not have been concerned with the precise appearance of forms and reports, only with which ones needed to exist and what their contents were. You may have distributed prototypes of forms and reports which emerged during analysis as a way to confirm requirements with users. Forms and reports are integrally related to various diagrams developed during requirements structuring. For example, every input form will be associated with a data flow entering a process on a DFD, and every output form or report will be a data flow produced by a process on a DFD. This means that the contents of a form or report correspond to the data elements contained in the associated data flow. Further, the data on all forms and reports must consist of data elements in data stores and on the E-R data model for the application, or must be computed from these data elements. (In rare instances, data simply go from system input to system output without being stored within the system.) It is common that, as you design forms and reports, you will discover flaws in DFDs and E-R diagrams; the project dictionary or CASE tool repository, therefore, continues to be the central and constantly updated source of all project information.

If you are unfamiliar with computer-based information systems, it will be helpful to clarify exactly what we mean by a form or report. A **form** is a business document containing some pre-defined data and often includes some areas where additional data are to be filled in. Most forms have a stylized format and are usually not in a simple row and column format. Examples of business forms are product order forms, employment applications, and class registration sheets. Traditionally, forms have been displayed on a paper medium, but today video display technology allows us to duplicate the layout of almost any printed form, including an organizational logo or any graphic, on a video display terminal. Forms displayed on a video display may be used for data display or data entry. Additional examples of forms are an electronic spreadsheet, computer sign-on or menu, and an ATM transaction layout.

Form: A business document that contains some pre-defined data and may include some areas where additional data are to be filled in. An instance of a form is typically based on one database record.

A **report** is a business document containing only pre-defined data; it is a passive document used solely for reading or viewing. Examples of reports are invoices, weekly sales summaries by region and salesperson, and a pie chart of population by age categories. We usually think of a report as printed on paper, but it may be printed to a computer file, a visual display screen, or some other medium such as microfilm. Often a report has rows and columns of data, but a report may consist of any format—for example, mailing labels. Frequently, the differences between a form and a report are subtle. A report is only for reading and often contains data about multiple unrelated records in a computer file. On the other hand, a form typically contains data from only one record or is, at least, based on one record, such as data about one customer, one order, or one student. The guidelines for the design of forms and reports are very similar.

Report: A business document that contains only pre-defined data; it is a passive document used solely for reading or viewing. A report typically contains data from many unrelated records or transactions.

The Process of Designing Forms and Reports

Designing forms and reports is a user-focused activity that typically follows a prototyping approach (see Figure 1-7). First, you must gain an understanding of the intended user and task objectives by collecting initial requirements during requirements determination. During this process, several questions must be answered. These questions attempt to answer the "who, what, when, where, and how" related to the creation of all forms or reports (see Table 13-1). Gaining an understanding of these questions is a required first step in the creation of any form or report.

For example, understanding who the users are—their skills and abilities—will greatly enhance your ability to create an effective design. In other words, are your users experienced computer users or novices? What is their educational level,

TABLE 13-1 Fundamental Questions when Designing Forms and Reports

1. Who will use the form or report?
2. What is the purpose of the form or report?
3. When is the form or report needed and used?
4. Where does the form or report need to be delivered and used?
5. How many people need to use or view the form or report?

business background, and task-relevant knowledge? Answers to these questions will provide guidance for both the format and content of your designs. Also, what is the purpose of the form or report? What task will users be performing and what information is needed to complete this task? Other questions are also important to consider. Where will the users be when performing this task? Will users have access to on-line systems or will they be in the field? Also, how many people will need to use this form or report? If, for example, a report is being produced for a single user, the design requirements and usability assessment will be relatively simple. A design for a larger audience, however, may need to go through a more extensive requirements collection and usability assessment process.

After collecting the initial requirements, you structure and refine this information into an initial prototype. Structuring and refining the requirements are completed independently of the users, although you may need to occasionally contact users in order to clarify some issue overlooked during analysis. Finally, you ask users to review and evaluate the prototype. After reviewing the prototype, users may accept the design or request that changes be made. If changes are needed, you will repeat the construction-evaluate-refinement cycle until the design is accepted. Usually, several iterations of this cycle occur during the design of a single form or report. As with any prototyping process, you should make sure that these iterations occur rapidly in order to gain the greatest benefits from this design approach.

The initial prototype may be constructed in numerous environments. The obvious choice is to use a CASE tool or the standard development tools used within your organization. Often, initial prototypes are simply mock screens that are not working modules or systems. Mock screens can be produced from a word processor, computer graphics design package, or electronic spreadsheet. It is important to remember that the focus of this phase within the SDLC is on the *design*—content and layout. How specific forms or reports are implemented (for example, the programming language or screen painter code) is left to later phases. Nonetheless, tools for designing forms and reports are rapidly evolving. In the past, inputs and outputs of all types were typically designed by hand on a coding or layout sheet. For example, Figure 13-2 shows the layout of a disk operating system (DOS) shell program using a coding sheet.

Although coding sheets are still used, their importance has diminished due to significant changes in system operating environments and the evolution of automated design tools. Prior to the creation of graphical operating environments, for example, analysts designed many inputs and outputs that were 80 columns (characters) by 25 rows, the standard dimensions for most video displays. These limits in screen dimensions are radically different in graphical operating environments such as Microsoft's Windows® where font sizes and screen dimensions can often be changed from user to user. Consequently, the creation of new tools and develop-

Figure 13-2
Designing a screen layout using a coding sheet

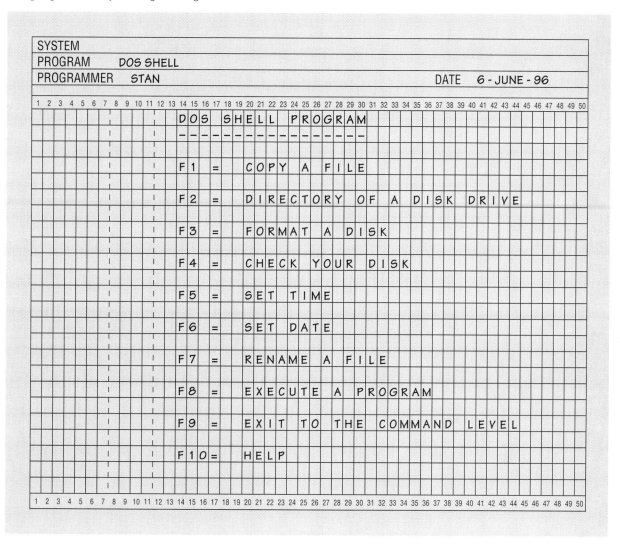

ment environments was needed to help analysts and programmers develop these graphical and flexible designs. Figure 13-3 shows an example of the same DOS shell screen as designed in Microsoft's Visual Basic®. Note the variety of fonts, sizes, and highlighting that was used. Needless to say, on-line graphical tools for designing forms and reports are rapidly becoming the de facto standard in most professional development organizations.

Deliverables and Outcomes

Each SDLC phase helps you to construct a system. In order to move from phase to phase, each phase produces some type of deliverable that is used in a later phase or activity. For example, within the project initiation and planning phase of the SDLC, the Baseline Project Plan serves as input to many subsequent SDLC activities. In the case of designing forms and reports, design specifications are the major

Figure 13-3
Designing a form using
Microsoft's Visual Basic®

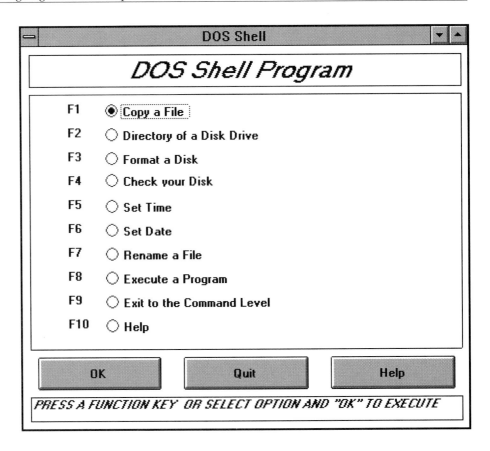

deliverables and are inputs to the system implementation phase. Design specifications have three sections:

1. Narrative overview

2. Sample design

3. Testing and usability assessment

The first section of a design specification contains a general overview of the characteristics of the target users, tasks, system, and environmental factors in which the form or report will be used. The purpose is to explain to those who will actually develop the final form why this form exists and how it will be used so that they can make the appropriate implementation decisions. In this section, you list general information and the assumptions that helped shape the design. For example, Figure 13-4 shows an excerpt of a design specification for a Customer Account Status form for Pine Valley Furniture. The first section of the specification, Figure 13-4a, provides a narrative overview containing the relevant information to developing and using the form within PVF. The overview explains the tasks supported by the form, where and when the form is used, characteristics of the people using the form, the technology delivering the form, and other pertinent information. For example, if the form is delivered on a visual display terminal, this section would describe the capabilities of this device, such as whether it has a touch screen and whether color and a mouse are available.

In the second section of the specification, Figure 13-4b, a sample design of the form is shown. This design may be hand-drawn using a coding sheet although, in most instances, it is developed using CASE or standard development tools. Using actual development tools allows the design to be more thoroughly tested and as-

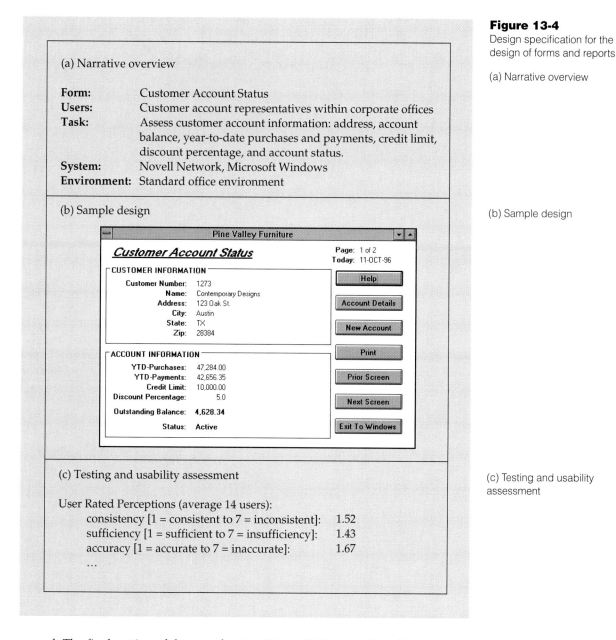

Figure 13-4
Design specification for the design of forms and reports

(a) Narrative overview

(b) Sample design

(c) Testing and usability assessment

sessed. The final section of the specification, Figure 13-4c, provides all testing and usability assessment information. Procedures for assessing designs are described later in the chapter. Some specification information may be irrelevant when designing some forms and reports. For example, the design of a simple Yes/No selection form may be so straightforward that no usability assessment is needed. Also, much of the narrative overview may be unnecessary unless intended to highlight some exception that must be considered during implementation.

FORMATTING FORMS AND REPORTS

A wide variety of information can be provided to users of information systems and, as technology continues to evolve, a greater variety of data types will be used. Unfortunately, a definitive set of rules for delivering every type of information to users has yet to be defined and these rules are continuously evolving along with the rapid changes in technology. Nonetheless, a large body of human-computer interaction research has provided numerous general guidelines for formatting information. Many of these guidelines will undoubtedly apply to the formatting of all evolving information types on yet-to-be-determined devices. Keep in mind that the mainstay of designing usable forms and reports requires your active interaction with users. If this single and fundamental activity occurs, it is likely that you will create effective designs.

Types of Information

Within the context of building information systems, there are two general types of information: internal information and external information. **Internal information** refers to information that is collected, generated, or consumed within an organization. Examples include lengthy reports summarizing business activity, such as salesperson or machine productivity, production scrap rates, forecasting trends, and so on. Internal information can also consist of simple informational reports summarizing daily activities, such as a sales contact report that a salesperson might produce for his or her boss or the forms used to collect daily travel expenditures.

Internal information: Information that is collected, generated, or consumed within an organization.

External information refers to information that is collected from or created for customers, suppliers, stockholders, competitors, or the general public. Examples include receipts, invoices, packing slips, checks, quarterly reports, product documentation, and so on. Like internal information, external information can be delivered in a variety of formats and media. For example, an automated teller machine (ATM) produces external information on both paper (e.g., a transaction receipt) and on a computer display. Many organizations have very specific standards for the formatting of external information, for example, to convey a consistent image to their customers. These standards cover use of company logos, colors, fonts, and paper size and quality.

External information: Information that is collected from or created for individuals and groups external to an organization.

A hybrid type of information that is both internal and external is called a **turnaround document.** A turnaround document is produced by a system as an output that may also "turn around" and return to the organization as an input. An example of a turnaround document is a warranty card for a Pine Valley Furniture product, shown in Figure 13-5. The bottom half of the card explains the product warranty to the customer. The top portion is a detachable registration card that must be returned to Pine Valley to register a purchase. This form has pre-printed system-generated information that a customer must verify and return to PVF. The customer also adds new information (e.g., name and address). The verified and new information then serves as an input to PVF's customer warranty tracking system. Other common examples of turnaround documents are credit card and utility bills.

Turnaround document: Information that is delivered to an external customer as an output that can be returned to provide new information as an input to an information system.

Some turnaround documents within high-volume transaction-based systems can contain computer-readable information such as a bar code (Figure 13-6a) or computer-readable text and numbers that are read using an optical or magnetic ink character reader. For example, the bottom of most bank checks contain computer-readable account information (Figure 13-6b) that is read by an optical character reader as input into a check-processing system. The keyless processing of computer-readable information not only speeds processing and customer service but also significantly reduces data entry errors.

You can format internal and external information in a variety of ways. For example, when producing a document using a pre-printed form, only the unique data for the form is printed onto the paper. This is similar to how information can be typed onto a pre-printed form using a typewriter (see Figure 13-7 for an example of a pre-printed form). Alternatively, laser and page printers now make it possible to

Figure 13-5

An example of a turnaround document (Pine Valley Furniture)

Pine Valley Furniture
Product Registration Warranty Card

URGENT
RETURN CARD
WITHIN 10 DAYS

Product Model: CAB-OO3-A
Registration Number: 95-17-3A-007
Date of Purchase: 15-DEC-96

Please indicate (√) type of use:
☐ Personal Use/Gift
☐ Commercial Use

NEW ADDRESS/OWNER:

Last Name   MI   First Name

Address

City   State   Zip

PURCHASER ON RECORD:
Larry Mueller
701 S. 6th East
Missoula, MT 59801

Tear here. Your name will not be added to other mailing lists.

Protect Your Investment!

Congratulations on choosing a Pine Valley Furniture Product. It's an intelligent decision that's sure to reward you for many years to come.

To receive all the privileges your purchase entitles you to, please be sure to complete and mail your Product Registration Card to us at once.

Return the attached card within 10 days to ensure:
√ Warranty Confirmation
√ Owner Verification
√ Model Registration

Warranty is outlined on the back of this card.

generate the headings and labels of a form or report along with the data when displayed or printed. This alternative allows you to print on normal paper stock, and special forms do not have to be kept in inventory only to become obsolete when requirements or design change. Different methods of producing standard forms and reports will, of course, influence the cost of producing information. Cost concerns are beyond the scope of our discussion, but these costs are primarily influenced by the type of media used. The types of media for displaying system-generated information is also beyond the scope of our discussion; see Alter (1995) for a review of this topic.

A variety of media can be used to deliver system information. From a design perspective, you must determine whether users will interact directly with the system—the information will be available on-line—or whether information will be collected and provided on paper—that is, on hard copy. Another distinction, and

Figure 13-6
Keyless data entry options

(a) Bar code

3700 62811

(b) Optical character
recognition

JOHN DOE   0101
123 MAIN STREET
ANYWHERE, USA 00000   19_____   12-345
                                                                   00

PAY TO THE
ORDER OF_____ $_____

_____ D O L L A R S

First National Bank
Downtown Branch
P.O. Box 456
Anywhere, USA 00000
MEMO_____

the one we are focusing on here, is whether system information is processed on-line or in batch.

On-line processing refers to the collection and delivery of the most recent available information, typically through an on-line workstation. On-line processing is preferred under the following conditions:

On-line processing: The collection and delivery of the most recent available information, typically through an on-line workstation.

1. Access to or capture of the information occurs randomly.

2. The format and type of information are not consistent (for example, an ad hoc query).

3. Information is continuously changing and the most current information is needed for proper processing and decision making.

4. Users are in locations where there is easy access to an information system or a remotely connected terminal.

Any one or all of these criteria suggest that information be processed on-line. There are many other factors—user characteristics, task, technological, and physical environment—that can also influence your design decisions (see below).

Batch processing, on the other hand, refers to the input or output of system information on a pre-determined and specific time interval. This means that information collected or generated through batch processing has a shelf life of a limited

Batch processing: Information that is collected or generated at some pre-determined time interval that can be accessed via hard copy or on-line devices.

Figure 13-7
Example of a pre-printed form

Application For Employment	Pine Valley Furniture

Personal Information	
Name:	Date:
Social Security Number:	
Home Address:	
City, State, Zip	
Home Phone:	Business Phone:
U.S. Citizen?	If Not Give Visa No. & Expiration:

Position Applying For	
Title:	Salary Desired:
Referred By:	Date Available:

Education	
High School (Name, City, State):	
Graduation Date:	
Business or Technical School:	
Dates Attended:	Degree, Major:
Undergraduate College:	
Dates Attended:	Degree, Major:
Graduate School:	
Dates Attended:	Degree, Major:

References

Form #2019
Last Revised: 9/15/95

time period whereas on-line processing always represents the most recent available information. Note, however, that batch processing can be collected and delivered both on-line or on hard copy. For example, when viewing an *on-line* batch processing report, users would access a static textual or graphical report from a terminal or workstation. In this case, the display is on-line, but the data are not current. Alternatively, a *hard copy* batch processing report would simply be a system output produced on some type of media such as a printer. For batch output, it is important to know both the date (and possibly time) the output is produced as well as when the data were current. For example, a system might produce a month-end sales report on the 10th of the following month.

Batch processing is preferred under the following conditions:

1. Access to information occurs regularly.

2. The format and type of information are consistent.

3. Information is stable for some time period and/or the users' informational needs do not require the most up-to-date information for effective decision making or task performance.

4. Users are in locations where access to an on-line information system or remotely connected terminal is not possible or extremely difficult.

Other factors to consider when choosing on-line versus batch processing relate to information control—assuring that only approved users have access to specific information. The most secure method of delivering information is using on-line delivery that requires a password for viewing. For hard copy information, routing slips and internal and external mail systems are used to help control unauthorized access to confidential outputs. Now that you understand how information can be processed and delivered, we will next focus on information formatting issues.

General Formatting Guidelines

Over the past several years, industry and academic researchers have spent considerable effort investigating how information formatting influences individual task performance and perceptions of usability. Through this work, several guidelines for formatting information have emerged (see Table 13-2). These guidelines reflect some of the general truths that apply to the formatting of most types of information (for more information, the interested reader should see the books by Shneiderman, 1992; Carroll, 1991; and Norman, 1991).

The differences between a well-designed form or report and one that is poorly designed will often be obvious to you. For example, Figure 13-8a shows a poorly designed form for viewing a current account balance for a PVF customer. Figure 13-8b is a better design incorporating several general guidelines from Table 13-2.

The first major difference between the two forms has to do with the title. The title on Figure 13-8a is ambiguous whereas the title on Figure 13-8b clearly and specifically describes the contents of the form. The form in Figure 13-8b also includes the date on which the form was generated so that, if printed, it will be clear to the reader when this occurred. Figure 13-8a displays information that is extraneous to the intent of the form—viewing the current account balance—and provides information that is not in the most useful format for the user. For example, Figure 13-8a provides all account transactions and a summary of year-to-date purchases and payments. The form does not, however, provide the current outstanding balance of the account without making a manual calculation. The layout of information

TABLE 13-2 General Guidelines for the Design of Forms and Reports
(Adapted from Dumas, 1988)

Meaningful Titles:

Clear and specific titles describing content and use of form or report

Revision date or code to distinguish a form or report from prior versions

Current date which identifies when the form or report was generated

Valid date which identifies on what date (or time) the data in the form or report were accurate

Meaningful Information:

Only needed information should be displayed

Information should be provided in a manner that is usable without modification

Balanced Layout:

Information should be balanced on the screen or page

Adequate spacing and margins should be used

All data and entry fields should be clearly labeled

Easy Navigation:

Clearly show how to move forward and backward

Clearly show where you are (e.g., page 1 of 3)

Notify user when on the last page of a multi-paged sequence

between the two forms also varies in balance and information density. Gaining an understanding of the skills of the intended system users and the tasks they will be performing is invaluable when constructing a form or report. By following these general guidelines, your chances of creating effective forms and reports will be enhanced. In the next sections we will discuss specific guidelines for highlighting information, using color, displaying text, and presenting numeric tables and lists.

Highlighting Information

As display technologies continue to improve, there will be a greater variety of methods available to you for highlighting information. Table 13-3 provides a list of the most commonly used methods for highlighting information. Given this vast array of options, it is more important than ever to consider how highlighting can be used to enhance an output and not prove a distraction. In general, highlighting should be used sparingly to draw the user to or away from certain information and to group together related information. There are several situations when highlighting can be a valuable technique for conveying special information:

- Notifying users of errors in data entry or processing

- Providing warnings to users regarding possible problems such as unusual data values or an unavailable device

- Drawing attention to keywords, commands, high priority messages, and data that have changed or gone outside normal operating ranges

Figure 13-8

Contrasting customer
information forms (Pine Valley
Furniture)

(a) Poorly designed form

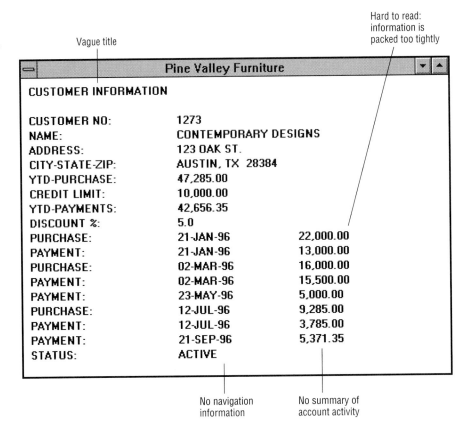

(b) Improved design for form

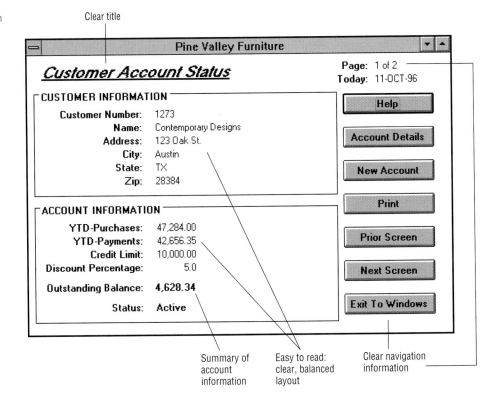

TABLE 13-3 Methods of Highlighting
(Adapted from Dumas, 1988)

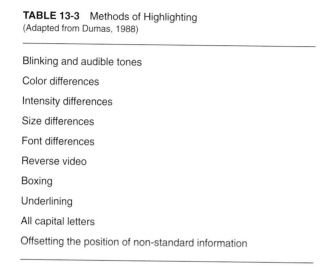

Blinking and audible tones

Color differences

Intensity differences

Size differences

Font differences

Reverse video

Boxing

Underlining

All capital letters

Offsetting the position of non-standard information

Additionally, many highlighting techniques can be used singularly or in tandem, depending upon the level of emphasis desired by the designer. Figure 13-9 shows a form where several types of highlighting are used. In this example, boxes clarify different categories of data; capital letters and different fonts distinguish labels from actual data; and intensity is used to draw attention to important data.

Much research has focused on the effects of varying highlighting techniques on task performance and user perceptions. A general guideline resulting from this research is that highlighting should be used conservatively. For example, blinking

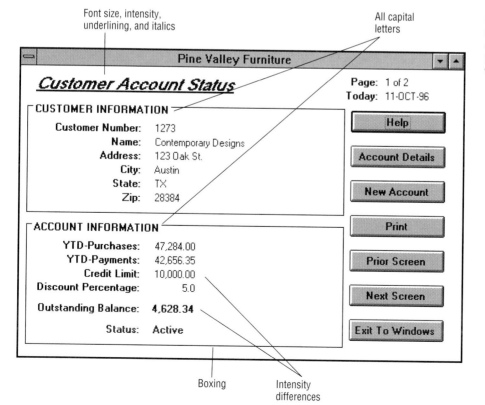

Figure 13-9
Customer account status display using various highlighting techniques (Pine Valley Furniture)

and audible tones should only be used to highlight critical information requiring an immediate response from the user. Once a response is made, these highlights should be turned off. Additionally, highlighting methods should be consistently used and selected based upon the level of importance of the emphasized information. It is also important to examine how a particular highlighting method appears on all possible output devices that could be used with the system. For example, some color combinations may convey appropriate information on one display configuration but wash out and reduce legibility on another.

Recent advances in the development of graphical operating environments have provided designers with some standard highlighting guidelines. However, these guidelines are often quite vague and are continuously evolving, leaving a great deal of control in the hands of the systems developer. Therefore, in order for organizations to realize the benefits of using standard graphical operating environments—such as reduced user training time and interoperability among systems—you must be disciplined in how you use highlighting.

Color versus No-Color

Color is a powerful tool for the designer in influencing the usability of a system. When applied appropriately, color provides many potential benefits to forms and reports, summarized in Table 13-4. As the use of color displays became widely available during the 1980s, a substantial amount of color versus no-color research was conducted. The objective of this research was to gain a better understanding of the effects of color on human task performance (see, for example, Benbasat, Dexter, and Todd, 1986).

The general findings from this research were that the use of color had positive effects on user task performance and perceptions when the user was under time constraints for the completion of a task. Color was also beneficial for gaining greater understanding from a display or chart. An important conclusion from this research

TABLE 13-4   Benefits and Problems from Using Color
(Adapted from Shneiderman, 1992)

Benefits from Using Color:

Soothes or strikes the eye.

Accents an uninteresting display.

Facilitates subtle discriminations in complex displays.

Emphasizes the logical organization of information.

Draws attention to warnings.

Evokes more emotional reactions.

Problems from Using Color:

Color pairings may wash out or cause problems for some users (e.g., color blindness).

Resolution may degrade with different displays.

Color fidelity may degrade on different displays.

Printing or conversion to other media may not easily translate.

was that color was *not* universally better than no-color. *The benefits of color only seem to apply if the information is first provided to the user in the most appropriate presentation format.* That is, if information is most effectively displayed in a bar chart, color can be used to enhance or supplement the display. If information is displayed in an inappropriate format, color has little or no effect on improving understanding or task performance.

There are also several problems associated with using color, also summarized in Table 13-4. Most of these dangers are related more to the technical capabilities of the display and hard copy devices than misuse. However, color blindness is a particular user issue that is often overlooked in the design of systems. Shneiderman (1992) reports that approximately eight percent of the males in the European and North American communities have some form of color blindness. He suggests that you first design video displays for monochrome and allow color (or better yet, a flexible palette of colors) to be a user-activated option. He also suggests that you limit the number of colors and where they are applied, using color primarily as a tool to assist in the highlighting and formatting of information.

Displaying Text

In business-related systems, textual output is becoming increasingly important as text-based applications such as electronic mail, bulletin boards, and information services (e.g., Dow Jones) are more widely used. The display and formatting of system help information, which often contains lengthy textual descriptions and examples, is one example of textual data which can benefit from following a few simple guidelines that have emerged from past research (see Table 13-5). The first guideline is simple: you should display text using common writing conventions such as mixed upper and lower case and appropriate punctuation. If space permits, text should be double-spaced and, minimally, a blank line should be placed between each paragraph. You should also left-justify text with a ragged right margin—research shows that a ragged right margin makes it easier to find the next line of text when reading than when text is both left- and right-justified.

When displaying textual information, you should also be careful not to hyphenate words between lines or use obscure abbreviations and acronyms. Users may not know whether the hyphen is a significant character if it is used to continue words across lines. Information and terminology that are not widely understood by the intended users may significantly influence the usability of the system. Thus, you should use abbreviations and acronyms only if they are significantly shorter than the full text and are commonly known by the intended system users. Figure 13-10 shows

TABLE 13-5 Guidelines for Displaying Text
(Adapted from Dumas, 1988)

Case	Display text in mixed upper and lower case and use conventional punctuation.
Spacing	Use double spacing if space permits. If not, place a blank line between paragraphs.
Justification	Left-justify text and leave a ragged right margin.
Hyphenation	Do not hyphenate words between lines.
Abbreviations	Use abbreviations and acronyms only when they are widely understood by users and are significantly shorter than the full text.

Figure 13-10

Contrasting the display of textual help information

(a) Poorly designed form

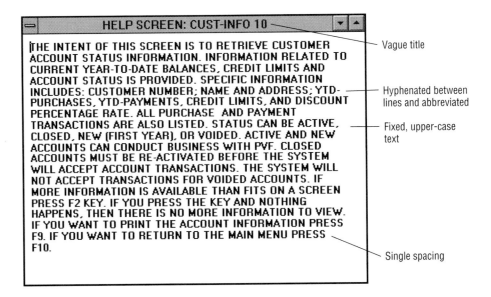

(b) Improved design for form

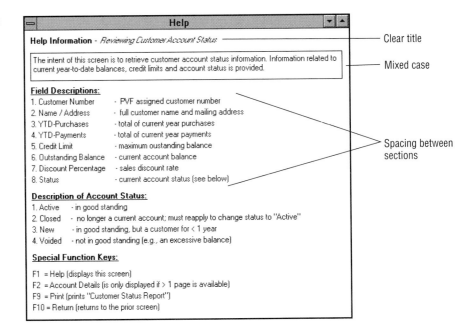

two versions of a help screen from an application systems at PVF. Figure 13-10a shows many violations of the general guidelines for displaying text whereas 13-10b shows the same information but follows the general guidelines for displaying text. Formatting guidelines for the *entry* of text and alphanumeric data is also a very important topic. These guidelines are presented in Chapter 14, *Designing Interfaces and Dialogues,* where we focus on issues of human-computer interaction.

Designing Tables and Lists

Unlike textual information, where context and meaning are significantly derived through reading, the context and meaning of tables and lists are significantly derived from the format of the information. Consequently, the usability of information displayed in tables and alphanumeric lists is likely to be much more influenced by effective layout than most other types of information display. As with the display of textual information, tables and lists can also be greatly enhanced by following a few simple guidelines. These are summarized in Table 13-6.

Figure 13-11 displays two versions of a form design from a Pine Valley Furniture application system that displays customer year-to-date transaction information in a table format. Figure 13-11a displays the information without consideration of the guidelines presented in Table 13-6 and Figure 13-11b displays this information after consideration of these guidelines.

One key distinction between these two display forms relates to labeling. The information reported in Figure 13-11b has meaningful labels that more clearly stand out as labels compared to the display in Figure 13-11a. Transactions are sorted by date and numeric data are right-justified and aligned by decimal point in Figure 13-11b, which helps to facilitate scanning. Adequate space is left between columns, and blank lines are inserted after every five rows in Figure 13-11b to help ease the finding and reading of information. Such spacing also provides room for users to annotate data that catch their attention. Taken together, using a few simple rules significantly enhanced the layout of the information for the user.

Most of the guidelines in Table 13-6 are rather obvious, but this and other tables serve as a quick reference to validate that your form and report designs will be usable. It is beyond our scope here to discuss each of these guidelines, but you should read each carefully and think about why each is appropriate. For example,

TABLE 13-6   General Guidelines for Displaying Tables and Lists
(Adapted from Dumas, 1988)

Use meaningful labels:

All columns and rows should have meaningful labels.

Labels should be separated from other information by using highlighting.

Re-display labels when the data extend beyond a single screen or page.

Formatting columns, rows, and text:

Sort in a meaningful order (e.g., ascending, descending, or alphabetic).

Place a blank line between every five rows in long columns.

Similar information displayed in multiple columns should be sorted vertically (that is, read from top to bottom, not left to right).

Columns should have at least two spaces between them.

Allow white space on printed reports for user to write notes.

Use a single typeface, except for emphasis.

Use same family of typefaces within and across displays and reports.

Avoid overly fancy fonts.

Formatting numeric, textual, and alphanumeric data:

Right-justify *numeric data* and align columns by decimal points or other delimiter.

Left-justify *textual data*. Use short line length, usually 30–40 characters per line (this is what newspapers use, and it is easier to speed read).

Break long sequences of *alphanumeric data* into small groups of three to four characters each.

Figure 13-11

Contrasting the display of
tables and lists (Pine Valley
Furniture)

(a) Poorly designed form

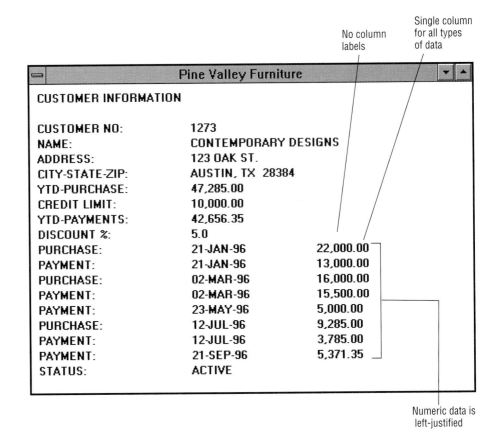

why should labels be repeated on subsequent screens and pages (the third guideline in Table 13-6)? One explanation is that pages may be separated or copied and the original labels will no longer be readily accessible to the reader of the data. Why should long alphanumeric data (see the last guideline) be broken into small groups? (If you have a credit card or bank check, look at how your account number is displayed.) Two reasons are that the characters will be easier to remember as you read and type them, and there will be a natural and consistent place to pause when you speak them over the phone; for example, when you are placing a phone order for products in a catalog.

A fundamental choice when you design the display of numeric information is to determine whether a table or a graph should be used. A considerable amount of research focusing on this topic has been conducted (see, for example, Jarvenpaa and Dickson, 1988, for very specific guidelines on the use of tables and graphs). In general, this research has found that tables are best when the user's task is related to finding an individual data value from a larger data set whereas line and bar graphs are more appropriate for gaining an understanding of data changes over time (see Table 13-7). For example, if the marketing manager for Pine Valley Furniture needed to review the actual sales of a particular salesperson for a particular quarter, a tabular report like the one shown in Figure 13-12 would be most useful. This report has been annotated to emphasize good report design practices. The report has both a printed date as well as a clear indication, as part of the report title, of the period over which the data apply. There is also sufficient white space to provide some room for users to add personal comments and observations. Often, to provide such white space, a report must be printed in landscape, rather than portrait, orientation. Alternatively, if the marketing manager wished to compare the

Clear and separate
column labels for
each data type

(b) Improved design for form

Pine Valley Furniture			
Pine Valley Furniture		Page:	2 of 2
Detail Customer Account Information		Today:	11-Oct-96
Year-to-Date Summary			

Customer Number: 1273
Name: Contemporary Designs

DATE	PURCHASE	PAYMENT	CURRENT BALANCE
01-Jan-96			0.00
21-Jan-96	22,000.00		22,000.00
21-Jan-96		13,000.00	9,000.00
02-Mar-96	16,000.00		25,000.00
02-Mar-96		15,500.00	9,500.00
23-May-96		5,000.00	4,500.00
12-Jul-96	9,285.00		13,785.00
12-Jul-96		3,785.00	10,000.00
21-Jul-96		5,371.65	4,628.35
YTD-SUMMARY	47,285.00	42,656.65	4,628.35

Help	Prior Screen	Exit to Windows

Numeric data is
right-justified

TABLE 13-7 Guidelines for Selecting Tables versus Graphs
(Adapted from Jarvenpaa and Dickson, 1988)

Use Tables for

Reading individual data values

Use Graphs for

Providing a quick summary of data

Detecting trends over time

Comparing points and patterns of different variables

Forecasting activities

Reporting vast amounts of information when relatively simple
impressions are to be drawn

overall sales performance of each sales region, a line or bar graph would be more
appropriate (see Figure 13-13). As with other formatting considerations, the key determination as to when you should select a table or a graph is the task being
performed by the user.

Figure 13-12

Tabular report illustrating numerous design guidelines (Pine Valley Furniture)

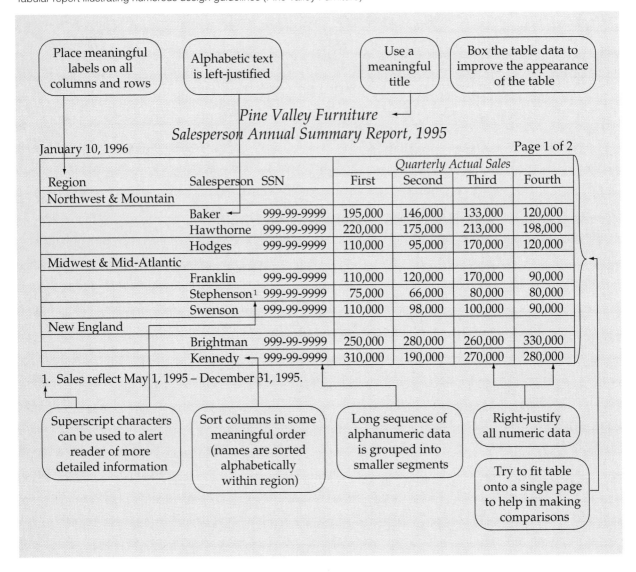

Place meaningful labels on all columns and rows

Alphabetic text is left-justified

Use a meaningful title

Box the table data to improve the appearance of the table

Pine Valley Furniture
Salesperson Annual Summary Report, 1995

January 10, 1996

Page 1 of 2

Region	Salesperson	SSN	Quarterly Actual Sales			
			First	Second	Third	Fourth
Northwest & Mountain						
	Baker	999-99-9999	195,000	146,000	133,000	120,000
	Hawthorne	999-99-9999	220,000	175,000	213,000	198,000
	Hodges	999-99-9999	110,000	95,000	170,000	120,000
Midwest & Mid-Atlantic						
	Franklin	999-99-9999	110,000	120,000	170,000	90,000
	Stephenson[1]	999-99-9999	75,000	66,000	80,000	80,000
	Swenson	999-99-9999	110,000	98,000	100,000	90,000
New England						
	Brightman	999-99-9999	250,000	280,000	260,000	330,000
	Kennedy	999-99-9999	310,000	190,000	270,000	280,000

1. Sales reflect May 1, 1995 – December 31, 1995.

Superscript characters can be used to alert reader of more detailed information

Sort columns in some meaningful order (names are sorted alphabetically within region)

Long sequence of alphanumeric data is grouped into smaller segments

Right-justify all numeric data

Try to fit table onto a single page to help in making comparisons

Figure 13-13

Graphs for comparison

(a) Line graph

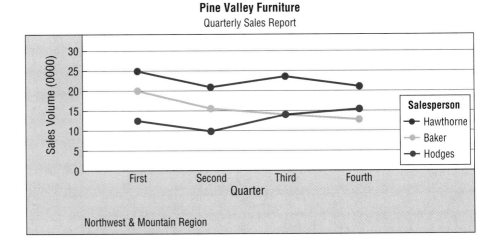

Pine Valley Furniture
Quarterly Sales Report

Northwest & Mountain Region

FORMATTING INFORMATION TO AVOID BIAS

A goal of this chapter is to help you understand how to select appropriate formats for conveying information on forms and reports. Unfortunately, this understanding is not widely held. For example, there are countless examples of unintended (or intended) biasing of information displayed on television or reported in the popular press. For example, in many automobile advertisements, bar charts are used with biased scales to project greater (or fewer) differences (depending upon their objective!) among products. As a designer of information systems, how you format information can have a tremendous influence on how information is perceived by users. In this section, we discuss information bias as it relates to the design of forms and reports. For a more general and somewhat humorous examination of information bias, see Huff's classic book *How to Lie with Statistics*.

There are several ways in which the format of output can bias user perceptions. Information biasing is often unavoidable and not necessarily a bad thing. It is simply the result of how the format of data can have intentional or unintentional influence on users. As a designer of system output, your role is to understand the sources of bias so that you can design outputs to either minimize their effects on users or to provide users with the ability to review information in order to gain an objective understanding of their data.

Bias can be introduced by both the type and format of the information provided. Sources of bias can include

1. Providing information (table or graph) that does not match the user's task

2. Providing charts that contain too many items

3. Using colors and highlighting improperly

4. Providing charts that use improper scaling

Providing tables and graphs that do not directly match the user's task is a common source of bias. An example of this is to provide a user with information in a table that might best be delivered as a graph. A more subtle example of introducing bias would be through the use of sorting. For example, suppose that a manager makes customer support assignments for support personnel based upon criteria such as hours worked or years of experience. If a report is produced listing support personnel sorted on some other criteria (for example, alphabetically by

(b) Bar graph

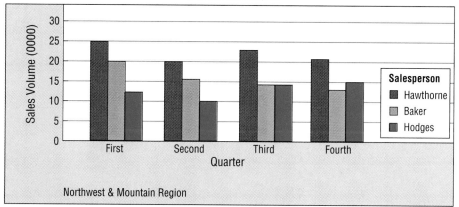

last name) it may be difficult for the manager to make the proper assignments.

A related source of bias is to include too much information on a graph or report so that finding the important information is too difficult. For example, too many lines, bars, data items, labels, and fonts can obscure the task-relevant information from the user. In these instances, users can experience information overload which often puts unwanted variance into a user's task performance. On the other hand, information can be summarized to such a point that distinctions between items are impossible to make or are irrelevant. Additionally, the improper or inconsistent use of color and highlighting can draw a user's attention away from important task-relevant information.

Another source of bias that is regularly found in output from the mass media, as well as from information systems, relates to information scaling. Information scaling refers to the range of values reported on a graph or table. For example, the scales for graphs should typically begin at zero and end with some value near, but greater than, the maximum value on the graph (see Figure 13-14a). To bias users' perceptions that there are greater differences between items on a graph than there really are can be done by setting the bottom of the scale to greater than zero (see Figure 13-14b). Alternatively, less variance between items can be implied by setting the top end of the scale much greater than any of the items' top end on the graph maximum value (see Figure 13-14c). Although the same data were used to construct the three graphs in the figure, the use of different scales suggests very different meanings to users. Making unusual adjustments to the high and low scales for graphs and tables is a commonly used biasing technique when advertising many products. Such techniques can also be used to bias system output.

Figure 13-14

How scaling can bias the meaning of information

(a) No bias in scaling

Normal Scale

Bottom of Scale = 0

Top of Scale = near maximum value

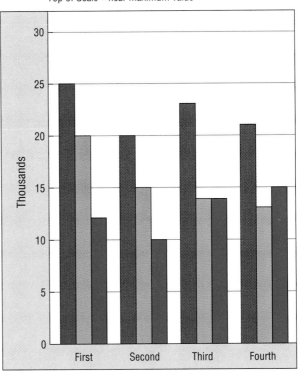

Biased Scale
Bottom of Scale = 10,000
Top of Scale = near maximum value

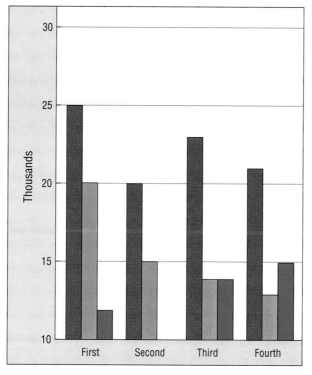

Biased Scale
Bottom of Scale = 0
Top of Scale = too much greater than maximum value

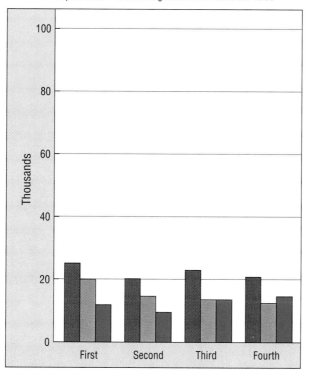

ASSESSING USABILITY

Usability: An overall evaluation of how a system performs in supporting a particular user for a particular task.

There are many factors to consider when you design forms and reports. The objective for designing forms, reports, and all human-computer interactions is **usability.** Usability typically refers to the following three characteristics:

1. *Speed.* Can you complete a task efficiently?

2. *Accuracy.* Does the output provide what you expect?

3. *Satisfaction.* Do you like using the output?

In other words, usability means that your designs assist, not hinder, user performance. In the remainder of this section, we describe numerous factors that influence usability and several techniques for assessing the usability of a design.

Usability Success Factors

Research and practical experience have found that design consistency is the key ingredient in designing usable systems. Consistency significantly influences users' ability to gain proficiency when interacting with a system. Consistency means, for example, that titles, error messages, menu options, and other design elements appear in the same place and look the same on all forms and reports. Consistency also means that the same form of highlighting has the same meaning each time it is used and that the system will respond in roughly the same amount of time each time a particular operation is performed. Other important factors found to be important include efficiency, ease (or understandability), format, and flexibility. Each of these usability factors, with associated guidelines, is described in more detail in Table 13-8.

TABLE 13-8   General Design Guidelines for Usability of Forms and Reports

Usability Factor	Guidelines for Achievement of Usability
Consistency	Consistent use of terminology, abbreviations, formatting, titles, and navigation within and across outputs. Consistent response time each time a function is performed.
Efficiency	Formatting should be designed with an understanding of the task being performed and the intended user. Text and data should be aligned and sorted for efficient navigation and entry. Entry of data should be avoided where possible (e.g., computing rather than entering totals).
Ease	Outputs should be self-explanatory and not require users to remember information from prior outputs in order to complete tasks. Labels should be extensively used, and all scales and units of measure should be clearly indicated.
Format	Information format should be consistent between entry and display. Format should distinguish each piece of data and highlight, not bury, important data. Special symbols, such as decimal places, dollar signs, and +/– signs should be used as appropriate.
Flexibility	Information should be viewed and retrieved in a manner most convenient to the user. For example, users should be given options for the sequence in which to enter or view data and for use of shortcut keystrokes, and the system should remember where the user stopped during the last use of the system.

When designing outputs, you must also consider the context in which the screens, forms, and reports will be used. As mentioned, numerous characteristics play an important role in shaping a system's usability. These characteristics are related to the intended users and task being performed in addition to the technological, social, and physical environment in which the system and outputs are used. Table 13-9 lists several factors in which variations in any item may influence the usability of a design. Your role is to gain a keen awareness of these factors so that your chances of creating highly usable designs are increased.

Measures of Usability

User friendliness is a term often used, and misused, to describe system usability. Although the term is widely used, it is too vague from a design standpoint to provide adequate information because it means different things to different people. Consequently, most development groups use several methods for assessing usability, including these considerations (Shneiderman, 1992):

- Time to learn
- Speed of performance
- Rate of errors
- Retention over time
- Subjective satisfaction

In assessing usability, you can collect information by observation, interviews, keystroke capturing, and questionnaires. Time to learn simply reflects how long it takes the average system user to become proficient using the system. Equally important is the extent to which users remember how to use inputs and outputs over time. The manner in which the processing steps are sequenced and the selection of one set of keystrokes over others can greatly influence learning time, the user's task performance, and error rates. For example, the most commonly used functions should be quickly accessed in the fewest number of steps (for example, pressing one key to save your work). Additionally, the layout of information should be consistent, both *within and across* applications, whether the information is delivered on a screen display or hard copy report.

TABLE 13-9 Characteristics for Consideration When Designing Forms and Reports
(Adapted from Norman, 1991)

Characteristic	Consideration for Form and Report Design
User	Issues related to experience, skills, motivation, education, and personality should be considered.
Task	Tasks differ in amount of information that must be obtained from or provided to the user. Task demands such as time pressure, cost of errors, and work duration (fatigue) will influence usability.
System	The platform on which the system is constructed will influence interaction styles and devices.
Environment	Social issues such as the users' status and role should be considered in addition to environmental concerns such as lighting, sound, task interruptions, temperature, and humidity. The creation of usable forms and reports may necessitate changes in the users' physical work facilities.

SUMMARY

We present this chapter first in a sequence of three chapters on logical system design topics since it focuses on a primary product of information systems: designing forms and reports. As organizations move into more complex and competitive business environments with greater diversity in the workforce, the quality of the business processes will determine success. One key to designing quality business processes is the delivery of the right information to the right people, in the right format, at the right time. The design of forms and reports concentrates on this goal. A major difficulty of this process comes from the great variety of information-formatting options available to designers.

We saw in this chapter that information can take three general forms: internal—for use within the organization; external—for distribution to customers, suppliers, and other parties outside the organization; and turnaround—documents which are produced for outside parties and returned as input to an internal system. Information processing can occur at pre-determined and regular times after sufficient data are collected, called batch processing, or it can be created or entered at user request, called on-line processing.

Specific guidelines should be followed when designing forms and reports. These guidelines, proven over years of experience with human-computer interaction, help you to create professional, usable systems. The chapter presented a variety of guidelines covering use of titles, layout of fields, navigation between pages or screens, highlighting data, use of color, format of text and numeric data, appropriate use and layout of tables and graphs, avoiding bias in information display, and achieving usable forms and reports.

Form and report designs are created through a prototyping process. Once created, designs may be stand-alone or integrated into actual working systems. The purpose, however, is to show users what a form or report will look like when the system is implemented. The outcome of this activity is the creation of a specification document where characteristics of the users, tasks, system, and environment are outlined along with each form and report design. Performance testing and usability assessments may also be included in the design specification.

The goal of form and report design is usability. Usability means that users can use a form or report quickly, accurately, and with high satisfaction. To be usable, designs must be consistent, efficient, self-explanatory, well-formatted, and flexible. These objectives are achieved by applying a wide variety of guidelines concerning such aspects as navigation, use of highlighting and color, display of text, tables, and lists.

CHAPTER REVIEW

KEY TERMS

Batch processing	Internal information	Turnaround document
External information	On-line processing	Usability
Form	Report	

REVIEW QUESTIONS

1. Define each of the following terms:
   a. batch processing
   b. external information
   c. internal information
   d. on-line processing
   e. turnaround document

2. Contrast the following terms:
   a. batch processing, on-line processing
   b. external information, internal information, turnaround document
   c. form, report

3. Describe the prototyping process of designing forms and reports. What deliverables are produced from this process? Are these deliverables the same for all types of system projects? Why or why not?

4. To which initial questions must the analyst gain answers in order to build an initial prototype of a system output?

5. Describe what is meant by the term internal information and provide at least three examples (not already used in this chapter) in your answer.

6. Describe what is meant by the term external information and provide at least three examples (not already used in this chapter) in your answer.

7. Describe what is meant by the term turnaround document and provide at least three examples (not already used in this chapter) in your answer.

8. What factors make the delivery of on-line information most practical? What factors make the delivery of batch processing most practical?

9. When can highlighting be used to convey special information to users?

10. Discuss the benefits, problems, and general design process for the use of color when designing system output.

11. How should textual information be formatted on a help screen?

12. What type of labeling can you use in a table or list to improve its usability?

13. What column, row, and text formatting issues are important when designing tables and lists?

14. Describe how numeric, textual, and alphanumeric data should be formatted in a table or list.

15. Using examples, discuss four ways in which the type and format of information can bias its meaning.

16. Provide some examples where variations in user, task, system, and environmental characteristics might impact the design of system forms and reports.

PROBLEMS AND EXERCISES

1. Match the following terms to the appropriate definitions.

   _____ form

   _____ balance

   _____ report

   _____ highlighting

   a. a way to draw attention to or away from data in output

   b. type of output that typically lists and summarizes data from several unrelated database records

   c. a characteristic of output which includes minimal density of information in the output

   d. type of output that typically lists data associated with one database record

2. Imagine that you are to design a budget report for a colleague at work using a spreadsheet package. Following the prototyping discussed in the chapter (see also Figure 1-7), describe the steps you would take to design a prototype of this report.

3. Consider a system that produces budget reports for your department at work. Alternatively, consider a registration system that produces enrollment reports for a department at a university. For whichever system you choose, answer the following design questions. Who will use the output? What is the purpose of the output? When is the output needed and when is the information that will be used within the output available? Where does the output need to be delivered? How many people need to view the output?

4. Imagine the worst possible reports from a system. What is wrong with them? List as many problems as you can. What are the consequences of such reports? What could go wrong as a result? How does the prototyping process help guard against each problem?

5. How many turnaround documents have you encountered in the past 30 days at work, home, school, or from any other organization? List what you like and dislike about these documents. Using one of these documents, describe how it could be improved.

6. Imagine a moderately sized flower shop. List as many internal and external informational items as you can for this business. Which of the internal information items should be delivered to users as on-line and/or hard copy reports? Will there be any turnaround documents? If so, describe.

7. Consider your answer for the previous question about the flower shop. Use the relevant criteria in this chapter to determine whether the output should be processed on-line or in batch.

8. Imagine an output display form for a hotel registration system. Using a software package for painting and/or drawing, such as MacPaint, Paintbrush, or Corel Draw, follow the design suggestions in this chapter and design this form entirely in black and white. Save the file and then, following the color design suggestions in this chapter, redesign the form using color. Based on this exercise, discuss the relative strengths and weaknesses of each output form.

9. Consider reports you might receive at work (e.g., budgets or inventory reports) or at a university (e.g., grade reports or transcripts). Evaluate the usability of these reports in terms of speed, accuracy, and satisfaction. What could be done to improve the usability of these outputs?

10. List the PC-based software packages you like to use. Describe each package in terms of the following usability characteristics: time to learn, speed of performance, rate of errors by users, retention over time, and subjective satisfaction. Which of these characteristics has led to your wanting to continue to use this package?

11. Given the guidelines presented in this chapter, identify flaws in the design of the Report of Customers shown below. What assumptions about users and tasks did you make in order to assess this design? Redesign this report to correct these flaws.

Report of Customers – *26-Oct-96*	
Cust-ID	**Organization**
AC-4	A.C. Nielson Co.
ADTRA-20799	Adran
ALEXA-15812	Alexander & Alexander, Inc.
AMERI-1277	American Family Insurance
AMERI-28157	American Residential Mortgage
ANTAL-28215	Antalys
ATT-234	AT&T Residential Services
ATT-534	AT&T Consumer Services
. . .	
DOLE-89453	Dole United, Inc.
DOME-5621	
DO-67	Doodle Dandies
. . .	
ZNDS-22267	Zenith Data System

12. Review the guidelines for attaining usability of forms and reports in Table 13-8. Consider an on-line form you might use to register a guest at a hotel. For each usability factor, list two examples of how this form could be designed to achieve that dimension of usability. Use examples other than those mentioned in Table 13-8.

FIELD EXERCISES

1. Find your last grade report. Given the guidelines presented in this chapter, identify flaws in the design of this grade report. Redesign this report to correct these flaws.

2. As stated in the chapter, most forms and reports are designed for contemporary information systems by using software to prototype output. Packages like Visual Basic, PowerBuilder, and Paradox for Windows have very sophisticated output design modules. Gain access to such a tool at your university or where you work and study all the features the software provides for the design of printed output. Write a report that lists and explains all the features for layout, highlighting, summarizing data, etc.

3. Investigate the display uses in another field (e.g., aviation). What types of forms and reports are used in this field? What standards, if any, are used to govern the use of these outputs?

4. Interview a variety of people you know about the different types of forms and reports they use in their jobs. Ask to examine a few of these documents and answer the following questions for each one:

   a. Do they provide internal or external information (or both)? Are any turnaround documents?
   b. Which are processed on-line, which are processed in batch?
   c. What types of tasks does each support and how is it used?
   d. What types of technologies and devices are used to deliver each one?
   e. Assess the usability of each form or report. Is each usable? Why or why not? How could each be improved?

5. Scan the annual reports from a dozen or so companies for the past year. These reports can usually be obtained in a university library. Describe the types of information and the ways that information has been presented in these reports. How have color and graphics been used to improve the usability of information? Describe any instances where formatting has been used to hide or enhance the understanding of information.

6. Choose a PC-based package you like to use and choose one that you don't like to use. Interview other users to determine their evaluations of these two packages. Ask each individual to evaluate each package in terms of the usability characteristics described in the previous question. Is there a consensus among these evaluations, or do the respondents' evaluations differ from each other or from your own evaluations? Why?

REFERENCES

Alter, S. 1995. *Information Systems: A Management Perspective.* Redwood City, CA: Benjamin/Cummings.

Benbasat, I., Dexter, A. S., & Todd, P. 1986. "The Influence of Color and Graphical Information Presentation in a Managerial Decision Simulation." *Human-Computer Interaction* 2: 65–92.

Carroll, J. M. 1991. *Designing Interaction.* Cambridge: Cambridge University Press.

Dumas, J. S. 1988. *Designing User Interfaces for Software.* Englewood Cliffs, NJ: Prentice-Hall.

Huff, D. 1954. *How to Lie with Statistics.* New York, NY: W. W. Norton.

Jarvenpaa, S. L. and G. W. Dickson. 1988. "Graphics and Managerial Decision Making: Research Based Guidelines." *Communications of the ACM* 31 (6): 764–74.

Norman, K. L. 1991. *The Psychology of Menu Selection.* Norwood, NJ: Ablex.

Shneiderman, B. 1992. *Designing the User Interface: Strategies for Effective Human-Computer Interaction.* Reading, MA: Addison-Wesley.

Logical Design: Designing a Customer Reservation Report

INTRODUCTION

In this case in the design of the Customer Activity Tracking System (CATS), we have been following the activities of Frank Napier, a systems analyst on the project. It is at this point that system outputs—screens, forms and reports—are being designed. Even though BEC has chosen pre-written software for CATS, outputs must be designed to verify that those in the package are adequate and to satisfy those requirements not provided in the package. The focus of this case is the customer-product assignment report, one of several reservation reports that is integral to the successful implementation of CATS. The customer-product assignment report is sent electronically to each store over BEC's private communications network and lists which customers have reserved which products. The report is stored on the file server at a store and a notification that a new report was sent is automatically displayed by the daily start-up routine from the store manager's personal computer. The report can be printed or used on-line. A store manager or other store personnel uses the report to notify customers that the product has arrived at the store and is available for purchase.

Initiating the Design

During a recent meeting, Jordan Pippen explained Frank's next project assignment.

"Frank, the customer-product assignment report, or what I like to call the CPA report, is a very important component in the system. The CPA report is sent over the BEC network to each store from BEC headquarters and contains the customer-product reservation information generated by CATS. The report is sent to arrive the day the store receives a new product shipment and lists each reservation in a format that makes it easy for the store to call customers to let them know the products they reserved are in the store. Other

reservation reports serve other purposes, like helping the store set aside the right quantity of product to cover reservations. The information in the CPA report must be delivered to stores in a timely manner and in a format that makes it easy for personnel to do their job. The store must be able to track who needs to be called and who has been contacted. The CPA report will be an important way stores will interface with CATS. How our stores view the system will be significantly influenced by the usability of the CPA report."

"Wow," said Frank, "this *is* an important report. Is there anything special you want me to do during the design?"

"No, not really," said Jordan. "The strength of our design methodology is that it works fine for the design of both routine and high-profile outputs, such as the CPA report. Yet, you should give some thought to which clients you want to work with during the design and usability assessment process. We must get the right people involved in this one and then thoroughly understand how they will use the report."

"OK, I will get on it," replied Frank as he returned to his office.

Designing the Customer-Product Assignment Report

After arriving back at his office, Frank reviewed the BEC output design methodology and was reminded that his first task was to answer several questions, questions that form the basis of the design. A summary of his work is shown in Figure 1. These questions and associated answers, along with his understanding of the information needed in the report, oriented Frank to his task. The next step was to identify individuals to help him design and assess the report. He remembered that Jordan had recently met with the managers of seven BEC retail stores with the highest sales figures. He thought, "Maybe those should be the users that I work with on this design?" After checking with

Figure 1
Fundamental questions and answers to the design of the customer reservation report

QUESTIONS	ANSWERS
Who will use the output?	All store managers and sales representatives
What is the purpose of the output?	Provide a list of customers with reservations Matching customers with inventory Provide a way for store to track contacts with customers concerning reservations
When is the output needed and used?	Needed (used) on (after) product delivery to store
Where does the output need to be delivered?	To each BEC store
How many people need to view the output?	The store manager and several sales representatives at each of several hundred stores in U.S.

Jordan for approval, he crafted an intra-company memo to these managers requesting their involvement in the design of the CPA report (Figure 2). The memo was routed to each manager via BEC's electronic mail system. He also asked Nancy Chen (v-p of U.S. Operations) and Wendy Yoshimuro (marketing manager responsible for CATS) to serve as design and usability reviewers. Nancy, Wendy, and all store managers agreed to participate.

The process Frank followed in the design of the CPA report is based on BEC's prototyping methodology (see Figure 3). At this stage, users will not be working with the report since it will be delivered electronically to the stores, but they will see copies of sample reports on which they can base reactions. Having several different users actually interact with a system before implementation can be awkward and misleading for the users which is why a passive prototyping process is used. Assuring that the report is delivered conveniently to users will be done during final system acceptance testing after all CATS modules have been implemented.

Using the available information from analysis and talking with design participants, an initial design was developed (step 1) from the ObjectStar CASE tool used at BEC (see Figure 4). Frank did this for two main reasons. First, the report could be linked to the rest of the CATS project repository for consistency among process, data, and interface models that show structured requirements. Second, from the report design the CASE tool could generate the necessary program code for the report for a variety of development platforms. The CPA report is one variation of the Reservation Report shown on DFDs for CATS (see Figure 4 in the BEC case after Chapter 9). Thus, the repository information about the composition of the Reservation Report and

other subflows on more detailed DFDs was a good starting point for Frank.

Most of the fields used to construct the prototype were already stored in the ObjectStar CASE repository, so standards for the formatting of fields and their expected length were already established. An exception to this was the reservation code field; this was a new field and not yet represented in the repository. The intent of the field is to uniquely identify customers, products, and stores and, in essence, this field could be the concatenation of three fields—customer, product, and store identifiers. This would, however, be an extremely long field. Since the customer must report the contents of this field to a phone operator in order to make a reservation, Frank decided to make this field a relatively short five-character, alphanumeric sequence number. The actual selection of the field format is a physical design and implementation issue that will be finalized later in the project.

The initial design was sent to all design participants with clear assessment instructions (step 2) and a firm date for critiquing the design. To ease the review process, each participant was also given a design review form with a check-box for either approving the design or recording change requests (Figure 5). In this way, during the second step of the prototyping methodology, each participant has the opportunity to critique the design and suggest changes. During step 3, Frank's role is to collect all participant critiques and consolidate the feedback into an overall evaluation report. The conclusions from this report can lead to either of the following choices:

1. Refine the design based upon the collective feedback of the participants and return to step 1.

2. Finalize the design specification (step 4).

Figure 2
Memorandum requesting participation in the design of the customer-product assignment report

Memorandum

To:	Selected BEC Store Managers, Nancy Chen, Wendy Yoshimuro
CC:	Jordan Pippen
From:	Frank Napier
Date:	June 21, 1995
Subject:	Request for help with an aspect of the Customer Activity Tracking System

As most of you are aware, BEC is actively working on the design of a new system, CATS, that will revolutionize the way in which we serve our customers. An important component of this system is a report summarizing the store-level reservations of customers and products. This report is called the customer-product assignment (CPA) report.

Due to your keen interest in this project and your awareness of the needs of the retail stores, you are being asked to participate in the design of this report. In order to move as quickly as possible, we must have your agreement to participate within one week (sooner if possible!).

See the attached document (an embedded object), The BEC Methodology for Designing Interfaces, Forms, and Reports, to get a better understanding of your role in this design process. I expect that this entire project will require less than 2 hours of your time.

Please respond to this request by either return e-mail, or by simply marking your decision below and faxing this sheet back to me at headquarters.

Check one:
    ☐ I agree to participate on the CPA report design team.
    ☐ I am sorry, but I won't be able to participate at this time.

_____   _____
Name (print)   Name (sign)

enclosure: The BEC Methodology for Designing Interfaces, Forms, and Reports

Figure 3

BEC's prototyping methodology for designing interfaces, forms, and reports

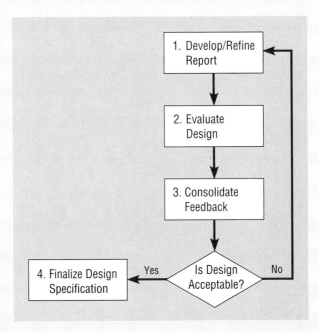

Figure 4

Initial design of the customer-product assignment report

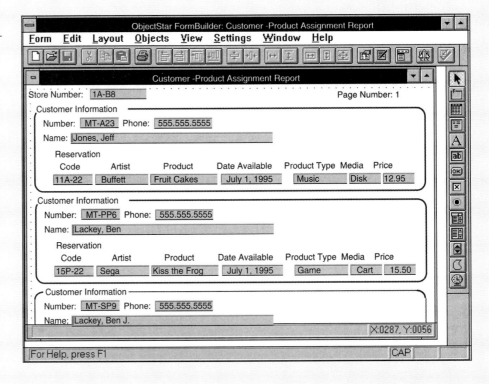

This means that if participants suggest changes, Frank must consolidate these changes into a revised design specification, sometimes resolving differences between conflicting suggestions. Using this feedback, he can then refine the design and repeat the evaluation and consolidation process until the design is accepted.

After three rounds, the design was accepted by the panel of users. During these iterations as well as during the usability assessment process, several changes to the original design were made. A final version of the report is shown in Figure 6 which highlights the differences between the initial and final versions.

Figure 5

Design review form for the customer-product assignment report

Broadway Entertainment Company
Report Assessment Form

Name: _____ Bret Williams _____

Location: _____ Norman, OK – store #4 _____

Report: _____ CPA _____ Version: _____ 1 _____

☐ I accept the current design.
☒ I request that the following changes be considered.

- -

Importance (circle one)   *Description of Change*

Low   Med   High
1 ------ (2) ------ 3   1. __Need more descriptive wording for the media field – I suggest__
                        __using word 'format' instead__

1 ------ 2 ------ 3   2. _____
                        _____
                        _____

1 ------ 2 ------ 3   3. _____
                        _____
                        _____

1 ------ 2 ------ 3   4. _____
                        _____
                        _____

1 ------ 2 ------ 3   5. _____
                        _____
                        _____

1 ------ 2 ------ 3   6. _____
                        _____
                        _____

1 ------ 2 ------ 3   7. _____
                        _____
                        _____

1 ------ 2 ------ 3   8. _____
                        _____
                        _____

1 ------ 2 ------ 3   9. _____
                        _____
                        _____

1 ------ 2 ------ 3   10. _____
                        _____
                        _____

Figure 6
Final design of the customer-product assignment report highlighting changes between the initial and final design

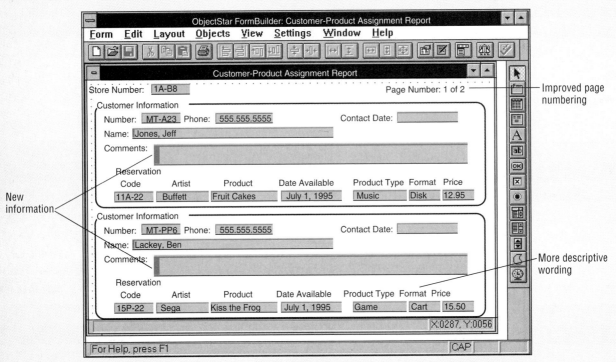

Frank's next activity was to assess the usability of the report.

Assessing Design Usability

The last major activity in the design of the CPA report was to assess its usability. When assessing design usability, BEC's methodology suggests that the following factors be considered:

- Time to learn
- Speed of performance
- Rate of errors
- Retention over time
- Subjective satisfaction

Since the report did not take users a significant amount of time to learn to use and since the report would be used daily, Frank determined that the most reasonable usability factors to assess for this report would be speed of performance, rate of errors, and subjective satisfaction. To make this assessment, he constructed a new sample report with fictitious customers and data. This report contained products of every category and

combination. The length of the report was determined by examining information collected during analysis. Frank's objective was to construct a report representative of those likely to be created by the actual system.

Next, he constructed a series of report evaluation questions about the content. These questions were recorded on a formal questionnaire that included questions related to three areas: background information on the person doing the assessment, objective questions of report comprehension, and user perceptions (see Figure 7 for questions excerpted from these three sections). Using both the sample report and evaluation questionnaire, Frank was now ready to assess the usability of the CPA report design. To do this, he identified a representative cross section of potential system users (both store managers and sales representatives, as identified in Figure 1). He then gave both the report and evaluation questions to these users, recording the number of errors and the speed with which each could complete the series of questions. To get a representative sample of potential system users, he visited several BEC stores across the country to collect usability data. Frank did not take special trips for these assessments only but conducted the assessment of the CPA report design along with other system components during these visits. In addition to questions

Figure 7
Sections of the usability questionnaire to assess the customer-product assignment report

Report Evaluation Questionnaire

Background Information:

1. Name (print): _____ 2. Location (store number): _____
3. How long have you worked for BEC? _____
4. Your current title:
   ☐ Store Manager
   ☐ Sales Representative
   ☐ Other _____

Report Evaluation: Use the information on the attached report to answer the following
questions.

1. Record current time now: _____ am/pm (circle one)
2. Check the customer names that have reserved a copy of Jimmy Buffet's "Fruit Cakes"
   compact disk.
   _____ Bill Jones   _____ Dana Merry
   _____ Janet Jones   _____ Billy Dolan
   _____ Anita Rosenburg   _____ Larry Mueller

8. Match customers to the products they reserved.
   _____ Babe Harnett   a. Revenge of the Nerds XI—video
   _____ Tina Albright   b. Yawni—"I Need to Take a Nap"—music CD
   _____ Bill Bixby   c. Cruise Control—Game
                               d. Snoop Doggy Dogg—music CD

Perceptions of Report Format:

1. Rate the consistency and dependability of the information on the report.
   inconsistent   1   2   3   4   5   6   7   consistent
   insufficient   1   2   3   4   5   6   7   sufficient
2. Rate the ability of the report to accurately transmit information.
   inaccurate   1   2   3   4   5   6   7   accurate
   insufficient   1   2   3   4   5   6   7   sufficient

6. Please note any other thing you think we should know about using this report (e.g.,
   were all fields clearly labeled?):
   _____
   _____
   _____
   _____

related to report information, he also asked a standard set of questions related to user perceptions about the reports. These questions were used to collect qualitative information and to assess overall user satisfaction. Additionally, background information was used to segment the data in order to compare the evaluations of long-term employees with new employees, managers with sales representatives, and so on.

SUMMARY

Here we described how the customer-product assignment report was designed and assessed within the Customer Activity Tracking System at Broadway En-

tertainment Company. It should be clear from this description that the design of even a single report requires both technical and interpersonal skills in addition to a lot of hard work. In the course of the design and assessment process, the systems analyst not only managed the prototyping development process but also interacted with many BEC employees from numerous levels and functions. The success of this design effort likely resulted from a combination of strong interpersonal and technical skills.

14

Designing Interfaces and Dialogues

L E A R N I N G   O B J E C T I V E S

After studying this chapter, you should be able to:

■ Explain the process of designing interfaces and dialogues and the deliverables for their creation.

■ Contrast and apply several methods for interacting with a system.

■ List and describe various input devices and discuss usability issues for each in relation to performing different tasks.

■ Describe and apply the general guidelines for designing interfaces and specific guidelines for layout design, structuring data entry fields, providing feedback, and system help.

■ Choose between various mechanisms for providing system security through a system's interface.

■ Design human-computer dialogues, including the use of dialogue diagramming.

■ Design graphical user interfaces.

INTRODUCTION

In this chapter, you learn about system interface and dialogue design. Interface design focuses on how information is provided to and captured from users; dialogue design focuses on the sequencing of interface displays. Dialogues are analogous to a conversation between two people. The grammatical rules followed by each person during a conversation are analogous to the interface. Thus, the design of interfaces and dialogues is the process of defining the manner in which humans and computers exchange information. A good human-computer interface provides a uniform structure for finding, viewing, and invoking the different components of a system. This chapter complements Chapter 13 which addressed design guidelines for the content of forms and reports. In Chapter 14, you will learn about navigation between forms, alternative ways for users to cause forms and reports to appear, and how to supplement the content of forms and reports with user help and error messages, among other topics.

In the next section, the process of designing interfaces and dialogues and the deliverables produced during this activity are described. This is followed by a section that describes interaction methods and devices. Next, interface design is described. This discussion focuses on layout design, data entry, providing feedback, and designing help. Since system security is an important part of interface design, a brief discussion of how to build user access controls follows. The last section examines techniques for designing human-computer dialogues.

DESIGNING INTERFACES AND DIALOGUES

This is the second chapter that focuses on logical design within the systems development life cycle (see Figure 14-1). In Chapter 13, you learned about the design of forms and reports. As you will see, the guidelines for designing forms and reports also apply to the design of human-computer interfaces. Chapter 15 focuses on the logical design of system files and databases. The specific ordering of when you *should* design forms and reports, interfaces and dialogues, or databases is open to some debate. Our sequence was chosen because each topic builds on prior ones. Yet, within organizations, the order of these activities is not sequential but likely to be done in an iterative or parallel fashion.

The Process of Designing Interfaces and Dialogues

Similar to designing forms and reports, the process of designing interfaces and dialogues is a user-focused activity. This means that you follow a prototyping

Figure 14-1
Systems development life cycle with logical design phase highlighted

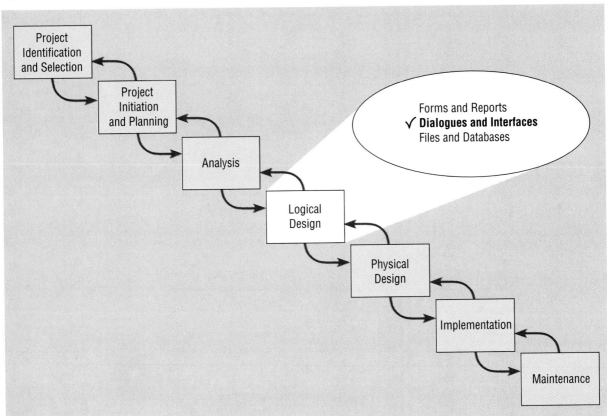

methodology of iteratively collecting information, constructing a prototype, assessing usability, and making refinements. To design usable interfaces and dialogues, you must answer the same who, what, when, where, and how questions used to guide the design of forms and reports (see Table 13-1). Thus, this process parallels that of designing forms and reports.

Deliverables and Outcomes

The deliverable and outcome from system interface and dialogue design is the creation of a design specification. This specification is also similar to the specification produced for form and report designs—with one exception. Recall that design specifications in Chapter 13 had three sections (see Figure 13-4):

1. Narrative overview

2. Sample design

3. Testing and usability assessment

For interface and dialogue designs, one additional subsection is included: a section outlining the dialogue sequence—the ways a user can move from one display to another. Later in the chapter you will learn how to design a dialogue sequence by using dialogue diagramming and state-transition diagramming. An outline for a design specification for interfaces and dialogues is shown in Figure 14-2.

Figure 14-2
Specification outline for the design of interfaces and dialogues

Design Specification

1. Narrative overview
    a.   Interface/Dialogue Name
    b.   User Characteristics
    c.   Task Characteristics
    d.   System Characteristics
    e.   Environmental Characteristics

2. Interface/Dialogue Designs
    a.   Form/Report Designs
    b.   Dialogue Sequence Diagram(s) and Narrative Description

3. Testing and Usability Assessment
    a.   Testing Objectives
    b.   Testing Procedures
    c.   Testing Results
        i)   Time to Learn
        ii)   Speed of Performance
        iii)   Rate of Errors
        iv)   Retention Over Time
        v)   User Satisfaction and Other Perceptions

INTERACTION METHODS AND DEVICES

Interface: A method by which users interact with information systems.

The human-computer **interface** defines the ways in which users interact with an information system. All human-computer interfaces must have an interaction style and use some hardware device(s) for supporting this interaction. In this section we therefore describe various interaction methods and guidelines for designing usable interfaces.

Methods of Interacting

When designing the user interface, the most fundamental decision you make relates to the methods used to interact with the system. Given that there are numerous approaches for designing the interaction, we briefly provide a review of those most commonly used.[1] Our review will examine the basics of five widely used styles: command language, menu, form, object, and natural language. We will also describe several devices for interacting, focusing primarily on their usability for various interaction activities.

Command language interaction: A human-computer interaction method where users enter explicit statements into a system to invoke operations.

Command Language Interaction   In **command language interaction,** the user enters explicit statements to invoke operations within a system. This type of interaction requires users to remember command syntax and semantics. For example, to copy a file named PAPER.DOC from one storage location (C:) to another (A:) using Microsoft's disk operating system (DOS), a user would type

COPY C:PAPER.DOC A:PAPER.DOC

Command language interaction places a substantial burden on the user to remember names, syntax, and operations. Most newer or large-scale systems no longer rely entirely on a command language interface. Yet, command languages are good for experienced users, for systems with a limited command set, and for rapid interaction with the system.

A relatively simple application such as a word processor may have hundreds of commands for such operations as saving a file, deleting words, canceling the current action, finding a specific piece of data, or switching between windows. Some of the burden of assigning keys to actions has been taken off your shoulders through the development of user interface standards such as the Macintosh, Microsoft Windows, and IBM's Common User Access (CUA) (see Apple, 1987; IBM, 1991). For example, Figure 14-3 shows the mapping of system functions within WordPerfect 6.0 for Windows. This mapping is based upon the CUA function key assignment standard. Note that designers still have great flexibility in how they interpret and implement these standards. This means that you still need to pay attention to usability factors and conduct formal assessments of designs. Any application may adopt the CUA function key standard, as appropriate for the functionality of that application.

Menu interaction: A human-computer interaction method where a list of system options is provided and a specific command is invoked by user selection of a menu option.

Menu Interaction   A significant amount of interface design research has stressed the importance of a system's ease of use and understandability. **Menu interaction** is a means by which many designers have accomplished this goal. A menu is simply a list of options; when an option is selected by the user, a specific command is invoked, or another menu is activated. Menus have become the most widely used interface method because the user only needs to understand simple signposts and route options to effectively navigate through a system.

---

[1]Readers interested in learning more about interaction methods are encouraged to see the books by Dumas (1988), Norman (1991), or Shneiderman (1992).

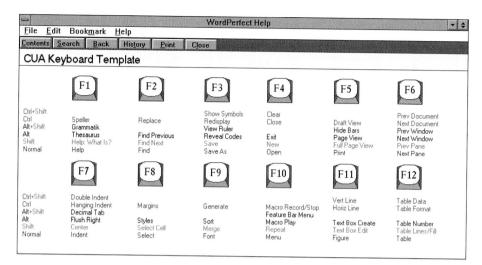

Figure 14-3
Function key mapping of commands within WordPerfect for Windows 6.0

Menus can differ significantly in their design and complexity. The variation of their design is most often related to the capabilities of the development environment, the skills of the developer, and the size and complexity of the system. For smaller and less complex systems with limited system options, you may use a single menu or a linear sequence of menus. A single menu has obvious advantages over a command language but may provide little guidance beyond invoking the command. A single menu for a DOS shell program is shown in Figure 14-4.

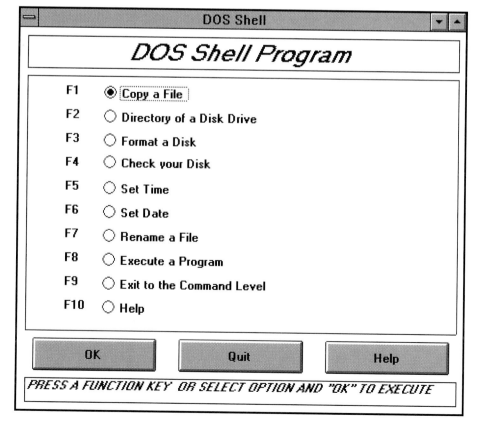

Figure 14-4
Single-level menu for disk operating system application

For large and more complex systems, you can use menu hierarchies to provide navigation between menus. These hierarchies can be simple tree structures or variations wherein children menus have multiple parent menus or which allow multi-level traversal. Variations as to how menus are arranged can greatly influence the usability of a system. Figure 14-5 shows a variety of ways in which menus can be structured and traversed. An arc on this diagram signifies the ability to move from one menu to another. Although more complex menu structures provide greater user flexibility, they may also confuse users about exactly where they are in the system. Structures with multiple parent menus also require the application to remember which path has been followed so that users can correctly backtrack.

There are two common methods for positioning menus. In a **pop-up menu** (also called a dialogue box), menus are displayed near the current cursor position so users don't have to move the position or their eyes to view system options (Figure 14-6a). A pop-up menu has a variety of potential uses. One is to show a list of commands relevant to the current cursor position (for example, delete, clear, copy, or validate current field). Another is to provide a list of possible values (from a look-up table) to fill in for the current field. For example, in a customer order form, a list of current customers could pop up next to the customer number field so the

Pop-up menu: A menu positioning method that places a menu near the current cursor position.

Figure 14-5
Various types of menu configurations (Adapted from Shneiderman, 1992)

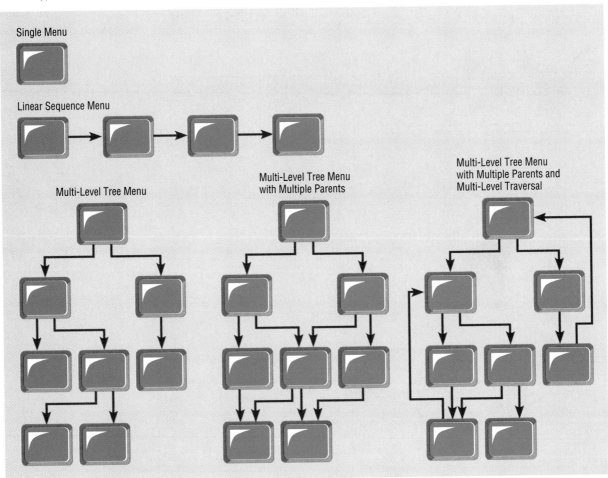

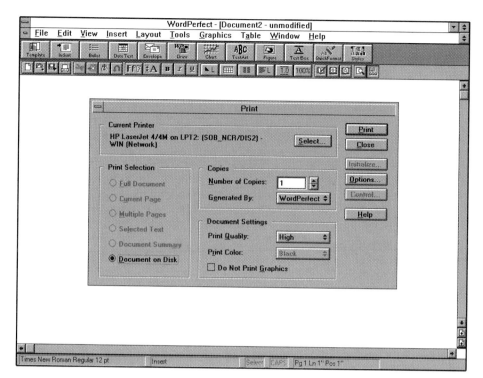

Figure 14-6
Menus from WordPerfect for
Windows 6.0

(a) Pop-up menu

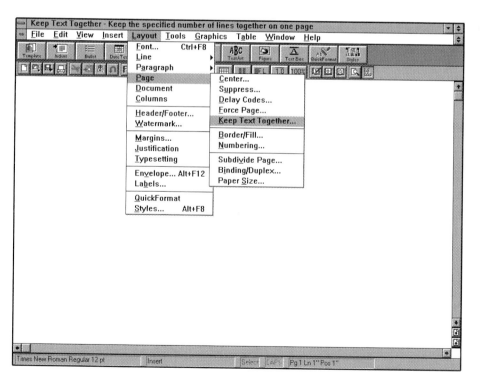

(b) Drop-down menu

Drop-down menu: A menu positioning method that places the access point of the menu near the top line of the display; when accessed, menus open by dropping down onto the display.

user can select the correct customer without having to know the customer's identifier. In a **drop-down menu,** menus drop down from the top line of the display (Figure 14-6b). Drop-down menus have become very popular in recent years because they provide consistency in menu location and operation among applications and efficiently use display space. Most advanced operating environments such as Microsoft Windows, Macintosh, and the Unix graphical operating environment called X-Windows provide a combination of both pop-up and drop-down menus.

When designing menus, there are several general rules that should be followed and these are summarized in Table 14-1. For example, each menu should have a meaningful title and be presented in a meaningful manner to users. A menu option of Quit, for instance, is ambiguous—does it mean return to the previous screen, exit the program, or exit to DOS? To more easily see how to apply these guidelines, Figure 14-7 contrasts a poorly designed menu with a menu following the menu design guidelines. Annotations on the two parts of this figure highlight poor and improved menu interface design features.

Many advanced programming environments provide powerful tools for designing menus. For example, Microsoft's Visual Basic allows you to quickly design a menu structure using a menu design facility. Figure 14-8a shows the design form, Menu Design Window, in which a menu structure is defined. Figure 14-8b shows how the menu looks to a user within the actual information system. To build a menu you only need to type the words that will represent each item on the menu. With the use of a few easily invoked options, you can also assign shortcut keys to menu items, connect help screens to individual menu items, define submenus, and set usage properties. Usage properties, for example, include the ability to dim the color of a menu item while a program is running to indicate that a function is currently unavailable. Menu building tools allow a designer to quickly and easily prototype a design that will look exactly as it will in the final system.

Form interaction: A highly intuitive human-computer interaction method whereby data fields are formatted in a manner similar to paper-based forms.

Form Interaction   The premise of **form interaction** is to allow users to fill in the blanks when working with a system. Form interaction is effective for both the input

TABLE 14-1   Guidelines for Menu Design
(Adapted from Dumas, 1988)

Wording	• Each menu should have a meaningful title
	• Command verbs should clearly and specifically describe operations
	• Menu items should be displayed in mixed upper- and lower-case letters and have a clear, unambiguous interpretation
Organization	• A consistent organizing principle should be used that relates to the tasks the intended users perform; for example, related options should be grouped together and the same option should have the same wording and codes each time it appears
Length	• The number of menu choices should not exceed the length of the screen
	• Submenus should be used to break up exceedingly long menus
Selection	• Selection and entry methods should be consistent and reflect the size of the application and sophistication of the users
	• How the user is to select each option and the consequences of each option should be clear (e.g., whether another menu will appear)
Highlighting	• Highlighting should be minimized and used only to convey selected options (e.g., a check mark) or unavailable options (e.g., dimmed text)

Figure 14-7
Contrasting menu designs

(a) Poor menu design

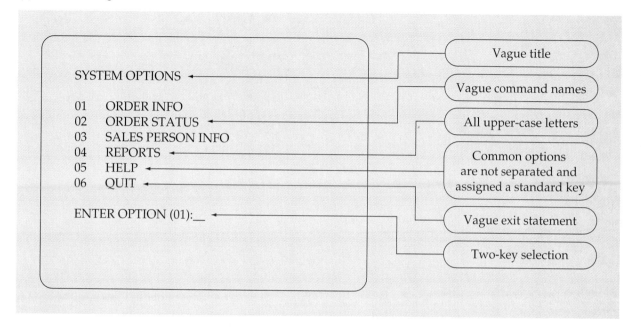

(b) Improved menu design

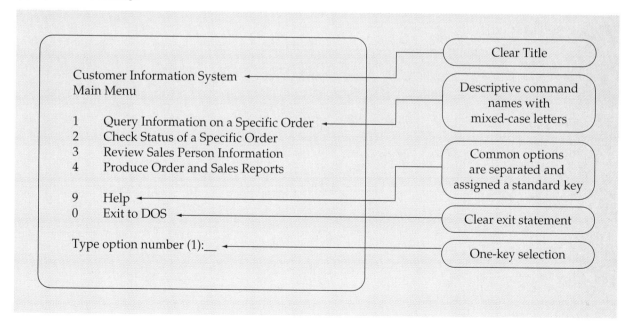

and presentation of information. An effectively designed form includes a self-explanatory title and field headings, has fields organized into logical groupings with distinctive boundaries, provides default values when practical, displays data in appropriate field lengths, and minimizes the need to scroll windows (Shneiderman, 1992). You saw many other design guidelines for forms in Chapter 13. Form interaction is the most commonly used method for data entry and retrieval in

Figure 14-8
Menu building within a
graphical user interface
environment

(a) Menu design window

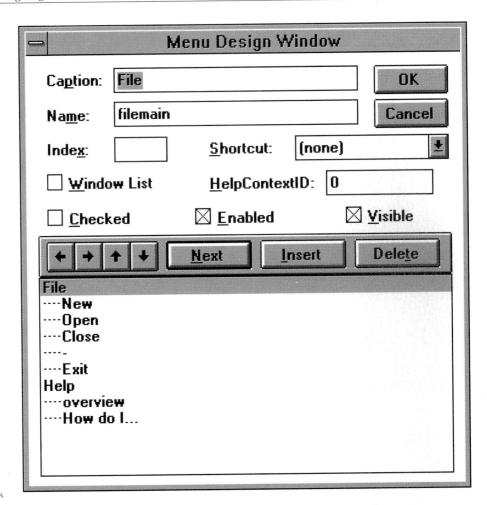

(b) Menu design as viewed
by user

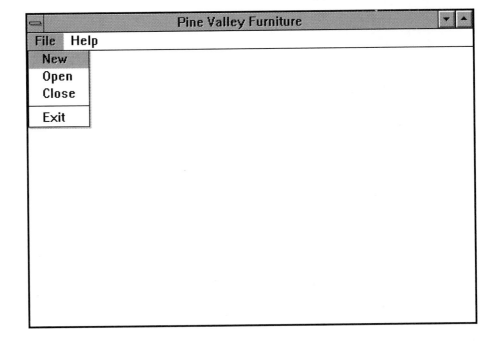

business-based systems. Figure 14-9 shows a form from Novell's World Wide Web site on the Internet. Using this interactive form, Web surfers can easily request Novell product information using Internet tools such as Netscape.

Object-Based Interaction   The most common method for implementing **object-based interaction** is through the use of icons. **Icons** are graphic symbols that look like the processing option they are meant to represent. Users select operations by pointing to the appropriate icon with some type of pointing device. The primary advantages to icons are that they take up little screen space and can be quickly understood by most users. An icon may also look like a button which, when selected or depressed, causes the system to take an action relevant to that form, such as cancel, save, edit a record, or ask for help. For example, Figure 14-10 illustrates an icon-based interface for configuring user preferences within WordPerfect for Windows 6.0.

Natural Language Interaction   A branch of artificial intelligence research studies techniques for allowing systems to accept inputs and produce outputs in a conventional language like English. This method of interaction is referred to as **natural language interaction.** Presently, natural language interaction is not as viable an interaction style as other methods presented. Current implementations can be tedious, frustrating, and time-consuming for the user and are often built to accept input in narrowly constrained domains (e.g., database queries). Natural language interaction is being applied within both keyboard and voice entry systems.

Hardware Options for System Interaction

In addition to the variety of methods used for interacting with a system, there is also a growing number of hardware devices employed to support this interaction (see Table 14-2 for a list of interaction devices along with brief descriptions of the

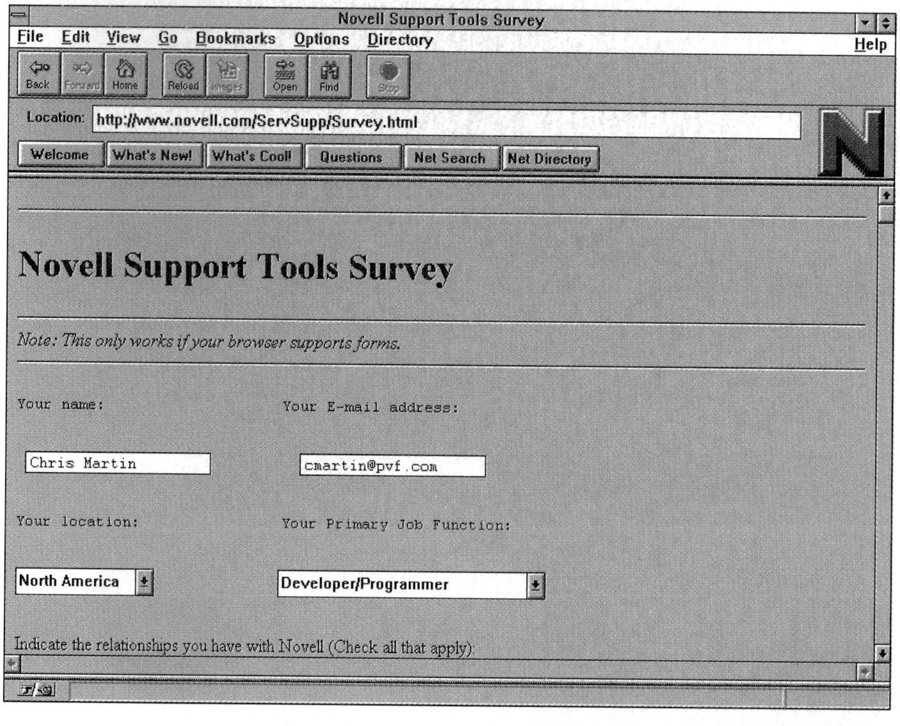

Object-based interaction: A human-computer interaction method where symbols are used to represent commands or functions.

Icon: Graphical pictures that represent specific functions within a system.

Natural language interaction: A human-computer interaction method whereby inputs to and outputs from a computer-based application are in a conventional speaking language such as English.

Figure 14-9

Example of form interaction in the Netscape™ World Wide Web browser

Figure 14-10
Icon-based interface for configuring user preferences within WordPerfect for Windows 6.0

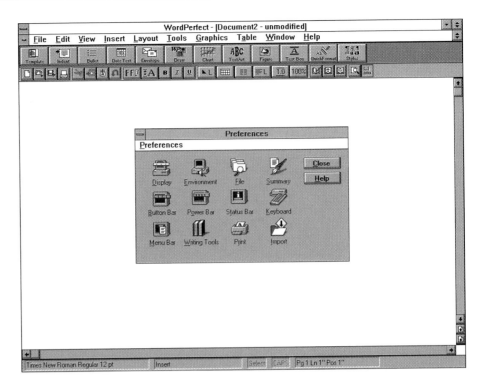

TABLE 14-2   Common Devices for Interacting with an Information System

Device	Description and Primary Characteristics or Usage
Keyboard	Users push an array of small buttons that represent symbols which are then translated into words and commands. Keyboards are widely understood and provide considerable flexibility for interaction.
Mouse	A small plastic box that users push across a flat surface and whose movements are translated into cursor movement on a computer display. Buttons on the mouse tell the system when an item is selected. A mouse works well on flat desks but may not be practical in dirty or busy environments, such as a shop floor or check-out area in a retail store. Newer pen-based mice provide the user with more of the feel of a writing implement.
Joystick	A small vertical lever mounted on a base that steers the cursor on a computer display. Provides similar functionality to a mouse.
Trackball	A sphere mounted on a fixed base that steers the cursor on a computer display. A suitable replacement for a mouse when work space for a mouse is not available.
Touch Screen	Selections are made by touching a computer display. This works well in dirty environments or for users with limited dexterity or expertise.
Light Pen	Selections are made by pressing a pen-like device against the screen. A light pen works well when the user needs to have a more direct interaction with the contents of the screen.
Graphics Tablet	Moving a pen-like device across a flat tablet steers the cursor on a computer display. Selections are made by pressing a button or by pressing the pen against the tablet. This device works well for drawing and graphical applications.
Voice	Spoken words are captured and translated by the computer into text and commands. This is most appropriate for users with physical challenges or when hands need to be free to do other tasks while interacting with the application.

typical usage of each). The most fundamental and widely used is the keyboard, which is the mainstay of most computer-based applications for the entry of alphanumeric information. Keyboards vary, from the typewriter kind of keyboards used with personal computers to special-function keyboards on point-of-sale or shop floor devices. The growth in graphical user environments, however, has spurred the broader use of pointing devices such as mice, joysticks, trackballs, and graphics tablets. The creation of notebook and pen-based computers which have trackballs, joysticks, or pens attached directly to the computer has also brought renewed interest to the usability of these various devices.

Research has found that each device has its strengths and weaknesses which must guide your selection of the appropriate devices to aid users in their interaction with an application. The selection of devices users will use for interaction must be made during logical design since different interfaces require different devices. Table 14-3 summarizes much of the usability assessment research by relating each device to various types of human-computer interaction problems. For example, for many applications keyboards do not give users a precise feel for cursor movement, do not provide direct feedback on each operation, and can be a slow way to enter data (depending on the typing skill of the user). Another means to gain an understanding of device usability is to highlight which devices have been found most useful for completing specific tasks. The results of this research are summarized in Table 14-4. The rows of this table list common user-computer interaction tasks, and the columns show three criteria for evaluating the usability of the different devices. After reviewing these three tables, it should be evident that no device is perfect and

TABLE 14-3 Summary of Interaction Device Usability Problems
(Adapted from Blattner & Schultz, 1988)

Device	Problem						
	Visual Blocking	User Fatigue	Movement Scaling	Durability	Adequate Feedback	Speed	Pointing Accuracy
Keyboard	□	□	■	□	■	■	□
Mouse	□	□	■	□	■	□	□
Joystick	□	□	■	□	■	□	■
Trackball	□	□	■	■	■	□	□
Touch Screen	■	■	□	■	□	□	■
Light Pen	■	■	□	□	□	□	■
Graphics Tablet	□	□	■	□	■	□	□
Voice	□	□	■	□	■	□	■

Key:

□ = little or no usability problems
■ = potentially high usability problems for some applications
Visual Blocking = extent to which device blocks display when using
User Fatigue = potential for fatigue over long use
Movement Scaling = extent to which device movement translates to equivalent screen movement
Durability = lack of durability or need for maintenance (e.g., cleaning) over extended use
Adequate Feedback = extent to which device provides adequate feedback for each operation
Speed = cursor movement speed
Pointing Accuracy = ability to precisely direct cursor

TABLE 14-4 Summary of General Conclusions from Experimental Comparisons of Input Devices in Relation to Specific Task Activities
(Adapted from Blattner & Schultz, 1988)

Task	Most Accurate	Shortest Positioning	Most Preferred
Target Selection	trackball, graphics tablet, mouse, joystick	touch screen, light pen, mouse, graphics tablet, trackball	touch screen, light pen
Text Selection	mouse	mouse	—
Data Entry	light pen	light pen	—
Cursor Positioning	—	light pen	—
Text Correction	light pen, cursor keys	light pen	light pen
Menu Selection	touch screen	—	keyboard, touch screen

Key:

Target Selection =	moving the cursor to select a figure or item
Text Selection =	moving the cursor to select a block of text
Data Entry =	entering information of any type into a system
Cursor Positioning =	moving the cursor to a specific position
Text Correction =	moving the cursor to a location to make a text correction
Menu Selection =	activating a menu item
— =	no clear conclusion from the research

that some are more appropriate for performing some tasks than others. To design the most effective interfaces for a given application, you should understand the capabilities of various interaction methods and devices.

DESIGNING INTERFACES

Building on the information provided in Chapter 13 on the design of content for forms and reports, in this section we discuss issues related to the design of interface layouts. This discussion provides guidelines for structuring and controlling data entry fields, providing feedback, and designing on-line help. Effective interface design requires that you gain a thorough understanding of each of these concepts.

Designing Layouts

To ease user training and data recording, you should use standard formats for computer-based forms and reports similar to paper-based forms and reports for recording or reporting information. A typical paper-based form for reporting customer sales activity is shown in Figure 14-11. This form has several general areas common to most forms:

- Header information
- Sequence and time-related information
- Instruction or formatting information
- Body or data details

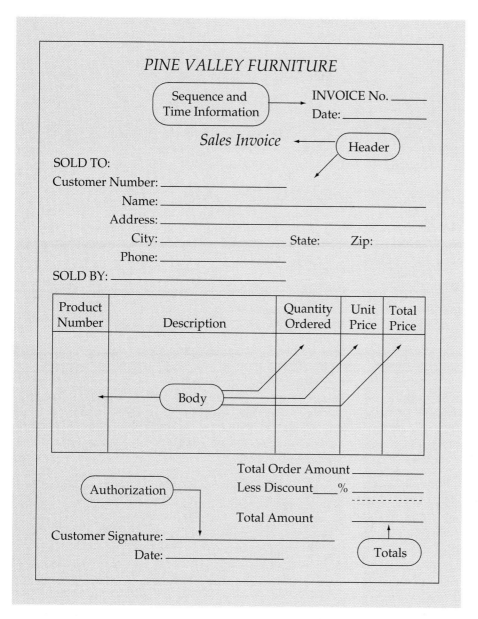

Figure 14-11
Paper-based form for reporting customer sales activity (Pine Valley Furniture)

- Totals or data summary

- Authorization or signatures

- Comments

In many organizations, data is often first recorded on paper-based forms and then later recorded within application systems. When designing layouts to record or display information on paper-based forms, you should try to make both as similar as possible. Additionally, data entry displays should be consistently formatted across applications to speed data entry and reduce errors. Figure 14-12 shows an equivalent computer-based form to the paper-based form shown in Figure 14-11.

Another concern when designing the layout of computer-based forms is the design of between-field navigation. Since you can control the sequence for users to

Figure 14-12

Computer-based form reporting customer sales activity (Pine Valley Furniture)

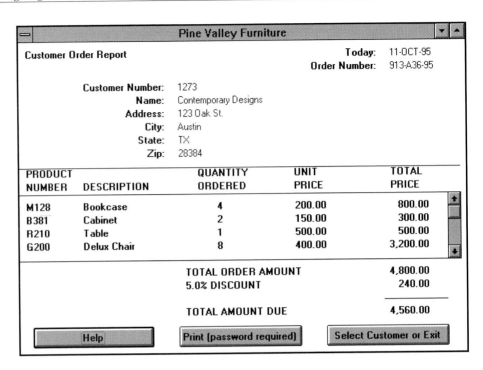

		Pine Valley Furniture		
Customer Order Report			**Today:**	11-OCT-95
			Order Number:	913-A36-95

Customer Number:	1273
Name:	Contemporary Designs
Address:	123 Oak St.
City:	Austin
State:	TX
Zip:	28384

PRODUCT NUMBER	DESCRIPTION	QUANTITY ORDERED	UNIT PRICE	TOTAL PRICE
M128	Bookcase	4	200.00	800.00
B381	Cabinet	2	150.00	300.00
R210	Table	1	500.00	500.00
G200	Delux Chair	8	400.00	3,200.00

TOTAL ORDER AMOUNT	4,800.00
5.0% DISCOUNT	240.00
TOTAL AMOUNT DUE	4,560.00

Help	Print (password required)	Select Customer or Exit

move between fields, standard screen navigation should flow from left to right and top to bottom just as when you work on paper-based forms. For example, Figure 14-13 contrasts the flow between fields on a form used to record business contacts. Figure 14-13a uses a consistent left to right, top to bottom flow. Figure 14-13b uses a flow that is non-intuitive. When appropriate, you should also group data fields into logical categories with labels describing the contents of the category. Areas of the screen not used for data entry or commands should be inaccessible to the user.

When designing the navigation procedures within your system, flexibility and consistency are primary concerns. Users should be able to freely move forward and backward or to any desired data entry fields. Users should be able to navigate each form in the same way or in as similar a manner as possible. Additionally, data should not *usually* be permanently saved by the system until the user makes an explicit request to do so. This allows the user to abandon a data entry screen, back up, or move forward without adversely impacting the contents of the permanent data.

Consistency extends to the selection of keys and commands. Each key or command should have only one function and this function should be consistent throughout the entire system and across systems, if possible. Depending upon the application, various types of functional capabilities will be required to provide smooth navigation and data entry. Table 14-5 provides a list of the functional requirements for providing smooth and easy navigation within a form. For example, a functional and consistent interface will provide common ways for users to move the cursor to different places on the form, editing characters and fields, moving among form displays, and obtaining help. These functions may be provided by keystrokes, mouse or other pointing device operations, or menu selection or button activation. It is possible that, for a single application, all functional capabilities listed in Table 14-5 may not be needed in order to create a flexible and consistent user interface. Yet, the capabilities that are used should be consistently applied to provide an optimal user environment. As with other tables in Chapters 13 and 14, Table 14-5 can serve as a checklist for you to validate the usability of user interface designs.

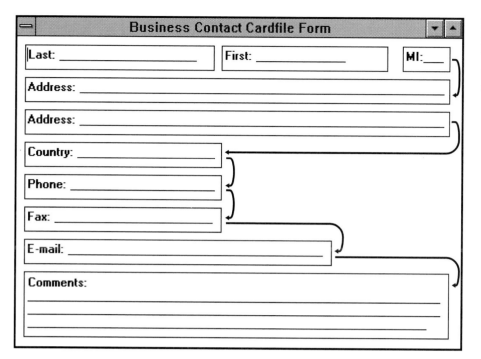

Figure 14-13
Contrasting the navigation
flow within a data entry form

(a) Proper flow between data
entry fields

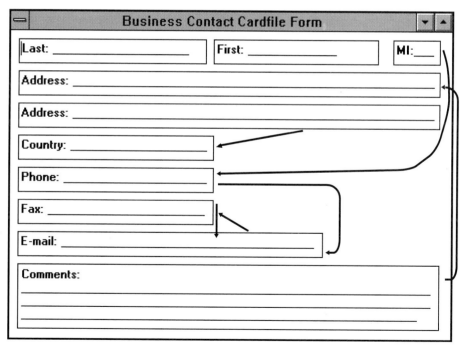

(b) Poor flow between data
entry fields

Structuring Data Entry

Several rules should be considered when structuring data entry fields on a form (see Table 14-6). The first is simple, but often violated by designers. To minimize data entry errors and user frustration, *never* require the user to enter information which is already available within the system or information that can be easily

TABLE 14-5 Data Entry Screen Functional Capabilities
(Adapted from Dumas, 1988)

Cursor Control Capabilities:

Move the cursor forward to the next data field

Move the cursor backward to the previous data field

Move the cursor to the first, last, or some other designated data field

Move the cursor forward one character in a field

Move the cursor backward one character in a field

Editing Capabilities:

Delete the character to the left of the cursor

Delete the character under the cursor

Delete the whole field

Delete data from the whole form (empty the form)

Exit Capabilities:

Transmit the screen to the application program

Move to another screen/form

Confirm the saving of edits or go to another screen/form

Help Capabilities:

Get help on a data field

Get help on a full screen/form

TABLE 14-6 Guidelines for Structuring Data Entry Fields
(Adapted from Dumas, 1988)

Entry	Never require data that is already on-line or that can be computed; for example, do not enter customer data on an order form if that data can be retrieved from the database, and do not enter extended prices which can be computed from quantity sold and unit prices.
Defaults	Always provide default values when appropriate; for example, assume today's date for a new sales invoice, or use the standard product price unless overridden
Units	Make clear the type of data units requested for entry; for example, indicate quantity in tons, dozens, pounds, etc.
Replacement	Use character replacement when appropriate; for example, allow the user to look up the value in a table or automatically fill in the value once the user enters enough significant characters.
Captioning	Always place a caption adjacent to fields; see Table 14-7 for caption options.
Format	Provide formatting examples when appropriate; for example, automatically show standard embedded symbols, decimal points, credit symbol, or dollar sign.
Justify	Automatically justify data entries; numbers should be right-justified and aligned on decimal points, and text should be left-justified.
Help	Provide context-sensitive help when appropriate; for example, provide a hot key, such as the F1 key, that opens the help system on an entry that is most closely related to where the cursor is on the display.

computed by the system. For example, never require the user to enter the current date and time, since each of these values can be easily retrieved from the computer system's internal calendar and clock. By allowing the system to do this, the user simply confirms that the calendar and clock are working properly.

Other rules are equally important. For example, suppose that a bank customer is repaying a loan on a fixed schedule with equal monthly payments. Each month when a payment is sent to the bank, a clerk needs to record that the payment has been received into a loan processing system. Within such a system, default values for fields should be provided whenever appropriate. This means that *only* in the instances where the customer pays *more or less* than the scheduled amount should the clerk have to enter data into the system. In all other cases, the clerk would simply verify that the check is for the default amount provided by the system and press a single key to confirm the receipt of payment.

When entering data, the user should also not be required to specify the dimensional units of a particular value. For example, a user should not be required to specify that an amount is in dollars or that a weight is in tons. Field formatting and the data entry prompt should make clear the type of data being requested. In other words, a caption describing the data to be entered should be adjacent to each data field. Within this caption, it should be clear to the user what type of data is being requested. As with the display of information, all data entered onto a form should automatically justify in a standard format (e.g., date, time, money). Table 14-7 illustrates a few options appropriate for printed forms. For data entry on video display terminals, you should highlight the area in which text is entered so that the exact number of characters per line and number of lines are clearly shown. You can also use check boxes or radio buttons to allow users to choose standard textual responses. And, you can use data entry controls to ensure that the proper type of data (alphabetic or numeric, as required) are entered. Data entry controls are discussed next.

Controlling Data Input

One objective of interface design is to reduce data entry errors. As data are entered into an information system, steps must be taken to ensure that the input is valid. As a systems analyst, you must anticipate the types of errors users may make and design features into the system's interfaces to avoid, detect, and correct data entry mistakes. Several types of data errors are summarized in Table 14-8. In essence, data errors can occur from appending extra data onto a field, truncating characters

TABLE 14-7   Options for Entering Text

Options	Example
Line caption	Phone Number ( ) -
Drop caption	( ) - Phone Number
Boxed caption	Phone number
Delimited characters	( ) - Phone Number
Check-off boxes	Method of payment (check one) ☐ Check ☐ Cash ☐ Credit card: Type

TABLE 14-8   Sources of Data Errors

Data Error	Description
Appending	Adding additional characters to a field
Truncating	Losing characters from a field
Transcripting	Entering invalid data into a field
Transposing	Reversing the sequence of one or more characters in a field

off a field, transcripting the wrong characters into a field, or transposing one or more characters within a field. Systems designers have developed numerous tests and techniques for catching invalid data before saving or transmission, thus improving the likelihood that data will be valid (see Table 14-9 which summarizes these techniques). These tests and techniques are often incorporated into both data entry screens and inter-computer data transfer programs.

Practical experience has also found that correcting erroneous data is much easier to accomplish before it is permanently stored in a system. On-line systems can notify a user of input problems as data are being entered. When data are processed on-line as events occur, it is much less likely that data validity errors will occur and not be caught. In an on-line system, most problems can be easily identified and resolved before permanently saving data to a storage device using many of the techniques described in Table 14-9. However, in systems where inputs are stored and entered (or transferred) in batch, the identification and notification of errors is

TABLE 14-9   Validation Tests and Techniques to Enhance the Validity of Data Input

Validation Test	Description
Class or Composition	Test to assure that data are of proper type (e.g., all numeric, all alphabetic, alphanumeric)
Combinations	Test to see if the value combinations of two or more data fields are appropriate or make sense (e.g., does the quantity sold make sense given the type of product?)
Expected Values	Test to see if data is what is expected (e.g., match with existing customer names, payment amount, etc.)
Missing Data	Test for existence of data items in all fields of a record (e.g., is there a quantity field on each line item of a customer order?)
Pictures/Templates	Test to assure that data conform to a standard format (e.g., are hyphens in the right places for a student ID number?)
Range	Test to assure data are within proper range of values (e.g., is a student's grade point average between 0 and 4.0?)
Reasonableness	Test to assure data are reasonable for situation (e.g., pay rate for a specific type of employee)
Self-Checking Digits	Test where an extra digit is added to a numeric field in which its value is derived using a standard formula (see Figure 14-14)
Size	Test for too few or too many characters (e.g., is social security number exactly nine digits?)
Values	Test to make sure values come from set of standard values (e.g., two-letter state codes)

more difficult. Batch processing systems can, however, reject invalid inputs and store them in a log file for later resolution.

Most of the tests and techniques shown in Table 14-9 are widely used and straightforward. Some of these tests can be handled by data management technologies, such as a database management system (DBMS), to ensure that they are applied for all data maintenance operations. If a DBMS cannot perform these tests, then you must design the tests into program modules. An example of one item that is a bit sophisticated, self-checking digits, is shown in Figure 14-14. The figure provides a description and an outline of how to apply the technique as well as a short example. The example shows how a check digit is added to a field before data entry or transfer. Once entered or transferred, the check digit algorithm is again applied to the field to "check" whether the check digit received obeys the calculation. If it does, it is likely (but not guaranteed, since two different values could yield the same check digit) that no data transmission or entry error occurred. If not equal, then some type of error occurred.

In addition to validating the data values entered into a system, controls must be established to verify that all input records are correctly entered and that they are only processed once. A common method used to enhance the validity of entering batches of data records is to create an audit trail of the entire sequence of data entry, processing, and storage. In such an audit trail, the actual sequence, count, time, source location, human operator, and so on are recorded into a separate transaction log in the event of a data input or processing error. If an error occurs, corrections can be made by reviewing the contents of the log. Detailed logs of data inputs are not only useful for resolving batch data entry errors and system audits but also serve as a powerful method for performing backup and recovery operations in the case of a catastrophic system failure. These types of file and database controls are discussed further in Chapter 16.

Figure 14-14
Using check digits to verify data correctness

Description	Techniques where extra digits are added to a field to assist in verifying its accuracy
Method	1. Multiply each digit of a numeric field by a weighting factor (e.g., 1,2,1,2,...). 2. Sum the results of weighted digits. 3. Divide sum by modulus number (e.g., 10). 4. Subtract remainder of division from modulus number to determine check digit. 5. Append check digits to field.
Example	Assume a numeric part number of: 12473 1–2. Multiply each digit of part number by weighting factor from right to left and sum the results of weighted digits: $$\begin{array}{ccccc} 1 & 2 & 4 & 7 & 3 \\ \times 1 & \times 2 & \times 1 & \times 2 & \times 1 \\ \hline 1 \;+ & 4 \;+ & 4 \;+ & 14 \;+ & 3 \;=\; 26 \end{array}$$ 3. Divide sum by modulus number. $26/10 = 2$ remainder 6 4. Subtract remainder from modulus number to determine check digit. check digit $= 10 - 6 = 4$ 5. Append check digits to field. Field value with appended check digit $= 124734$

Providing Feedback

When talking with a friend, you would be concerned if he or she did not provide you with feedback by nodding and replying to your questions and comments. Without feedback, you would be concerned that he or she was not listening, likely resulting in a less than satisfactory experience. Similarly, when designing system interfaces, providing appropriate feedback is an easy method for making a user's interaction more enjoyable; not providing feedback is a sure way to frustrate and confuse. System feedback can consist of three types:

1. Status information

2. Prompting cues

3. Error or warning messages

Status Information   Providing status information is a simple technique for keeping users informed of what is going on within a system. For example, relevant status information such as displaying the current customer name or time, placing appropriate titles on a menu or screen, or identifying the number of screens following the current one (e.g., Screen 1 of 3) all provide needed feedback to the user. Providing status information during processing operations is especially important if the operation takes longer than a second or two. For example, when opening a file you might display "Please wait while I open the file" or, when performing a large calculation, flash the message "Working . . ." to the user. Further, it is important to tell the user that besides working, the system has accepted the user's input and the input was in the correct form. Sometimes it is important to give the user a chance to obtain more feedback. For example, a function key could toggle between showing a "Working . . ." message and giving more specific information as each intermediate step is accomplished. Providing status information will reassure users that nothing is wrong and make them feel in command of the system, not vice versa.

Prompting Cues   A second feedback method is to display prompting cues. When prompting the user for information or action, it is useful to be specific in your request. For example, suppose a system prompted users with the following request:
  READY FOR INPUT:_____
With such a prompt, the designer assumes that the user knows exactly what to enter. A better design would be specific in its request, possibly providing an example, default values, or formatting information. An improved prompting request might be as follows:
  Enter the customer account number (123-456-7):_____-_____-__

Errors and Warning Messages   A final method available to you for providing system feedback is using error and warning messages. Practical experience has found that a few simple guidelines can greatly improve their usefulness. First, messages should be specific and free of error codes and jargon. Additionally, messages should never scold the user and should attempt to guide the user toward a resolution. For example, a message might say "No Customer Record Found for That Customer ID. Please verify that digits were not transposed." Messages should be in user, not computer, terms. Hence, such terms as "end of file," "disk I/O error," or "write protected" may be too technical and not helpful for many users. Multiple messages can be useful so that a user can get more detailed explanations if wanted or needed. Also, error messages should appear in roughly the same format and placement each time so that they are recognized as error messages and not as some other information. Examples of good and bad messages are provided in Table 14-10. Using

TABLE 14-10   Examples of Poor and Improved Error Messages

Poor Error Messages	Improved Error Messages
ERROR 56 OPENING FILE	The file name you typed was not found. Press F2 to list valid file names.
WRONG CHOICE	Please enter an option from the menu.
DATA ENTRY ERROR	The prior entry contains a value outside the range of acceptable values. Press F9 for list of acceptable values.
FILE CREATION ERROR	The file name you entered already exists. Press F10 if you want to overwrite it. Press F2 if you want to save it to a new name.

these guidelines, you will be able to provide useful feedback in your designs. A special type of feedback is answering help requests from users. This important topic is described next.

Providing Help

Designing how to provide help is one of the most important interface design issues you will face. When designing help, you need to put yourself in the user's place. When accessing help, the user likely does not know what to do next, does not understand what is being requested, or does not know how the requested information needs to be formatted. A user requesting help is much like a ship in distress, sending an SOS. In Table 14-11, we provide our SOS guidelines for the design of system help: Simplicity, Organize, and Show. Our first guideline, *simplicity,* suggests that help messages should be short, to the point, and use words that enable understanding. This leads to our second guideline, *organize,* which means that help messages should be written so that information can be easily absorbed by users. Practical experience has found that long paragraphs of text are often difficult for people to understand. A better design organizes lengthy information in a manner easier for users to digest through the use of bulleted and ordered lists. Finally, it is often useful to explicitly *show* users how to perform an operation and the outcome of procedural steps. Figure 14-15 contrasts the designs of two help screens, one employing our guidelines and one that does not.

Many commercially available systems provide extensive system help. For example, Table 14-12 lists the range of help available in a popular electronic spreadsheet. Many systems are also designed so that users can vary the level of detail provided. Help may be provided at the system level, screen or form level, and individual field level. The ability to provide field level help is often referred to as "context-sensitive" help. For some applications, providing context-sensitive help for all system options is a tremendous undertaking that is virtually a project in itself. If you do decide to design an extensive help system with many levels of detail,

TABLE 14-11   Guidelines for Designing Usable Help

Guideline	Explanation
Simplicity	Use short, simple wording, common spelling, and complete sentences. Give users only what they need to know, with ability to find additional information.
Organize	Use lists to break information into manageable pieces.
Show	Provide examples of proper use and the outcomes of such use.

Figure 14-15
Contrasting help screens

(a) Poorly designed help
display

(b) Improved design for help
display

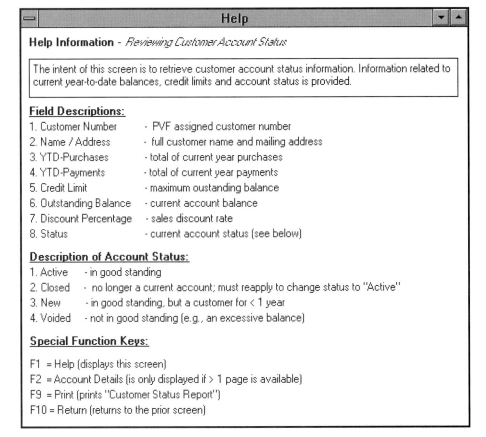

TABLE 14-12   Types of Help
(Adapted from Dumas, 1988)

Type of Help	Example of Question
Help on Help	How do I get help?
Help on Concepts	What is a customer record?
Help on Procedures	How do I update a record?
Help on Messages	What does "Invalid File Name" mean?
Help on Menus	What does "Graphics" mean?
Help on Function Keys	What does each Function key do?
Help on Commands	How do I use the "Cut" and "Paste" commands?
Help on Words	What do "merge" and "sort" mean?

you must be sure that you know exactly what the user needs help with, or your efforts may confuse users more than help them. After leaving a help screen, users should always return back to where they were prior to requesting help. If you follow these simple guidelines, you will likely design a highly usable help system.

As with the construction of menus, many programming environments provide powerful tools for designing system help. For example, Microsoft's Help Compiler allows you to quickly construct hypertext-based help systems. In this environment, you use a text editor to construct help pages that can be easily linked to other pages containing related or more specific information. Linkages are created by embedding special characters into the text document that make words hypertext buttons— that is, direct linkages—to additional information. The Help Compiler transforms the text document into a hypertext document. For example, Figure 14-16 shows a

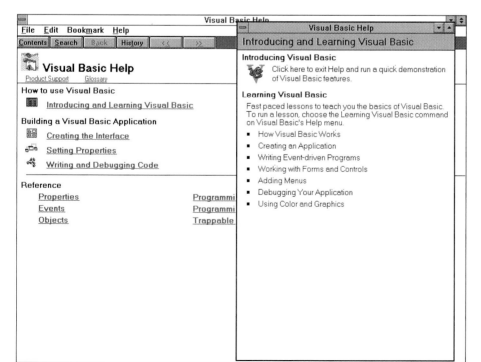

Figure 14-16
Hypertext-based help system from Microsoft's Visual Basic

hypertext-based help screen from Microsoft's Visual Basic. Hypertext-based help systems have become the standard environment for most commercial Windows-based applications. This has occurred for two primary reasons. First, standardizing system help across applications eases user training. Second, hypertext allows users to selectively access the level of help they need, making it easier to provide effective help for both novice and experienced users within the same system.

CONTROLLING USER ACCESS

Corporations and government agencies are putting more and more sensitive data on computers. As the number and sophistication of user access paths to the data increase, those data become vulnerable to unwanted access or corruption. A recent survey of computer security professionals estimates the annual losses from computer abuse at over $500,000,000 nationally. Among the survey respondents, the average installation experienced losses of over $100,000 and 365 person-hours per year (Bloombecker, 1989). Part of your job during interface design is to decide which forms of user access controls are necessary and make your choice or advise other system developers on which mechanisms will be most effective in securing the application.

You have several techniques available to control user access:

1. Views or subschemas: Restrict user views of a database

2. Authorization rules: Restrict user access and activities within a system

3. Encryption procedures: Encode data in an unrecognizable form

4. Authentication schemes: Identify users with certainty so that other security mechanisms can work

In essence, since an interface defines how users are able to create, read, update, or delete data, security controls should be designed into an interface when it is initially designed. Experience has shown that if designed separately from the rest of the interface, security procedures are usually cumbersome and less effective.

Views

View: A subset of the database that is presented to one or more users.

One form of user access control for system interfaces is to provide a customized version of the systems data to the user. A **view** is a subset of the database that is presented to one or more users. As long as the user can access data only through the system, they will be able to use only the data, raw or summarized, available in the view supported by the parts of the system they are authorized to use. Many database management systems include this concept. If not, it is possible to achieve the same effect through controls on access to specific programs and use of other mechanisms described below. Since views are commonly covered in database management texts (McFadden and Hoffer, 1994), we do not illustrate this control here.

Although views promote security by restricting user access to limited data, views are not adequate security measures, as unauthorized persons may gain knowledge of or access to a particular view. Also, several persons may share a particular view; all may have authority to read the data but only a restricted few may be authorized to update the data. Finally, with high-level query languages, an unauthorized person may gain access to data through simple experimentation. As a result, more sophisticated security measures are normally required.

Authorization Rules

Controls incorporated in an application or system software that restrict access to data and also restrict the actions that people may take when data are accessed are called **authorization rules.** For example, a person who can supply a particular password may be authorized to gain access to a system and read any record in a database but cannot necessarily modify any of those records.

Fernandez, Summers, and Wood (1981) have developed a conceptual model for authorization rules within database environments. Their model expresses authorization rules in the form of a table (or matrix) that includes subjects, objects, actions, and constraints. Each row of the table indicates that a particular subject is authorized to take a certain action on an object in the database, perhaps subject to some constraint. An example of such an authorization matrix is shown in Figure 14-17. This table contains several entries pertaining to records in an accounting database. The first row in the table indicates that anyone in the Sales Department is authorized to insert a new customer record in the database, provided that the customer's credit limit does not exceed $5,000. The last row indicates that the program AR4 is authorized to modify order records without restriction. Owners of data along with data and system administration are responsible for determining authorization rules.

Authorization rules: Controls incorporated to restrict access to systems and data and also to restrict the actions that people may take once in the system.

Encryption Procedures

For highly sensitive data (such as company financial data), data encryption can be used. Encryption is the coding (or scrambling) of data so that they cannot be read by humans (see Chapter 16). Special hardware and software can be purchased to encrypt data so that users cannot read stored data except through the authorized use of decoding mechanisms. Some DBMS products include encryption routines that automatically encode sensitive data when they are stored or transmitted over communications channels. For example, encryption is commonly used in electronic funds transfer (EFT) systems. Any system that provides encryption facilities must also provide complementary routines for decoding the data. These decoding routines must, of course, be protected by adequate security, or else the advantages of encryption are lost.

Authentication Schemes

A longstanding problem in computer circles is how to positively identify persons who are trying to gain access to a computer or its resources. Passwords cannot, of themselves, ensure the security of a computer and its databases because they give no indication of who is trying to gain access. For systems that involve highly sensitive data or perform functions initiated by a select group of people, it is important

SUBJECT	OBJECT	ACTION	CONSTRAINT
Sales Dept.	Customer record	Insert	Credit limit ≤ $5,000
Order Trans.	Customer record	Read	None
Terminal 12	Customer record	Modify	Balance due only
Accounting Dept.	Order record	Delete	None
Luke Skywalker	Order record	Insert	Order amt < $2,000
Program AR4	Order record	Modify	None

Figure 14-17
Authorization matrix

Biometric device: An instrument that detects personal characteristics such as fingerprints, voice prints, retina prints, or signature dynamics.

Smart card: A thin plastic card the size of a credit card with an embedded microprocessor and memory.

to design mechanisms into the system interface to ensure that only authorized people are using the system.

The computer industry is developing devices and techniques to positively identify any prospective user. The most promising of these appear to be **biometric devices** which measure or detect personal characteristics such as fingerprints, voice prints, retina prints, or signature dynamics. To implement this approach, several companies have developed a **smart card**—a thin plastic card, the size of a credit card, with an embedded microprocessor and memory. An individual's unique biometric data (such as fingerprints) are stored permanently on the card. To access a computer, the user inserts the card into a device that reads the person's unique biometric data from the smart card. A second device reads the user's actual characteristics and compares this data with the data stored on the smart card. The two must match to gain computer access. A lost or stolen card would be useless to another person, since the biometric data would not match.

In this and the prior sections, we have described a wide variety of issues— from how to structure data entry fields to how to be sure human-system interactions are legitimate—relevant to the design of system interfaces. Designing useful interfaces requires that you effectively apply numerous guidelines and pay careful attention to issues of usability. Doing this will ensure that information is clearly laid out, data entry fields are effectively designed, and appropriate amounts of feedback and help are provided. After completing interface design, you will have available individual displays to support specific tasks, such as user sign-on, entry of a customer order, or selection of a particular reporting option. The next step is to design the sequencing of, or navigation between, these displays. This process is the topic of the next section—designing dialogues.

DESIGNING DIALOGUES

Dialogue: The sequence of interaction between a user and a system.

The process of designing the overall sequences that users follow to interact with an information system is called dialogue design. A **dialogue** is the sequence in which information is displayed to and obtained from a user. As the designer, your role is to select the most appropriate interaction methods and devices (described above) and to define the conditions under which information is displayed and obtained from users. There are three major steps:

1. Designing the dialogue sequence

2. Building a prototype

3. Assessing usability

There are a few general rules that should be followed when designing a dialogue and they are summarized in Table 14-13. For a dialogue to have high usability, it must be consistent in form, function, and style. All other rules regarding dialogue design are mitigated by the consistency guideline. For example, the effectiveness of how well errors are handled or feedback is provided will be significantly influenced by consistency in design. If the system does not consistently handle errors, the user will often be at a loss as to why certain things happen.

One example of these guidelines concerns removing data from a database or file (see the Reversal entry in Table 14-13). It is good practice to display the information that will be deleted before making a permanent change to the file. For example, if the customer service representative wanted to remove a customer from the database, the system should ask only for the customer ID in order to retrieve the correct customer account. Once found, and before allowing the confirmation of the

TABLE 14-13 Guidelines for the Design of Human-Computer Dialogues
(Adapted from Shneiderman, 1992)

Guideline	Explanation
Consistency	Dialogues should be consistent in sequence of actions, keystrokes, and terminology (e.g., the same labels should be used for the same operations on all screens, and the location of the same information should be the same on all displays).
Shortcuts and Sequence	Allow advanced users to take shortcuts using special keys (e.g., CTRL-C to copy highlighted text). A natural sequence of steps should be followed (e.g., enter first name before last name, if appropriate).
Feedback	Feedback should be provided for every user action (e.g., confirm that a record has been added, rather than simply putting another blank form on the screen).
Closure	Dialogues should be logically grouped and have a beginning, middle, and end (e.g., the last in the sequence of screens should indicate that there are no more screens).
Error Handling	All errors should be detected and reported; suggestions on how to proceed should be made (e.g., suggest why such errors occur and what user can do to correct the error). Synonyms for certain responses should be accepted (e.g., accept either "t," "T," or "TRUE").
Reversal	Dialogues should, when possible, allow the user to reverse actions (e.g., undo a deletion); data should not be destructed without confirmation (e.g., display all the data for a record the user has indicated is to be deleted).
Control	Dialogues should make the user (especially an experienced user) feel in control of the system (e.g., provide a consistent response time at a pace acceptable to the user).
Ease	Dialogues should be simple for users to enter information and navigate between screens (e.g., provide means to move forward, backward, and to specific screens, such as first and last).

deletion, the system should display the account information. For actions making permanent changes to system data files and when the action is not commonly performed, many system designers use the *double-confirmation* technique where the users must confirm their intention twice before being allowed to proceed.

Designing the Dialogue Sequence

Your first step in dialogue design is to define the sequence. In other words, you must first gain an understanding of how users might interact with the system. This means that you must have a clear understanding of user, task, technological, and environmental characteristics when designing dialogues. Suppose that the marketing manager at Pine Valley Furniture (PVF) wants sales and marketing personnel to be able to review the year-to-date transaction activity for any PVF customer. After talking with the manager, you both agree that a typical dialogue between a user and the Customer Information System for obtaining this information might proceed as follows:

1. Request to view individual customer information

2. Specify the customer of interest

3. Select the year-to-date transaction summary display

4. Review customer information

5. Leave system

As a designer, once you understand how a user wishes to use a system, you can then transform these activities into a formal dialogue specification.

Dialogue diagramming: A formal method for designing and representing human-computer dialogues using box and line diagrams.

A formal method for designing and representing dialogues is **dialogue diagramming.** Dialogue diagrams have only one symbol, a box with three sections; each box represents one display (which might be a full screen or a specific form or window) within a dialogue (see Figure 14-18). The three sections of the box are used as follows:

1. *Top:* Contains a unique display reference number used by other displays for referencing it

2. *Middle:* Contains the name or description of the display

3. *Bottom:* Contains display reference numbers that can be accessed from the current display

All lines connecting the boxes within dialogue diagrams are assumed to be bi-directional and thus do not need arrowheads to indicate direction. This means that users are allowed to always move forward and backward between adjacent displays. If you desire only uni-directional flows within a dialogue, arrowheads should be placed on one end of the line. Within a dialogue diagram, you can easily represent the sequencing of displays, the selection of one display over another, or the repeated use of a single display (e.g., a data entry display). These three concepts—sequence, selection, and iteration—are illustrated in Figure 14-19.

Continuing with our PVF example, Figure 14-20 shows a partial dialogue diagram for processing the marketing manager's request. In this diagram, the analyst placed the request to view year-to-date customer information within the context of the overall Customer Information System. The user must first gain access to the system through a log-on procedure (item 0). If log-on is successful, a main menu is displayed that has four items (item 1). Once the user selects the Individual Customer Information (item 2), control is transferred to the Select Customer display (item 2.1). After a customer is selected, the user is presented with an option to view customer information four different ways (item 2.1.1). Once the user views the customer's year-to-date transaction activity (item 2.1.1.2), the system will allow the user to back up to select a different customer, back up to the main menu, or exit the system (see bottom of item 2.1.1.2).

Building Prototypes and Assessing Usability

Building dialogue prototypes and assessing usability are often optional activities. Some systems may be very simple and straightforward. Others may be more complex but are extensions to existing systems where dialogue and display standards have already been established. In either case, you may not be required to build prototypes and do a formal assessment. However, for many other systems, it is critical that you build prototype displays and then assess the dialogue; this can pay numerous divi-

Figure 14-18

Sections of a dialogue diagramming box

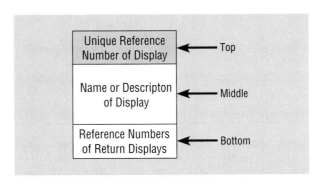

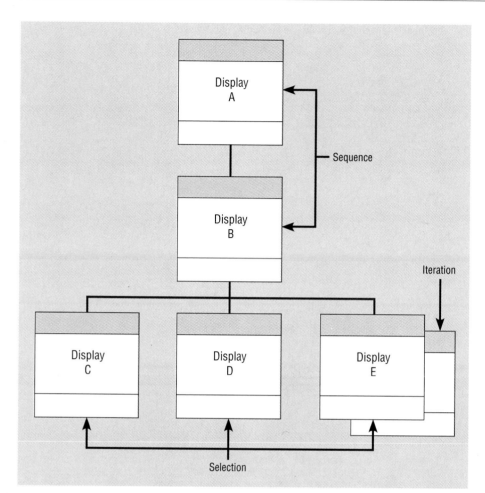

Figure 14-19
Dialogue diagram illustrating sequence, selection, and iteration

dends later in the systems development life cycle (for example, it may be easier to implement a system or train users on a system they have already seen and used).

Building prototype displays is often a relatively easy activity for you to perform when using CASE tools or many of the graphical development environments such as Microsoft's Visual Basic or Visual C. Recall that Chapter 4 extensively describes the use of CASE for the design of system displays and interfaces. Some systems development environments include easy-to-use input and output (form, report, or window) design utilities. There are also several tools called "Prototypers" or "Demo Builders" that allow you to quickly design displays and show how an interface will work within a full system. These demo systems allow users to enter data and move through displays as if using the actual system. Such activities are not only useful for you to show how an interface will look and feel, they are also useful for assessing usability and for performing user training long before actual systems are completed. In the next section, we extend our discussion of interface and dialogue design to consider issues specific to graphical user interface environments.

Designing Interfaces and Dialogues in Graphical Environments

Graphical user interface (GUI) environments are rapidly growing in popularity. Although all of the interface and dialogue design guidelines presented previously apply to designing GUIs, there are additional issues that are unique to these environments that must be considered. Here, we briefly discuss some of these issues.

Figure 14-20

Dialogue diagram for the Customer Information System (Pine Valley Furniture)

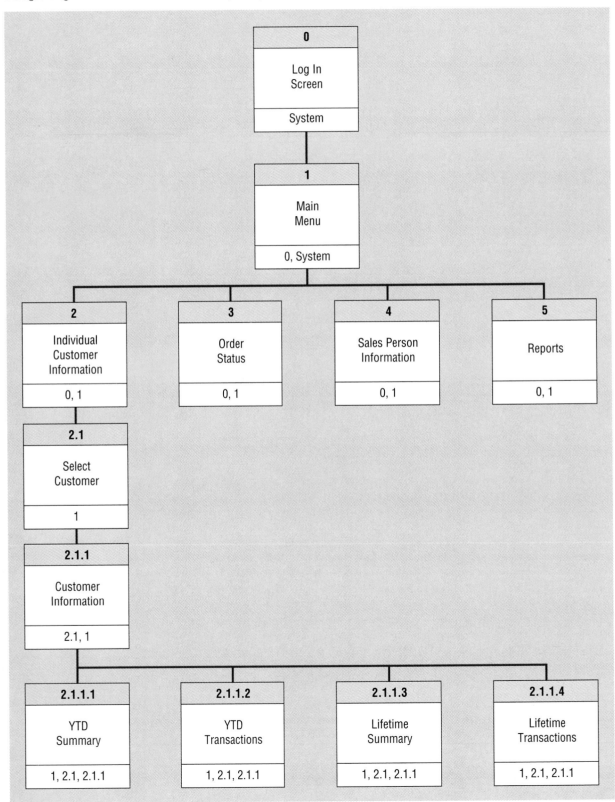

Graphical Interface Design Issues   When designing GUIs for an operating environment such as Microsoft Windows or for the Macintosh, there are numerous factors to consider. Some factors are common to all GUI environments while others are specific to a single environment. We will not, however, discuss the subtleties and details of any single environment. Instead, our discussion will focus on a few general truths that experienced designers mention as critical to the design of usable GUIs (for more information on GUI design, see the books by Porter, 1993, and May, 1991). In most discussions of GUI programming, two rules repeatedly emerge as comprising the first step to becoming an effective GUI designer:

1. *Become an expert user of the GUI environment.*

2. *Understand the available resources and how they can be used.*

The first step should be an obvious one. The greatest strength of designing within a standard operating environment is that *standards* for the behavior of most system operations have already been defined. For example, how you cut and paste, set up your default printer, design menus, or assign commands to functions have been standardized both within and across applications. This allows experienced users of one GUI-based application to easily learn a new application. Thus, in order to design effective interfaces in such environments, you must first understand how other applications have been designed so that you will adopt the established standards for "look and feel." Failure to adopt the standard conventions in a given environment will result in a system that will likely frustrate and confuse users.

The second rule—gaining an understanding of the available resources and how they can be used—is a much larger undertaking. For example, within Windows you can use menus, forms, and boxes in many ways. In fact, the flexibility with which these resources *can be used* versus the established standards for how most designers *actually use* these resources makes design especially challenging. For example, you have the ability to design menus using all upper-case text, putting multiple words on the top line of the menu, and other non-standard conventions. Yet, the standards for menu design require that top-level menu items consist of one word and follow a specific ordering. Numerous other standards for menu design have also been established (see Figure 14-21 for illustrations of many of these standards). Failure to follow standard design conventions will likely prove very confusing to users.

In GUIs, information is requested by placing a window (or form) on the visual display screen. Like menu design, forms can also have numerous properties that can be mixed and matched (see Table 14-14). For example, properties about a form determine whether a form is resizable or movable after being opened. Since properties define how users can actually work with a form, the effective application of properties is fundamental to gaining usability. This means that, in addition to designing the layout of a form, you must also define the "personality" of the form with its characteristic properties. Fortunately, numerous GUI design tools have been developed that allow you to "visually" design forms and interactively engage properties. Interactive GUI design tools have greatly facilitated the design and construction process.

In addition to the issues related to interface design, the sequencing of displays turns out to be a bit more challenging in graphical environments. This topic is discussed next.

Dialogue Design Issues in a Graphical Environment   When designing a dialogue, your goal is to establish the sequence of displays (full screens or windows) that users will encounter when working with the system. Within many GUI environments, this process can be a bit more challenging due to the GUI's ability to suspend activities (without resolving a request for information or exiting the application

Figure 14-21

Highlighting graphical user interface design standards

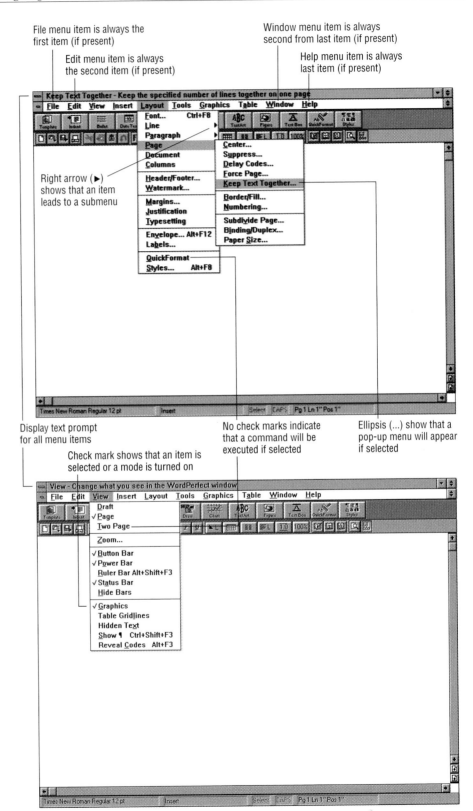

TABLE 14-14 Common Properties of Windows and Forms in a Graphical User Interface Environment that can be Active or Inactive
(Adapted from Wagner, 1994)

Property	Explanation
Modality	Requires users to resolve the request for information before proceeding (e.g., need to cancel or save before closing a window).
Resizable	Allows users to resize a window or form (e.g., to make room to see other windows also on the screen).
Movable	Allows users to move a window or form (e.g., to allow another window to be seen).
Maximize	Allows users to expand a window or form to a full screen in size (e.g., to avoid distraction from other active windows or forms).
Minimize	Allows users to shrink a window or form to an icon (e.g., to get the window out of the way while working on other active windows).
System Menu	Allows a window or form to also have a system menu to directly access system-level functions (e.g., to save or copy data).

altogether) and switch to another application or task. For example, within Word-Perfect for Windows 6.0, the spell checker executes independently from the general word processor. This means that you can easily jump between the spell checker and word processor without exiting either one. Conversely, when selecting the print operation, you must either initiate printing or abort this request before returning to the word processor. This is an example of the concept of "modality" described in Table 14-14. Thus, Windows-type environments allow you to create forms that either *require* the user to resolve a request before proceeding (print example) or *selectively choose* to resolve a request before proceeding (the spell checker). Creating dialogues that allow the user to jump from application to application or from module to module within a given application requires that you carefully think through the design of dialogues.

One easy way to deal with the complexity of designing advanced graphical user interfaces is to require users to *always* resolve all requests for information before proceeding. For such designs, the dialogue diagramming technique is an adequate design tool. This, however, would make the system operate in a manner similar to a traditional non-GUI environment where the sequencing of displays is tightly controlled. The drawback to such an approach would be the failure to capitalize on the task-switching capabilities of these environments. Thus, if you want to allow users to selectively perform some functions or to selectively switch between tasks, a more powerful design tool may be needed. One tool that is often applied to this problem is the state-transition diagram (STD), introduced in Chapter 10. STDs can be used to model dialogues where displays are selectively suspended and activated by the user. For example, Figure 14-22 shows a simple STD that models the various states the WordPerfect spell checker encounters when in use.

Recall that the first step when designing a STD is to identify all possible states for a given object or process. In the case of the spell checker, it can have the following states:

1. Closed (not loaded)

2. Open, but idle (waiting to check)

3. Open and checking

Figure 14-22

State-transition diagram for spell checker

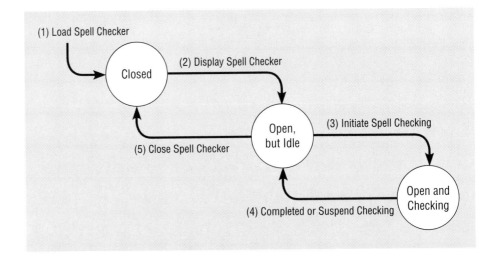

Further, several events or transitions occur that move the spell checker from one state to another. The first transition (labeled 1) loads the spell checker. In other words, this represents a user's request to activate the spell-checking system by selecting the appropriate command within WordPerfect. As the spell checker can be activated at virtually any time within the system, it would be difficult if not impossible to model all possible display sequences using normal dialogue diagramming techniques.

The second transition occurs when the spell checker moves from the "Closed" state to the "Open, but Idle" state, referred to as the "Display spell checker" transition (labeled 2). At this point, the spell checker can be requested to "Initiate Spell Checking" (labeled 3), left idle, or closed (labeled 5). If spell checking is initiated, the spell checker moves into its final unique state, "Open and Checking." As before, the user has the option to suspend checking or wait for the spell checking to be completed (labeled 4). When either of these transitions occur, the spell checker will return to the "Open, but Idle" state. From there, the spell checker can be closed (labeled 5) or re-initiated. For each state, or transformations between states, a specific display, dialogue box, or request must be designed. The sequencing of these displays and dialogue boxes will be guided by the STD. This means that it is only through a clear understanding of all possible states, and the transitions that make them change, that a complete dialogue specification can be designed for most multi-windowing GUI applications.

In summary, designing dialogues in environments where the sequence between displays cannot be pre-determined offers significant challenges to the designer. In fact, the focus of object-oriented systems analysis and design, a much more advanced design approach, is to provide designers with tools and techniques for modeling systems employing these concepts (see Appendix D).

SUMMARY

In this chapter, our focus was to acquaint you with the process of designing human-computer interfaces and dialogues. Understanding the characteristics of various interaction methods (command language, menu, form, object, natural language) and devices (keyboard, mouse, joystick, trackball, touch screen, light pen, graphics tablet, voice) is a fundamental skill you must master. No single interaction style or device is the most appropriate in all instances: each has its strengths and weaknesses. You must consider characteristics of the intended users, the tasks being performed, and various technical and environmental factors when making design decisions.

The chapter also reviewed design guidelines for computer-based forms. You learned that most forms have a header, sequence or time-related information, instructions, a body, summary data, authorization, and comments. Users must be able to move the cursor position, edit data, exit with different consequences, and obtain help. Techniques for structuring and controlling data entry were presented along with guidelines for providing feedback, prompting, and error messages. A simple, well-organized help function that shows examples of proper use of the system should be provided. A variety of help types were reviewed. A final topic of interface design focused on controlling user access to systems and data. This discussion presented four techniques: views, authorization rules, encryption procedures, and authentication schemes.

Next, guidelines for designing human-computer dialogues were presented including a description of dialogue diagramming. These guidelines are consistency, allowing for shortcuts, providing feedback and closure on tasks, handling errors, allowing for operations reversal, giving the user a sense of control, and ease of navigation. Designing dialogues and procedures for assessing their usability were also reviewed. Finally, several interface and dialogue design issues were described within the context of designing graphical user interfaces. These included the need to follow standards to provide the capabilities of modality, resizing, moving, maximizing and minimizing windows, and to offer a system menu choice. This discussion highlighted how concepts presented earlier in the chapter can be applied or augmented in these emerging environments.

Our goal was to provide you with a foundation for building highly usable human-computer interfaces. As more and more development environments provide rapid prototyping tools for the design of interfaces and dialogues, many complying to common interface standards, the difficulty of designing usable interfaces will be reduced. However, you still need a solid understanding of the concepts presented in this chapter in order to succeed. Learning to use a computer system is like learning to use a parachute—if a person fails on the first try, odds are he or she won't try again (Blattner & Schultz, 1988). If this analogy is true, it is important that a user's first experience with a system be a positive one. By following the design guidelines outlined in this chapter, your chances of providing a positive first experience to users will be greatly enhanced.

C H A P T E R   R E V I E W

K E Y   T E R M S

Authorization rules
Biometric device
Command language
   interaction
Dialogue
Dialogue diagramming

Drop-down menu
Form interaction
Icon
Interface
Menu interaction

Natural language
   interaction
Object-based interaction
Pop-up menu
Smart card
View

REVIEW QUESTIONS

1. Define each of the following terms:

   a. dialogue
   b. drop-down menu
   c. icon
   d. pop-up menu

   e. authorization rules
   f. smart card
   g. view

2. Contrast the following terms:

   a. dialogue, interface
   b. command language interaction, form interaction, menu interaction, natural language interaction, object-based interaction
   c. drop-down menu, pop-up menu
   d. authorization rules, biometric device, view

3. Describe the process of designing interfaces and dialogues. What deliverables are produced from this process? Are these deliverables the same for all types of system projects? Why or why not?

4. Describe five methods of interacting with a system. Is one method better than all others? Why or why not?

5. Describe several input devices for interacting with a system. Is one device better than all others? Why or why not?

6. Describe the general guidelines for the design of menus. Can you think of any instances when it would be appropriate to violate these guidelines?

7. List and describe the general sections of a typical business form. Do computer-based and paper-based forms have the same components? Why or why not?

8. List and describe the functional capabilities needed in an interface for effective entry and navigation. Which capabilities are most important? Why? Will this be the same for all systems? Why or why not?

9. Describe the general guidelines for structuring data entry fields. Can you think of any instances when it would be appropriate to violate these guidelines?

10. Describe four types of data errors.

11. Describe the methods used to enhance the validity of data input.

12. Describe the types of system feedback. Is any form of feedback more important than the others? Why or why not?

13. Describe the general guidelines for designing usable help. Can you think of any instances when it would be appropriate to violate these guidelines?

14. Describe four ways to control user access to a system or data.

15. What steps do you need to follow when designing a dialogue? Of the guidelines for designing a dialogue, which is most important? Why?

16. Describe the properties of windows and forms in a graphical user interface environment. Which property do you feel is most important? Why?

PROBLEMS AND EXERCISES

1. Match the following terms to the appropriate definitions.

   _____ interface

   _____ command language interaction

   _____ menu interaction

   _____ object-based interaction

   _____ natural language interaction

   _____ dialogue

   _____ drop-down menu

   a. human-computer interaction method where a list of system options is provided and, if selected, each option invokes a command
   b. menu positioning method that places the access point of the menu near the top line of the display; when accessed, menus open by dropping down onto the display
   c. human-computer interaction method where symbols are used to represent commands or functions
   d. method by which users interact with information systems
   e. human-computer interaction method where explicit statements are entered into a system to invoke operations
   f. sequence of interaction between a user and a system
   g. human-computer interaction method where inputs to and outputs from a computer-based application are in a conventional speaking language such as English

2. Consider the design of a registration system for a hotel. Following design specification items in Figure 14-2, briefly describe the relevant users, tasks, and displays involved in such a system.

3. Consider software applications that you regularly use that have menu interfaces, whether they be PC- or mainframe-based applications. Evaluate these applications in terms of the menu design guidelines outlined in Table 14-1.

4. Imagine the design of a system used to register students at a university. Discuss the user, task, system, and environmental characteristics (see Table 13-9) that should be considered when designing the interface for such a system.

5. For each of the following interaction methods recall a software package that you have used recently: command language, menus, and objects. What did you like and dislike about each package? What were the strengths and weaknesses of each? Which type do you prefer for which circumstances? Which type do you believe will become most prevalent? Why?

6. Briefly describe several different business tasks that are good candidates for form-based interaction within an information system.

7. List the physical input devices described in this chapter that you have seen or used. For each device, briefly describe your experience and provide your personal evaluation. Do your personal evaluations parallel the evaluations provided in Tables 14-3 and 14-4?

8. Propose some specific settings where natural language interaction would be particularly useful and explain why.

9. Examine the help systems for some software applications that you use. Evaluate each using the general guidelines provided in Table 14-11.

10. Design one sample data entry screen for a hotel registration system using the data entry guidelines provided in this chapter (see Table 14-6). Support your design with arguments for each of the design choices you made.

11. Describe some typical dialogue scenarios between users and a hotel registration system. For hints, reread the section in this chapter that provides sample dialogue between users and the Customer Information System at Pine Valley Furniture.

12. Represent the dialogues from the previous question through the use of dialogue diagrams.

FIELD EXERCISES

1. Research the topic "natural language" at your library. Determine the status of applications available with natural language interaction. Forecast how long it will be before natural language capabilities are prevalent in information systems use.

2. Examine two PC-based graphical user interfaces (e.g., Microsoft's Windows and IBM's OS/2). If you do not own these interfaces, you are likely to find them at your university, workplace, or at a computer retail store. You may want to supplement your hands-on evaluation with recent formal evaluations published in magazines such as *PC Computing*. In what ways are these two interfaces similar and different? Are these interfaces intuitive? Why or why not? Is one more intuitive than the other? Why or why not? Which interface seems easier to learn? Why? What types of system requirements does each interface have, and what are the costs of each interface? Which do you prefer? Why?

3. Interview a variety of people you know about the various ways they interact, in terms of inputs, with systems at their workplaces. What types of technologies and devices are used to deliver these inputs? Are the input methods and devices easy to use and do they help these people complete their tasks effectively and efficiently? Why or why not? How could these input methods and devices be improved?

4. Interview systems analysts and programmers in an organization where graphical user interfaces are used. Describe the ways that these interfaces are developed and used. How does the use of such interfaces enhance or complicate the design of interfaces and dialogues?

REFERENCES

Apple. 1987. *Human Interface Guidelines: The Apple Desktop Interface.* Reading, MA: Addison-Wesley.

Blattner, M. and E. Schultz. 1988. "User Interface Tutorial." Presented at the 1988 Hawaii International Conference on System Sciences, Kona, Hawaii, January, 1988.

Bloombecker, J. J. 1989. "Short-Circuiting Computer Crime." *Datamation* 55 (Oct. 1): 71–72.

Dumas, J. S. 1988. *Designing User Interfaces for Software.* Englewood Cliffs, NJ: Prentice-Hall.

Fernandez, E. B., R. C. Summers, and C. Wood. 1981. *Database Security and Integrity.* Reading, MA: Addison-Wesley.

IBM. 1991. *Systems Application Architecture: Common User Access Guide to User Interface.* IBM Document SC34-4289-00. October.

May, J. C. 1991. *Extending the Macintosh Toolbox: Programming Menus, Windows, Dialogues, and More.* Reading, MA: Addison-Wesley.

McFadden, F. R. and J. A. Hoffer. 1994. *Modern Database Management.* Redwood City, CA: Benjamin/Cummings.

Norman, K. L. 1991. *The Psychology of Menu Selection.* Norwood, NJ: Ablex.

Porter, A. 1993. *C++ Programming for Windows.* Berkeley, CA: Osborne McGraw-Hill.

Shneiderman, B. 1992. *Designing the User Interface: Strategies for Effective Human-Computer Interaction.* Reading, MA: Addison-Wesley.

Wagner, R. 1994. "A GUI Design Manifesto." *Paradox Informant* 5 (June): 36–42.

Logical Design: Designing the Customer Reservation Interface and Dialogue

INTRODUCTION

In this Broadway Entertainment Company (BEC) case, we review the activities that occurred during the design of an important custom interface to the Customer Activity Tracking System (CATS). Specifically, we focus on the activities of one systems analyst, Frank Napier, during the creation of the on-line customer-product reservation confirmation module within CATS. This interface occurs where a BEC telephone reservation clerk confirms a customer reservation request after the customer receives a postcard notifying him or her of a new product release.

DESIGNING THE CUSTOMER RESERVATION INTERFACE AND DIALOGUE

After completing the design of the customer-product assignment report (see BEC case after Chapter 13), Frank Napier was assigned to develop an initial design of the interface and dialogue for recording customer product reservations within CATS. This custom dialogue, not provided in the purchased software, was to be only a very small piece of CATS, but it must be consistent with the overall style for CATS. CATS uses a graphical user interface with a fairly strict navigation path. Buttons on each on-line form rather than general menus are used to move from form to form. Forms are used to capture data to enter into the CATS database. Each reservation clerk will work from a personal computer equipped with a color monitor, keyboard, and mouse.

BEC's system development methodology requires that Frank first gain an understanding of the data and process requirements for making reservations. To do this, he reviewed the data flow diagrams, entity-relationship models, data element list, and designs of system outputs. In essence, he reviewed all relevant analysis and design information concerning customers placing reservations for products. From this review, he concluded that the dialogue flow for recording the reservation within CATS would take the following five steps:

1. A customer receives a notification postcard containing a product description, date of availability, and reservation code (the product / customer / store code).

2. If a customer desires to make a product reservation, he or she calls BEC to confirm the reservation.

3. To record the reservation, a BEC telephone clerk requests the reservation code from the customer; this code is entered into a CATS module.

4. After entering the reservation code, the information system displays customer and product information; this information will be visually inspected for correctness by the clerk through questions and answers with the customer.

5. After verifying the information correctness, the clerk will confirm the reservation, thank the customer, and hang up the telephone call.

After outlining the flow and sequence of the reservation recording process, Frank outlined the sequence of on-line forms using dialogue diagramming (Figure 1). The dialogue diagram contained three forms in its sequence. Form 1, Reservation Code Entry, equates to step 3. Form 2, Information Verification, equates to step 4 while Form 3, Reservation Confirmation, equates to step 5 in the sequence. For each form, the dialogue diagram shows that the user is able to back up to the previous form. Also, after Reservation Confirmation (Form 3), the user is taken back to Form 1, Reservation Code Entry. In other words, after a reservation confirmation, the clerk will be ready to take another reservation.

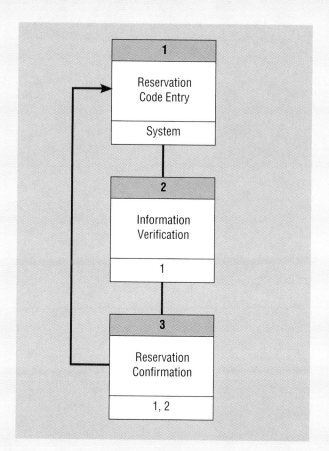

Figure 1
Dialogue diagramming for the
customer product reservation
sequence

After designing the dialogue flow, Frank's next step was to design prototypes of each form in the sequence. The BEC systems development group uses a variety of tools for prototyping including CASE and visual interface design tools. The CATS design team decided to use Microsoft's Visual Basic for Windows to design the prototype because display forms can be easily constructed and compiled into a working prototype. These prototypes can then be distributed to users and others within the design team to assess system usability and completeness.

Figure 2 shows the initial prototype design for Form 1, Reservation Code Entry. This form contains a single field for entering the reservation code and three selection buttons: OK, Help, and Quit. Frank also defined a picture statement for the reservation code field to restrict it to a five-character, alphanumeric value with a dash between the third and fourth characters. Using a picture statement on data entry fields is one easy way to help users enter correct values.

Until any of the buttons is pushed, the reservation clerk can edit the reservation code using normal Windows field editing functions. Selecting the OK button

(or pressing the Enter key) will advance the sequence of the dialogue to Form 2. In this initial design, Frank does not deal with data entry errors (in this case, an improper reservation code); these will be designed after evaluating the usability of the basic design. Selecting Help will invoke the help system (also not yet designed), and Quit will terminate the execution of the reservation system. Under the OK button, Frank added two lines of Basic language programming code to "unload" the current form (Cats1) and to "show" the second form in the sequence (Cats2) when executing the prototype (see Figure 3). Thus, when the application is compiled, users can use a mouse to press the OK button to move to Form 2, Information Verification. This means that the prototype can be easily made to behave in a manner consistent with the actual system implementation, greatly improving the usefulness of any usability assessment activities.

As described, Form 2, Information Verification, contains both customer and product information (Figure 4). All of this information is retrieved from the CATS database based on the reservation code. The customer data fields may be updated (except for the Customer

Figure 2
Reservation code entry form
prototype

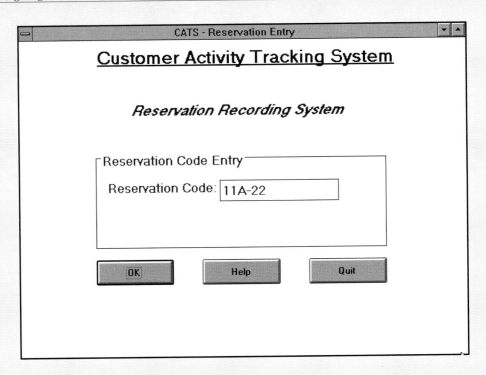

Figure 3
Visual Basic programming
code attached to the OK
button to unload Form 1 and
display Form 2

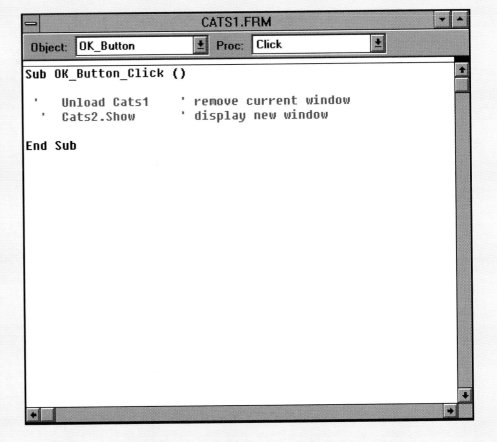

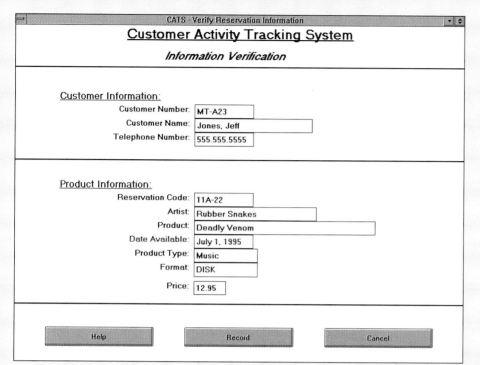

Figure 4
Information verification form
prototype

Number field), if necessary. This form also has three buttons: Record, Help, and Cancel. Choosing Record or pressing the Enter key advances the sequence of the dialogue to Form 3 while pressing the Cancel button returns the user to Form 1. As with the prior form, Frank programmed the Record button to load Form 3, Reservation Confirmation. This time, however, he did not program the system to unload Form 2 but simply to overlay Form 3 on top of Form 2 (Figure 5). After Form 3 is displayed, a user can confirm the reservation

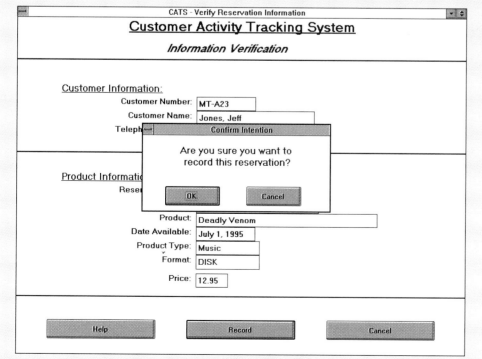

Figure 5
Reservation confirmation form
prototype

Figure 6
One-to-one mapping of dialogue diagramming sequence to prototype form designs

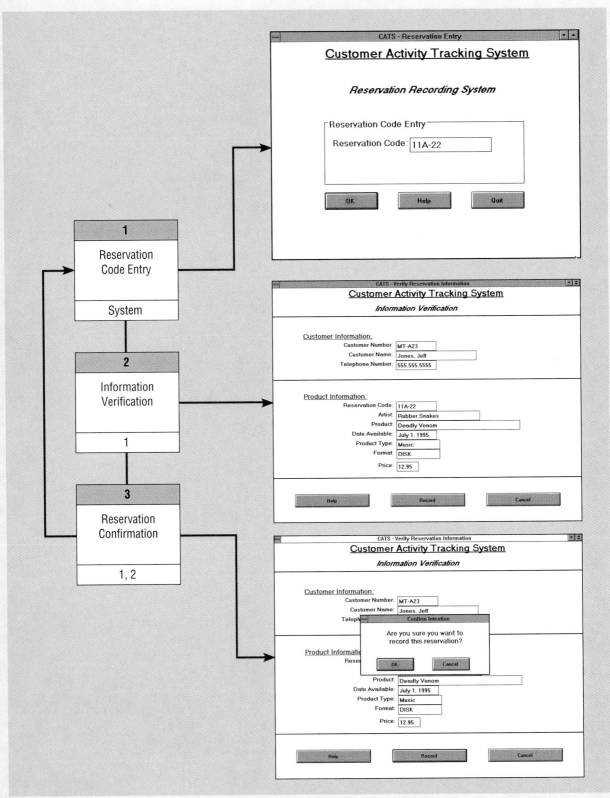

by selecting the OK button, or cancel the reservation request and return to Form 2 by selecting the Cancel button. Since Form 3 is a window, the reservation clerk can move this confirmation box to anywhere on the display screen. Again, Frank programmed the selection buttons so that the prototype would behave in a manner consistent with the design represented in the dialogue diagram.

In summary, the customer-product reservation dialogue sequence contains three forms. Each form has a one-to-one mapping with dialogue sequence and movement between each form controlled by selection buttons (see Figure 6).

SUMMARY

After completing the dialogue and form design process, Frank assessed the design's usability. As he had done with system outputs, Frank first contacted several store managers and arranged to have them use the dialogue on the PCs in their stores. After consolidating some comments from these potential users, he developed new form layouts and a hard copy script of sample forms which he distributed to a cross section of potential users. He prepared a questionnaire that captured user background data, responses to comprehension questions, and satisfaction perceptions (see Figure 7 from the BEC installment after Chapter 13). Comprehension tasks involved responding to which buttons to press to perform a certain navigation activity, the meaning of fields on the form, and how to undo data already entered.

From this survey Frank discovered several improvements were needed:

- More descriptive button labels

- More specific form titles

- Color coding fields which are read only

- Changes to the information verification and confirmation forms to handle declining a reservation

Frank realized that these were initial form designs since he would have to add additional functions (like pop-up lists to look up valid field values) to support modifying and deleting reservations (inevitable when customers changed their minds). Also, he would have to design such elements as help displays, error messages, shortcut keystrokes, and more methods to undo data entries before the dialogue and interface were complete. In addition, the reservation interface must be integrated into the total interface available to each telephone clerk. However, a reaction to an initial design was important before more detailed work was completed. Fortunately, both potential users and development colleagues enthusiastically praised Frank's efforts at initial dialogue and interface designs. Nevertheless, Frank knew that much more work was required to translate the form prototypes into an operational system. He was, however, reassured that BEC's system development methodology would help guide him through the additional work that remains to be done.

Designing Databases:
Logical Data Modeling

After studying this chapter, you should be able to:

■ Concisely define each of the following key logical data modeling terms: *relation, functional dependency, foreign key,* and *normalization.*

■ Explain the role of logical data modeling in the overall analysis and design of an information system.

■ Describe four steps in logical data modeling.

■ Be able to represent unary, binary, and ternary relationships in the relational data model.

■ Give concise definitions of first, second, and third normal forms.

■ Merge normalized relations from separate user views into a consolidated set of normalized relations.

■ Transform an entity-relationship (E-R) diagram into an equivalent set of normalized relations.

INTRODUCTION

In Chapter 11 you learned how to describe the data requirements of an information system using a graphical notation, the entity-relationship (E-R) data model. You used E-R data models to represent a conceptual view of organizational data independent of any particular database processing technology. Further, you learned how to use E-R diagramming to represent data about an organization based on business rules and policies rather than on specific information processing requirements. Although we concentrated on representing the nature of data in an organization, we did not concern ourselves with guidelines on what were good or poor data models. We address this important topic in this chapter.

In Chapter 15 you learn more about structuring the description of data through logical data modeling. Logical data modeling has three purposes. First, in logical data modeling you use the *normalization* process which results in a structure for data that has some very desirable properties. These properties help you represent data in more stable structures that are not likely to change over time and that have minimal redundancy. If you were to implement the database for an information system from normalized data, database processing would be simple, generally efficient, and less susceptible to problems, or *anomalies,* during maintenance.

The second purpose of logical data modeling is to develop a data model from which we can do physical database design. Since most information systems today use relational database management systems, logical data modeling usually produces a relational database model. It is possible to produce network, hierarchical, or object-oriented data models from logical data modeling, but we cover only the most typical situation in this chapter. In fact, we suggest that even if network or hierarchical technologies are used, you still should do logical data modeling to produce relations. You can then translate these relations into the target data model during physical database design. Here we will present the basic principles of the relational database model and will show you how to develop relational models from E-R models.

The final purpose of logical data modeling is to develop a data model that reflects the actual data requirements that exist in the forms (hard copy and computer displays) and reports of an information system. Thus, logical data modeling occurs after or concurrent with the design of system externals and outputs. Logical data modeling includes a *bottom-up* process of deriving relations from specific forms and reports, or what are called user views of an information system. We will show how to do this bottom-up logical design and then describe how to integrate or merge separate relations for individual forms and reports into a consolidated logical data model. Then we will illustrate how to translate an E-R model into relational form so that the conceptual data model can be compared with the logical data model and differences resolved.

Familiarity with the concepts and terminology introduced in Chapter 11 is essential for reading and understanding the contents of Chapter 15 (you may want to review Chapter 11 now). Once you master the conceptual and logical data modeling techniques explained in Chapters 11 and 15, you will likely design normalized data into entity-relationship diagrams. Thus, although we introduce conceptual and logical data modeling as separate topics, sometimes these project activities are combined into one data modeling project step.

LOGICAL DATA MODELING

Logical database model: A description of data using a notation which corresponds to a data organization used by a database management system.

A **logical data model** describes data using a notation which corresponds to a data organization used by a database management system. The most common style for a logical data model is the *relational database model.*

You typically do logical data modeling in parallel with other systems design steps (see Figure 15-1). Thus, you collect the detailed specifications of data necessary for logical data modeling as you design system inputs and outputs. Logical data modeling is driven not only from the previously developed E-R data model for the application but also from form and report layouts. You study data elements on these system inputs and outputs and identify inter-relationships among the data. As with conceptual data modeling, the work of all systems development team members is coordinated and shared through the project dictionary or repository. Increasingly, this repository is maintained by a common CASE tool although some organizations still use manual documentation. In either case, it is essential that the process, logic, system input, system output, and data model descriptions of a system be consistent and complete since each describes different views of the same information system.

Figure 15-1
Systems development life cycle with logical design phase highlighted

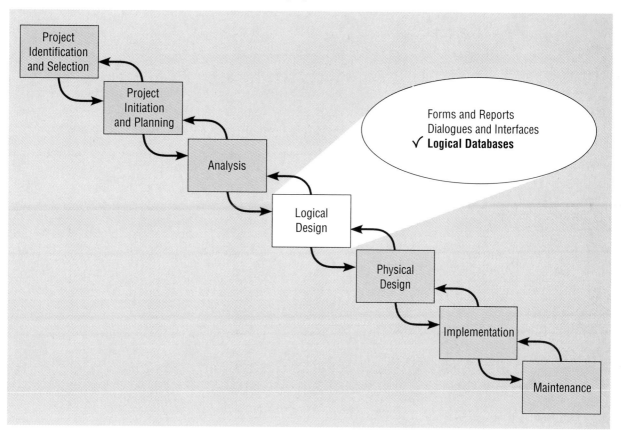

The Process of Logical Data Modeling

Figure 15-2 shows that data modeling occurs in the project planning, analysis, and logical design steps of the systems development process. In this chapter we discuss methods which help you finalize a data model during the logical design phase. Data modeling during project planning and analysis is done using entity-relationship modeling which uses a special notation to represent as much meaning about data as possible. In logical design you use a process called normalization, which is a way to build a data model that has some desirable properties of simplicity, non-redundancy, and maintenance. We will discuss normalization later within the context of developing a relational database model. We develop only the fundamentals of logical data modeling in this chapter. The interested reader is referred to McFadden and Hoffer (1994) for a more thorough treatment of techniques for logical data modeling.

There are four key steps in logical data modeling:

1. Develop a logical data model for each known user view (form and report) for the application using normalization principles.

2. Combine all normalized user views into one consolidated logical data model; this step is called *view integration.*

Figure 15-2

Relationship between data modeling and the systems development life cycle

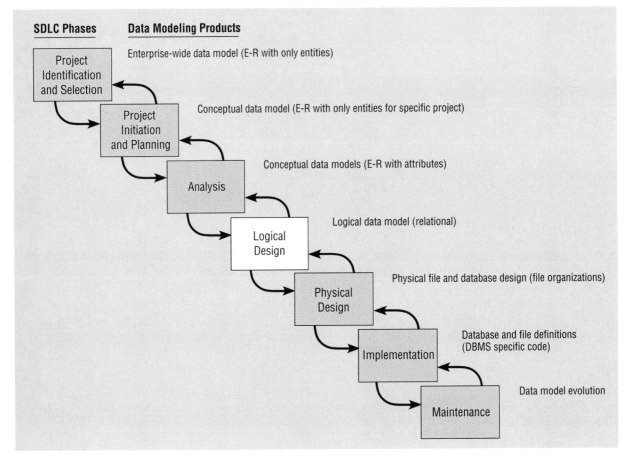

3. Translate the conceptual E-R data model for the application, developed without explicit consideration of specific user views, into normalized relations.

4. Compare the consolidated logical data model with the translated E-R model and produce, through view integration, one final logical data model for the application.

We will show how to do each of these steps in this chapter.

Deliverables and Outcomes

During logical data modeling, you must account for every data element on a system input or output—form or report—and on the E-R model in the logical data model. Each data element (like customer name, product description, or purchase price) must be a piece of raw data kept in the system's database or else, in the case of a data element on a system output, the element can be derived from data in the database. Figure 15-3 (on pages 558–559) illustrates the outcomes from the four-step logical data modeling process listed above. Figures 15-3a and 15-3b (step 1) contain two sample system outputs for a customer order processing system in Pine Valley

Furniture. Normalized relations are listed below each output diagram. Each relation is named and its attributes are listed within parentheses. The primary key attribute is indicated by an underline, and an attribute of a relation which is the primary key of another relation is indicated by a dashed underline. Figure 15-3c (step 2) shows the result of integrating these two separate sets of normalized relations. Figure 15-3d (step 3) shows an E-R diagram for a customer order processing application that might be developed during conceptual data modeling along with equivalent normalized relations. Finally, Figure 15-3e (step 4) shows a set of normalized relations that would result from reconciling the logical data models of Figures 15-3c and 15-3d. Normalized relations like those in Figure 15-3e are the primary deliverable from logical data modeling.

Updating the CASE repository or project dictionary is the second output from logical data modeling. Figure 11-12 listed the typical project dictionary entries from conceptual data modeling and Figure 15-4 (on page 560) shows the additional entries made during logical data modeling. Since some of this information is collected during design of system externals and outputs, you make some of these entries during other logical design steps. A key difference between conceptual and logical data modeling is that you have structured data requirements during logical data modeling into relations, rather than entities. Because of normalization, there is not necessarily a one-to-one correspondence between entities and relations. Also, as we will see, relationships are not explicitly represented in relations but are shown indirectly by using common attributes *(foreign keys)* in associated relations. Thus, you do not add new relationship information during logical database design. Any new relationships discovered during logical data modeling, however, would be added into the conceptual data model and the project dictionary would be updated. Relationship information is useful in physical database and computer program design, which we will illustrate in later chapters.

It is important to remember that relations do not correspond to computer files. The project dictionary still contains a logical description of data at this point. In physical database design (Chapter 16), we show how relations are transformed into the design for physical computer files.

Since the relational data model is central to our discussion of logical data modeling, the next section introduces this data model.

RELATIONAL DATABASE MODEL

The **relational database model** (Codd, 1970) represents data in the form of tables or relations. You need only a few simple concepts to describe the relational model, and it is easily understood and used by those unfamiliar with the underlying theory.

A **relation** is a named, two-dimensional *table* of data. Each relation (or table) consists of a set of named *columns* and an arbitrary number of unnamed *rows*. Each column in a relation corresponds to an attribute of that relation. Each row of a relation corresponds to a record that contains data values for an entity.

Figure 15-5 shows an example of a relation named EMPLOYEE1. This relation contains the following attributes describing employees: EMPID, NAME, DEPT, and SALARY. There are five rows in the table, corresponding to five employees.

You can express the *structure* of a relation by a shorthand notation in which the name of the relation is followed (in parentheses) by the names of the attributes in the relation. The primary key attribute is underlined. For example, you would express EMPLOYEE1 as follows:

EMPLOYEE1(<u>EMPID</u>,NAME,DEPT,SALARY)

Relational database model: A data model that represents data in the form of tables or relations.

Relation: A named, two-dimensional table of data. Each relation consists of a set of named columns and an arbitrary number of un-named rows.

Figure 15-3

Simple example of logical
data modeling

(a) Highest volume customer
query screen

```
HIGHEST VOLUME CUSTOMER

ENTER PRODUCT NO.: M128
START DATE:        11/01/93
END DATE:          12/31/93
- - - - - - - - - -
CUSTOMER NO.:      1256
NAME:              Commonwealth Builder
VOLUME:            30
```

This inquiry screen shows the customer with the largest volume total
sales of a specified product during an indicated time period.

Relations:
      CUSTOMER (CUSTOMER NO., NAME)
      ORDER (ORDER NO., CUSTOMER NO., ORDER DATE)
      PRODUCT (PRODUCT NO.)
      LINE ITEM (ORDER NO., PRODUCT NO., ORDER QUANTITY)

(b) Backlog summary report

```
                                                    PAGE 1

                     BACKLOG SUMMARY REPORT
                          11/30/93

                                    BACKLOG
                  PRODUCT NO.       QUANTITY
                     B381              0
                     B975              0
                     B985              6
                     E125             30
                      ⋮
                     M128              2
                      ⋮
```

This report shows the unit volume of each product that has been ordered
less that amount shipped through the specified date.

Relations:
      PRODUCT (PRODUCT NO.)
      LINE ITEM (PRODUCT NO., ORDER NO., ORDER QUANTITY)
      ORDER (ORDER NO., ORDER DATE)
      SHIPMENT (PRODUCT NO., INVOICE NO., SHIP QUANTITY)
      INVOICE (INVOICE NO., INVOICE DATE)

```
CUSTOMER (CUSTOMER NO., NAME)
PRODUCT (PRODUCT NO.)
INVOICE (INVOICE NO., INVOICE DATE)
ORDER (ORDER NO., CUSTOMER NO., ORDER DATE)
LINE ITEM (ORDER NO., PRODUCT NO., ORDER QUANTITY)
SHIPMENT (PRODUCT NO., INVOICE NO., SHIP QUANTITY)
```

(c) Integrated set of relations

(d) Conceptual data model
and transformed relations

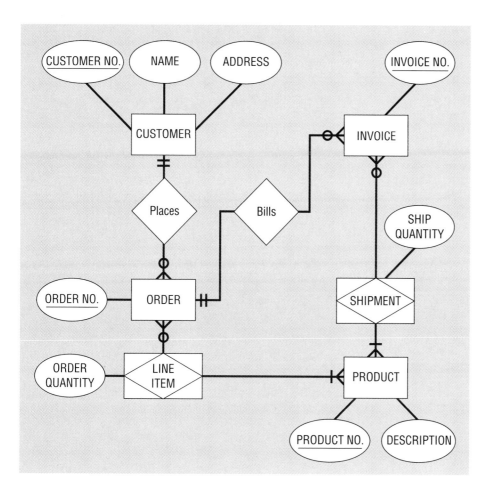

```
Relations:
    CUSTOMER (CUSTOMER NO., NAME, ADDRESS)
    PRODUCT (PRODUCT NO., DESCRIPTION)
    ORDER (ORDER NO., CUSTOMER NO.)
    LINE ITEM (ORDER NO., PRODUCT NO., ORDER QUANTITY)
    INVOICE (INVOICE NO., ORDER NO.)
    SHIPMENT (INVOICE NO., PRODUCT NO., SHIP QUANTITY)
```

```
CUSTOMER (CUSTOMER NO., NAME, ADDRESS)
PRODUCT (PRODUCT NO., DESCRIPTION)
ORDER (ORDER NO., CUSTOMER NO., ORDER DATE)
LINE ITEM (ORDER NO., PRODUCT NO., ORDER QUANTITY)
INVOICE (INVOICE NO., ORDER NO., INVOICE DATE)
SHIPMENT (INVOICE NO., PRODUCT NO., SHIP QUANTITY)
```

(e) Final set of normalized
relations

Figure 15-4

Typical data model elements added to a project dictionary during logical data modeling

RELATION	
Name	A label for the relation; may correspond to an entity name if the meaning of the relation is the same as some entity from the conceptual data model
Description	Explanation so that it is clear which objects are covered by this relation; clarifies relationship of this relation to any entities from the conceptual data model
Attributes and null restrictions	List of attributes included in this relation and for each a statement of whether the attribute may be null (i.e., have no value) for this relation
Primary key	Name(s) of attribute(s) that form the unique identifier for each row of this relation
Foreign keys and referential integrity restrictions	Name of each attribute that is a cross reference to a primary key in another relation; for each foreign key, a statement of the relationship between a value of the foreign key attribute in this relation and values for this attribute found in the associated relation in which this attribute is the primary key
ATTRIBUTE	
Name	A name that uniquely labels the attribute; this would be an attribute name already entered from conceptual data modeling, and the other information about this attribute (like description and domain) would be updated, if necessary
Data type	A specification of the natural format of the attribute (such as alphanumeric, integer, date) as well as a maximum length, if appropriate

Figure 15-5

EMPLOYEE1 relation with sample data

EMPLOYEE1

EMPID	NAME	DEPT	SALARY
100	Margaret Simpson	Marketing	42,000
140	Allen Beeton	Accounting	39,000
110	Chris Lucero	Info Systems	41,500
190	Lorenzo Davis	Finance	38,000
150	Susan Martin	Marketing	38,500

Properties of Relations

We have defined relations as two-dimensional tables of data; however, not all tables are relations. Relations have several properties that distinguish them from nonrelational tables:

1. *Entries in columns are atomic.* An entry at the intersection of each row and column is *atomic* (or single-valued).

2. *Entries in columns are from the same domain.* In a relation, all entries in a given column are drawn from the same domain of values.

3. *Each row is unique.* No two rows in a relation are identical. Uniqueness is guaranteed since the relation has a nonnull primary key.

4. *The sequence of columns (left to right) is insignificant.* The columns of a relation can be interchanged without changing the meaning or use of the relation.

5. *The sequence of rows (top to bottom) is insignificant.* As with columns, the rows of a relation may be interchanged or stored in any sequence. Users may view the rows of a relation in different logical sequences.

Well-Structured Relations

To prepare for our discussion of normalization, we need to address the following question: What constitutes a well-structured relation? Intuitively, a **well-structured relation** contains a minimum amount of redundancy and allows users to insert, modify, and delete the rows in a table without errors or inconsistencies. EMPLOYEE1 (Figure 15-5) is such a relation. Each row of the table contains data describing one employee, and any modification to an employee's data (such as a change in salary) is confined to one row of the table.

In contrast, EMPLOYEE2 (Figure 15-6) contains data about employees and the courses they have completed. Each row in this table is unique for the combination of EMPID and COURSE, which becomes the primary key for the table. This is not a well-structured relation. If you examine the sample data in the table, you will notice a considerable amount of redundancy. For example, the EMPID, NAME, DEPT, and SALARY appear in two separate rows for employees 100, 110, and 150. Consequently, if the salary for employee 100 changes, we must record this fact in two rows (or more, for some employees).

Redundancies in a table may result in errors or inconsistencies (called **anomalies**) when a user attempts to update the data in the table. Three types of anomalies are possible: insertion, deletion, and modification.

1. *Insertion anomaly.* Suppose that you need to add a new employee to EMPLOYEE2. Since the primary key for this relation is the combination of EMPID and COURSE you must supply values for both EMPID and COURSE (since primary key values cannot be null or nonexistent). This is an anomaly, since the user should be able to enter employee data without supplying course data.

2. *Deletion anomaly.* Suppose that the data for employee number 140 is deleted from the table. This will result in losing the information that this employee completed a course (Tax Acc) on 12/8/9x. In fact, it results in losing all the information about this course.

3. *Modification anomaly.* Suppose that employee number 100 gets a salary increase. You must record the increase in each of the rows for that employee (two occurrences in Figure 15-6), otherwise the data will be inconsistent.

Well-structured relation: A relation that contains a minimum amount of redundancy and allows users to insert, modify, and delete the rows in a table without errors or inconsistencies.

Anomalies: Errors or inconsistencies that may result when a user attempts to update a table that contains redundant data. There are three types of anomalies: insertion, deletion, and modification anomalies.

Figure 15-6
Relation with redundancy

EMPLOYEE2

EMPID	NAME	DEPT	SALARY	COURSE	DATE COMPLETED
100	Margaret Simpson	Marketing	42,000	SPSS	6/19/9X
100	Margaret Simpson	Marketing	42,000	Surveys	10/7/9X
140	Alan Beeton	Accounting	39,000	Tax Acc	12/8/9X
110	Chris Lucero	Info Systems	41,500	SPSS	1/12/9X
110	Chris Lucero	Info Systems	41,500	C++	4/22/9X
190	Lorenzo Davis	Finance	38,000	Investments	5/7/9X
150	Susan Martin	Marketing	38,500	SPSS	6/19/9X
150	Susan Martin	Marketing	38,500	TQM	8/12/9X

These anomalies indicate that EMPLOYEE2 is not a well-structured relation. The problem with this relation is that it contains data about two entities: EMPLOYEE and COURSE. You will use normalization to divide EMPLOYEE2 into two relations. One of the resulting relations is EMPLOYEE1 (Figure 15-5). The other we will call EMP COURSE, which appears with sample data in Figure 15-7. The primary key of this relation is the combination of EMPID and COURSE, and we underline these attribute names in Figure 15-7 to highlight this fact. Examine Figure 15-7 to verify that EMP COURSE is free of the types of anomalies described previously and is therefore well-structured.

CONCEPTS OF NORMALIZATION

Normalization: The process of converting complex data structures into simple, stable data structures.

We have presented an intuitive discussion of well-structured relations; however, we need formal definitions of such relations, together with a process for designing them. **Normalization** is a process for converting complex data structures into simple, stable data structures (see Dutka and Hanson, 1989, for a thorough explanation of normalization and a wide variety of normal forms). For example, we used the principles of normalization to convert the EMPLOYEE2 table with its redundancy to EMPLOYEE1 (Figure 15-5) and EMP COURSE (Figure 15-7).

Steps in Normalization

Normal form: A state of a relation that can be determined by applying simple rules regarding dependencies to that relation.

Normalization can be accomplished in stages, each of which corresponds to a normal form (see Figure 15-8). A **normal form** is a relation state that can be determined by applying simple rules regarding dependencies (or relationships between attributes) to that relation. Many normal forms, each of which rids a data model of anomalies, have been identified. Space does not permit a thorough treatment of all commonly used normal forms (see McFadden and Hoffer, 1994). We concentrate here on the three most frequently used:

1. *First normal form (1NF).* Any repeating data have been removed, so there is a single value at the intersection of each row and column of the table.

2. *Second normal form (2NF).* Nonkey attributes require the whole key for identification (what we will call *full functional dependency*).

3. *Third normal form (3NF).* Nonkey attributes do not depend on other nonkey data elements (what we will call no *transitive dependencies*).

Figure 15-7
EMP COURSE relation

EMP COURSE

EMPID	COURSE	DATE COMPLETED
100	SPSS	6/19/9X
100	Surveys	10/7/9X
140	Tax Acc	12/8/9X
110	SPSS	1/22/9X
110	C++	4/22/9X
190	Investments	5/7/9X
150	SPSS	6/19/9X
150	TQM	8/12/9X

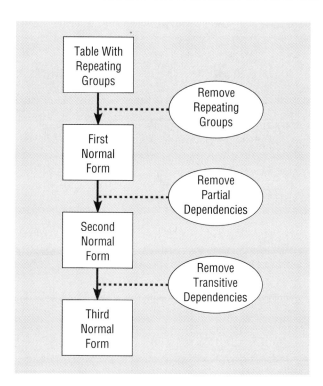

Figure 15-8
Steps in normalization

Thus, the result of normalization is that every nonkey attribute depends upon the whole primary key and nothing but the primary key.

Functional Dependence and Keys

Normalization is based on the analysis of functional dependence. A **functional dependency** is a particular relationship between two attributes. For any relation R, attribute B is *functionally dependent* on attribute A if, for every valid instance of A, that value of A uniquely determines the value of B (Dutka and Hanson, 1989). The functional dependence of B on A is represented by an arrow, as follows: A→B (for example, EMPID→NAME in the relation of Figure 15-5). Functional dependence does not imply mathematical dependence—that the value of one attribute may be computed from the value of another attribute; rather, functional dependence of B on A means that there can be only one value of B for each value of A. Thus, for a given EMPID value, there can be only one NAME value associated with it; the value of NAME, however, cannot be derived from the value of EMPID.

An attribute may be functionally dependent on two (or more) attributes, rather than on a single attribute. For example, consider the relation EMP COURSE (EMPID,COURSE,DATE COMPLETED) shown in Figure 15-7. We represent the functional dependency in this relation as follows: EMPID,COURSE→DATE COMPLETED.

Common examples of functional dependencies are the following:

1. SSN→NAME,ADDRESS,BIRTHDATE: A person's name, address, and birth date are functionally dependent on that person's Social Security number.

2. VIN→MAKE,MODEL,COLOR: The make, model, and color of a vehicle are functionally dependent on the vehicle identification number.

Functional dependency:
A particular relationship between two attributes. For any relation R, attribute B is functionally dependent on attribute A if, for every valid instance of A, that value of A uniquely determines the value of B. The functional dependence of B on A is represented as A→B.

3. ISBN→TITLE: The title of a book is functionally dependent on the book's international standard book number (ISBN).

You should be aware that the instances (or sample data) in a relation do not prove that a functional dependency exists. Only knowledge of the problem domain, obtained from a thorough requirements analysis, is a reliable method for identifying a functional dependency. However, you can use sample data to demonstrate that a functional dependency does *not* exist between two or more attributes. For example, consider the sample data in the relation EXAMPLE(A,B,C,D), shown in Figure 15-9. The sample data in this relation proves that attribute B is not functionally dependent on attribute A since A does not uniquely determine B (two rows with the same value of A have different values of B). You should examine Figure 15-9 to eliminate other possible functional dependencies based on the sample data.

First Normal Form

First normal form (1NF):
A relation that contains no repeating data.

A relation is in **first normal form (1NF)** if it contains no repeating data. Recall that the first property of a relation is that the value at the intersection of each row and column is atomic. A table that contains multivalued attributes or repeating data is not a relation.

Figure 15-10 shows a table with repeating data. Figure 15-6 shows one simple way of removing the repeating data from the table in Figure 15-10. A table with repeating data can be converted to a relation in first normal form by extending the data in each column to fill cells that are empty because of the repeating data structures. Another, and better, way of removing the repeating data in Figure 15-10 is to separate the employee data from the employee-course data, which has been done in Figures 15-5 and 15-7. This second approach also happens to address the characteristic of second normal form, which we define next.

Figure 15-9
EXAMPLE relation

EXAMPLE

A	B	C	D
X	U	X	Y
Y	X	Z	X
Z	Y	Y	Y
Y	Z	W	Z

Figure 15-10
Table with repeating data

EMPLOYEE2

EMPID	NAME	DEPT	SALARY	COURSE	DATE COMPLETED
100	Margaret Simpson	Marketing	42,000	SPSS	6/19/9X
				Surveys	10/7/9X
140	Alan Beeton	Accounting	39,000	Tax Acc	12/8/9X
110	Chris Lucero	Info Systems	41,500	SPSS	1/12/9X
				C++	4/22/9X
190	Lorenzo Davis	Finance	38,000	Investments	5/7/9X
150	Susan Martin	Marketing	38,500	SPSS	6/19/9X
				TQM	8/12/9X

Second Normal Form

A relation is in **second normal form (2NF)** if it is in first normal form and every nonkey attribute is fully functionally dependent on the primary key. Thus no non-key attribute is functionally dependent on part, but not all, of the primary key. A relation that is in first normal form will be in second normal form if any one of the following conditions apply:

1. The primary key consists of only one attribute (such as the attribute EMPID in relation EMPLOYEE1).

2. No nonkey attributes exist in the relation.

3. Every nonkey attribute is functionally dependent on the full set of primary key attributes.

EMPLOYEE2 (Figure 15-6) is an example of a relation that is not in second normal form. The shorthand notation for this relation is
EMPLOYEE2(<u>EMPID</u>,NAME,DEPT,SALARY,<u>COURSE</u>,DATE COMPLETED).
The functional dependencies in this relation are the following:
EMPID→NAME,DEPT,SALARY
EMPID,COURSE→DATE COMPLETED
The primary key for this relation is the composite key EMPID,COURSE. Therefore, the nonkey attributes NAME, DEPT, and SALARY are functionally dependent on only EMPID but not on COURSE. A **partial functional dependency** exists when one or more nonkey attributes (such as NAME) are functionally dependent on part, but not all, of the primary key.

The partial functional dependency in EMPLOYEE2 creates redundancy in that relation, which results in anomalies when the table is updated. We described the anomalies in EMPLOYEE2 in the previous section on Well-Structured Relations, so we do not repeat the discussion here.

To convert a relation to second normal form, you decompose the relation into new relations using the attributes, called determinants, that determine other attributes; the determinants are the primary keys of these relations. EMPLOYEE2 is decomposed into the following two relations:

1. EMPLOYEE1(<u>EMPID</u>,NAME,DEPT,SALARY) This relation satisfies the first second normal form condition (sample data is shown in Figure 15-5).

2. EMP COURSE(<u>EMPID,COURSE</u>,DATE COMPLETED) This relation satisfies second normal form condition three (sample data appear in Figure 15-7).

Examine these new relations to verify that they are free of the anomalies associated with EMPLOYEE2.

Third Normal Form

A relation is in **third normal form (3NF)** if it is in second normal form and no transitive dependencies exist. A **transitive dependency** in a relation is a functional dependency between two (or more) nonkey attributes. For example, consider the relation
SALES(<u>CUST NO.</u>,CUST NAME,SALESPERSON,REGION) (sample data for this relation is shown in Figure 15-11a).

Second normal form (2NF): A relation is in second normal form if it is in first normal form and every nonkey attribute is fully functionally dependent on the primary key. Thus no nonkey attribute is functionally dependent on part (but not all) of the primary key.

Partial functional dependency: A dependency in which one or more nonkey attributes are functionally dependent on part, but not all, of the primary key.

Third normal form (3NF): A relation is in third normal form if it is in second normal form and no transitive dependencies exist.

Transitive dependency: A functional dependency between two (or more) nonkey attributes in a relation.

Figure 15-11

Removing transitive
dependencies

(a) Relation with transitive
dependency

SALES

CUST NO.	NAME	SALESPERSON	REGION
8023	Anderson	Smith	South
9167	Bancroft	Hicks	West
7924	Hobbs	Smith	South
6837	Tucker	Hernandez	East
8596	Eckersley	Hicks	West
7018	Arnold	Faulb	North

(b) Relations in 3NF

SALES1

CUST NO.	NAME	SALESPERSON
8023	Anderson	Smith
9167	Bancroft	Hicks
7924	Hobbs	Smith
6837	Tucker	Hernandez
8596	Eckersley	Hicks
7018	Arnold	Faulb

SPERSON

SALESPERSON	REGION
Smith	South
Hicks	West
Hernandez	East
Faulb	North

The following functional dependencies exist in the SALES relation:

1. CUST NO.→CUST NAME,SALESPERSON,REGION (CUST NO. is the primary key.)

2. SALESPERSON→REGION (Each salesperson is assigned to a unique region.)

Notice that SALES is in second normal form since the primary key consists of a single attribute (CUST NO.). However, there is a transitive dependency: REGION is functionally dependent on SALESPERSON and SALESPERSON is functionally dependent on CUST NO. As a result, there are update anomalies in SALES.

1. *Insertion anomaly.* A new salesperson (Robinson) assigned to the North region cannot be entered until a customer has been assigned to that salesperson (since a value for CUST NO. must be provided to insert a row in the table).

2. *Deletion anomaly.* If customer number 6837 is deleted from the table, we lose the information that salesperson Hernandez is assigned to the East region.

3. *Modification anomaly.* If salesperson Smith is reassigned to the East region, several rows must be changed to reflect that fact (two rows are shown in Figure 15-11a).

These anomalies arise as a result of the transitive dependency. The transitive dependency can be removed by decomposing SALES into the two relations, based on the two determinants, shown in Figure 15-11b. These relations are the following:
SALES1(CUST NO.,CUST NAME,SALESPERSON)
SPERSON(SALESPERSON,REGION).

Foreign key: An attribute that appears as a nonkey attribute in one relation and as a primary key attribute (or part of a primary key) in another relation.

Note that SALESPERSON is the primary key in SPERSON. SALESPERSON is also a foreign key in SALES1. A **foreign key** is an attribute that appears as a nonkey attribute in one relation (such as SALES1) and as a primary key attribute (or part of a primary key) in another relation. You designate a foreign key by using a dashed underline.

One business rule we did not discuss earlier is called referential integrity. **Referential integrity** specifies that the value of an attribute in one relation depends on the value of the same attribute in another relation. A foreign key must satisfy referential integrity, as must any part of a composite key that refers to a primary key in another table. Referential integrity is one of the most important types of business rules implied by the relational model.

Referential integrity: An integrity constraint specifying that the value (or existence) of an attribute in one relation depends on the value (or existence) of an attribute in the same or another relation.

The relations in Figure 15-11b are now in third normal form, since no transitive dependencies exist. You should verify that the anomalies that exist in SALES are not present in SALES1 and SPERSON.

TRANSFORMING E-R DIAGRAMS INTO RELATIONS

As mentioned earlier, normalization is a bottom-up process. Normalization produces a set of well-structured relations that contain all of the data mentioned in system externals and outputs. Normalization is grounded in specific information requirements developed in other logical design steps. Since these specific information requirements may not represent all future information needs or all business rules, the E-R diagram you developed in conceptual data modeling is another source of insight into possible information requirements for a new application system. To compare the conceptual data model and the normalized relations developed so far, your E-R diagram must be transformed into relational notation, normalized, and then merged with the existing normalized relations.

Transforming an E-R diagram into normalized relations and then merging all the relations into one final, consolidated set of relations can be accomplished in four steps. These steps are summarized briefly here, and then steps 1, 2, and 4 are discussed in detail in the remainder of this chapter.

1. *Represent entities.* Each entity type in the E-R diagram is represented as a relation in the relational data model. The identifier of the entity type becomes the primary key of the relation, and other attributes of the entity type become nonkey attributes of the relation.

2. *Represent relationships.* Each relationship in an E-R diagram must be represented in the relational model. How we represent a relationship depends on its nature. For example, in some cases we represent a relationship by making the primary key of one relation a foreign key of another relation. In other cases, we create a separate relation to represent a relationship.

3. *Normalize the relations.* The relations that are created in Steps 1 and 2 may have unnecessary redundancy and may be subject to anomalies when they are updated. So, we need to normalize these relations to make them well-structured.

4. *Merge the relations.* So far in logical data modeling we have created various relations from both a bottom-up normalization of user views and from transforming one or more E-R diagrams into sets of relations. Across these different sets of relations, there may be redundant relations (two or more relations that describe the same entity type) that must be merged and renormalized to remove the redundancy.

Represent Entities

Each regular entity type in an E-R diagram is transformed into a relation. The primary key (or identifier) of the entity type becomes the primary key of the corresponding relation. You should check to make sure that this key satisfies the following two properties:

1. The value of the key must uniquely identify every row in the relation.

2. The key should be nonredundant; that is, no attribute in the key can be deleted without destroying its unique identification.

Some entities may have keys that include the primary keys of other entities. For example, an EMPLOYEE DEPENDENT may have a NAME for each dependent but, to form the primary key for this entity, you must include the EMPLOYEE ID attribute from the associated EMPLOYEE entity. Such an entity whose primary key depends upon the primary key of another entity is called a weak entity.

Each nonkey attribute of the entity type becomes a nonkey attribute of the relation. The relations that are formed from entity types may be modified as relationships are represented.

Representation of an entity as a relation is straightforward. Figure 15-12a shows the CUSTOMER entity type for Pine Valley Furniture Company. The corresponding CUSTOMER relation is represented as follows:

CUSTOMER(CUSTOMER NO.,NAME,ADDRESS,CITY STATE ZIP,DISCOUNT)

In this notation, the entity type label is translated into a relation name. The identifier of the entity type is listed first and underlined. All nonkey attributes are listed after the primary key. This relation is shown as a table with sample data in Figure 15-12b.

Represent Relationships

The procedure for representing relationships depends on both the degree of the relationship—unary, binary, ternary—and the cardinalities of the relationship.

Binary 1:*N* and 1:1 Relationships   A binary one-to-many (1:*N*) relationship in an E-R diagram is represented by adding the primary key attribute (or attributes) of

Figure 15-12

Transforming an entity type to a relation

(a) E-R diagram

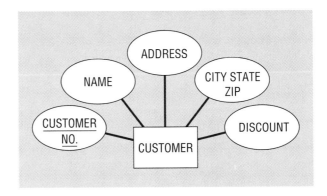

(b) Relation

CUSTOMER

CUSTOMER NO.	NAME	ADDRESS	CITY STATE ZIP	DISCOUNT
1273	Contemporary Designs	123 Oak St.	Austin, TX 38405	5%
6390	Casual Corner	18 Hoosier Dr.	Bloomington, IN 45821	3%

the entity on the one-side of the relationship as a foreign key in the relation that is on the many-side of the relationship.

Figure 15-13a, an example of this rule, shows the Places relationship (1:*N*) linking CUSTOMER and ORDER at Pine Valley Furniture Company. Two relations, CUSTOMER and ORDER, were formed from the respective entity types. CUSTOMER NO., which is the primary key of CUSTOMER (on the one-side of the relationship) is added as a foreign key to ORDER (on the many-side of the relationship).

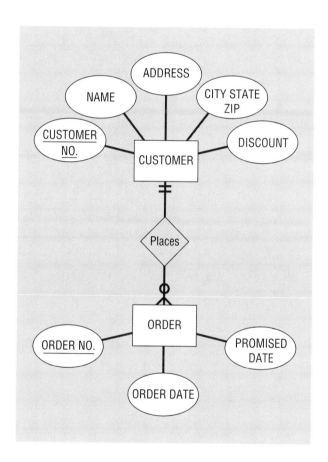

Figure 15-13
Representing a 1:*N* relationship

(a) E-R diagram

(b) Relations

CUSTOMER

CUSTOMER NO.	NAME	ADDRESS	CITY STATE ZIP	DISCOUNT
1273	Contemporary Designs	123 Oak St.	Austin, TX 38405	5%
6390	Casual Corner	18 Hoosier Dr.	Bloomington, IN 45821	3%

ORDER

ORDER NO.	ORDER DATE	PROMISED DATE	CUSTOMER NO.
57194	3/15/9X	3/28/9X	6390
63725	3/17/9X	4/01/9X	1273
80149	3/14/9X	3/24/9X	6390

One special case under this rule was mentioned in the previous section. If the entity on the many-side needs the key of the entity on the one-side as part of its primary key (this is a so-called weak entity), then this attribute is added not as a nonkey but as part of the primary key.

For a binary or unary one-to-one (1:1) relationship between two entities A and B (for a unary relationship, A and B would be the same entity type), the relationship can be represented by any of the following choices:

1. Adding the primary key of A as a foreign key of B

2. Adding the primary key of B as a foreign key of A

3. Both of the above

Binary and Higher Degree *M:N* Relationships   Suppose that there is a binary many-to-many (*M:N*) relationship or a gerund between two entity types A and B. For such a relationship or gerund, we create a separate relation C. The primary key of this relation is a composite key consisting of the primary key for each of the two entities in the relationship. Any nonkey attributes that are associated with the *M:N* relationship or gerund are included with the relation C.

Figure 15-14a, an example of this rule, shows the Requests relationship (*M:N*) between the entity types ORDER and PRODUCT for Pine Valley Furniture Company. Figure 15-14b shows the three relations (ORDER, PRODUCT, and ORDER LINE) that are formed from the entity types and the Requests relationship. First, a relation is created for each of the two entity types in the relationship (ORDER and PRODUCT). Then a relation (called ORDER LINE in Figure 15-14b) is created for the Requests relationship. The primary key of ORDER LINE is the combination (ORDER NO.,PRODUCT NO.), which are the respective primary keys of ORDER and PRODUCT. The nonkey attribute QUANTITY ORDERED also appears in ORDER LINE.

Occasionally, the relation created from an *M:N* relationship requires a primary key that includes more than just the primary keys from the two related relations. Consider, for example, the following situation:

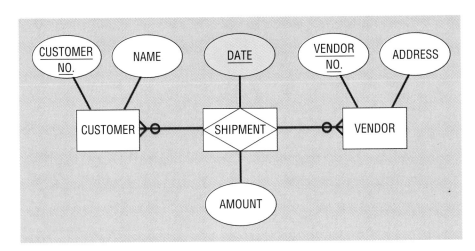

In this case, DATE must be part of the key for the SHIPMENT relation to uniquely distinguish each row of the SHIPMENT table, as follows:

SHIPMENT(<u>CUSTOMER NO.</u>,<u>VENDOR NO.</u>,<u>DATE</u>,AMOUNT)

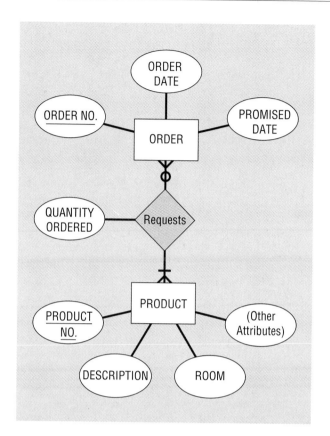

Figure 15-14
Representing an *M:N*
relationship

(a) E-R diagram

ORDER

(b) Relations

ORDER NO.	ORDER DATE	PROMISED DATE
61384	2/17/9X	3/01/9X
62009	2/13/9X	2/27/9X
62807	2/15/9X	3/01/9X

ORDER LINE

ORDER NO.	PRODUCT NO.	QUANTITY ORDERED
61384	M128	2
61384	A261	1

PRODUCT

PRODUCT NO.	DESCRIPTION	(OTHER ATTRIBUTES)
M128	Bookcase	——
A261	Wall unit	——
R149	Cabinet	——

If each shipment has a separate nonintelligent key, say a shipment number, then DATE becomes a nonkey and CUSTOMER NO. and VENDOR NO. become cross reference keys, as follows:

SHIPMENT(<u>SHIPMENT NO.</u>,CUSTOMER NO.,VENDOR NO.,DATE,AMOUNT)

In some cases, there may be a relationship among three or more entities. In such cases, we create a separate relation that has as a primary key the composite of the primary keys of each of the participating entities (plus any necessary additional key elements). This rule is a simple generalization of the rule for a binary *M:N* relationship.

Unary Relationship   To review, a *unary* relationship is a relationship between the instances of a single entity type, which are also called *recursive relationships.* Figure 15-15 shows two common examples. Figure 15-15a shows a one-to-many relationship named Manages that associates employees of an organization with another employee who is their manager. Figure 15-15b shows a many-to-many relationship that associates certain items with their component items. This relationship, introduced in Chapter 11, is called a *bill-of materials structure.*

For a unary 1:*N* relationship, the entity type (such as EMPLOYEE) is modeled as a relation. The primary key of that relation is the same as for the entity type. Then a foreign key is added to the relation that references the primary key values. A **recursive foreign key** is a foreign key in a relation that references the primary key values of that same relation. We can represent the relationship in Figure 15-15a as follows:

EMPLOYEE(<u>EMP ID</u>,NAME,BIRTHDATE,<u>MANAGER ID</u>)

> Recursive foreign key: A foreign key in a relation that references the primary key values of that same relation.

In this relation, MANAGER ID is a recursive foreign key that takes its values from the same domain of worker identification numbers as EMP ID.

For a unary *M:N* relationship, we model the entity type as one relation. Then we create a separate relation to represent the *M:N* relationship. The primary key of this new relation is a composite key that consists of two attributes (which need not have the same name) that both take their values from the same primary key domain. Any attribute associated with the relationship (such as QUANTITY in Figure 15-15b) is included as a nonkey attribute in this new relation. We can express the result for Figure 15-15b as follows:

ITEM(<u>ITEM NO.</u>,NAME,COST)
ITEM-BILL(<u>ITEM NO.</u>,<u>COMPONENT NO.</u>,QUANTITY)

IS-A Relationship (Class-Subclass)   IS-A relationships are explained in Appendix C. The relational data model does not directly support class-subclass (or IS-A) relationships. We use the following rules to represent IS-A relationships:

1. Create a separate relation for the class and for each of the subclasses.

2. The table (relation) for the class consists only of the primary key and the attributes that are common to all of the subclasses.

3. The table for each subclass contains only its primary key and the columns unique to that subclass.

4. The primary keys of the class and each of the subclasses are from the same domain, but do not have to have the same names.

An IS-A relationship (with two subclasses for different types of rental properties) appears in the E-R diagram of Figure 15-16a, and the relations that are derived by applying the rules are shown in Figure 15-16b. Notice that there are three relations: PROPERTY, BEACH, and MOUNTAIN. The primary key of each of these

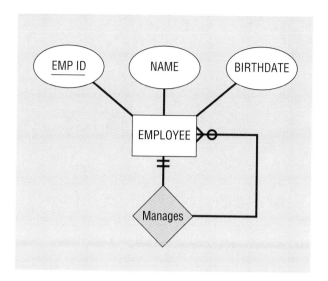

Figure 15-15
Two unary relationships

(a) EMPLOYEE with Manages
relationship (1:*N*)

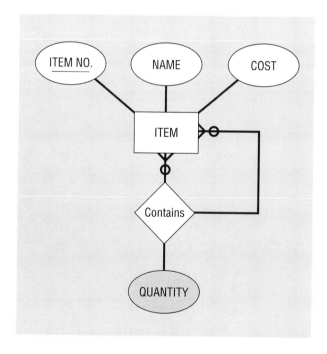

(b) Bill-of-materials structure
(*M:N*)

relations is the composite key (STREET ADDRESS,CITY STATE ZIP). Although the primary key of a subclass does not have to have the same name as the primary key of the class as in this example, they must be from the same domain.

The PROPERTY relation contains those descriptors (or nonkey attributes) that are common to both subclasses: NO. ROOMS and TYPICAL RENT. The relations BEACH and MOUNTAIN (representing the subclasses) have the same primary key as PROPERTY. Each relation contains an attribute, however, that is unique to the subclass (BLOCKS TO BEACH for BEACH, SKIING for MOUNTAIN). We can form a subclass with its inherited attributes by combining the subclass relation with the class relation. This is called joining in relational terminology.

Figure 15-16

Representing IS-A relationships

(a) E-R diagram

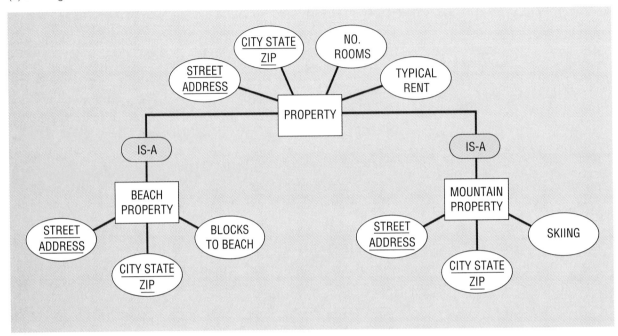

(b) Relations

PROPERTY

STREET ADDRESS	CITY STATE ZIP	NO. ROOMS	TYPICAL RENT
120 Surf Dr.	Honolulu, HI 99987	3	500
100 Mogul Dr.	Jackson, WY 89204	3	250

BEACH

STREET ADDRESS	CITY STATE ZIP	BLOCKS TO BEACH
120 Surf Dr.	Honolulu, HI 99987	2

MOUNTAIN

STREET ADDRESS	CITY STATE ZIP	SKIING
100 Mogul Dr.	Jackson, WY 89204	N

Summary of Transforming E-R Diagrams to Relations

We have now described how to transform E-R diagrams to relations. Table 15-1 lists the rules discussed in this section for transforming entity-relationship diagrams into equivalent relations. After this transformation, you should check the resulting relations to determine whether they are in third normal form and, if necessary, perform normalization as described earlier in the chapter.

TABLE 15-1   E-R to Relational Transformation

E-R Structure	Relational Representation
Regular entity	Create a relation with primary key and nonkey attributes
Weak entity	Create a relation with a composite primary key (which includes the primary key of the entity on which this entity depends) and nonkey attributes
Binary or unary 1:1 relationship	Place the primary key of either entity in the relation for the other entity or do this for both entities
Binary 1:*M* relationship	Place the primary key of the entity on the one-side of the relationship as a foreign key in the relation for the entity on the many-side
Binary or unary *M:N* relationship or gerund	Create a relation with a composite primary key using the primary keys of the related entities, plus any nonkey attributes of the relationship or gerund
Binary or unary *M:N* relationship or gerund with additional key(s)	Create a relation with a composite primary key using the primary keys of the related entities and additional primary key attributes associated with the relationship or gerund, plus any nonkey attribute of the relationship or gerund
Binary or unary *M:N* relationship or gerund with its own key	Create a relation with the primary key associated with the relationship or gerund, plus any nonkey attributes of the relationship or gerund and the primary keys of the related entities (as nonkey attributes)
IS-A relationship	Create a relation for the superclass which contains the primary key and all nonkey attributes in common with all subclasses, plus create a separate relation for each subclass with the same primary key (with the same or a local name) but with only the nonkey attributes related to that subclass

MERGING RELATIONS

As part of the logical data modeling process, normalized relations likely have been created from a number of separate E-R diagrams and various user views. Some of the relations may be redundant—they may refer to the same entities. If so, you should merge those relations to remove the redundancy. This section describes merging relations or *view integration,* which is the last step in the logical data modeling process defined earlier in the chapter.

An Example of Merging Relations

Suppose that modeling a user view or transforming an E-R diagram results in the following 3NF relation:

EMPLOYEE1(<u>EMPLOYEE NO.</u>,NAME,ADDRESS,PHONE)

Modeling a second user view might result in the following relation:

EMPLOYEE2(<u>EMPLOYEE NO.</u>,NAME,ADDRESS,JOBCODE,NO. YEARS)

Since these two relations have the same primary key (EMPLOYEE NO.) and describe the same entity, they should be merged into one relation. The result of merging the relations is the following relation:

EMPLOYEE(<u>EMPLOYEE NO.</u>,NAME,ADDRESS,PHONE,JOBCODE,NO. YEARS)

Notice that an attribute that appears in both relations (such as NAME in this example) appears only once in the merged relation.

View Integration Problems

When integrating relations, you must understand the meaning of the data and must be prepared to resolve any problems that may arise in that process. In this section, we describe and briefly illustrate four problems that arise in view integration: *synonyms, homonyms, transitive dependencies,* and *class/subclass relationships.*

Synonyms In some situations, two or more attributes may have different names but the same meaning, as when they describe the same characteristic of an entity. Such attributes are called **synonyms.** For example, EMPLOYEE ID and EMPLOYEE NO. may be synonyms.

> **Synonyms:** Two different names that are used to refer to the same data item (for example, *car* and *automobile*).

When merging the relations that contain synonyms, you should obtain, if possible, agreement from users on a single standardized name for the attribute and eliminate the other synonym. Another alternative is to choose a third name to replace the synonyms. For example, consider the following relations:

STUDENT1(<u>STUDENT ID</u>,NAME)
STUDENT2(<u>MATRICULATION NO.</u>,NAME,ADDRESS)

In this case, the analyst recognizes that both the STUDENT ID and MATRICULATION NO. are synonyms for a person's social security number and are identical attributes. One possible resolution would be to standardize on one of the two attribute names, such as STUDENT ID. Another option is to use a new attribute name, such as SSN, to replace both synonyms. Assuming the latter approach, merging the two relations would produce the following result:

STUDENT(<u>SSN</u>,NAME,ADDRESS)

> **Alias:** An alternative name given to an attribute.

Often when there are synonyms, there is a need to allow some system users to refer to the same data by different names. Users may need to use familiar names that are consistent with terminology in their part of the organization. An **alias** is an alternative name used for an attribute. Many database management systems and CASE repositories allow the definition of an alias that may be used interchangeably with the primary attribute label.

Homonyms In other situations, a single attribute, called a **homonym,** may have more than one meaning or describe more than one characteristic. For example, the term *account* might refer to a bank's checking account, savings account, loan account, or other type of account; therefore, *account* refers to different data, depending on how it is used.

> **Homonym:** A single name that is used for two or more different attributes (for example, the term *invoice* to refer to both a customer invoice and a supplier invoice).

You should be on the lookout for homonyms when merging relations. Consider the following example:

STUDENT1(<u>STUDENT ID</u>,NAME,ADDRESS)
STUDENT2(<u>STUDENT ID</u>,NAME,PHONE-NO.,ADDRESS)

In discussions with users, the systems analyst may discover that the attribute ADDRESS in STUDENT1 refers to a student's campus address while in STUDENT2 the same attribute refers to a student's home address. To resolve this conflict, we would probably need to create new attribute names and the merged relation would become

STUDENT(<u>STUDENT ID</u>,NAME,PHONE NO.,CAMPUS ADDRESS,
    PERMANENT ADDRESS)

Transitive Dependencies When two 3NF relations are merged to form a single relation, *transitive dependencies* may result. For example, consider the following two relations:

STUDENT1(<u>STUDENT ID</u>,MAJOR)
STUDENT2(<u>STUDENT ID</u>,ADVISOR)

Since STUDENT1 and STUDENT2 have the same primary key, the two relations may be merged:

STUDENT(<u>STUDENT ID</u>,MAJOR,ADVISOR)

However, suppose that each major has exactly one advisor. In this case, ADVISOR is functionally dependent on MAJOR:

MAJOR→ADVISOR

If the above dependency exists, then STUDENT is 2NF but not 3NF, since it contains a transitive dependency. The analyst can create 3NF relations by removing the transitive dependency (MAJOR becomes a foreign key in STUDENT):

STUDENT(<u>STUDENT ID</u>,MAJOR)

MAJOR ADVISOR(<u>MAJOR</u>,ADVISOR)

Class/Subclass (IS-A)   Class/subclass relationships (defined in Appendix C) may be hidden in user views or relations. Suppose that we have the following two hospital relations:

PATIENT1(<u>PATIENT NO.</u>,NAME,ADDRESS)

PATIENT2(<u>PATIENT NO.</u>,ROOM NO.)

Initially, it appears that these two relations can be merged into a single PATIENT relation. However, suppose that there are two different types of patients: inpatients and outpatients. PATIENT1 actually contains attributes common to *all* patients. PATIENT2 contains an attribute (ROOM NO.) that is a characteristic only of inpatients. In this situation, you should create *class/subclass (IS-A)* relationships for these entities:

PATIENT(<u>PATIENT NO.</u>,NAME,ADDRESS)

INPATIENT(<u>PATIENT NO.</u>,ROOM NO.)

OUTPATIENT (<u>PATIENT NO.</u>,DATE TREATED)

For an extended discussion of view integration in database design, see Navathe, et al. (1986).

LOGICAL DATA MODELING FOR HOOSIER BURGER

In Chapter 11 we developed a conceptual data model, an E-R model, for a new inventory control system at Hoosier Burger. That E-R diagram is repeated here as Figure 15-17. In this section we show how this E-R model is translated into normalized relations and how to normalize and then merge the relations for a new report with the relations from the E-R model.

In this E-R model, four entities exist independently of other entities: SALE, PRODUCT, INVOICE, and INVENTORY ITEM. Given the attributes shown in Figure 15-17, we can represent these entities in the following four relations:

SALE(<u>Receipt No.</u>,Sale Date)

PRODUCT(<u>Product No.</u>,Product Description)

INVOICE(<u>Vendor No.,Invoice No.</u>,Invoice Date,Paid)

INVENTORY ITEM(<u>Item No.</u>,Item Description,Quantity in Stock,Minimum Order Quantity,Type of Item)

The entities ITEM SALES and INVOICE ITEM as well as the gerund RECIPE each have composite primary keys taken from the entities to which they relate, so we can represent these three entities in the following three relations:

ITEM SALE(<u>Receipt No.,Product No.</u>,Quantity Sold)

INVOICE ITEM(<u>Vendor No.,Invoice No.,Item No.</u>,Quantity Added)

RECIPE(<u>Product No.,Item No.</u>,Quantity Used)

Since there are no many-to-many, one-to-one, IS-A, or unary relationships, we have now represented all the entities and relationships from the E-R model. Also, each of

Figure 15-17
Final E-R diagram for Hoosier Burger's inventory control system

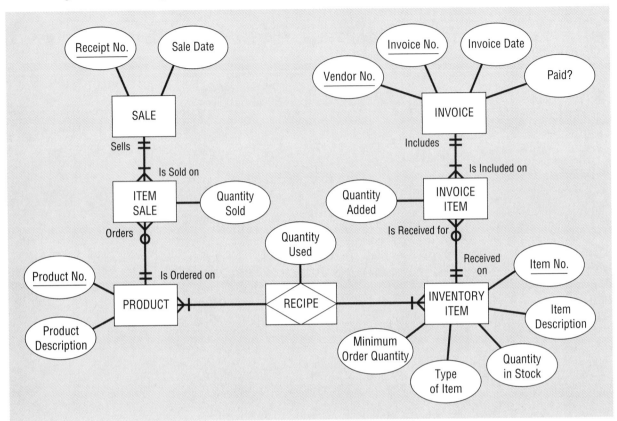

the above relations is in 3NF since all attributes are simple, all nonkeys are fully dependent on the whole key, and there are no dependencies between nonkeys in the INVOICE and INVENTORY ITEM relations.

Now suppose that Bob Mellankamp wanted an additional report which was not previously known by the analyst who designed the inventory control system for Hoosier Burger. A rough sketch of this new report, listing usage of inventory items in both regular item sales and Hoosiers in a Hurry meal sales, appears in Figure 15-18. A Hoosiers in a Hurry meal is a special, pre-defined package of several products, such as a hamburger, regular fries, and a medium drink. The combination of products in a Hoosiers in a Hurry meal is not the same concept as a RECIPE; in fact, each product within A Hoosiers in a Hurry meal has a recipe.

This report contains data about several relations already known to the analyst, including

SALE(<u>Receipt No.</u>,Sale Date)
PRODUCT(<u>Product No.</u>,Product Description)
INVENTORY ITEM(<u>Item No.</u>,Item Description)
ITEM SALE(<u>Receipt No.</u>,<u>Product No.</u>,Quantity Sold)
RECIPE(<u>Product No.</u>,<u>Item No.</u>,Quantity Used)

Even though data such as Receipt No. and Quantity Used are not on the report, the analyst realizes that these data are needed to identify relations or to compute data

INVENTORY USAGE REPORT
for Sales *Date 1–Date 2*

Page x of n
Date Printed

Figure 15-18
Hoosier Burger inventory
usage report

REGULAR PRODUCT NO.	DESCRIPTION	DATE USED	CONSUMPTION		
			REGULAR	HOOSIER	TOTAL
xxx		aaa	—	—	—
		bbb	—	—	—
		ccc	—	—	—
		⋮	⋮	⋮	⋮
yyy		—	—	—	—
		—	—	—	—
		—	—	—	—
		⋮	⋮	⋮	⋮
⋮					

on the report. Since the five relations show no new relations or attributes, the current normalized logical data model is so far sufficient to cover the new report.

The analyst recognizes, however, that Hoosiers in a Hurry meals do not actually have recipes as only individual products have recipes. Thus, the RECIPE relation does not really help to compute the amount of inventory used for Hoosiers in a Hurry consumption, whereas these are products with a product number. Consequently, the analyst realizes that there are really two kinds (subclasses) of products: regular products and Hoosiers in a Hurry products, and that there is a many-to-many relationship between Hoosiers in a Hurry and regular products. Thus, the analyst defines two subclass PRODUCT relations and adds a new relation to represent the relationship between them, as follows:

REGULAR PRODUCT(<u>Regular Product No.</u>)
HOOSIER PRODUCT(<u>Hoosier Product No.</u>)
MEAL(<u>Hoosier Product No.</u>,<u>Regular Product No.</u>,Meal Quantity)

where Meal Quantity is the number of a particular regular product used in a particular Hoosiers in a Hurry meal.

The analyst then recognizes that the RECIPE relation should be changed to capture the fact that only regular products have a recipe, as follows:

RECIPE(<u>Regular Product No.</u>,<u>Item No.</u>,Quantity Used)

The final set of relations, after considering this new report and merging relations for it with those identified before, look like this:

SALE(<u>Receipt No.</u>,Sale Date)
PRODUCT(<u>Product No.</u>,Product Description)
REGULAR PRODUCT(<u>Regular Product No.</u>)
HOOSIER PRODUCT(<u>Hoosier Product No.</u>)
MEAL(<u>Hoosier Product No.</u>,<u>Regular Product No.</u>,Meal Quantity)
INVOICE(<u>Vendor No.</u>,<u>Invoice No.</u>,Invoice Date,Paid)
INVENTORY ITEM(<u>Item No.</u>,Item Description,Quantity in Stock,Minimum Order Quantity,Type of Item)
INVOICE ITEM(<u>Vendor No.</u>,<u>Invoice No.</u>,<u>Item No.</u>,Quantity Added)
ITEM SALE(<u>Receipt No.</u>,<u>Product No.</u>,Quantity Sold)
RECIPE(<u>Regular Product No.</u>,<u>Item No.</u>,Quantity Used)

The updated E-R model for these relations appears in Figure 15-19.

When comparing this list of ten relations with the E-R diagram, we note that the ten relations correspond to the ten entities and gerunds in Figure 15-19. If there had been any *M:N* relationships on the E-R diagram, these, too, would have been

Figure 15-19

E-R diagram corresponding to normalized relations of Hoosier Burger's inventory control system

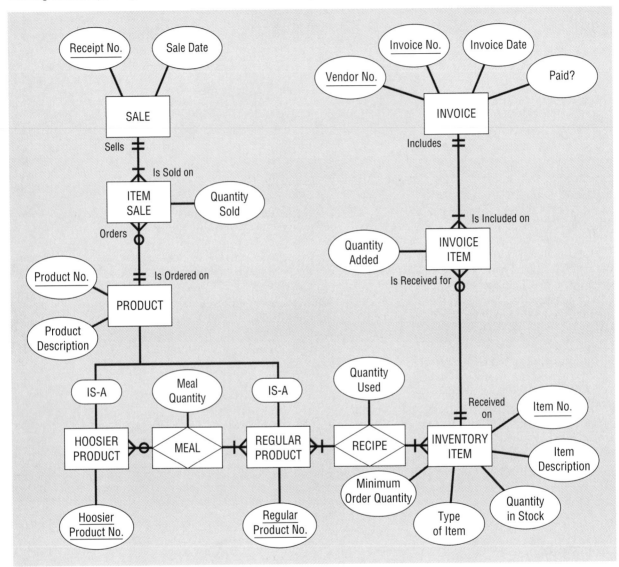

represented by a relation, which would contain only the primary keys of the related entities. Also note in Figure 15-19 that there is a relationship, the MEAL gerund, connecting two subclass entities. The list of ten relations, with the E-R diagram as a visual representation, now becomes input to physical file and database design, which is the topic of the next chapter.

SUMMARY

In this chapter, we have presented the process and techniques used to structure the logical data requirements of an information system. Logical data modeling uses the relational data model to represent data requirements as tables of data in which every cell has an atomic value. We also showed how to transform conceptual data models between entity-relationship and relational notations.

The data normalization process transforms unnormalized relations identified from E-R diagrams into first, second, and third normal forms. Normalization develops a structure for data requirements that has desirable properties which avoid insertion, deletion, and maintenance anomalies in data. Normalization is based on functional dependencies between attributes, in which a relation in third normal form has each nonkey attribute dependent on the full primary key, and nothing but this key, of its relation.

In addition, we showed how to merge relations developed from separate E-R diagrams or normalization activities into consolidated, normalized relations. We discussed how to handle issues of synonyms, homonyms, transitive dependencies, and IS-A relationships during relation merging. Such a consolidated set of normalized relations, along with updating the project dictionary, are the deliverables from logical data modeling.

Logical data modeling is based on certain constructs about the structure of data. These constructs include relation, functional dependency, primary key, foreign key, and referential integrity. Logical data modeling is grounded in specific data requirements (forms and reports), but the final steps reconcile the structure of data found in specific system externals and outputs with the view of data portrayed in a conceptual data model.

This chapter completes our coverage of the logical design of an information system. The chapters in the following section address the techniques necessary for the physical design of an information system. Chapter 16 begins by discussing physical database design, which follows directly from the results demonstrated in this chapter on logical data modeling.

C H A P T E R   R E V I E W

K E Y   T E R M S

Alias	Normal form	Relational database model
Anomalies	Normalization	Second normal form (2NF)
First normal form (1NF)	Partial functional	Synonyms
Foreign key	dependency	Third normal form (3NF)
Functional dependency	Recursive foreign key	Transitive dependency
Homonym	Referential integrity	Well-structured relation
Logical data model	Relation	

R E V I E W   Q U E S T I O N S

1. Define each of the following terms:

a. functional dependency

b. normalization

c. relation

d. primary key

e. foreign key

f. anomalies

2. What is the purpose of normalization?

3. What is the role of a conceptual data model during logical database design?

4. List five properties of relations.

5. What do you need to know about an information system to do logical data modeling?

6. What are the two major deliverables from logical data modeling?

7. What problems can arise during view integration or merging relations?

8. How are relationships between entities represented in the relational data model?

9. How do you represent a 1:*M* unary relationship in a relational database model?

10. How do you represent a *M:N* ternary relationship in a relational database model?

11. What is the relationship between the primary key of a relation and the functional dependencies among all attributes within that relation?

PROBLEMS AND EXERCISES

1. Match the following terms to the appropriate definitions.

_____ view integration	a. named two-dimensional table of data
_____ third normal form	b. state of a relation that does not have transitive dependencies
_____ referential integrity	c. functional dependency between two (or more) non-key attributes in a relation
_____ homonym	d. process of merging relations developed from a conceptual data model with those from different user views
_____ relation	e. particular relationship between two attributes
_____ functional dependency	f. name used for several different attributes
_____ transitive dependency	g. requirement that the value of an attribute come from the domain of another attribute

2. Assume that in Pine Valley Furniture products are comprised of components, products are assigned to salespersons, and components are produced by vendors. Also assume that in the relation
PRODUCT(PRODNAME,SALESPERSON,COMPNAME,VENDOR)
VENDOR is functionally dependent on COMPNAME and COMPNAME is functionally dependent on PRODNAME. Eliminate the transitive dependency in this relation and form 3NF relations.

3. Draw an E-R diagram for the situation described in Problem 2. Transform the E-R diagram into a set of 3NF relations.

4. In this chapter we emphasize the importance of well-structured relations and normalization of data. In practice, however, information systems designers often choose to unnormalize relations when creating files. Why do you think this is done? (*Hint:* Remember, normalization is done during logical data modeling and files are designed during physical file and database design.)

5. Transform the E-R diagram of Figure 11-3 into a set of 3NF relations. Make up a primary key and one or more nonkeys for each entity.

6. Transform the following E-R diagram into a set of 3NF relations.

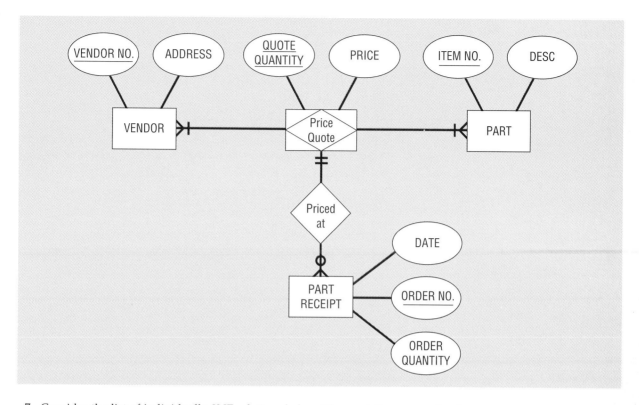

7. Consider the list of individually 3NF relations below. These relations were developed from several separate normalization activities.
   PATIENT(<u>PATIENT NO.</u>,ROOM NO.,ADMIT DATE,ADDRESS)
   ROOM(<u>ROOM NO.</u>,PHONE,DAILY RATE)
   PATIENT(<u>PATIENT NO.</u>,TREATMENT DESCRIPTION,ADDRESS)
   TREATMENT(<u>TREATMENT ID</u>,DESCRIPTION,COST)
   PHYSICIAN(<u>PHYSICIAN ID</u>,NAME,DEPARTMENT)
   PHYSICIAN(<u>PHYSICIAN ID</u>,NAME,SUPERVISOR ID)

   a. Merge these relations into a consolidated set of 3NF relations. Make and state whatever assumptions you consider necessary to resolve any potential problems you identify in the merging process.
   b. Draw an E-R diagram for your answer to part (a).

8. Develop a set of 3NF relations for the customer account information screen display shown in Figure 13-11b.

9. Consider the following 3NF relations about a sorority or fraternity:
   MEMBER(<u>MEMBER ID</u>,NAME,ADDRESS,DUES OWED)
   OFFICE(<u>OFFICE NAME</u>,OFFICER ID,TERM START DATE,BUDGET)
   EXPENSE(<u>LEDGER NO.</u>,OFFICE NAME,EXPENSE DATE,AMT OWED)
   PAYMENT(<u>CHECK NO</u>,EXPENSE LEDGER NO.,AMT PAID)
   RECEIPT(<u>MEMBER NO.</u>,<u>RECEIPT DATE</u>,DUES RECEIVED)
   COMMITTEE(<u>COMMITTEE ID</u>,OFFICER-IN-CHARGE)
   WORKERS(<u>COMMITTEE ID</u>,<u>MEMBER NO.</u>)

   a. Foreign keys are not indicated in these relations. Decide which attributes are foreign keys and justify your decisions.
   b. Draw an E-R diagram for these relations, using your answer to part (a).
   c. Explain the assumptions you made about cardinalities in your answer to part (b). Explain why it is said that the E-R data model is more expressive or more semantically rich than the relational data model.

10. Consider the following functional dependencies:
    APPLICANT ID→APPLICANT NAME
    APPLICANT ID→APPLICANT ADDRESS
    POSITION ID→POSITION TITLE
    POSITION ID→DATE POSITION OPENS
    POSITION ID→DEPARTMENT
    APPLICANT ID + POSITION ID→DATE APPLIED
    APPLICANT ID + POSITION ID + DATE INTERVIEWED→

    a. Represent these attributes with 3NF relations. Provide meaningful relation names.
    b. Represent these attributes using an E-R diagram. Provide meaningful entity and relationship names.

11. Develop 3NF relations for the E-R diagram you drew for Problem and Exercise 6 in Chapter 11.

12. Develop 3NF relations for the E-R diagram you drew for Problem and Exercise 9 in Chapter 11.

FIELD EXERCISES

1. Find in the literature discussions of additional normal forms other than first, second, and third normal forms. Describe each of these additional normal forms and give examples of each. How are these additional normal forms different from those presented in this chapter? What additional benefit does their use provide?

2. Find a form or report from a business organization. Draw an E-R diagram from the entities presented in the form or report. Transform the diagram into a set of third normal form relations.

3. Interview data analysts in organizations where normalization is done. Describe the ways that they conduct normalization. What problems do they encounter when conducting normalization? Do they ever unnormalize relations when creating files? Why or why not?

4. Interview systems builders in a few different business organizations. Is logical data modeling done? Why or why not? If so, what are the steps taken and how closely do these fit with the steps described in this chapter? If not, what are the potential disadvantages of skipping or shortening this process?

5. Consider a recent interaction with an organization (e.g., paying a utility bill, purchasing a ticket for a game or movie). Draw an E-R diagram for all the entities relevant to this transaction (even indirectly involved entities). Transform the diagram into a set of third normal form relations.

REFERENCES

Codd, E. F. 1970. "A Relational Model of Data for Large Relational Databases." *Communications of the ACM* 13 (June): 77–87.

Dutka, A. F., and H. H. Hanson, 1989. *Fundamentals of Data Normalization.* Reading, MA: Addison-Wesley.

McFadden. F. R., and J. A. Hoffer. 1994. *Modern Database Management.* Redwood City, CA: Benjamin/Cummings Publishing.

Navathe. S., R. Elmasri, and J. Larson. 1986. "Integrating User Views in Database Design." *Computer* (Jan.): 50–62.

Logical Design:
Designing the Logical Database

INTRODUCTION

In this last BEC case for the logical design phase of the Customer Activity Tracking System (CATS) project, we follow the activities of Jorge Lopez, a systems analyst on the project, and Buffy Jarvis, a data analyst on loan to the team. In addition, we track the overall progress of the CATS project as the team makes the transition from logical to physical design. In particular, this BEC case shows how Jorge and Buffy developed the logical data model for the CATS project. This consolidated relational database model shows all functional aspects of the CATS databases, except for those dimensions which relate to the features of the database management system to be used for CATS. Jorge and Buffy developed the normalized relations from the final CATS E-R model derived during conceptual database design and supplemented this with more data details discovered during other logical design steps. These normalized relations will be input to physical file and database design when the CATS team decides if and how they will change the data structures of the purchased system.

TRANSFORMING THE E-R MODEL

"Jorge, you are really catching on to data modeling," observes Buffy Jarvis as she reviews the preliminary work Jorge Lopez has done to prepare normalized relations. "The relations appear to be normalized, but why don't you review for me how you developed these relations."

"The CASE tool was a big help," responded Jorge. "Since the E-R model we prepared was in the repository (Figures 1 and 2), I simply had the CASE tool use its built-in rules for transforming an E-R model into relations—you know, rules like each entity becomes a relation, each many-to-many relationship becomes a relation, and how foreign keys are assigned. You'll remember, we updated the E-R model since conceptual

design when we discovered that the 'Superior product category name' attribute represented a unary relationship between product profiles (see the Has Superior relationship in Figure 1). I thought this E-R diagram would be a good starting point."

Jorge continued, "Then I looked over the relations that were generated to see if they made sense (see Figure 3). You can see all the work the CASE tool did for me by comparing the E-R model attributes (Figure 2) with the generated relations (Figure 3). For example, it added the foreign key STORE_ID in the CUSTOMER relation, represented the unary relation on the PRODUCT PROFILE entity as a foreign key in the PRODUCT_PROFILE relation, and created a relation PRODUCT_FOR_CUSTOMER with a compound primary key for the many-to-many Products for Customer relationship. The CASE tool also forced discipline in creating field names, such as each field in the PRODUCT relation has a PROD prefix. As you can see, there are some relations, like the ones for the subclasses of products and the one for customer profile, with only primary key attributes; I still need to investigate why we don't have any nonkey attributes for those relations yet.

"I didn't worry about field data types yet, although I understand we'll need to make some basic decisions on data types before the end of logical design. And the primary and foreign keys imply referential integrity constraints, which we will also have to decide how to implement during physical design.

"Next, since an E-R model is not necessarily normalized, I checked the functional dependencies within each relation. Since there were no multivalued attributes on the E-R model, I knew we were in first normal form. I then used that guideline you told me one day: make sure that each nonkey is 'fully dependent on the whole key and nothing but the key, so help me Codd.' This guideline covers both second and third normal forms. Generally, I found no normal form violations, but Buffy, should we do something about the relationship between zip code and city and state in the CUSTOMER relation? Clearly there is a relationship here."

Figure 1

Updated CATS E-R model
from conceptual data
modeling

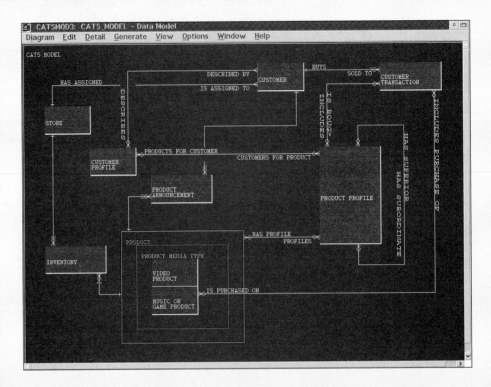

Buffy replied, "I think just about every project that deals with addresses faces this question. You are right, zip code determines state and city, and even restricts valid values for street. However, as a company, we have not implemented a standard zip code table for all systems, so I suggest we ignore this functional dependency for CATS.

"As you said, our CASE tool created VIDEO_ PRODUCT and MUSIC_OR_GAME_PRODUCT relations for the IS-A relationships in the E-R diagram. This is fine for now. If these subclass relations do not carry any data in addition to what any PRODUCT does, we may not actually create tables for these relations, but we'll decide that in physical design, too. Jorge, what did you do next?"

CREATING NORMALIZED RELATIONS FOR SYSTEM INPUTS AND OUTPUTS

"Next I took each report and on-line form in the system and made sure that each field was contained in or could be computed from data in the E-R model. Not surprisingly, I found numerous examples of data in these system inputs and outputs, or what you call user views, that we had not anticipated during conceptual design. For example, here is the first analysis I've

done—the customer-product assignment (CPA) report that indicates which customers have reserved which products (see Figure 4) for pick-up at a particular store. You may recall, Buffy, that this report is generated from central CATS and sent as an electronic document to the store CATS. Thus, the system design does not require that any of this data be part of the store CATS database. Consequently, a field like Contact Date is really just a label printed on the report for use by the store personnel, not a database field. What I did was create a list of fields on this report and the functional dependencies between them (see Figure 5) to see if the central CATS database was sufficient to produce this report. I used, where possible, the field names generated from the CASE tool to avoid creating synonyms. I'll bet that when I look at other reports and forms I'll find some nonkey attributes for the relations with only primary key attributes."

"This looks good, Jorge," said Buffy. "It looks like the CUSTOMER relation generated from the E-R model is complete since all of the attributes dependent on CUST_ID are in this relation. And the necessary changes to the PRODUCT relation seem pretty straightforward because we can just add the PROD_ARTIST and PROD_PRICE attributes to the relation since these attributes also depend on PROD_ID. But what does the last functional dependency mean? RESERVATION_ CONFIRMATION was not in the E-R model, and RESERVATION_DATE was."

Figure 2

Data elements for entities in updated CATS E-R data model

CATS_MODEL - Updated Data Model Element List

Type	Name	
Subject Area	┌CATS_MODEL	
Entity Type	├CUSTOMER	
Attribute	I CUSTOMER_NUMBER	(Number, 9, Mandatory, Basic)
Attribute	CUSTOMER_NAME	(Text, 30, Mandatory, Basic)
Attribute	CUSTOMER_STREET	(Text, 30, Mandatory, Basic)
Attribute	CUSTOMER_CITY	(Text, 20, Mandatory, Basic)
Attribute	CUSTOMER_STATE	(Text, 2, Mandatory, Basic)
Attribute	CUSTOMER_ZIP	(Number, 9, Mandatory, Basic)
Attribute	TELEPHONE_NUMBER	(Text, 10, Mandatory, Basic)
Attribute	CREDIT_CARD_TYPE	(Text, 2, Mandatory, Basic)
Attribute	CREDIT_CARD_NUMBER	(Text, 16, Mandatory, Basic)
Attribute	CREDIT_CARD_EXPIRATION_DATE	(Date, 8, Mandatory, Basic)
Attribute	SOCIAL_SECURITY_NUMBER	(Number, 9, Optional, Basic)
Relationship	Sometimes DESCRIBED_BY One or More CUSTOMER_PROFILE	
Relationship	Sometimes HAVE One or More PRODUCT_ANNOUNCEMENT	
Relationship	Sometimes BUYS One or More CUSTOMER_TRANSACTION	
Relationship	Always IS_ASSIGNED_TO One STORE	
Entity Type	┌CUSTOMER_PROFILE	
Attribute	I CUSTOMER_PROFILE_CATEGORY	(Text, 2, Mandatory, Basic)
Attribute	I CUSTOMER_CATEGORY_LEVEL	(Text, 2, Mandatory, Basic)
Relationship	Sometimes PRODUCTS_FOR_CUSTOMER One or More PRODUCT_PROFILE	
Relationship	Always DESCRIBES One or More CUSTOMER	
Entity Type	┌CUSTOMER_TRANSACTION	
Attribute	I TRANSACTION_DATE	(Date, 8, Mandatory, Basic)
Relationship	Sometimes INCLUDES_PURCHASE_OF One or More MUSIC_OR_GAME_PRODUCT	
Relationship	Sometimes IS_BOUGHT_ON One or More PRODUCT_PROFILE	
Relationship	I Always SOLD_TO One CUSTOMER	
Entity Type	┌INVENTORY	
Attribute	QUANTITY_IN_STOCK	(Number, 4, Mandatory, Basic)
Relationship	I Always HAVE One STORE	
Relationship	I Always HAVE One PRODUCT	
Entity Type	┌PRODUCT	
Attribute	I PRODUCT_NUMBER	(Number, 6, Mandatory, Basic)
Attribute	PRODUCT_TITLE	(Text, 30, Mandatory, Basic)
Attribute	PRODUCT_MEDIA_TYPE	(Text, 2, Mandatory, Basic)
Attribute	DATE_PRODUCT_IS_AVAILABLE	(Date, 8, Mandatory, Basic)
Relationship	Sometimes HAS_PROFILE One or More PRODUCT_PROFILE	
Relationship	Sometimes HAVE One or More INVENTORY	
Relationship	Sometimes HAVE One or More PRODUCT_ANNOUNCEMENT	
Partitioning	┌PRODUCT_MEDIA_TYPE	
Subtype	├MUSIC_OR_GAME_PRODUCT	
Relationship	Sometimes IS_PURCHASED_ON One or More CUSTOMER_TRANSACTION	
Subtype	VIDEO_PRODUCT	
Entity Type	┌PRODUCT_ANNOUNCEMENT	
Attribute	I DATE_ANNOUNCEMENT_SENT	(Date, 8, Mandatory, Basic)
Attribute	CATS_RUN_NUMBER_FOR_ANNOUNCEMENT	(Number, 4, Mandatory, Basic)
Attribute	DATE_CUSTOMER_RESERVED_PRODUCT	(Date, 8, Mandatory, Basic)
Attribute	PCS_CODE	(Text, 2, Mandatory, Basic)
Relationship	I Always HAVE One PRODUCT	
Relationship	I Always HAVE One CUSTOMER	
Entity Type	┌PRODUCT_PROFILE	
Attribute	I PRODUCT_CATEGORY_NAME	(Text, 20, Mandatory, Basic)
Relationship	Sometimes HAS_SUBORDINATE One or More PRODUCT_PROFILE	
Relationship	Sometimes HAS_SUPERIOR One PRODUCT_PROFILE	
Relationship	Sometimes PROFILES One or More PRODUCT	
Relationship	Sometimes CUSTOMERS_FOR_PRODUCT One or More CUSTOMER_PROFILE	
Relationship	Sometimes INCLUDES One or More CUSTOMER_TRANSACTION	
Entity Type	┌STORE	
Attribute	I STORE_NUMBER	(Number, 4, Mandatory, Basic)
Attribute	STORE_MANAGER_NAME	(Text, 30, Mandatory, Basic)
Relationship	Sometimes HAVE One or More INVENTORY	
Relationship	Always HAS_ASSIGNED One or More CUSTOMER	

Figure 3

Normalized relations from E-R model

```
STORE (STORE_ID, STORE_MRG_NAME)
PRODUCT (PROD_ID, PROD_TITLE, PROD_MEDIA, PROD_DATE_AVAILABLE)
VIDEO_PRODUCT (VIDEO_PROD_ID)
MUSIC/GAME_PRODUCT (MUSIC/GAME_PROD_ID)
CUSTOMER (CUST_ID, STORE_ID, CUST_NAME, CUST_STREET, CUST_CITY, CUST_STATE,
    CUST_ZIP, CUST_PHONE, CUST_CARD_TYPE, CUST_CARD_NO, CUST_CARD_EXPIRES,SSN)
INVENTORY (STORE_ID, PROD_ID, QTY_IN_STOCK)
PRODUCT_ANNOUNCEMENT (PROD_ID, CUST_ID, ANNOUNCE_DATE, RUN_NUMBER,
    RESERVATION_DATE, PCS_CODE)
CUSTOMER_TRANSACTION (CUST_TRAN_ID, CUST_ID, CUST_TRAN_DATE)
PURCHASE (CUST_TRAN_ID, MUSIC/GAME_PROD_ID)
CUSTOMER_PROFILE (CUST_PROF_CATEGORY, CUST_PROF_CAT_LEVEL)
PRODUCT_PROFILE (PROD_PROF_CATEGORY, PROD_PROF_SUPERIOR_CAT)
PRODUCT_FOR_CUSTOMER (CUST_PROF_CATEGORY, CUST_PROF_CAT_LEVEL, PROD_PROF_CATEGORY)
```

(*Note:* Primary keys are underlined and foreign keys are in italics.)

Figure 4

Customer-product assignment (CPA) report

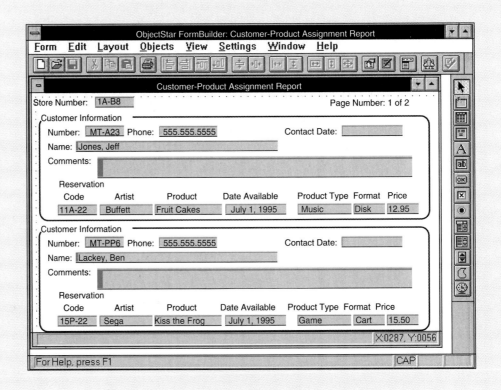

Figure 5

Functional dependencies for customer-product assignment (CPA) report

```
CUST_ID → CUST_NAME, CUST_PHONE, STORE_ID
PROD_ID → PROD_ARTIST, PROD_TITLE, PROD_MEDIA, PROD_DATE_AVAILABLE, PROD_PRICE
PCS_CODE → RESERVATION_CONFIRMATION, CUST_ID, PROD_ID
```

"Yes, that's the one area I've been wondering about, too," replied Jorge. "First, let me explain what this last functional dependency represents. I included it since only announcements that have been confirmed are on the report, so CATS has to know the confirmation status for each product announcement to a customer. The PCS_CODE stands for this announcement.

"What I further concluded is that until a customer confirms an announcement, the RESERVATION_DATE is null (empty). Once customers call to confirm the reservation, the date they call is automatically put in this field. Thus, RESERVATION_CONFIRMATION represents this logical status of the RESERVATION_DATE field. What I propose we do is simply use the RESERVATION_DATE attribute, treating a null value as an unconfirmed reservation.

"Further, what this functional dependency says is that if we know a PCS_CODE, which is unique for each announcement of a new product release to a customer, then there is only one possible value for whether

```
STORE (STORE_ID, STORE_MRG_NAME)
PRODUCT (PROD_ID, PROD_TITLE, PROD_MEDIA, PROD_DATE_AVAILABLE, PROD_ARTIST,
    PROD_PRICE)
VIDEO_PRODUCT (VIDEO_PROD_ID)
MUSIC/GAME_PRODUCT (MUSIC/GAME_PROD_ID)
CUSTOMER (CUST_ID, STORE_ID, CUST_NAME, CUST_STREET, CUST_CITY, CUST_STATE,
    CUST_ZIP, CUST_PHONE, CUST_CARD_TYPE, CUST_CARD_NO, CUST_CARD_EXPIRES,SSN)
INVENTORY (STORE_ID, PROD_ID, QTY_IN_STOCK)
PRODUCT_ANNOUNCEMENT (PCS_CODE, PROD_ID, CUST_ID, RUN_NUMBER, RESERVATION_DATE)
RUN (RUN_NUMBER, ANNOUNCE_DATE)
CUSTOMER_TRANSACTION (CUST_TRAN_ID, CUST_ID, CUST_TRAN_DATE)
PURCHASE (CUST_TRAN_ID, MUSIC/GAME_PROD_ID)
CUSTOMER_PROFILE (CUST_PROF_CATEGORY, CUST_PROF_CAT_LEVEL)
PRODUCT_PROFILE (PROD_PROF_CATEGORY, PROD_PROF_SUPERIOR_CAT)
PRODUCT_FOR_CUSTOMER (CUST_PROF_CATEGORY, CUST_PROF_CAT_LEVEL, PROD_PROF_CATEGORY)
```

Figure 6
Revised normalized relations based on functional dependency analysis of the CPA report

(*Note:* Primary keys are underlined and foreign keys are in italics.)

that customer has confirmed the reservation of that product associated with that PCS_CODE. The PCS_CODE was included in the PRODUCT_ANNOUNCE-MENT relation as a nonkey. I think that PCS_CODE may also be a candidate key for this relation, and maybe we ought to make it the primary key instead of the compound key previously chosen. Then PROD_ID and CUST_ID will be foreign keys. Remember, the PCS_CODE stands for a product/customer/store combination. The store is the one to which the customer is assigned, so we don't need to keep that attribute in this relation. What do you think about these changes?"

Buffy thought for a moment before replying. "What you say makes sense, but I don't think you've gone far enough. What do we do with ANNOUNCE_DATE in the PRODUCT_ANNOUNCEMENT relation? It seems that ANNOUNCE_DATE may actually depend on RUN_NUMBER, since this is the date CATS produces an announcement. Maybe there is a transitive dependency here since the conceptual data modeling did not identify a RUN entity. All of the announcements generated from the same run will have the same ANNOUNCE_DATE, won't they? Is it possible that CATS might announce the same product to the same customer for the same store on more than one occasion, so PCS_CODE cannot identify a unique RUN_NUMBER? Jorge, am I getting confused here?"

"No, I think you are asking some very good and very tough questions, some of which I asked myself," said Jorge. "I went back to the on-line interview notes to see what was said about PCS_CODE and RUN_NUMBER. It really helped that we decided to store those interview notes as word processing documents indexed by AskSAM, a text DBMS. That made it a lot easier to search through all the notes for certain keywords. I couldn't find any insights, however. Let's see if we can get hold of Wendy Yoshimuro (the marketing manager responsible for CATS) on the phone and try to resolve this."

Jorge used a programmed speed dial button on his speaker phone to call Wendy. Fortunately, Wendy was in and confirmed to Jorge and Buffy that her intent was that BEC would give a customer only one announcement for a given product, as the product was being released. There would be no second announcement.

Jorge summarized his conclusions after this phone call. "Well, I think we can now finalize the changes to the normalized relations due to the CPA report. We'll make PCS_CODE the primary key of the PRODUCT_ANNOUNCEMENT relation; we'll use the ANNOUNCE_DATE attribute to know whether to generate an entry for the CPA report; and we'll create a RUN relation to hold the ANNOUNCE_DATE, using RUN_NUMBER as a foreign key in PRODUCT_ANNOUNCEMENT to tie these data together. Does this seem right, Buffy?" (Figure 6 shows the new set of normalized relations.)

"As I said earlier, you really are getting the hang of data modeling," responded Buffy. "That seems to me to be exactly what we should do. Next, you can have the CASE tool reverse engineer a new E-R data model from the revised normalized relations. If I had more time we could go through all the other outputs and inputs you've analyzed, but it looks like you know what you are doing. I'll take a look at your final product when we do a team-level walkthrough of the logical design for the interim report Jordan said she wants to produce for the Systems Priority Board. Call me or send me an e-mail if you want me to look at any particular changes."

THE TRANSITION TO PHYSICAL DESIGN

Buffy continued, "I really like Jordan's idea of this status report, even though we don't owe the SPB anything until the end of physical design. I think it gives the SPB a chance to informally check our work and gives them a chance to execute their overseers' role. Based on what you've done so far, Jorge, do you see any need to recommend to Jordan any changes in the project schedule?"

"No, but I was wondering what else we'll need to know for physical database design. I want to make sure that the CASE repository and other notes contain all the information we'll need when we have to make final, physical design decisions."

"I'm glad you are looking ahead. I wanted to talk about exactly this point at our next weekly project team meeting. Briefly, you'll want to review interview notes and other documentation to make sure we know attribute formats and lengths, validation rules on each attribute (like a range of permitted values), the number of instances of each relation, what relations are used together for each report and form, and how frequently different processes that update or retrieve from the database occur. DFDs will tell us which data stores are used together, and these data stores must be mapped to relations. Hopefully the description of processes tells us how frequently certain activities occur. We may need to put together a questionnaire to send to selected CATS users to capture some of this information if we don't already have these details. The whole point is that we want to match the capabilities of the DBMS with the characteristics of the CATS operating environment. So far we have concentrated on what we wanted CATS to do; now it's time to consider how efficiently CATS has to work.

"It's so important that we deal with issues in the incremental fashion dictated by the systems develop-ment methodology. Otherwise, we have more to deal with than we can handle. The methodology allows us to deal with issues step-by-step, delaying costly decisions as long as possible. But, I think we are just about ready to make the critical physical design decisions, including database design."

SUMMARY

The CATS project team is in the final stages of the logical design phase of BEC's systems development process. As indicated in this case, new functional specifications have been developed during logical design, but the previous analysis information is still referenced and will continue to be useful during physical design. The amount of information about the CATS project is becoming overwhelming. Without automated tools, like a CASE tool and project management software, coordination across team members and across models and specifications would be too laborious and would likely become inaccurate. The team must ensure that every report, form, relation, process, data store, and so forth are complete and consistent with one another before physical system design can begin. Physical design is the last point at which CATS will be abstract, as implementation will begin to make CATS real and very costly to change if design errors are found. Thus, the work done by Jorge Lopez, Buffy Jarvis, and other team members will be carefully coordinated before moving into the next systems development phase.

Consensys Group

Design Issues for a Mobile LAN

Clay Tyler, owner/manager of Consensys Group, in San Diego, California, specializes in systems integration. He believed that Group Support System (GSS) services would be attractive to his customers. He now uses a completely mobile GSS as one of many consulting services that his company offers.

Consensys' mobile GSS consists of twelve Winbook XP, 486 DX4/75 MHz, color notebooks, each with 8 MB of RAM and a 150 MB hard drive. The GSS software is GroupSystems V for DOS and for Windows, running on Novell Netware 3.12. Each PC has a Xircom, full duplex, PCMCIA ethernet card. Tyler investigated wireless technology for networking but found that it still is not quite fast enough for his needs. Instead, the sys-

tem runs on coaxial cable in a bus configuration and has the capability to run on RJ45 twisted-pair wire in a star configuration. For projection Tyler uses a high-end Proxima DP 2800 color projection system.

Tyler has had to address some unique problems and concerns with his mobile multi-user LAN that are not typically seen with traditional stand-alone applications, or even with a permanent LAN. For example, concurrency control is a major issue. Also, for now, Tyler has all the same computers and network interface cards attached to his network. However, once he introduces new computers and cards, a likely scenario, his Novell 3.12 server will be able to load and bind all the necessary protocols (i.e., IEEE 802.3 & 802.2). Tyler plans ahead by loading and binding all protocols that he can foresee incorporating into his LAN.

Tyler is concerned about viruses on his network. Often he must do quick file transfers from floppy disks. He explains:

> Most times I don't have time to scan everything, but I need to change my policy. I recently fell victim to the 'Form' and 'Anti-Exe' viruses that prevented me from using Windows at a crucial moment. I got the viruses earlier in the day from a floppy disk that someone handed me during a time-critical situation (I didn't make time to scan and clean it). Although it didn't prevent me from executing GroupSystems V, it could have been a more 'fatal' situation.

As with any LAN, controlling access is important for his mobile LAN. He argues that simple solutions are often overlooked and leave a network vulnerable. Tyler has a password on the supervisor account and has made sure that accounts have only the access their users need and nothing more. He also has a login script common to all users and, on an "as needed" basis, he will add additional permissions to certain people who need them, being sure to document what permissions he has given out.

Because Tyler's LAN is mobile, wear and tear is his biggest concern, particularly on the file server. He uses polyethelene travel cases with high density foam to pack his notebooks. Although these cases are durable, Tyler stills plans for the worst.

> I have a file server with an 80 MB volume (plenty for what I need) and the remaining 45 MB of the hard disk is a DOS partition upon which I've loaded Windows for Workgroups 3.11. I have also constructed another notebook in a similar fashion with all the critical applications that I use. This provides me with a comfortable level of redundancy, because if my primary server has a problem, I can reboot it as a DOS machine (a network node) and reboot the backup as the server (all in about 3 minutes).

Another important factor, given that the LAN is mobile, is which type of network adapter to use. Tyler claims that using PCMCIA is advantageous compared to parallel port pocket adapters. Although the pocket adapters can be less expensive, he believes that the extra $50 or so per internal PCMCIA card is worth the price for the gain in speed. Thus, they never have to be disconnected and are invisible to the user. The pocket adapters are relatively cumbersome and have to be packed separately during travel.

After providing GSS services for two months, Tyler concludes that the GSS services are a nice fit with the process consulting he does for systems integration. Tyler has expanded his customer base and reached clients he had "never dreamed of reaching" before adding GSS to his list of services. As a result, he does a great deal of GSS-related consulting and even resells GSS-related equipment.

For the future, Tyler plans to continue providing GSS services and to find new ways to integrate GSS services with the other consulting services he provides. He will move to wireless technology when it is ready, and he plans to use Lotus Notes®. He hopes to pioneer applications that will blend Notes capabilities with those of GroupSystems.

An Overview of Part VI:

Physical Design

The project team often expands during the physical design phase to include a variety of specialists in such areas as databases, computer security, data networks, systems controls and audits, programming, and development environments. The purpose of physical design is to specify all the technological characteristics of the system so that those involved in the implementation phase can concentrate on building the system, not on deciding how the system should be built. Whereas prior phases dealt with what the system should do and how the system should look to the user, physical design specifies the structures for data and programs that will make the system work efficiently and securely, including considerations for the location of data and data processing on a computer network.

As depicted in Figure 1, physical design includes these steps:

- *Designing physical files and databases:* Describes how data will be stored and accessed in secondary computer memory and how the quality of data will be insured.

- *Designing system and program structures:* Describes the various programs and program modules which correspond to the data flow diagrams and other documentation developed in prior life cycle stages.

- *Designing distributed processing strategies:* Describes how your system will make data and processing available to users on computer networks within the capabilities of existing computer networks.

Physical design is tightly linked to previous systems development phases, especially analysis and logical design. For example, data flow diagrams outline all the various processing steps required in the system. These steps must be transformed into programs and program modules, traditionally described as hierarchies of processing units rather than in flow diagrams. Programs will accept or produce data through the forms and reports designed during logical design. In physical design, the functional capabilities of system inputs, outputs, and dialogues are specified within the constraints of the implementation tools. Finally, physical file and database design uses the normalized relations developed in logical data modeling.

As with prior life cycle phases, it is possible that during physical design you will discover inadequacies or errors in the logical design. So, iteration back to previous phases to gather additional information or to fill in gaps in the design is quite possible. Since physical design is the last life cycle phase before implementation,

Figure 1

Steps in physical design

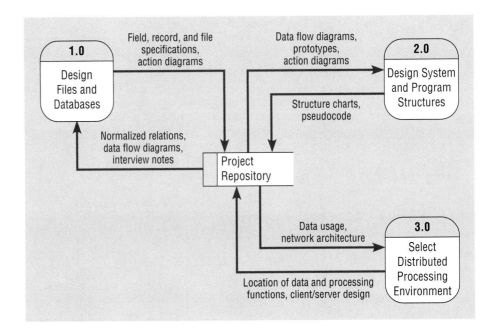

this is your last chance to make sure the system design is complete and consistent—before the cost to correct mistakes escalates. You will also generate new information to be stored in the CASE repository or project dictionary.

As with most life cycle phases and sub-phases, the three sub-phases of physical design are not necessarily sequential. The physical design of files and databases and the physical design of distributed processing methods interact to identify flaws and missing elements in each other. Further, the references to physical data elements in program design charts and the physical structure of files and databases must be consistent. Finally, for a program module which accesses files, the logic of how the data will be accessed must match the storage and file structure capabilities provided in the file design. Thus, the project dictionary or CASE repository becomes an active and evolving component of the management of system development during physical design as the central source of all design specifications.

During the design of physical files and databases, the subject of Chapter 16, you will learn about the wide variety of decisions that you must make in order to create efficient data storage and accessing methods. Your overall goal in designing files and databases is to balance the costs to store, access, and maintain data in secondary computer memory such as hard disks. These costs are affected by the format of each stored attribute or field, how data are divided into separate records, and the type of arrangement of records within a computer file. Because there are so many options available, and the options you have depend on the environment in which the system will be built, this chapter reviews only the general decisions you will have to make.

File and database design is presented in a bottom–up fashion in Chapter 16. First you learn about how to translate attributes, or columns, from normalized relations into physical data field specifications; the choice of data types and coding schemes are emphasized. Next, we review issues in the formation of records, including reasons why you might want to violate data normalization in order to improve data processing performance. Then, you will study the pros and cons of various methods for arranging records within a computer file, called file organizations. This is followed by an overview of the database design choices you must make, including the selection of the appropriate architecture for a database and the associated database management system. One of the most important decisions made during the design of files and databases is the choice of indexes for providing rapid access to data based on content; the chapter provides guidelines for choosing indexes.

Central to the design of data storage and access is controlling the integrity of data as they are captured and stored. Chapter 16 covers controls on fields, such as range checks and default values, which help to create accurate data. Contamination and even destruction of data are inevitable in complex, modern information systems; you will therefore also learn about methods to recover data after a failure has occurred.

Chapter 17 addresses techniques and principles you should use in the structuring of systems and programs that give programs easily understood and maintainable form. Central to this is how to transform the sequential flow of data between processes in data flow diagrams, a conceptual design notation, into hierarchical structures that can guide programmers in building programs. Program structure is typically represented as a hierarchy of main and iteratively subordinate modules. A chart showing such a hierarchical structure is called a structure chart, and Chapter 17 reviews the notation used in these charts.

Chapter 17 shows you how to translate data flow diagrams into structure charts. The process of transforming DFDs into structure charts is slightly different for different types of systems, and we illustrate these differences. Once program modules are shown in their hierarchical relationship with one another, the processing logic in each module must be specified. This processing logic is outlined in a generic language called pseudocode, which can be translated easily into any programming language. Thus, the physical system and program specifications developed in this step can be translated into a variety of development environments, or even a mixed environment if parts of the system will be implemented in different platforms. Nassi–Shneiderman diagramming, a popular alternative notation, is also introduced, since some organizations prefer this notation over pseudocode.

Developing well-structured programs is guided by two principles, coupling and cohesion. In Chapter 17 you will learn more about these principles and how to identify different forms of coupling and cohesion, including the desirable and undesirable properties of these different forms. Chapter 17 gives you both the notations and the principles to apply as you transform the logical design of the system into its physical, program structure.

Most modern information systems are shared systems: they are used by several or many people who work at geographically dispersed locations. Special issues arise in such distributed processing environments not present in stand-alone or single-user systems, and Chapter 18 addresses these special issues. However, we do not cover the design of computer networks. In most cases, the design of an individual information system does not change the basic design of computer networks in an organization. As the system's designer, you must be aware of the options you have within existing networks so that you can design your system for efficiency and integrity. Thus, Chapter 18 concentrates on the issues involved in designing an information system in a distributed processing environment, not on the design of telecommunications networks.

A common platform for modern systems is local area networks. Chapter 18 reviews several forms of networks, including different types of client/server networks, and highlights how systems are built differently for each network environment. Since many organizations support both file server and client/server networks, the chapter provides guidelines on when each is applicable so that you can choose the right network for your application.

Distributed systems mean that potentially both data and programs are handled on multiple linked computers. A fundamental question is where to locate data in a network of computers. Options for distributing data are compared in Chapter 18 so that you know which factors to consider when addressing a data distribution decision. In data distribution, you allocate whole or parts of files to secondary memory media at different computers. Each computer may primarily service certain business functions or geographical regions. The goal is to locate data as close as possible to where they are needed, with a controlled amount of redundancy. As usual, you are trying to balance data storage, accessing, and maintenance costs.

The chapter concludes with an overview of the evolving alternatives for distribution of data processing functions in a computer network. It is shown that three general components of systems—data presentation, data analysis, and data management—can be placed in various combinations on local PCs, server computers, or both. The features of six different combinations of locating these system functions are compared so that you can understand the relative merits and risks of each.

Each chapter in the physical design section of the book concludes with a case from Broadway Entertainment Company (BEC). These cases illustrate the development of the Customer Activity Tracking System (CATS) as physical files, databases, and programs are designed. The BEC case after Chapter 18 raises issues related to data privacy and ethics confronting organizations which store extensive data about customers in their information systems.

Designing Physical Files
and Databases

After studying this chapter, you should be able to:

- Define each of the following key physical file and database terms: *null value, pointer, record, page, file, file organization, index, database,* and *database management system.*

- Define six types of data files maintained by information systems.

- Choose storage formats (data types, coding schemes, display patterns, etc.) for attributes from a logical data model.

- Select controls on fields and files that will improve data accuracy and protect data from loss, contamination, or abusive use.

- Translate a relational data model into efficient storage structures for database processing, including knowing when to denormalize the logical data model.

- Choose among alternative methods of representing and handling missing data.

- Explain when to use different types of file organizations to store computer files.

- Calculate file sizes and estimate the amount of secondary computer memory required to store an application's data.

- Describe the purpose of indexes and the important considerations in selecting attributes to be indexed.

INTRODUCTION

In Chapters 11 and 15 you learned how to describe and model organizational data during the analysis and logical design phases of the systems development process. The processes and notations shown in those chapters were technology independent. That is, we use the processes and notations of E-R diagrams and normalization to develop abstractions of organizational data that capture the meaning of data; however, these notations do not explain *how* data will be processed or stored. The purpose of physical file and database design is to translate the logical description of data into the technical specifications for storing data. Systems analysts develop these specifications jointly with programmers, database analysts and administrators, data security staff, and other IS technical specialists. Beginning with this chapter, we turn our attention to issues of processing and storage efficiency, since prior development stages have structured the business information requirements as thoroughly and accurately as possible.

Physical file and database design has two main purposes:

1. You must translate relations from a logical data model into a technical design. This design includes choosing storage formats for each attribute, grouping attributes from one or more relations into one or more physical computer records, choosing organizations for the contents of each file of records, and designing methods to access data in and between files.

2. You must choose data storage technologies that will be used to help manage data. These technologies include various operating system functions, called access methods, and database management systems. These different technologies follow certain architectures, each of which is best suited for different data processing situations.

It is important to remember that physical file and database design does not include implementing (i.e., creating and loading data into) files and databases. Physical design produces the technical specifications that programmers and others involved in system construction will use during the implementation phase of systems development to create and populate files, file organizations, and databases.

PHYSICAL FILE AND DATABASE DESIGN

Physical file and database design is a natural, succeeding step after logical data modeling, which is why these are sequential chapters (see Figures 16-1 and 16-2). As in logical design, the three steps of the physical design phase—physical file and database design, program and system structure design, and distributed system design—are typically conducted in parallel because of the overlap of issues addressed. For example, in Chapter 18 we discuss the design of distributed databases. Thus, you will use together the concepts and skills you learn in Chapters 16–18.

The Process of Physical File and Database Design

In most situations, many physical data storage decisions are implicit or eliminated when you choose the data management technologies to use with the application you are designing. Since many organizations have standards for operating systems, database management systems, and data access languages, you must deal only with those choices not implicit in the given technologies. For example, many relational database management systems (DBMSs) give you little if any choice on the way to arrange records within computer files. In addition, computer center staff or systems programmers may have selected such parameters as virtual memory page sizes, which will affect data retrieval performance; these cannot be changed for individual applications. Thus, we will cover only those decisions you will make most frequently, plus a few selected decisions fundamental to your interactions with other systems developers making these decisions.

The goal of physical design, in general, and physical file and database design, in particular, is data processing efficiency. In other words, prior SDLC stages concen-

Figure 16-1
Systems development life cycle with physical design phase highlighted

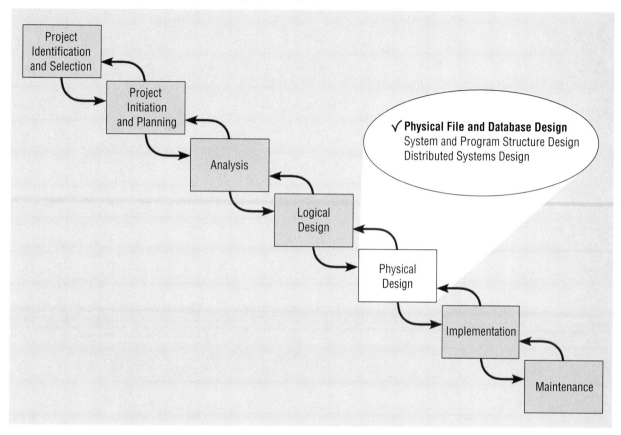

trated on "doing the right things," whereas physical design focuses on "doing the right things right." Today, with ever-decreasing costs for computer technology per unit of measure (both speed and space measures), it is typically very important for you to develop a physical system design that will minimize the time required by users to interact with the system. This means shortening data retrieval time and computer job turnaround time. Thus, we will concentrate on how to make processing of physical files and databases efficient, with less attention on efficient use of space.

Designing physical files and databases requires certain information that should have been collected and produced during prior SDLC phases. The information needed for physical file and database design includes these requirements:

- Normalized relations, including volume estimates

- Definitions of each attribute

- Descriptions of where and when data are used: entered, retrieved, deleted, and updated (including frequencies)

- Expectations or requirements for response time and data integrity

- Descriptions of the technologies used for implementing the files and database so that the range of required decisions and choices for each is known

Figure 16-2
Relationship between data modeling and the systems development life cycle

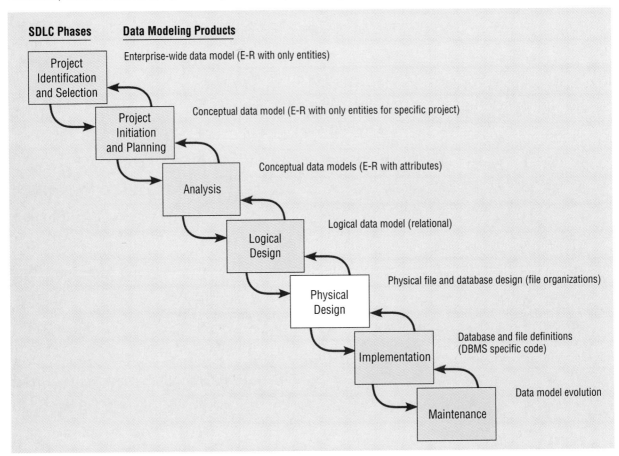

In physical file and database design you will make a number of critical decisions that will affect the integrity and performance of the application system. These key decisions include the following:

1. Choosing the storage format (called *data type*) for each attribute from the logical data model; the format is chosen to minimize storage space and to maximize data integrity. Data type involves choosing length, coding scheme, number of decimal places, minimum and maximum values, and potentially many other parameters for each attribute.

2. Grouping attributes from the logical data model into physical records (in general, this is called selecting a *stored record, or data, structure*).

3. Arranging related records in secondary memory (hard disks and tapes) so that individual and groups of records can be stored, retrieved, and updated rapidly (called *file organizations*). Consideration must be given also to protecting data and recovering data after errors are found.

4. Selecting media and structures for storing data to make access more efficient. The choice of media affects the utility of different file organizations. If you are using a database management system, this step involves selecting options for

random and sequential data retrieval as well as implementing relationships between entities and relations. The primary structure used today to make access to data more rapid is primary and secondary key indexes.

In this chapter we will address the factors that must be considered in making each of these decisions.

Deliverables and Outcomes

The deliverables from physical file and database design are a set of specifications that a programmer or database analyst will use to define the format and structure of data in secondary computer storage (usually hard disks). You are defining the structure of files recognized by system software, such as database management systems and operating systems. For many modern systems, these specifications would contain all the information needed to write or generate SQL (a standard relational database language) data definition statements. In other cases, these specifications would allow a COBOL, C, or C++ programmer to write data definitions. The specifications may be in the form of a pseudocode, but often the specifications are provided in the target language. The statements are not, however, executed to allocate space in secondary computer memory. Since data analysts and programmers often have to enter the code and verify specifications provided from several analysts or even several projects, specifications are often combined across different projects or major parts of a project before file and database implementation occurs.

The CASE repository or manual project dictionary is the natural place to store all of these specifications. Figure 16-3 lists the typical elements added to a project dictionary or repository during physical file and database design. For example, you would enter a specific data type for each attribute (called a field in physical design) object. Although a general data type may have been entered during logical design (for example, alphanumeric), it is during physical design that you finally decide whether an alphanumeric field is of fixed or variable length, what its maximum length is, and which unit of measure numeric values will represent. You would also enter specifications for how a field might be coded or formatted to more compactly represent longer, sparse sets of values. You group fields to form physical records and state rules—called referential integrity—to specify which fields must have matching values in different records (for example, the customer number in an order record must match a customer number field in a customer record). For each file, you must specify on what medium and device the file will be stored and the organization of records within the file. And you must indicate which files constitute each database and what structures will be used to link together the different files in a database.

We take a bottom-up approach to reviewing physical file and database design. Thus, we begin the physical design phase by addressing the design of physical fields for each attribute in a logical data model.

DESIGNING FIELDS

A **field** is the smallest unit of application data recognized by system software, such as a programming language or database management system. An attribute from a logical data model may be represented by several fields. For example, a student name attribute in a normalized student relation might be represented as three fields: last name, first name, and middle initial. Each field requires a separate definition when the application system is implemented.

Field: The smallest unit of named application data recognized by system software.

Figure 16-3

Typical data specifications added to a project dictionary during physical file and database design

FIELD	
Name	A name, in the format accepted by the DBMS, that uniquely labels the field; logical data model attribute and field names must be matched. *Example*: QTY_ON_HAND for the quantity in stock of an inventory item.
Data type	Computer storage format, including length. *Example*: Two-byte numeric.
Unit of measure	The units represented by the value of the field. *Example*: Inches, pounds, dollars.
Coding	Abbreviations used for each possible real-world field value. *Example*: States represented by standard two-character abbreviation.
Data integrity rules	Specifications of the limitations on what constitutes a legitimate value. *Example*: A default value of 0 for QTY_ON_HAND or a range from 0 to .25 for any price discount value.
Maintenance controls	Indication if value is allowed to change once it is entered. *Example*: Invoice data cannot change once the invoice is created.
Missing data procedures	Specifications of how to handle a missing value for the field when records in which the field appear are processed. *Example*: Ignore the field or use an average of other instances in the field.
Formula	For a calculated field, the equation used to compute the value for the field. *Example*: Extended price is unit price times quantity sold.
Check digit	If used, specifications for how certain digits within the field are mathematically related to other digits. *Example*: Adding to the end of a field the remainder after dividing the field value by 43.
Referential integrity	Specifications if values for the field must match values in a field in another record. *Example*: The customer number field value in a customer order must match a customer number field value in a customer record.
Ownership	Who is responsible for the meaning of this field and methods which access and update its value. *Example*: Storeroom manager is responsible for the accuracy of QTY_ON_HAND.
RECORD TYPE	
Fields	List of fields that appear in record. *Example*: ITEM_NO, DESCRIPTION, and QTY_ON_HAND for an inventory item record.
Structural data	Specification of any structural data necessary to interpret the format of a record. *Example*: Pointers and other data about the location of variable-length fields within the record.
Retention	Specification of how long the record is to be kept in a data file before it is purged or archived. *Example*: Retain an inactive customer record for two years after the most recent sales transaction.

In general, you will represent each attribute from each normalized relation as one or more fields. The basic decisions you must make in specifying each field concern the type of data (or storage type) used to represent the field, data integrity controls, and how to handle missing values for the field.

FILE

Name and location	Label for file and the medium and device to which the file is allocated. *Example*: CUSTOMER file located on hard disk drive D:.
Record(s)	Specification of which record types are stored in the file; if only certain records of a given record type are included, then the logic of which records are included is provided. *Example*: The file contains records for two types, both IN_PATIENT and OUT_PATIENT; and the first field of each record indicates which type record it is.
Primary key	The field or combination of fields that will serve as the unique identifier of each record in the file. *Example*: ITEM_NO for an inventory item file.
Indexes	Which fields are indexed and whether the index is a primary or secondary key index. *Example*: Inventory items are indexed by the FINISH field which is a secondary key (since more than one item can have the same finish).
Record blocking factor	Number of records per page or record block. *Example*: Ten ITEM records will be stored in each secondary memory page, and records cannot span page boundaries.
Retention and backup	How long the file is kept before it is purged or archived; procedures for file backup. *Example*: The ITEM file is backed up daily and the daily sales transaction file is kept for seven days, in case item data must be restored due to damage.
File organization	Operating system access method and file or database system method for arranging records within the file. *Example:* ITEM file will be stored as a hashed file based on ITEM_NO.
Controls	Specifications of audit trail or image files and encryption method. *Example:* File contents will be scrambled using a standard encryption procedure.

DATABASE

Files	The files which are included in the database and their location in secondary memory. *Example:* The ITEM, CUSTOMER, ORDER, and LINE_ITEM files form the order entry database, stored in subdirectory F:\ENTRY.
Architecture	Type of database structure used to organize files. *Example:* The order entry database is managed by a relational DBMS.
Relationships	Mechanisms used to link files together. *Example:* In a network or hierarchical database, what types of pointers are used to link records in different files.

Choosing Data Types and Data Representations

A **data type** is a detailed coding scheme recognized by system software for representing organizational data. The bit pattern of the coding scheme is usually transparent to you, but the space to store data and the speed required to access data are of consequence in the physical file and database design. The specific file or database management software you will use with your system will dictate which choices are available to you. For example, Table 16-1 lists the data types available in many SQL systems. Additional data types might be available for currency, variable-length

Data type: A detailed coding scheme recognized by system software for representing organizational data.

TABLE 16-1 Typical SQL Data Types

Data Type	Description
DECIMAL(m,n)	Signed numbers, where m is the total number of digits and n is the number of digits to the right of the decimal point
INTEGER	Large (up to 11 digits) positive or negative whole numbers
SMALLINT	Small (5 or 6 digits) positive or negative whole numbers (requires less storage space than INTEGER)
FLOAT(m,n)	Whole and fractional numbers represented in scientific notation, where m is the total number of digits and n is the number of digits to the right of the decimal point
CHAR(n)	Alphanumeric, where n is the maximum length
DATE	Calendar date, using several different styles
LOGICAL	True or false

alphanumeric data where only the required space is used (often called a memo field), and time.

Selecting a data type involves four objectives which will have different relative importances for different applications:

1. Minimize storage space

2. Represent all possible values

3. Improve data integrity

4. Support all data manipulations

You want to choose a data type for a field that can, in the minimal amount of space, represent every possible value for the associated attribute and for which you can do the required data manipulations. For example, suppose a quantity sold field can be represented by a SMALLINT data type. Some SQL systems will automatically assign the SMALLINT data type to the temporary field which is the sum of the quantity sold values. Since it is possible that the sum may have more digits than each component field, SMALLINT may not be of sufficient size for this field. Further, you want a data type that will restrict users from entering inappropriate values. For example, a numeric data type is stronger than an alphanumeric one and will reject entry of alphabetic values. In SQL, a SMALLINT data type might be sufficient for a quantity on hand attribute for most organizations but it does not, by itself, restrict negative numbers, which may be invalid.

Be careful—the data type must be suitable for the life of the application, otherwise maintenance will be required. Thus, it is advisable that you choose data types for future needs, anticipating growth.

To illustrate supporting data manipulations, consider an invoice date attribute. Two possible data types for this field would be DATE and CHAR(6); in CHAR(6) the date would be 6 digits as MMDDYY. Besides evaluating how much space is required for a DATE versus a CHAR(6) data type, for example, you would also want to know if there will ever be any date arithmetic done on invoice date. Most systems that support a date data type will allow subtraction and addition of dates with each other and with constants.

Some implementation platforms will support more advanced data types. Newer data types include voice, image or graphic, video, and user-defined. There is growing recognition that business applications require a more diverse set of data rather than just numbers and characters, so the data types supported are significantly improving.

Calculated Fields   It is common that a logical data model will include attributes which are mathematically related to other data. For example, an invoice may include a total due field, which represents the sum of the amount due on each item on the invoice. A field which can be derived from other database fields is called a **calculated field** (or a computed or derived field). Some implementation environments allow you to explicitly define calculated fields as a special data type. You would then usually be prompted to enter the formula for the calculation; the formula can involve other fields from the same record and possibly fields from records in related files.

Calculated field: A field which can be derived from other database fields. Also called computed or derived field.

In general, data storage technologies provide three ways to handle a calculated field:

1. You cannot represent the calculated field, but you must compute it on a record-by-record basis in each program that needs the computed value.

2. You may store the calculated field in the file as you would any other field so that it need not be recalculated every time it is needed (it must be recalculated only if the data it depends on changes), but use "extra" storage space to save the computational time.

3. You may virtually store the calculated field, but the data management technology actually stores only the formula in the data definitions and not the computed value. The data management technology transparently recalculates the field value at run time, giving you the impression it was retrieved.

The third option, if available, is usually the most desirable. It requires the consumption of the least storage space and the computational time is usually negligible. The second option allows the calculated field to be processed like any other field, which is desirable, but it does require redundant storage. For some systems like dBASE IV version 2.0, you define calculated fields not as part of record definitions but as fields on forms or reports. This is similar to the first option, but dBASE generates the code to do the calculations.

Coding and Compression Techniques   Some attributes have very sparse domains. For example, although a six-digit field (five numbers plus sign) can represent numbers –99999 to 99999, maybe only 100 positive values within this range will ever exist. Thus, a SMALLINT data type does not adequately restrict the permissible values, and storage space for five digits plus sign is wasteful. To more efficiently use space, you can define a field for an attribute so that the possible attribute values are not represented literally but rather are abbreviated. For example, suppose in Pine Valley Furniture each product has a finish attribute, with possible values Birch, Walnut, Oak, and so forth. To store this attribute as a character string might require 12, 15, or even 20 bytes to represent the longest finish character string value. Suppose that even a liberal estimate is that Pine Valley Furniture will never have more than 25 finishes. Thus, a single alphabetic or alphanumeric character would be more than sufficient. In the case of such coding, a separate code look-up table is created which matches each code value with the associated real-world value, as illustrated in Figure 16-4. On all user interfaces the real-world values are shown, but the internal storage is minimized to the length of the coded value in each product record, at the expense of minimal space for a look-up table and the (negligible) cost for the look-up. This look-up operation may be directly supported by the data

Figure 16-4

Example code look-up table (Pine Valley Furniture)

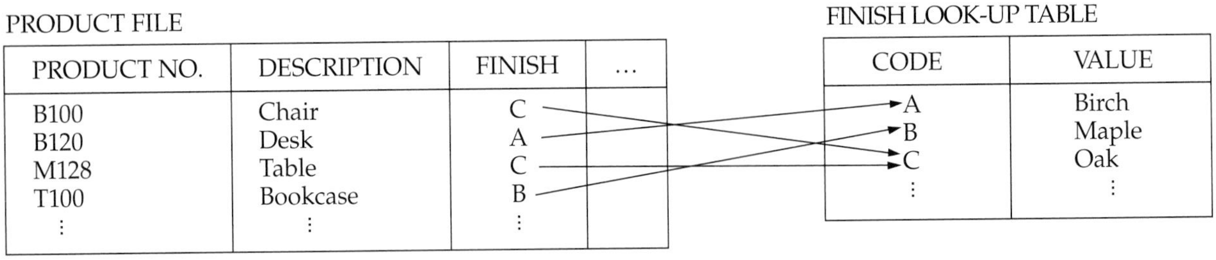

PRODUCT FILE

PRODUCT NO.	DESCRIPTION	FINISH	...
B100	Chair	C	
B120	Desk	A	
M128	Table	C	
T100	Bookcase	B	
⋮	⋮	⋮	

FINISH LOOK-UP TABLE

CODE	VALUE
A	Birch
B	Maple
C	Oak
⋮	⋮

management software or may have to be handled by application software. A further advantage is that the dense code (no wasted values) reduces the likelihood of incorrect data entry since each entered value must correspond with an entry in the code table. Thus, we not only reduce storage space but also increase integrity, which helps to achieve two of the physical file and database design goals. Codes also have disadvantages. If used in system inputs and outputs, they can be more difficult for users to remember, and programs must be written to decode fields if codes will not be displayed.

Data compression techniques reduce data storage requirements by looking for patterns in data and then coding frequently appearing patterns with fewer bits. A popular technology that uses data compression techniques is Stacker® from Stac Electronics, which compresses MS-DOS files.

> **Data compression technique:**
> Pattern matching and other methods which replace repeating strings of characters with codes of shorter length.

Selection of Primary Key Values   Sometimes proper primary keys have not been chosen prior to the physical design phase. The relations resulting from logical data modeling may have primary keys that violate the principles for formulating primary keys. The result of applying the guidelines for key design described in Chapter 11 is typically a value that has no real-world meaning, a so-called nonintelligent key. For example, a product relation in third normal form may have a primary key of product number; if this attribute changes value over time, is null, or indicates location or classification of products, then it is not a good selection for the primary key. Consider an intelligent product number of M128 (where M indicates the line of product, for example, modern). In this case, it would be better to create a PRODUCT_ID primary key, with a new system assigned value for each record as it is added to the file. The intelligent product number would also be stored, but only as a regular attribute.

Controlling Data Integrity

We have already explained that data typing helps control data integrity by limiting the possible range of values for a field. In this section, we discuss several additional physical file and database design options you might use to ensure higher-quality data. This discussion supplements the coverage of validation tests and techniques from Chapter 14 (see Table 14-9). Although these controls can be imposed within application code, it is better to include these as part of the file and database definitions so that the controls are guaranteed to be applied all the time as well as uniformly for all programs. There are five popular data integrity control methods: default values, picture statement, range control, referential integrity, and null value controls.

Default Value   Some attributes are such that a large percentage of the instances of that attribute have a common value. For example, the city and state of most customers for a particular retail store will likely be the same as the store's city and state. A **default value** is the value a field will assume unless an explicit value is entered for the field. Assigning a default value to a field can reduce data entry time (the field can simply be skipped during data entry) and helps to reduce data entry errors, such as typing 'IM' instead of 'IN' for Indiana.

Default value: A value a field will assume unless an explicit value is entered for that field.

Picture Control   Some data must follow a specified pattern. A **picture (or template)** is a pattern of codes that restricts the width and possible values for each position within a field. For example, a product number at Pine Valley Furniture is four alphanumeric characters—the first is alphabetic and the next three are numeric—defined by a picture of A999 where A means that only alphabetic characters are accepted and 9 means that only numeric digits are accepted. Thus, M128 is an acceptable value but 3128 or M12H would be unacceptable.

Picture (or template): A pattern of codes that restricts the width and possible values for each position of a field.

A picture may specify that certain special characters appear in certain positions within a field during display. For example, for the bank account example, a template might specify a hyphen between the second and third and between the sixth and seventh digits. A picture helps to not only visually format data for easy inspection of correctness but also allows the computer software to check that individual digits are not skipped during entry. Consider a picture of

$$\$999,999.00$$

for an integer field. This picture could be used during data entry to confirm to the user that the entered data is in dollars, the proper placement of digits is within a limit of six significant digits (even though an integer data type might support a much larger number), and that values must be in whole dollars. Other types of picture controls can be used to specify the following choices:

- How to show negative numbers

- Whether to suppress showing leading zeros

- Whether a floating currency symbol should appear immediately to the left of the most significant digit

- Justification of the value within the space for the field display

Range Controls   Both numeric and alphabetic data may have a limited set of permissible values. For example, a field for the number of product units sold may have a lower bound of 0 and a field which represents the month of a product sale may be limited to the values JAN, FEB, and so forth.

Although a significant help in ensuring data integrity, range and other data integrity controls must be used with caution because the world changes and such restrictions may themselves change. For example, should the value PR be accepted for a U.S. state? Or must all years begin with 19? Many systems that support range and other controls will allow you to update these controls as long as no field values violate the new restrictions. Furthermore, if you have implemented integrity controls as "hard code" in programs rather than through data definitions, it may be very cumbersome and expensive to update the restrictions.

Referential Integrity   The term referential integrity was defined in Chapter 15 as limiting values for a given attribute to the existing values of another attribute in the data model. The most common example of this is cross referencing between relations. For example, consider the pair of relations in Figure 16-5a. In this case, the values for the foreign key CUSTOMER_ID field within a customer order must be limited to the set of CUSTOMER_ID values from the customer relation; we would not want to accept an order for a nonexisting or unknown customer. Referential

Figure 16-5

Examples of referential
integrity field controls

(a) Referential integrity
between relations

(b) Referential integrity within
a relation

CUSTOMER (**CUSTOMER_ID**, CUST_NAME, CUST_ADDRESS,...)

CUST_ORDER (ORDER_ID, <u>**CUSTOMER_ID**</u>, ORDER_DATE,...)

    and CUSTOMER_ID may not be null since every order must be for some
    existing customer

EMPLOYEE (**EMPLOYEE_ID**, <u>**SUPERVISOR_ID**</u>, EMPL_NAME,...)

    and SUPERVISOR_ID may be null since not all employees have supervisors

integrity may be useful in other instances. Consider the employee relation example in Figure 16-5b. In this example, the employee relation has a field of SUPERVISOR_ID. This field refers to the EMPLOYEE_ID of the employee's supervisor and should have referential integrity on the EMPLOYEE_ID field within the same relation. Note in this case that since some employees do not have supervisors, this is a weaker referential integrity constraint because the value of a SUPERVISOR_ID field may be empty.

Referential integrity acts like a dynamic table (in this case, file) look-up. Whereas a range control limits a field to those values found in a table defined when the files and databases were created, referential integrity says that a field is limited to those values currently in some field in the database. Since it is cross references between relations that hold a relational database together, it is usually critical that all possible referential integrity controls be defined in relational databases.

Note that referential integrity guarantees that only some existing cross reference value is used, not that it is the correct one. This is why visual checks are still usually applied. For example, when entering a customer order, it is advisable to not only enforce referential integrity on the CUSTOMER_ID field but also to retrieve the associated customer data and display them for visual verification by the data entry person.

Null Value Control   It is not uncommon that when it is time to enter data—for example, a new customer—you might not know the customer's phone number. The question is whether a customer, to be valid, must have a value for this field. The answer for *this* field is probably no, since most data processing can continue without knowing the customer's phone number. It is possible that a customer's phone number might be unknown when the customer record is created, but it may not be allowed to be unspecified when you are ready to ship product to the customer. On the other hand, you must always know a value for the CUSTOMER_ID field. Due to referential integrity, you cannot enter any customer orders for this new customer without knowing an existing CUSTOMER_ID value, and customer name is essential for visual verification of correct data entry.

Null value: A special field value, distinct from 0, blank, or any other value, that indicates that the value for the field is missing or otherwise unknown.

A **null value** is a special field value, distinct from 0, blank, or any other value, that indicates that the value for the field is missing or otherwise unknown. Usually when you apply a referential integrity restriction on a field, it implies that the field may not be null or empty; the example in Figure 16-5b illustrates that this is not always the case. In addition, you should also specify whether nonkey fields such as the customer name field of a customer record may be null.

Handling Missing Data

When a field may be null, there is a question of how to handle null values in processing data. For example, suppose a customer zip code field is null and a report summarizes total sales by month and zip code. How should sales to customers with unknown zip codes be handled? Or what if monthly total sales for some product is missing for a given month? How should a sales forecasting routine cope with this missing data? During physical file and database design you must specify how your system is to handle missing data for each field.

Two options for handling or preventing missing data have already been mentioned: using default values and not permitting missing (null) values. Both of these methods bias some results since neither may accurately depict the true state of the organization. Certainly preventing missing data to the extent possible minimizes the ramifications of data inaccuracies. But some missing data is inevitable. According to Babad and Hoffer (1984), other possible methods for handling missing data are the following:

1. Substitute an estimate of the missing value. For example, for monthly product sales, use a formula involving the mean of monthly sales for that product indexed by total sales for that month across all products. Such estimates must be marked so that users know that these are not actual values.

2. Track missing data so that special reports and other system elements cause people to quickly resolve unknown values. This method allows data to be entered promptly in a system yet calls attention to the missing data in order to expedite finding the missing values.

3. Perform sensitivity testing so that missing data are ignored unless knowing a value might significantly change results; if, for example, total monthly sales for a particular salesperson are almost over a threshold that would make a difference in that person's compensation. This is the most complex of the methods mentioned and hence requires the most sophisticated programming to implement.

Missing data are a fact of information systems. Yet, in practice, explicit methods for handling missing data are often overlooked during physical file and database design. Ignoring missing data may, however, cause misrepresentation of the organization's state and result in bad decisions.

DESIGNING PHYSICAL RECORDS

In logical data modeling you grouped into a relation those attributes which concern some unifying business concept, such as a customer, product, or employee. In contrast, a **physical record** is a group of fields stored in adjacent memory locations and retrieved together as a unit. The design of a physical record involves choosing the sequence and grouping of fields into adjacent storage locations to achieve two goals: efficient use of secondary storage and data processing speed.

Physical record: A group of fields stored in adjacent memory locations and retrieved together as a unit.

One consideration when designing physical data records is conserving secondary memory space. Computer operating systems actually read data from hard disks in units called pages. A **page** is the amount of data read or written in one secondary memory input or output operation. (With magnetic tapes, the equivalent concept is a *record block*.) The page size is fixed by system programmers and is selected to most efficiently use RAM across all applications. Depending on the computer system, a physical record may or may not be allowed to span two pages. Thus, if page length is not an integer multiple of the physical record length, wasted space may occur at the end of a page. For example, consider a physical record containing fields whose total length is 1,034 bytes. If a page is 2,000 bytes and the file management software does not allow physical records to span pages, then there would be only one record per page, with 966 wasted bytes per page. The number

Page: The amount of data read or written in one secondary memory (disk) input or output operation. For I/O with a magnetic tape, the equivalent term is *record block*.

Blocking factor: The number of physical records per page.

Record partitioning: The process of splitting logical records into separate physical segments based on affinity of use.

of records per page is called the **blocking factor.** If storage space is scarce and physical records cannot span pages, you should consider creating multiple physical records from one logical relation so that wasted storage space is minimized.

A second and often more important consideration when selecting a physical record design is efficient data processing. Data are most efficiently processed when the needed data are stored close to one another in secondary memory, thus minimizing the number of input/output (I/O) operations that must be performed. **Record partitioning** is the process of splitting normalized relations into separate physical segments based on affinity of use. Each segment is a separate physical record, stored with other like segments in a separate physical file. For example, consider patient data in Figure 16-6. In this figure, the logical record has been broken into two segments. The primary segment contains fields used in almost every retrieval of patient data. The secondary segment contains data used on an intermittent basis. The result is very short primary segments, so there will be many primary segments per page. Since many primary segments can be read in one I/O operation, total I/O time for many processing functions is minimized. Other examples of partitioning might include splitting a product record into separate segments with each containing only engineering, accounting or marketing data, or dividing customer account data by geographical region.

Partitioning is a deliberate form of denormalization—defining physical records not in third or higher normal forms. Denormalization also includes joining attributes from several relations together to avoid the cost of accessing several files instead of one. For example, at the cost of redundantly storing the customer data, you could store customer and customer order data in one physical record to avoid retrieving two separate records. Since the relative benefits of denormalization may change with new technology, you must maintain the normalized relations in the project dictionary so that the decision to denormalize can be readdressed as new technology is introduced.

Denormalization can create some of the problems (anomalies) that normalization, as discussed in Chapter 15, is designed to avoid. As Finkelstein (1988) points out, denormalization can increase the chance of errors and inconsistencies and can force reprogramming systems if business rules change. Further, denormalization optimizes certain data processing at the expense of others, so if the frequencies of different processing activities change, the benefits of denormalization may no longer exist.

Various forms of denormalization can be done and there are no hard-and-fast rules for deciding when to denormalize data. Rodgers (1989) discusses the general trade-offs. You should consider denormalization in the following situations (Fig-

Figure 16-6
Example of record partitioning

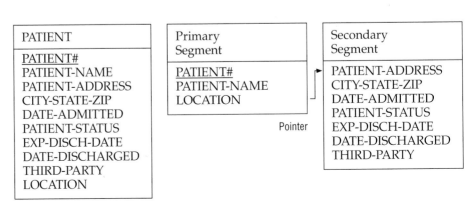

Logical record   Physical record segments

ure 16-7 shows examples of normalized and denormalized relations for each of these three situations):

1. *Two entities with a one-to-one relationship.* Even if one of the entities is an optional participant, if the matching entity exists most of the time, then it may be wise to combine these two relations into one record definition. Figure 16-7a shows student data with optional data from a standard scholarship application a student may complete. In this case, one record could be formed with four fields from the STUDENT and SCHOLARSHIP APPLICATION FORM normalized relations. (*Note:* In this case, fields from the optional entity must have null values allowed.)

2. *A many-to-many relationship (gerund) with nonkey attributes.* Rather than joining three files to extract data from the two entities in the relationship, it may be advisable to combine attributes from one of the entities into the record representing the many-to-many relationship, thus avoiding one join in many data access modules. Figure 16-7b shows price quotes for different items from different vendors. In this case, fields from ITEM and PRICE QUOTE relations might be combined into one record to avoid having to join all three files together. (*Note:* This may create considerable duplication of data—in the example, the ITEM fields, such as DESCRIPTION, would repeat for each price quote—and excessive updating if duplicated data changes.)

3. *Reference data.* Reference data exists in an entity on the one-side of a one-to-many relationship, and this entity participates in no other database relationships. You should seriously consider merging the two entities in this situation into one record definition when there are few instances of the entity on the many-side for each entity instance on the one-side. See Figure 16-7c in which several ITEMs have the same STORAGE INSTRUCTIONS and STORAGE INSTRUCTIONS only relate to ITEMs. In this case, the storage instruction data could be stored in the ITEM records creating, of course, redundancy and the potential for extra data maintenance.

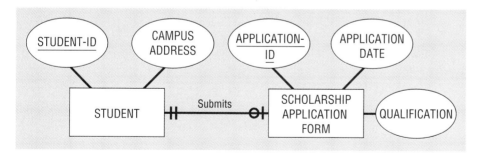

Figure 16-7
Possible denormalization situations

(a) Two entities with one-to-one relationship

Normalized relations:
    STUDENT (STUDENT-ID, CAMPUS-ADDRESS, APPLICATION-ID)
    APPLICATION (APPLICATION-ID, APPLICATION DATE, QUALIFICATIONS, STUDENT-ID)

Denormalized relation:
    STUDENT (STUDENT-ID, CAMPUS-ADDRESS, APPLICATION DATE, QUALIFICATIONS)

    and APPLICATION DATE and QUALIFICATIONS may be null

(continues)

(Note: We assume APPLICATION-ID is not necessary when all fields are stored in one record, but this field can be included if it is required application data.

Figure 16-7 *(continued)*

(b) A many-to-many relation-ship with nonkey attributes

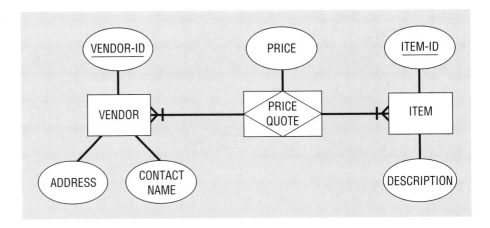

```
Normalized relations:
    VENDOR (VENDOR-ID, ADDRESS, CONTACT NAME)
    ITEM (ITEM-ID, DESCRIPTION)
    PRICE QUOTE (VENDOR-ID, ITEM-ID, PRICE)

Denormalized relations:
    VENDOR (VENDOR-ID, ADDRESS, CONTACT NAME)
    ITEM-QUOTE (VENDOR-ID, ITEM-ID, DESCRIPTION, PRICE)
```

(c) Reference data

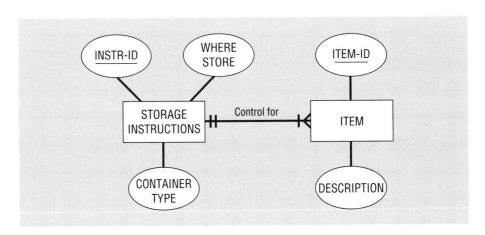

```
Normalized relations:
    STORAGE (INSTR-ID, WHERE STORE, CONTAINER TYPE)
    ITEM (ITEM-ID, DESCRIPTION, INSTR-ID)

Denormalized relation:
    ITEM (ITEM-ID, DESCRIPTION, WHERE STORE, CONTAINER TYPE)
```

Figure 16-8
Positional field format
technique

ELWAYbbbbbbbJbRbbQBbbb83

Handling Fixed-Length Fields

Designing a physical record is easier if every field is of fixed length. In the case of fixed-length fields, a positional technique is most often used to place and find fields. This technique is illustrated in Figure 16-8. With this technique, character fields have trailing blanks and numeric fields have leading zeros to make all instances of a given field the same length. In this scheme, we know that the fifth field is always offset from the beginning of the record by a known displacement, which is the sum of the lengths of the preceding fields. Further, since every record has the same fields, each record has the same length. Thus, it is easy to find the position of the mth field in the nth record in this way:

address of beginning of file + (n-1)*record length + sum (length$_i$)
$$i = 1 \text{ to } m - 1$$

where length$_i$ is the length of field i. Other techniques are used and the interested reader can refer to database and data structure texts for explanations of alternatives.

Handling Variable-Length Fields

A variable-length field, such as a memo field, makes the location of a particular field and of a particular record irregular. That is, depending on which records exist and the precise values for fields, different records and fields will be in different locations. A common way you can handle variable-length fields is to break the relation into a fixed-length physical record containing all of the fixed-length fields and one or more variable-length physical records. You design the fixed-length record as described previously. You can then allocate each variable-length field to a separate file. This is the automatic physical record design technique used by most microcomputer DBMSs that support a memo field, such as Paradox and dBASE IV. Other techniques for handling variable-length fields are addressed in many data structure texts.

DESIGNING PHYSICAL FILES

A **physical file** is a named portion of secondary memory (a magnetic tape or hard disk) allocated for the purpose of storing physical records. Some computer operating systems allow a physical file to be split into separate pieces. If this occurs, it will be transparent to you as the file designer and is the result of the operating system trying to use every page of secondary memory. In subsequent sections, we will assume that a physical file is not split and each record has the same structure.

Physical file: A named set of contiguous records.

Before we present the principles and options for designing files, it is worthwhile reviewing the various types of files that might exist in a system, each of which requires a design.

Types of Files

Table 16-2 lists and defines the six different types of files associated with a typical information system, along with considerations for the design of each. During physical file and database design you must anticipate the need for each of these files and design the structure of each. The nuances of each file are beyond the scope of our discussion, but the following sections review general principles of file design that can apply to each file type. In subsequent examples we will exclusively use data files since these are the types of files on which you typically concentrate most attention during physical file and database design.

TABLE 16-2 Types of Files

Type of File	Description
Data File (also called Master File)	A file that contains business data related to the conceptual and logical data models for applications. The file is permanent but the contents change. Data files should be designed for both random and sequential processing speed and storage efficiency and to accommodate changes in size. *Example:* customer file or invoice file.
Look-Up Table File	A list of reference data used to validate field values in one or more data files. Such files are usually static and should be designed for rapid data retrieval of individual entries. *Example:* list of valid product finishes.
Transaction File	A temporary file of data about the day-to-day activities of an organization. This file may hold data until they can be processed to update data files or the file may be retained in case the data must be reused. Transaction files are usually designed for rapid sequential processing. *Example:* all the ATM transactions for a given day.
Work File	A temporary file in which intermediate results are held. A temporary file is automatically deleted once no longer needed and recreated when again needed. Work file design is unique to the uses of that file. *Example:* a sorted list of customer and customer order data needed to produce a report.
Protection File (also called an Audit Trail)	A file used to restore other files in the case that the other files are damaged. This type of file may include before or after images of data file records so old values can be restored. This type of file is usually designed for rapid sequential processing. *Example:* a backup copy of a customer file.
History File	A file which contains data that are no longer current but which are needed to show what the state of the organization was at some prior point in time. History files are usually designed for efficient storage space. *Example:* a bank account file showing account balances at the end of the prior business day.

Pointer

All files are organized by using two basic constructs to link one piece of data with another piece of data: sequential storage and pointers. With sequential storage one field or record is stored right after another field or record. As illustrated in Figure 16-6, various circumstances can occur when the sequential storage of data must be broken and we need to know where the "next" field of data is. A **pointer** is a field of data that can be used to locate a related field or record of data. In most cases, a pointer contains the address of the associated data. For example, in Figure 16-6, a pointer is used to link each primary segment to its associated secondary segment. Pointers are used in many different contexts in physical file and database design.

Pointer: A field of data that can be used to locate a related field or record of data.

Access Methods

All I/O operations are ultimately handled by the data management portion of the computer's operating system. Each operating system supports one or more different algorithms for storing and retrieving data and these algorithms are called **access methods.** Your understanding of the access methods on the target computer system is fundamental to designing data files for an application.

Access method: An operating system algorithm for storing and locating data in secondary memory.

There are basically two types of access methods: relative and direct. A relative access method supports accessing data as an offset from the most recently referenced point in secondary memory. A sequential access method is a special case of this type since the "next" record begins the distance of one record from the beginning of the current record. In general, a relative access method supports finding the nth record from the current position or from the beginning of the file. A direct access method uses some calculation to generate the beginning address of a record. The simplest form of a direct method is to tell the access method to go to a particular disk address. Another variation is to provide a record's primary key and the direct access method determines where this record should be located. We will discuss this access method under the more inclusive concept of file organizations in the next section.

It is rare that you will directly use the operating system's access methods. Rather, you will more likely use the data organizations supported by a programming language or database or file management system. These data arrangements are called file organizations, which we discuss next.

File Organizations

A **file organization** is a technique for physically arranging the records of a file on secondary storage devices. In choosing a file organization for a particular file, you should consider seven important factors:

1. Fast data retrieval

2. High throughput for processing transactions

3. Efficient use of storage space

4. Protection from failures or data loss

5. Minimizing need for reorganization

6. Accommodating growth

7. Security from unauthorized use

Often these objectives conflict, and you must select an organization for each file that provides a reasonable balance among the criteria within the resources available.

Literally hundreds of different file organizations and variations have been created but we will outline the basics of three families of file organizations used in most file management environments: sequential, indexed, and hashed (see Figure 16-9). You will need to understand the particular variations of each method available in the environment for which you are designing files.

Sequential File Organizations   In a **sequential file organization,** the records in the file are stored in sequence according to a primary key value (see Figure 16-9a). To locate a particular record, a program must normally scan the file from the beginning until the desired record is located. A common example of a sequential file is the alphabetical list of persons in the white pages of a phone directory (ignoring any index that may be included with the directory). Sequential files are very fast if you want to process records sequentially but they are essentially impractical for random record retrievals. Deleting records can cause wasted space or the need to compress the file. Adding records requires rewriting the file, at least from the point of insertion. Updating a record may also require rewriting the file, unless the file organization supports rewriting over the updated record only. Moreover, only one sequence can be maintained without duplicating the records.

Indexed File Organizations   In an **indexed file organization,** the records are either stored sequentially or non-sequentially and an index is created that allows the application software to locate individual records (see Figure 16-9b). Like a card catalog in a library, an **index** is a table that is used to determine the location of rows in a file that satisfy some condition. Each entry matches a key value with one or more records. An index can point to unique records (a primary key index, such as on the

File organization: A technique for physically arranging the records of a file on secondary storage devices.

Sequential file organization: The records in the file are stored in sequence according to a primary key value.

Indexed file organization: The records are either stored sequentially or non-sequentially and an index is created that allows software to locate individual records.

Index: A table or other data structure used to determine the location of rows in a file that satisfy some condition.

Figure 16-9

Comparison of file organizations

(a) Sequential

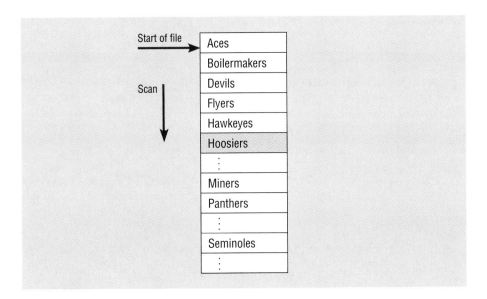

(b) Indexed

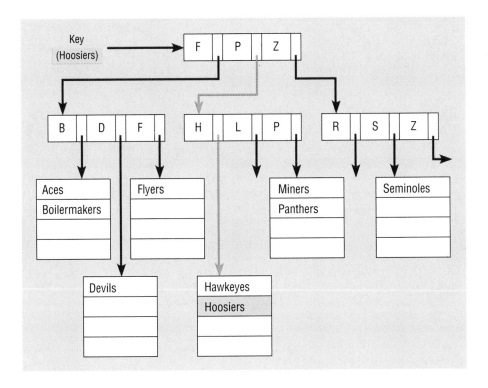

Secondary key: One or a combination of fields for which more than one record may have the same combination of values.

PRODUCT_ID field of a product record) or to potentially more than one record. An index that allows each entry to point to more than one record is called a **secondary key** index. Secondary key indexes are important for supporting many reporting requirements and for providing rapid ad hoc data retrieval. An example would be an index on the FINISH field of a product record.

The example in Figure 16-9b illustrates that indexes can be built on top of indexes, creating a hierarchical set of indexes. This may be desirable since an index itself is a file and, if it is very large, it too can be organized by indexing the index.

(c) Hashed

Each index entry in Figure 16-9b has a key value and a pointer to another index or to a data record. For example, to find the record with key 'Hoosiers', the file organization would start at the top index and take the pointer after the entry P, which points to another index for all keys that begin with the letter G through P in the alphabet. Then the software would follow the pointer after the H in this index, which represents all those records with keys that begin with the letters G through H. Eventually, the search through the indexes either locates the desired record or indicates that no such record exists. See McFadden and Hoffer, 1994, for a thorough treatment of the most popular indexing methods.

The main disadvantages to indexed file organizations are the extra space required to store the indexes and the extra time required to access and maintain indexes. Usually these disadvantages are more than offset by the advantages. Since the index is kept in sequential order, both random and sequential processing are practical. Also, since the index is separate from the data, you can build multiple index structures on the same data file (just as in the library where there are multiple indexes on author, title, subject, and so forth). With multiple indexes, software may rapidly find records that have compound conditions, such as finding books by Tom Clancy on espionage. Also, since the raw data can be stored nonsequentially, data addition, deletion, and modification may be easy. Depending on how the indexes are managed, however, the file may need to be reorganized after many updates. Two popular variations on indexed file organizations are distinguished by this last factor: indexed sequential access method, ISAM, files must be periodically reorganized and virtual storage access method, VSAM, files are dynamically reorganized with negligible overhead.

Hashed File Organizations   In a **hashed file organization,** the address of each record is determined using a hashing algorithm (see Figure 16-9c). A **hashing algorithm** is a routine that converts a primary key value into a record address. Although there are several variations of hashed files, in most cases the records are located nonsequentially as dictated by the hashing algorithm. Thus, sequential data processing is impractical. On the other hand, retrieval of random records is very fast. There are issues in the design of hashing file organizations, such as how to handle two primary keys that translate into the same address, but again, these issues are beyond our scope (see McFadden and Hoffer, 1994, for a thorough discussion).

Hashed file organization: The address for each record is determined using a hashing algorithm.

Hashing algorithm: A routine that converts a primary key value into a relative record number (or relative file address).

Summary of File Organizations

The three families of file organizations cover most of the file organizations you will have at your disposal as you design physical files and databases. Other more basic data structures, such as linked lists and chains, are typically used by operating systems and file management software. Unless you are building a system without the aid of modern system software, you will not deal with these structures directly.

Table 16-3 summarizes the comparative features of sequential, indexed, and hashed file organizations. You can use this table to help choose a file organization by matching the file characteristics and file processing requirements with the features of the file organization.

Designing Controls for Files

When you design files and databases, it is important to decide how you will protect them from destruction or contamination. Certainly the field data integrity controls discussed earlier in the chapter help, but you can use other techniques to protect the whole file. Two frequently used techniques are backup procedures and encryption.

File Backup Procedures   Before the system is written, you should decide how and how often each file will be backed up. The easiest approach is to specify that all files will be periodically copied onto a separate electronic medium and stored off site from the current versions. Since a total backup of all files may be time-consuming and each file has a different volatility, some files may need to be backed up more frequently than others. Although frequent backups can be expensive in lost computer time and media, you have to reconstruct less work to restore a file if it is damaged. You must also decide how many backup versions need to be retained.

TABLE 16-3   Comparative Features of Different File Organizations

Factor	File Organization		
	Sequential	**Indexed**	**Hashed**
Storage Space	No wasted space	No wasted space for data, but extra space for index	Extra space may be needed to allow for addition and deletion of records
Sequential Retrieval on Primary Key	Very fast	Moderately fast	Impractical
Random Retrieval on Primary Key	Impractical	Moderately fast	Very fast
Multiple Key Retrieval	Possible, but requires scanning whole file	Very fast with multiple indexes	Not possible
Deleting Records	Can create wasted space or require reorganizing	If space can be dynamically allocated, this is easy, but requires maintenance of indexes	Very easy
Adding Records	Requires rewriting file	If space can be dynamically allocated, this is easy, but requires maintenance of indexes	Very easy, except multiple keys with same address require extra work
Updating Records	Usually requires rewriting file	Easy, but requires maintenance of indexes	Very easy

Legal reasons or an estimate of how long it might take to discover an error in a file may dictate the age of the oldest file.

Recovery from file failure can be done by two methods, only one of which, forward recovery, requires a file backup (see Figure 16-10). In **forward recovery** (Figure 16-10a) you restore the most recent correct copy of the file and rerun all maintenance (and possibly display and reporting) activity against the file input since the time of the backup copy. In **backward recovery** (Figure 16-10b) you take the current state of the file, assuming the file is usable, and backout (undo) each transaction to a point at which the file is correct, then rerun all the transactions up to the current point. A database administrator can be very helpful in deciding which of these approaches makes sense for the files involved in the application you are designing. For a shared database managed by a database management system, backup procedures may already be specified, and you have no decision to make in support of your individual application.

The file which contains a chronological history of changes and accesses to a file or database is called an **audit trail.** An audit trail (and related files like a transaction log) might contain images of records before or after modification, copies of transaction data, and other identifying information useful in tracing problems found in files and databases (see Figure 16-11). An audit trail can be used to rebuild damaged files and can also be used to test whether file update procedures are working correctly. The audit trail can be inspected by computer auditors to see if the results of sampled transactions are correct or to trace a chain of data updates to discover where erroneous data were introduced when contaminated data are discovered.

Figure 16-11a illustrates a simple situation of maintenance and retrieval of ON_HAND inventory data in an ITEM file. An analysis of the audit trail entries indicates that some type of data maintenance error occurred during the second file update operation, since the ON_HAND after image of 60 and the transaction operation do not agree. In this case, the audit trail not only helps us identify where the

Forward recovery (roll-forward): An approach to rebuilding a file in which one starts with an earlier version of the file and either reruns prior transactions or replaces a record with its image after each transaction.

Backward recovery (roll-back): An approach to rebuilding a file in which before images of changed records are restored to the file in reverse order until some earlier state is achieved.

Audit trail: A list of changes to a data file which allows business transactions to be traced. Both the updating and use of data should be recorded in the audit trail, since the consequences of bad data should be discovered and corrected.

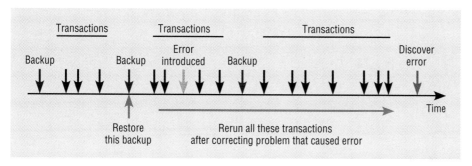

Figure 16-10
File recovery approaches

(a) Forward recovery

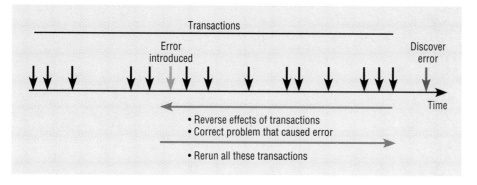

(b) Backward recovery

problem occurred but it may also be used to reprocess transactions once the problem is corrected. Further, the audit trail also identifies retrieval activity (transaction R1) after the error is introduced. We could use this information to notify the system user associated with this retrieval that he or she is working with erroneous data. Figure 16-11b lists and defines some possible contents for each audit trail entry. The information in each audit trail entry is used to diagnose the possible source of the data contamination as well as which business areas might be affected by bad data. Since software is written by humans (even the logic of generated code was designed by a programmer), application software is fallible. Thus, systems should be tested periodically for accuracy, and an audit trail is an effective mechanism to help in this testing and in restoring data to an uncontaminated state.

Encryption: The coding (or scrambling) of data so that they cannot be read by humans.

Encryption   Security of file contents is an important issue for some applications, often when financial, personal, or proprietary data are involved. Although various schemes for security may be in place, one security mechanism for an individual file is encryption. **Encryption** is the coding or scrambling of data so that they cannot be

Figure 16-11
Audit trail controls on a file

(a) Sample audit trail

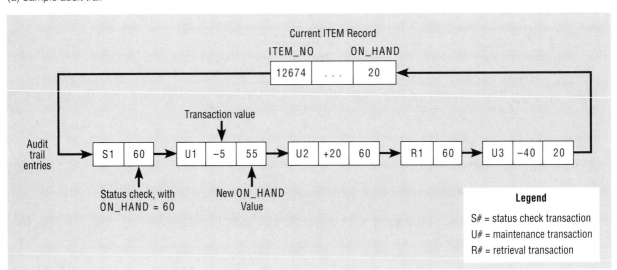

(b) Possible audit trail contents

AUDIT TRAIL DATA	EXPLANATION
Transaction identifier	Primary key value for each audit trail file record
Transaction value	The value in the transaction. For a numeric field, this may be the value by which the existing field value is changed. For an alphanumeric field, this will be the new field value
Before or after image	The value of the field before or after the change is made
Type of operation	The type of data operation logged: retrieval, insertion, deletion, modification
Date/Time stamp	A unique indication of when the transaction occurred
Program name	The name and version number of the computer program initiating the transaction
User/Device	The user name or computer device from which the transaction was initiated

read by humans. A person or a software program must know how to descramble the data to make them readable. Since the encryption scheme is usually more difficult to subvert than other security measures, encryption is considered the soundest form of data security. Either the operating system or data management software may provide an encryption algorithm and encryption can be done by computer chips; if not, you can build an encryption routine yourself using a function like a hashing algorithm or random number generator. Of course, the encryption algorithm itself must be secured. The U.S. federal government is attempting to standardize an encryption chip, the Clipper chip, so that federal authorities will have a way to decode any computer-encrypted data. Encryption is not always allowed. For example, since some foreign governments (e.g., France) forbid the transmission of encrypted data across national borders, companies doing business in these countries cannot use encryption to protect data files transported in and out of the country.

Calculating File Size

Sometimes it is necessary to calculate the potential sizes of data files. There are two purposes for calculating the total amount of space a file might consume. First, in some computer environments, you must request a fixed amount of space for each file, and you want to ask for enough but not too much. Second, there may be an issue of whether enough space exists, or you may need to determine the size of a hard disk needed for an application.

The size of a file depends on the following specifications:

- Data type of each field and structure of records

- Media (magnetic tape, floppy diskette, or hard disk)

- Page sizes (and, hence, blocking factors)

- File organization

The data types and record structures determine how much raw data space is required. The medium dictates how data is generally laid out. For example, floppy disks come in different capacities and use a sector as the basic physical storage unit (similar to a page). The page size and whether records can span pages for hard disks, as discussed earlier, influence the use and waste of space. Finally, the file organization determines the need for overhead data besides the raw data.

Because there are so many options, it is not feasible to illustrate all possible situations you might face. To understand all parameters, you should refer to reference manuals for the hardware and operating system that you are designing for.

As a hypothetical example of the general approach to calculating file size, consider the hashed file in Figure 16-12. This hard disk file was designed to accommodate a maximum of 172 records of 240 bytes each. Assume that space for the file cannot be dynamically allocated, so the space for the maximum file size must be assigned at once. The records are grouped into blocks of 4 records each, for a total of 43 blocks (172/4). This is done to create a number of blocks which corresponds to a prime number (a typical basis for a hashing algorithm). The prime number, 43 in this case, is used in the hashing algorithm called division hashing. In division hashing, the primary key is divided by the prime number and the remainder is the record's address (block number, from 0 to 42). Assuming that records (and also blocks) may span page boundaries, we are not concerned with page size in this example. But, hard disks have tracks, and let us assume that records (and also blocks) may not span track boundaries. In this example, a track is 4,000 bytes. Visually we

Figure 16-12

Example hash file layout and size

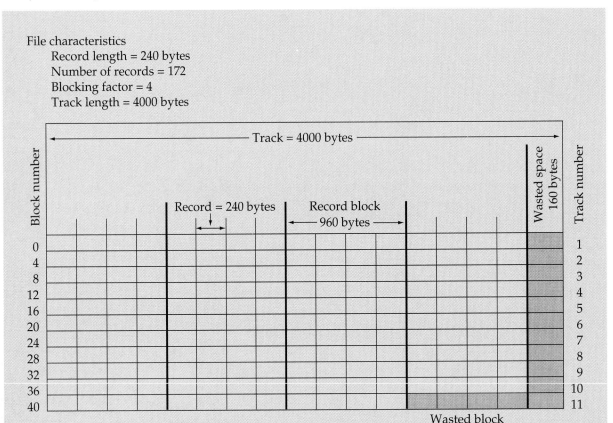

see from the layout that we can fit 16 records per track and need 11 tracks, or 44,000 bytes, for this file. Within each track, 160 bytes are wasted due to the restriction on record spanning tracks, and an additional record block of 960 bytes is not needed in the last track. Thus, there are 1,600 (10*160) plus 960 bytes wasted, or 1,760 bytes, or roughly 4 percent of the allocated space is wasted.

DESIGNING DATABASES

Database management system (DBMS): Software that is used to create, maintain, and provide controlled access to user databases.

Most modern information systems utilize database technology, called database management systems (DBMSs), for data storage and retrieval. A **database management system (DBMS)** is the software that sits between application programs and the operating system and extends the capabilities of the operating system for organizing data. Technologies such as IMS and DB2 on IBM mainframes, Ingres and Oracle on minicomputers (and other environments), Sybase SQL Server on client/server networks, and Paradox for Windows on personal computers provide a wide range of data storage and accessing options.

Recall that a database is a collection of logically related data, designed to meet the information needs of multiple users in an organization. A more operational definition for our purpose is that a database is a set of inter-related files. The relationship between files is due to relationships identified in the conceptual and logical data models. The relationships imply access paths between data. For example, consider Figure 16-13 for Hoosier Burger's inventory control system (you may also want to refer to the normalized relations located in Chapter 15 on page 580). To discover whether you have enough inventory in stock to fill a large order, this data model says that you could actually trace an item sale through product data, then through recipe data to inventory item data.

Choosing Database Architectures

There are different styles of database management systems, each characterized by the way data are defined and structured, called database architectures. Deciding

Figure 16-13

Final E-R diagram for Hoosier Burger's inventory control system

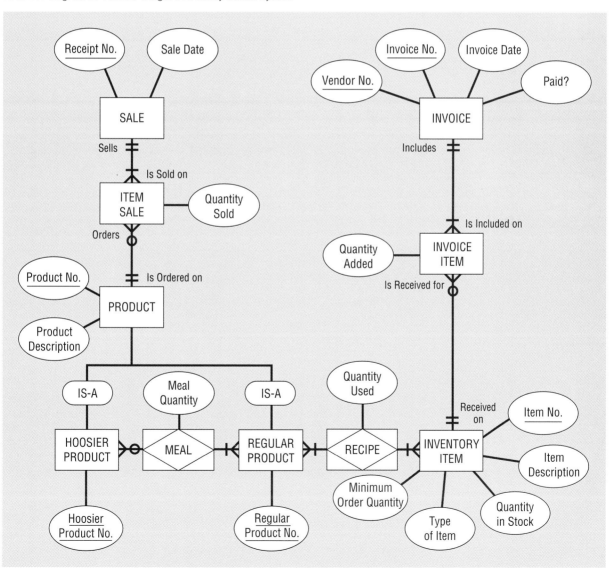

which database architecture best suits your database is fundamental to database design. A particular database management system supports one of four different architectures. Figure 16-14 compares these four architectures:

1. *Hierarchical database model:* In this model, files are arranged in a top-down structure that resembles a tree or genealogy chart. The top file is called the root, the bottom files are called leaves, and intermediate files have one parent, or owner, file and one or several children files. Among the oldest of the database architectures, many hierarchical databases exist in larger organizations. This technology is best applied when the conceptual data model also resembles a tree and when most data access begins with the same file. Hierarchical database technology is used for high-volume transaction processing and MIS applications. Few new databases are developed with hierarchical DBMSs since newer applications tend to have broader needs than simply transaction processing or summarization of transaction data.

2. *Network database model:* In this model, each file may be associated with an arbitrary number of files. Although very flexible as any relationships can be implemented, the form of implementation, usually using pointers between related records in different files, creates significant overhead in storage space and maintenance time. Network model systems are still popular on powerful mainframes and for high-volume transaction processing applications. Since the database designer has such detailed control over data organizations, it is possible to design highly optimized databases with network systems. Network systems support a wider variety of processing requirements than do hierarchical database systems, but network systems still require significant programming and database design knowledge and time, and hence are used primarily in those organizations with significant expertise with such technologies.

3. *Relational database model:* The most common database model for new systems, this model defines simple tables for each relation and many-to-many relationship. Cross-reference keys link the tables together, representing the relationships between entities. Primary and secondary key indexes provide rapid access to data based upon qualifications. Most new applications are built using relational DBMSs, and many relational DBMS products exist.

4. *Object-oriented database model:* In this model, attributes and methods that operate on those attributes are encapsulated in structures called object classes. Relationships between object classes are shown, in part, by nesting or encapsulating one object class within another. New object classes are defined from more general object classes. A major advantage of this data model is that complex data types like graphics, video, and sound are supported as easily as simpler data types. This is the newest DBMS technology and larger organizations are gaining experience with it by selectively using it when complex data or event-driven programming is appropriate for the application.

Since relational databases are the most common kind of database used today, we will cover the most significant relational database design issue, the selection of indexes.

Selecting Indexes

During physical file and database design you must choose which fields to use to create indexes. As mentioned earlier in the chapter, indexes can be created for both primary and secondary keys. When using indexes, there is a trade-off between

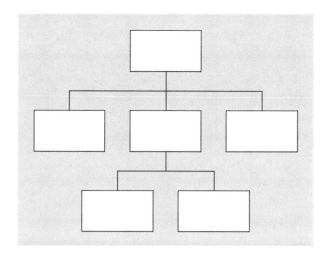

Figure 16-14
Database architectures

(a) Hierarchical

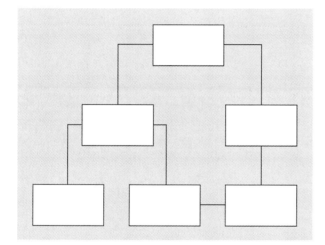

(b) Network

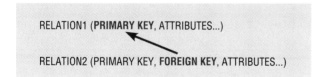

(c) Relational

RELATION1 (**PRIMARY KEY**, ATTRIBUTES...)

RELATION2 (PRIMARY KEY, **FOREIGN KEY**, ATTRIBUTES...)

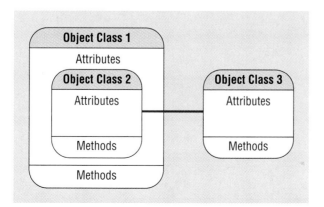

(d) Object-Oriented

improved performance for retrievals and degrading performance for inserting, deleting, and updating the records in a file. Thus, indexes should be used generously for databases intended primarily to support data retrievals, such as for decision support applications. Because they impose additional overhead, indexes should be used judiciously for databases that support transaction processing and other applications with heavy updating requirements (Gibson et al., 1989).

Following are some guidelines for choosing indexes for relational databases.

1. Specify a unique index for the primary key of each table (file). This selection ensures the uniqueness of primary key values and speeds retrieval based on those values. Random retrieval based on primary key value is common for answering multi-table queries and for simple data maintenance tasks.

2. Specify an index for foreign keys used in joining tables. As in point 1, this speeds processing multiple-table queries.

3. Specify an index for nonkey fields that are referred to in qualification, sorting, and grouping commands for the purpose of retrieving data.

 To illustrate the use of these rules, consider the following relations for Pine Valley Furniture Company:

PRODUCT (<u>PRODUCT_NO</u>, DESCRIPTION, FINISH, ROOM, PRICE)
ORDER (<u>ORDER_NO</u>, <u>PRODUCT_NO</u>, QUANTITY)

You would normally specify a unique index for each primary key: PRODUCT_NO in PRODUCT and ORDER_NO in ORDER. Other indexes would be assigned based on how the data are used. For example, suppose that there is a system module that requires PRODUCT and PRODUCT_ORDER data for products with a price below $500, grouped by PRODUCT_NO. To speed up this retrieval, you could consider specifying indexes on the following nonkey attributes:

1. PRICE in PRODUCT since it satisfies rule 3.

2. PRODUCT_NO in ORDER since it satisfies rule 2.

Since users may direct a potentially large number of different queries against the database, especially for a system with a lot of ad hoc queries, you will probably have to be selective in specifying indexes to support the most common or frequently used queries.

Selecting indexes is arguably the most important physical database design decision, but it is not the only way you can improve the performance of a database. Other ways address such issues as reducing the costs to relocate records, optimizing the use of extra or so-called free space in files, and optimizing query processing algorithms (Viehman, 1994).

Representing Data Volume

For making some database design decisions (such as calculating file sizes) and for developing data access strategies, it is advisable to create diagrams that show the volume of data. As an analyst, you may not be responsible for finalizing database designs; rather, this may be the job of a database analyst who has specialized skills and intimate knowledge of the DBMS features used for the application. In either case, you are responsible for outlining the need for data based on your knowledge of the application. Thus, you are in a better position than database specialists to determine the volume of data your application must handle.

To size files and to choose physical database structures like file organizations, you need to know how many records to expect for each file. A starting point for this is illustrated in Figure 16-15, which shows a variation of an E-R diagram, indicating

Figure 16-15
Data volume chart

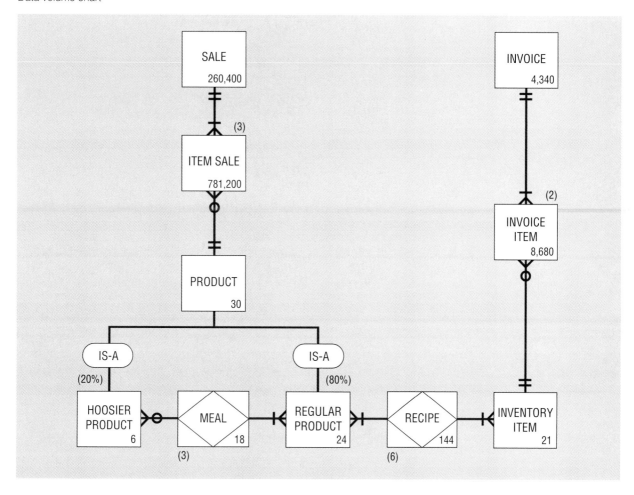

the number of occurrences of each entity with ratios that help you estimate the number of occurrences. Similar to Figure 16-13, this figure omits relationship names and attributes and the number of instances of each entity is added. You might develop such a chart by starting with any entity for which you know the volume of instances. Using information collected during analysis, you can then estimate ratios and other volumes. For example, if you knew that Hoosier Burger has 30 products on the menu board, and that roughly 20 percent of the products are Hoosier products, then there are approximately 6 Hoosier products and 24 Regular products. Also, since your study of past accounting records shows that Hoosier Burger generates approximately 1,200 sales a day for seven days a week, and one month of sales data will be kept on-line, then there will be approximately 1200*7*31, or 260,400 SALE records. With an average of three items on each sale, there will be roughly 781,200 ITEM SALE records.

Your role as a systems analyst is to provide as much information as possible about data volume and usage in a useful format so that database specialists (analysts and administrators) can create the final, technical designs which will make database processing and storage most efficient. As with telecommunications, detailed database design is typically done by these specialists, but your role of translating the system requirements into clear specifications for data is invaluable in the total process. In organizations that do not have database specialists, the final decisions on database design will rest with you. Thus, you should study the field of database design to either communicate effectively with specialists or to do database design when specialists are not available.

CASE IN DATABASE DESIGN

There are integrated and stand-alone CASE tools which can be useful during physical file and database design in four ways:

1. By storing the results of file and database design in a repository which shows how logical data attributes, relations, and entities map into physical data structures.

2. By generating file (like COBOL Data Division) or database (like SQL) definition code. This function saves considerable time in the implementation phase of the life cycle and ensures that the data definitions comply with the design. Figure 16-16 shows a typical example of generated SQL database defintion code from an E-R diagram and attribute definitions in the repository of a CASE tool. Figure 16-16a shows a simple E-R diagram and Figure 16-16b shows SQL code generated from this E-R diagram and the descriptions (not shown) of the entities and attributes associated with this E-R model. Often a CASE tool can generate to several different SQL dialects associated with

Figure 16-16
CASE tool SQL database definition code generation

(a) Sample E-R diagram

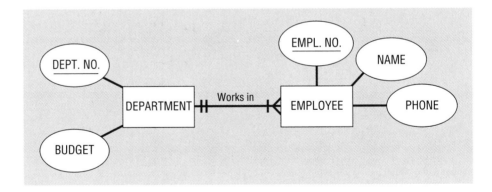

(b) Generated SQL code

```
CREATE TABLE DEPARTMENT
(
   DEPT_NO      SMALLINT          NOT NULL,
   BUDGET       DECIMAL (8,2)            ,
);

CREATE UNIQUE INDEX DEPT_INDEX ON DEPARTMENT (DEPT_NO);

CREATE TABLE EMPLOYEE
(
   EMPL_NO      SMALLINT          NOT NULL,
   NAME         CHAR(20)                 ,
   PHONE        INT                      ,
   DEPT_NO      SMALLINT          NOT NULL,

   FOREIGN KEY (DEPT_NO) REFERENCES DEPARTMENT
);

CREATE UNIQUE INDEX EMPL_INDEX ON EMPLOYEE (EMPL_NO);
```

different SQL-based DBMSs. As the database evolves, analysts can simply change the E-R model and attribute repository entries, and the CASE tool will generate the new database definition code.

3. By reverse engineering existing file and database definitions for elements of current systems that will be reused in the new application. Often the structure of data is not obvious in existing, or legacy, systems. Reverse data engineering tools are very sophisticated (akin to an expert system), and can analyze the data definitions as well as program code to surmise functional dependencies and other information necessary to develop a data model of existing data. These descriptions are then combined with the descriptions of the new data in the CASE repository for generation of the new data structures.

4. By assisting in making file and database design decisions. There are some CASE tools and stand-alone analysis tools, like the IBM Database Design Aid, that will analyze data characteristics (sizes and volumes) along with usage patterns and suggest a good or optimal set of secondary key indexes. Such tools can be of great assistance when the database has tens or hundreds of files and hundreds of database access modules—which would not be atypical of major application systems.

SUMMARY

We have presented a wide range of concepts and techniques you will need for designing computer files and databases during the physical design phase of the systems development life cycle. Because of the technical and often technology-specific nature of physical design, you will often assist or provide information to specialists who will actually design the physical file and database.

Physical file and database design takes, as input, the normalized relations from logical data modeling as well as other models that explain how data in files and databases will be accessed. You must have a basic understanding of the capabilities of the data management technology used with the system since your design options may be limited by the technology's capabilities. The deliverables are specifications that can be given to programmers and data management specialists who will write the code to define file and database structures.

Physical file and database design involves designing fields, records, files, and databases. Designing a field includes selecting a data type, coding and compression methods to reduce storage space and improve data accuracy, the format for primary keys, and data integrity controls to ease data entry and reduce errors. The chapter also discussed several techniques for handling missing fields, a common situation for many application systems.

Designing physical records involves grouping fields together based on a shared primary key and an affinity of usage. Records are also designed to minimize wasted space in secondary memory storage units, called pages. You also learned that record partitioning and denormalization are two approaches to creating efficient record structures. The techniques for handling fixed-length and variable-length records use constructs such as sequential storage and pointers to arrange fields within a record.

There are six types of physical files: data, look-up table, transaction, work, temporary, and history. Most of the effort in physical file and database design focuses on data files, but the need for and structure of each of these files must be handled during this life cycle step. Operating systems handle the physical input/output operations with files using relative and direct access methods. File and database management software use these basic access methods to create file organizations that can support a variety of data accessing needs. File organizations are grouped into three general families: sequential, indexed, and hashing. The most common file organization in use today is indexed, so the selection of primary and secondary keys for indexes is one of the major physical file and database design choices. The performance of different file organizations varies by such factors as storage space consumed, speed of sequential or random record retrieval, and speed for adding, deleting, and updating records. Some file organizations are specifically designed to rapidly find data based on multiple-key selection statements, such as are supported in most relational database query languages.

Designing files also includes selecting strategies for file backup and security (such as data encryption). Another file design task is estimating file size so that suitable secondary memory space can be allocated for the application you are designing. The estimation of file size depends on the storage medium, field data types, record structures, volume of data, secondary memory page or record block sizes, and file organizations you choose.

Designing databases presents additional physical file and database design issues. Files must be designed within the capabilities of the database environment, including the database management system, the data definition and data manipulation languages used in the application, and the data dictionary. Four database architectures are available to you: hierarchical, network, relational, and object-oriented, with relational being the norm for modern systems. The database design decision of greatest import in relational environments is selecting primary and secondary key indexes.

Physical file and database design also includes documenting the number of instances of each entity from the logical and conceptual data models. Information about data volume and usage are important parameters for making a variety of file and database design decisions.

CASE tools can help in physical file and database design by storing the results of file and database design in a CASE repository, generating file (like COBOL Data Division) or database (like SQL) definition code, reverse engineering existing file and database definitions for elements of current systems that will be reused in the new application, and assisting in making file and database design decisions (such as index selection).

Physical file and database design is one of three parallel steps in the physical design phase of the systems development life cycle. Program and distributed system design are the other two steps and are the subjects of the following two chapters.

CHAPTER REVIEW

KEY TERMS

Access method	Default value	Null value
Audit trail	Encryption	Page
Backward recovery	Field	Physical file
Blocking factor	File organization	Physical record
Calculated field	Forward recovery	Picture (or template)
Data compression technique	Hashed file organization	Pointer
Data type	Hashing algorithm	Record partitioning
Database management system (DBMS)	Index	Secondary key
	Indexed file organization	Sequential file organization

R E V I E W   Q U E S T I O N S

1. Define each of the following terms:
   a. data type
   b. file organization
   c. index
   d. check digit
   e. page
   f. hashing algorithm

2. List and describe six types of files found in information systems.

3. Describe the inputs to physical file and database design.

4. Describe the deliverables from physical file and database design.

5. In what way does the choice of a data type for a field help to control the integrity of that field?

6. Explain the term referential integrity. Do only foreign keys have referential integrity? Why or why not?

7. How can missing data be handled for a field?

8. What is the difference between how a range control statement and a referential integrity control statement are handled by a file management system?

9. What is the purpose of record partitioning and denormalization? Why might you not want to create one file for each relation in a logical data model?

10. What techniques can be used to keep track of where variable-length fields are in a physical record?

11. What factors influence the decision to create an index on a field?

P R O B L E M S   A N D   E X E R C I S E S

1. Match the following terms to the appropriate definitions.

   _____ backward recovery

   _____ forward recovery

   _____ data compression technique

   _____ sequential file organization

   _____ indexed file organization

   _____ hashed file organization

   a. an approach to locating and storing records based on searching a table of keys and locations for the associated record(s) with that key value
   b. use of a backup file and a transaction log to restore a damaged file to an uncontaminated state
   c. a way to get rid of unnecessary and redundant data in a file or field
   d. use of an algorithm to convert a key value into the location where a record is stored
   e. use of a log of records before images to restore an erroneous file to a prior state before reprocessing transactions
   f. location of records one right after another

2. On page 580 of Chapter 15 a list of normalized relations for Hoosier Burger's inventory control system appears. Choose data types for the fields in these relations based on figures about this database shown throughout the book.

3. Explain the four criteria that influence the choice of a data type for a field and give examples of how each criterion is applied.

4. Suppose you were designing a file of student records for your university's placement office. One of the fields that would likely be in this file is the student's major. Develop a coding scheme for this field that achieves the objectives outlined in this chapter for field coding.

5. In Problem and Exercise 6 in Chapter 15 you developed normalized relations for a particular E-R data model. Choose primary keys for the files that would hold the data for these relations. Did you use attributes from the relations for primary keys or did you design new fields? Why or why not?

6. Suppose you created a file for each relation in your answer to Problem and Exercise 6 from Chapter 15. If the following queries represented the complete set of accesses to this database, suggest and justify what primary and secondary key indexes you would build.

   a. For each PART in ITEM NO. order list in VENDOR NO. sequence all the vendors and their associated prices for that part.

   b. List all PART RECEIPTs including related PART fields for all the parts received on a particular day.

   c. For a particular VENDOR, list all the PARTs and their associated prices that VENDOR can supply.

7. The following questions involve the construction of check digits for the account numbers with a local telephone company. The account number is the telephone number, not including area code.

   a. What are the first seven prime numbers?

   b. Suppose that the check digits on the account number are calculated by summing the multiplications of each telephone number digit by the associated prime number starting with the right-most digit (that is, in the telephone number 555-6789 the 9 would be multiplied by the first prime number, the 8 by the second prime number, and so forth). The check digit is the three least significant digits in the result of this formula. What would be the check digits for the above telephone number?

8. Suppose you were designing a default value for the age field in a student record at your university. What possible values would you consider and why? How might the default vary by other characteristics about the student, such as school within the university or degree sought?

9. The following questions concern the use of null values for a field.

   a. When a student has not chosen a major at a university, the university often enters a value of "Undecided" for the major field. Is "Undecided" a default value or a way to represent the null value?

   b. What problems might allowing a null value for the field PERMANENT ADDRESS cause for processing student records at a university?

10. Assume that you have a hard disk file designated to accommodate a maximum of 1,000 records of 240 bytes each and all other factors are similar to those described in the example related to Figure 16-12. How many bytes will be needed for this file?

11. Consider the E-R diagram for Hoosier Burger in Figure 16-13.

    a. Represent this data model as a set of 3NF relations.

    b. Suggest physical record structures for these relations. Which relations might be denormalized in order to speed up frequent data-processing steps, such as entering sales transactions, generating an order to the kitchen, and handling the receipt of new inventory?

FIELD EXERCISES

1. Find out what database management systems are available at your university for student use. Investigate which data types these DBMSs support. Compare these DBMSs based upon data types supported and suggest which types of applications each DBMS is best suited for based on this comparison.

2. Contact someone in the data center at your school, university, or place of work. Find out how frequently they back up files on different computers or file servers. Why do they back up with that frequency? Where do they keep these files and for how long are the backup files kept?

3. Investigate data compression products, such as Stacker from Stac Electronics, and PKZIP by PKWARE. Gather information from popular press articles found in the library, directly from the software writers, or from hands-on testing. What products are available? Are they useful? Are they reliable? How much do they cost? Do they seem to be worthwhile given the costs and benefits? Why or why not? What is the extent of use in the market of these data compression products?

4. Investigate data encryption techniques presented in the academic research literature. What is the state of the art in data encryption? What seem to be the strengths and weaknesses of these techniques? What applications do you foresee for data encryption in the future?

5. Investigate object-oriented database products. Gather information from popular press articles found in the library, directly from the software writers, or from hands-on testing. What products are available? Are they useful? How much do they cost? Do they seem to be worthwhile given the costs and benefits? Why or why not? For what types of applications are object-oriented databases being used? Do you foresee that object-oriented databases will replace other types of databases discussed in this chapter? Why or why not?

REFERENCES

Babad, Y. M. and J. A. Hoffer. 1984. "Even No Data Has a Value." *Communications of the ACM* 27 (Aug.):748–56.

Finkelstein, R. 1988. "Breaking the Rules Has a Price." *Database Programming & Design* 1 (June):11–14.

Gibson, M., C. Hughes, and W. Remington. 1989. "Tracking the Trade-Offs with Inverted Lists." *Database Programming & Design* 2 (Jan.):28–34.

McFadden, F. R. and J. A. Hoffer. 1994. *Modern Database Management.* Redwood City, CA: Benjamin/Cummings Publishing.

Rodgers, U. 1989. "Denormalization: Why, What, and How?" *Database Programming & Design* 2(12) (Dec.):46–53.

Viehman, P. 1994. "24 Ways to Improve Database Performance." *Database Programming & Design* 7(2) (Feb.):32–41.

Physical Design: Designing Files and Databases

INTRODUCTION

As with other elements of the logical and physical design of the Customer Activity Tracking System (CATS), the project team continues with a complete design effort even though the chosen strategy is to use packaged software as much as possible. The design effort is important to identify what BEC wants from CATS. In the final stages of design, the team will decide if the package needs to be modified to meet the detailed needs of CATS or if the specifications for CATS need to be changed to fit the capabilities of the package or if a combination of both of these modifications is required. Frank Napier, an analyst on the CATS team, and Buffy Jarvis, a data analyst assigned to help the CATS team, are reviewing several issues related to tuning the performance of the physical files and database and identifying how the data in CATS relate to data in other BEC systems. These are some of the issues the CATS team is addressing during file and database design.

CATS RELATIONSHIP WITH EXISTING FILES AND DATABASES

"Hi, Buffy, come on in," Frank Napier greets Buffy Jarvis at the entrance to Frank's cubicle. "Thanks for stopping by between appointments. Please sit down. I'm anxious to see what you discovered from your analysis of existing BEC files." Buffy has been studying the attributes in the normalized relations developed from the logical data modeling step in the CATS project (see Figure 6 in the BEC case at the end of Chapter 15). The purpose of this analysis is to determine what data identified for CATS already exist in other BEC files and databases. This will be used to decide what data CATS must capture itself and what will simply be fed from other systems.

"Frank, I think I've discovered everything we'll need to know to make some decisions on data sourc-

ing," responds Buffy. "In many ways, we got lucky, Frank. I called the head of information systems at TwoCat Petfood, the supplier of the package we intend to use for CATS. I found out that their system is written in Paradox for Windows, a popular PC relational DBMS package, and all of the data is stored in Paradox tables. It's written as a simple LAN application, but they were smart enough to build a lot of controls into it. We don't currently use Paradox, but we knew we'd have to get a LAN license for it.

"The TwoCat package is written to use the LAN server as a shared hard disk, with each workstation addressing the server just like it would a local hard drive. This is about the simplest kind of LAN database environment, what people call a file server environment. It allows data to be shared by several users, but each user is running a separate copy of the database management system. As you know, BEC decided several years ago to use an SQL database system called Sybase SQL Serve™, which is a lot more sophisticated than Paradox. We've been building all our new systems under Sybase and converting older systems as it made sense. Sybase allows us to design programs to be run on workstations for all data capture, analysis, and presentation functions, but all data management functions are done at the server. This has proven to be a great way for us to decouple various components of systems, giving us greater flexibility to use the latest technology for each application we build yet still have central management of data. It has also allowed us to minimize network traffic, a big plus for a company our size.

"At first I was concerned that we'd have to write a lot of bridge routines to regularly transfer data between our current BEC files managed by SQL Server and CATS, but then I saw an article in *PC Week* about a product called SQL Link® from Borland, the vendor for Paradox. What SQL Link would allow us to do is access the SQL files in Sybase as if they were Paradox files. Thus, whenever CATS needs data we already have under SQL Server, we'll just need to define the path to it and a Paradox application can access it without much modification. This is going to save us a lot of time in installing the package since the source code can generally stay the same and we won't have to convert data; this will make it easier to get final approval from management for the implementation phase.

"What we will end up with is a database environment that will look something like this (Buffy quickly sketches a drawing on the whiteboard on Frank's office wall; see Figure 1). As far as the Paradox programs are concerned, all the data appear to be in Paradox tables on the server. When a query requires data from a Sybase table, SQL Link will send a request to the server which will return a table to the PC which Paradox will then join with Paradox tables. Although CATS will be able to use and maintain data housed in Sybase databases, CATS will still run as a Paradox application. This means that the PC workstations will do all the work. From the analysis I've done, most of the CATS data are in existing Sybase tables. Eventually we'll want to recode the package into a language that can allow Sybase to do as much work as possible, but at least we can make the package work now in our environment. As you can imagine, I was quite relieved to reach this conclusion. This has really simplified satisfying our data sourcing requirements in order to get CATS up and running soon."

"This sounds like one of your advertisements for using standards, Buffy." Frank said with just a hint of sarcasm. "Okay, so we can get to existing data, but where are the data?"

Buffy continued, "Then what I did was prepare a table listing all of the relations and selected attributes from the normalized relations for CATS and where these items are in our Sybase corporate database. I also included what I would suggest we do in terms of data sourcing for CATS (Buffy hands Frank a copy of Table 1). For example, let's look at the STORE relation.

This relation is a simple subset of the standard BEC Sybase database table of the same name. I don't see where CATS will interfere with this Sybase table in any way, so I think that CATS ought to use the existing database table. We ought to make this a read-only table to CATS, and we can disable the STORE data maintenance functions in the package. The same is true for PRODUCT, CUSTOMER, INVENTORY, CUSTOMER_ TRANSACTION, and PURCHASE. In each of these cases, CATS just wants to use a subset of attributes in existing Sybase tables.

"The PURCHASE table does require some minor modification. Implicitly this relation in CATS means a purchase generated from a reservation. Since the idea of a reservation is new, our Point of Sale (POS) system and the corporate sales tracking system don't have any field to record this. I've prepared a change request for these systems to add this field and to modify the POS to record a reservation sale.

"I've looked at the statistics you, Jorge, and Jordan gave me about the frequency of various CATS accesses to the data, and I see no reason to separate the CATS processing from the other activity against this database. The bulk of the activity in CATS is handling reservations, and that will require real-time access to the Sybase CUSTOMER and PRODUCT tables; the rest of CATS is a minimally extra load on Sybase. Thus, it would not be wise to try to maintain two copies of these files, with one copy strictly for CATS use. So, what I am recommending is that we use existing Sybase tables, appropriately modified, when a table exists."

Frank interjected, "Now, what about the relations and fields that are new for CATS? Can we just use the Paradox tables found in the TwoCat package?"

Buffy replied, "To some degree, but it depends on the relation. Let's take a look at the PRODUCT_ ANNOUNCEMENT relation (see Table 1). The TwoCat package has such a table, but it does not use a code like our PCS_CODE. There is, of course, a primary key

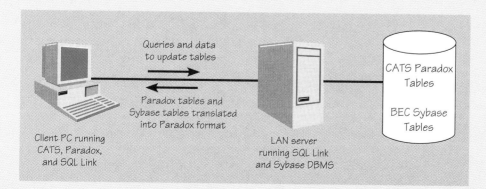

Figure 1
BEC database environment with Paradox and Sybase

Queries and data to update tables

Paradox tables and Sybase tables translated into Paradox format

Client PC running CATS, Paradox, and SQL Link

LAN server running SQL Link and Sybase DBMS

CATS Paradox Tables

BEC Sybase Tables

TABLE 1 Sources of CATS Data

CATS Data	Source(s)	Recommendation
STORE	standard STORE relation	Use standard relation
PRODUCT	standard PRODUCT relation	Use standard relation
VIDEO_PRODUCT	standard VIDEO_PRODUCT relation	Use standard relation
MUSIC/GAME_PRODUCT	standard MUSIC/GAME_PRODUCT relation	Use standard relation
CUSTOMER	standard CUSTOMER relation	Use standard relation
INVENTORY	standard INVENTORY relation	Use standard relation

PRODUCT_ANNOUNCEMENT
  PCS_CODE
 	
  RUN_NUMBER
  ANNOUNCE_DATE
	package
	  8 alphanumerics
	    (99XXXXXX)
	  S (small integer)
	  DATE
		Modify package to accept our PCS code;
		use Paradox table, but add indexes on
		RUN_NUMBER and ANNOUNCE_DATE

RUN	does not exist	Combine into PRODUCT_ANNOUNCEMENT table
CUSTOMER_TRANSACTION	standard CUSTOMER_ TRANSACTION relation	Use standard relation
PURCHASE	standard PURCHASE relation	Use standard relation; modify standard relation to indicate reservation purchase

CUSTOMER_PROFILE
  CUST_PROF_CATEGORY
  CUST_PROF_CAT_LEVEL
	package	Modify package with new data integrity controls for
		these two fields; use Paradox table

PRODUCT_PROFILE
  PROD_PROF_CATEGORY
  PROD_PROF_SUPERIOR_CAT
	package	Modify package with new data integrity controls for
		these two fields; use Paradox table

PRODUCT_FOR_CUSTOMER	package	No changes to package; use Paradox table

for the file; we'll just need to change the field's definition, and my contact at TwoCat said this is pretty easy to do. What I suggest for a primary key is an 8-character alphanumeric, with the first two characters being the last two digits of the year in which we send the announcement to the customer. This will give us 26^6, or roughly 309 million, announcements per year. From the estimates that you sent me on transaction volumes, this should be sufficient for at least five years. By then, CATS will need a major overhaul, and the format of the PCS_CODE could be addressed then.

"The CUSTOMER_PROFILE, PRODUCT_PROFILE, and PRODUCT_FOR_CUSTOMER tables can be used straight out of the package. The only thing we'll need to do is modify what Paradox calls validity checks on some of the fields. Validity checks control the range of possible values and table look-ups for lists of legal values, like the allowable values for CUST_PROF_CATEGORY. Not surprisingly, TwoCat uses different categories and levels than we plan to use, but the

concept is the same. The format of these fields in the package will be fine."

"What about the RUN_NUMBER field in PRODUCT_ANNOUNCEMENT and the RUN relations?" asked Jorge. "Your table says these are new for us?"

"Yes, interestingly, TwoCat designed their system to run daily, so they did not need a run concept. They just store the date of the announcement as part of the PRODUCT_ANNOUNCEMENT table. I think we ought to do the same. This violates normalization, but I think this would be an appropriate situation to denormalize. Almost every time any PRODUCT_ANNOUNCEMENT data are used, so is the ANNOUNCE_DATE. Thus, the system would be doing a lot of joining of these two tables. Also, ANNOUNCE_DATE is not a very large field, so even though there are potentially a very large number of PRODUCT_ANNOUNCEMENT records each with the same date, we won't be storing much extra data. We should create a secondary key on ANNOUNCE_DATE as well as RUN_NUMBER, which TwoCat did not do for their system. We do summaries of announcements by both RUN_NUMBER and ANNOUNCE_DATE, so these indexes will speed those processes."

"Okay, that makes sense. Did you look at all the integrity rules, file organizations, backup procedures, and indexes used throughout the package—do we need to make any other modifications for CATS?" asked Frank. "We seem to be a lot bigger operation than TwoCat, so we might need more tuning of the database."

"Good questions, Frank. That must have been a fine database course you took at the University of Western Florida. You new guys think you have all the answers," chided Buffy. "Actually, I have looked over those aspects of the TwoCats system and I have only a few changes to recommend, but several are significant. First, Paradox, as opposed to Sybase, does not do any automatic logging of transactions or keep audit trails so files can be rebuilt in case of damage. Since CATS will be installed as a Paradox application running on each client PC, we'll need to modify the package to write these additional files. Second, TwoCats put a manual backup option into the main system menu. I think we'll want to create a batch procedure that will automatically run every night to back up the PRODUCT_ANNOUNCEMENT table, and every week back up the whole database. Third, Paradox does not allow encryption of tables, and I don't see a need for that in this system. Finally, I looked at the statistics on data usage you sent me and I have identified some additional fields for indexing, since we have designed

some new on-line and batch reports that the TwoCat system does not produce. The report programs will be easy to write, but production of the reports will go faster with some additional indexes. I've placed these changes in the CATS repository as notes, so whoever defines the Paradox tables in our environment will know to create these other indexes."

Frank then said, "Buffy, another fine job. I'll brief Jorge on your ideas and we'll check them against the program module plan we are currently developing. Again, we want to see how well our methods for system design match with the package's design. You've helped on the data side, and we need to do some work on the processing side. Thanks again for your help. Thanks for stopping by on your way to the Sybase training session you are leading for end users. See you at next Monday's team meeting."

SUMMARY

This Broadway Entertainment Company case has shown a part of the analysis of physical file and database issues during the CATS project. Some unique issues arose since CATS will adopt a software package where possible, and this package needs to fit into the overall Sybase LAN data environment at BEC. Besides the data aspects of CATS, the project team must match the data processing requirements of CATS with those of the package. We will see an example of program design in the next BEC case.

Designing the Internals:
Program and Process Design

L E A R N I N G   O B J E C T I V E S

After studying this chapter, you should be able to:

- Draw and explain a structure chart.

- Develop both transform-centered and transaction-centered information system designs.

- Turn a data flow diagram into a structure chart using transform analysis.

- Refine an initial system design into a more detailed design.

- Explain and apply the guidelines for good program design.

- Differentiate between the five types of coupling and apply these in program design.

- Differentiate between the seven types of cohesion and apply these in program design.

- Represent the processing logic inside program modules using pseudo-code and Nassi-Shneiderman diagrams.

INTRODUCTION

Once you know what the input, output, interface, dialogue, and database for an information system are, you can design the inside of the system, the part that will make the interface operable, generate output, and access and update the organization's databases. In this chapter, you will learn about designing a system's "internals." We will begin at an abstract level, taking what we already know about a system's processes, in the form of documented data flow diagrams, and converting them to *structure charts*. Structure charts graphically represent system designs. You will learn how to draw structure charts and how to derive them from data flow diagrams. Just as most CASE tools support the creation and revision of data flow diagrams and screen and report design, most CASE tools also support the creation and revision of structure charts.

The structure charts you create will form the basis for the structure of the system you design and build. Decisions you make at this point will heavily influence the overall design and implementation of your information system. Therefore, you need to be aware of what constitutes a *good* system design. You will learn about good design and some guidelines that will help you achieve it. The primary goal behind good design is to make your system easy to read and easy to maintain. The primary way the structured systems development techniques you have been studying promote good design is through the division of problem solutions into smaller and smaller pieces. The smaller the piece, or *module,* the easier it is to program, to read, and to revise due to changing business conditions. Modularization, then, promotes ease of coding and maintenance. On the other hand, modularization is not simply a reflection of size—it is also a reflection of function, that is, what a particular piece of a system is supposed to do. One guideline for good design, then, is to maximize *cohesion,* the extent to which a part of the system is designed to perform one and only one function. Modules that perform a single task are easier to write and maintain than those that perform many varied tasks. As modularization also involves how different parts of the system work in conjunction with each other, another guideline for good design is to minimize *coupling,* the extent to which different parts of the system are dependent on each other. The processing logic represented by each system module needs to become an explicit part of the system specifications you will pass on to programmers for coding, so you will also learn how to represent processing logic as *pseudocode* and as Nassi-Shneiderman diagrams.

DESIGNING THE INTERNALS

Chapters 16, 17, and 18 all focus on the physical design phase in the systems development life cycle (see Figure 17-1). In this chapter, you will learn how to design computer programs and processes contained in information systems. Processes take incoming data and transform them or route them to other locations in the information system. In designing programs and processes, however, you will not immediately begin to write computer code. Coding doesn't begin until the implementation phase. The focus in designing the internals is to turn the previously defined relationships among data, processing, and output into a detailed blueprint programmers can use as a guide to writing programs once implementation begins.

The Process of Designing the Internals

You begin designing internals armed with a great deal of information about the system chosen for development and implementation. With the exception of the detailed physical file and database design, most of that information is at a logical level. For the processes that have been identified as part of the system, the information takes the form of data flow diagrams and mostly textual descriptions of what each process is supposed to accomplish. Your job now is to take the logical information and turn it into a blueprint for the physical information system that will be programmed.

Deliverables and Outcomes

There are two primary outcomes of the physical design process for the system's internals: a set of structure charts and a set of physical design specifications for each separate part of the system. In addition to providing the functional descriptions for each part of the system, you must also provide information about input received and output generated for each program and its component parts. Table 17-1 provides more detailed information on the deliverables from designing the system internals.

Figure 17-1

The systems development life cycle with physical design phase highlighted

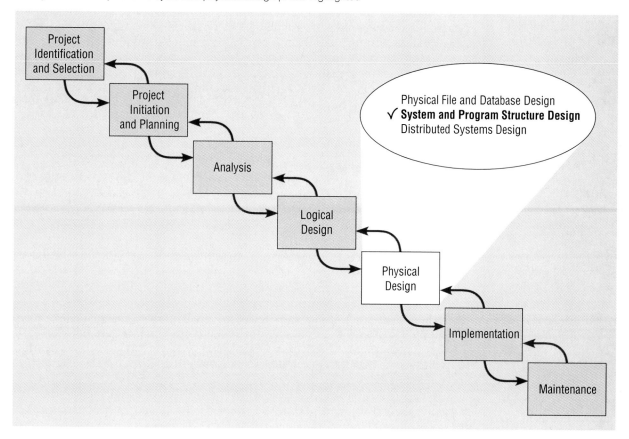

TABLE 17-1 Deliverables from the Design of System Internals

1. Structure Charts
   a. Fully factored structure charts
   b. Complete descriptions of data couples and flags

2. Module Specifications
   a. Input specifications
      1. Database specifications
      2. Other input specifications
         a. On-line or batch
         b. Files
         c. Other modules
   b. Processing specifications
      1. Pseudocode
      2. Nassi-Shneiderman charts
   c. Output specifications
      1. Database and file update specifications
      2. Print or on-line specifications
      3. Other modules

The specifications for input to and output from the database will come from the physical database description, often in the form of pseudocode or a special pseudocode called action diagrams. All of the data that appear in the structure charts should have already been defined in the database or in input form designs. Specifications for modules that capture system inputs or produce system outputs may be in narrative form since the processing logic of these modules is usually quite simple. If input modules also include procedures to validate data, handle errors in the data, or capture different data depending on the data that are input, then pseudocode can also be used to explain these system modules. These specifications will be supplemented by the actual form, report, and dialogue designs.

STRUCTURE CHARTS

Just as you can diagram how information flows through a system and how data entities are related to each other, you can also diagram how the various program parts of an information system are physically organized. The most common architecture for representing the physical structure of a system is hierarchical. A hierarchy looks like an inverted tree or organization chart with one root or main routine at the top and multiple levels of other modules nested underneath. A **structure chart** shows how an information system is organized in a hierarchy of components, called modules. The purpose of a structure chart is to show graphically how the parts of a system or program are related to each other, in terms of passing data and in terms of the basic components of structured programming: sequence, selection, and repetition. A structure chart redefines the flow and processing of data from data flow diagrams into a structure of system components that follow certain principles of good program design.

Structure chart: Hierarchical diagram that shows how an information system is organized.

Structure charts are used to show the breakdown of a system into programs and the internal structure of programs written in third- and fourth-generation languages. The structure of programs written in newer object-oriented or event-driven programming languages is usually depicted by state-transition diagrams and Structured English, which we discussed in Chapter 10.

Module: A self-contained component of a system, defined by function.

A **module** is a relatively small unit of a system that is defined by its function. Modules are self-contained system components. As much as possible, all of the computer instructions contained in a module should contribute to the same function. Modules are executed as units and, in most instances, have a single point of entry and a single point of exit. Standard programming terms used to identify modules include COBOL sections, paragraphs, or subprograms, and subroutines in BASIC or FORTRAN. For object-oriented programming languages, a module would roughly be a method. A computer program is typically made up of several modules. Modules may also represent separately compiled programs, subprograms, or identifiable internal procedures.

In a structure chart, there is a single coordinating module at the root and on the next level are modules that the coordinating module calls. For on-line systems, you can think of the root as the main menu and each of the subordinate modules as the main menu options. Each of the modules that report to the coordinating module may call additional modules in the next row down. A module calls another, lower-level module when the subordinate module is needed to perform its function. Modules at the lowest levels do not call any other modules; instead they only perform specific tasks. Middle-level modules act as coordinating modules for those lower-level modules they control but may perform some processing as well. In a structure chart, each module is represented as a rectangle containing a descriptive name of its function (see Figure 17-2). Each module name should concisely and ac-

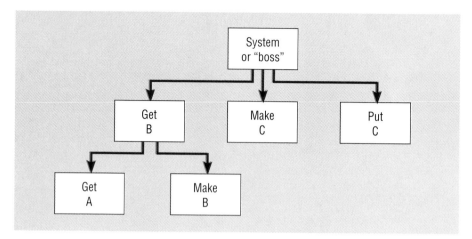

Figure 17-2
An illustration of the hierarchy
of a structure chart

curately reflect what that module does. When naming modules, you should avoid names that include conjunctions as conjunctions indicate that the module performs more than a single function. Modules that seem to need compound names would probably be better represented as multiple modules. Modules are called in order from left to right.

Modules in a structure chart communicate with each other through passing parameters. These parameters take the form of data, represented as data couples and flags. A **data couple** shows data being exchanged between two modules. Data couples are drawn as circles with arrows coming out of them (see Figure 17-3a). The arrow indicates the direction of movement of the data couple between modules. A data couple's circle is not filled in. A data couple is usually a single data element, although it can also be a data structure or even an entire record. A **flag** shows control data, or a message, being passed between modules. You represent a flag (also called a control flag) using the same symbol as a data couple except that you fill in the circle. A flag represents information the system needs for processing; a flag itself is never processed. For example, a flag may carry a message such as "End of file" or "Value out of range." A flag should never represent one module telling another module what to do, such as "Write EOF message."

You use special symbols with modules to indicate specific types of processing or special types of modules. A diamond shape at the bottom of a module means that only one of the subordinates attached to the diamond will be called; other modules, called each time, are not attached to the diamond. The diamond indicates there is a conditional statement in the module's code that determines which subordinate module to call (see Figure 17-3b). The diamond is how we show selection in structure charts. If a module's subordinates are called over and over again until some terminal condition, a curved line is drawn through the arrows connecting the module to these subordinates to indicate repetition (see Figure 17-3c). Again, other subordinates may be called only once and, hence, the curved line does not intersect the connection to these modules. If a module is pre-defined, meaning that its function is dictated by some pre-existing part of the system, then you represent the module with a vertical bar drawn down each side (see Figure 17-3d). For example, a pre-defined module may already exist because it is part of the operating system or because its function is dictated by some piece of hardware necessary for the system to work. Pre-defined modules are most commonly found on the very bottom layer of a structure chart because the types of functions they perform deal with such things as reading raw data from the original source or printing to a specific type of printer.

Data couple: A diagrammatic representation of the data exchanged between two modules in a structure chart.

Flag: A diagrammatic representation of a message passed between two modules.

Figure 17-3

Special symbols used in structure charts

(a) Data couples and control flag

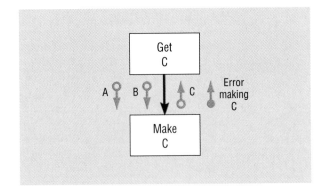

(b) Conditional call of subordinates

(c) Repetitive calls of subordinates

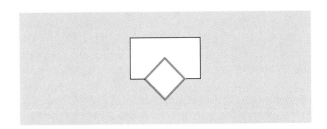

(d) Pre-defined module

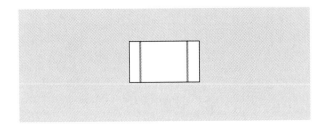

(e) Embedded module

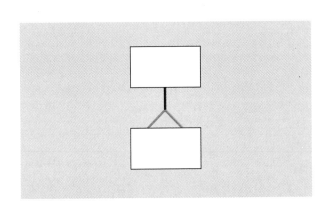

In Figure 17-3e, the special symbol connecting the two modules is called a "hat." A hat means that the function in the subordinate module is important logically to the system, but so few lines of code are needed to perform the function that the code itself is actually contained in the superior module. For example, a university registrar's system will definitely contain code for calculating a student's grade point average (GPA). A module that calculates GPA will be a central transform to a registrar's system, so it will appear on a structure chart as a separate module. However, to calculate GPA takes only a few lines of code, not enough to really merit a separate module. The Calculate GPA module, then, appears on the structure chart as a separate module logically, but the actual code needed for the function will be contained in the boss module.

Structure charts are read from left to right, and the order in which modules are called is determined not so much by the placement of the modules as it is by the placement of the arrows connecting the modules. Look at Figure 17-4a. The arrows connecting the modules do not overlap. The system module first calls the Get Valid A module. Get Valid A first calls the module Read A which executes its function and returns the data couple A to its coordinating module, often called the boss module. A is then passed to Validate A which returns the data couple Valid A. Control is returned to Get Valid A which then returns the data couple Valid A and control to the boss module. A similar set of steps occurs in connection with the data couple Valid B. Next, Valid A and Valid B are passed to Make C which returns the data couple C and control to the system module. Finally, the system module calls Put C and passes C to the module.

Now look at Figure 17-4b. Here the system designer has decided that the Validate B module is redundant and that the validation process implemented in the module Validate A works just as well for validating B. Therefore, there is only one validation module, called Validate data. Processing continues here as in the structure chart described previously. Notice, however, that the arrows leaving the module Get Valid B now cross over each other. The left-most arrow that leaves the module

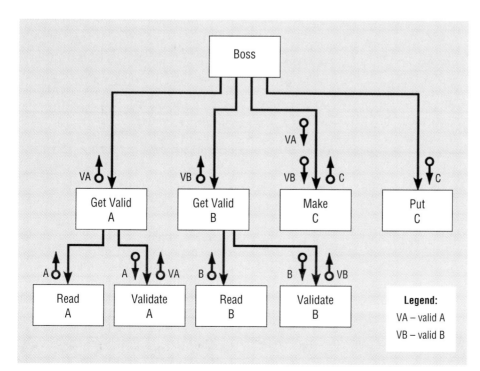

Figure 17-4
How to read a structure chart

(a) Non-overlapping arrows

Figure 17-4 *(continued)*
How to read a structure chart

(b) Overlapping arrows

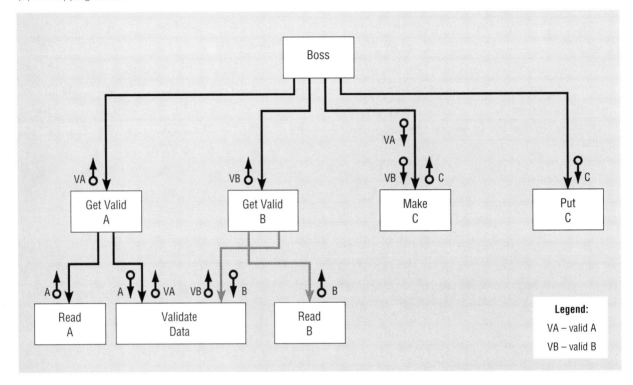

points to Read B, and the right-most arrow points to Validate data. Read B is called first, as the arrow that points to it is the left-most arrow leaving Get Valid B. Validate data is called next. You can also tell the order in which the modules are called by reading the labels—you can't validate B until you have read it into the system.

Once you understand all of the different symbols that make up structure charts, you can begin to use them to represent program designs. When you begin drawing structure charts, you can actually begin with data flow diagrams, converting DFDs to structure charts. *Transform analysis* and *transaction analysis* are two structured methods you can use to convert DFDs to structure charts. Transform analysis and transaction analysis are discussed in more detail later. But before you can use transaction analysis or transform analysis, you have to be able to distinguish between transaction-centered systems and transform-centered systems.

TRANSACTION-CENTERED AND TRANSFORM-CENTERED DESIGNS

Transaction-centered system:
An information system that has as its focus the dispatch of data to their appropriate locations for processing.

Information systems (or parts of an IS) are typically either transaction-centered or transform-centered. In a **transaction-centered system,** the system's primary function is to send data to their proper destinations within a more general system. Data come into the central module of the system, the transaction center, and they are dispatched to their proper locations based on their data type. An example of a transaction-centered system is a bank system designed to process banking transactions (see Figure 17-5). Coming into the transaction center would be all kinds of banking transactions: deposits to checking, deposits to savings, withdrawal from checking, withdrawal from savings, car loan payments, and so on. Within the trans-

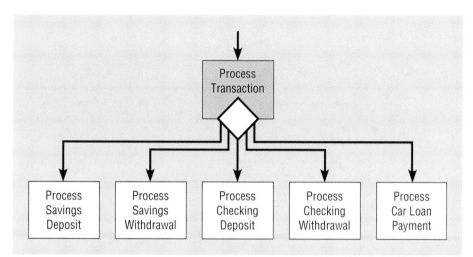

Figure 17-5
A transaction-centered
system design

action center, the data are evaluated and dispatched, depending on their type. Deposits to checking would take one path, withdrawals from checking another path, car loan payments yet another path. Each path leads to modules designed for processing that particular type of transaction. For on-line systems, each path is a menu choice leading to the part of the system handling that type of transaction. Processes along a transaction path often have user interactions.

A **transform-centered system** has as its central function the derivation of new data values from existing data values. An example is converting student grades and class hours to grade point average. Another example is calculating a mortgage payment from an interest rate, compounding period, and principal. In transform-centered systems, the derivation of new data, or the transformation, tends to be the core of the system. As such, transformations are often transparent to users; once the transformation is initiated, there is little if any interaction with users until the transformation is complete. The structure charts in Figure 17-4 are transform-centered designs. The module or modules that represent the core, in this case Make C, are called the **central transform.** The modules that perform the task of bringing data into the system are called **afferent** modules. Afferent modules are arranged in groups referred to as afferent branches. The modules that perform tasks associated with the output of the transformed data are called **efferent** modules, which are arranged in efferent branches. Transaction-centered systems would also have afferent and efferent modules and branches.

As should be obvious from Figures 17-4 and 17-5, the type of design determines how a system's structure chart is constructed. You should be able to look at a data flow diagram and determine if what is represented is transform-centered or transaction-centered. In a transform-centered design, there are one or more data flows that converge on a single area of the system where they are "transformed." The area of transformation may consist of one or several processes. The transformed or derived data then flow away in one or more streams (see Figure 17-6). In a transaction-centered design, the data flows converge on a single process and then flow away along many different paths (see Figure 17-7). The single point of distribution in the system is the transaction center. It is also certainly possible to have a transaction-based system that has a transform center and a transform-based system that includes a transaction center. In the next section, you will learn how to start with a data flow diagram and proceed to a refined structure chart which reflects the system design as contained in the DFD. For transform-centered designs, this process is called **transform analysis.**

Transform-centered system: An information system that has as its focus the derivation of new information from existing data.

Central transform: The area of a transform-centered information system where the most important derivation of new information takes place.

Afferent module: A module of a structure chart related to input to the system.

Efferent module: A module of a structure chart related to output from the system.

Transform analysis: The process of turning data flow diagrams of a transform-centered system into structure charts.

Figure 17-6
A central transform in a data flow diagram

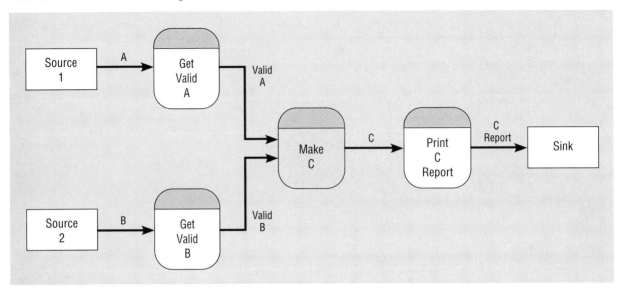

Figure 17-7
A transaction center in a data flow diagram

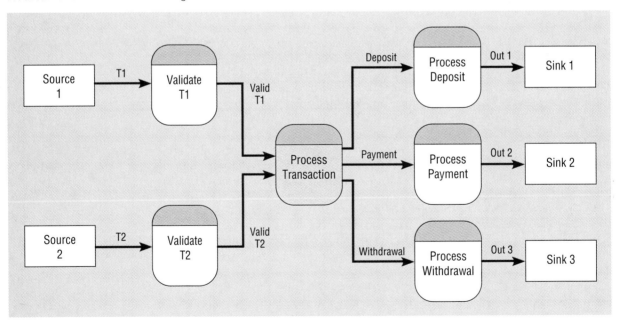

TRANSFORM ANALYSIS

The goal of transform analysis is to convert a transform-centered data flow diagram into a structure chart that remains faithful to the system description contained in the DFD. Transform analysis is not an algorithm: it is a structured process, but you are left with some discretion. One choice you make in the process is which level data flow diagram to convert to a structure chart. In some cases, you will want to start with a level-0 diagram; in other situations, you will want to start at a more detailed level. If you find as you proceed that you need more detail, then switch to a more detailed DFD (Page-Jones, 1980). The transform analysis process has two main parts: top-level design and detailed design.

Top-Level Design

The goal of top-level design is to create a structure chart that captures the essence of the afferent central transform and efferent processes in a DFD. Let's begin with Figure 17-6. We already know that the system here is transform-centered. The first thing we need to do is find the central transform. One way to find the central transform is to trace each afferent flow forward until it disappears and to trace each efferent flow backward until it disappears. The point at which all the flows disappear is the central transform. Sometimes this point consists of more than one process. In such a case, each process should become a separate high-level module in your initial structure chart. You want to identify the central transform because it contains the essential elements of your system where data input streams are transformed into data output streams. In the DFD in Figure 17-6, we see that the data flows Valid A and Valid B converge on the process Make C. This process would seem to be the central transform: the afferent data flows converge on it and the name indicates new data will be derived from existing data. To provide added evidence that Make C is the central transform, we can also look at the process that follows, Print C Report. A process that prints a report would most likely not derive any new data as its primary purpose is to format the information necessary for the report and print it. It seems, then, that Make C is indeed our central transform. Figure 17-8 shows the data flow diagram with the central transform circled.

The next step is to find a boss or coordinating module. Generally, coordinating modules are not evident in a data flow diagram. DFDs map the flow of information through a system, and while such a mapping is very useful in identifying key data and processes in a system, there is no one-to-one correspondence between the processes in a DFD and the modules in a structure chart. Many modules necessary for an information system must be created by the analyst. The coordinating module must almost always be created. For the DFD in Figure 17-8, we must create a coordinating module.

Now that we have identified a central transform and added a coordinating module, it is a straightforward process to identify the afferent and efferent flows. In Figure 17-8, we see there are two afferent flows, from Source 1 through Get Valid A to the central transform, and from Source 2 through Get Valid B to the central transform. The data flowing into the central transform are Valid A and Valid B. There is only one efferent branch, leaving Make C and ending at the Sink. Now that we have identified the central transform and the data flowing in and out of it, we have enough information to draw a top-level structure chart that represents the system's initial design (see Figure 17-9).

Notice that the structure chart in Figure 17-9 has two levels in its hierarchy, one for the coordinating module, and one for the main modules in each of the system's primary branches: the two afferent branches, the central transform, and the efferent branch. We have now defined the basic structure of our information system. Now we want to refine the design. To do that, we need to go back to the original data flow diagram.

Detailed Design

In order to refine the structure chart we created for Figure 17-9, we need to further develop the afferent and efferent branches and, if necessary, the central transform. In this particular example, the central transform needs no further elaboration as it

Figure 17-8
The data flow diagram from Figure 17-6 with the central transform circled

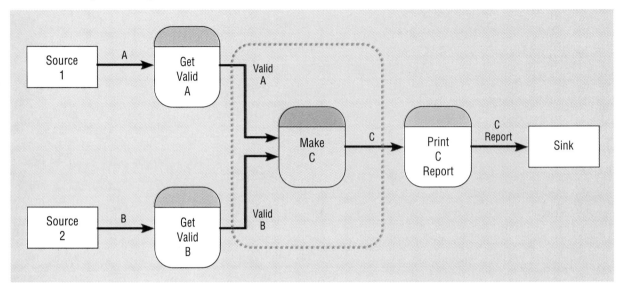

Figure 17-9
The top-level structure chart derived from the data flow diagram in Figure 17-8

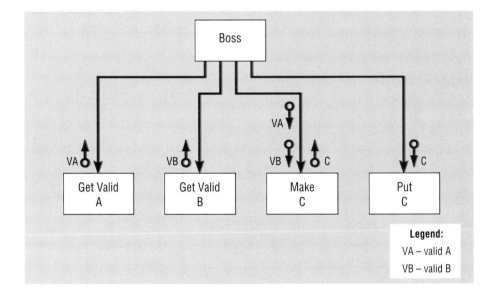

performs a single function, namely making C from Valid A and Valid B. We must, however, account for the data flows that bring A and B into the system and validate them. In both cases, the data are read into the system from outside sources and validated. We can represent these actions with two modules for each afferent flow, one module for reading the data and the other for validating them. Figure 17-10 shows how the fully developed afferent flows would look.

Now we turn our attention to the efferent branch. We see from Figure 17-8 that the derived data C are processed by the module Print C Report. The C Report then flows to a sink. The print process implies two modules, one for formatting the report and one for printing the report. The module for formatting must be inferred, much as the coordinating module was. We know that to turn data into a report, the

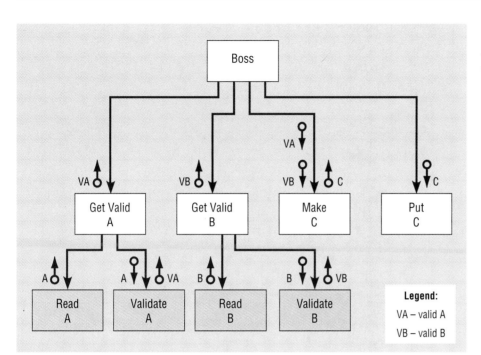

Figure 17-10
First refinement—the structure chart from Figure 17-9 with both afferent branches refined

data must be formatted in a particular way so that they are printed in the right places, and the report will need to have headings and page numbers and other information. On the structure chart, therefore, we will want to include a separate module for formatting the report. Once we add the formatting and printing modules subordinate to the Put C module, we have finished our first refinement (see Figure 17-11).

Any additional refinements to this particular structure chart will be in the form of design improvements. For example, we might want to consider having the modules Get Valid A and Get Valid B share the same validation module, much as we did in Figure 17-4b. Later in this chapter, you will learn about additional design guidelines that will affect how you further refine and improve your structure charts. (For an excellent example of how refining a structure chart improves its design and dramatically changes how it looks, see Chapter 10 in Yourdon & Constantine, 1979.)

The example we have presented is a simple one. System and program design has many more complexities than we have presented here, so we refer you to software engineering and other systems development texts which treat transform analysis and program design in much more detail (see, for example, Adams, Powers, and Owles, 1985). The example we present next, from Pine Valley Furniture, is also relatively simple but it introduces transform analysis in a more typical business context.

An Example: PVF's Purchasing Fulfillment System

In Chapter 2, you were introduced to Pine Valley Furniture's Purchasing Fulfillment System. The data flow diagram for the system is reproduced in Figure 17-12. Using the Purchasing Fulfillment System as an example, we will demonstrate how to use transform analysis to convert the data flow diagram into a structure chart useful for beginning the physical design process.

The first step is to identify the area of central transformation. If we look at the data flow diagram for the Purchasing Fulfillment System, we see that some

Figure 17-11
Complete first refinement of the structure chart from Figure 17-9

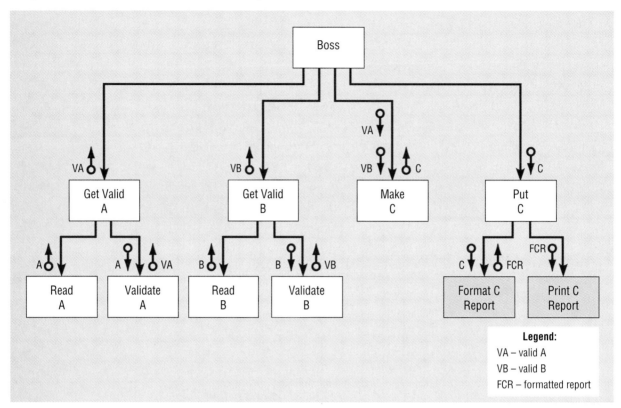

transformation actually occurs in each process. What we want to do, however, is identify the most important or central transformations. Process 1.0 is a transform, where production capacities are turned into material forecasts. However, Process 1.0 is not a central transform as it involves a single input and a single output, and the data generated by Process 1.0 must be transformed again before they are useful to the ordering process. Process 6.0 is not considered the central transform because it primarily uses existing data to prepare orders and derives very little new data. Process 6.0 is an assembly point for much of the data created by the system. Process 4.0 is not the central transform either. In fact, very little transformation takes place in Process 4.0, as the primary function of this process is to select a preferred supplier based on information about suppliers and criteria that guide the selection. Process 5.0 is a transform, as it takes product designs and converts them to bills of materials but, like Process 1.0, Process 5.0 has only one input and one output. That leaves us with Processes 2.0 and 3.0 as potential central transforms. Notice that, in determining the central transform, we have worked our way inward from the periphery of the DFD. We reasoned through each process' function and eliminated those that were either not transforms or were simple one-input, one-output processes.

For this system, by process of elimination, the area of central transformation encompasses the remaining two processes: 2.0: Plan Purchase Agreements, and 3.0: Develop Purchased Goods Specs. Process 2.0 has three inputs, making it central to processing early in the system. Process 3.0 has only one input but it generates two outputs that are necessary for Processes 4.0 and 5.0 to perform their functions. Why did we choose both processes as the central transform? Why wasn't it just Process

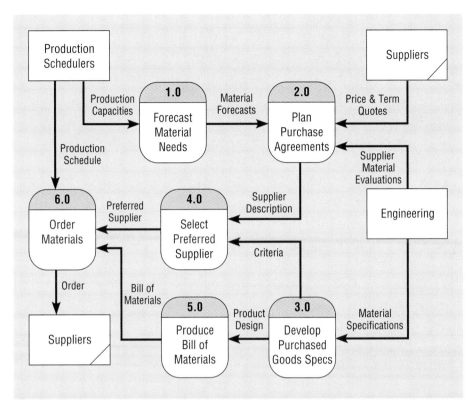

Figure 17-12
PVF Purchasing Fulfillment
System data flow diagram
(from Figure 2-4)

2.0 or Process 3.0? The answer lies in the structure of the DFD. Processes 2.0 and 3.0 are independent of each other—notice there are no direct data flows between them. Both processes serve as starting points for streams of transformed data that eventually become inputs to the assembly point for all of the system's tranformed data, Process 6.0. Figure 17-13 shows the data flow diagram for the Purchasing Fulfillment System with the area of central transform circled.

Now that we have identified the central transforms, we have to create a coordinating module, as no coordinating process is evident on the DFD. In our top-level structure chart, the coordinating module will be at the root of the hierarchy. Each process of the central transform will be a coordinating module at the next level. We have two central transform modules in this example because the processes we identified are independent of each other. Now that we have created a coordinating module and established the central transform in the top-level structure chart (see Figure 17-14), we can focus on the main afferent and efferent flows in the system. The efferent flow is the easiest to identify here as it consists solely of outputting the materials order. To identify the afferent branches, we have to look carefully at the data flowing into Processes 2.0 and 3.0 on the data flow diagram. There are three inflows of data into Process 2.0: Material Forecasts from Process 1.0, Price & Term Quotes from Suppliers, and Supplier Material Evaluations from Engineering. Process 3.0 has a single input: Material Specifications from Engineering. As the inputs to Processes 2.0 and 3.0 are either about the suppliers or the material, we can identify two afferent branches, organized by entity: Get Supplier Data, and Get Material Data. Now that we have identified all of the highest-level modules in the chart, we need to identify the data couples that are input to and output from each of these modules. The now complete top-level structure chart for the Purchasing Fulfillment System is shown in Figure 17-14. Notice that Process 1.0 is not shown on the

Figure 17-13
PVF Purchasing Fulfillment System data flow diagram with central transform circled

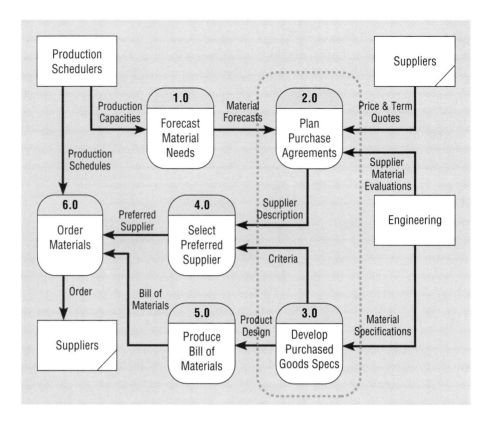

Figure 17-14
Top-level structure chart for the PVF Purchasing Fulfillment System

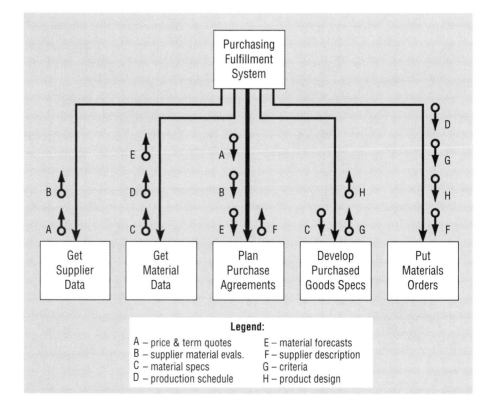

structure chart at this level because it is part of the afferent branch that gets material data. Process 1.0 will be shown in our refinements of the structure chart. Similarly, Processes 4.0 and 5.0 are not shown on the structure chart because they are all downstream from the central transforms and so are actually part of the efferent branch that produces orders. Both of these processes will also be shown in further refinements of the structure chart.

The next step is to refine the top-level design so that we have a detailed structure chart. As the data flow diagram in Figure 17-13 is more complicated than the DFD in Figure 17-8, the refinement process here will be more complicated as well. Refining a structure chart is an iterative process. You continue to improve the structure chart based on how complete, internally consistent, and correct it is. You also rely on the principles of good design we list later in the chapter.

Our starting point is our top-level structure chart. Working left to right, we begin with the afferent branches. We have identified two data couples representing supplier data and three data couples representing material data. We need to account for each data couple in our elaboration of the afferent branches. Figure 17-15 shows one possible refinement of the afferent branches. Notice that we have added two modules that are subordinate to Get Supplier Data. Each module is responsible

Figure 17-15
Refined afferent branches for the PVF Purchasing Fulfillment System structure chart

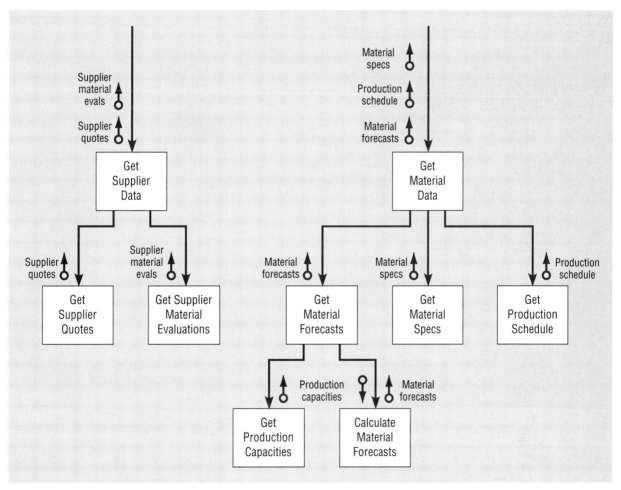

for the task of obtaining a data couple that has to do with suppliers. The same logic is used for further elaboration of the Get Material Data module. Three subordinate modules have been introduced, each given the task of obtaining a particular data couple associated with material. One of these subordinate modules, Get Material Forecasts, has been assigned two subordinate modules of its own, one for obtaining production capacities and one for transforming production capacities into material forecasts. This latter module effectively represents the action of Process 1.0 in the data flow diagram in Figure 17-13.

Our next task is to refine the central transforms. Neither central transform needs any further elaboration. Both are shown in Figure 17-16. Notice that all the data couples the transforms use have already been obtained by the system and that each transform returns the same data couples as are on the corresponding DFD.

We are now ready to focus on the efferent branch Put Materials Orders. As you can see from the DFD in Figure 17-13, Process 6.0 needs three inputs, Bill of Materials, Preferred supplier, and Production schedule. We have already obtained Production schedule in the Get Material Data afferent branch. We need to get the other two inputs to produce and send orders, but we notice from the DFD that the other inputs are generated by Processes 4.0 and 5.0, respectively. We have decided

Figure 17-16
Refined central transforms and efferent branch for the PVF Purchasing Fulfillment System structure chart

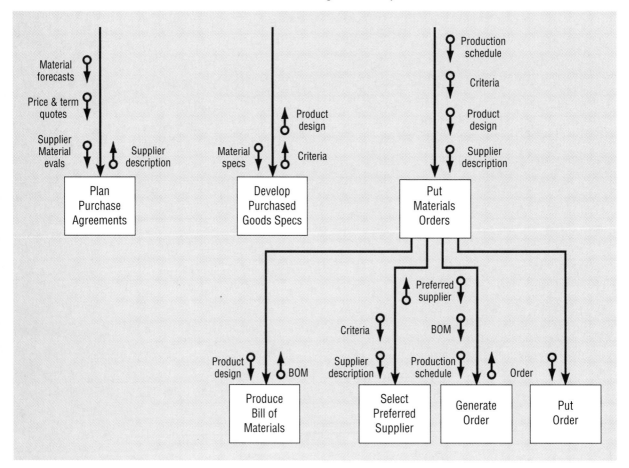

to make the modules that correspond to Processes 4.0 and 5.0 subordinate to the Put Materials Orders module. The inputs needed by Processes 4.0 and 5.0 are passed down to Put Materials Orders, which calls the modules corresponding to each process and passes down the needed data. Process 5.0 becomes the Produce Bill of Materials module and Process 4.0 becomes the Select Preferred Supplier module. Produce Bill of Materials generates the bill of materials (BOM) data couple, and Select Preferred Supplier generates the Preferred Supplier data couple. Now we have all the inputs we need for our Process 6.0, Order Materials. That part of Process 6.0 that generates orders is represented by the Generate Order module subordinate to Put Materials Orders. The finished order is returned to Put Materials Order from where it is passed to the final module, Put Order. Figure 17-16 shows the refinement of the efferent branch as we have described it, and Figure 17-17 shows the overall shape of the entire structure chart we have constructed.

Figure 17-17

Complete refined structure chart for the PVF Purchasing Fulfillment System

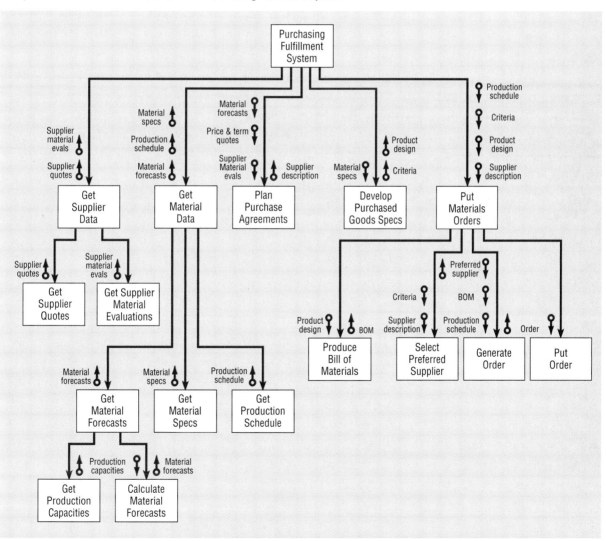

Figure 17-18
Generic transaction-centered
structure chart

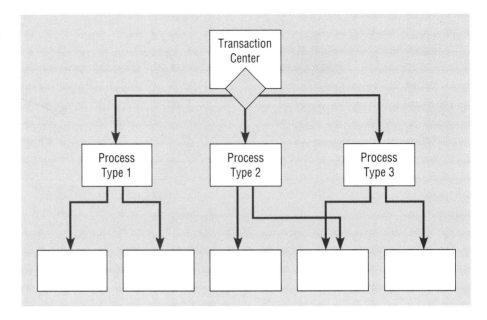

TRANSACTION ANALYSIS

Transaction analysis: The
process of turning data flow
diagrams of a transaction-
centered system into structure
charts.

Transaction analysis is similar to transform analysis in that you begin with a DFD and use it as the basis for a structure chart. The primary difference in the procedures is that, as the name implies, you focus on the transaction center in transaction analysis, instead of focusing on the central transform. The transaction center is shown as a high-level module with a diamond-shaped symbol to indicate choice (see Figure 17-3b). Each transaction type is represented as a module at the next level in the structure chart. The transaction center module does nothing more than recognize the type of transaction coming in and route it to the appropriate transaction module at the next level. The actual process of transactions is done at the third level, where specific processing actions are performed. A generic transaction-centered structure chart is shown in Figure 17-18. In creating a top-level structure chart for a transaction-centered system, you would identify and represent the transaction center and each major transaction type. During refinement, you would identify and represent the modules at the third and subsequent levels.

GUIDELINES OF GOOD DESIGN

As we said earlier in this chapter, further refinement of a structure chart is influenced by a set of guidelines for good design (see Page-Jones, 1980, for a much more extensive discussion of good design guidelines). The goal of good design is to produce a system that is easy to read, easy to code, and easy to maintain. The system designs that result from adherence to the guidelines are considered good because these systems meet the overall goals of systems design:

1. The system should be modular; that is, the system should be organized hierarchically into several smaller units.

2. Each module should control the functions of a reasonable number of subordinate modules at the next hierarchical level.

3. Modules should be relatively independent of each other in that no module's function should rely on the internal workings of other modules; therefore, the amount of communication between modules can be kept to a minimum.

4. Each module should be of reasonable size.

5. To the greatest extent possible, each module should perform one and only one function.

6. The code within each module should be as generic as possible so that each module can be used as often as possible within a system.

Each of these system design goals corresponds to a guideline of good design, as follows:

1. *Factoring:* Factor, or decompose, a system into smaller and smaller parts. (Use the guidelines for Reasonable Size and Cohesion to help you decide when to stop factoring.)

2. *Span of Control:* In general, no superordinate module should control more than seven subordinate modules.

3. *Coupling:* Minimize the extent to which modules are dependent on each other and, by so doing, minimize the amount of communication between modules. In the best case, communication should only be through passing data elements and informational flags.

4. *Reasonable Size:* In general, modules should be limited to between 50 and 100 lines of code. The more reasonable the size, the easier the module is to read.

5. *Cohesion:* To the greatest extent possible, all instructions within a module should pertain to the same function. You should be able to accurately describe the function that a module performs in just a few words, with no "ands" or "ors" in the name.

6. *Shared Use:* Especially at the lower levels of a system design, subordinate modules should be called by multiple superordinate modules.

We have already seen several examples of good design guidelines in this chapter as we have been creating and refining structure charts. Although we did not call it factoring, we were factoring when we took a module in a structure chart and decomposed it into several more modules at the next level. Look, for example, at Figures 17-11 and 17-15 through 17-17 where we were refining afferent and efferent branches in structure charts. In refining our structure charts, we also adhered to the guideline of span of control. The highest number of modules any module controlled in Figure 17-17 was five, and that was the coordinating module. We used the guideline of shared use back in Figure 17-4b when we decided to have a single module, Validate Data, called by two different superordinate modules, Get Valid A and Get Valid B.

Reasonable size is not something we can show in a structure chart. The idea behind reasonable size is to make a module small enough so that how it works can be readily understood by the programmer who has to modify it. Coupling can be shown in a structure chart through the passing of data couples and flags between modules. Coupling is evident throughout the structure charts we have created in this chapter. Whether or not the coupling you have seen is good coupling is a different question. In fact, there are actually five different types of coupling, some good, some not so good (Page-Jones, 1980), and we will provide definitions and examples of all of them in the next section. Similarly, cohesion can be shown in a structure chart through the way modules are named. Note that there are good and not so good forms of cohesion, and to distinguish between good and not so good requires a little more information. We will discuss the seven types of cohesion after we finish with coupling.

FIVE TYPES OF COUPLING

The extent to which modules are interdependent is called *coupling*. Ideally, you want to minimize interdependence among modules. The more dependent one module is on another, the more chance there is for error, as a mistake in one module is no longer confined to that module but also affects another module. Also, programmers who modify information systems have a much more difficult job to do if the code inside one module is dependent on the code in other modules for performing its function. Changing the code in one module may then cause unwanted or unexpected changes in other modules that are dependent on it. Keeping modules as independent as possible makes life easier for everyone involved and makes for a better information system.

Despite its desirability, it is not always possible in practice to minimize coupling. Five different types of coupling have been identified (Page-Jones, 1980), and we present them in order from best to worst.

1. *Data Coupling:* This is the best kind of coupling, as the modules in question are so independent of each other that they can communicate only through passing data elements or informational flags. Neither module has any idea about what goes on inside the other module, which is as it should be. Look at Figure 17-19. The superordinate module has no need to know what goes on inside Calculate New Balance. All the superordinate module needs to know is what data to pass to its subordinate. Similarly, the subordinate module has no need to know what all goes on inside its superordinate module—it only needs to know what data it requires and what data it returns.

2. *Stamp Coupling:* In stamp coupling, data are passed in the form of data structures or entire records. Stamp coupling is not quite as good as data coupling because using data structures instead of data elements tends to make the system more complicated than it needs to be. A change in a data structure will affect all modules that use it, even if the modules do not use the same parts of

Figure 17-19
Example of data coupling

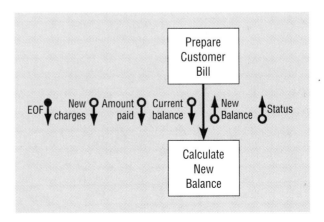

the structure. Stamp coupling makes modules more dependent on each other since, in order to avoid errors, modules must have some knowledge of the internal workings of other modules that use the same data structure. Stamp coupling exposes modules to more data than they need, which can result in problems for the system. Figure 17-20 presents an example of stamp coupling. The example is similar to Figure 17-19, except that we have substituted the data structure CUSTOMER RECORD for the data element Current balance. Clearly, the entire customer record is much more data than the module Calculate New Balance needs. Data structure CUSTOMER RECORD is then used again by the process Format Customer Bill. Once again, CUSTOMER RECORD contains too much data for the process. It would be better for both subordinate modules, and for the system, if only the relevant data elements were passed, instead of the entire record.

3. *Control Coupling:* When one module passes control information to another module, the two are said to be control-coupled. The control information, which may appear as a flag on a structure chart, tells the receiving module what actions it should perform. Obviously, for such control information to be passed, the sending module must know a great deal about the inner workings of the receiving module. In such a case, the modules in question are far too interdependent. An example of control coupling is in Figure 17-21. Notice that the flag's label starts with a verb, indicating some control information is being

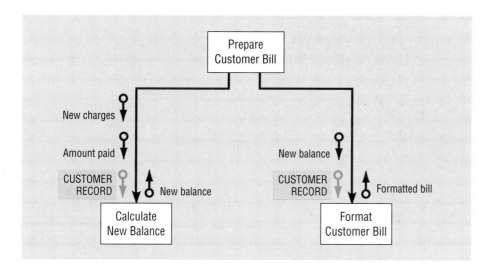

Figure 17-20
Example of stamp coupling

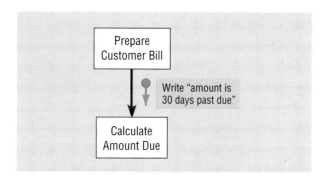

Figure 17-21
Example of control coupling

passed. Control information may also be passed from a subordinate to a superordinate module. Such a practice is referred to as an inversion of authority because the subordinate is telling the boss what to do.

4. *Common Coupling:* When two modules refer to the same global data area, they are common-coupled. Global data areas are possible in several computer languages, such as FORTRAN (common block) and COBOL (the data division is global to any paragraph in the procedure division). The main reason common coupling is undesirable is because of the tremendous opportunity for errors to spread throughout the system. An error in any module using the global data area can show up in any other module using the same area. The level of module interdependence becomes quite high.

5. *Content Coupling:* By far the worst type of coupling, content coupling, involves one module directly referring to the inner workings of another module. For example, one module may alter data in a second module or change a statement coded into another module. In content coupling, modules are tightly intertwined. There is no semblance of independence. Fortunately, most higher-level languages have no provisions for creating content coupling.

SEVEN TYPES OF COHESION

The extent to which all instructions in a module relate to a single function is called *cohesion.* In a truly cohesive module, all of the instructions in the module pertain to performing a single, unified task. You want to maximize cohesion in modules. Maximally cohesive modules also tend to be the most loosely coupled, so achieving high levels of cohesion in your system design also helps you minimize coupling. If a module is designed to perform one and only one function, then it has no need to know about the interior workings of other modules. The cohesive module only needs to take the data it is passed, act on them, and pass its output on to its superordinate module. Seven types of cohesion have been identified (Page-Jones, 1980) and we will discuss them in order of best to worst.

1. *Functional Cohesion:* A functionally cohesive module is the most desirable type in that all instructions contained in the module pertain to a single function or task. Many times, the name of a module will indicate that it is functionally cohesive, e.g., Maintain proper temperature for steel furnace, Calculate interest rate, or Select supplier.

2. *Sequential Cohesion:* If a module is sequentially cohesive, the instructions inside of it are related to each other through the data that are input rather than through the task being performed. In sequential cohesion, the first instruction acts on the data that are passed in, and the second instruction uses the output of the first instruction as its input. The output of the second instruction then becomes input for the third instruction, the output of the third instruction becomes input for the fourth, and so on. Sequence, or the ordering of events, is very important in sequential cohesion. Let's look at a non-systems example.
   CUT DOWN TREE
   CUT TREE INTO PLANKS
   PLANE PLANKS

SAND PLANKS
SAW PLANKS TO SPECIFICATIONS
PUT PLANKS TOGETHER TO MAKE DOOR

A living tree is the initial input for this set of instructions. The cut-down tree is the input to the second instruction where the tree is made into planks. The raw planks are the input for the next instruction where the planks are planed. Planed planks are the input for the next instruction where the planks are sanded. This pattern of processing continues until we end up with a wooden door. Information system modules follow a similar pattern of logic.

3. *Communicational Cohesion:* In a module that has communicational cohesion, the activities are also related to each other by the data that the module uses, but here sequence is not important. Each instruction in a communicational module acts on the same input data or is concerned with the same output data. An example of communicational cohesion is in Figure 17-22a. The module Find Part Details is so vague that it tips you off to the module as not being functionally cohesive. The module is designed to use part # as input to find a part's name, cost, and supplier. Sequence is not important as it does not matter whether name, cost, or supplier are found first or last. Page-Jones (1980) states that a communicationally cohesive module is easier to understand and maintain if it is split into two functionally cohesive modules, as is the case in Figure 17-22b.

4. *Procedural Cohesion:* As we progress further in the list of the different types of cohesion, we get further away from functional cohesion. Modules exhibiting these kinds of cohesion become more and more difficult to maintain. In

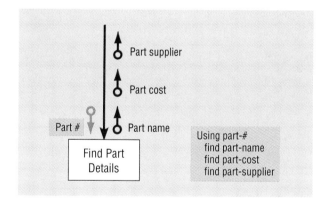

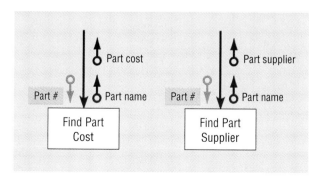

Figure 17-22

Examples of communicational cohesion

(a) Example of a communicational cohesion module

(b) Example of a communicationally cohesive module split into two functionally cohesive modules

procedural cohesion, the instructions in a module are related to each other through flow of control. Instructions have to be carried out in a specific order, so sequence is important. The instructions in a procedurally cohesive module are generally more related to activities in other modules than they are to each other. A non-systems example of procedural cohesion would be the following set of instructions:

PICK UP THE NEWSPAPER
CHECK MAILBOX FOR YESTERDAY'S MAIL
PUT NEWSPAPER AND MAIL IN BOX
WATER PLANTS
CHECK ON WATER AND FOOD FOR DOGS
MAKE SURE DOORS ARE LOCKED BEFORE LEAVING

You might leave a set of instructions like this for someone who is watching your house while you are away on vacation. Although it might seem that the order of some of the instructions can be interchanged, there is a logic to the order as presented. The instructions are written in the order they would be followed if the person watching your house started at the street, entered the house, checked on the dogs in the backyard, and then locked up before leaving.

5. *Temporal Cohesion:* In a temporally cohesive module, the instructions are also related to each other through the flow of control, but sequence does not matter. The only reason the instructions are in the module at all is that they occur at about the same point in time, hence the name temporal. The classic example of a temporally cohesive module is one that contains instructions for initializing a whole host of variables, counters, switches, and so on, throughout the system. Such a module is related to several other modules, making it very difficult to change the timing of a particular initialization step without affecting other modules that rely on the other instructions in the initialization module.

6. *Logical Cohesion:* In a logically cohesive module, the instructions are hardly related to each other at all. A logically cohesive module consists of several sets of instructions, but the particular set that is executed is determined from outside the module. Typically, a flag that specifies what is to be done is passed in from the outside. A non-systems example is the following:

EAT AT A RESTAURANT
EAT AT YOUR DESK
EAT AT HOME
SKIP LUNCH

According to Page-Jones, people write logically cohesive modules to overlap parts of functions that have the same lines of code or the same buffers. Maintenance is very difficult.

7. *Coincidental Cohesion:* Coincidental cohesion is the worst type because the instructions have no relationship to each other at all. As with logical cohesion, a flag from outside is generally sent in to tell the coincidentally cohesive module what to do. Such modules are the result of haphazard factoring, attempts to save time in design or programming, or bad fixes to existing code. Fortunately, modules that suffer from coincidental cohesion are rare.

Page-Jones (1980) has developed a useful diagram for keeping the different types of cohesion straight (Figure 17-23). Using his decision tree, first ask whether the module is performing a single function. If so, the module is functionally cohe-

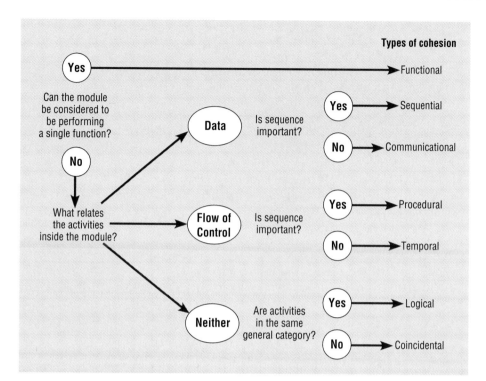

Types of cohesion

Yes ——————————————→ Functional

Can the module
be considered to
be performing
a single function?

Data — Is sequence important? — Yes → Sequential / No → Communicational

No

What relates
the activities
inside the module? — Flow of Control — Is sequence important? — Yes → Procedural / No → Temporal

Neither — Are activities in the same general category? — Yes → Logical / No → Coincidental

Figure 17-23
A decision tree that helps
distinguish among the
seven types of cohesion
(Adapted from Page-Jones,
1980)

sive. If not, ask what connects the activities inside the module. If the connection is data, ask whether sequence is important. If sequence is important, you have sequential cohesion; if not, you have communicational. If it is the flow of control that connects the activities in a module, ask whether sequence is important. If it is, you have procedural cohesion; if not, you have temporal. Finally, if neither data nor control relates the activities in a module to each other, ask whether the activities are in the same general category. If they are, you have logical cohesion; if not, you have coincidental cohesion, the worst type, and the farthest from functional cohesion you can get.

We should make two more points before leaving the topic of cohesion. The first point relates to naming conventions. If a module is truly functionally cohesive, its function can be summed up in a name made up of a single imperative verb and a concise direct object, such as READ STUDENT MASTER FILE, CALCULATE GPA, and PRINT DEAN'S LIST REPORT. Modules with sequential or communicational cohesion will usually include "and" in their names, indicating that they perform more than one function. As you go further down the list of types of cohesion, names that capture what modules do become more general and less understandable. The second point we should make is that the functions modules perform differ according to the level you are considering in your structure chart. As we said previously, coordinating modules, at the root and at the first level of modules in the structure chart, are primarily concerned with decision functions, that is, deciding which modules to call and which parameters to pass and when. The modules at the very bottom of the chart have no decision responsibilities. They simply take inputs, perform their functions, and return output to the modules that called them. Modules in the middle of the chart, then, have a mixture of decision responsibility and task performance responsibility. How you show the details of what a module does is the subject of the next section.

SPECIFYING THE CONTENTS OF MODULES

So far, we have concentrated on depicting a system's hierarchical design in the form of a structure chart. We have focused on transforming a DFD into a structure chart and on refining the initial design using factoring and five other guidelines for good design. If you were a practicing systems analyst, you would have a pretty good overall design at this point. You already know which modules to include in your system and the specifics of how they communicate with each other through the passing of data and informational flags. You have a pretty good idea of what goes on inside each module, but now you must turn your attention to describing the specific instructions contained in each module. We will look at two ways to describe the contents of each module: pseudocode and Nassi-Shneiderman diagrams. Both methods rely on representing the three basic parts of structured programming: sequence, conditional statements, and iterative statements (Bohm and Jacopini, 1966). Pseudocode and Nassi-Shneiderman diagrams both perform the same task, which is representing the contents of system modules. A variation of pseudocode, developed for showing the logic of file and database accesses, is action diagramming. Action diagramming is described in some database management texts and is used with some CASE tools.

Pseudocode

Pseudocode: A method for representing the instructions in a module with language very similar to computer programming code.

In Chapter 10, you learned about representing the logic of a data flow diagram process with Structured English. In design, a more exact representation of processing logic is required, as programmers will use these representations as specifications for coding. Representing program code as text is the purpose of **pseudocode.** Think of pseudocode as "almost code," a personalized way to describe a computer program and all of its steps using a language that is not quite a programming language but not quite English either. Pseudocode serves two functions:

1. It helps an analyst think in a structured way about the task a module is designed to perform.

2. It acts as a communication tool between the analyst, who understands what the module needs to do, and the programmer, who must write the code to make the module perform appropriately.

Every analyst develops his or her own version of pseudocode, so there is no one right way to write it. However, some organizations have standardized on a version of pseudocode and require its use. Whether its origin is personal or standard, all pseudocode must have a way to represent at least sequence, conditional statements, and iteration. The actual form pseudocode takes will depend on the programming languages the analyst knows best. For example, a COBOL-based representation of sequence might be as follows:

```
OPEN INPUT CUSTOMER-MASTER-FILE
OUTPUT CUSTOMER-BILLING-FILE
PERFORM PROCEDURE-INITIALIZE-FIELDS
READ CUSTOMER-MASTER-FILE
AT END MOVE "Y" TO WS-END-OF-FILE-SWITCH
```

A conditional statement based on Pascal might be represented in this way:

```
if A>B
then
     writeln('A is greater')
else writeln('B is greater')
```

Iteration can be represented as either do-while or do-until loops, as illustrated with the following, again based on Pascal:

```
begin
    read(k);
    while k > 0
        do k:= k + 1;
    writeln(k)
end.
```

```
begin
    read(k);
    repeat
        k := k + 1
    until k <= 0;
    writeln(k)
end.
```

The format and phrasing of pseudocode is not as important as the information it conveys. The point in using pseudocode in this phase of the life cycle is to provide the programmers who will write the code with a specific description of the contents of each module. Figure 17-24 shows what the pseudocode description of the module Calculate New Balance from Figure 17-19 could be like.

Notice in Figure 17-24 that the name of the module is specified, as are the data passed in and out of the module. Notice also that indentation is used to call attention to the nested conditional statements and the contents of the do-until loop. Finally, notice that the data couple Status is assigned an integer value. The meanings of the integers are explained in a note at the end of the pseudocode.

Nassi-Shneiderman Diagrams

Dissatisfied with logic flow charting techniques and the problems of using them in structured programming, Nassi and Shneiderman developed their own flow

```
Module name: Calculate New Balance
Receives: Current-balance, New-charges, Amount-paid, EOF
Returns: New-balance, Status

Set New-balance, Status to 0
Read Current-balance, New-charges, Amount-paid, EOF
Repeat
   New-balance = Current-balance + New-charges - Amount-paid
   Beginif
      If New-balance < 0
      Then Status = 1
      Else
         Beginif
         If New-balance = 0
         Then Status = 2
         Else Status = 3
         Endif
   Endif
   Return New-balance, Status
Until EOF = YES
```

Figure 17-24

Pseudocode description of Calculate New Balance module

(*Note:* The data couple Status has three values, 1, 2, and 3. The value 1 corresponds to "credit," 2 to "nothing due," 3 to "amount due.")

charting techniques for structured programming (1973). Rather than depend on multiple flow charting symbols to represent the basic constructs of structured programming, Nassi and Shneiderman developed their own symbols, shown in Figure 17-25. The diagrams that result from using these symbols are called Nassi-Shneiderman charts. As the figure shows, there are specific symbols for conditional statements, do-while loops, do-until loops, and for sequential statements. There is also a special symbol for case statements.

There are a few rules for using Nassi-Shneiderman charts (Schneyer, 1984):

1. The charts are always rectangular in form.

2. Nassi-Shneiderman charts always start at the top.

3. Movement through the chart is always top to bottom, except at the end of a loop. If the end of a loop has been reached but the exit condition has not been satisfied, movement begins again at the beginning of the same loop. Loops can only be exited through the horizontal bar of the L-shape.

4. Never cross vertical lines.

5. A rectangle can only be exited in one direction at any given time.

6. A rectangle that contains no other lines is either a single instruction (sequence) or another Nassi-Schneideman chart.

Using these rules, follow the flow of instructions through Figure 17-26, which is the pseudocode from Figure 17-24 converted to a Nassi-Shneiderman chart. Once you have learned the different symbols, one glance at a Nassi-Shneiderman chart should give you a more immediate idea of what the diagrammed program does than a similar glance at a pseudocode listing. For example, looking at Figure 17-26, you can tell immediately that there is a do-until loop and two conditional statements in the module.

Figure 17-25

Basic symbols in Nassi-Shneiderman charts

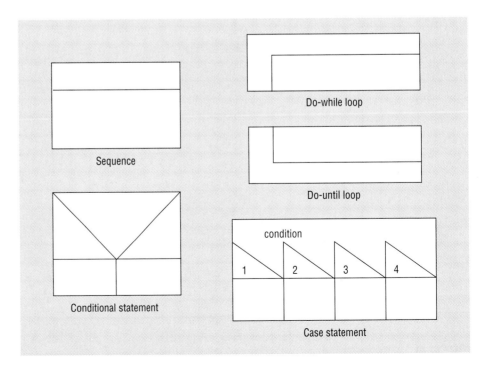

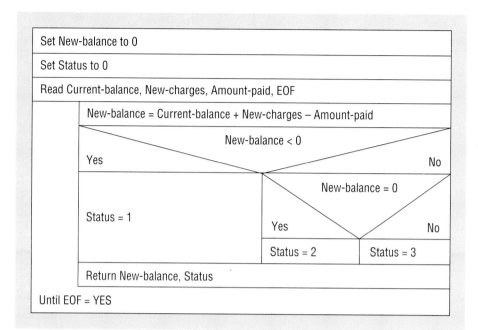

Figure 17-26
The contents of the module
Calculate New Balance in the
form of a Nassi-Shneiderman
chart

CASE TOOLS IN DESIGNING PROGRAMS

As shown in this chapter, the design of programs through structure charts, pseudo-code, and Nassi-Shneiderman diagrams is an analyst-led process of tranforming DFDs into program designs. During this manual process, you can use CASE tools for their diagramming capabilities, like an intelligent drawing package, and for maintaining the program designs as part of the project repository. A CASE drawing tool will contain the special symbols of structure charts, for example, and will know to place data couples and control flags along the lines connecting modules. As with other diagrams, each symbol on a structure chart (module, data couple, and control flag) is an object in the repository, and the relationships between objects shown on the diagrams are also maintained in words within the repository, for reporting purposes. Analysis functions of the CASE tool can check that diagrams are complete, consistent, and that they follow rules for drawing that type of diagram.

Figure 17-27 shows an excerpt of a sample system design adapted from an example repository included with the educational version of the Visible Analyst Workbench (VAW) CASE tool. These diagrams represent part of a simple library information system. Figure 17-27a contains the level-0 DFD for the entire system and Figure 17-27b shows a simple level-1 diagram for process 2, Checkout and Return Books. The five processes of this DFD were transformed into the structure chart in Figure 17-27c.

The Main module in Figure 17-27c represents the initial, sign-on module of the system. This structure chart shows that a sign-on screen would be displayed and the user would enter a UserID to be validated as a legitimate user of the system. The User Menu Lib1.8 connector points to another structure chart not shown. Finally, a Library Main Menu transaction module routes the user to the function he or she wants to perform. Each function is shown on another structure chart, with the connector pointing to that chart. Two of these other structure charts correspond to processes 2.1 (Checkout Books 1.6) and 2.2 (Return Books Lib1.7) from Figure 17-27b. The structure chart for Checkout Books 1.6 appears in Figure 17-27d.

Figure 17-27

Example CASE tool usage for program structure design

(a) Level-0 DFD for library system

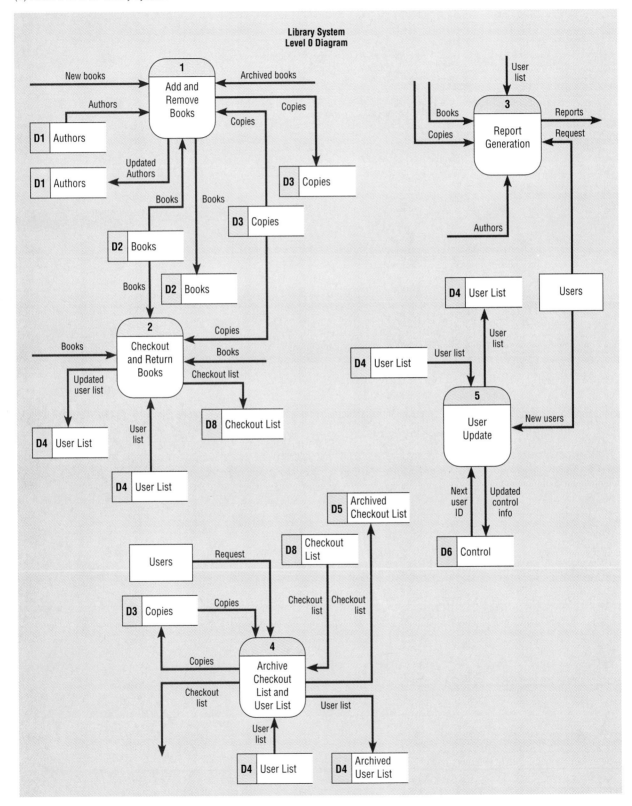

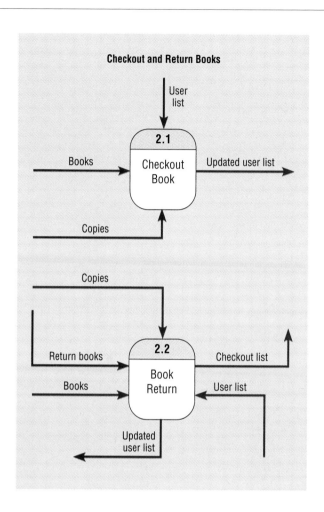

The Checkout_Books module coordinates the display of a pertinent screen and then controls calling other modules for the various functions necessary to check out a book.

Figures 17-27e, 17-27f, and 17-27g show the repository entries for the Main, Library Main Menu, and Checkout_Books modules, respectively. In these figures you can see how pseudocode is used to describe the logic within that module. The Location information within each module entry lists what other modules that module is called by or calls and the data couples and control flags passed between pairs of modules. The Date Last Altered at the end of the entry indicates which version of this entry we are viewing. If you are using VAW, DFDs and structure charts are separate, unrelated structures, so there is no information in the repository about the relationship of processes and data flows on DFDs to modules and data couples in structure charts.

SUMMARY

In physical information systems design, designers must be concerned with the system's input, output, databases, and programs' physical specifications. Designers must consider what each component will look like and how it will be physically

Figure 17-27 *(continued)*
Example CASE tool usage for program structure design

(c) Structure chart for level-0 DFD

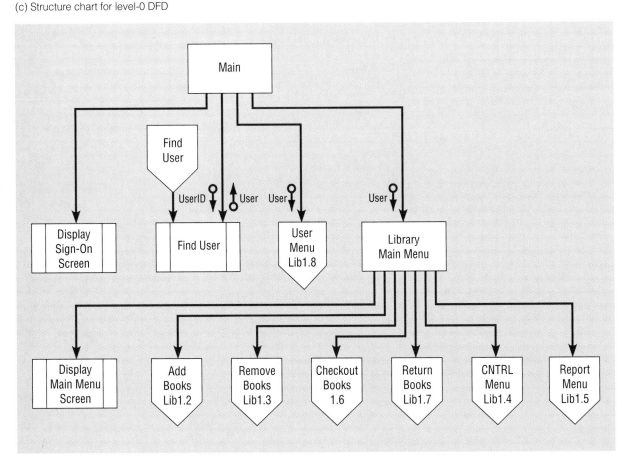

implemented in the new system. In this chapter, we have focused on physical program design. You have learned how to draw structure charts, and you have learned how structure charts are a useful diagramming tool for representing a system's program components and their inter-relations.

Structure charts can represent both transform-centered and transaction-centered system designs. In a transform-centered design, the primary activity of the system is to derive new data from existing data or to transform existing data into new data. In a transaction-centered design, the primary activity is to dispatch transactions to their proper destinations, where they are further processed. Transform-centered designs can be derived from data flow diagrams using a process called transform analysis.

The first step in transform analysis is to find the area of central transform, where the most important transformations of data take place. Once the central transforms have been identified, the second step is to either find or infer a coordinating module for the top of the structure chart. The third step is to identify the primary afferent (input) and efferent (output) data flows. The fourth step in transform analysis is to draw a top-level structure chart, using the information you uncovered in the first three steps. The top-level structure chart has only two hierarchical levels, one for the coordinating module and one for the primary modules for the afferent and efferent branches and for the central transforms. The fifth step is to further re-

(d) Structure chart for Checkout Books section of system

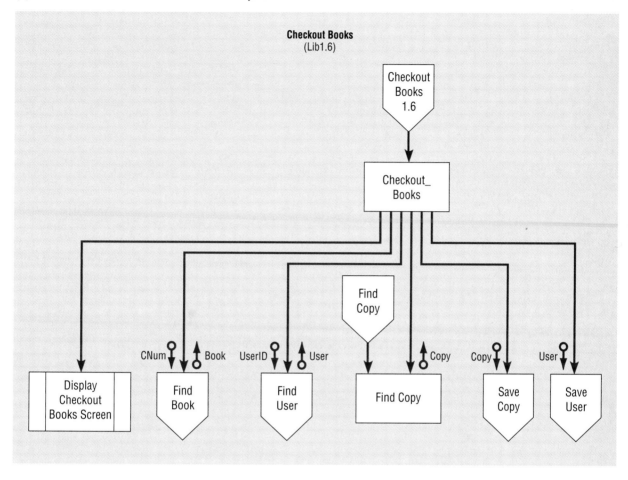

fine the structure chart until the origin of data entering the system, the functions of the central transforms, and the final disposition of the system's outputs have all been satisfactorily accounted for. Transform analysis is an iterative process, so the fifth step may well be the most involved and take the most effort.

The factoring of modules that you undertake to refine a structure chart is the first of six guidelines you can use to help ensure good systems designs. Other guidelines are to limit the number of subordinate modules to no more than seven per boss module; to minimize the interdependence of modules (coupling); to limit the number of lines of code in a module to a reasonable number; to maximize the extent to which all instructions in a module are related to a single function or task (cohesion); and to design some modules, especially at lower levels, so that they may be called by more than one superordinate module. There are five types of coupling and seven types of cohesion. Analysts should be able to identify the various types of coupling and cohesion so that they can recognize the less desirable forms and understand what it takes to convert the less desirable to the preferred forms.

Once the analyst has determined the overall design of an information system, he or she must describe the contents of the individual modules that make up the system. Two methods for describing module contents are pseudocode, a textual way of representing program instructions that is similar to a programming language, and Nassi-Shneiderman charts, a form of program logic flow charting for

Date: 10/3/94	*Project:* LIB	*Page:* 1
Time: 10:56 PM		

Single Entry Listing
Structure Charts

Main   Module

Description:

    Allow a user to sign on to the Library System

Module Description:

    (Sign_On module)
        Display_Screen(SignOn)
        While(1 not = 0) & (not Exit)
            Accept_ UserID(User.User_ID)
            Find_User(User.User_ID)
            Uf (not User_Found)
                Display_Error_Message("Invalid User ID")
            Else
                Accept_Password(Password)
                If(User.Password not = Password)
                    Display_Error_Message("Incorrect password")
                Else
                    If(User.Staff = "Y")
                        Call "Main_Menu"
                    Else
                        Call "User_Menu"
                    End_If
                End_If
            End_If
        End_While
        Return

Location:

    LIB.0

            Calls: *** Nothing or Unnamed Module***
                Passed Couples:
                    User (Data ITR)

            Calls: Library Main Menu (Module)
                Passed Couples:
                    User (Data ITR)

            Calls: Display Sign-On Screen (Library Module)

            Calls: Find User (Library Module)
                Passed Couples:
                    UserID (Data Couple)
                Returned Couples:
                    User (Data ITR)

Date Last Altered: 1/15/93   *Date Created:* 1/15/93

(f) Repository entry for Library Main Menu module

Single Entry Listing
Structure Charts

Library Main Menu   Module

   Description:

        The Library (Staff) main menu.

   Module Description:

        (Staff_Menu module)
             Display_Screen(mmenu)
             While (not Exit)
                 Get_ Menu_Option(Menu_Option)
                 If (not Exit)
                     Case Menu_Option = "1"
                         Call "Add_Books"
                             Menu_Option = "2"
                         Call "Remove_Books"
                             Menu_Option = "3"
                         Call "Checkout_Books"
                             Menu_Option = "4"
                         Call "Return_Books"
                             Menu_Option = "5"
                         Call "Control_Menu"
                             Menu_Option = "6"
                         Call "Report_Menu"
                     End_Case
                 End_If
             End_While
             Return

   Location:

        LIB.0

                 Called by: Main (Module)
                     Passed Couples:
                         User (Data ITR)

                 Calls: *** Nothing or Unnamed Module ***

                 Calls: *** Nothing or Unnamed Module ***

                 Calls: *** Nothing or Unnamed Module ***

                 Calls: *** Nothing or Unnamed Module ***

                 Calls: Display Main Menu Screen (Library Module)

                 Calls: *** Nothing or Unnamed Module ***

                 Calls: *** Nothing or Unnamed Module ***

Date Last Altered: 1/15/93   Date Created: 1/15/93

Figure 17-27 *(continued)*
Example CASE tool usage for program structure design

(g) Repository entry for Checkout_Books module

Date: 10/3/94	*Project:* LIB	*Page:* 1
Time: 10:55 PM		

Single Entry Listing
Structure Charts

  Checkout_Books   Module

    Description:

      Checkout a book for a user.

    Module Description:

```
        (Checkout_Book module)
            Display_Screen(CheckOutBooks)
        Get_UserID:
            Accept(User.UserID)   {Required}
            Find_User
            If (not Valid_User)
                Display_Error_Message("User ID not on file")
            Else
                Display(User.NumberCheckedOut)
                Display(User.BorrowLimit)
                If (User.NumberCheckedOut < User.BorrowLimit)
                    Accept(Book.CallNumber) {Required}
                    Find_Book Using CallNumber
                    If (not Already_in_Library)
                        Display_Error_Message("Book is not on file")
                    Else
                        Display(Book.Title)
                        Accept(Copy.CopyNumber) {Required}
                        Find Copy
                        Accept(Copy.DateOut) {Required}
                        Calc (Copy.DateOut + 14) to Copy.DateDueBack
                        Display(Copy,DateDueBack)
                        Accept(Continue) {Required, "Y/N"}
                        If (Continue not = "Y")
                            Goto Get_UserID
                        Else
                            Save_Copy
                            Calc (User.NumberCheckedOut + 1) to User.NumberChe
                            Save_User
                        End_If
                    End_If
                End_If
            End_If
            Return
```

(continues)

(g) Repository entry for Checkout_Books module *(continued)*

Date: 10/3/94	*Project:* LIB	*Page:* 2
Time: 10:55 PM		

Single Entry Listing
Structure Charts

Location: Module

    LIB1.6

        Calls: Display Checkout Books Screen (Library Module)
            Passed Couples:
                CNum (Data Couple)
            Returned Couples:
                Book (Data ITR)

        Calls: *** Nothing or Unnamed Module ***
            Passed Couples:
                User ID (Data Couple)
            Returned Couples:
                User (Data ITR)

        Calls   *** Nothing or Unnamed Module ***
            Passed Couples:
                Copy (Data Couple)

        Calls: *** Nothing or Unnamed Module ***
            Passed Couples:
                User (Data ITR)

        Called by: *** Nothing or Unnamed Module ***

        Calls: Find Copy (Module)
            Returned Couples:
                Copy (Data Couple)

Date Last Altered: 1/15/93   *Date Created:* 1/15/93

structured programming. Analysts must also provide specifications for the data input by each module and their sources, as well as specifications for the output the system generates.

CASE tools help analysts in program structuring. The drawing component of a CASE tool will maintain the structure chart diagram and analyze it for completeness, consistency, and structural errors. Each object on a structure chart is linked to a repository entry. An entry for a module can contain pseudocode to explain the logic which must be programmed to implement that module.

C H A P T E R   R E V I E W

K E Y   T E R M S

Afferent module	Module	Transaction-centered
Central transform	Pseudocode	system
Data couple	Structure chart	Transform analysis
Efferent module	Transaction analysis	Transform-centered
Flag		system

R E V I E W   Q U E S T I O N S

1. Define each of the following terms:
   a. structure chart
   b. module
   c. coordinating module
   d. central transform
   e. transform analysis
   f. transaction-centered system

2. What is a structure chart's role in physical information system design?

3. What is the difference between a data couple and a flag?

4. What does a diamond mean in a structure chart? What does a hat in a structure chart mean? What is a pre-defined module? How do you represent repetition in a structure chart?

5. What is the difference between a transform-centered system and a transaction-centered system? Give an example of each.

6. What is transform analysis? What is its goal?

7. Explain how to use transform analysis to create a top-level structure chart from a data flow diagram.

8. Explain how to refine a top-level structure chart.

9. Explain the six guidelines to good design.

10. Distinguish among the five types of coupling. How are five categories of coupling useful to a systems analyst?

11. Distinguish among the seven types of cohesion. How are seven categories of cohesion useful to a systems analyst?

12. What are the key differences between pseudocode and Nassi-Shneiderman charts?

PROBLEMS   AND   EXERCISES

1. Match the following terms to the appropriate definitions.

   _____ afferent

   _____ efferent

   _____ flag

   _____ coupling

   _____ cohesion

   _____ pseudocode

   a. diagrammatic representation of a message passed between two modules
   b. extent to which modules are interdependent
   c. output
   d. input
   e. representing the instructions in a module with language similar to code
   f. extent to which all instructions in a module relate to a single function

2. Using transform analysis, convert the level-0 data flow diagram for Hoosier Burger's food ordering system (see Figure 9-5) to a refined structure chart.

3. Using transform analysis, convert the level-0 data flow diagram for Hoosier Burger's inventory control system (see Figure 9-16) to a refined structure chart.

4. Study Figure 17-17, the complete refined structure chart for Pine Valley Furniture's Purchasing Fulfillment System. How many different types of coupling can you identify in the structure chart? If you find evidence of more types of coupling than simple data coupling, how can you convert all types of coupling to data coupling? How many different types of cohesion are represented by the structure chart's functions? If you find evidence of more types than simple functional cohesion, what can you suggest to convert all types of cohesion to functional cohesion?

5. Represent the processing logic of each module in Figure 17-15 (or 17-16) using pseudocode.

6. Represent the processing logic of each module in Figure 17-15 (or 17-16) using Nassi-Shneiderman charts.

7. Compare the pseudocode and Nassi-Shneiderman charts you created in Questions 5 and 6 (for either Figure 17-15 or Figure 17-16). Which type of representation is the clearest? Which works best for you? Explain why.

8. Starting with Figure 17-12 (Pine Valley Furniture's Purchasing Fulfillment System), use a CASE tool to convert the data flow diagram to a complete, refined structure chart.

FIELD   EXERCISES

1. Interview a systems analyst about the physical design process where the analyst works. What specific design methodology does the analyst's organization follow? What are the roles of structure charts and transform analysis in the methodology? Does the methodology prescribe pseudocode or Nassi-Shneiderman charts?

2. Look for an Internet USENET discussion group on systems design. Follow the discussion for several weeks and document what you have learned in a journal to be turned in to your instructor.

3. Research the principles of good design for information systems and for some other artifact that people design and build, such as bridges, computer peripherals,

furniture, or cars. Prepare a report on the principles of good design for information systems and for whatever else you have chosen to study. How are the principles different? How are they similar?

4. Research the physical information design process in Japan and in Europe. How do these processes compare to what you have learned here about North American practice? How are the processes similar? How are they different?

REFERENCES

Adams, D. R., M. J. Powers, and V. A. Owles. 1985. "Computer Information Systems Development: Design and Implementation." Cincinnati, OH: South-Western Publishing Co.

Bohm, C. and I. Jacopini. 1966. "Flow Diagrams, Turing Machines, and Languages with Only Two Formation Rules." *Communications of the ACM* 9 (May): 366–71.

Nassi, I. and B. Shneiderman. 1973. "Flowchart Techniques for Structured Programming." *ACM SIGPLAN Notices* 8(8) (Aug.): 12–26.

Page-Jones, M. 1980. *The Practical Guide to Structured Systems Design.* New York, NY: Yourdon Press.

Schneyer, R. 1984. *Modern Structured Programming: Program Logic, Style, and Testing.* Santa Cruz, CA: Mitchell Publishing Inc.

Yourdon, E. and L. L. Constantine. 1979. *Structured Design.* Englewood Cliffs, NJ: Prentice-Hall.

Designing the Internals: Program and Process Design

INTRODUCTION

The Customer Activity Tracking System (CATS) team is progressing on schedule. Considerable documentation has been generated about what CATS should do, but not much has been specified on how CATS will be structured. Since BEC will purchase a large portion of CATS, the purchased functions only need to be checked against BEC's unique requirements to determine whether the structure of the package is sufficient. The in-store component of CATS, however, is not in the package and must be added on. Thus, the in-store processes, identified in analysis, must be transformed into a software design so the BEC programmers can build this portion. This case includes a discussion between Jordan Pippen, the CATS project leader, and Frank Napier, a systems analyst on the CATS team, as they review structure charts generated from the ObjectStar CASE tool for the in-store portion of CATS.

TRANSFORM ANALYSIS FOR IN-STORE CATS

"Everything is coming together," Jordan Pippen told Frank Napier. "I am really pleased at the progress we have been making. Using the TwoCats system as the basis for our corporate Customer Activity Tracking System (CATS) was a good choice."

"It doesn't help that much with our in-store component, though," Frank replied. "We still have to pretty much grow our own on that one."

"So," Jordan answered. "How much progress have you made in coming up with a program design for the in-store component?"

"I've done a rough draft on the structure chart. Want to see it? I thought you might ask about it, so I brought my work with me."

"Yeah, now would be a good time—I have a couple of minutes before my next meeting."

"OK, good," Frank said, pulling out some printouts. "Here's what I've been working on. I started with the data flow diagram we drew for the in-store component of CATS (Figure 1). You can see that there are four processes, one for updating store inventory, one for updating product reservations, one for recording customer purchases, and one for generating the CATS success report.

"I had to figure out which one of these processes was the central transform. That wasn't a straightforward decision. But after analyzing the data flows coming into the system, it seemed to me that everything pointed to the fourth process, Produce Success Evaluation Report. Start at Process 4.0 and trace back those two data flows that enter it, New Reservations, and Store Transactions. All of the activity in the system, except for the activity surrounding Update Store Inventory, is focused on populating and updating the two data stores that are the sources of New Reservations and Store Transactions."

"Hmmmm, I'm not so sure," Jordan replied. "Are you sure this process is the central transform? All Process 4.0 does is produce a report. There's not much transformation going on there. And look at these other processes. They aren't really transforms either. Processes 1.0 and 2.0 update data stores; Process 3.0 simply records store transactions. In fact, it doesn't even look like Process 1.0 is part of this system at all. The in-store CATS system never uses any of the data in the store inventory records. Isn't there already a system in place that updates the inventory for each store as shipments arrive? And now that I think about it, Process 3.0 shouldn't be in our system either. The point-of-sale system we run at all of our stores automatically captures each transaction."

Looking slightly discouraged, Frank responded, "Yes, I think you're right. I guess I didn't look at the DFD that carefully, I was so focused on transform analysis. Processes 1.0 and 3.0 don't really belong in our system. We included Process 1.0 in the in-store CATS system because the process will have to take

Figure 1

DFD of CATS in-store
component

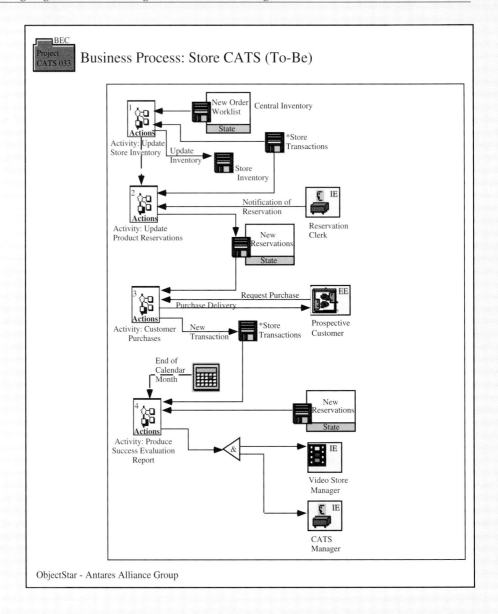

reservation purchases into consideration for updating
inventory, but that inventory update system is already
in place. I need to find out who is responsible for maintaining the store inventory update system. I'll have to
work with that person so that we're sure the reservation purchases our in-store CATS system generates are
taken into consideration."

"You're going to have to do the same thing with
Process 3.0, Frank," Jordan added. "That process is
just not in our system. But this DFD does point to
some changes that will have to be made. The point-of-
sale system will have to be modified so that it can
record a purchase as a reservations purchase. Also, we
need a process on our DFD that can extract reservation purchase data from the Store Transactions data

store, even if that process is really just a single query.
Why don't you redraw the DFD so that the inventory
update system is an external entity that receives the
reservation purchases data flow and so that Process
3.0 is replaced with an extraction process?"

"How about this?" Frank asked, redrawing the
DFD as he spoke (see Figure 2).

"Meanwhile," Jordan said, "You've still got a DFD
with no real central transform. So this is a transaction-
centered system, even though there is no obvious
transaction center. The system is linked together by
the data stores."

"Another reason to see this as transaction-centered
is timing." Frank stated. "Is there any reason to believe
these three processes have to be performed in order or

Figure 2
Revised DFD of CATS in-store component

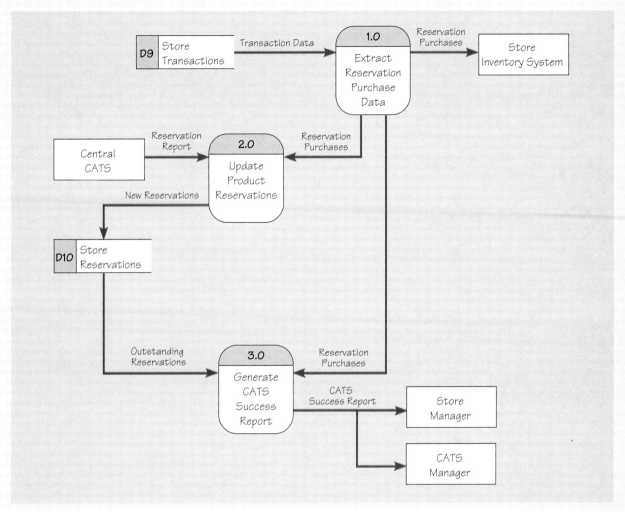

at the same time? When is the CATS success report generated?"

"It's projected to be a daily report," Jordan responded, "Although I guess it could be generated on demand. Look at this process—Update Product Reservations. One of the events that triggers it is the reservation report from Central CATS. We need to find out how often and when the Central CATS reservation report is generated. I bet it's out of synch with generating the CATS success report. And how often do we want to run this purchase data extraction process?"

"I think your idea of making this system transaction-centered is right. Because of the timing differences, this system seems ideally suited to be menu-based. We end up with one transaction module for two of the processes on the revised DFD, Update Product Reservations and Generate CATS

Success Report. That pretty much gives us a first-cut transaction-centered design," Frank said (see Figure 3).

"What about Process 1.0, Extract Reservations Purchase Data?" Jordan asked. "Did you forget that?"

"No," Frank said, "I didn't forget it. Both of the other processes need the reservations purchases data in order to function, so it seems to me that we should make Process 1.0 subordinate to both of the other modules. When a user calls for a product reservation update or calls for the CATS report to be generated, then the main module for whatever the user chooses calls Process 1.0. So we'll show Process 1.0 on the refined structure chart."

"OK, I see. Now let's refine this," Jordan said.

"First we'll add modules that we need for a transaction-based system," Frank replied. "So we have a module called Display Function Menu and another

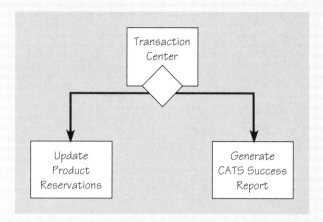

Figure 3

First-cut transaction-centered structure chart for the CATS in-store component

called Accept Valid Choice. Then under Update Product Reservations we'll need a module that corresponds to Process 1.0 and another for getting the Central CATS Reservation Report, like this (Figure 4)."

"Add another subordinate module for adding new reservations," Jordan said, "and we're done at the subordinate level for updating product reservations. Now let's draw an arrow from Generate

Figure 4

First refinement of the transaction-centered structure chart for the CATS in-store component

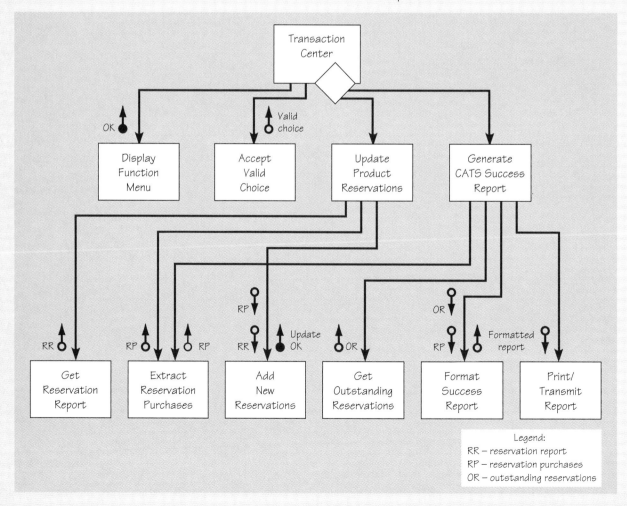

CATS Success Report back to Extract Reservation Purchase. . . ."

"Good—reuse."

"And then add a module responsible for getting the reservations still outstanding. Now we add a couple of modules for producing the report, one for formatting and one for printing or transmitting it," continued Jordan.

"Hey, this is good, I like how this ended up," said Frank.

"Now go back to your office and refine this some more. And then you're ready to create the Nassi-Schneiderman charts," Jordan smiled.

"I want to work with some other team members on that," Frank said. "You know how big a chore putting together all of the physical specs for a system can be."

"Oh, I know, all right," Jordan replied. "Sometimes I wish I could do more of that kind of work and less managing, but hey, that's my job. Speaking of which, I'm late for my meeting—gotta run. Keep me posted, Frank."

And Jordan practically ran out of her office.

SUMMARY

The specifications for the in-store portion of CATS are taking shape. The project team is in a transition within the project. They are moving from a functional to a technical perspective and soon will be turning work over to new members of the team responsible for building CATS. Thus, whereas the functional specifications were for communications within the group and with clients, the program specifications are guidelines to the programmers. The programmers may not have sufficient time to review all of the repository, so the structure charts need to capture just what they need to know in order to write program modules. Frank's work is very detailed and critical to the continued success of the project. We will see in the next BEC case that, even with all the analysis and study done during the project, unanticipated issues can arise. Each project is unique and presents new challenges to systems developers and to the organization.

18

Designing Distributed Systems

After studying this chapter, you should be able to:

■ Define the key terms client/server architecture, local area network, and distributed database.

■ Choose between file server and other client/server architectures for an application, based on the relative advantages of each.

■ Distinguish between file server and client/server environments, and contrast how each is used in a local area network.

■ Describe and apply four major strategies for distributing data in distributed systems.

■ Describe alternative designs for distributed systems and their trade-offs.

INTRODUCTION

The advances in personal computer technology and the rapid evolution of graphical user interfaces, networking, and communications are changing the way today's computing systems are being used to meet ever more demanding business needs. In many organizations, previous stand-alone personal computers are being linked together to form networks that support workgroup computing (this process is sometimes called *upsizing*). At the same time, other organizations (or even the same organization) are migrating mainframe applications to personal computers, workstations, and networks (this process is sometimes called *downsizing*) to take advantage of the greater cost-effectiveness of these environments. As technology continues to evolve, users are increasingly capable of interconnecting various types of computers, efficiently and transparently, to distribute data and applications within and across organizations.

A variety of new opportunities and competitive pressures are driving the trend toward these technologies. Corporate restructuring—mergers, acquisitions, consolidations—makes it necessary to connect or replace existing stand-alone applications. Similarly, corporate downsizing has caused individual managers to have a broader span of control, thus requiring access to a wider range of data, applications, and people. Applications are being downsized from expensive mainframes to networked microcomputers and workstations (possibly with a mainframe as a server) that are much more cost-effective and scalable, and can also be more user-friendly.

However, data traffic congestion on traditional local area networks (LANs) with a simple file server architecture can cause serious performance problems. Thus, how distributed systems are designed and configured can have obvious ramifications for day-to-day organizational activities and the success of an information system.

In this chapter we describe several technologies that are being used to upsize, downsize, and distribute information systems and data. These technologies are LAN-based DBMSs, client/server DBMSs, and distributed databases. The capabilities and issues surrounding these technologies are the foundation for understanding how to migrate single processor applications and designs into a multiprocessor, distributed computing environment. As you will learn, deciding to design a new information system as a distributed system must be made prior to implementation. For example, in order to effectively share data across locations, you must consider trade-offs between various designs for distributing data. Choosing a design without considering how data will be created, shared, updated, and destroyed across locations may significantly reduce overall system throughput and organizational effectiveness.

DESIGNING DISTRIBUTED SYSTEMS

In this section we will briefly discuss the process and deliverables when designing distributed systems. Given the direction of organizational change and technological evolution, it is likely that most future systems development efforts will need to consider the issues surrounding the design of distributed systems.

The Process of Designing Distributed Systems

This is the last chapter in the text dealing with physical design issues within the systems development life cycle (see Figure 18-1). In the previous chapters on physical design, specific techniques for representing and refining modules of processes and data were presented. In this chapter, however, no specific techniques will be presented on how to design distributed systems since no generally accepted techniques exist. Alternatively, we will focus on increasing your awareness of common environments for deploying distributed systems and the issues you will confront surrounding their design and implementation. You will learn about the trade-offs between different ways to deal with distributed systems design, and should be able to apply this understanding when you design multi-user, distributed systems.

Designing distributed systems is much like designing single-location systems. The primary difference is that because a system will be distributed over two or more locations, numerous design issues must be considered due to their influence on the reliability, availability, and survivability of the system when it is implemented. Because distributed systems have more components than a single-location system—that is, more processors, networks, locations, data, and so on—there are more potential places for a failure to occur. Consequently, various strategies can be used when designing and implementing distributed systems. For example, replicating all data elements across all distributed sites will enhance the *reliability* that a given piece of information will be available at a given time, and enhance the overall system *survivability* if a single node has a catastrophic failure. Yet, replicating all data elements at all sites may result in not having the latest information at a given site since it may not be possible to simultaneously update changes to all locations. Such a design is somewhat low on the *availability* dimension.

Thus, when designing distributed systems, you will need to consider numerous trade-offs that will influence reliability, survivability, and availability. To create effective designs, you need to understand the characteristics of the commonly used

Figure 18-1
Systems development life cycle with physical design phase highlighted

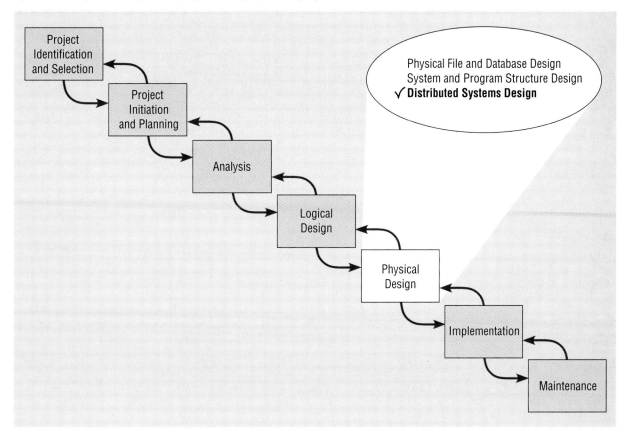

architectures for supporting distributed systems and sharing data. In other words, you must decide whether you want to partition data and processing across single or multiple sites and, if you do, what trade-offs to make.

Deliverables and Outcomes

When designing distributed systems, the deliverable is a document that will consolidate the information that must be considered when implementing a physical distributed system design. Figure 18-2 lists the types of information that must be considered when implementing a distributed system. In general, the information that must be considered when designing a distributed system is the site, processing, and data information for *each* location (or processor) in a distributed environment. Specifically, information related to physical distances between locations, counts and usage patterns by users, building and location infrastructure issues, personnel capabilities, data usage (use, create, update, or destroy), and local organizational processes must be described. Additionally, the pros and cons of various implementation solutions for each location should be reviewed. The collection of this information, in conjunction with the physical design information already developed, will provide the basis for implementing the information system in the distributed environment. Note, however, that our discussion assumes that any required network infrastructure is already in place. In other words, we focus only on those issues in which you will likely have a choice.

Figure 18-2

Outcomes and deliverables from designing distributed systems

1. Description of Site (for each site)
   a. geographical information
   b. physical location
   c. infrastructure information
   d. personnel characteristics (education, technical skills, etc.)
   e. ...

2. Description of Data Usage (for each site)
   a. data elements used
   b. data elements created
   c. data elements updated
   d. data elements deleted

3. Description of Business Process (for each site)
   a. list of processes
   b. description of processes

4. Contrasts of Alternative IS Architectures for Site, Data, and Process Needs (for each site)
   a. pros and cons of no technological support
   b. pros and cons of non-networked, local system
   c. pros and cons of various distributed configurations
   d. ...

DESIGNING SYSTEMS FOR LOCAL AREA NETWORKS

Personal computers and workstations can be used as stand-alone systems to support local applications. However, organizations have discovered that if data are valuable to one employee, they are probably also valuable to other employees in the same workgroup or in other workgroups. By interconnecting their computers, workers can exchange information electronically and can also share devices such as laser printers that are too expensive to be used by only a single user.

Local area network (LAN): The cabling, hardware, and software used to connect workstations, computers, and file servers located in a confined geographical area (typically within one building or campus).

A **local area network (LAN)** supports a network of personal computers, each with its own storage, that are able to share common devices and software attached to the LAN. Each PC and workstation on a LAN is typically within 100 feet of the others, with a total network cable length of under 1 mile. At least one computer (a microcomputer or larger) is designated as a file server, where shared databases and applications are stored. The LAN modules of a DBMS, for example, add concurrent access controls, possibly extra security features, and query or transaction queuing management to support concurrent access from multiple users of a shared database.

File Servers

File server: A device that manages file operations and is shared by each client PC attached to a LAN.

In a basic LAN environment (see Figure 18-3), all data manipulation occurs at the workstations where data are requested. One or more file servers are attached to the LAN. A **file server** is a device that manages file operations and is shared by each client PC that is attached to the LAN. In a file server configuration, each file server acts as an additional hard disk for each client PC. For example, your PC might recognize a logical F: drive, which is actually a disk volume stored on a file server on the LAN. Programs on your PC refer to files on this drive by the typical path specification, involving this drive and any directories, as well as the file name.

When using a DBMS in a file server environment, each client PC is authorized to use the DBMS application program on that PC. Thus, there is one database on the file server and many concurrently running copies of the DBMS on each active PC. The primary characteristic of a client-based LAN is that all data manipulation is performed at the client PC, not at the file server. The file server acts simply as a shared data storage device and is an extension of a typical PC. It also provides additional resources (e.g., disk drives, shared printing), collaborative applications (e.g., electronic mail), in addition to the shared data. Software at the file server only queues access requests; it is up to the application program at each client PC, working with the copy of the DBMS on that PC, to handle all data management functions. This means that in an application that wants to view a single customer account record in a database stored on the server, the file containing all customer account records will be sent over the network to the PC. Once at the PC, the file will be searched to find the desired record. Additionally, data security checks and file and record locking are done at the client PCs in this environment, making multi-user application development a relatively complex process.

Limitations of File Servers

There are three limitations when using file servers on local area networks:

1. Excessive data movement

2. Need for powerful client workstation

3. Decentralized data control

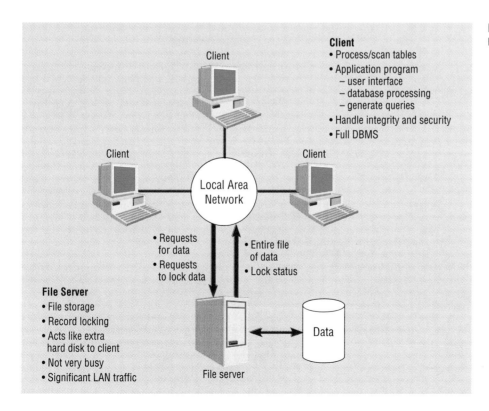

Figure 18-3
File server model

Figure 18-4

File servers transfer entire files when data are requested from a client

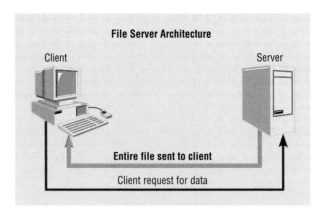

File Server Architecture

Client

Server

Entire file sent to client

Client request for data

First, when using a file server architecture, considerable data movement is generated across the network. For example, when an application program running on a client PC in Pine Valley Furniture wants to access the Birch products, the whole Product table is transferred to the client PC; then, the table is scanned at the client to find the few desired records. Thus, the server does very little work, the client is busy with extensive data manipulation, and the network is transferring large blocks of data (see Figure 18-4). Consequently, a client-based LAN places a considerable burden on the client PC to carry out functions that have to be performed on all clients and creates a high network traffic load.

Second, since each client workstation must devote memory to a full version of the DBMS, there is less room on the client PC to rapidly manipulate data in high-speed random access memory (RAM). Often, data must be swapped between RAM and a relatively slower hard disk when processing a particularly large database. Further, because the client workstation does most of the work, each client must be rather powerful to provide a suitable response time. File server-based architectures also benefit from having a very fast hard disk and cache memory in both clients and the server to enhance their ability to transfer files to and from the network, RAM, and hard disk.

Third, and possibly most important, the DBMS copy in each workstation must manage the shared database integrity. In addition, each application program must recognize, for example, locks on data and take care to initiate the proper locks. A lock is necessary to stop users from accessing data that are in the process of being updated. Thus, application programmers must be rather sophisticated to understand various subtle conditions that can arise in a multiple-user database environment. Programming is more complex since you have to program each application with the proper concurrency, recovery, and security controls.

DESIGNING SYSTEMS FOR A CLIENT/SERVER ARCHITECTURE

Client/server architecture: A LAN-based computing environment in which a central database server or engine performs all database commands sent to it from client workstations, and application programs on each client concentrate on user interface functions.

A recent improvement in LAN-based systems is the **client/server architecture** in which application processing is divided (not necessarily evenly) between client and server. The client workstation is most often responsible for managing the user interface, including presenting data, and the database server is responsible for database storage and access such as query processing. The typical client/server architecture is illustrated in Figure 18-5. Variations on the basic client/server design are discussed later in the chapter.

In the typical client/server architecture, all database recovery, security, and concurrent access management is centralized at the server, whereas this is the re-

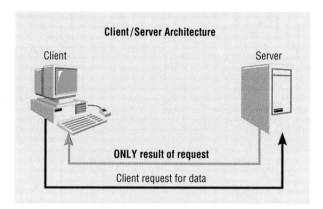

Figure 18-5
Client/server architecture transfers only the required data after a request from a client

sponsibility of each user workstation in a simple LAN. These central DBMS functions are often called a **database engine** in a client/server environment. Some people refer to the central DBMS functions as the back-end functions whereas client-based delivery of applications to users using PCs and workstations are called front-end applications. Further, in the client/server architecture, the server executes all requests for data so that only data that match the requested criteria are passed across the network to client stations. This is a significant advantage of client/server over simple file server-based designs. Since the server provides all shared database services, this leaves the **client** software to concentrate on user interface and data manipulation functions. The trade-off is that the server must be more powerful than the server in a file server environment.

An application built using the client/server architecture is also different from a centralized database system on a mainframe. The primary difference is that each client is an *intelligent* part of the application processing system. In other words, the application program executed by a user is running on the client, not on the server. The application program handles all interactions with the user and local devices (printer, keyboard, screen, etc.). Thus, there is a division of duties between the server (database engine) and the client: the database engine handles all database access and control functions and the client handles all user interaction and data manipulation functions. The client PC sends database commands to the database engine for processing. Alternatively, in a mainframe environment, all parts of the information system are managed and executed by the central computer.

Another advantage of client/server architectures is the ability to decouple the client environment from the server environment. Clients can consist of multiple types (for example, different computers, operating systems, and application programs) which means that the client can be running any application system that can generate the proper commands (often SQL) to request data from the server. For example, the application program might be written in Quattro Pro, dBASE IV, a report writer, a sophisticated screen painter, or any fourth-generation language that has an **application program interface,** or API, for the database engine. The database engine might be DB2 on an IBM mainframe or mid-range computer, or Sybase or Oracle running on a variety of platforms. An API calls library routines that transparently route SQL commands from the front-end client application to the database server. An API might work with existing front-end software, like a third-generation language or custom report generator, and it might include its own facilities for building applications. When APIs exist for several program development tools, then you have considerable independence to develop client applications in the most convenient front-end programming environment, yet still draw from a common server database. With some APIs, it is possible to access data from both the client and server in one database operation, as if the data were in one location managed

Database engine: The (back-end) portion of the client/server database system running on the server and providing database processing and shared access functions.

Client: The (front-end) portion of the client/server database system that provides the user interface and data manipulation functions.

Application program interface (API): Software which allows a specific front-end program development platform to communicate with a particular back-end database engine, even when the front-end and back-end were not built to be compatible.

by one DBMS. Figure 18-6 reviews several popular API environments that are used to provide this functionality.

Client/Server Advantages and Cautions

There are several significant benefits that can be realized by adopting a client/server architecture:

1. It allows companies to leverage the benefits of microcomputer technology. Today's workstations deliver impressive computing power at a fraction of the costs of mainframes.

2. It allows most processing to be performed close to the source of processed data, thereby improving response times and reducing network traffic.

3. It facilitates the use of graphical user interfaces (GUIs) and visual presentation techniques commonly available for workstations.

4. It allows for and encourages the acceptance of open systems.

Many other vendors of relational DBMSs and other LAN-based technologies are attempting to migrate their products into the client/server environment. However, products which were not designed from the beginning under a client/server architecture may have problems adapting to this new environment (see Radding, 1992, for a discussion of such issues). This is because new issues and new spins on old issues arise in this new environment. These issues and areas include compatibility of data types, query optimization, distributed databases, data administration of distributed data, CASE tool code generators, cross operating system integration, and more. In general, there is a lack of tools for systems design and performance monitoring in a client/server environment. As versions of different front- and back-end tools change, problems may arise with compatibility, until the API evolves, and these problems must be handled directly by the programmer, not by development tools.

Figure 18-6
SQL-based application program interface (API) environments for client/server-based applications development

- SQL Link from Borland, which allows Paradox programs to access SQL Server (from both Microsoft and Sybase), Oracle Server, IBM's DB2/2, and several other database engines
- DataEase SQL, which allows programs written in the DataEase language to access SQL Server, Database Manager, and Oracle Server databases
- Q + E add-in to the Excel electronic spreadsheet system, which allows access to SQL Server, Database Manager, and Oracle Server databases from within Excel
- @SQL add-in to Lotus 1-2-3 versions 2.01 and 2.2, which supports access to SQL Server databases
- SQL Windows version 3.1 from Gupta Technologies provides screen painters, a report writer, and context-sensitive help. It features front-end facilities to SQLBase, SQL Server, Oracle, IBM's Data Manager, DB2, Btrieve, Teradata, and other servers
- PowerBuilder from Powersoft Corporation, which provides a variety of object-oriented programming tools to build client applications which access Sybase and Microsoft SQL Server, SQL Base, Oracle, DB2, and other database servers
- Object/1 version 3.0 from mdbs, Inc., which is an object-oriented programming environment for an OS/2 client and SQL Server, Oracle, OS/2 Data Manager, and mdbs' MDBS IV server product

Now that you have an understanding of the general differences between file server and client/server architectures, we will next discuss how data can be managed within a distributed environment. After discussing data management options, we will present several design alternatives for distributed systems. All LAN-based distributed system designs are implemented using some configuration of the general file server or client/server architectures and data management options.

MANAGING DATA IN DISTRIBUTED SYSTEMS

When an organization is geographically dispersed, it may choose to store its databases on a central computer and make this data available over telecommunications lines to multiple users. More recently, organizations are choosing to distribute data among a network of local computers. A **distributed database** is a single logical database that is physically spread across computers in multiple locations that are connected by a data communications network. We emphasize that a distributed database is truly a database, not a loose collection of files. The network must allow the users to share the data; thus a user (or program) at location A must be able to access and perhaps update data at location B. The sites of a distributed system may be distributed over a large area such as the United States or the world, or over a small area such as a building or campus. The computers may range from micros to large-scale computers or even supercomputers.

Distributed database: A single logical database that is spread across computers in multiple locations which are connected by a data communications link.

Objectives and Trade-Offs

A major objective of distributed databases is to provide easy access to data for users at many different locations. To meet this objective, the distributed database system must provide what is called location transparency. **Location transparency** means that a user (or user program) requesting data need not know at which site these data are located. Any request to retrieve or update data at a nonlocal site is automatically forwarded by the system to that site. Ideally, the user is unaware of the distribution of data, and all data in the network appear as a single logical database. Distributed databases have several advantages and disadvantages that are summarized in Table 18-1. Advantages include greater system reliability, data integrity and control modularity, and faster user response. Disadvantages include greater cost, complexity, processing overhead, and increased security concerns. Whether distributed databases are appropriate for a particular organization or application will depend upon the balancing of these factors.

Location transparency: A design goal for a distributed database which says that a user (or user program) requesting data need not know at which site those data are located.

Options for Distributing a Database

Given that there are various options for distributing databases, a fundamental question is how should you distribute data for one or several applications among the sites (or nodes) of a distributed network? Research and practical experience have found that there are four basic strategies you can follow for distributing databases:

1. Data replication

2. Horizontal partitioning

3. Vertical partitioning

4. Combinations of 1, 2, and 3

We will explain and illustrate each approach using relational databases.

TABLE 18-1   Advantages and Disadvantages of Distributed Databases

Advantages of Distributed Databases:

1. Increased system reliability due to redundancy

2. Local control of data, which tends to promote improved data integrity and administration

3. Modular growth of applications and databases without disruption to existing users

4. Lower communication costs by reducing communication traffic

5. Faster response since most applications use data at a local site

Disadvantages of Distributed Databases:

1. Software cost and complexity, since more complex software is required for a distributed environment

2. Processing overhead to exchange messages among the sites

3. Data integrity, which is more difficult to control with multiple and dispersed copies of data

4. Slow response if the data and application software are not distributed properly according to their usage

---

Suppose that a bank has numerous branches located throughout a state. One of the relations in the bank's database is the Customer relation. The format for an abbreviated version of this relation is shown in Figure 18-7. For simplicity, the sample data in the relation apply to only two of the branches (Lakeview and Valley). The primary key in this relation is account number (ACCT NO.). BRANCH NAME is the name of the branch where customers have opened their accounts and, therefore, where they presumably perform most of their transactions.

Data Replication   One option for data distribution is to store a separate copy of the database at each of two or more sites. The Customer relation in Figure 18-7 could be stored at both Lakeview and Valley, for example. If a copy is stored at every site, we have the case of full replication. The advantages and disadvantages of data replication are summarized in Table 18-2. In short, data replication enhances system reliability and system response but requires increased data storage, processing complexity, and cost. These issues are especially acute when each copy of the data must be constantly identical. Newer technologies, called replication database server systems, are reducing the cost of replicated data when users can tolerate some lack of synchronization across copies of data. Given its trade-offs, full data replication is favored for data where most transactions are read-only and where the data are relatively static, such as catalogs, telephone directories, train schedules, and so on. CD-ROM storage technology has promise as an economical medium for replicated databases.

Figure 18-7

Customer relation for a bank

ACCT NO.	CUSTOMER NAME	BRANCH NAME	BALANCE
200	Jones	Lakeview	1000
324	Smith	Valley	250
153	Gray	Valley	38
426	Dorman	Lakeview	796
500	Green	Valley	168
683	McIntyre	Lakeview	1500
252	Elmore	Lakeview	330

TABLE 18-2   Advantages and Disadvantages of Data Replication

Advantages of Data Replication:

1. Reliability: If one of the sites containing the relation (or database) fails, a copy can always be found at another site.

2. Fast response: Each site that has a full copy can process queries locally, so queries can be processed rapidly.

Disadvantages of Data Replication:

1. Storage requirements: Each site that has a full copy must have the same storage capacity that would be required if the data were stored centrally.

2. Complexity and cost of updating: Whenever a relation is updated, it must be updated at each of the sites that holds a copy, which requires careful coordination.

Horizontal Partitioning   With **horizontal partitioning,** some of the rows of a table (or relation) are put into a base relation at one site and other rows are put into a base relation at another site. More generally, the rows of a relation are distributed to many sites. The advantages and disadvantages of horizontal partitioning are summarized in Table 18-3. Horizontal partitioning has the advantages of greater processing efficiency and security but is susceptible to inconsistent data access speeds and backup vulnerability due to the need to join data from multiple sites.

Horizontal partitioning: Distributing the rows of a table into several separate tables.

An example of horizontal partitioning within the Customer relation is shown in Figure 18-8. Each row is now located at its "home" branch. If a customer, in fact, conducts most of his or her transactions at the home branch, such transactions are processed locally and response times are minimized. When a customer initiates a transaction at another branch, the transaction must be transmitted to the home branch for processing and the response transmitted back to the initiating branch (this is the normal pattern for persons using automated teller machines, or ATMs). If a customer's usage pattern changes (perhaps because of a move), the system may be able to detect this change and dynamically move the record to another location where most transactions are being initiated. Thus, horizontal partitions are usually used when an organizational function is distributed, but each site is concerned with only a subset of the entity instances (frequently based on geography).

TABLE 18-3   The Advantages and Disadvantages of Horizontal Partitioning

Advantages of Horizontal Partitioning:

1. Efficiency: Data are stored close to where they are used and separate from other data used by other users or applications.

2. Local optimization: Data can be stored to optimize performance for local access.

3. Security: Data not relevant to usage at a particular site are not made available.

Disadvantages of Horizontal Partitioning:

1. Inconsistent access speed: When data from several partitions are required, the access time can be significantly different from local-only data access.

2. Backup vulnerability: Since data are not replicated, when data at one site become inaccessible or damaged, usage cannot switch to another site where a copy exists; data may be lost if proper backup is not performed at each site.

Figure 18-8
Horizontal partitions

ACCT NO.	CUSTOMER NAME	BRANCH NAME	BALANCE
200	Jones	Lakeview	1000
426	Dorman	Lakeview	796
683	McIntyre	Lakeview	1500
252	Elmore	Lakeview	330
(a) Lakeview Branch			

ACCT NO.	CUSTOMER NAME	BRANCH NAME	BALANCE
324	Smith	Valley	250
153	Gray	Valley	38
500	Green	Valley	168
(b) Valley Branch			

Vertical partitioning: Distributing the columns of a table into several separate tables.

Vertical Partitioning   With the **vertical partitioning** approach, some of the columns of a relation are moved into a base relation at one of the sites while other columns are moved into a base relation at another site (more generally, columns may be moved to several sites). The relations at each of the sites must share a common key so that the original table can be reconstructed.

To illustrate vertical partitioning, we use an application for a manufacturing company. A Part relation with PART NO. as the primary key is shown in Figure 18-9. Some of these data are used primarily by manufacturing while others are used mostly by engineering. The data are distributed to the respective departmental computers using vertical partitioning, as shown in Figure 18-10. Each of the partitions shown in Figure 18-10 is obtained by taking selected columns of the original relation. The original relation in turn can be obtained by joining the partitions into a single table.

Combinations for Data Distribution   To complicate matters further, there are almost unlimited combinations of the preceding strategies. Some data may be stored centrally while other data are replicated at the various sites. Also, for a given relation, both horizontal and vertical partitions may be desirable for data distribution. Figure 18-11 is an example of a combination strategy which shows the following:

1. Engineering parts, Accounting, and Customer data are each centralized at different locations.

2. Standard parts data is partitioned horizontally among the three locations.

3. The Standard price list is replicated at all three locations.

Figure 18-9
Part relation

PART NO.	NAME	COST	DRAWING NO.	QTY ON HAND
P2	Cap	100	123-7	20
P7	Lead	550	621-0	100
P3	Spring	48	174-3	0
P1	Clip	220	416-2	16
P8	Body	16	321-0	50
P9	Gripper	75	400-1	0
P6	Eraser	125	129-4	200

PART NO.	DRAWING NO.
P2	123-7
P7	621-0
P3	174-3
P1	416-2
P8	321-0
P9	400-1
P6	129-4

Figure 18-10
Vertical partitioning of Part relation

(a) Engineering

PART NO.	NAME	COST	QTY ON HAND
P2	Cap	100	20
P7	Lead	550	100
P3	Spring	48	0
P1	Clip	220	16
P8	Body	16	50
P9	Gripper	75	0
P6	Erase	125	200

(b) Manufacturing

Figure 18-11
Data distribution strategies

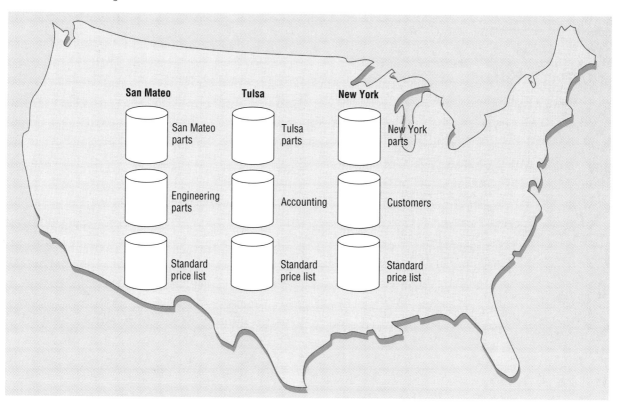

The overriding principle in distributed database design as part of designing distributed applications is that data should be stored at the sites where they will be accessed most frequently. Other considerations, such as security, data integrity, and cost, are also likely to be important.

Distributed systems, because they permit multiple users to concurrently use data, raise special issues on data integrity controls. With concurrent processing involving updates, a database without **concurrency control** will be compromised. Concurrency controls refer to techniques that prevent loss of data integrity due to interference between users in a multi-user environment. There are many techniques that database administrators can utilize to enforce concurrency control. Since this topic involves issues beyond general system design, the interested reader is encouraged to see McFadden and Hoffer (1994) for a detailed discussion of concurrency control techniques and related issues.

> **Concurrency control:** A method for preventing loss of data integrity due to interference between users in a multi-user environment.

ALTERNATIVE DESIGNS FOR DISTRIBUTED SYSTEMS

A clear trend in physical systems design is to move away from central mainframe systems and stand-alone PC applications to some form of system that distributes data and processing across multiple computers. Figure 18-12 represents these trends, including the trend that simple file server use has flattened or slightly declined in recent years. The excerpt in the box "Can a Client/Server Environment Replace a Mainframe?" characterizes what many organizations are attempting to do and the success of these efforts with existing technology.

One result of this increased interest in distributing system functions is that new forms of distributed processing are being developed, increasing your choices as a systems analyst for the design of new systems. In this section, we briefly review the major differences between two common types of network computing environments introduced earlier, file servers and database servers. In the following section, we discuss the trade-offs among newer ways to separate processing between clients and servers that are giving you greater options. Remember, development tools, such as CASE environments, are still relatively immature in generating code for systems in the wide variety of distributed processing options available to you, so be prepared to have significant portions of systems coded by hand if you choose some of the more advanced options.

Choosing Between File Server and Client/Server Architectures

Both file server and client/server architectures use personal computers and workstations and are interconnected using a LAN. Yet, a file server architecture is very different from a client/server architecture. A file server architecture supports only the distribution of data whereas the client/server architecture supports

---

Can a Client/Server Environment Replace a Mainframe?

Chase Manhattan Bank conducted a test for the possible conversion of a mainframe decision support application, the Affluent Database, to a Sybase SQL Server Unix application. This database contains 3 gigabytes of storage, with data on two million customers, 2.9 million accounts, and 1.6 million households. The server in the test was a Sun SPARC Server 10. Clients were AT&T Global Information Solution PCs, some running DOS and some running OS/2. Microsoft Access®, QBE Vision® from Coromandel Industries, and Lotus Approach® were chosen as the GUI front ends. Data were transferred from the IBM DB2 mainframe database using Platinum Technology's Platinum Migration Utility. A side-by-side comparison was done of identical queries running on the previous mainframe and the new client/server systems. Queries in the client/server environment were tuned to achieve the greatest possible performance, given that the existing mainframe application had already been similarly tuned. The results showed that simple queries and on-line transactions processed equally as quickly on the server as on the mainframe, but that complex queries took much longer in the client/server. Part of this can be explained by the newness of the technology. The front-end and server software are not as mature, and have not been optimized over repeated updates as has the mainframe software. Also, more problems arose in the client/server environment, in part because of the difficulties of getting software from multiple vendors to work smoothly together. The test showed that the migration of a DSS from a mainframe to a client/server environment was possible, but that performance is less predictable and reliable in a client/server environment.

(Adapted from Rideout, 1994)

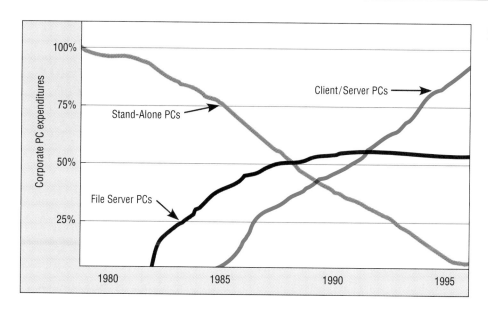

Figure 18-12
Trends in distributed systems

both the distribution of data and the distribution of processing. This is an important distinction that has ramifications for systems design.

Table 18-4 summarizes some of the key differences between file server and client/server architectures. Specifically, a file server architecture is the simplest method for interconnecting PCs and workstations. In this architecture, the file server simply acts as a shared storage device for all clients on the network. Entire programs and databases must be transferred to each client when accessed. This means that a file server architecture is most appropriate for applications that are relatively small in size with little or no concurrent data access by multiple users. The file server architecture has been found to be very effective for systems that do not require the simultaneous sharing of the same data and when the size of the applications "served" to workstations are relatively small in size and infrequently used. For example, numerous LAN-based electronic mail packages run exceptionally well in a file server environment.

Alternatively, a client/server architecture overcomes many of the limitations of the file server architecture since both client and server share the processing workload of a task and only transfer needed information. This characteristic has resulted

TABLE 18-4 Several Differences between File Server and Client/Server Architectures

Characteristic	File Server	Client/Server
Processing	Client only	Both client and server
Concurrent Data Access	Low—managed by each client	High—managed by server
Network Usage	Large file and data transfers	Efficient data transfers
Database Security and Integrity	Low—managed by each client	High—managed by server
Software Maintenance	Low—software changes just on server	Mixed—some new parts must be delivered to each client
Hardware and System Software Flexibility	Client and server decoupled and can be mixed	Need for greater coordination between client and server

in having many organizations migrate very large applications with extensive data sharing requirements to client/server environments. In fact, client/server computing has become the workhorse architecture for many organizations where multiple clients are likely to be concurrently working with the same data. Also, if the systems and databases are relatively large in size, the client/server architecture is preferred because of the client and server's ability to distribute the work and to transfer only needed information (e.g., only a record if that is all that is needed rather than an entire database as in a file server environment).

Advanced Forms of Client/Server Architectures

The client/server vendors are evolving into architectures that represent variations in the way different application system functions are distributed between client and server computers. These variations are based on the concept that there are three general components to any information system:

1. *Data management:* These functions manage all interaction between software and files and databases, including data retrieval/querying, updating, security, concurrency control, and recovery.

2. *Data presentation:* These functions manage just the interface between system users and the software, including the display and printing of forms and reports and possibly validating system inputs.

3. *Data analysis:* These functions transform inputs into outputs, including simple summarization to complex mathematical modeling like regression analysis.

Different client/server architectures distribute each of these functions to one or both of the client or server computers. Thus, in theory, there are 27 different client/server architectures with each of the three components distributed to either the client, server, or both. In practice, only six architectures have emerged (see Figure 18-13). Technology exists to allow you to develop an application using any of these six architectures, although CASE tools do not have equal code-generation capabilities for each.

Distributed Presentation   The distributed presentation form of client/server (Figure 18-13a) is used to freshen up the delivery of existing server-based applications to distributed clients. Often the server is a mainframe, and the existing mainframe code is not changed. Without the client presentation support, users must interact directly with mainframe programs, often using character-based interfaces and whatever response time the server supports. In this client/server architecture, technologies called "screen scrappers" work on the client machines to simply reformat the same data handled by the server forms and reports. The result is more appealing and has easier-to-use interfaces without destroying or having to rewrite legacy systems. Network traffic is limited to transmitting data only on the forms and reports. Distributed presentation is limited to the functionality of existing forms and reports and, when these change, the presentation modules on both the clients and server must change, so there is double maintenance. Thus, this approach may be very costly if the user interface designs frequently change. Client environments (e.g., Microsoft Windows, Apple, and UNIX X-Windows) can be mixed, which may open access to mainframe data to a wider group of users.

Remote Presentation   The remote presentation style of client/server architecture (Figure 18-13b) places all data presentation functions on the client machine so that software on the client has total responsibility for formatting data. This architecture gives you greater flexibility compared to the distributed presentation style since the presentation on the client will not be constrained by having to be compatible with that on the server. There will be less maintenance since only software on the client needs to change when users want forms and reports to be reorganized or new content added. There may be more network traffic than with distributed presentation since the client must extensively interact with the server to receive all the needed data.

Remote Data Management   The remote data management form of client/server architecture (Figure 18-13c) places all software on the client except for the data management functions. This form is the closest to what we have called client/server earlier in the chapter. Network traffic is higher than in the previous two options since all data needed for analysis, not just for presentation, must be transferred to clients. However, you now have the flexibility to use any software, like PC-based spreadsheets, for analysis. Further, the analysis processing is done on the less expensive client machine. The database is still shared but centrally controlled, which is simpler than some of the subsequent architectures we will discuss. If you want to change the analysis code, you have to make sure that each copy is changed. A mixed client environment may be more difficult to support than in the previous architectures since you must learn multiple analysis programming environments, not just those for presentation tools.

Distributed Function   The distributed function client/server architecture (Figure 18-13d) splits analysis functions between the client and server, leaving all presentation on the client and all data management on the server. This architecture

FUNCTION	CLIENT	SERVER
Data management		All data management
Data analysis		All data analysis
Data presentation	Data for presentation on server are reformatted for presentation to user	Data delivered to client using server presentation technologies

Figure 18-13
Types of client/server architectures

(a) Distributed presentation

FUNCTION	CLIENT	SERVER
Data management		All data management
Data analysis		All data analysis
Data presentation	Data from analysis on server are formatted for presentation to user	

(b) Remote presentation

FUNCTION	CLIENT	SERVER
Data management		All data management
Data analysis	Raw data from server are retrieved and analyzed	
Data presentation	All data presentation	

(c) Remote data management

Figure 18-13
(continued)
Types of client/server
architectures

(d) Distributed function

FUNCTION	CLIENT	SERVER
Data management		All data management
Data analysis	Selective data from server retrieved and analyzed	Selective data from server retrieved and analyzed, then transmitted to client
Data presentation	All data presentation, from analyses on both server and client	

(e) Distributed database

FUNCTION	CLIENT	SERVER
Data management	Local data management	Shared management of data on server
Data analysis	Data retrieved from both client and server for analysis	
Data presentation	All data presentation	

(f) Distributed processing

FUNCTION	CLIENT	SERVER
Data management	Local data management	Shared management of data on server
Data analysis	Data retrieved from both client and server for analysis	Data retrieved from server for analysis, then sent to client for further analysis and presentation
Data presentation	All data presentation	

allows analysis to be located on the most cost-efficient machine; for example, analysis that requires superior numerical calculations can be done on the server, which might be a supercomputer. Also, analysis functions that require extensive data access can be located on the server in order to reduce network traffic. This is a very difficult environment in which to develop, test, and maintain software due to the potential for considerable coordination between analysis functions on both client and server.

Distributed Database The distributed database client/server architecture (Figure 18-13e) places all functionality on the client, except data storage and management which is divided between client and server. Technologies like Paradox or dBASE on a client running SQL Link into an Oracle or Sybase server support this client/server architecture. In this case, data can be located at an optimal site but, given today's technologies, usually only the data on the server is shared across all clients. Client programs can mix calls to local and server data, possibly even within the same query. Concepts of vertical and horizontal partitioning of data can be used to decide what data to locate on each client or on the server. Although possible today, this is a very unstable architecture since it requires considerable compatibility and communication between software on the client and server, which may never have been meant to be compatible. Thus, organizations attempting to implement this architecture tend to standardize on a limited set of development tools to simplify distributed system design and implementation. Also, a decision to locate data on a client may make it difficult or impossible to share this data with remote or mobile computers.

Distributed Processing   The distributed processing client/server architecture (Figure 18–13f) combines the best features of distributed function and distributed database by splitting both of these across client and server, with presentation functions under the exclusive responsibility of the client machine. This permits even greater flexibility since analysis functions and data both can be located wherever it makes the most sense. Complexity is actually less than with the distributed database architecture since analysis and database functions on the same machine can be very compatible.

As the designer of information systems, you have more choices available to you today than ever before. You must weigh the factors discussed above and outlined in Table 18-4 to determine a distributed system design that will be most beneficial to the organization. As with other physical design decisions, organizational standards may limit your choices, and you will make such application design decisions in cooperation with other system professionals. You, however, are in the best position to understand user requirements and to be able to estimate the ramifications of distributed system design decisions on response time and other factors for the user.

SUMMARY

We have covered in this chapter various issues and technologies involved in the sharing of systems and data by multiple people across space and time in distributed systems. You learned about the client/server architecture which is being used in both network personal computers and workstations (upsizing) and to replace older mainframe applications (downsizing). Components of the client/server architecture, including local area networks, database servers, application programming interfaces, and application development tools were described.

Two common types of local area network-based architectures—file server and client/server—were compared. It was shown that the newer client/server technologies have significant advantages over the older file servers. These advantages include less network traffic, greater flexibility to develop applications in convenient environments, and a more sensible distribution of queries to form a cooperative computing situation.

We saw that a distributed database is a single logical database spread across computers in multiple locations connected by a data communications network. The network must allow the users to share the data as transparently as possible. There are numerous advantages to distributed databases. The most important of these are the following: increased reliability and availability of data, local control by users over "their" data, modular (or incremental) growth, reduced communication costs, and faster response to requests for data. There are also several costs and disadvantages of distributed databases: software is more costly and complex, processing overhead often increases, maintaining data integrity is often more difficult, and if data is not distributed properly, response to requests for data may be very slow. Both the advantages and disadvantages should be weighed by an organization that is considering distributed database management.

There are several options for distributing data in a network: data replication, horizontal partitioning, vertical partitioning, and combinations of these approaches.

With data replication, a separate copy of the database (or part of the database) is stored at each of two or more sites. Data replication can result in improved reliability and faster response; however, additional storage capacity is required and it may be difficult to keep the data updated at each of the sites. With horizontal partitioning, some of the rows of a relation are placed at one site while other rows are placed in a relation at another site (or several sites). On the other hand, with vertical partitioning, the columns of a relation are distributed among different sites. The objectives of data partitioning include improved performance and security.

We also outlined in this chapter the evolution of distributed systems and newer forms of client/server technologies that are giving you more options for distributed system design. These client/server architectures—distributed presentation, remote presentation, remote data management, distributed function, distributed database, and distributed processing—are supported by some CASE tool code generators, but CASE tool support may be limited. Thus, the decision to use very sophisticated distributed processing environments must consider the potential for considerable human support for both programming and training.

We did not have the space in this chapter to address several additional issues concerning distributed systems. Many of these issues are handled by other systems professionals, such as database administrators, telecommunications experts, and computer security specialists. Systems analysts must work closely with other professionals to build sound distributed systems.

CHAPTER REVIEW

KEY TERMS

Application program interface (API)	Concurrency control	Horizontal partitioning
Client	Database engine	Local area network (LAN)
Client/server architecture	Distributed database	Location transparency
	File server	Vertical partitioning

REVIEW QUESTIONS

1. Define each of the following terms:
   a. application program interface (API)
   b. client
   c. client/server architecture
   d. concurrency control
   e. database engine
   f. distributed database
   g. file server
   h. horizontal partitioning
   i. local area network (LAN)
   j. location transparency
   k. vertical partitioning

2. Contrast the following terms.
   a. file server, client/server architecture, local area network
   b. horizontal partitioning, vertical partitioning, data replication
   c. distributed database, database engine
   d. data management, data analysis, data presentation

3. Describe the limitations of a file server architecture.

4. Describe the advantages of a client/server architecture.

5. Explain the relative advantages and disadvantages of distributed databases.

6. What are the advantages and disadvantages of replicated databases?

7. What are the advantages and disadvantages of partitioned databases?

8. What are the major advantages and issues of the client/server architecture?

9. Summarize the six possible architectures for client/server systems.

PROBLEMS AND EXERCISES

1. Match each of the following terms with the most appropriate definition:

_____	file server	a. rows of a table are distributed to nodes
_____	concurrency control	b. user is unaware that data are distributed to several nodes
_____	horizontal partition	c. technique used to prevent data integrity errors
_____	location transparency	d. one logical database allocated to several physical nodes
_____	distributed database	e. allows PCs to share data and resources on a LAN

2. Under what circumstances would you recommend that a file server approach, as opposed to a client/server approach, be used for a distributed information system application? What warnings would you give the prospective user for this file server approach? What factors would have to change for you to recommend the move to a client/server approach?

3. A supplier of water pumps and filtration systems has one warehouse and sales outlets in several towns in the region. The warehouse is used strictly to store the various products and ship them to the sales outlets in predetermined monthly deliveries and upon specialized requests. In response to the owners' wish to employ a distributed database, recommend one of the options discussed in this chapter (e.g., data replication, horizontal or vertical partitioning, combination) and provide your rationale.

4. A bank has one main office and five branch office locations. Design a series of database tables that employ data replication, horizontal, and/or vertical partitioning for sharing data between the main office and branches. Define at least 10 separate data attributes for these tables. Provide a definition for your attributes that describes their meaning and list all assumptions about the bank and your data to help clarify your design.

5. Develop a table that summarizes the comparative capabilities of the six client/server architectures. You might start with Table 18-4 for some ideas.

6. Suppose you are responsible for the design of a new order entry and sales analysis system for a national chain of auto part stores. Each store has a PC that supports office functions. The company also has regional managers who travel from store to store working with the local managers to promote sales. There are four national offices for the regional managers, who each spend about one day a week in their office and four on the road. Stores place orders to replenish stock on a daily basis, based on sales history and inventory levels. The company uses high-speed dial-in lines and modems to attach store PCs into the company's main computer. Each

regional manager has a laptop computer with a modem and a network connection for times when the manager is in the office. Would you recommend a client/server distributed system for this company and, if so, which architecture would you recommend? Why?

7. The Internet is a network of networks. Using the terminology of this chapter, what type of distributed network architecture is used on the Internet?

8. Suppose you were designing applications for a standard file server environment. One issue discussed in the chapter for this distributed processing environment is that the application software on each client PC must share in the responsibilities for data management. One data management problem that can arise is that applications running concurrently on two clients may want to update the same data at the same time. What could you do to manage this potential conflict? Is there any way this conflict might result in both PCs making no progress (in other words, going into an infinite loop)? How might you avoid such problems?

FIELD EXERCISES

1. Visit an organization that has installed a local area network (LAN). Explore the following questions:

   a. Inventory all application programs that are delivered to client PCs using a file server architecture. How many users use each application? What are their professional and technical skills? What business processes are supported by the application? What data are created, read, updated, or destroyed in each application? Could the same business processes be performed without using technology? If so, how? If not, why not?

   b. Inventory all application programs that are delivered to client PCs using a client/server architecture. How many users use each application? What are their professional and technical skills? What business processes are supported by the application? What data are created, read, updated, or destroyed in each application? Could the same business processes be performed without using technology? If so, how? If not, why not?

2. Visit an organization that has installed a distributed database. Explore the following questions:

   a. Is this truly a distributed database? If so, how are the data distributed: replication, horizontal partitioning, vertical partitioning?

   b. What commercial distributed DBMS software products are used? What are the advantages, disadvantages, and problems with this system?

   c. To what extent does this system provide location transparency?

   d. What are the organization's plans for future evolution of this system?

3. Scan the literature and determine the various local area network operating systems available. Describe the relative strengths and weaknesses of these systems. Do these systems seem to be adequate for distributed information system needs in organizations? Why or why not? Determine the current sales volume and approximate market shares for these systems. Why are they selling so well?

4. In this chapter file servers were described as one way of providing information to users of a distributed information system. What file servers are available and what are their relative strengths and weaknesses and costs? What other types of servers are available and/or for what other uses are file servers employed (e.g., print servers)?

5. Examine the capabilities of a client/server API environment. List and describe the types of client-based operations that you can perform with the API. List and describe the types of server-based operations that you can perform with the API. How are these operations the same/different?

REFERENCES

McFadden, F. R., and J. A. Hoffer. 1994. *Modern Database Management.* 4th ed., Redwood City, CA: Benjamin/Cummings Publishing.

Radding, A. 1992. "DBAs Find Tools Gap in C/S." *Software Magazine.* Client/Server Special Section 12 (16) (November), 33–38.

Rideout, R. 1994. "Banking on Client/Server." *DBMS* 7(9) (August): 57–62.

Data Privacy Within CATS

INTRODUCTION

In a recent interview with *Video Entrepreneur Magazine*, Nigel Broad, chairman of BEC, described his road to success through the creation of his video and music empire. He stressed BEC's ability to be more innovative than their competition as their major competitive advantage. This spurred the interviewer to ask Nigel for an example of an innovative project that was currently underway at BEC so that she could provide the reader with an example of such leading-edge thinking. After pausing a moment to collect his thoughts, Nigel described the Customer Activity Tracking System (CATS).

DATA PRIVACY ISSUES ARISE

During this exchange Nigel explained, "CATS will allow us to really understand and, consequently, serve our customers. We will know what they like and don't like. Although such information on general consumer buying patterns has been readily available for years, CATS will be the first such system for home entertainment." He continued, "With appropriate modeling and analysis, we will be able to very accurately predict the likelihood of a given customer to purchase or rent a specific movie, music CD or cassette, or video game. This will ensure that we have in stock exactly what people want when they enter our stores. Again, we will have this for *each* customer, not just an aggregate view of the overall patterns at a particular store. This will be very powerful information for our marketing people. To take this a bit further, we could combine the CATS data with general consumer buying information on a particular customer and be able to construct a very accurate understanding of each customer. We will know what they drive, where they live, what they eat, how much they earn, and, most impor-

tantly for our business, what they like to do with their leisure time."

The interviewer found this to be a very interesting development in the conversation so she followed up the description of CATS with a few questions. "Mr. Broad, could you discuss some of the ethical ramifications of constructing such profiles you have considered and the steps BEC is going to follow to assure the accuracy of this data and to protect the privacy of your customers?"

Nigel was lucky that this was a phone interview, due to the crimson tint his face took on while listening to the interviewer's question. This topic was not something he, nor to his knowledge anyone at BEC, had yet considered. Caught off guard, Nigel took a moment before answering. He began by first assuring the interviewer that BEC had a fundamental obligation to protect the privacy of their customers and that federal laws prohibited tracking the actual movie titles rented by each customer. He continued, "CATS is being created only to provide better service to our customers. In no way will we invade the privacy of anyone. Precautions will be taken to assure information validity. Additionally, each customer's profile will be guarded like the 'Crown Jewels'."

After the interview, Nigel felt a bit uneasy, not really knowing what was being done to assure data validity and to protect the privacy of individuals. But, he believed what he had told the interviewer; BEC *did* have an obligation to protect the privacy of their customers. To follow through on his belief, his next action was to contact Karen Gardner, director of MIS.

"Karen, I have been thinking about CATS," Nigel began. "How are we going to protect the privacy of each individual's profile from falling into the wrong hands? I mean, analyzing a customer's long-term buying patterns is going to be very revealing about how a given customer spends his or her leisure time. I understand that this is exactly why we want CATS, and what

TABLE 1   Data Privacy Questions Regarding Customer Profiles

1. Can the details of individual customer profiles be viewed or can data only be viewed in aggregate?

2. Who can access customer profile information? Do we need security access levels for viewing less aggregate data as we do in personnel databases?

3. What security mechanisms should be followed to secure the customer profile information?

4. Can customers view their profiles and, if so, what mechanisms should be followed to allow customers to access, validate, or even modify their data?

5. Does BEC have an obligation to tell customers that a profile has been created on them?

6. Should BEC combine CATS data with other consumer databases (such as shopping data or even driver's license data) to get a more detailed and revealing profile of customers?

7. Should BEC sell customer profile data to other companies and, if so, in what form should the data be made available? Should customers be informed about this?

we need it to do, but are we taking adequate steps to protect the privacy of each individual?"

Karen responded, "Nigel, your questions are certainly good ones. In fact, Dwight Ford, our legal counsel, raised a similar query when we initially reviewed the CATS proposal. What I've discovered is that most BEC information systems employees are members of one of the industry's professional societies. As a member, you are asked to follow a code of ethics that includes protecting the privacy of others [see Figure 3-10 for the code of ethics for one such society]. However, other than that, we haven't talked about securing the contents of customer profiles beyond normal security mechanisms. But I think that this is an important issue. How should we proceed?"

Over the next few days, Nigel and Karen met and outlined several questions that both felt needed to be resolved before implementing CATS. Nigel intended to pose these questions to BEC managers and the Board of Directors (see Table 1). The answers to these

questions would help shape how customer profiles would be accessed and secured, clearly important issues for both BEC and their customers.

SUMMARY

Strong customer relationships begin with trust, and Nigel Broad believes that trust is fundamental to BEC's continued growth. Furthermore, Nigel feels a social responsibility to maintain the highest ethical standards in the face of intense competition. Confronting ethical issues is relatively new in the systems development field, and BEC has little experience to draw on to decide how to handle consumer data privacy in CATS. The questions in Table 1 are only a start at clarifying the obligation BEC has for protecting the rights of customers, employees, suppliers, and other stakeholders.

PART

VII

Implementation and Maintenance

Albertson's, Inc.

Organizational and Human Issues in the Implementation of Business Re-Engineering

Albertson's, Inc., is the nation's fourth largest retail food and drug chain with well over 700 retail stores in 19 states throughout the west, midwest, and southern United States. For an organization as large as Albertson's, operating effectively and efficiently is not a trivial task. To be more productive, they develop information systems that satisfy real user needs, solve real problems, and support business processes well.

As have other forward-thinking organizations, Albertson's realized that their information systems solutions were too frequently only incremental changes to current business processes. They wanted to help systems development people and users stop thinking about their work in traditional, routine ways. They wanted to have employees be able to step back from their work, critically analyze it, and figure out how to do it better, even if it meant doing it in dramatically different ways. They decided to learn more about business process re-engineering by visiting other organizations. Albertson's management has adapted some of the re-engineering techniques they saw and launched their own re-engineering efforts, but in a unique way.

Albertson's houses their re-engineering effort in specially created meeting rooms called "labs." These labs are equipped with PCs and printers connected to the organization's network, electronic whiteboards, conference tables and chairs; the walls are nearly covered with regular whiteboards. There are currently six different labs in a variety of sizes. One lab is a spacious 22' × 22', though there are other smaller labs.

Albertson's brings together in a lab a core group of two to six relevant stakeholders for a particular project, including systems developers, users, and managers, and dedicates them to the project for anywhere from four to eight weeks. This is their only task and their primary place of work; they must temporarily reassign their normal duties to other employees. Albertson's commonly brings in additional members to the teams as they are needed. The teams use various methodologies such as interviewing, Joint Application Design, and group problem-solving techniques to elicit information and solve problems.

Albertson's believes that a key ingredient of the lab approach is bringing together the key people for a project, even if it means bringing in people direct from the retail stores, a district manager, a foreman, or a store manager. Once in the lab, hierarchy, job titles, and status are abandoned as the team slowly breaks down hierarchical and discipline-based barriers and builds camaraderie and trust. In addition, they are free from distractions and interruptions in the lab.

The physical and structural changes resulting from the lab approach lead to psychological changes in the participants. The lab approach helps them to achieve their goals, to develop real solutions to real problems, to think outside the box, and to imagine, if they could do anything to solve their particular problem or improve their particular process, what they could do.

During Albertson's first lab, senior managers addressed the fundamental strategic question, "What should the company really be doing and where should it really be heading?" They critically analyzed the Hammer and Champy book, *Reengineering the Corporation,* and applied it to their company. In this trial experience with the lab they learned a lot about the company and how to refine and best use the lab approach. It worked so well and was liked so much that they quickly built more labs and applied the approach to other problems.

Hadley Wagner, vice-president of Information Systems, says that as a result of using the lab approach, they are now much closer to providing real solutions to real problems than they were before using the labs. He admits that, like so many other companies, they were simply "automating" traditional business processes when building information systems. He adds that they now garner much higher levels of commitment to new systems by users and managers because they are now very heavily involved and invested in the process of designing and building the new system and enhancing the business process.

Craig Olsen, senior vice-president and chief financial officer, says that the return on investment in the labs is huge. First, the investment in the labs themselves is minuscule. Then considering the salaries of a half dozen people working for two to three months and add in their travel expenses, the total investment is still relatively small. Further, this investment is far outweighed by the savings and revenue from the solutions that teams generate and implement. For example, one team determined a much more cost-effective way to transfer goods to stores. This also will ultimately result in better service, better product, improved customer satisfaction, and increased revenue. Surprisingly, teams frequently hit home runs like this. In Olsen's mind there is no question about the payback on the labs.

The labs do have some potential challenges. It is difficult to break down the barriers between people and build trust among the team members. In addition, because the participants are so motivated, there are differences of opinion and even arguments at times. Participants can also become frustrated when the rest of the organization cannot move as quickly and easily as the team would like it to. To address these and other concerns, Albertson's will temporarily bring into a new team a person from a previous team who will serve as a mentor. An additional potential challenge is that there often are some language differences among people from different levels and functional areas within the organization. As a result, part of the methodology is to build a project term dictionary. This helps everyone to better interpret and understand each other.

Albertson's learns more about the labs and improves the process with each new team. Given their success thus far, they plan to build more labs in other locations and to apply the lab approach to other types of business problems and processes. They like what the labs have helped them to do thus far and they want to empower even more employees to develop real solutions to solve real problems, to break out of tradition and incremental change, and to effect revolutionary, purposeful, strategic change.

Implementation
and Maintenance

Implementation and maintenance are the last two phases of the systems development life cycle. The end of implementation typically brings your responsibilities for a system to a close, but others will continue to work with the system to support and maintain it until it is replaced by another system. The purpose of implementation is to build a properly working system and to install it in the organization, replacing old systems and work methods. Implementation also includes finalizing all system and user documentation, thoroughly training users and others to effectively use the new system, and preparing support systems to assist users as they encounter difficulties. The purpose of maintenance is to fix and enhance the system to respond to problems and changing business conditions. Maintenance work also includes activities from all systems development phases and usually eventually leads to a decision to abandon a system and build a replacement, starting the cycle over again.

As depicted in Figure 1, implementation includes the following phases:

- *Writing computer software:* for a systems analyst, this may mean actually writing code or monitoring coding done by programmers to insure that programs meet design specifications

- *Testing software:* involves using test data and scenarios to verify that each component and the whole system work under normal and abnormal circumstances

- *Converting from the old to the new system:* includes not only installing the new system in organizational sites but also dealing with personal and organizational resistance to the change the new system causes

- *Documenting the system:* includes reviewing all project dictionary or CASE repository entries for completeness as well as finalizing all user documentation, such as user guides, reference cards, and tutorials

- *Training users and others:* may include a variety of human and computer-assisted sessions as well as tools to explain the purpose and use of the system

- *Designing support procedures:* ensures that users can obtain the assistance they need as questions and problems arise

Implementation also includes the close-down of the project, including evaluating personnel, reassigning staff, assessing the success of the project, and turning all

Figure 1
Steps in implementation

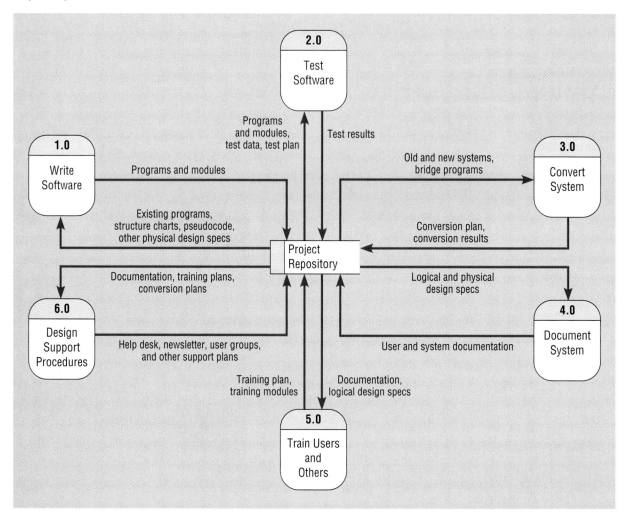

resources over to those who will support and maintain the system. Thus, implementation is tightly linked to the next phase—maintenance—which is why we combine these two phases into one section of the book.

As depicted in Figure 2, maintenance includes the following procedures:

- *Responding to requests to change the system:* the backlog of change requests must be prioritized and assigned to maintenance staff and then grouped into batches of work for each new release of the system, a step similar to the project identification and selection phase of the SDLC

- *Transforming requests into changes:* assessing if the change can technically be made and if so how the change will affect the existing system and determining what work must be done to make the requested changes, a step similar to the project initiation and analysis phases

- *Designing the changes:* may include activities such as those found in logical and physical design phases

Figure 2
Steps in maintenance

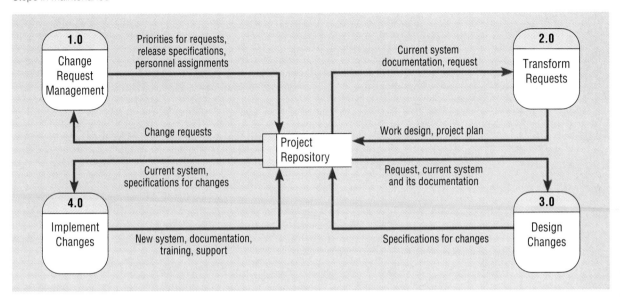

- *Implementing changes:* rewriting code, documentation, training plans, and so forth, similar to the work done during the implementation phase

Maintenance is tightly linked to all phases of the SDLC since it draws upon and adds to information about system requirements, design, and implementation. In fact, maintenance projects are small systems development projects. Maintenance must also keep all documentation, including the CASE repository, up-to-date so that any version of the system can be reconstructed at any time.

Because of the variety of work done during system implementation, we address this rather comprehensive phase in two chapters, 19 and 20. In Chapter 19 you will learn about your role in coding a system. Since other courses in your curriculum likely address the skills required in coding, Chapter 19 concentrates on the management of coding, in which you as an analyst will be directly involved. You will learn about testing systems and system components and methods to ensure and measure software quality.

As you know from your own programming experience, coding must be tested. Your role as a systems analyst is to develop a plan for testing, including all the test data required to exercise every part of the system. You start developing the test plan earlier in the project, usually during analysis, since testing requirements are highly related to system functional requirements. The testing plan you design will usually follow a bottom-up approach, beginning with small modules, extensive testing by the programming group, beta testing with users, and final acceptance testing. A complete history of test data and runs must be kept so that future versions of the system can be compared to demonstrate that discovered problems were fixed and new features do not cause additional problems. Testing insures the quality of software, and you can use certain measures and methods such as structured walkthroughs to assess the quality of software. We illustrate an exemplary software quality assurance plan used for the IBM AS/400 system.

Once built, a system must be installed. For effective installation, existing data must be converted and you must follow a process to "unplug" the existing system and "plug in" the new system. In Chapter 19 we review four popular conversion strategies you can employ and adapt: direct, parallel, single location (pilot), and phased. The comparative advantages and disadvantages of each are discussed so that you can gain insight into the selection of the most appropriate or hybrid strategy for a given system.

Installing a new system involves more than making technical changes to computer systems. Information systems include the people who supply and consume data and information as well as those interested in the organizational units supported by the system. Thus, managing installation involves managing organizational changes as much as it does technical changes. We review several frameworks you can use to anticipate and control human and organizational resistance to change. You will learn about factors, such as organizational politics, that can affect the successful implementation of a new system.

In Chapter 20 we continue with other implementation activities: documenting the system, training, and supporting users. Documentation is extensive for any system. You have already been developing most of the system documentation needed by system maintenance staff by keeping a thorough project workbook or CASE repository. You now need to finalize user documentation: user's guide, release description, system administrator's guide, and reference manuals. In Chapter 20 you will see a generic outline for a user's guide as well as a wide range of guidelines you can use to develop high-quality user documentation. Remember, documentation is part of the system and it too must be tested for completeness, accuracy, and readability.

While documentation is being finalized, user support activities also need to be designed and implemented. First, you must develop a training plan. You may choose to conduct training face-to-face, via computer-based tutorials, or through vendors. New electronic performance support systems are becoming a popular way of delivering on-demand training, and Chapter 20 discusses this and other training methods. Once trained, however, users will still encounter difficulties. So, ongoing support from help desks, newsletters, user groups, on-line bulletin boards, and other mechanisms must be designed, tested, and implemented. You will likely be responsible for designing these support elements, or at least assessing whether such elements exist. You may have to train central information or data center staff on how to handle user questions, how to back up files, and how to recover the system in case of failure. Putting system support operations in place can involve considerable coordination with other system professionals.

One professional who can be of extensive help during implementation is a technical communicator. This person has the skills of a technical writer, but is also able to review all documentation, training, and support literature. This leaves you free to concentrate on the right content and on coordinating all the professionals involved in system implementation. We conclude Chapter 20 with a brief review of project close-down activities, since the end of implementation means the end of the project.

Both Chapters 19 and 20 are followed by case segments for the CATS project at Broadway Entertainment Company. The segment after Chapter 19 deals with conversion problems that have arisen for the in-store part of CATS, and the segment after Chapter 20 deals with creating support procedures for CATS and closing down the project.

The end of the project does not mean that work on the system ceases. In fact, the work is only beginning. Today, as much as 80 percent of the life cycle cost of a system occurs *after* implementation. Because systems can live for so long and must be constantly updated to correct flaws and changed to work with new technologies and to meet new business conditions and regulations, systems maintenance is a major part of the total systems analysis and design process. In Chapter 21, the last chapter, you will learn about your role in systems maintenance.

There are four kinds of maintenance: corrective, adaptive, perfective, and preventive. You can help control the potentially monumental maintenance cost for a system by making systems maintainable. Various factors affect the maintainability of a system, most of which you can influence, such as the number of defects, the number and skill of users, the quality of documentation, and software structure.

You may also be involved in establishing a maintenance group for a system. You will learn about different organizational structures for maintenance personnel and the reasons for each. You will also learn how to measure maintenance effectiveness, fundamental to controlling the cost and quality of maintenance activities. Configuration management, deciding how to handle changes, is at the heart of managing maintenance. You will learn how a systems librarian keeps track of baseline software modules, checks these out to maintenance staff as required, and then how the librarian rebuilds systems. CASE can play a major role in managing documentation, structuring poorly designed system components, and regenerating system modules to implement changes.

In Chapter 21 we discuss one of the major sources of system changes and new systems today, Business Process Re-engineering (BPR). In contrast to the incremental improvements from Total Quality Management (TQM) methods that might drive many adaptive, perfective, and preventive change requests, BPR causes radical changes to systems. New, or disruptive technologies can make current business processes obsolete, requiring extensive system maintenance, or even new systems. We show how BPR relates to corporate and information systems planning and how maintenance makes the SDLC a true life cycle.

Chapter 21 concludes with a discussion of a hypothetical maintenance situation with the Purchasing Fulfillment System at Pine Valley Furniture. In this discussion you will see how a maintenance programmer/analyst applies maintenance procedures to deal with a crisis situation.

System Implementation: Coding, Testing, and Installation

After studying this chapter, you should be able to:

- Describe the process of coding, testing, and converting an organizational information system and outline the deliverables and outcomes of the process.

- Prepare a test plan for an information system.

- Discuss system quality assurance strategies.

- Explain organization change using the Lewin and Schein models of change.

- Apply four installation strategies: direct, parallel, single location, and phased installation.

- Explain why systems implementation sometimes fails.

- Compare the factor and political models of the implementation process.

INTRODUCTION

After maintenance, the implementation phase of the systems development life cycle is the most expensive and most time-consuming phase of the entire life cycle. Implementation is expensive because so many people are involved in the process; it is time-consuming because of all the work that has to be completed during implementation. Physical design specifications must be turned into working computer code, the code must be tested until most of the errors have been uncovered and corrected, the system must be installed, user sites must be prepared for the new system, and users must come to rely on the new system rather than the existing one to get their work done.

Implementing a new information system into an organizational context is not a mechanical process. The organizational context has been shaped and reshaped by the people who work in the organization. The work habits, beliefs, inter-relationships, and personal goals of an organization's members all affect the implementation process. Although factors important to successful implementation have been identified, there are no sure recipes you can follow. During implementation, you must be attuned to key aspects of the organizational context, such as history, politics, and environmental demands—aspects that can contribute to implementation failures if ignored.

Here and in Chapter 20, you will learn about the many activities that comprise the implementation phase. Here we focus on installation and implementation success, although we will also discuss coding, testing, and quality assurance. (Chapter 20 is about documentation, user training, and support for a system after it is installed.) Our intent is not to teach you how to program and test systems—most of you have already learned about writing and testing programs in the courses you took before this one. Rather, this chapter shows you where coding and testing fit in the overall scheme of implementation and stresses the view of implementation as an organizational change process that is not always successful.

After a brief overview of the coding, testing, and installation process and the deliverables and outcomes from this process, we will talk about software application testing. Using the development of IBM's OS/400 operating system as a case in point, we will then list and explain some strategies you can employ to assure high levels of software quality. The following section presents the four types of installation: direct, parallel, pilot, and phased. The final section of the chapter portrays implementation as an organizational change process and discusses the organizational and people issues involved in the implementation effort.

CODING, TESTING, AND INSTALLATION

Coding, testing, and installation (or conversion) are three separate processes in the implementation phase of the systems development life cycle (see Figure 19-1) The purpose of these steps is to convert the final physical system specifications into working and reliable software and hardware. These steps are often done by other project team members besides analysts, although analysts may do some programming. In any case, analysts are responsible for ensuring that coding, testing, and installation are properly planned and executed.

The Process of Coding, Testing, and Installation

Coding, as we mentioned before, is the process whereby the physical design specifications created by the analysis team are turned into working computer code by the programming team. Depending on the size and complexity of the system, coding can be an involved, intensive activity. Once coding has begun, the testing process can begin and proceed in parallel. As each program module is produced, it can be tested individually, then as part of a larger program, and then as part of a larger system. You will learn about the different strategies for testing later in the chapter. We should emphasize that although testing is done during implementation, you must begin planning for testing earlier in the project. Planning involves determining what needs to be tested and collecting test data. This is often done during the analysis phase because testing requirements are related to system requirements.

Finally, installation is the process during which the current system is replaced by the new system. This includes conversion of existing data, software, documentation, and work procedures to those consistent with the new system. Users must give up their old ways of doing their jobs, whether manual or automated, and adjust to accomplishing the same tasks with the new system. Users will sometimes resist the change to the new system and you must help users adjust. Unfortunately, there are only so many aspects of the installation process that you can affect.

Deliverables and Outcomes

Table 19-1 shows the deliverables from the coding, testing, and installation processes. The most obvious outcome is the code itself, but just as important as the

Figure 19-1

The systems development life cycle with the implementation phase highlighted

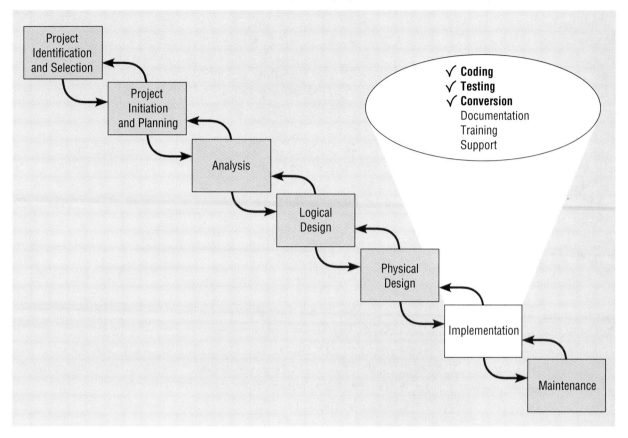

TABLE 19-1   Deliverables for Coding, Testing, and Installation

1. Coding
   a. Code
   b. Program Documentation

2. Testing
   a. Test Scenarios and Test Data
   b. Results of Program and System Testing

3. Installation
   a. User Guides
   b. User Training Plan
   c. Installation and Conversion Plan
      i. Software and Hardware Installation Schedule
      ii. Data Conversion Plan
      iii. Site and Facility Remodeling Plan

code is documentation of the code. Some programming languages, such as COBOL, are said to be largely self-documenting because the language itself spells out much about the program's logic, the labels used for data and variables, and where data are accessed and output. But even COBOL code can be mysterious to maintenance programmers who must maintain the system for years after the original system was written and the original programmers have moved on to other jobs. Therefore, clear, complete documentation for all individual modules and programs is crucial to the system's continued smooth operation. Increasingly, CASE tools are used to maintain the documentation needed by systems professionals. The results of program and system testing are important deliverables from the testing process, as they document the tests as well as the test results. For example, what type of test was conducted? What test data were used? How did the system handle the test? The answers to these questions can provide important information for system maintenance as changes will require retesting and similar testing procedures will be used during the maintenance process.

The next two deliverables, user guides and the user training plan, result from the installation process. User guides provide information on how to use the new system, and the training plan is a strategy for training users so they can quickly learn the new system. The development of the training plan probably began earlier in the project and some training, on the concepts behind the new system, may have already taken place. During early stages of implementation, the training plans are finalized and training on the use of the system will begin. Similarly, the installation plan lays out a strategy for moving from the old system to the new, from the beginning to the end of the process. Installation includes installing the system (hardware and software) at central and user sites. The installation plan answers such questions as when the new system will be installed, which installation strategies will be used, who will be involved, what resources are required, which data will be converted and cleansed, and how long the installation process will take. It is not enough that the system is installed; users must actually use it.

As an analyst, your job is to ensure that all of these deliverables are produced and are done well. You may produce some of the deliverables, such as test data, user guides, and an installation plan; for other deliverables, like code, you may only supervise or simply monitor their production or accomplishment. The extent of your implementation responsibilities will vary according to the size and standards of the organization you work for, but your ultimate role includes ensuring that all the implementation work leads to a system that meets the specifications developed in earlier project phases. See the box "The Future Programmer" for some insights into the changing role of programming and the nature of systems implementation in systems development.

The Future Programmer

Where and by whom programming is done continues to change as the nature of programming languages evolves, resulting in improved programming productivity and the opening of programming to less highly skilled personnel. One prediction suggests that future programmers can be grouped into four categories:

- *IS department programmers:* these people who work for the IS function are in a clear decline, from nearly 2 million in 1994 to several hundred thousand by 2010; some people believe that these jobs are really being distributed out of the central IS function into business units, possibly under different titles

- *Software company programmers:* these programmers work for consulting and packaged software companies, and the number will likely rise from roughly 600,000 in 1994 to several million by 2010

- *Embedded software programmers:* these programmers produce code that is embedded in other products, like cars, office equipment, and consumer electronics; this group will likely dramatically increase from several million in 1994 to over 10 million by 2010

- *Occasional programmers:* these include professionals and technicians (accountants, engineers, managers, and so forth) who program as part of their main duties; this group should rise from roughly 20 million in 1994 to over 100 million by 2010

One theory is that standard business system components (objects in some terminologies) can be assembled into new systems by less skilled programmers. Thus, the job of what we call today a programmer will be to build these components and to ensure the quality of assembled systems. Although the number of occasional programmers (likely not trained in information systems) is exploding, the need for highly skilled programmers and programming work is far from diminishing.

Adapted from Bloor, 1994

SOFTWARE APPLICATION TESTING

As we mentioned previously, analysts prepare system specifications that are passed on to programmers for coding. Although coding takes considerable effort and skill, the practices and processes of writing code do not belong in this text.

However, as software application testing is an activity analysts plan and sometimes supervise, depending on organizational standards, you need to understand the essentials of the testing process.

Testing software begins earlier in the systems development life cycle, even though much of the actual testing activities are carried out in implementation. During analysis, you develop a master test plan. During design, you develop a unit test plan, an integration test plan, and a system test plan. During implementation, these various plans are implemented and the actual testing is performed.

The purpose of these written test plans is to improve communication among all of the people involved in testing the application software. Everyone learns from the plan what their roles are during testing. The test plans also serve as checklists you can use to determine whether all of the master test plan has been completed. The master test plan is not just a single document but a collection of documents. Each of the component documents represents a complete test plan for one part of the system or for a particular type of test. Presenting a complete master test plan is far beyond the scope of this book. We refer you to Mosley's *The Handbook of MIS Software Application Testing* for a complete test plan, which comprises a 101-page appendix. To give you an idea of what a master test plan involves, we present in Table 19-2 an abbreviated table of contents.

A master test plan is a project within the overall system development project. Since at least some of the system testing will be done by people who have not been involved in the system development so far, the Introduction provides general information about the system and the needs for testing. The Overall Plan and Testing Requirements sections are like a baseline project plan for testing, with a schedule of events, resource requirements, and standards of practice outlined. Procedure Control explains how the testing is to be conducted, including how changes to fix errors will be documented. The fifth and final section explains each specific test necessary to validate that the system performs as expected.

Some organizations have specially trained personnel who supervise and support testing. Testing managers are responsible for developing test plans, establishing testing standards, integrating testing and development activities in the life cycle, and ensuring that test plans are completed. Testing specialists help develop test plans, create test cases and scenarios, execute the actual tests, and analyze and report test results.

Seven Different Types of Tests

Software application testing is an umbrella term that covers several types of tests. Mosley (1993) organizes the types of tests according to whether they employ static or dynamic techniques and whether the test is automated or manual. Static testing means that the code being tested is not executed. The results of running the code are not an issue for that particular test. Dynamic testing, on the other hand, does involve execution of the code. Automated testing means the computer conducts the test while manual means that people do. Using this framework, we can categorize types of tests as shown in Table 19-3.

Let's examine each type of test in turn. **Inspections** are formal group activities where participants manually examine code for occurrences of well-known errors. Each programming language lends itself to certain types of errors that programmers make when coding in it, and these common errors are well-known and documented (for an example of common coding errors in COBOL, see Litecky and Davis, 1976). Code inspection participants compare the code they are examining to a checklist of the well-known errors for that particular language. Exactly what the code does is not investigated in an inspection. It has been estimated that code

Inspections: A testing technique in which participants examine program code for predictable language-specific errors.

TABLE 19-2 Table of Contents of a Master Test Plan
(Adapted from Mosley, 1993)

1. Introduction
   a. Description of system to be tested
   b. Objectives of the test plan
   c. Method of testing
   d. Supporting documents

2. Overall Plan
   a. Milestones, schedule, and locations
   b. Test materials
      1. Test plans
      2. Test cases
      3. Test scenarios
      4. Test log
   c. Criteria

3. Testing Requirements
   a. Hardware
   b. Software
   c. Personnel

4. Procedure Control
   a. Test initiation
   b. Test execution
   c Test failure
   d. Access/change control
   e. Document control

5. Test Specific or Component Specific Test Plans
   a. Objectives
   b. Software description
   c. Method
   d. Milestones, schedule, progression, and locations
   e. Requirements
   f. Criteria
   g. Resulting test materials
   h. Execution control
   i. Attachments

TABLE 19-3 A Categorization of Test Types
(Adapted from Mosley)

	Manual	**Automated**
Static	Inspections	Syntax Checking
Dynamic	Walkthroughs	Unit Test
	Desk Checking	Integration Test
		System Test

inspections have been used by organizations to detect from 60 to 90 percent of all software defects as well as to provide programmers with feedback that enables them to avoid making the same types of errors in future work (Fagan, 1986). The inspection process can also be used for such things as design specifications.

Unlike inspections, what the code does is an important question in a *walkthrough.* Structured walkthroughs are a very effective method of detecting errors in code. As you saw in Chapter 7, structured walkthroughs can be used to review many systems development deliverables, including logical and physical design specifications as well as code. Whereas specification walkthroughs tend to be formal reviews, code walkthroughs tend to be informal. Informality tends to make programmers less apprehensive about walkthroughs and helps increase their frequency. According to Yourdon (1989), code walkthroughs should be done frequently when pieces of work reviewed are relatively small and before the work is formally tested. If walkthroughs are not held until the program is being tested, the programmer has already spent much time looking for errors which the programming team could have found much more quickly. The programmer's time has been wasted, and the other members of the team may become frustrated because they will not find as many errors as they would have if the walkthrough had been conducted earlier. Further, the longer a program goes without being subjected to a walkthrough, the more defensive the programmer becomes when the code is reviewed. Although each organization that uses walkthroughs conducts them differently, there is a basic structure that you can follow that works well (see Figure 19-2).

It should be stressed that the purpose of a walkthrough is to detect errors, not to correct them. It is the programmer's job to correct the errors uncovered in a walkthrough. Sometimes it can be difficult for the reviewers to refrain from suggesting ways to fix the problems they find in the code, but increased experience with the process can help change a reviewer's behavior.

What the code does is also important in **desk checking,** an informal process where the programmer or someone else who understands the logic of the program works through the code with a paper and pencil. The programmer executes each instruction, using test cases that may or may not be written down. In one sense, the reviewer acts as the computer, mentally checking each step and its results for the entire set of computer instructions.

> **Desk checking:** A testing technique in which the program code is sequentially executed mentally by the reviewer.

Among the list of automated testing techniques in Table 19-3, there is only one technique that is also static, syntax checking. Syntax checking is typically done by a compiler. Errors in syntax are uncovered but the code is not executed. For the other three automated techniques, the code is executed.

The first such technique is unit testing, sometimes called module testing. In unit testing, each module is tested alone in an attempt to discover any errors that may exist in the module's code. But since modules co-exist and work with other modules in programs and systems, modules must be tested together in larger groups. Combining modules and testing them is called **integration testing.** Integration

> **Integration testing:** The process of bringing together all of the modules that comprise a program for testing purposes. Modules are typically integrated in a top-down, incremental fashion.

GUIDELINES FOR CONDUCTING A CODE WALKTHROUGH
1. Have the review meeting chaired by the project manager or chief programmer, who is also responsible for scheduling the meeting, reserving a room, setting the agenda, inviting participants, and so on.
2. The programmer presents his or her work to the reviewers. Discussion should be general during the presentation.
3. Following the general discussion, the programmer walks through the code in detail, focusing on the logic of the code rather than on specific test cases.
4. Reviewers ask to walk through specific test cases.
5. The chair resolves disagreements if the review team cannot reach agreement among themselves and assigns duties, usually to the programmer, for making specific changes.
6. A second walkthrough is then scheduled if needed.

Figure 19-2
Steps in a typical code walkthrough (Adapted from Yourdon, 1989)

testing is gradual. You first test the coordinating module (the root module in a structure chart tree) and only one of its subordinate modules. After the first test, you add one or two other subordinate modules from the same level. Once the program has been tested with the coordinating module and all of its immediately subordinate modules, you add modules from the next level and then test the program. You continue this procedure until the entire program has been tested as a unit. **System testing** is a similar process, but instead of integrating modules into programs for testing, you integrate programs into systems. System testing follows the same incremental logic that integration testing does. Under both integration and system testing, not only do individual modules and programs get tested many times, so do the interfaces between modules and programs.

Current practice calls for a top-down approach to writing and testing modules. Under a top-down approach, the coordinating module is written first. Then the modules at the next level in the structure chart are written, followed by the modules at the next level, and so on, until all of the modules in the system are done. Each module is tested as it is written. Since top-level modules contain many calls to subordinate modules, you must wonder how they can be tested if the lower level modules haven't been written yet. The answer is **stub testing.** Stubs are two or three lines of code written by a programmer to stand in for the missing modules. During testing, the coordinating module calls the stub instead of the subordinate module. The stub accepts control and then returns it to the coordinating module.

Figure 19-3 illustrates stub and integration system testing. Stub testing is depicted as the innermost oval. Here the Get module is being written and tested, but

System testing: The bringing together of all the programs that comprise a system for testing purposes. Programs are typically integrated in a top-down, incremental fashion.

Stub testing: A technique used in testing modules, especially where modules are written and tested in a top-down fashion, where a few lines of code or stubs are used to substitute for subordinate modules.

Figure 19-3
Comparing stub and integration testing

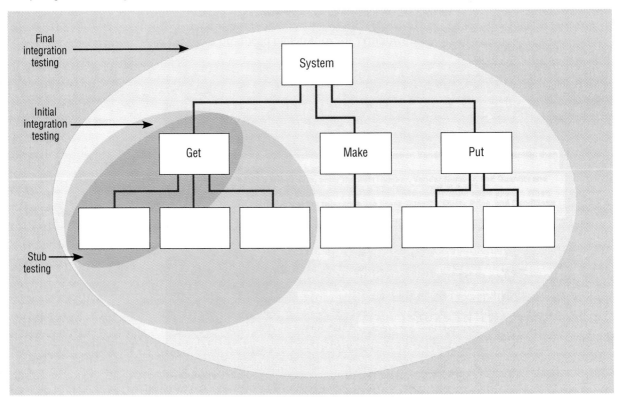

as none of its subordinate modules have been written yet, each one is represented by a stub. In the stub testing illustrated by Figure 19-3, Get is tested with only one stub in place, for its left-most subordinate module. You would of course write stubs for all of the Get module's subordinate modules, just as you would for the Make and Put modules. Once all of the subordinate modules are written and tested, you would conduct an integration test of Get and its subordinate modules, as represented by the second oval. As stated previously, the focus of an integration test is on the inter-relationships among modules. You would also conduct integration tests of Make and its subordinate modules, of Put and its subordinates, and of System and its subordinates, Get, Make, and Put. Eventually your integration testing would include all of the modules in the large oval that encompasses the entire program.

System testing is more than simply expanded integration testing where you are testing the interfaces between programs in a system rather than testing the interfaces between modules in a program. System testing is also intended to demonstrate whether a system meets its objectives. This is not the same as testing a system to determine whether it meets requirements—that is the focus of acceptance testing, discussed later. To verify that a system meets its objectives, system testing involves using non-live test data in a non-live testing environment. Non-live means the data and situation are artificial, developed specifically for testing purposes, although both data and environment are similar to what users would encounter in everyday system use. The system test is typically conducted by information systems personnel led by the project team leader, although it can also be conducted by users under MIS guidance. The scenarios that form the basis for system tests are prepared as part of the master test plan.

The Testing Process

Up to this point, we have talked about the master test plan and seven different types of tests for software applications. We haven't said very much about the process of testing itself. There are two important things to remember about testing information systems:

1. The purpose of testing is confirming that the system satisfies requirements, including finding errors.

2. Testing must be planned.

These two points have several implications for the testing process, regardless of the type of test being conducted. First, testing is not haphazard. You must pay attention to many different aspects of a system, such as response time, response to boundary data, response to no input, response to heavy volumes of input, and so on. You must test anything (within resource constraints) that could go wrong or be wrong about a system. At a minimum, you should test the most frequently used parts of the system and as many other paths through the system as time permits. Planning gives analysts and programmers an opportunity to think through all the potential problem areas, list these areas, and develop ways to test for problems. As indicated previously, one part of the master test plan is creating a set of test cases, each of which must be carefully documented (see Figure 19-4 for an outline of a test case description).

A test case is a specific scenario of transactions, queries, or navigation paths that represent a typical, critical, or abnormal use of the system. A test case should

Figure 19-4

Test case description form
(Adapted from Mosley, 1993)

Pine Valley Furniture Company
Test Case Description

Test Case Number:
Date:
Test Case Description:

Program Name:
Testing State:
Test Case Prepared By:

Test Administrator:

Description of Test Data:

Expected Results:

Actual Results:

be repeatable, so that it can be rerun as new versions of the software are tested. Even though analysts often do not do the testing, systems analysts, because of their intimate knowledge of the application, often make up or find test data. The people who create the test cases should not be the same people as those who coded and tested the system. In addition to a description of each test case, there must also be a description of the test results, with an emphasis on how the actual results differed from the expected results (see Figure 19-5). This description will indicate why the results were different and what, if anything, should be done to change the software. This description will then suggest the need for retesting, possibly introducing new tests necessary to discover the source of the differences.

One important reason to keep such a thorough description of test cases and results is so that testing can be repeated for each revision of an application. Although

Figure 19-5
Test case results form
(Adapted from Mosley, 1993)

Pine Valley Furniture Company
Test Case Results

Test Case Number:
Date:

Program Name:
Module Under Test:

Explanation of difference between actual and expected output:

Suggestions for next steps:

new versions of a system may necessitate new test data to validate new features of the application, previous test data usually can and should be reused. Results from use of the test data with prior versions are compared to new versions to show that changes have not introduced new errors and that the behavior of the system, including response time, is no worse. A second implication for the testing process is that test cases must include illegal and out-of-range data. The system should be able to handle any possibility, no matter how unlikely; the only way to find out is to test.

If the results of a test case do not compare favorably to what was expected, the error causing the problem must be found and fixed. Programmers use a variety of debugging tools to help them locate and fix errors. A sophisticated debugging tool called a symbolic debugger allows the program to be run on-line, even one

Automating Testing

Software testing tools provide the following functions which improve the quality of testing:

- Record or build scripts of data entry, menu selections and mouse clicks, and input data, which can be replayed in exact sequence for each test run
- Compare the results of one test run with those from prior test cases to identify errors or to highlight the results of new features
- Supported unattended script playing to simulate high-volume or stress situations

Such tools can reduce the time for software testing by almost 80 percent.

instruction at a time if the programmer desires, and allows the programmer to observe how different areas of data are affected as the instructions are executed. This cycle of finding problems, fixing errors, and rerunning test cases continues until no additional problems are found. There are specific testing methods that have been developed for generating test cases and guiding the test process. Discussion of these methods is beyond the scope of this book, but you are referred to Mosley's 1993 book for an excellent single source on several different applications software testing methods. See the box "Automating Testing" for an overview of tools to assist you in testing software.

Acceptance Testing by Users

Acceptance testing: The process whereby actual users test a completed information system, the end result of which is the users' acceptance of it.

Alpha testing: User testing of a completed information system using simulated data.

Beta testing: User testing of a completed information system using real data in the real user environment.

Once the system tests have been satisfactorily completed, the system is ready for **acceptance testing,** testing the system in the environment where it will be used by the people who will eventually be using it. Acceptance refers to the fact that users typically sign off on the system and "accept" it once they are satisfied with it. As we said previously, the purpose of acceptance testing is for users to determine whether the system meets their requirements. The extent of acceptance testing will vary with the organization and with the system in question. The most complete acceptance testing will include **alpha testing,** where simulated but typical data are used for system testing; **beta testing,** in which live data are used in the users' real working environment; and a system audit conducted by the organization's internal auditors or by members of the quality assurance group.

During alpha testing, the entire system is exercised to discover whether or not the system is overtly destructive to itself or to the rest of the environment. The types of tests performed during alpha testing include the following:

- *Recovery testing*—forces the software (or environment) to fail in order to verify that recovery is properly performed;

- *Security testing*—Verifies that protection mechanisms built into the system will protect it from improper penetration;

- *Stress testing*—tries to break the system (for example, what happens when a record is written to the database with incomplete information or what happens under extreme on-line transaction loads or a large number of concurrent users);

- *Performance testing*—determines how the system performs on the range of possible environments in which it may be used (for example, different hardware configurations, networks, operating systems, and so on); often the goal is to have the system perform with similar response time and other performance measures in each environment.

In beta testing, a subset of the intended users exercise the system in their own environments using their own data. The intent of the beta test is to determine whether the software, documentation, technical support, and training activities work as intended. In essence, beta testing can be viewed as a rehearsal of the installation phase. Problems uncovered in alpha and beta testing in any of these areas must be corrected before users will accept the system for daily use. There are many stories systems analysts can tell about long delays in final user acceptance due to system bugs (see box "Bugs in the Baggage" for one famous incident).

QUALITY ASSURANCE

Regardless of how competent and careful analysts and programmers are in designing and coding an information system, there will always be mistakes made in the design and in the code. Software defects can be classified into six general categories: requirements, design, coding, documentation, administrative, and bad fixes (Jones, 1986). Requirements, design, coding, and documentation defects are created during the analysis, design, and implementation stages. Administrative defects consist of software maintenance and service problems occurring after delivery. Bad fixes are new defects incorrectly or accidentally injected into software when programmers try to fix other defects. The common metric for measuring software defects is the rate of defects per thousand lines of code (KLOC). Figure 19-6 shows the number of defects per KLOC to be expected in a typical software development project (Phan, 1990; Phan, George, and Vogel, 1994). This figure shows that even after beta testing, we can expect one percent of the system to be faulty! This result suggests that we must be diligent in testing, that even thoroughly tested systems will fail, and that we must be prepared to recover and support systems after their inevitable failure.

Although the number of defects in in-house and commercial software may seem high, they are actually low compared to historical standards. The primary reason that defect rates are as low as they are is that software developers have come to rely more and more on quality assurance measures.

What is quality assurance? Software quality assurance covers a range of procedures and techniques used during the software development process to enhance the inputs to the process, to improve the process itself and, in so doing, to prevent problems. The focus of quality assurance, then, is the prevention of problems during the production process itself. Testing is not so much quality assurance as it

Bugs in the Baggage

Testing a complex software system can be long and frustrating. A case in point was the software to control 4,000 baggage cars at the Denver International Airport. Errors in the software put the airport's opening on hold for months, costing taxpayers $500,000 a day and turning airport bonds into junk status. (Note, as of the writing of this book, the airport was still not open more than a year after the discovery of the baggage problems; so, the rest of this story is left for you to investigate.) Various causes of the delay were identified, including last minute design change requests from airport officials and mechanical problems. "The bottom-line lesson is that system designers must build in plenty of test and debugging time when scaling up proven technology into a much more complicated environment."

Adapted from Scheier, 1994

Development Activities	Defects per KLOC				
	Begin	Created	Detected & Removed	Bad Fixes	Remaining
Design & Review	0.0	34.0	(23.8)	4.8	15.0
Coding & Review	15.0	16.0	(21.7)	4.3	13.6
Unit Test	13.6	0.0	(9.5)	1.9	6.0
Integration Test	6.0	0.0	(4.2)	0.8	2.6[1]
Beta Test	2.6	0.0	(1.8)	0.3	1.1[2]

[1] Defect rate at delivery for in-house software development
[2] Defect rate at delivery for commercial software development

Figure 19-6
Estimated software defects in each stage of development and testing

is quality control, since testing acts as a last line of defense where problems are discovered and corrected before the final product is delivered to customers. To introduce you to quality assurance methods in the systems development process and to show you how quality assurance methods work, let's look at quality assurance in a recent large-scale development effort, the development of OS/400 R.1 (release 1).

Assuring Quality in the OS/400 Operating System

OS/400 R.1 is the operating system for the IBM Corporation's AS/400 mid-range business computer system. IBM at Rochester, Minnesota, developed OS/400 R.1 during 26 months in 1986–1988. Over 50 million lines of high-level code, including scaffolding, were written. **Scaffolding** consists of modules and data written for debugging purposes but never intended to be part of the finished product. The code was then refined to over 3.6 million lines of actual product code, 1.2 million lines of which were reused (Phan, 1990). The challenge for IBM Rochester was to identify and satisfy the many needs of customers in the mid-range computer marketplace.

Scaffolding: Program modules and data written for debugging purposes but never intended to be part of the final product delivered to users or customers.

IBM Rochester employed several different quality assurance measures to improve the software development process and to reduce the number of problems with the software as it was being produced. These measures included (1) rigorous requirements definition (2) aggressive resource control techniques (3) multiple-level code reviews (4) high level of code reuse and (5) rigorous control of the change request process.

IBM Rochester began by rigorously defining customer requirements and by defining and implementing resource control strategies. The OS/400 project team undertook a careful study of customer requirements and market opportunities, using customer surveys and interviews, literature searches, and conference attendance. Customer requirements were grouped and converted into project requirements and estimates were made of the resources needed to develop key requirements.

As might be expected for any such project, IBM's efforts were constrained by limited resources. Deviating from the traditional IBM culture and structural inertia that had made it difficult for divisions to open up to outsiders and move quickly to new market niches, IBM headquarters gave IBM Rochester the independence to control resources. This allowed Rochester to employ such resource control strategies as collaboration, coordination, merging and distributing of workload, reorganization, contraction, and use of outside software partners.

Throughout the development, code written for OS/400 R.1 had to go through various inspections, including reviews of high-level design, low-level design, and code. The high-level design review examined functions, interface, and data structures of the software. The low-level design review focused on the logic flow, pseudocode, and error conditions in the design. The code reviews consisted of structured program walkthroughs conducted by experienced programmers, and problems and defects were logged by the review moderators. All defects discovered had to be resolved before proceeding to the next step. Many of these reviews and tests were performed jointly with customers and end users. User involvement and a good feedback process helped to detect and fix problems at early stages. IBM Rochester business executives and project team members set up regular meetings with users to discuss what users disliked about the system. Customers were invited to review and validate project requirements and test results to make sure that the delivered products would be trouble-free at customer IS sites. User satisfaction levels, feedback, and system performance evaluations were gathered frequently by surveys conducted by IBM and independent consulting firms. The IBM customer support team phoned every customer 90 days after installation to collect feedback and call attention to additional problems that might have been encountered.

During the development period, the Integrated Development Support System (IDSS), an internal IBM CASE tool, played a major role in supporting OS/400 development activities and in maintaining an inventory of millions of lines of good quality reusable code. The IDSS software development environment provides tools for various development life cycle stages and includes CASE tools for designing, coding, compiling, testing and debugging, integrating, building, and maintenance. Supporting the IDSS data management are application databases and a library control system database. In the OS/400 project, the application databases were applied primarily to project management and office management functions; the library control system database was applied to software development. In addition, IDSS collects information about activities performed in the development process and makes the information available for query, reporting, and analysis.

During the development process, there was a constant flow of requests, suggestions for improvement, and new requirements from various sources. To control the quality of the development process, the changes that took place required a thorough understanding of the requirements and the use of a controlled approach for changes. Project management chose two tools to help control and improve software quality: a design change request (DCR), which controlled the entire life cycle of the project; and a problem tracking report (PTR), which controlled defects during development.

DCR was used in the OS/400 project to record and track design changes, beginning with the initial investigation of an idea through the development phases to the final integration necessary for building an operational product. It was also used to control the implementation of new functions for incremental improvements. Whenever there was a need to create or to change system capabilities, features, or design, a DCR inspection meeting was conducted. If the DCR passed the inspection, an initial estimate of needed resources and efforts was made and sent to management for approval.

On the other hand, PTR was used to record and track suspected problems and their solutions. It also provided management with a means of gathering defect removal data, identifying error-prone components, judging the workload, and certifying software quality. The problem tracking report was an on-line form used to record problems found during development and to route those problems to the responsible development team for fixes. Whenever a user discovered a defect, a PTR was created and logged into the PTR subsystem. Then the system notified the developer responsible for providing answers and fixes. Each defect had a specific timetable for being corrected, with the allocated time dependent on the seriousness of the defect.

The code was subjected to testing at four levels: stub, integration, system, and beta. Experience from prior systems development projects helped developers build test plans and test cases early, even before coding started. In all test phases, defects were recorded into a database for on-line query and monitoring. Project documentation was maintained on-line using the IDSS library.

The OS/400 R.1 project was very successful. The operating system came in on time, within budget, and met user requirements. In 1990, the Rochester facility was awarded the Malcolm Baldrige National Quality Award for its OS/400 development efforts. The software quality assurance techniques IBM Rochester used were key contributors to the project's success and national recognition (details on this project are derived largely from Phan, 1990, and Phan, George, and Vogel, 1994).

Measuring Software Quality

Part of quality assurance is measuring how good the code is at different stages of testing and when in operation. Based on the fundamental Total Quality Management principle that you cannot control something you do not measure, you and quality assurance specialists need metrics to gauge the quality of the code in the coding process. Metrics are also at the heart of encouraging and tracking improvements and determining productivity. Users will expect that the system you design is of high quality; unless you can measure the quality of software, you will not be able to deliver what users expect.

What you measure to determine and manage software quality will determine what improvements are made, so the choice of measures is critical. A general rule is not to use just one or two measures since these tend to provide too simplistic a view and focus attention on only limited aspects of coding, at the sacrifice of other aspects. Another general rule is to measure both the product (the code) and the process (the coding) since, as we saw previously, improvement in the product will come from improvements in the process.

Measures used for evaluating code and coding should be selected based on the following criteria:

- *Understandable:* Metrics must be understood by both systems managers and programmers to know what to do to improve performance.

- *Field-tested:* Measurements should not be used until they are fully tested to confirm that improvements in the measurements actually lead to better software.

- *Economical:* Metrics should be relatively easy to capture and compute.

- *High leverage:* Measurements that lead to only minute improvements are less attractive than those that result in major improvements.

- *Timely:* Metrics should be captured in time to fix problems before the situation becomes much worse.

Measuring software quality is based on measuring defects, their nature, source, severity, and when the defect was discovered. Severity could be measured on a numeric scale or simply by broad categories such as major, moderate, and minor. Errors may be categorized by such characteristics as the following:

- Problems in controlling the number of times through a loop

- Incomplete edit of data

- Incorrect interface between modules

- Subscript out of range or

- Many other technical coding features

Identifying the source of the error is a matter of assigning responsibility for the cause. In general, the more skilled programmer would create the fewest number of errors (possibly weighted by severity) per unit of code produced (amount of code produced can be measured by number of lines of code or number of elemental units of code called function points). The source of a defect can be in particular program modules, interfaces between modules, or in system design specifications.

The point of measuring when errors are detected is to assess the overall quality assurance process. Errors detected early in testing versus later during system operation imply a better quality assurance process. To judge such numbers requires capturing metrics over time or comparing measures to industry benchmarks. Consulting organizations and several industry trade groups can be a source of benchmark statistics on software quality.

Yourdon (1993) summarizes findings by various practitioners and researchers on possible software quality metrics. Some of the metrics listed by Yourdon include the following:

- Defect rate by hour, day, week, or month

- Defect rate per function point or similar unit of software size

- Mean time between failures (MTBF)

- Mean time to repair a defect (MTTR)

- Size of defect backlog

- Number of clean compilations or system component runs on first attempt

- Cumulative defects per version

- Timeliness of response to defect or time to fix defect

- Customer satisfaction with quality of work done to fix defect

As an analyst, you are responsible to your client and other stakeholders for the system you are building. The quality of software, whether measured quantitatively or subjectively by the attitude of your clients, reflects on the quality of the project and your work. Thus, although you may not do any or all of the coding, you should be concerned that the coding is done with high quality.

INSTALLATION

Together, quality assurance techniques and a sound test plan help produce high-quality information systems which can then be turned over to organizational members for daily use. The process of moving from the old information system to the new one is called **installation.** All employees who use a system, whether they were consulted during the development process or not, must give up their reliance on the current system and begin to rely on the new system. Four different approaches to installation have emerged over the years: direct, parallel, single location, and phased (Figure 19-7). The approach an organization decides to use will depend on the scope and complexity of the change associated with the new system and the organization's risk aversion.

Installation: The organizational process of changing over from the current information system to a new one.

Direct Installation

The **direct** or abrupt approach to **installation** (also called "cold-turkey") is as sudden as the name indicates: the old system is turned off and the new system is turned on (Figure 19-7a). Under direct installation, users are at the mercy of the new system. Any errors resulting from the new system will have a direct impact on the users and how they do their jobs and, in some cases—depending on the centrality of the system to the organization—on how the organization performs its

Direct installation: Changing over from the old information system to a new one by turning off the old system as the new one is turned on.

Figure 19-7

Comparison of installation strategies

(a) Direct installation

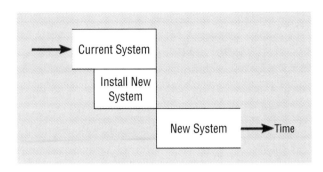

(b) Parallel installation

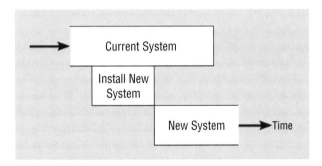

business. If the new system fails, considerable delay may occur until the old system can again be made operational and business transactions reentered to make the database up-to-date. For these reasons, direct installation can be very risky. Further, direct installation requires a complete installation of the whole system. For a large system, this may mean a long time until the new system can be installed, thus delaying system benefits or even missing the opportunities which motivated the system request. On the other hand, it is the least expensive installation method and it creates considerable interest in making the installation a success. Sometimes, a direct installation is the only possible strategy since there is no way for the current and new systems to coexist, which happens in some way in each of the other installation approaches.

Parallel Installation

Parallel installation: Running the old information system and the new one at the same time until management decides the old system can be turned off.

Parallel installation is as riskless as direct installation is risky. Under **parallel installation,** the old system continues to run alongside the new system until users and management are satisfied that the new system is effectively performing its duties and the old system can be turned off (Figure 19-7b). All of the work done by the old system is concurrently performed by the new system. Outputs are compared (to the extent possible) to help determine whether the new system is performing as well as the old system. Errors discovered in the new system do not cost the organization much, if anything, as errors can be isolated and the business can be supported from the old system. Since all work is essentially done twice, a parallel installation can be very expensive, as running two systems also implies employing (and paying) two staffs to not only operate both systems but also to maintain them. A parallel approach can also be confusing to users since they must deal with both systems. As with direct installation, the whole new system is installed, so there can be a considerable delay until the full system is ready for installation. A parallel ap-

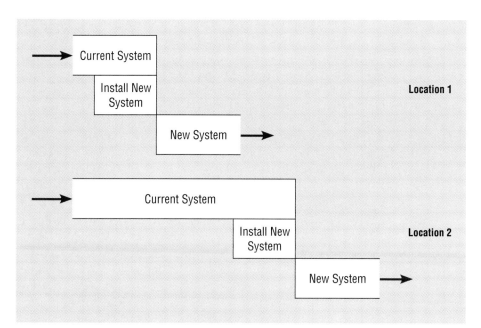

(c) Single location installation (with direct installation at each location)

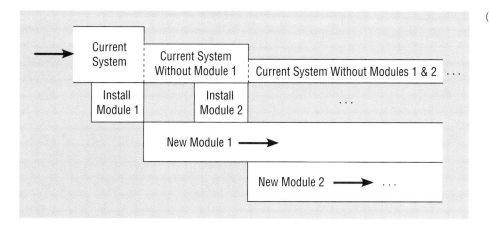

(d) Phased installation

proach may not be feasible, especially if the users of the system (such as customers) cannot tolerate redundant effort.

Single Location Installation

Single location installation, also known as location and pilot installation, is a middle-of-the-road approach compared to direct and parallel installation. Rather than convert all of the organization at once, **single location installation** involves changing from the old to the new system in only one place or in a series of separate sites over time (Figure 19-7c depicts this approach for a simple situation of two locations). The single location may be a branch office or a single factory or one department, and the actual approach used for installation in that location may be any of the other approaches. The key advantage to single location installation is that it limits potential damage and the potential cost (as in the case of parallel installation) by

Single location installation: Trying out a new information system at one site and using the experience to decide if and how the new system should be deployed throughout the organization.

limiting the effects to a single site. Once management has determined that installation has been successful in one location, the new system may be deployed in the rest of the organization, possibly continuing with installation in one location at a time. Success at the pilot site can be used to convince reluctant sites that the system can be worthwhile for them as well. Problems with the system (the actual software as well as documentation, training, and support) can be resolved before deployment to other sites. The single location installation approach is simpler for users since they are working with only one system, but it still places a large burden on IS staff to support two versions of the system. On the other hand, because problems are isolated at one site at a time, IS staff can devote all their efforts to the success at the pilot site. Also, if different locations require sharing of data, extra programs will need to be written to synchronize the current and new systems; although this will happen transparently to users, it is extra work for IS staff. As with each of the prior approaches, the whole system is installed, and not all parts of the organization will get the help of the new system, delaying benefits even longer for some sites.

Phased Installation

Phased installation:
Changing over from the old information system to a new one incrementally, starting with one or a few functional components and then gradually extending the installation to cover the whole new system.

Phased installation (also called staged installation) is an incremental approach to implementing a new information system. Under **phased installation,** the new system is brought on-line in functional components; different parts of the old and new systems are used in cooperation until the whole new system is installed (Figure 19-7d shows the phase-in of the first two modules of a new system). Phased installation, like single or a series of location installations, is an attempt to limit the organization's exposure to risk, whether in terms of cost or in terms of disruption to the business. By converting gradually, the organization's risk is spread out over time and place. Also, a phased installation allows for some benefits from the new system to be achieved before the whole system can be ready. For example, new data capture methods can be used before all reporting modules are ready. For a phased installation, the new and replaced system must be able to coexist and probably share data. Thus, bridge programs connecting old and new databases and programs often must be built. Sometimes, the new and old systems are so incompatible (built using a totally different structure) that pieces of the old system cannot be incrementally replaced, so this strategy becomes infeasible. A phased installation is akin to bringing out a sequence of releases of the system. Thus, a phased approach requires careful version control, repeated conversions at each phase, and a long period of change, which may be frustrating and confusing to users. On the other hand, each phase of change is smaller and more manageable for all involved.

Planning Installation

Each installation strategy involves converting not only software but also data and potentially hardware, documentation, work methods, job descriptions, office and other facilities, training materials, business forms, and all other aspects of the system. For example, it is necessary to recall or replace all the old system documentation and business forms, which suggests that the IS department must keep track of who has these items so that they can be notified and receive replacement items. In practice you will rarely choose a single strategy to the exclusion of all others; most installations will rely on a combination of two or more approaches. For example, if you choose a single location strategy, you have to decide how installation will proceed there and at subsequent sites. Will it be direct, parallel, or phased?

Of special interest in the installation process is the conversion of data. Since existing systems usually contain data required in the new system, current data must be rid of errors, unloaded from current files, combined with new data, and loaded into new files. Data may need to be reformatted to be consistent with more advanced data types supported by newer technology used to build the new system. New data fields may have to be entered in large quantities so that every record copied from the current system has all the new fields populated. Manual tasks, such as taking a physical inventory, may need to be done in order to validate data before it is transferred to the new files. The total data conversion process can be tedious. Further, this process may require current systems to be shut off while the data is extracted so that updates to old data, which would contaminate the extract process, cannot occur.

Any decision that requires the current system to be shut down, in whole or in part, before the replacement system is in place must be done with care. Typically, off hours are used for installations that require a lapse in system support. Whether a lapse in service is required or not, the installation schedule should be announced to users well in advance to let them plan their work schedules around outages in service and periods when their system support might be erratic. Successful installation steps should also be announced, and special procedures put in place so that users can easily inform you of problems they encounter during installation periods. You should also plan for emergency staff to be available in case of system failure so that business operations can be recovered and made operational as quickly as possible. Another consideration is the business cycle of the organization. Most organizations face heavy workloads at particular times of year and relatively light loads at other times. A well-known example is the retail industry where the busiest time of year is the fall, right before the year's major gift-giving holidays. You wouldn't want to schedule installation of a new point-of-sale system to begin December 1 for a department store. Make sure you understand the cyclical nature of the business you are working with before you schedule installation.

Planning for installation may begin as early as analysis, when you study the organization supported by the system. Some installation activities, such as buying new hardware, remodeling facilities, validating data to be transferred to the new system, and collecting new data to be loaded into the new system must be done before the software installation can occur. Often the project team leader will be responsible for anticipating all installation tasks and will assign responsibility for each to different analysts.

Each installation process involves getting workers to change the way they work. As such, installation should be looked at not as simply installing a new computer system, but as an organizational change process. More than just a computer system is involved—you are also changing how people do their jobs and how the organization operates.

MANAGING ORGANIZATIONAL CHANGE

As the previous discussion of installation approaches implied, implementation is an organizational change process. As such, implementation should be handled like other organizational changes. Perhaps the best known prescription for handling permanent social change was offered by Lewin in the 1940s: unfreezing the current situation and circumstances, moving to the new level of operations, and refreezing the situation at the new level (Lewin, 1951). Unfreezing involves breaking current habits, getting people to stop doing things the way they are used to, essentially persuading workers to *unlearn* the way they do things. Once old processes and operations have been unfrozen, new processes and operations can be introduced, and once the new ways take effect, they can be frozen into place as the accepted methods for working. Without unfreezing, organizational change becomes a process of pushing new methods to workers while they continue to think and act in the old ways. Workers may change for a short duration but, eventually, if the old ways are still frozen in place, workers will return to the old ways without regard for the new methods to which they have been introduced.

Edgar Schein, after years of helping organizations make permanent changes, offers an updated version of the Lewin model (see Figure 19-8). Schein writes that for unfreezing to occur successfully, three conditions must be met. First, employee expectations about outcomes for a given action are consistently not being met; this is what Schein calls disconfirmation. Second, this lack of expected outcomes must generate anxiety about missed goals or guilt about the violation of some ideal. Third, employees must feel a sense of psychological safety so they can recognize that past ways of operating were inadequate. In other words, workers must feel psychologically that making a particular change is safe. Unfreezing typically takes the largest proportion of time in the overall organizational change process. If unfreezing has been successful, change itself becomes much easier. According to Schein, motivation and readiness to change are essential to the process, but organizational members will also need to identify with someone who is spearheading the change, such as a manager, an analyst, or a key user. For refreezing to be effective, workers have to accept and act on the new ways of operating, and the local group of people they interact with in the organization must also be receptive and supportive of the change.

The Lewin and Schein models are general representations of any organizational change process. How can these models be applied to information systems implementation? There are several guidelines you can glean from the Lewin and Schein models. First, analysts should understand the importance of refreezing. Users have been known to enthusiastically use a new system for a short while after installation, only to fall back on old methods later. In some cases, users have used the new system just enough to keep from getting into trouble, but they have depended on paper versions of the old system they created and keep going. For example, a factory worker may refer to old hard copy blueprints taped to the side of the toolbox rather than the new blueprint that accompanies the job. Workers may begin to rely on the old system if the new system behaves other than as expected, especially if the worker senses the possibility of errors in the new system. In other cases, users have used the new system in ways they were not designed for in order to make the new system seem more like the old one.

Figure 19-8

Schein's three-stage model of the change process (Adapted from Schein, 1987)

Stage 1: *Unfreezing*	Creating motivation and readiness to change through disconfirmation; creation of guilt or anxiety; provision of psychological safety
Stage 2: *Changing Through Cognitive Restructuring*	Helping employees to see things, judge things, feel things, and react to things differently based on a new point of view
Stage 3: *Refreezing*	Helping employees to integrate the new point of view into their personal view and into their local organizational relationships

Second, you should be aware that unfreezing may take a great deal of time. Third, for change to be successful, you may have to act as an advocate for the new system and be willing to be the person with whom the users identify. If you cannot assume this role, you may have to find someone in the user ranks to perform it. Finally, for refreezing, you must consider how those who interact with users will react to the new system.

Seeing implementation as an organizational change process, with the three phases of unfreezing, changing, and refreezing, will not guarantee that an implementation effort is successful. However, such a perspective will allow the analyst team to plan and manage the change process involved in implementation better than if the process were only viewed as a technical exercise.

ORGANIZATIONAL ISSUES IN SYSTEMS IMPLEMENTATION

Despite the best efforts of the systems development team to design and build a quality system and to manage the change process in the organization, the implementation effort sometimes fails. Sometimes employees will not use the new system that has been developed for them or, if they do use the system, their level of satisfaction with it is very low. Why do systems implementation efforts fail? This question has been the subject of information systems research for the past 25 years. In this section, we will try to provide some answers, first by looking at what conventional wisdom says are important factors related to implementation success, then by investigating factor-based models and political models of systems implementation.

Why Implementation Sometimes Fails

The conventional wisdom that has emerged over the years is that there are at least two conditions necessary for a successful implementation effort: management support of the system under development and the involvement of users in the development process (Ginzberg, 1981b). Conventional wisdom holds that if both of these conditions are met, you should have a successful implementation. Yet, despite the support and active participation of management and users, information systems implementation sometimes fails (see Figure 19-9 for examples).

The importance of management support and user involvement may be overrated. Research has shown that the link between user involvement and implementation success is weak, if it exists at all (Ives and Olson, 1984). User involvement can help reduce the risk of failure when the system is complex, but user participation in the development process only makes failure more likely when there are financial and time constraints in the development process (Tait and Vessey, 1988). Information systems implementation failures are too common, and the implementation process is too complicated, for the conventional wisdom to be correct and complete. The search for better explanations for implementation success and failure has led to two alternative approaches: factor models and political models.

Factor Models of Implementation Success

Several research studies have found other factors that are important to a successful implementation process. Ginzberg (1981b) found three additional important factors: commitment to the project, commitment to change, and extent of project definition and planning. Commitment to the project involves managing the system

Figure 19-9
Spectacular information systems implementation failures

In 1985, the New Jersey Motor Vehicles Division implemented a new vehicle registration system. The new system had been developed by a major accounting firm in Applied Data Research's Ideal, a 4GL. Although appropriate for management information systems or decision support systems, 4GLs are not well-suited for high-volume transaction processing systems. As might have been predicted, the vehicle registration system was a disaster. Using the system entailed as much as a five minute response time, instead of the two second time the Division requested. System use resulted in a large number of incorrect vehicle registrations: over one million drivers in New Jersey had problems with their car registrations. The critical parts of the system had to be reprogrammed in COBOL at a cost of over two million dollars (which the accounting firm paid for) (V. Zwass, *Management Information Systems*, Wm. C. Brown, 1992).

Another state system has had an even more colorful history. In 1988, the Department of Health and Rehabilitative Services (HRS) began work on the development of the Florida On-Line Recipient Integrated Data Access system, or FLORIDA for short. The purpose of FLORIDA was to provide a single point of eligibility testing for the welfare services administered by the state, including food stamps, Aid to Families with Dependent Children (AFDC), and Medicaid. Access to FLORIDA would be possible from state welfare offices throughout the state, but FLORIDA and all of its files would reside in Tallahassee. In 1989, EDS Federal Corporation won the bid for primary contractor for FLORIDA. One of the contract's stipulations was that the system be developed and implemented within 29 months.

FLORIDA was handed over to the state in June 1992. By the time the system came on-line, it had changed in at least two key aspects from the original design. First, the system was now running on a single mainframe system instead of three. Second, whereas the original design had called for eligibility rulings to be generated overnight, the final system allowed instant, on-line eligibility determinations.

From the time it was first handed over to the state, FLORIDA was a disaster. Response time often approached eight minutes. System crashes, lasting as long as eight hours, were common. Adding to the problems was the fact that, from July 1990 to July 1993, Florida's food stamps, AFDC, and Medicaid enrollments doubled, from 800,000 to 1.6 million people. FLORIDA was pushed to the limits of its capacity. IBM was commissioned to bring in a new, powerful mainframe system for FLORIDA to run on. As of September 1993, over $173 million has been spent on FLORIDA's development and implementation, almost double the original estimate, and the system is still not complete. Consultants estimate the system has only two more years left to operate, given FLORIDA's current capacity and projections for the state's welfare enrollments.

The human and financial toll from FLORIDA has easily surpassed those that resulted from New Jersey's vehicle registration system. FLORIDA's use has resulted in at least $263 million in welfare over- and underpayments. EDS has sued HRS for $46 million in payments EDS says it never received. Throughout 1992 and 1993, several key information systems personnel in HRS were fired or forced to resign. In the spring of 1993, the Governor demoted the Secretary of HRS and sent the Lieutenant Governor to HRS to manage the FLORIDA disaster. In August, 1993, a grand jury indicted the former project leader and former deputy secretary of HRS for information systems, for filing false status reports and for falsifying payment reports. In the spring of 1994, all charges were dropped against the former deputy secretary of HRS. The FLORIDA system's former project manager went to trial in May, 1994, and was convicted of two misdemeanor counts of making false statements. The judge in the case overthrew the verdict in June. (Compiled from 1993–94 stories in the *Tallahassee Democrat* and *Information Week*.)

development project so that the problem being solved is well understood and the system being developed to deal with the problem actually solves it. Commitment to change involves being willing to change behaviors, procedures, and other aspects of the organization. The extent of project definition and planning is a measure of how well planned the project is. The more extensive the planning effort, the less likely is implementation failure. In other research, Ginzberg (1981a) uncovered another important factor related to implementation success: user expectations. The more realistic a user's early expectations about a new system and its capabilities, the more likely it is that the user will be satisfied with the new system and actually use it.

Lucas, who has extensively studied information systems implementation, identified five factors related to implementation success (1989):

1. *Technical characteristics:* the extent to which the information technology is standard, reliable, available, matches the functionality required for the system, and so on;

2. *Management actions:* management must make their support and the reasons for their system-related decisions widely known;

3. *Attitudes toward the system and the information systems staff:* positive user attitudes towards the system and the IS staff increase the chance of a successful implementation;

4. *Decision style:* since people make decisions in different ways, the more aligned an information system is to how people decide, the better the chance for success;

5. *Personal and situational factors:* individuals with a vested interest in the new system, as well as the organizational context in which the system is being developed, constitute this important factor.

Figure 19-10 illustrates how these five factors are related to successful implementation and to each other. Notice that in constructing this model, Lucas hypothesizes that each of the five factors can have a direct effect on the implementation process. In other words, each individual factor is necessary, though not sufficient, to implementation success. You must also see that appropriate information technology is employed, that management takes the appropriate actions, that user attitudes toward the system and the IS staff are managed, that the new system matches existing decision styles, and that personal and situational factors are not neglected.

Figure 19-10
Lucas' model of implementation success (1989)

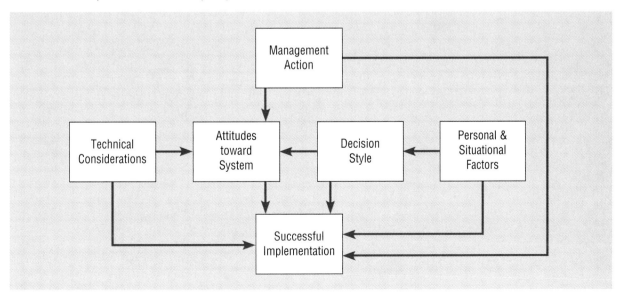

The model also hypothesizes indirect relationships among the factors and success. The effects of management action, technical characteristics, and decision style can all be mediated by user attitudes toward the system. Similarly, personal and situational factors can be mediated by decision style, which can in turn be mediated by attitudes. To see how these indirect relationships work, imagine a company where there is a history of mistrust and miscommunication between users and the IS staff and where users are generally antagonistic toward information systems. The president of this company decides a new automated inventory system is needed. She goes out of her way to express her support for the system, and her management team publicizes that support and the rationale behind the new system. The IS staff works carefully to choose the appropriate information technology. Yet, despite the best efforts of management and the IS staff, the chances for a successful systems implementation are not good, because the attitudes of users toward IS and the IS staff overwhelm the other factors.

How much control do you have over Lucas' five implementation factors? As you might expect, you will have more control over some factors than over others (see Figure 19-11). You have the most control over a system's technical characteristics, followed by control over management's actions (although analysts may not have all that much control over the extent of management's support). On the other hand, psychologists tell us that attitudes are very difficult to change and decision styles almost impossible to influence. You have virtually no influence at all over personal and situational factors. Having little or no influence over the latter three factors does not mean that you should ignore them, however. If anything, knowing that there is little you can do to influence attitudes, decision style, or context should only make these factors that much more critical to identify and understand.

Political Implementation Models

Factor models of the implementation process have helped analysts better understand implementation and why it may or may not be successful. Certainly, taking the different factors into account and realizing how they work together to influence implementation is important to success. But just as the conventional wisdom about implementation could not explain the whole story, neither can factor models hope to completely explain the implementation process. Political models have been proposed as another perspective to help you understand how a systems development project can succeed. We will use two examples from the implementation literature in MIS to illustrate the usefulness of political models.

Political models assume that individuals who work in an organization have their own self-interested goals which they pursue in addition to the goals of their departments and the goals of their organizations. Political models also recognize that power is not distributed evenly in organizations. Some workers have more power than others. People may act to increase their own power relative to that of

Figure 19-11

The extent to which analysts can influence Lucas' five factors

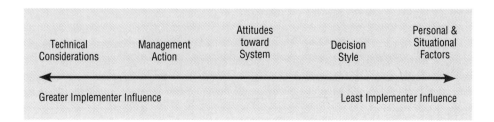

their co-workers and, at other times, people will act to prevent co-workers with more power (such as bosses) from using that power or from gaining more.

Markus (1981) tells the story of a division of an organization where implementation appears to have succeeded. Workers in two manufacturing plants, called Athens and Capital City, were using the new work-in-progress (WIP) system, which made it possible for management to use a planning and forecasting system based in part on the output of the WIP system. Workers from both plants had been involved in the systems development process, especially workers from the Capital City plant, and when workers at Athens had resisted using the new WIP system, extensive management pressure seemed to have forced people at Athens to give in and begin using the new system. It appeared that the conventional wisdom about systems development had prevailed: workers had participated in development, and management had been forcefully supportive. Implementation was a success.

Markus presents a political interpretation of the story that provides another explanation of events (see Figure 19-12). She begins by examining the history and power relationships within the division and the two plants. Although Athens was the plant that resisted using the new WIP system, they actually had superior information systems support at the beginning of the WIP system development process. Athens had once been a separate company and, until the division head decided new WIP and new planning systems were needed, Athens had been allowed a large amount of autonomy in how it operated. Further, the work performed at the two plants was tightly coupled. The Athens plant manufactured airplane parts which were then refined and finished at Capital City. Athens' manufacturing process was unpredictable and unreliable and resulted in a high scrap rate. Capital City never knew where Athens was in the manufacturing process for any particular part, and this uncertainty complicated Capital City's efforts to complete its work and finish parts in time for promised delivery dates.

According to Markus, when the opportunity arose to develop a system that would lessen the dependency of Capital City on Athens, people at Capital City were anxious to participate in the system's development. With a new WIP system containing data on Athens' production process, data which Capital City could access, the Capital City plant could greatly reduce the uncertainty associated with its own work. Understandably, the people at Athens were not so enthusiastic, and resisted participating in the development process and using the new system. After

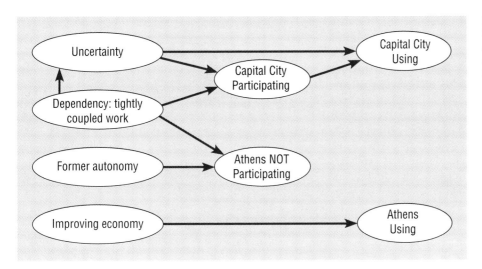

Figure 19-12
Illustrating the political explanation of system success at Athens and Capital City

the WIP system was completed and installed in both plants, the people at Athens continued to use their old WIP system, much to the chagrin of division management and Capital City. Division management applied a great deal of pressure, until finally Athens began to use the system. Markus states, however, that the people at Athens began using the new system not because of management pressure but because the economy began to improve, and managing the work at Athens became too difficult without the new system.

One lesson to be learned from Markus' case study is that factor-based analyses can only go so far in explaining successful implementation in organizations. History and power relationships are also important considerations. Sometimes political interpretations provide better explanations for the implementation process and why events took the course they did.

An example of how political considerations affect information systems after implementation is provided by Kling and Iacono (1984). The company they describe, which they call PRINTCO, successfully implemented a material requirements planning (MRP) system to support their manufacturing process for computer printers. PRINTCO was a very successful company and its business grew dramatically: the MRP system became inadequate for the company's needs. After six months of investigation, a group of top managers at PRINTCO, dominated by managers from manufacturing, found a new MRP system they wanted. Installation of the new system, which required buying new hardware as well, was supposed to take one year but, after 18 months, the company was not enjoying any benefits from the new MRP system. PRINTCO fired its data processing manager and hired a new one from outside the company. The new manager declared the installation effort a failure and concentrated on improving PRINTCO's existing MRP system which, though inadequate, had been operating all this time. The new manager was fired after 10 months, and management promoted the manager of engineering services to be the new data processing manager. Top management immediately purchased another new mainframe system. During the years of focus on the MRP installation, many other legitimate information systems needs throughout the company had been ignored. Users began to obtain microcomputers to take care of their own needs. By the time top management had become aware of the proliferation of microcomputers throughout PRINTCO, over $1 million had been invested in the machines. Top management immediately formed a committee to oversee the company's many microcomputers.

A political interpretation helps explain how the information systems turned out the way they did at PRINTCO. It is important to realize that MRP was very important to the manufacturing group at PRINTCO, and that manufacturing therefore dominated systems development at PRINTCO. Manufacturing managers dominated the group of managers who determined that a new MRP system would be found and implemented, that two data processing managers would be fired for not bringing the manufacturing vision to fruition, and that other legitimate information systems needs would be ignored in favor of MRP. Users did not openly rebel at PRINTCO because they had been educated about the benefits of MRP and its importance to PRINTCO in seminars provided by the manufacturing group. People throughout the company believed in MRP, and they supported Manufacturing's efforts to improve MRP but, in the meantime, workers had to make do with an inadequate MRP system and little or no support for their other data processing needs. Workers' efforts to take care of their own systems needs were brought under the control of top management when the managers created the committee to oversee the microcomputers. As Kling and Iacono (1984) point out, management took action to build support for its efforts (e.g., education) and to quiet any opposition (e.g., the microcomputer committee). The politics involved in the post-implementation developments for PRINTCO's MRP system provide a useful framework for understanding those developments.

SUMMARY

This chapter presented an overview of the coding, testing, and installation aspects of the systems implementation process. In the chapter, you studied seven different types of testing: (1) code inspections, where code is examined for well-known errors; (2) walkthroughs, where a group manually examines what the code is supposed to do; (3) desk checking, where an individual mentally executes the computer instructions; (4) syntax checking, typically done by a compiler; (5) unit or module testing; (6) integration testing, where modules are combined and tested together until the entire program has been tested as a whole; and (7) system testing, where programs are combined to be tested as a system and where the system's meeting of its objectives is examined. You also learned about acceptance testing, where users test the system for its ability to meet their requirements, using live data in a live environment. Different ways to assure software quality were illustrated by the example of IBM's OS/400 software development process.

You also read about four types of installation: (1) direct, where the old system is shut off just as the new one is turned on; (2) parallel, where both old and new systems are run together until it is clear the new system is ready to be used exclusively; (3) single location, where one site is selected to test the new system; and (4) phased, where the system is installed bit by bit. You saw how implementation is an organizational change process, which must be managed as such. And you saw how information systems researchers have been trying to explain what constitutes a successful implementation, using conventional wisdom, factor models, and political models.

If there is a single main point in this chapter, it is that implementation is a complicated process, from managing programmer teams to the politics that influence what happens to a system after it has been successfully implemented. Whether the concern is applying a particular quality assurance measure or understanding the subtle resistance of politically motivated users to adopting a new system, analysts have many factors to identify and manage for a successful systems implementation. Successful implementation rarely happens by accident or occurs in a totally predictable manner. The first step in a successful implementation effort may be realizing just that fact.

As noted in the chapter, system implementation includes not only coding, testing, and installation, but also documentation, training, and support. We address these other aspects of implementation in the next chapter.

CHAPTER REVIEW

KEY TERMS

Acceptance testing
Alpha testing
Beta testing
Desk checking
Direct installation

Inspections
Installation
Integration testing
Parallel installation
Phased installation

Scaffolding
Single location installation
Stub testing
System testing

REVIEW QUESTIONS

1. Define each of the following terms:
   a. walkthrough
   b. stub testing
   c. scaffolding
   d. phased installation
   e. alpha testing
   f. beta testing

2. What are the deliverables from coding, testing, and installation?

3. Explain the testing process for code.

4. What are structured walkthroughs for code? What is their purpose? How are they conducted? How are they different from code inspections?

5. List and describe the software quality assurance techniques used by IBM in their development of OS/400 R.1.

6. What are the four approaches to installation? Which is the most expensive? Which is the most risky? How does an organization decide which approach to use?

7. Explain the Lewin/Schein model of organizational change.

8. What is the conventional wisdom about implementation success?

9. List and define the factors that are important to successful implementation efforts.

10. Explain Lucas' model of implementation success.

11. How would you characterize political models of the implementation (and post-implementation) process? What can political models do that factor models don't?

PROBLEMS AND EXERCISES

1. Match the following terms to the appropriate definitions.

   _____ acceptance testing

   _____ direct installation

   a. bringing together all of the programs that comprise a system for testing purposes. Programs are typically integrated in a top-down, incremental fashion

   b. process whereby actual users test a completed information system, the end result of which is the users' acceptance of the system

_____ parallel installation

c. bringing together all of the modules that comprise a program for testing purposes

_____ system testing

d. changing over from the old system to a new one by turning off the old system as the new one is turned on

_____ integration testing

e. running the old system and the new one at the same time until the old system can be turned off

2. Prepare a testing strategy or plan for Pine Valley Furniture's Purchasing Fulfillment System (see Chapter 17 for details).

3. Develop a plan to assure the quality of the code being written to implement Pine Valley Furniture's Purchasing Fulfillment System. What specific approaches do you recommend? Why?

4. Which installation strategy would you recommend for PVF's Purchasing Fulfillment System? Which would you recommend for Hoosier Burger's Inventory Control System? If you recommended different approaches, please explain why. How is PVF's case different from Hoosier Burger's?

5. Develop a table that compares the four installation strategies, showing the pros and cons of each. Try to make a direct comparison when a pro of one is a con of another.

6. One of the difficult aspects of using the single location approach to installation is choosing an appropriate location. What factors should be considered in picking such a pilot site?

7. You have been a user of many information systems including, possibly, a class registration system at your school, a bank account system, a word processing system, and an airline reservation system. Pick a system you have used and assume you were involved in the beta testing of that system. What would be the criteria you would apply to judge whether this system was ready for general distribution?

8. What is the purpose of scaffolding code? What functions might scaffolding code perform in order to improve system testing?

9. Why is it important to keep a history of test cases and the results of these test cases even after a system has been revised several times?

FIELD EXERCISES

1. Interview someone you know or have access to who works for a medium to large organization. Ask for details on a specific instance of organizational change: What changed? How did it happen? Was it well planned or ad hoc? How were people in the organization affected? How easy was it for employees to move from the old situation to the new one? When you have finished your interview, analyze the situation from the perspective of the Lewin and Schein models of organizational change. How well do the models fit the situation? Can you explain why?

2. Re-examine the data you collected in the interview in Field Exercise 1. This time, analyze the data from a political perspective. How well does the political model explain how the organization dealt with change? Explain why the political model does or does not fit.

3. Ask a systems analyst you know or have access to about implementation. Ask what the analyst believes is necessary for a successful implementation. Try to determine whether the analyst believes in factor models or political models of implementation.

4. Prepare a research report on successful and unsuccessful information system implementations. After you have found information on two or three examples of both successful and unsuccessful system implementations, try to find similarities and differences among the examples of each type of implementation. Do you detect any patterns? Can you add to either the factor or political models you read about in the chapter?

REFERENCES

Bloor, R. 1994. "The Disappearing Programmer." *DBMS* 7(9) (Aug.): 14–16.

Brooks, F. P. 1978. *The Mythical Man-Month.* Reading, MA: Addison-Wesley.

Fagan, M. E. 1986. "Advances in Software Inspections." *IEEE Transactions on Software Engineering* SE-12(7) (July): 744–51.

Ginzberg, M. J. 1981a. "Early Diagnosis of MIS Implementation Failure: Promising Results and Unanswered Questions." *Management Science* 27(4): 459–78.

Ginzberg, M. J. 1981b. "Key Recurrent Issues in the MIS Implementation Process." *MIS Quarterly* 5(2) (June): 47–59.

Ives, B. and M. H. Olson. 1984. "User Involvement and MIS Success: A Review of Research." *Management Science* 30(5): 586–603.

Jones, C. 1986. *Programming Productivity.* New York, NY: McGraw-Hill.

Kling, R. and S. Iacono. 1984. "The Control of Information Systems After Implementation." *Communications of the ACM* 27(12): 1218–26.

Lewin, K. 1951. *Field Theory in Social Science.* New York, NY: Harper and Brothers.

Litecky, C. R. and G. B. Davis. 1976. "A Study of Errors, Error Proneness, and Error Diagnosis in COBOL." *Communications of the ACM* 19(1): 33–37.

Lucas, H. C. 1989. *Managing Information Services.* New York, NY: Macmillan.

Markus, M. L. 1981. "Implementation Politics: Top Management Support and User Involvement." *Systems/Objectives/Solutions* 1(4): 203–15.

Mosley, D. J. 1993. *The Handbook of MIS Application Software Testing.* Englewood Cliffs, NJ: Yourdon Press.

Phan, D. 1990. *Information Systems Project Management: An Integrated Resource Planning Perspective Model.* Unpublished doctoral dissertation, University of Arizona.

Phan, D., J. F. George, and D. R. Vogel. 1994. "Lessons Learned from Modeling Software Quality in a Very Large Software Development Project." Working paper, St. Cloud State University, St. Cloud, Minnesota.

Scheier, R. L. 1994. "Software Snafu Grounds Denver's High-Tech Airport." *PC Week* 11(19) (May 16): 1, 11.

Schein, E. H. 1987. *Process Consultation,* vol. 2. Reading, MA: Addison-Wesley.

Tait, P. and I. Vessey. 1988."The Effect of User Involvement on System Success: A Contingency Approach." *MIS Quarterly* 12(1) (March): 91–108.

Yourdon, E. 1989. *Managing the Structured Techniques.* 4th ed. Englewood Cliffs, NJ: Prentice-Hall.

Yourdon, E. 1993. *Decline & Fall of the American Programmer.* Englewood Cliffs, NJ: Yourdon Press.

Zwass, V. 1992. *Management Information Systems.* Des Moines, IA: William C. Brown.

Coding, Testing, and Installation

INTRODUCTION

System testing ensures that software work according to specifications. No matter how thorough the testing, however, you may never really know how well a system will perform until it is used on the job in a real setting. Beta testing will help you discover many potential problems, but often behavioral issues do not surface until the system is installed and used in live operation for a while. The period shortly after conversion to the new system can be a very revealing time for a new system. After the novelty of the new system wears off and until users totally give up work methods suited for the old system, you may discover design deficiencies. This was the situation with the Customer Activity Tracking System (CATS). In this case we listen in on a conversation between several principles in Broadway Entertainment as they try to explain problems that have arisen soon after CATS is installed.

PROBLEMS SURFACE WITH IN-STORE CATS

Jordan Pippen, CATS (Customer Activity Tracking System) project leader, was working in her office with Frank Napier, a systems analyst on the CATS team, and Wendy Yoshimuro, a marketing manager responsible for CATS. They were reviewing the final stages of the CATS system implementation.

"We used our initial test plan as a checklist to determine if we had accomplished all we set out to do," Frank said. "We had to make a few changes to the plan when we decided to go with the TwoCats customer preference system as the basis for corporate CATS, as the original test plan presumed we would build the system in-house. For example, we looked at how TwoCats had tested the package and we eliminated testing elements that we thought TwoCats had tested thoroughly. Other than that, testing went the way we expected—unit testing, integration testing, system testing, all went according to plan. We've been using corporate CATS for several weeks now, using live data in the live environment as part of acceptance testing. And Wendy's group is ready to sign off."

"That's right," replied Wendy. "The system meets all of our expectations."

"So now we're involved in pilot testing, including the in-store component of CATS," Frank continued. "The system is being tested right now in seven different stores, with live data in the workday environment."

"How did you decide to do a pilot test?" Wendy asked. "Why not a direct conversion using all the stores? But given the decision to pilot test, how did you choose these seven stores?"

"This is a new system," Jordan replied. "It provides new capabilities we didn't have before, plus we wrote this in-store system ourselves. I didn't think it would be a good idea to install the in-store CATS at every location all at once. Also, I didn't see a way to do a phased implementation since the system is totally new and different modules do not stand on their own. I don't anticipate any major problems with the system, you understand, but I wanted to be sure.

"And as for how we chose the seven pilot sites, we wanted the stores with the highest sales revenues—the best stores, the ones with the most customers, the ones most likely to have customers who would receive lots of merchandise notifications. Plus, many of these stores worked closely with us during development. Some provided substantial insight into design issues, some provided test cases, and others helped us generate test scenarios for our system tests. Because of their involvement and excitement about CATS, some of the managers from these stores insisted on being involved in the pilot. Their enthusiasm has been really great.

If the pilot installations go well, they will be great spokespersons as we roll out CATS. We may want to do a second pilot with a representative mix of other stores, maybe some with high sales to teenagers or stores that serve college students. You know, you'd be surprised when something unexpected can occur. We should learn more after receiving the pilot survey forms from the managers and clerks in these stores and then the debriefing sessions we will hold."

Just then the phone in Jordan's office rang. It was Karen Gardner, v-p of Information Systems, calling from Miami.

"Jordan, we have a problem with CATS." Jordan quickly turned on the speaker phone so Frank and Wendy could join in the conversation.

"What do you mean by a problem with CATS?" Jordan responded. Frank, who had been responsible for many of the design details for the store component of CATS, shifted a little in his chair.

"I'm not talking about the corporate CATS," Karen replied. "The problem is with the store component of the system. The corporate version seems to be working fine, at least to the extent we have been using it so far. Frank, deciding to base the corporate system on the TwoCats customer preference system was a sound decision. No, the problem is with the store component. We're seeing some user resistance at Miami store number 23."

"What exactly is the problem?" Jordan asked. "Store 23 is one of our strongest stores. It's the flagship store for the Metro Dade area, and management at Store 23 has been on-board since the beginning of this development project. I myself have met with Julio Mejias and his staff many times since we started the CATS project. That's why I insisted that Store 23 be one of the seven stores in our initial installation effort."

"Yeah, I met with Mr. Mejias and his staff and his clerks myself when I was working on the interface for the in-store CATS," Frank added. "I used a lot of their suggestions. I don't understand why that store would be the one to give us trouble. Like Jordan said, they've been on-board since the beginning of this project. People there have been heavily involved in the entire design process."

"Maybe it's something else, then," Wendy offered. "Karen, what seems to be the specific problem? What are they doing? Or not doing?"

"The biggest problem," Karen replied, "is that very few clerks in Store 23 are using CATS. Customers of 23 are receiving the notification cards from corporate CATS. And I must say, we seem to be right on target. People are getting notification of the release of products they

are really interested in. And customers are calling the toll-free number to make reservations for the products they want."

"We've been monitoring the generation of notifications and the traffic on the 800-number line," Jordan said. "Karen's right, notification volume is meeting our expectations, and so is the traffic on the 800-number line. Volumes are right at what we projected for the seven stores in our pilot."

"And the customers are coming in to 23 to fill the reservations," Karen continued. "The clerks are filling the orders but not all of them are using CATS to record the sales as reservation sales. We know customers are coming in to get the products their reservations guarantee them, but we can't track the specific activity. Reservation records are not being updated properly. Remember, the point-of-sale system was modified to include an in-store CATS module to handle reservation items on a sale, and the clerks seem to be skipping over this function."

"Can't Mr. Mejias make the clerks use CATS?" Frank asked impatiently. "They work for him, after all. He's the manager. Can't he make them use the system?"

"Julio can oversee the activities at the cash registers," Jordan answered, "But he can't ride herd on everybody all the time. This is a new function for the clerks, and apparently we just didn't design the system to make it easy for them to record the reservations fulfillment activity."

"I don't agree at all," Frank said. "The user interface we designed for the in-store CATS is about the easiest to use I've ever seen. And it's simple. User friendly. I can't believe usability is an issue. It must be something else. Maybe they're just not motivated properly. I mean these are people making little better than minimum wage. How committed to Broadway are they?"

"No, I don't believe that," Jordan said. "I've been to Store 23. It's a very busy place. On Fridays and Saturdays, that place is packed. It looks like Christmas Eve shopping in there sometimes. I've worked with a lot of those people on this project. I think most of them are motivated and committed. You've worked with them, too, Frank. You know Julio has hired good people."

"Yeah, you're right. The people I worked with there were good workers. But I can't speak for all of them," Frank replied.

"I think it must be something else," Jordan said. "Karen, what's the status of the implementation at the other six stores?"

"Everything seems to be going well at the other stores," Karen said. "Store 23 is the only one with a user resistance problem. I do tend to agree with Frank in that it is Julio's responsibility to make using CATS part of routine sales activities. Maybe if we pressured him more . . ."

"I just don't think that will work very well," Jordan said.

"It sounds to me as if the problem might be related to how busy they are," said Wendy. "I've never been to Store 23, so I can't offer any opinions about how dedicated the workers are, but when Jordan talked about how the store is like Christmas Eve shopping on the weekends, it made me think. Maybe the problem has to do with the fact that the clerks are so busy they just don't have time to deal with yet another new demand on their time. Maybe the problem is not so much the user interface as it is just plain old business volume. Maybe it's all the clerks can do to take care of the customers' demands for service. I mean, the clerks are moving product. Isn't that their main responsibility? They've never had to track special sales before, have they?"

"That could be it," Karen replied. "Jordan's right, this is an awfully busy store. Confidentially, Julio believes he has the volume to justify another expansion, and he has asked top management for the opportunity to expand."

"Why don't we program a workaround so that the clerks don't have to access CATS directly?" Frank suggested. "That way they would record the reservation sales in CATS and they wouldn't have to depart from their usual routine."

"What's a workaround?" asked Wendy.

"A workaround is a special procedure you create for using a system that's needed when a user resists using the system," replied Frank. "A workaround doesn't change the functionality of a system, but it can allow a user to interact with a new system the same way he or she did with the old system. That way, the users are happy because they don't have to learn a brand new way of doing things, and management is happy because they are still getting the benefits the new system was designed to deliver. Sometimes a workaround is nothing more than a simple change in the interface. Other times it may be a redesigned procedure that makes it easier for users to incorporate a new system or new system features in their daily work routines. Lots of times, users create workarounds on their own, especially when a new system doesn't deliver some of the functionality they want. Users will figure out a way to make a system do something it was

not quite designed to do. In this case, maybe we can let the clerks indicate an item is a reservation sale, but not have them match the sale with a reservation record. It's possible we could write a batch program to do the matching, although it is more reliable to be done by the clerk at the time of the sale."

"Your explanation makes it seem that a workaround reduces some of the benefits we've designed into the system?" Jordan asked. "Plus, I don't like the idea of CATS having one interface in some stores and a different interface in others. Besides, Karen says CATS is working fine in the other six pilot locations."

"The workaround is a good idea, Frank," Karen said, "But I agree with Jordan on this one. I've never liked workarounds, and I don't want CATS to be inconsistently applied. On the other hand, we do need the clerks in Store 23 to use the system. If Wendy is right, once Julio has finished his expansion—I'm sure the board will approve Julio's request—then the volume per sales clerk should be close to our store average, and the clerks should start using CATS the way we intended. In the meantime, we need the clerks to use CATS. So, Frank, see what you can do about a temporary fix for Store 23 and other very high-volume stores."

"Right," said Frank, "I'll get right on it. You know, though, on Christmas Eve, every store will look like Christmas Eve. I mean, every store will be very busy. It may be that we need to rethink how CATS works when sales volume is seasonally high for every store."

"Good point, Frank," Karen said. "Maybe this particular problem extends beyond Store 23. Maybe this is serious enough to think more in terms of some design changes rather than just a temporary workaround. Well, I've got to go. This was a good meeting—better than I expected. Jordan, keep me posted. Bye."

"Well," Jordan said to Frank and Wendy, "We did accomplish more than I thought we would. Good insight, Wendy."

"More work for me," Frank said. "Sales volume is one thing I didn't anticipate having an effect on CATS."

"Hey," Jordan said, "That's why we chose to go with a pilot conversion, so we could discover any serious problems with CATS before a more general roll-out."

SUMMARY

No matter how thoroughly an organization is studied during systems development, installing a new system can cause unanticipated results. Sometimes the new system itself can so affect the business operation that issues arise that were not present before the introduction of the system. Organizations, and systems, are dynamic. The increased sales volume due to the success of CATS created unexpected operational and behavioral issues for the CATS team. Fortunately, however, they recognized that they likely could not anticipate all of the effects of CATS and deployed this new system in a limited number of pilot sites where they could concentrate attention and limit any diverse consequences. Their lack of overconfidence made the team members more receptive to making needed changes in their new child—CATS.

Simply installing CATS is not the end of the CATS project. Other important work in implementing CATS needs to be finalized. We look at these other aspects of the CATS implementation in the next BEC case.

System Implementation: Documenting the System, Training, and Supporting Users

After studying this chapter, you should be able to:

■ List the deliverables for documenting the system and for training and supporting users.

■ Distinguish between system and user documentation and determine which types of documentation are necessary for a given information system.

■ Write task-oriented user documentation.

■ Compare the many modes available for organizational information system training, including self-training and electronic performance support systems.

■ Prepare a training plan for an information system.

■ Discuss the issues of providing support for computing end users.

■ Describe the role of the technical communicator.

INTRODUCTION

In Chapter 19, you learned about implementing information systems in organizations both from a technical and an organizational perspective. Implementation is not complete just because the new system has been installed, conversion is complete, and user resistance to the new system has been analyzed and resolved. As an analyst you must also consider providing documentation about the new system for the information systems personnel who will maintain the system and for the system's users. These same users must be trained how to use what you have developed and installed in their workplace. Once training has ended and the system has become institutionalized, users will have questions about the system's implementation and how to use it effectively. You must provide a means for users to get answers to these questions and to identify needs for further training.

As a member of the system development team that developed and implemented the new system, your job is winding down now that installation and conversion is complete. At this point, you must focus on issues of documentation, training, and supporting users, the topics of this chapter. These activities complete your study of the life cycle's implementation phase. Chapter 21 completes the life cycle by focusing on maintenance and business process re-engineering. The end of implementation also marks the time for you to begin the process of project close-down. You read about project close-down in Chapter 4 when you learned about project management. At the end of this chapter, we will return to the topic of formally ending the systems development project.

In this chapter, you will read about the process of documenting systems and training and supporting users as well as the deliverables from these processes. You will also learn about the various types of documentation and numerous methods available for delivering training and support services. Finally, the chapter describes the role of technical communicator, a person who can provide assistance throughout the documentation, training, and support processes.

DOCUMENTING THE SYSTEM, TRAINING, AND SUPPORTING USERS

Documenting the system, training, and supporting users are three processes that comprise part of the implementation stage of the systems development life cycle (Figure 20-1). Although you may give more attention to documentation during implementation, documentation is actually relevant throughout the systems development life cycle. The process of documenting the system began with the first document created as part of the life cycle and with the first entry made into the system repository. Training and support, on the other hand, tend to receive the most focus from the analysis team during implementation. Once the system has been successfully coded and tested and the conversion process begins, the analysis team turns its attention to the related tasks of training and supporting users.

The Processes of Documenting the System, Training, and Supporting Users

Although the process of documentation proceeds throughout the life cycle, it receives formal attention during the implementation phase because the end of implementation largely marks the end of the analysis team's involvement in system development. As the team is getting ready to move on to new projects, you and the other analysts need to prepare documents that reveal all of the important information you have learned about this system during its development and implementation. There are two audiences for this final documentation: (1) the information systems personnel who will maintain the system throughout its productive life and (2) the people who will use the system as part of their daily lives, the users. The analysis team in a large organization can get help in preparing documentation from specialized staff in the information systems department.

Larger organizations also tend to provide training and support to computer users throughout the organization. Some of the training and support is very specific to particular application systems while the rest is general to particular operating systems or off-the-shelf software packages. For example, it is common to find courses on Microsoft Windows™ and WordPerfect™ in organization-wide training facilities. Analysts are mostly uninvolved with general training and support, but they do work with corporate trainers to provide training and support tailored to particular computer applications they have helped develop. Centralized information system

Figure 20-1
The systems development life cycle with the implementation phase highlighted

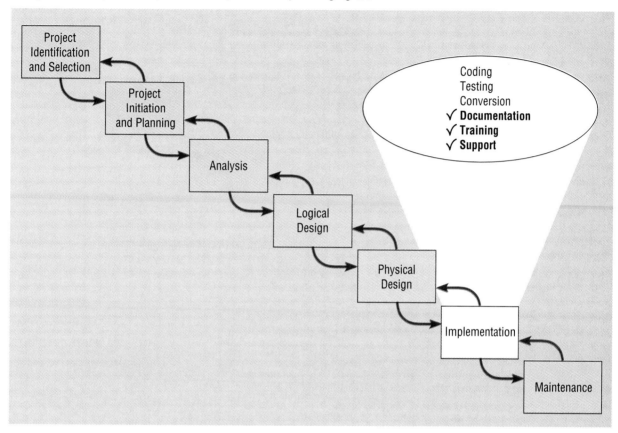

training facilities tend to have specialized staff who can help you with training and support issues. In smaller organizations that cannot afford to have well-staffed centralized training and support facilities, fellow users are the best source of training and support users have, whether the software is customized or off-the-shelf (Nelson and Cheney, 1987).

Deliverables and Outcomes

Table 20-1 shows the deliverables from documenting the system, training, and supporting users. At the very least, the development team must prepare a set of user documentation. The documentation can be paper-based, but it should also include computer-based modules. For modern information systems, this documentation includes any on-line help designed as part of the system interface. The development team should think through the user training process: Who should be trained? How much training is adequate? What do users need to learn during training? The training plan should be supplemented by actual training modules, or at least outlines of such modules, that at a minimum address the three questions stated previously. If the organization is sufficiently staffed, a technical communicator can help with the preparation of documentation and training modules. Finally, the development team should also deliver a user support plan that addresses such issues as how users will be able to find help once the information system has become

TABLE 20-1 Deliverables for Documenting the System, Training, and Supporting Users

1. Documentation
   a. system documentation
   b. user documentation

2. User training plan
   a. classes
   b. tutorials

3. User training modules
   a. training materials
   b. computer-based training aids

4. User support plan
   a. help desk
   b. on-line help
   c. bulletin boards and other support mechanisms

integrated into the organization. The development team should consider a multitude of support mechanisms and modes of delivery. Each deliverable is addressed in more detail later in the chapter.

DOCUMENTING THE SYSTEM

In one sense, every information systems development project is unique and will generate its own unique set of documentation. In another sense, though, system development projects are probably more alike than they are different. Each project shares a similar systems development life cycle which dictates that certain activities will be undertaken and that each of those activities will be or should be documented. Bell and Evans (1989) illustrate how a generic systems development life cycle maps onto a generic list of systems development documentation (Table 20-2). As you compare the generic life cycle in Table 20-2 to the life cycle presented in this book, you will see that there are differences, but the general structure of both life cycles is the same, as both include the basic phases of analysis, design, implementation, and project planning. Specific documentation will vary depending on the life cycle you are following and the format and content of the documentation may be mandated by the organization you work for, but the basic outline of documentation you will need to consider providing is contained in Table 20-2.

We can simplify the situation even more by dividing documentation into two basic types, system documentation and user documentation. **System documentation** records detailed information about a system's design specifications, its internal workings, and its functionality. In Table 20-2, all of the documentation listed, except for that listed under System Delivery, would qualify as system documentation. System documentation can further be divided into internal and external documentation (Martin and McClure, 1985). **Internal documentation** is part of the program source code or is generated at compile time. **External documentation** includes the outcome of all of the structured diagramming techniques you have studied in this book, such as data flow and entity-relationship diagrams. Although not part of the code itself, external documentation can provide useful information to the primary users of system documentation—maintenance programmers. For example, structure charts and Nassi-Shneiderman charts together provide a good overview of a system's larger structure and the details of its inner workings. Recall that Nassi-

System documentation: Detailed information about a system's design specifications, its internal workings, and its functionality.

Internal documentation: System documentation that is part of the program source code or is generated at compile time.

External documentation: System documentation that includes the outcome of such structured diagramming techniques as data flow and entity-relationship diagrams.

TABLE 20-2 SDLC and Generic Documentation Corresponding to Each Phase
(Adapted from Bell and Evans, 1989)

Generic Life Cycle Phase	Generic Document
Requirements Specification	System Requirements Specification
	Resource Requirements Specification
Project Control Structuring	Management Plan
	Engineering Change Proposal
System Development	
Architectural Design	Architecture Design Document
Prototype Design	Prototype Design Document
Detailed Design & Implementation	Detailed Design Document
Test Specification	Test Specifications
Test Implementation	Test Reports
System Delivery	User's Guide
	Release Description
	System Administrator's Guide
	Reference Guide
	Acceptance Sign-off

Shneiderman charts are themselves used to model specifications for writing code. In the past, external documentation was typically discarded after implementation, primarily because it was considered too costly to keep up-to-date, but today's CASE environment makes it possible to maintain and update external documentation as long as desired.

While system documentation is intended primarily for maintenance programmers (see Chapter 21), **user documentation** is intended primarily for users. An organization may have definitive standards on system documentation, often consistent with CASE tools and the system development process. These standards may include the outline for the project dictionary and specific pieces of documentation within it. Standards for user documentation are less likely.

User documentation: Written or other visual information about an application system, how it works, and how to use it.

User Documentation

User documentation consists of written or other visual information about an application system, how it works, and how to use it. An excerpt of on-line user documentation for Microsoft Access appears in Figure 20-2. Notice how the documentation first explains to the user what each important term means. The documentation then lists the steps necessary to actually perform the task the user inquired about. The "notes" section that follows explains specific restrictions and constraints that will affect what the user is attempting. You should also notice how some words in the documentation are fainter (in this printout) and are underlined with dotted lines. This notation signifies that these particular words are hypertext links to related material elsewhere in the documentation. Hypertext techniques, rare in on-line PC documentation five years ago, are now the rule rather than the exception.

Figure 20-2 represents the content of a reference guide, just one type of user documentation (there is also a quick reference guide). Other types of user documentation

Figure 20-2
Example of user documentation (from Microsoft Access™)

Creating a Calculated Control on a Form or Report

See Also   Examples

A calculated control on a form or report displays a value calculated using data from one or more fields from the underlying table or query or from other controls. The calculation is the result of an expression assigned to the ControlSource property for a control.

To create a calculated control
1. In Design view of a form or report, add a text box or other control to the form or report. Although text boxes are the most common control for displaying calculated values, you can use any control that has a ControlSource property.
2. While the control is selected, type the appropriate expression.
   -Or-
   Create the expression using the Expression builder.

Notes
- Use a calculated control when you want Microsoft Access to generate the value for a particular field automatically so you don't have to enter or calculate a new value each time you use the form or generate a report. For example, use a calculated control to compute the monthly sales total for each of your employees or to display the current date.
- For check boxes, option buttons, or option groups, you must set the ControlSource property in the property sheet rather than typing an expression in the control.
- If the property sheet is displayed and you click a control that's already selected, the property sheet will go blank. If you're entering a property setting in a control, press Enter to set the value and redisplay the property sheet. If you're not entering a setting, press the Esc key to redisplay the property sheet.
- In Form view, calculated fields are read-only.
- Microsoft Access doesn't store the result of a calculated field in a table but recomputes it each time the record is displayed.
- You can set the Format property to display data in either a standard format or a user-defined format.

---

include a user's guide, release description, system administrator's guide, and acceptance sign-off (Table 20-2). The reference guide consists of an exhaustive list of the system's functions and commands, usually in alphabetical order. Most on-line reference guides allow you to search by topic area or by typing in the first few letters of your key word. Reference guides are very good for very specific information (as in Figure 20-2) but not as good for the broader picture of how you perform all the steps required for a given task. The quick reference guide provides essential information about operating a system in a short, concise format. Where computer resources are shared and many users perform similar tasks on the same machines (as with airline reservation or mail order catalog clerks), quick reference guides are often printed on index cards or as small books and mounted on or near the computer terminal. An outline for a generic user's guide (from Bell and Evans, 1989) is shown in Table 20-3. The purpose of such a guide is to provide information on how users can use computer systems to perform specific tasks. The information in a user's guide is typically ordered by how often tasks are performed and how complex they are.

In Table 20-3, sections with an *n* and a title in square brackets mean that there would likely be many such sections, each for a different topic. For example, for an accounting application, sections 4 and beyond might address topics such as entering a transaction in the ledger, closing the month, and printing reports. The items in rounded brackets are optional, included as necessary. An index becomes more important the larger the user's guide. Although a generic user's guide outline is helpful in providing an overview for you of what a user's guide might contain, we have included outlines of user guides from several popular PC software packages in Figure 20-3. Notice how different they are. For example, the Access and Word-Perfect outlines are quite general and macro whereas the CA Simple Tax outline is very detailed. Some sections are unique due to features present in the package, such as Cue Cards for Access. WordPerfect Corporation includes a section on related products which hints at how the user can expand the functionality of WordPerfect. Since CA Simple Tax is task-oriented, rather than general-purpose software, the outline is organized around user-centered tasks rather than generic functions.

A release description contains information about a new system release, including a list of complete documentation for the new release, features and enhancements, known problems and how they have been dealt with in the new release, and information about installation. A system administrator's guide is intended primarily for those who will install and administer a new system and contains information about the network on which the system will run, software interfaces for peripherals such as printers, troubleshooting, and setting up user accounts. Finally, the acceptance sign-off allows users to test for proper system installation and then signify their acceptance of the new system with their signatures.

TABLE 20-3   Outline of a Generic User's Guide
(Adapted from Bell and Evans, 1989)

Preface

1. Introduction
    1.1 Configurations
    1.2 Function Flow

2. User Interface
    2.1 Display Screens
    2.2 Command Types

3. Getting Started
    3.1 Login
    3.2 Logout
    3.3 Save
    3.4 Error Recovery
    3.n [Basic Procedure Name]

n. [Task Name]
Appendix A—Error Messages
([Appendix])

Glossary
    Terms
    Acronyms

(Index)

Figure 20-3

Outlines of user's guides from various popular PC software packages

(a) Microsoft Access™

MICROSOFT ACCESS HELP CONTENTS

Help Features   *What's New*

Using Microsoft Access
Step-by-step instructions to help you complete your tasks.

Cue Cards
The online coach that helps you learn Microsoft Access as you do your work.

General Reference
Guides to menu commands, keyboard shortcuts, toolbars, and windows and answers to common questions.

Language and Technical Reference
Complete reference information about properties, actions, events, objects, and the Access Basic language.

Technical Support
Available support options so that you can get the most from Microsoft Access.

(b) WordPerfect™ for Windows 6.0

CONTENTS

Welcome to WordPerfect 6.0 for Windows Help

To find information, choose from the following items. To search for information and to move through Help, use the buttons along the top of the Help window.

Choose	For information about
Search (Index)	topics listed alphabetically
How Do I	performing tasks
Glossary	meanings of terms
Menu Commands	features by menus
WordPerfect Bars	topics by Feature Bar, Power Bar, Ruler Bar, Button Bar, and Status Bar
Keystrokes	keystrokes and templates
What's New	features new to WPWin
Other Products	other WPCorp products
Using Help	how to use Help

(c) CA Simple Tax 1993

CONTENTS

Preparing User Documentation

User documentation, regardless of its content or intended audience, was once provided almost exclusively in big, bulky paper manuals, and it was out-of-date almost as soon as it was printed. Most documentation is now delivered on-line, in hypertext format. Regardless of format, user documentation is an up-front investment that should pay off in reduced training and consultation costs later (Torkzadeh and Doll, 1993). As a future analyst, you need to consider the source of documentation, its quality, and whether its focus is on the information system's functionality or on the tasks the system can be used to perform.

The traditional source of user documentation has been the organization's information systems department. Even though information systems departments have always provided some degree of user documentation, for much of the history of data processing in organizations the primary focus of documentation was the system. In a traditional information systems environment, the user interacted with an analyst and computer operations staff for *all* of his or her computing needs (Figure 20-4). The analyst acted as intermediary between the user and all computing

Figure 20-4

Traditional information system
environment and its focus
on system documentation
(Adapted from Torkzadeh and
Doll, 1993)

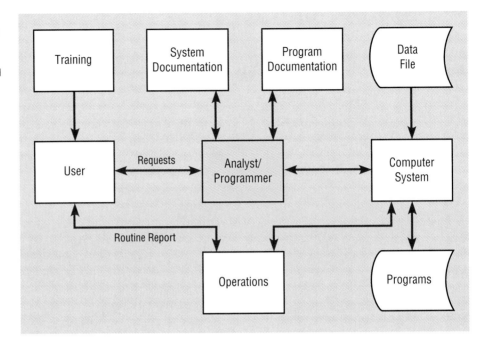

resources. Any reports or other output that went to the user were generated by the
operations staff, based on a regular reporting schedule. Since users were consumers and providers of data and information, most documentation developed during
a traditional information systems development project was system documentation
for the analysts and programmers who had to know how the system worked. Although some user documentation was generated, most documentation was intended
to assist maintenance programmers who tended to the system for years after the
analysis team had finished its work.

In today's end-user information systems environment, users interact directly
with many computing resources, users have many options or querying capabilities
from which to choose when using a system, and users are able to develop many
local applications themselves (Figure 20-5). Analysts often serve as consultants for
these local end-user applications. For end-user applications, the nature and purpose
of documentation has changed from documentation intended for the maintenance
programmer to documentation for the end user. Application-oriented documentation, whose purpose is to increase user understanding and utilization of the
organization's computing resources, has also come to be important. While some
of this user-oriented documentation continues to be supplied by the information
systems department, much of it now originates with vendors and with users
themselves.

Characteristics of Quality User Documentation

Simply providing user-oriented and application-oriented documentation, regardless
of source, may not be enough, however. Torkzadeh and Doll (1993) found that users
also need high-quality documentation. The higher the quality of the end-user documentation, the more satisfied users are with an information system application.

Even with high-quality documentation, how much documentation is enough?
Some believe documentation should be as brief, as graphical, and as to-the-point as
possible, available when needed but no sooner (Russell, 1994). Perhaps the real test

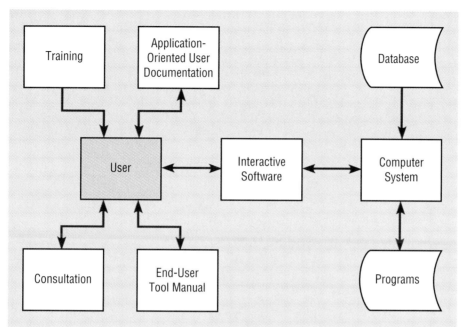

Figure 20-5
End user information system
environment and its focus
on user documentation
(Adapted from Torkzadeh
and Doll, 1993)

is not so much how much documentation has been created as it is how helpful the documentation is for users. Is the documentation complete? Is it easy to read and understand? Does it describe what the system can do? Does it describe how to use the system for your work?

Mathieson (1993) states that user documentation that only describes the system and its functionality may only go so far in supporting users in their jobs. He suggests that functionality-oriented documentation be supplemented with task-based user documentation, sometimes contained in a user's guide. Tasks can be broken down into a number of subtasks which can themselves be broken down into other subtasks. Many larger tasks share the same subtasks. Task-based documentation would provide detailed information on which steps to take to successfully perform a particular subtask. For example, if a secretary needed to send out 250 letters asking alumni for donations, task-based documentation would explain how to use the software to create the form letter, how to create a list of names and addresses, how to use the mail merge process, and how to generate addressed envelopes. Many off-the-shelf software products now include some task-based documentation, such as the "How Do I . . ." sections of help facilities.

It may be difficult for an information systems department to know about all of the possible tasks users perform and the best way to structure subtasks to complete each task. One possible solution is to have users create the task-based documentation themselves (Mathieson, 1993). The information systems department becomes the overseer and distributor of this user-generated documentation. User-generated documentation is probably a better reflection of what users actually do with the information system, and the IS department's costs of providing documentation are decreased.

Regardless of who writes the documentation or who the specific audience for it is, there are certain guidelines that researchers in this area have established for developing good documentation. These guidelines are listed in Table 20-4. As documentation of an information system may often contain both text and graphics, there are guidelines to use for each form of communication. Although following

TABLE 20-4   Guidelines for Improving Documentation Quality

Concern	Guideline
Document Organization	The document should be clearly structured. Organization should be reinforced through control of space by including such structures as a table of contents, section headings, chapter overviews, chapter summaries, and indexes.
Document Length	Make documents as short as possible: summarize, get rid of unnecessary information.
Reader Involvement	Involve the reader as soon as possible. Users want to use the system now; maintenance personnel want to fix a problem now.
Overall Appearance	Make the document look as professional as possible. Make text accessible: limit the use of italics and all upper-case text.
Style & Comprehension	Keep sentences short (no more than 25 words). Avoid abstract and difficult words. Use active voice most of the time, although passive voice does have its place. Be careful with conditional rules that have many conditions. Use short scenarios or narratives to illustrate concrete situations or actions. (*Note:* Grammar and style software, like RightWriter™ and Grammatik™, can be very helpful to check that text follows such rules.)
Representing Systems	Use pictures whenever possible to represent systems, components, equipment, and hardware.
Photographs	Use for presenting overall context.
Line Drawings	Use when the contour of an object is important and readers need the detail necessary to distinguish among components or to understand particular features.
Orientation in a Manual	Graphic organizers (e.g., color coding, icons) can help readers remember information within a manual and how it's organized.
Illustrating Processes	Graphic tools, like flow charts and hierarchy charts, can help readers complete abstract processes more quickly and with fewer errors.

these guidelines will not necessarily lead to perfect documentation, it will certainly lead to better documentation than would have been produced otherwise.

TRAINING AND SUPPORTING USERS

Computing infrastructure:
All the resources and practices required to help people adequately use computer systems to do their primary work.

Training and support are two aspects of an organization's computing infrastructure (Kling and Iacono, 1989). **Computing infrastructure** is made up of all the resources and practices required to help people adequately use computer systems to do their primary work (Kling and Scacchi, 1982; Gasser, 1986). It is analogous to the infrastructure of water mains, electric power lines, streets, and bridges that form the foundation for providing essential services in a city. Henderson and Treacy (1986) identify infrastructure as one of four fundamental issues IS managers must address. They suggest that training and support are most important in the early stages of end-user computing growth and less so later on. Rockart and Short (1989) cite the "development and implementation of a general, and eventually 'seamless,' information technology infrastructure" as a major demand on the information technology organization. They list the creation of an effective information technology infrastructure as one of the five key issues for senior organizational managers in the 1990s. Thus, training and support are critical for the success of an information system. As the person whom the user holds responsible for the new system, you and other analysts on the project team must ensure that high-quality training and support are available.

Although training and support can be talked about as if they are two separate things, in organizational practice the distinction between the two is not all that clear, as the two sometimes overlap. After all, both deal with learning about computing. It is clear that support mechanisms are also a good way to provide training, especially for intermittent users of a system (Eason, 1988). Intermittent or occasional system users are not interested in, nor would they profit from, typical user training methods. Intermittent users must be provided with "point of need support," specific answers to specific questions at the time the answers are needed. A variety of mechanisms, such as the system interface itself and on-line help facilities, can be designed to provide both training and support at the same time.

The value of support is often underestimated. Few organizations invest heavily in support staff, which can lead to users solving problems for themselves or somehow working around them (Gasser, 1986). Adequate user support may be essential for successful information system development, however. One study found that user satisfaction with support provided by the information systems department was the factor most closely related to overall satisfaction with user development of computer-based applications (Rivard and Huff, 1988).

Training Information System Users

Computer use requires skills, and training people to use computer applications can be expensive for organizations (Kling and Iacono, 1989). Training of all types is a major activity in American corporations (they spent $48 billion on training in 1993), but information systems training is often neglected. Many organizations tend to underinvest in computing skills training. It is true that some organizations institutionalize high levels of information system training, but many others offer no systematic training at all.

Some argue that information systems departments are similar to hospitals: both are high-technology environments, both are staffed by well-educated professionals, both are capital-intensive, and both have less than adequate "bedside manners" (Schrage, 1993). In this scenario, users, like patients, are seen as problems to be solved rather than as people. One response to this issue might be to more evenly spread the cost of computer training. Even though users come from all parts of the organization, the information systems department is often stuck paying for most of the computer training itself. Attitudes in information systems towards users and the availability of high-quality training might change if other organizational departments began to pay their share of the computer training bill.

Others argue that it is difficult to demonstrate any direct benefits from training once the training is complete, since managers are more likely to continually upgrade hardware and software (de Jager, 1994). Progress from upgrading is easy to measure—just look at all the empty cartons— but upgrading may not result in the increases in productivity you would expect. Training users to be effective with the systems they have now may be a more cost-effective way to increase productivity: "If the goal is maximum productivity at the lowest cost, then a day of training usually delivers more productivity per dollar than costly hardware and software upgrades" (de Jager, 1994: 86). Though not always taken seriously in practice, the value and cost-effectiveness of information systems training should not be underestimated.

For the success of an application, on what topics might you need to train system users? The type of necessary training will vary by type of system and expertise of users. The potential topics from which you must determine if training will be useful include the following:

- Use of the system (e.g., how to enter a class registration request)

- General computer concepts (e.g., computer files and how to copy them)

- Information system concepts (e.g., batch processing)

- Organizational concepts (e.g., FIFO inventory accounting)

- System management (e.g., how to request changes to a system)

- System installation (e.g., how to reconcile current and new systems during phased installation)

As you can see from this partial list, there are many potential topics that go beyond simply how to use the new system. It may be necessary for you to develop training for users in other topics so that users will be ready, conceptually and psychologically, to use the new system. Some training, like concept training, should begin early in the project since this training can assist in the unfreezing element of the organizational change process (see Chapter 19).

Each element of training can be delivered in a variety of ways. Table 20-5 lists the most common training methods used by information system departments a few years ago. Despite the importance and value of training, most of the methods listed in Table 20-5 are under-utilized in many organizations. Users primarily rely on just one of these delivery modes: more often than not, users turn to the resident expert and to fellow users for training, as shown in how often people use each mode listed in Table 20-5 (see Figure 20-6). One study reported that 89.4 percent of end users consulted their colleagues about how to use microcomputers, but only 48 percent consulted central information systems staff (Lee, 1986). Users are more likely to turn to local experts for help than to the organization's technical support staff because the local expert understands both the users' primary work and the computer systems they use (Eason, 1988). Given their dependence on fellow users for training, it should not be surprising that end users describe their most common mode of computer training as "self-training" (Nelson and Cheney, 1987). Self-training has been found to be associated with particular sets of user skills: using

TABLE 20-5   Seven Common Methods for Computer Training in 1987
(Adapted from Nelson and Cheney, 1987)

1. **Tutorial**—one person taught at one time

2. **Course**—several people taught at one time

3. **Computer-Aided Instruction**

4. **Interactive Training Manuals**—combination of tutorials and computer-aided instruction

5. **Resident Expert**

6. **Software Help Components**

7. **External Sources,** such as vendors

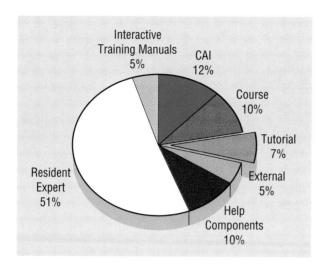

Figure 20-6
Frequency of use of
computer training methods
(Nelson & Cheney, 1987)

application development software, using packaged application software, data communication, using hardware, using operating systems, and graphics skills. The last four sets of skills are also highly associated with company-provided training. There appear to be areas of training best accomplished by centralized, company-provided training on the one hand and by self-training on the other.

One conclusion from the experience with user training methods is that an effective strategy for training on a new system is to first train a few key users and then organize training programs and support mechanisms which involve these users to provide further training, both formal and on demand. Often, training is most effective if you customize training to particular user groups, and lead trainers from these groups are in the best position to do this.

To improve information system training and increase the utilization of centralized training services, many researchers have studied training computing end users. The aim of this work has been to find ways to improve end-user training, to "... ensure that users acquire the necessary EUC (*end-user computing*) skills ... in the most efficient and effective ways possible" (Sein and Bostrom, 1990). Most of the conclusions from this body of research point to the importance of considering individual differences in designing training programs:

- End-user training should be tailored to match differences in individual learning modes (Sein and Bostrom, 1989; 1990; Bostrom et al., 1990).

- Lectures are better for learning computer skills than self-study, and level of education and individual learning style should also be taken into account when designing a training program (Davis and Davis, 1990).

- Experts in both task domain and a particular software package solved a problem differently from people who were experts in one or the other, or who were novices in both (Mackay and Lamb, 1991).

While individualized training is expensive and time-consuming for its providers, technological advances and decreasing costs have made individualized training more feasible. Similarly, the number of training modes used by information systems departments today has increased dramatically beyond what is listed in Table 20-5. Training modes now include videos, interactive television for remote

Electronic performance support system (EPSS): Component of a software package or application in which training and educational information is embedded. An EPSS can take several forms, including a tutorial, an expert system shell, and hypertext jumps to reference material.

training, multimedia training, on-line tutorials, and electronic performance support systems (EPSS).

Electronic performance support systems are on-line help systems that go beyond simply providing help—they embed training directly into a software package (Panepinto, 1993). An EPSS may take on one or more forms: they can be an on-line tutorial, provide hypertext-based access to context-sensitive reference material, or consist of an expert system shell that acts as a coach. The main idea behind the development of EPSSs is that the user never has to leave the application to gain the benefits of training. Users get to learn a new system or unfamiliar features on their own time and on their own machines, without having to lose work time to remote group training sessions. Further, this learning is on demand, when the user is most motivated to learn, since the user has a task to do. One example of an EPSS with which you may be familiar is Microsoft's Cue Cards™. Cue Cards (Figure 20-7) look like flash cards that come up on top of such applications as Access™ and Office™. They provide educational information, such as graphics, examples, and procedures as well as hypertext nodes for jumping to related help topics. Microsoft's Cue Cards communicate with the application you are running to see where you are, so you can determine, by reading the context-sensitive Cue Card, if what you want to do is possible from your present location (Galager, 1994). Some EPSS environments actually walk the user step-by-step through the task, coaching the user on what to do or allowing the user to get additional on-line assistance at any point.

As both training and support for computing are increasingly able to be delivered on-line in modules embedded in software packages and applications, the already blurred distinction between them blurs even more. Some of the issues most particular to computer user support are examined in the next section.

Figure 20-7
A Microsoft Cue Card™ for Microsoft Access™

CUE CARDS

Do your own work as you learn. Cue Cards can walk you through common Microsoft Access tasks step by step.

▶ What do you want to do?

[>] *Build a Database with Tables*
Build a database; create, import or attach tables.

[>] *Work with Data*
Add, view, edit, sort, or filter data in forms and datasheets.

[>] *Design a Query*
Create or troubleshoot queries.

[>] *Design a Form*
Create or customize forms.

[>] *Design a Report or Mailing Labels*
Create, customize, or print reports or mailing labels.

[>] *Write a Macro*
Design, run, or attach macros.

[>] *I'm Not Sure*
Show me what a database is and what I can do with it.

To read about Cue Cards and how to use them, see *About Cue Cards*.

Supporting Information System Users

Historically, computing **support** for users has been provided in one of a few forms: on paper, on-line versions of paper-based support, vendors, or by other people who work for the same organization. As we stated earlier, support, whatever its form, has often been inadequate for users' needs. Yet users consider support to be extremely important. A 1993 J. D. Power and Associates survey found user support to be the number one criterion contributing to user satisfaction with personal computing, cited by 26 percent of respondents as the most important factor (Schurr, 1993).

As organizations moved away from mainframe-based computing to increased reliance on personal computers in the 1980s, the need for support among users also increased. More support was made available on-line, but organizations also began to provide institutionalized user support in the form of information centers and *help desks.* An **information center** is a group of people who can answer questions and assist users with a wide range of computing needs, including the use of particular information systems. Information center staff might handle the following tasks:

- Install new hardware or software

- Consult with users writing programs in fourth-generation languages

- Extract data from organizational databases onto personal computers

- Set up user accounts

- Answer basic on-demand questions

- Provide a demonstration site for viewing hardware and software

- Work with users to submit system change requests

When you expect an organizational information center to help support a new system, you likely will want to train the information center staff as soon as possible on the system. Information center staff will need periodic additional training as new system features are introduced, new phases of a system are installed, or system problems and workarounds are identified. Even if new training is not required, information center staff should be aware of system changes, since such changes may result in an increase in demand for information center services.

Personnel in an information center or help desk (we discuss help desks in a later section) were typically drawn from the ranks of information systems workers or from the ranks of knowledgeable users in functional area departments. There was rarely any type of formal training people could take to learn how to work in the support area and, in general, this remains true today. As organizations move to client/server architectures, their needs for support have increased even more than during the introduction of PCs, and organizations find themselves in the position of having to rely more than ever on vendor support (From, 1993). This increased need for support comes in part from the lack of standards governing client/server products and the resulting need to make equipment and software from different vendors compatible. Vendors are able to provide the necessary support, but as they have shifted their offerings from primarily expensive mainframe packages to inexpensive off-the-shelf software, they find they can no longer bear the cost of providing the support for free. Most vendors now charge for support, and many have instituted 900 numbers or sell customers unlimited support for a given monthly or annual charge.

Automating Support In an attempt to cut the costs of providing support and to catch up with the demand for additional support services, vendors have automated

Support: Providing on-going educational and problem-solving assistance to information system users. For in-house developed systems, support materials and jobs will have to be prepared or designed as part of the implementation process.

Information center: An organizational unit whose mission is to support users in exploiting information technology.

much of their support offerings. Common methods for automating support include on-line support forums, bulletin board systems, on-demand fax, and voice-response systems (Schneider, 1993). On-line support forums provide users access to information on new releases, bugs, and tips for more effective usage. Forums are offered on such on-line services as Prodigy™, CompuServe™, and GEnie™. On-demand fax allows users to order support information through an 800 number and receive that information instantly over their fax machines. Finally, voice-response systems allow users to navigate option menus that lead to pre-recorded messages about usage, problems, and workarounds. Organizations have established similar support mechanisms for systems developed or purchased by the organization. Internal e-mail, group support systems, and office automation can be used to support such capabilities within an organization.

Other organizations can follow the lead of vendors in automating some of their support offerings in similar ways. Some software vendors will work with large accounts and corporations to set up or enhance their own support offerings. The cost to the corporation can vary from $5,000 to as much as $20,000. Offerings include access to knowledge bases about a vendor's products, electronic support services, single point of contact, and priority access to vendor support personnel (Schneider, 1993). Product knowledge bases include all of the technical and support information about vendor products and provide additional information for on-site personnel to use in solving problems. Some vendors allow users to access some of the same information that is available on on-line forums through bulletin boards which the vendors set up and maintain themselves. Many vendors now supply complete user and technical documentation on CD-ROM, including periodic updates, so that a user organization can provide this library of documentation, bug reports, workaround notices, and notes on undocumented features on-line to all internal users. Electronic support services include all of the vendor support services discussed earlier but tailored specifically for the corporation. The single point of contact is a system engineer who is often based on site and serves as a liaison between the corporation and the vendor. Finally, priority access means that corporate workers can always get help via telephone or e-mail from someone at the vendor, usually within a pre-specified response time of four hours or less.

Such vendor-enhanced support is especially appropriate in organizations where a wide variety of a particular vendor's products are in use, or where most in-house application development either utilizes vendor products as components of the larger system or where the vendor's products are themselves used as the basis for applications. An example of the former would be the case where an organization has set up a client/server architecture based on a particular vendor's SQL server and application programmer interfaces (APIs). Whichever applications are developed in-house to run under the client/server architecture depends heavily on the server and APIs, and direct vendor support dealing with problems in these components would be very helpful to the enterprise information systems staff. An example of the second would include order entry and inventory control application systems developed using Borland's dBASE IV™ or Lotus Development Corporation's Lotus 1-2-3™. In this case, the system developers and users, who are sometimes the same people for such package-based applications, can benefit considerably from directly questioning vendor representatives about their products.

Providing Support Through a Help Desk   Whether assisted by vendors or going it alone, the center of support activities for a specific information system in many organizations is the help desk. A **help desk** is an information systems department function, staffed by IS personnel, possibly part of the information center unit. The help desk is the first place users are to call when they need assistance with an in-

Help desk: A single point of contact for all user inquiries and problems about a particular information system or all users in a particular department.

formation system. The help desk staff either deal with the users' questions or refer the users to the most appropriate person.

For many years, a help desk was the dumping grounds for people IS managers did not know what else to do with. Turnover rates were high because the position was sometimes little more than a complaints department, the pay was low, and burnout rates were high. In today's information systems-dependent enterprises, however, help desks are gaining new respect as management comes to appreciate the special combination of technical skills and people skills needed to make good help desk staffers. In fact, a recent survey reveals that the top two valued skills for help desk personnel are related to communication and telephone and customer service (Crowley, 1993).

Help desk personnel (as well as the personnel in the more general information center) need to be good at communicating with users, listening to users' problems, and intelligently communicating potential solutions. These personnel also need to understand the technology they are helping users with. It is crucial, however, that help desk personnel know when new systems and releases are being implemented and when users are being trained for new systems. Help desk personnel themselves should be well-trained on new systems. One sure recipe for disaster is to train departments full of users on new systems but not train the help desk personnel these same new users will turn to for their support needs.

Help desks may be very small operations involving only a few people, or they may be large, fully staffed departments. Help desks may be integrated within the more general information center, or isolated for particular systems or particular departments. At their largest, help desks may employ five levels of professionals:

1. The first-level help desk employee takes calls, screens them, and routes them to the appropriate location. In some organizations, this first-level employee may have been replaced with interactive voice mail systems.

2. At the next level are the people who deal with most of the problems users have.

3. If the problems are too difficult, they are passed on to the third level of help desk employee, specialists who can solve almost any problem the organization's users may have.

4. The fourth level is occupied by supervisors, who have the day-to-day overseers' responsibilities for those who work directly with users.

5. The final level of employee is the help desk manager.

Call screeners make about $24,000 per year while managers make about $50,000 per year on average (Crowley, 1993). There are no information systems industry-wide standards for training help desk personnel, although there have been some discussions about certification programs. For now, much of the training comes from on-the-job training and experience.

Support Issues for the Analyst to Consider

Support is more than just answering user questions about how to use a system to perform a particular task or about the system's functionality. Support also consists of such tasks as providing for recovery and backup, disaster recovery, PC maintenance, providing newsletters and other types of proactive information sharing, and setting up user groups. It is the responsibility of the analysts for a new system to be sure that all forms of support are in place before the system is installed.

For medium to large organizations with active information system functions, many of these issues are dealt with centrally. For example, users may be provided with backup software by the central information systems unit and a schedule for routine backup. Policies may also be in place for initiating recovery procedures in case of system failure. Similarly, disaster recovery plans are almost always established by the central IS unit. Information systems personnel in medium to large organizations are also routinely responsible for PC maintenance, as the PCs belong to the enterprise. There may also be IS unit specialists in charge of composing and transmitting newsletters, or overseeing automated bulletin boards and organizing user groups.

When all of these (and more) services are provided by central IS, you must follow the proper procedures to include any new system and its users on the lists of those to whom support is provided. You must design training for the support staff on the new system, and you must make sure that system documentation will be available to them. You must make the support staff aware of the installation schedule. And you must keep these people informed as the system evolves. Similarly, any new hardware and off-the-shelf software has to be registered with the central IS authorities.

When there is no official IS support function to provide support services, you must come up with a creative plan to provide as many services as possible. You may have to write backup and recovery procedures and schedules, and the users themselves may have to purchase and be responsible for the maintenance of their hardware. In some cases, software and hardware maintenance may have to be outsourced to vendors or other capable professionals. In such situations, user interaction and information dissemination may have to be more informal than formal: informal user groups may meet over lunch or over a coffee pot rather than in officially formed and sanctioned forums.

Whether working with a vendor, external experts, or relying on their own information systems personnel, organizations can benefit from the skills of people specially trained to provide training and support services. Such individuals are called technical communicators.

THE ROLE OF TECHNICAL COMMUNICATOR

Technical communicator:
Skilled individual who is trained to write system and user documentation about mechanical and information systems.

A **technical communicator** is a person who has been specially trained to create system and user documentation about complex mechanical or information systems. Such a person used to be called a technical writer. Technical communicators are now being trained to do more than just write documentation—they also have roles in training and support and helping information systems department users create their own system documentation (Craig & Beck, 1993). Technical communicators typically work for the information systems unit in an enterprise and can play many roles in the development process (Table 20-6). They can provide documentation assistance throughout the systems development life cycle for both analysts and for users, and they can develop user training and support materials.

In our discussions about documentation and training and support materials, we have never paid much attention to the fact that someone has to do the actual writing. The writing does not happen by itself. Given their skills, technical communicators are better equipped to develop user documentation and training and support materials than programmers or analysts. They share some of the same skills as programmers and analysts, as they must also know about screen and interface design, but technical communicators also know about the latest research in learning and teaching. In designing training materials, the technical communicator's focus is on encouraging users to engage in exploratory, active learning at the keyboard rather than simply reading written documents.

TABLE 20-6   System Development Roles for Technical Communicators
(Adapted from Craig and Beck, 1993)

1. Documentation Writer

2. Trainer

3. Computer-Based Instruction Programmer

4. Usability Tester

5. Hypertext Developer

6. On-Line Help Writer

7. Quick Reference Guide Writer

8. Standards Writer

9. Front-End Analyst (for proposed documentation or training project)

10. Instructional Designer

Despite the usefulness of a technical communicator for the processes of documenting a system and for training and supporting users, not every organization is large enough or wealthy enough to afford such specialized personnel. In such cases, much of the documentation, training, and support work will be left to others, such as programmers, analysts and, as we have seen, to users themselves. It is also possible for an information systems department to outsource some of the documentation work or to hire free-lance technical communicators.

PROJECT CLOSE-DOWN

In Chapter 4 you learned about the various phases of project management, from project initiation to closing down the project. If you are the project manager and you have successfully guided your project through all of the phases of the systems development life cycle presented so far in this book, you are now ready to close down your project. Although the maintenance phase is just about to begin, the development project itself is over. As you will see in the next chapter, maintenance can be thought of as a series of smaller development projects, each with its own series of project management phases.

As you recall from Chapter 4, your first task in closing down the project involves many different activities, from dealing with project personnel to planning a celebration of the project's ending. You will likely have to evaluate your team members, reassign most to other projects, and perhaps terminate others. As project manager, you will also have to notify all of the affected parties that the development project is ending and that you are now switching to maintenance mode.

Your second task is to conduct post-project reviews both with your management and with your customers. In some organizations, these post-project reviews will follow formal procedures and may involve internal or EDP (electronic data processing) auditors. The point of a project review is to critique the project, its methods, its deliverables, and its management.

The third major task in project close-down is closing out the customer contract. Any contract that has been in effect between you and your customers during the project (or as the basis for the project) must be completed, typically through the consent of all contractually involved parties. This may involve a formal "signing-off" by the clients that your work is complete and acceptable. The maintenance phase will typically be covered under new contractual agreements. If your customer is outside of your organization, you will also likely negotiate a separate support agreement.

As an analyst member of the development team, your job on this particular project ends during project close-down. You will likely be reassigned to another project dealing with some other organizational problem. Maintenance on your new system will begin and continue without you. To complete our consideration of the systems development life cycle, however, we will cover the maintenance phase and its component tasks in Chapter 21.

SUMMARY

In this chapter you learned about four types of documentation: system documentation, which describes in detail the design of a system and its specifications; internal documentation, that part of system documentation which is included in the code itself or emerges at compile time; external documentation, that part of system documentation which includes the output of such diagramming techniques as data flow and entity-relationship diagramming; and user documentation, which describes a system and how to use it for the system's users. You also learned about task-oriented documentation, which focuses on how a user can perform a specific task with an information system instead of on the system's functionality.

Training and support are both part of an organization's computing infrastructure. A computing infrastructure is analogous to a city's infrastructure of streets, water mains, and electrical lines. The computing infrastructure allows an organization's computer users to continue to operate, just as a city's infrastructure allows the city to continue to operate. In many organizations, the computing infrastructure and the people charged with providing and maintaining it are underfunded.

Computer training has typically been provided in classes and tutorials. While there is some evidence lectures have their place in teaching people about computing and information systems, the current emphasis in training is on automated delivery methods, such as on-line reference facilities, multimedia training, and electronic performance support systems. The latter embed training in applications themselves in an attempt to make training a seamless part of using an application for daily operations. The emphasis in support is also on providing on-line delivery, including on-line support forums and bulletin board systems. As organizations move toward client/server architectures, they rely more on vendors for support. Vendors provide many on-line support services, and they work with customers to bring many aspects of on-line support in-house. An information center provides general support to all users, and a help desk provides aid to users in a particular department or for a particular system.

Technical communicators, formerly called technical writers, are specially trained personnel who can help information systems departments develop and maintain documentation and training and support materials. Finally, the steps necessary to close down a project include reassigning project personnel, conducting post-project reviews, and closing out the project contract with your customers.

C H A P T E R   R E V I E W

K E Y   T E R M S

Computing infrastructure
Electronic performance
   support system (EPSS)
External documentation

Help desk
Information center
Internal documentation
System documentation

Support
Technical communicator
User documentation

R E V I E W   Q U E S T I O N S

1. Define each of the following terms:
   a. computing infrastructure
   b. EPSS
   c. technical communicator
   d. support
   e. information center

2. What is the difference between system documentation and user documentation?

3. What is task-oriented documentation?

4. What were the common methods of computer training in 1987? What training methods have been added since then?

5. What is self-training?

6. What proof do you have that individual differences matter in computer training?

7. Why are corporations relying so heavily on vendor support as they move to client/server architectures?

8. Describe the delivery methods many vendors employ for providing support.

9. Describe the role in systems development of technical communicators.

10. Describe the various roles typically found in a help desk function.

P R O B L E M S   A N D   E X E R C I S E S

1. Match the following terms to the appropriate definitions.

   _____ system documentation

   _____ user documentation

   _____ internal documentation

   _____ external documentation

   a. detailed information about a system's design specifications, its internal workings, and its functionality

   b. system documentation that includes the outcome of such structured diagramming techniques as data flow and entity-relationship diagrams

   c. written or other visual information about an application system, how it works, and how to use it

   d. system documentation that is part of the program source code or is generated at compile time

2. How much documentation is enough?

3. What is the purpose of electronic performance support systems? How would you design one to support a word processing package? A database package?

4. Discuss the role of a centralized training and support facility in a modern organization. Given advances in technology and the prevalence of self-training and consulting among computing end users, how can such a centralized facility continue to justify its existence?

5. Is it good or bad for corporations to rely on vendors for computing support? List arguments both for and against reliance on vendors as part of your answer.

6. Suppose you were responsible for establishing a training program for users of Hoosier Burger's inventory control system described in previous chapters. Which forms of training would you use? Why?

7. Suppose you were responsible for establishing a help desk for users of Hoosier Burger's inventory control system described in previous chapters. Which support system elements would you create to help users be effective? Why?

8. Your university or school probably has some form of microcomputer center or help desk for students. What functions does this center perform? How do these functions compare to those outlined in this chapter?

9. Suppose you were responsible for organizing the user documentation for Hoosier Burger's inventory control system, described in previous chapters. Write an outline that shows the documentation you would suggest be created, and generate the table of contents or outline for each element of this documentation.

10. Find the documentation for the word processor you use. Evaluate this documentation using the guidelines in Table 20-4. Based on your evaluation, what specific improvements would you suggest need to be made to improve this documentation?

11. Compare the user's guide documentation for the spreadsheet package you use to that of your package's chief competitor. Then compare both to the generic outline in Table 20-3. To what extent does each user's guide resemble the generic outline? How do you explain any variation you see?

FIELD EXERCISES

1. Talk with people you know who use computers in their work. Ask them to get copies of the user documentation they rely on for the systems they use at work. Analyze the documentation. Would you consider it good or bad? Support your answer. Whether good or bad, how might you improve it?

2. Interview a technical communicator. How did this person move into this career? What is his or her background and training? What tasks does this person perform at work?

3. Volunteer to work for a shift at a help desk at your school's computer center. Keep a journal of your experiences. What kind of users did you have to deal with? What kinds of questions did you get? Do you think help desk work is easy or hard? What skills are needed by someone in this position?

4. Let's say your professor has asked you to help him or her train a new secretary on how to prepare class notes for electronic distribution to class members. Your professor uses word processing software and an e-mail package to prepare and distribute the notes. Assume the secretary knows nothing about either package. Prepare a user task guide that shows the secretary how to complete this task.

REFERENCES

Bell, P. and C. Evans. 1989. *Mastering Documentation*. New York: John Wiley & Sons.

Bostrom, R. P., L. Olfman, and M. K. Sein. 1990. "The Importance of Learning Style in End-User Training." *MIS Quarterly* 14 (March): 101–119.

Craig, J. S. and B. E. Beck. 1993. "A New Look at Documentation and Training: Technical Communicator as Problem Solver." *Information Systems Management* 10 (Summer): 47–55.

Crowley, A. 1993. "The Help Desk Gains Respect." *PC Week* 10 (November 15): 138.

Davis, D. L. and D. F. Davis. 1990. "The Effect of Training Techniques and Personal Characteristics on Training End Users of Information Systems." *Journal of MIS* 7 (Fall): 93–110.

de Jager, P. 1994. "Are We Just Plain Lazy?" *Computerworld* 28 (February 21): 85.

Eason, K. 1988. *Information Technology and Organisational Change*. London, Taylor & Francis.

From, E. 1993. "Shouldering the Burden of Support." *PC Week* 10 (November 15): 122, 144.

Galagar, P. 1994. "The Instructional Designer in a New Age." *Training & Development* 48 (March): 52–53.

Gasser, L. 1986. "The Integration of Computing and Routine Work." *ACM Transactions on Office Information Systems* 4 (July): 205–25.

Henderson, J. C. and M. E. Treacy. 1986. "Managing End-User Computing for Competitive Advantage." *Sloan Management Review* (Winter): 3–14.

Kling, R. and S. Iacono. 1989. "Desktop Computerization and the Organization of Work." In *Computers in the Human Context*, edited by Forester, T. Cambridge, MA, The MIT Press: 335–56.

Kling, R. and W. Scacchi. 1982. "The Web of Computing: Computer Technology as Social Organization." *Advances in Computers* 21: 1–90.

Lee, D. M. S. 1986. "Usage Pattern and Sources of Assistance for Personal Computer Users." *MIS Quarterly,* 10 (December): 313–25.

Mackay, J. M. and C. W. Lamb, Jr. 1991. "Training Needs of Novices and Experts with Referent Experience and Task Domain Knowledge." *Information & Management* 20 (March):183–89.

Martin, J. and C. McClure. 1985. *Structured Techniques for Computing*. Englewood Cliffs, NJ: Prentice-Hall.

Mathieson, K. 1993. "Effective User Documentation: Focusing on Tasks Instead of Systems." *Journal of Systems Management* 44 (May): 25–27.

Nelson, R. R. and P. H. Cheney. 1987. "Training End Users: An Exploratory Study." *MIS Quarterly* 11 (December): 547–59.

Panepinto, J. 1993. "Delivering Training." *Computerworld* 27 (November 8): 93.

Rivard, S. and S. L. Huff. 1988. "Factors of Success for End-User Computing." *Communications of the ACM* 31 (May): 552–61.

Rockart, J. F. and J. E. Short. 1989. "IT in the 1990s: Managing Organizational Interdependence." *Sloan Management Review* (Winter): 7–17.

Russell, D. L. 1994. "Just Say NO to Documentation." *Computerworld* 28 (January 17): 35.

Schneider, J. 1993. "Shouldering the Burden of Support." *PC Week* 10 (November 15): 123, 129.

Schrage, M. 1993. "Unsupported Technology: A Prescription for Failure." *Computerworld* 27 (May 10): 31.

Schurr, A. 1993. "Support is No. 1." *PC Week* 10 (November 15): 126.

Sein, M. and R. Bostrom. 1990. "An Experimental Investigation of the Role and Nature of Mental Models in the Learning of Desktop Systems." In *Desktop Information Technology,* edited by Kaiser, K. M. and H. J. Oppelland. Amsterdam, Elsevier Science Publishers, B.V.:253–76.

Sein, M. and R. Bostrom. 1989. "Individual Differences and the Training of Novice Users." *Human-Computer Interaction* 4(3): 197–229.

Torkzadeh, G. and W. J. Doll. 1993. "The Place and Value of Documentation in End-User Computing." *Information & Management* 24 (3): 147–58.

Supporting Users and Project Close-Down

INTRODUCTION

Frank Napier sighed. The CATS (Customer Activity Tracking System) project was just about over. Frank had two important meetings today. The first was a CATS project review, headed by Jordan Pippen, CATS project leader. The meeting was to last approximately two hours, about the right amount of time for a project of this complexity, Frank thought. The second meeting, to be held later this afternoon, would include only Frank, Jordan, and Karen Gardner, the CIO. The purpose of the second meeting was to review Frank's performance on CATS and to reassign him to a new project. There was a slight chance that Frank would be terminated—that was always one possible outcome of these personnel meetings—but as Frank mentally reviewed the project and his performance, he felt pretty sure that BEC would keep him on.

SUPPORTING CATS USERS

The CATS review meeting was scheduled from 10 a.m. until noon, so Frank had less than an hour to review his work on support arrangements for CATS. There were really two sets of arrangements, one for the corporate part of CATS and a second for the in-store part. Frank had long ago finished the arrangements for the corporate part. He had worked with Lina Wertmueller, one of BEC's best technical communicators, on formulating a systems administrator guide for corporate CATS. Frank had also worked with Central Operations to produce a backup schedule and to integrate CATS into Operations' recovery and disaster recovery plans.

Now Frank was finishing up the details on support arrangements for the in-store CATS. Frank and Lina had also worked together on a quick reference guide for the in-store version of the system. As CATS was not that complicated to understand or use at the store level, Lina suggested they only design a quick

reference guide. She argued the system was not complex enough to justify a more thorough reference guide or a complete user's guide. What was really needed, she said, was a reference guide any clerk at a point-of-sale terminal could pick up and quickly scan for help in completing a CATS transaction. Frank agreed, and after Jordan saw a prototype of the guide, she agreed as well.

The quick reference guide Lina developed was made up of a dozen or so colored index cards, joined together with a ring (Figure 1). The colors of the cards indicated certain functions that would be performed on CATS: "How do I enter CATS?" in grey; "How do I enter or update customer information?" in sky blue; "How do I record a reservation sale?" in yellow; error messages in red, and so on. The information on the cards tended to be presented graphically rather than textually, and most cards contained pictures of screens from CATS. There were two reasons for this:

1. Lina felt that this type of reference information was better presented graphically.

2. As BEC was becoming more of an international company, it was necessary to be sensitive to non-English speakers.

The cards fit into a customized clear plastic bag that could be attached to the terminals with Velcro strips.

The quick reference guide had tested well with BEC clerks and other potential users. Since the decision had been made to implement CATS in all North American stores, the quick reference guide was being produced for eventual distribution. Now that the quick reference guide had been finalized, Frank could turn his attention to other items on the support arrangements checklist (Figure 2).

Frank reviewed the checklist. In his mind, the most important component of the in-store CATS system was the quick reference guide, and that was all done. Many of the support issues were covered by

Figure 1

Quick reference guide for in-store CATS

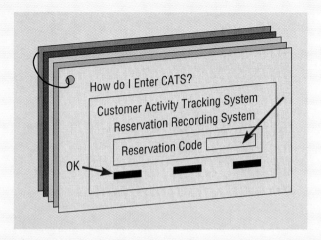

Figure 2

Support arrangements checklist for the in-store CATS component

BEC SUPPORT ARRANGEMENTS CHECKLIST	
System: In-Store CATS **Prepared by:** Frank Napier	
Date: July 1, 1995 **Approved by:**	
Issue:	**Arrangements:**
User guides	Quick reference guide designed by Lina Wertmueller, technical communicator; approved 6/1/95 by Jordan Pippen, CATS project manager; distribution to stores planned for 9/1/95.
Help Desk	Training session for help desk personnel scheduled for 8/1/95.
On-line help	No on-line help components currently exist; recommended for inclusion with next releases.
Hardware maintenance	Already covered by BEC maintenance agreements for store-based hardware (see ISD Policy MBX-234.5).
Backup & recovery	Already covered by standard BEC policies for backup and recovery of store-based data (see ISD Policy MBY-1.1).
Disaster recovery	Already covered by BEC store disaster recovery plans (see ISD Policy MBY-1.15).
Newsletters	None recommended.
User groups	None recommended.
New releases	Minor changes will trigger relevant adjustments to current arrangements. Major changes will be treated as new systems, with resultant support review and design.
Other (specify)	No other relevant issues identified.

existing centralized BEC policies, so Frank didn't have to worry about hardware maintenance or disaster recovery. All of the existing policies that covered the store-based systems would also cover the in-store component of CATS. Frank did need to carefully consider one other area of support. Sometimes clerks would call the BEC help desk when they couldn't figure out something about the transaction processing systems they used. BEC encouraged clerks to call rather than make a mistake, especially when they were unable to find a co-worker who might have the answer. As the in-store component of CATS was just about ready to be implemented throughout North America, help desk personnel needed to know how to answer the questions they might get about the system. Fortunately, Frank had already scheduled a training session for help desk personnel. The in-store component of CATS was not difficult to understand or use, and there probably wouldn't be that many questions to the help desk about it, but a little training for the help desk people was still a good idea. There were few things worse, help desk personnel had told Frank, than a new BEC system up and running that the help desk had little information about.

Frank looked up at the clock on his computer screen. Almost time to go to the review meeting. He looked over his support notes one last time and checked his portfolio to make sure he had everything he needed for the review. I wonder what project I'll be assigned to next, Frank thought, as he left his cubicle and headed toward the conference room.

SUMMARY

The CATS project is just about over for Frank, Jordan, Jorge, Wendy, and the others who have been part of one of the most creative projects ever conducted at Broadway Entertainment Company. Certainly CATS did not do everything imaginable of a customer tracking system (there has to be something, Frank thought, for the maintenance staff to do), and competitors were certain to react to CATS with similar systems of their own, so enhancements would be inevitable. In any case, Frank headed into the final CATS project review meeting with confidence and pride. CATS had been a wonderful experience and a tremendous responsibility for someone who had graduated college only a year ago. He sure would have a lot to talk about at the Computer Information Systems homecoming party this coming fall.

Maintaining Information Systems

After studying this chapter, you should be able to:

- Explain and contrast four types of maintenance.

- Describe several factors that influence the cost of maintaining an information system and apply these factors to the design of maintainable systems.

- Describe maintenance management issues including alternative organizational structures, quality measurement, processes for handling change requests, and configuration management.

- Explain the role of CASE when maintaining information systems.

- Describe business process re-engineering and explain why it can be a significant source for generating information system maintenance requests.

INTRODUCTION

In this final chapter, we discuss systems maintenance, the largest systems development expenditure for many organizations. In fact, more programmers today work on maintenance activities than work on new development. Your first job after graduation may very well be as a maintenance programmer/analyst. This disproportionate distribution of maintenance programmers is interesting since software does not wear out in a physical manner as do buildings and machines.

There is no single reason why software is maintained; however, most reasons relate to a desire to evolve system functionality in order to overcome internal processing errors or to better support changing business needs. Thus, maintaining a system is a fact of life for most systems. This means that maintenance can begin soon after the system is installed. As with the initial design of a system, maintenance activities are not limited only to software changes but include changes to hardware and business procedures. A question many people have about maintenance relates to how long organizations should maintain a system. Five years? Ten years? Longer? There is no simple answer to this question, but it is most often an issue of economics. In other words, at what point does it make financial sense to discontinue evolving an older system and build or purchase a new one? The focus of a great deal of upper IS management attention is devoted to assessing the trade-offs between maintenance and new development. In this chapter, we will provide you with a better understanding of the maintenance process and describe the types of issues that must be considered when maintaining systems.

In this chapter, we also briefly describe the systems maintenance process and the deliverables and outcomes from this process. This is followed by a detailed discussion contrasting the types of maintenance, an overview of critical management issues, and a description of the role of CASE technology in the maintenance process. We also describe business process re-engineering, a method many organizations are using to radically improve business performance. A result of many business process re-engineering efforts is identifying ways to alter how existing information systems process, store, and share information to better serve changing business needs. This means that business process re-engineering efforts often lead to the development of system maintenance requests. Finally, we describe the process of resolving a maintenance request at Pine Valley Furniture.

MAINTAINING INFORMATION SYSTEMS

Once an information system is installed, the system is essentially in the maintenance phase of the SDLC. When a system is in the maintenance phase, some person within the systems development group is responsible for collecting maintenance requests from system users and other interested parties, like system auditors, data center and network management staff, and data analysts. Once collected, each request is analyzed to better understand how it will alter the system and what business benefits and necessities will result from such a change. If the change request is approved, a system change is designed and then implemented. As with the initial development of the system, implemented changes are formally reviewed and tested before installation into operational systems.

The Process of Maintaining Information Systems

Throughout this book, we have drawn the systems development life cycle as the waterfall model where one phase leads to the next with overlap and feedback loops. As shown in Figure 21-1, the maintenance phase is the last phase of the SDLC. Yet, a life cycle, by definition, is circular in that the last activity leads back to the first. This means that the process of maintaining an information system is the process of returning to the beginning of the SDLC (see Figure 21-2) and repeating development steps until the change is implemented.

As shown in Figure 21-1, four major activities occur within maintenance:

1. Obtaining Maintenance Requests

2. Transforming Requests into Changes

3. Designing Changes

4. Implementing Changes

Obtaining maintenance requests requires that a formal process be established whereby users can submit system change requests. Earlier in the book, we presented a user request document called a systems service request (SSR) which is shown in Figure 21-3. Most companies have some sort of document like an SSR to

Figure 21-1
Systems development life cycle with maintenance phase highlighted

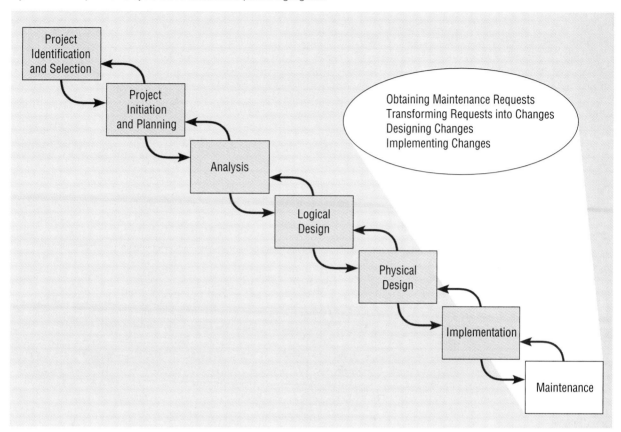

request new development, to report problems, or to request new system features with an existing system. When developing the procedures for obtaining maintenance requests, organizations must also specify an individual within the organization to collect these requests and manage their dispersal to maintenance personnel. The process of collecting and dispersing maintenance requests is described in much greater detail later in the chapter.

Once a request is received, analysis must be conducted to gain an understanding of the scope of the request. It must be determined how the request will affect the current system and the duration of such a project. As with the initial development of a system, the size of a maintenance request can be analyzed for risk and feasibility (see Chapter 7). Next, a change request can be transformed into a formal design change which can then be fed into the maintenance implementation phase. Thus, many similarities exist between the SDLC and the activities within the maintenance process. Figure 21-4 on page 796 equates SDLC phases to the maintenance activities described previously. The figure shows that the first phase of the SDLC—project identification and selection—is analogous to the maintenance process of obtaining a maintenance request (step 1). SDLC phases project initiation and planning and analysis are analogous to the maintenance process of transforming requests into a specific system change (step 2). The logical and physical design phases of the SDLC, of course, equate to the designing changes process (step 3). Finally, the SDLC phase implementation equates to step 4, implementing changes. This similarity

Figure 21-2
Maintenance phase makes the systems development process a life cycle

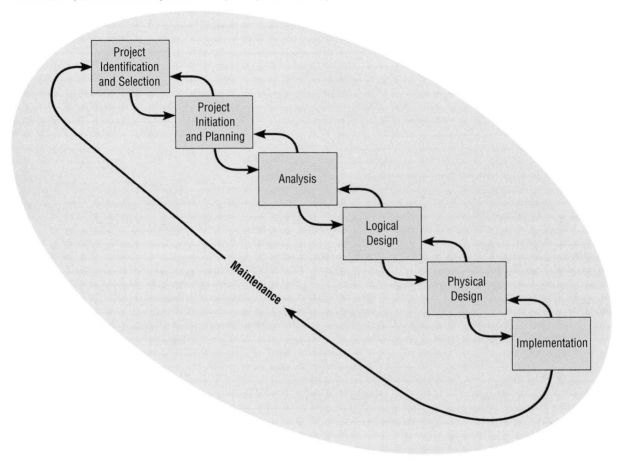

between the maintenance process and the SDLC is no accident. The concepts and techniques used to initially develop a system are also used to maintain it.

Deliverables and Outcomes

Since the maintenance phase of the SDLC is basically a subset of the activities of the entire development process, the deliverables and outcomes from the process are the development of a new version of the software and new versions of all design documents developed or modified during the maintenance process. This means that all documents created or modified during the maintenance effort, including the system itself, represent the deliverables and outcomes of the process. Those programs and documents that did not change may also be part of the new system. Since most organizations archive prior versions of systems, all prior programs and documents must be kept to ensure the proper versioning of the system. This enables prior versions of the system to be recreated if needed. A more detailed discussion of configuration management and change control is presented later in the chapter.

Because of the similarities between the steps, deliverables, and outcomes of new development and maintenance, you may be wondering how to distinguish

Figure 21-3
Systems Service Request for Purchasing Fulfillment System (Pine Valley Furniture)

Pine Valley Furniture
System Service Request

REQUESTED BY _____Juanita Lopez_____ DATE _____November 1, 1994_____

DEPARTMENT _____Purchasing, Manufacturing Support_____

LOCATION _____Headquarters, 1-322_____

CONTACT _____Tel: 4-3267   FAX: 4-3270   e-mail: jlopez_____

TYPE OF REQUEST   URGENCY

[X]   New System   [   ]   Immediate – Operations are impaired or
                                                              opportunity lost
[   ]   System Enhancement   [   ]   Problems exist, but can be worked around
[   ]   System Error Correction   [X]   Business losses can be tolerated until new
                                                              system installed

PROBLEM STATEMENT

Sales growth at PVF has caused greater volume of work for the manufacturing support unit within
Purchasing. Further, more concentration on customer service has reduced manufacturing lead times,
which puts more pressure on purchasing activities. In addition, cost-cutting measures force Purchasing
to be more agressive in negotiating terms with vendors, improving delivery times, and lowering our
investments in inventory. The current modest systems support for manufacturing purchasing is not
responsive to these new business conditions. Data are not available, information cannot be summarized,
supplier orders cannot be adequately tracked, and commodity buying is not well supported. PVF is
spending too much on raw materials and not being responsive to manufacturing needs.

SERVICE REQUEST

I request a thorough analysis of our current operations with the intent to design and build a completely
new information system. This system should handle all purchasing transactions, support display and
reporting of critical purchasing data, and assist purchasing agents in commodity buying.

IS LIAISON _Chris Martin (Tel: 4-6204   FAX: 4-6200   e-mail: cmartin)_____

SPONSOR _Sal Divario, Director, Purchasing_____

- TO BE COMPLETED BY SYSTEMS PRIORITY BOARD -

[   ]   Request approved   Assigned to _____
                                         Start date _____
[   ]   Recommend revision
[   ]   Suggest user development
[   ]   Reject for reason _____
                                     _____
                                     _____

Figure 21-4
Maintenance activities in relation to the SDLC

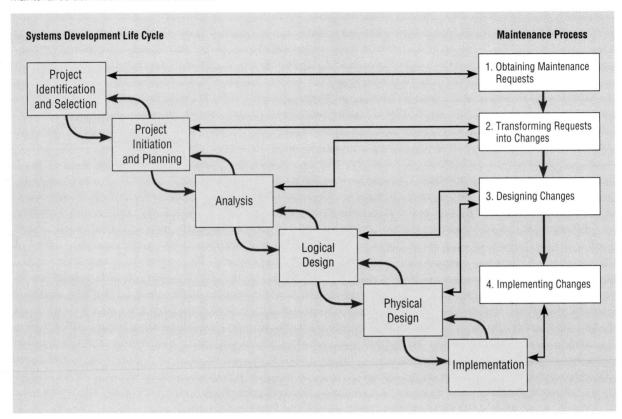

between these two processes. One difference is that maintenance reuses most existing system modules in producing the new system version. Other distinctions are that we develop a new system when there is a change in the hardware or software platform or when fundamental assumptions and properties of the data, logic, or process models change.

CONDUCTING SYSTEMS MAINTENANCE

A significant portion of the expenditures for information systems within organizations does not go to the development of new systems but to the maintenance of existing systems. We will describe various types of maintenance, factors influencing the complexity and cost of maintenance, alternatives for managing maintenance, and the role of CASE during maintenance. Given that maintenance activities consume the majority of information systems-related expenditures, gaining an understanding of these topics will yield numerous benefits to your career as an information systems professional.

Maintenance: Changes made to a system to fix or enhance its functionality.

Corrective maintenance: Changes made to a system to repair flaws in its design, coding, or implementation.

Types of Maintenance

There are several types of **maintenance** that you can perform on an information system (see Table 21-1). By maintenance, we mean the fixing or enhancing of an information system. **Corrective maintenance** refers to changes made to repair defects

TABLE 21-1 Types of Maintenance

Type	Description
Corrective	Repair design and programming errors
Adaptive	Modify system to environmental changes
Perfective	Evolve system to solve new problems or take advantage of new opportunities
Preventive	Safeguard system from future problems

in the design, coding, or implementation of the system. For example, if you had recently purchased a new home, corrective maintenance would involve repairs made to things that had never worked as designed, such as a faulty electrical outlet or misaligned door. Most corrective maintenance problems surface soon after installation. When corrective maintenance problems surface, they are typically urgent and need to be resolved to curtail possible interruptions in normal business activities. Of all types of maintenance, corrective accounts for as much as 75 percent of all maintenance activity (Andrews and Leventhal, 1993). This is unfortunate because corrective maintenance adds little or no value to the organization; it simply focuses on removing defects from an existing system without adding new functionality (see Figure 21-5).

Adaptive maintenance involves making changes to an information system to evolve its functionality to changing business needs or to migrate it to a different operating environment. Within a home, adaptive maintenance might be adding storm windows to improve the cooling performance of an air conditioner. Adaptive maintenance is usually less urgent than corrective maintenance because business and technical changes typically occur over some period of time. Contrary to corrective maintenance, adaptive maintenance is generally a small part of an organization's maintenance effort but does add value to the organization.

Adaptive maintenance: Changes made to a system to evolve its functionality to changing business needs or technologies.

Perfective maintenance involves making enhancements to improve processing performance, interface usability, or to add desired, but not necessarily required, system features ("bells and whistles"). In our home example, perfective maintenance would be adding a new room. Many system professionals feel that perfective maintenance is not really maintenance but new development.

Perfective maintenance: Changes made to a system to add new features or to improve performance.

Preventive maintenance involves changes made to a system to reduce the chance of future system failure. An example of preventive maintenance might be to increase the number of records that a system can process far beyond what is currently needed or to generalize how a system sends report information to a printer

Preventive maintenance: Changes made to a system to avoid possible future problems.

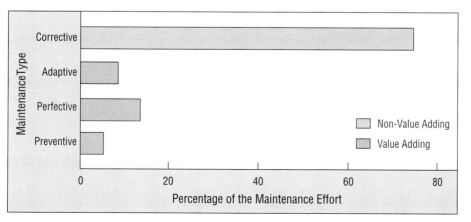

Figure 21-5
Types of maintenance
(Adapted from Andrews and
Leventhal, 1993)

so that the system can easily adapt to changes in printer technology. In our home example, preventive maintenance could be painting the exterior to better protect the home from severe weather conditions. As with adaptive maintenance, both perfective and preventive maintenance are typically a much lower priority than corrective maintenance. Over the life of a system, corrective maintenance is most likely to occur after initial system installation or after major system changes. This means that adaptive, perfective, and preventive maintenance activities can lead to corrective maintenance activities if not carefully designed and implemented.

The Cost of Maintenance

Information systems maintenance costs are a significant expenditure. For some organizations, as much as 80 percent of their information systems budget is allocated to maintenance activities (Pressman, 1992). Additionally, the proportion of systems expenditures has been rising due to the fact that many organizations have accumulated more and more older systems that require more and more maintenance. For example, Figure 21-6 shows that in the 1970s, most of an organization's information systems' expenditures were allocated to new development rather than to maintenance. However, over the years, this mix has changed so that the majority of expenditures are now earmarked for maintenance. This means that you must understand the factors influencing the **maintainability** of systems. Maintainability is the ease with which software can be understood, corrected, adapted, and enhanced. Systems with low maintainability result in uncontrollable maintenance expenses.

Maintainability: The ease with which software can be understood, corrected, adapted, and enhanced.

Numerous factors influence the maintainability of a system. These factors, or cost elements, will determine the extent to which a system has high or low maintainability. Table 21-2 shows numerous elements that influence the cost of maintenance, many of which you can influence as a systems analyst. Of these factors, three are most significant: number of latent defects, number of customers, and documentation quality.

The number of latent defects refers to the number of unknown errors existing in the system after it is installed. Because corrective maintenance accounts for most maintenance activity, the number of latent defects in a system influences most of the costs associated with maintaining a system. If there are no errors in the system after it is installed, then maintenance costs will be relatively low. If there are a large number of defects in the system when it is installed, maintenance costs will likely be high.

Figure 21-6
New development versus maintenance as a percent of software budget (Adapted from Pressman, 1987)

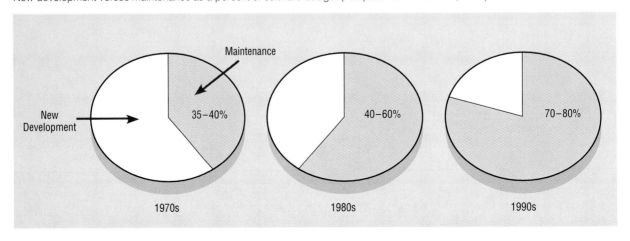

Maintenance

New Development

35–40%   40–60%   70–80%

1970s   1980s   1990s

TABLE 21-2   Cost Elements of Maintenance
(Adapted from Jones, 1986)

Element	Description
Defects	Number of unknown defects in a system when it is installed
Customers	Number of different customers that a maintenance group must support
Documentation	Quality of technical system documentation including test cases
Personnel	Number and quality of personnel dedicated to the support and maintenance of a system
Tools	Software development tools, debuggers, hardware, and other resources
Software Structure	Structure and maintainability of the software

A second factor influencing maintenance costs is the number of customers for a given system. In general, the greater the number of customers, the greater the maintenance costs. For example, if a system has only one customer, problem and change requests will come from only one source. A single customer also makes it much easier for the maintenance group to know how the system is actually being used and the extent to which users are adequately trained. If the system fails, the maintenance group is only notifying, supporting, and retraining a small group of people and updating their programs and documentation. On the other hand, for a system with thousands of users, change requests (possibly contradictory or incompatible) and error reports come from many places, and notification of problems, customer support, and system and documentation updating become a much more significant problem. For example, notifying all customers of a problem is fast and easy when you have a single customer whom you can contact using a telephone, fax, or electronic mail message. Yet, it will be difficult and expensive to quickly and easily contact thousands of users about a catastrophic problem. In sum, the greater the number of customers, the more critical it is that a system have high maintainability.

A third major contributing factor to maintenance costs is the quality of system documentation. Figure 21-7 shows that without quality documentation, maintenance

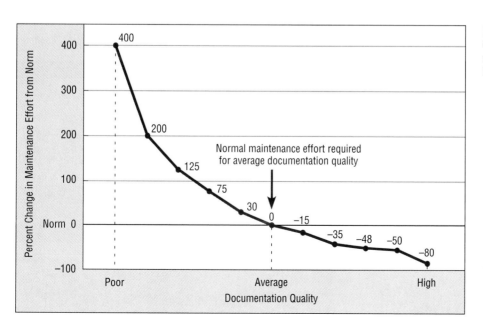

Figure 21-7
Quality documentation eases maintenance (Adapted from Hanna, 1992)

effort can increase exponentially. In conclusion, numerous factors will influence the maintainability and thus the overall costs of system maintenance. System professionals have found that the number of defects in the installed system drive all other cost factors. This means that it is important that you remove as many errors as possible before installation.

The quality of the maintenance personnel also contributes to the cost of maintenance. In some organizations, the best programmers are assigned to maintenance. Highly skilled programmers are needed because the maintenance programmer is typically not the original programmer and must quickly understand and carefully change the software. Tools, such as those that can automatically produce system documentation where none exists, can also lower maintenance costs. Finally, well-structured programs make it much easier to understand and fix programs.

Managing Maintenance

As maintenance activities consume more and more of the systems development budget, maintenance management has become increasingly important. For example, Table 21-3 shows the worldwide growth of programmers who are working on new development versus maintenance. These data show that the number of people working on maintenance has now surpassed the number working on new development. Maintenance has become the largest segment of programming personnel, and this implies the need for careful management. We will address this concern by discussing several topics related to the effective management of systems maintenance.

Managing Maintenance Personnel   One concern with managing maintenance relates to personnel management. Historically, many organizations had a "maintenance group" that was separate from the "development group." With the increased number of maintenance personnel, the development of formal methodologies and tools, changing organizational forms, end-user computing, and the widespread use of very high-level languages for the development of some systems, organizations have rethought the organization of maintenance and development personnel (Chapin, 1987). In other words, should the maintenance group be separated from the development group? Or, should the same people who build the system also maintain it? A third option is to let the primary end users of the system in the functional units of the business have their own maintenance personnel. The advantages and disadvantages to each of these organizational structures are summarized in Table 21-4.

TABLE 21-3   Worldwide Totals of Programmers Working on New Development versus Maintenance
(Adapted from Jones, 1986)

Year	Programmers on New Programs	Programmers on Maintenance
1950	90	10
1960	8,500	1,500
1970	65,000	35,000
1980	1,200,000	800,000
1990	3,000,000	4,000,000
2000	4,000,000	6,000,000

TABLE 21-4 Advantages and Disadvantages of Different Maintenance Organizational Structures

Type	Advantages	Disadvantages
Separate	Formal transfer of systems between groups improves the system and documentation quality	All things cannot be documented, so the maintenance group may not know critical information about the system
Combined	Maintenance group knows or has access to all assumptions and decisions behind the system's original design	Documentation and testing thoroughness may suffer due to a lack of a formal transfer of responsibility
Functional	Personnel have a vested interest in effectively maintaining the system and have a better understanding of functional requirements	Personnel may have limited job mobility and lack access to adequate human and technical resources

In addition to the advantages and disadvantages listed in Table 21-4, there are numerous other reasons why organizations should be concerned with how they manage and assign maintenance personnel. One key issue is that many systems professionals don't want to perform maintenance because they feel that it is more exciting to build something new rather than change an existing system (Martin, De-Hayes, Hoffer, and Perkins, 1994). In other words, maintenance work is often viewed as "cleaning up someone else's mess." Also, organizations have historically provided greater rewards and job opportunities to those performing new development, thus making people shy away from maintenance-type careers. As a result, no matter how an organization chooses to manage its maintenance group—separate, combined, or functional—it is now common to rotate individuals in and out of maintenance activities. This rotation is believed to lessen the negative feelings about maintenance work and to give personnel a greater understanding of the difficulties and relationships between new development and maintenance.

Measuring Maintenance Effectiveness   A second management issue is the measurement of maintenance effectiveness. As with the effective management of personnel, the measurement of maintenance activities is fundamental to understanding the quality of the development and maintenance efforts. To measure effectiveness, you must measure these factors:

- Number of failures
- Time between each failure
- Type of failure

Measuring the number and time between failures will provide you with the basis to calculate a widely used measure of system quality. This metric is referred to as the **mean time between failures (MTBF).** As its name implies, the MTBF measure shows the average length of time between the identification of one system failure until the next. Over time, you should expect the MTBF value to rapidly increase after a few months of use (and corrective maintenance) of the system (see Figure 21-8 for an example of the relationship between MTBF and age of a system). If the MTBF does not rapidly increase over time, it will be a signal to management that major problems exist within the system that are not being adequately resolved through the maintenance process.

A more revealing method of measurement is to examine the failures that are occurring. Over time, logging the types of failures will provide a very clear picture

Mean time between failures (MTBF): A measurement of error occurrences that can be tracked over time to indicate the quality of a system.

Figure 21-8

How the mean time between failures should change over time

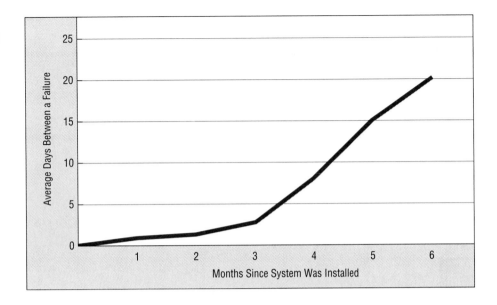

of where, when, and how failures occur. For example, knowing that a system repeatedly fails logging new account information to the database when a particular customer is using the system can provide invaluable information to the maintenance personnel. Were the users adequately trained? Is there something unique about this user? Is there something unique about an installation that is causing the failure? What activities were being performed when the system failed?

Tracking the types of failures also provides important management information for future projects. For example, if a higher frequency of errors occurs when a particular development environment is used, such information can help guide personnel assignments, training courses, or the avoidance of a particular package, language, or environment during future development. The primary lesson here is that without measuring and tracking maintenance activities, you cannot gain the knowledge to improve or know how well you are doing relative to the past. To effectively manage and to continuously improve, you must measure and assess performance over time.

Controlling Maintenance Requests   Another maintenance activity is managing maintenance requests. There are various types of maintenance requests—some correct minor or severe defects in the system while others improve or extend system functionality. From a management perspective, a key issue is deciding which requests to perform and which to ignore. Since some requests will be more critical than others, some method of prioritizing requests must be determined.

Figure 21-9 shows a flow chart that suggests one possible method you could apply for dealing with maintenance change requests. First, you must determine the type of request. If, for example, the request is an error—that is, a corrective maintenance request—then the flow chart shows that a question related to the error's severity must be asked. If the error is "very severe," then the request has top priority and is placed at the top of a queue of tasks waiting to be performed on the system. In other words, for an error of high severity, repairs to remove it must be made as soon as possible. If, however, the error is considered "non-severe," then the change request can be categorized and prioritized based upon its type and relative importance.

If the change request is not an error, then you must determine whether the request is to adapt the system to technology changes and/or business requirements or to enhance the system so that it will provide new business functionality. For

Figure 21-9
Flow chart of how to control maintenance requests (Adapted from Pressman, 1992)

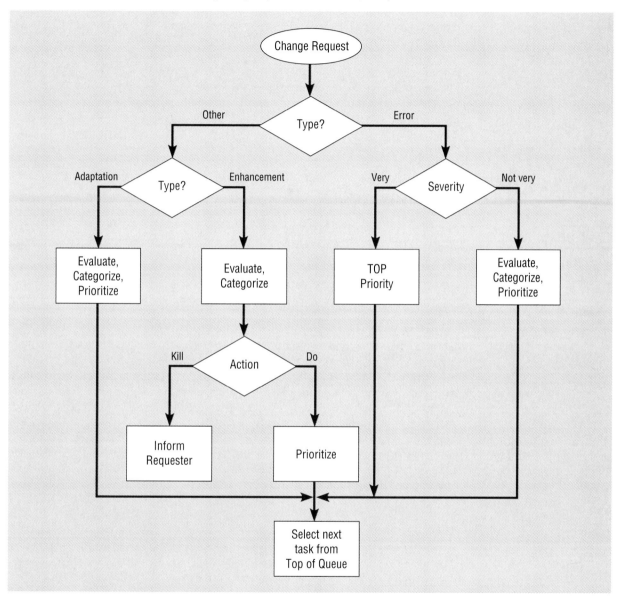

adaptation requests, they too will need to be evaluated, categorized, prioritized, and placed in the queue. For enhancement-type requests, they must first be evaluated to see whether they are aligned with future business and information systems' plans. If not, the request will be rejected and the requester will be informed. If the enhancement appears to be aligned with business and information systems plans, it can then be prioritized and placed into the queue of future tasks. Part of the prioritization process includes estimating the scope and feasibility of the change. Techniques used for assessing the scope and feasibility of entire projects should be used when assessing maintenance requests (see Chapter 7).

The queue of maintenance tasks for a given system is dynamic—growing and shrinking based upon business changes and errors. In fact, some lower-priority change requests may never be accomplished, since only a limited number of changes can be accomplished at a given time. In other words, changes in business needs between the time the request was made and when the task finally rises to the top of the queue may result in the request being deemed unnecessary or no longer important given current business directions. Thus, managing the queue of pending tasks is an important activity.

To better understand the flow of a change request, see Figure 21-10. Initially, an organizational group that uses the system will make a request to change the system. This request flows to the project manager of the system (labeled 1). The project manager evaluates the request in relation to the existing system and pending changes, and forwards the results of this evaluation to the System Priority Board (labeled 2). This evaluation will also include a feasibility analysis that includes estimates of project scope, resource requirements, risks, and other relevant factors. The board evaluates, categorizes, and prioritizes the request in relation to both the strategic and information systems plans of the organization (labeled 3). If the board decides to kill the request, the project manager informs the requester and explains the rationale for the decision (labeled 4). If the request is accepted, it is placed in the queue of pending tasks. The project manager then assigns tasks to maintenance personnel based upon their availability and task priority (labeled 5). On a periodic basis, the project manager prepares a report of all pending tasks in the change request queue. This report is forwarded to the priority board where they reevaluate the requests in light of the current business conditions. This process may result in removing some requests or re-prioritizing others.

Figure 21-10

How a maintenance request moves through an organization (Adapted from Pressman, 1987)

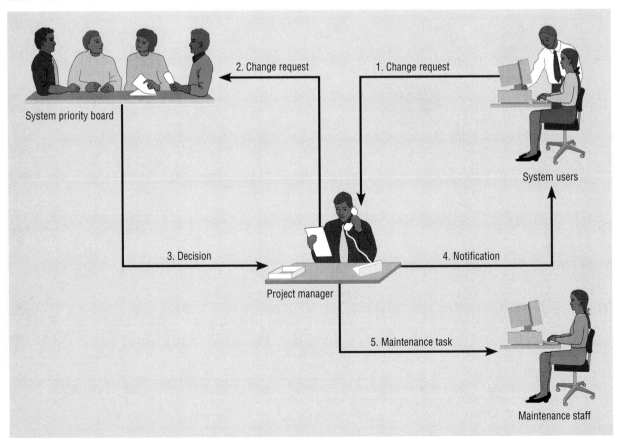

System priority board

2. Change request

1. Change request

System users

3. Decision

Project manager

4. Notification

5. Maintenance task

Maintenance staff

Although each change request goes through an approval process as depicted in Figure 21-10, changes are usually implemented in batches, forming a new release of the software. It is too difficult to manage a lot of small changes. Further, batching changes can reduce maintenance work when several change requests affect the same or highly related modules. Frequent releases of new system versions may also confuse users if the appearance of displays, reports, or data entry screens changes.

Configuration Management   A final aspect of managing maintenance is **configuration management,** which is the process of assuring that only authorized changes are made to a system. Once a system has been implemented and installed, the programming code used to construct the system represents the **baseline modules** of the system. The baseline modules are the software modules for the most recent version of a system where each module has passed the organization's quality assurance process and documentation standards. A **system librarian** controls the baseline source code modules. If maintenance personnel are assigned to make changes to a system, they must first check out a copy of the baseline system modules because no one is allowed to directly modify the baseline modules. Only those that have been checked out and have gone through a formal check-in process can reside in the library. Before any code can be checked back in to the librarian, the code must pass the quality control procedures, testing, and documentation standards established by the organization.

When various maintenance personnel working on different maintenance tasks complete each task, the librarian notifies those still working that updates have been made to the baseline modules. This means that all tasks being worked on must now incorporate the latest baseline modules before being approved for check-in. Following a formal process of checking modules out and in, a system librarian helps to assure that only tested and approved modules become part of the baseline system. It is also the responsibility of the librarian to keep copies of all prior versions of all system modules, including the **build routines** needed to construct *any version* of the system that *ever* existed. It may be important to reconstruct old versions of the system if new ones fail, or to support users that cannot run newer versions on their computer system.

Special software systems have been created to manage system configuration and version control activities (see box "Configuration Management Tools"). This software is increasingly necessary as the change control process is complicated in organizations deploying several different networks, operating systems, languages, and database management systems in which there may be many concurrent versions of an application, each for a different platform. One function of this software is to control access to libraried modules. Each time a module is checked out or in, this activity is recorded, after being authorized by the librarian. The software helps the librarian track that all necessary steps have been followed before a new module is released to production, including all integration tests, documentation updates, and approvals.

Configuration management: The process of assuring that only authorized changes are made to a system.

Baseline modules: Software modules that have been tested, documented, and approved to be included in the most recently created version of a system.

System librarian: A person responsible for controlling the checking out and checking in of baseline modules for a system when a system is being developed or maintained.

Build routines: Guidelines that list the instructions to construct an executable system from the baseline source code.

Configuration Management Tools

There are two general kinds of configuration management tools: revision control and source code control. With revision control tools, each system module file is "frozen" (unchangeable) to a specific version level—or "floating"— a programmer may check out, lock, and modify. Only the most recent version of a module is stored; previous versions are reconstructed when needed by applying changes in reverse order. Source code control tools extend the above description to address inter-related files. These tools also help in rebuilding any historic version of a system by recompiling the proper source code modules. They also trace an executable code module back to its source code version.

Role of CASE in Maintenance

In traditional systems development, much of the time is spent on coding and testing. When software changes are approved, code is first changed and then tested. Once the functionality of the code is assured, the documentation and specification documents are updated to reflect system changes. Over time, the process of keeping all system documentation "current" can be a very boring and time-consuming activity that is often neglected. This neglect makes future maintenance by the same or *different* programmers difficult at best.

A primary objective of using CASE for systems development and maintenance is to radically change the way in which code and documentation are modified and updated. When using an integrated CASE environment, analysts maintain design documents such as data flow diagrams and screen designs, not source code. In other words, design documents are modified and then code generators automatically create a new version of the system from these updated designs. Also, since the changes are made at the design specification level, most documentation changes such as an updated data flow diagram will have already been completed during the maintenance process itself. Thus, one of the biggest advantages to using CASE is its benefits during system maintenance.

In addition to using general CASE tools for maintenance, two special-purpose tools, reverse engineering and re-engineering tools (see Chapter 5), are primarily used to maintain older systems that have incomplete documentation or that were developed prior to CASE use. These tools are often referred to as *design recovery tools* since their primary benefit is to create high-level design documents of a program by reading and analyzing its source code.

Recall that reverse engineering tools are those that can create a representation of a system or program module at a design level of abstraction. For example, reverse engineering tools read program source code as input, perform an analysis, and extract information such as program control structures, data structures, and data flow. Once a program is represented at a design level using both graphical and textual representations, the design can be more effectively restructured to current business needs or programming practices by an analyst. Similarly, re-engineering tools extend reverse engineering tools by automatically, or interactively with a systems analyst, altering an existing system in an effort to improve its quality or performance.

In the next section, we describe business process re-engineering (BPR), a method that identifies business processes that can be "re-engineered" to radically improve business effectiveness and efficiency. Consequently, BPR is becoming a popular method for identifying how to evolve information systems to support new business processes.

BUSINESS PROCESS RE-ENGINEERING

Many information system change requests you will receive will ask for incremental improvements in systems, but will assume that the business will continue to operate as in the past. In fact, the user's request for a new system may simply be motivated by a desire to automate the current manual steps with which the user is comfortable, yielding cost efficiencies, fewer errors, better customer service, and a host of other benefits. If you take the user's request as submitted, you will miss the opportunity to study whether automation would support a radically different way of doing business, which would result in gains an order of magnitude greater than is possible by simply automating existing procedures.

Consider the following analogy. Suppose you are a successful European golfer who has tuned your game to fit the style of golf courses and weather in Europe. You have learned how to control the flight of the ball in heavy winds, roll the ball on wide open greens, putt on large and undulating greens, and aim to a target without the aid of the landscaping common on North American courses. When you come to the U.S. to make your fortune on the U.S. tour, you discover that incrementally improving your putting, driving accuracy, and sand shots will help but that the new competitive environment is simply not suited to your style of game. You need to re-engineer your whole approach, learning how to aim at targets, spin and stop a ball on the green, and manage the distractions of crowds and press. If you are good enough, you may survive, but without re-engineering, you will never be a winner.

Similarly, the competitiveness of our global economy has driven most companies into a mode of continuously improving the quality of their products and services (Dobyns and Crawford-Mason, 1991). Organizations have realized that creatively using information technologies can yield significant improvements in most business processes. Consequently, during the late 1980s and early 1990s, the concept of **Total Quality Management (TQM)** was embraced as a method for identifying ways to continuously and incrementally improve business products and services. TQM is based on the premise of continuous improvement: the identification of ways to continuously improve customer satisfaction, product quality, or customer service. Using this approach requires that all aspects of the business be measured so that performance over time can be understood. It is only through measurement that improvements can be made. TQM has generally been viewed as a successful initiative, but many believe that applying TQM restricts one's vision for dramatic gains.

Total Quality Management (TQM): A systematic approach to improving business processes through a philosophy of continuous improvement.

Although TQM still has its advocates, a new concept for viewing business improvement, **Business Process Re-engineering (BPR),** surfaced in the early 1990s. Many of the concepts of TQM and BPR are similar; however, one fundamental philosophy of BPR is to use information technologies to *radically* change or re-engineer business processes to yield dramatic improvements in products and services. Thus, the idea is not just to improve each business process but, in a systems modeling sense, to reorganize the complete flow of data in major sections of an organization to eliminate unneccessary steps, achieve synergies between previously separate steps, and become more responsive to future changes. Companies such as IBM, Procter & Gamble, Wal-Mart, and Ford are actively pursuing BPR efforts and have had great success. Yet, many other companies have found difficulty in applying TQM and BPR principles (Moad, 1994). Nonetheless, both TQM and BPR concepts are actively applied in both corporate strategic planning and information systems planning as a way to radically improve business processes (as described in Chapter 6). Figure 21-11 shows the relationship between TQM and BPR. As shown in this figure, BPR creates quantum improvements whereas TQM yields gradual, but steady, progress.

Business Process Re-engineering (BPR): The search for, and implementation of, radical change in business processes to achieve breakthrough improvements in products and services.

Both TQM and BPR assume that in order to provide better products and services, you must improve your business processes. Thus, over time, TQM advocates would expect an upward sloping line of linear process improvements. Such improvements would be the adaptive and perfective maintenance activities discussed earlier in the chapter. Alternatively, BPR advocates suggest that radical increases in the quality of business processes can be achieved through creative application of information technologies. BPR advocates also suggest that radical improvement cannot be achieved by tweaking existing processes but rather by using a clean sheet of paper and asking, "If we were a new organization, how would we accomplish this activity?" Changing the way work is performed also changes the way information is shared and stored, which means that the results of many BPR efforts are the development of information system maintenance requests or requests for system replacement. Due to this close relationship between BPR and system maintenance, we will describe BPR in greater detail. It is likely that you will encounter or have encountered BPR initiatives in your own organization. A recent survey of IS executives found that they view BPR to be a top IS priority for the coming years (Hayley, Plewa, and Watts, 1993).

Figure 21-11

How Total Quality Management differs from Business Process Re-engineering

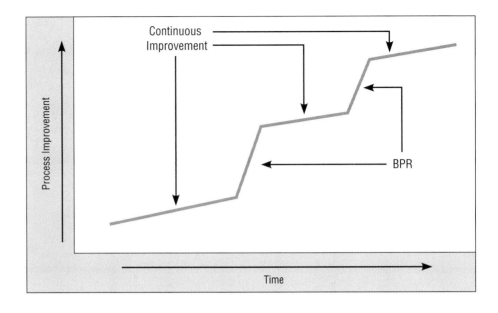

Identifying Processes to Re-engineer

Key business processes: The structured, measured set of activities designed to produce a specific output for a particular customer or market.

A first step in any BPR effort relates to understanding what processes to change. To do this, you must first understand which processes represent the **key business processes** for the organization. Key business processes are the structured set of measurable activities designed to produce a specific output for a particular customer or market. The important aspect of this definition is that key processes are focused on some type of organizational outcome such as the creation of a product or the delivery of a service. Key business processes are also customer-focused. In other words, key business processes would include all activities used to design, build, deliver, support, and service a particular product for a particular customer. BPR efforts, therefore, first try to understand those activities that are part of the organization's key business processes and then alter the sequence and structure of activities to achieve radical improvements in speed, quality, and customer satisfaction.

After identifying key business processes, the next step is to identify specific activities that can be radically improved through re-engineering. Hammer and Champy (1993) suggest that three questions be asked to identify activities for radical change:

1. How important is the activity to delivering an outcome?

2. How feasible is it that the activity can be changed?

3. How dysfunctional is the activity?

The answers to these questions provide guidance for selecting which activities to change. Those activities deemed important, changeable, yet dysfunctional, are primary candidates. To identify dysfunctional activities, they suggest you look for activities where there are excessive information exchanges between individuals, where information is redundantly recorded or needs to be rekeyed, where there are excessive inventory buffers or inspections, and where there is a lot of rework or complexity. Many of the tools and techniques for modeling data, processes, events, and logic within the IS development process are also being applied to model business processes within BPR efforts (see Davenport, 1993). Thus, the skills of a systems analyst are often central to many BPR efforts.

Disruptive Technologies

Once key business processes and activities have been identified, information technologies must be applied to radically improve business processes. To do this, Hammer and Champy (1993) suggest that organizations think "inductively" about information technology. Induction is the process of reasoning from the specific to the general which means that managers must *learn* the power of new technologies and *think* of innovative ways to alter the way work is done. This is contrary to deductive thinking where problems are first identified and solutions are then formulated.

Hammer and Champy suggest that managers especially consider **disruptive technologies** when applying inductive thinking. Disruptive technologies are those that enable the breaking of long-held business rules that inhibit organizations from making radical business changes. For example, Saturn is using production schedule databases and electronic data interchange (EDI) to allow its suppliers to operate as if they were one company. Suppliers do not wait until Saturn sends them a purchase order for more parts but simply monitor inventory levels and automatically send shipments as needed (Hammer and Champy, 1993: 90). Table 21-5 shows several long-held business rules and beliefs that constrain organizations from making radical process improvements. For example, the first rule suggests that information can only appear in one place at a time. However, the advent of distributed databases (see Chapter 18) has "disrupted" this long-held business belief.

In this section, we discussed how BPR is increasingly being used to identify ways to adapt existing information systems to changing organizational information needs and processes. It was our intent to provide a brief introduction to this topic, as the specific tools and techniques for performing BPR are still evolving. For more information on this exciting topic, the interested reader is encouraged to see the books by Davenport (1993) and Hammer and Champy (1993).

We now return to the general topic of system maintenance with an example from Pine Valley Furniture and the Purchasing Fulfillment System (see Chapter 2) to illustrate what may happen to you someday when you are supporting and maintaining systems.

> **Disruptive technologies:** Technologies that enable the breaking of long-held business rules that inhibit organizations from making radical business changes.

TABLE 21-5 Long-Held Organizational Rules that Are Being Eliminated through Disruptive Technologies

Rule	Disruptive Technology
Information can appear in only one place at a time	Distributed databases allow the sharing of information
Only experts can perform complex work	Expert systems can aid non-experts
Businesses must choose between centralization and decentralization	Advanced telecommunications networks can support dynamic organizational structures
Managers must make all decisions	Decision-support tools can aid non-managers
Field personnel need offices where they can receive, store, retrieve, and transmit information	Wireless data communication and portable computers provide a "virtual" office for workers
The best contact with a potential buyer is personal contact	Interactive communication technologies allow complex messaging capabilities
You have to find out where things are	Automatic identification and tracking technology know where things are
Plans get revised periodically	High-performance computing can provide real-time updating

MAINTAINING AN INFORMATION SYSTEM AT PINE VALLEY FURNITURE

Early on a Saturday evening, Juanita Lopez, head of the manufacturing support unit of the Purchasing Department at Pine Valley Furniture (PVF), was developing a new four-week production schedule to prepare purchase orders for numerous material suppliers. She was working on Saturday evening because she was leaving the next day for a long overdue two-week vacation to the Black Hills of South Dakota. Before she could leave, however, she needed to prepare purchase orders for all material requirements for the next four weeks so that orders could be placed during her absence. She was using the Purchasing Fulfillment System (described in detail in Chapter 2) to assist her with this activity.

Midway through the process of developing a new production schedule, the system failed and could not be restarted. When she tried to restart the program, an error message was displayed on the terminal:

Data Integrity Error: Corrupt or missing supplier file.

Given that her plane for Rapid City left in less than 12 hours, Juanita had to figure out some way to overcome this catastrophic system error. Her first thought was to walk over to the offices of the information systems development group within the same building. When she did, she found no one there. Her next idea was to contact Chris Ryan, the project manager for the development and maintenance of the system. She placed a call to Chris' home and found that he was at the grocery store but would be home soon. Juanita left a message for Chris to call her ASAP at the office.

Within 30 minutes, Chris returned the call and was on his way into the office to help Juanita. Although it is not a common occurrence, this is not the first time that Chris has gone into the office to assist users when systems have failed during off hours. Chris was looking forward to the day when he could handle all of these problems from home using a home PC and modem; he had been able to do this when all that was required was to scan data files for errors or issue a command to restore a database. Based on Juanita's explanation of the problem and a few quick inquiries from his home PC, Chris decided he had better make the trip to the office where he had all CASE and other tools available.

PVF's system development methodology for performing system maintenance is a formal process in which a user must first write a Systems Service Request (SSR) before maintenance is performed. After it is reviewed by the project manager, it is then forwarded to the Systems Priority Board (see Figure 21-10 for a description of a similar process). For catastrophic problems requiring instant correction so as not to delay normal business operations, the project manager has the discretion to circumvent the normal request process. After arriving on the scene, reviewing the error messages, and learning of Juanita's pending vacation, Chris believed that the failure with the Purchasing Fulfillment System was an instance where he would circumvent the normal maintenance process. His quick investigation suggested a failure in a new version of a system module which had been installed late on Friday afternoon. Chris noticed that the CASE tool records showed that this replacement module had not been tested against a standard test data set related to the type of work Juanita was doing, which made him suspect that this was the source of the problem. After patching the system to make it run, he would have to go back and document and test his changes so that they conformed to the development standards of PVF.

Over the next two hours, Chris used system backups to rebuild the supplier database. He reinstalled a previous version of the system's potentially faulty module (stored in the CASE library) that seemed more reliable, and then he quickly ran a test data set to check that the patches would hold the system together for now. He had to refresh himself on how to mount a tape cartridge on which the backup supplier data had been archived. Juanita was able to complete her task on time to easily make her flight the next morning. She thanked Chris for "going beyond the call of duty." Her appreciation made Chris feel good, but he was still uneasy. When making the "quick-fix" on the system, he did not perform carefully planned testing nor did he confirm what had caused the error. Thus, he knew that the system could fail at any time. He did, however, have a copy of all of Juanita's actions just prior to the system failure. He hoped that through a careful review of those actions he would be able to learn why the system failed. But that would be a job for Monday morning.

SUMMARY

Maintenance is the final phase in the systems development life cycle where systems are changed to rectify internal processing errors or to extend the functionality of the system. Maintenance is where a majority of the financial investment in a system occurs and can span more than 20 years. More and more information systems professionals have devoted their careers to systems maintenance and, as more systems move from initial development into operational use, it is likely that even more professionals will in the future.

It is during maintenance that the SDLC becomes a life cycle since requests to change a system must first be approved, planned, analyzed, designed, and then implemented. In some special cases where business operations are impaired due to an internal system error, quick-fixes can be made. This, of course, circumvents the normal maintenance process. After quick-fixes are made, maintenance personnel must back up and perform a thorough analysis of the problem to make sure that the correction conforms to normal systems development standards for design, programming, testing, and documentation.

Maintenance requests can be one of four types: corrective, adaptive, perfective, and preventive. Corrective maintenance is used to repair design and programming errors. Adaptive maintenance is used to modify the system to changes in business informational or processing needs, or to migrate the system onto a different technology platform. Perfective maintenance is used to evolve the system so that it can solve new problems or take advantage of new opportunities; in other words, to extend the capabilities of the system beyond those initially intended for the system. Preventive maintenance is used to safeguard the system from future problems.

How a system is designed and implemented can greatly impact the cost of performing maintenance. The number of unknown errors in a system when it is installed is a primary factor in determining the cost of maintenance. Other factors, such as the number of separate customers and the quality of documentation, significantly influence maintenance costs.

Organizations can choose three general methods for managing maintenance personnel. One method is to have a separate maintenance group. A second approach is to have the same people who construct the system also maintain it. A third option is to have maintenance personnel housed within the functional areas of the business that use the system on a day-to-day basis. Each approach has its advantages and disadvantages and no approach is universally best. A second maintenance management issue relates to understanding how to measure the quality of the maintenance effort. Most organizations track the frequency, time, and type of each failure and compare performance over time. Since limited resources preclude organizations from performing all maintenance requests, some formal process for reviewing requests must be established to make sure that only those requests deemed consistent with organizational and information systems plans be performed. A central source, usually a project manager, is used to collect maintenance requests. When requests are submitted, this person forwards each request to a committee charged to assess its merit. Once assessed, the project manager assigns higher-priority activities to maintenance personnel.

Maintenance personnel must be controlled from making unapproved changes to a system. To do this, most organizations assign one member of the maintenance staff, typically a senior programmer or analyst, to serve as system librarian. The librarian controls the checking out and checking in of system modules to assure that appropriate procedures for performing maintenance such as adequate testing and documentation are adhered to.

CASE technology is actively employed during maintenance. The primary benefit to using CASE is its ability to enable maintenance to be performed on design documents rather than low-level source code. Reverse engineering and re-engineering CASE tools are used to recover design specifications of older systems that were not constructed using CASE or for systems with inadequate design specifications. Once recovered into a design specification, these older systems can then be changed at the design level rather than the source code level, yielding a significant improvement in maintenance personnel productivity.

Business process re-engineering is a major source for maintenance requests. BPR is often part of the formal strategic organizational and information systems planning effort where key business processes are redesigned using information technologies to radically improve business performance. Many of the tools and techniques used to design information systems are also used in the BPR process.

CHAPTER REVIEW

KEY TERMS

Adaptive maintenance
Baseline modules
Build routines
Business Process Re-
 engineering (BPR)
Configuration
 Management

Corrective maintenance
Disruptive technologies
Key business processes
Maintainability
Maintenance
Mean time between
 failures (MTBF)

Perfective maintenance
Preventive maintenance
System librarian
Total Quality Management
 (TQM)

REVIEW QUESTIONS

1. Define the following terms:
   a. baseline modules
   b. configuration management
   c. maintainability
   d. mean time between failures
   e. build routines
   f. disruptive technologies

2. Contrast the following terms:
   a. Business Process Re-engineering, maintenance, Total Quality Management
   b. adaptive maintenance, corrective maintenance, perfective maintenance, preventive maintenance
   c. baseline modules, build routines, system librarian
   d. maintenance, maintainability

3. List the steps in the maintenance process and contrast them with the phases of the systems development life cycle.

4. What are the different types of maintenance and how do they differ?

5. Describe the factors that influence the cost of maintenance. Are any factors more important? Why?

6. Describe three ways for organizing maintenance personnel and contrast the advantages and disadvantages of each approach.

7. What types of measurements must be taken to gain an understanding of the effectiveness of maintenance? Why is tracking mean time between failures an important measurement?

8. What managerial issues can be better understood by measuring maintenance effectiveness?

9. Describe the process for controlling maintenance requests. Should all requests be handled in the same way or are there situations when you should be able to circumvent the process? If so, when and why?

10. What is meant by configuration management? Why do you think organizations have adopted the approach of using a systems librarian?

11. How is CASE used in the maintenance of information systems?

12. What is the difference between reverse engineering and re-engineering CASE tools?

13. When conducting a business process re-engineering study, what should you look for when trying to identify business processes to change? Why?

14. What are disruptive technologies and how do they enable organizations to radically change their business processes?

PROBLEMS AND EXERCISES

1. Match the following terms to the appropriate definitions.

   _____ maintenance
   _____ Business Process Re-engineering
   _____ adaptive maintenance
   _____ perfective maintenance
   _____ configuration management
   _____ Total Quality Management

   a. process that assures that only authorized changes are made to the proper software modules
   b. changing a system to take advantage of new opportunities
   c. changing a system in response to environmental changes
   d. a process of continuous improvements to business processes
   e. changing or fixing a system
   f. a process of radical changes to business processes

2. Maintenance has been presented as both the final stage of the SDLC (see Figure 21-1) and as a process similar to the SDLC (see Figure 21-4). Why does it make sense to talk about maintenance in both of these ways? Do you see a conflict in looking at maintenance in both ways?

3. In what ways is a request to change an information system handled differently from a request for a new information system?

4. According to Figure 21-5, corrective maintenance is by far the most frequent form of maintenance. What can you do as a systems analyst to reduce this form of maintenance?

5. Is systems maintenance more like continuous improvement or Business Process Re-engineering? Why?

6. Briefly discuss how a systems analyst can manage each of the six Cost Elements of Maintenance listed in Table 21-2.

7. Suppose you were a system librarian. Using entity-relationship diagramming notation, describe the database you would need to keep track of the information necessary in your job. Consider operational, managerial, and planning aspects of the job.

8. Review the material in Chapter 6 on corporate and information systems strategic planning. How are these processes different from Business Process Re-engineering? What new perspective might BPR bring that classical strategic planning methods may not have?

9. Software configuration management is similar to configuration management in any engineering environment. For example, the product design engineers for a refrigerator need to coordinate dynamic changes in compressors, power supplies, electronic controls, interior features, and exterior designs as innovations to each occur. How do such product design engineers manage the configuration of their products? What similar practices do systems analysts and librarians have to follow?

FIELD EXERCISES

1. Study an information systems department with which you are familiar or to which you have access. How does this department organize for maintenance? Has this department adopted one of the three approaches outlined in Table 21-4 or does it use some other approach? Talk with a senior manager in this department to discover how well this maintenance organization structure works.

2. Study an information systems department with which you are familiar or to which you have access. How does this department measure the effectiveness of systems maintenance? What specific metrics are used and how are these metrics used to effect changes in maintenance practices? If there is a history of measurements over several years, how can changes in the measurements be explained?

3. With the help of other students or your instructor, contact a system librarian in an organization. What is this person's job description? What tools does this person use to help him or her in the job? To whom does this person report? What previous jobs did this person hold, and what jobs does this person expect to be promoted into in the near future?

4. With the help of other students or your instructor, contact someone in an organization that has carried out a BPR study. What effects did this study have on information systems? In what ways did IT (especially disruptive technologies) facilitate making the radical changes discovered in the BPR study?

REFERENCES

Andrews. D. C. and N. S. Leventhal. 1993. *Fusion: Integrating IE, CASE, JAD: A Handbook for Reengineering the Systems Organization.* Englewood Cliffs, NJ: Prentice-Hall.

Chapin, N. 1987. "The Job of Software Maintenance." *Proceedings of the Conference on Software Maintenance.* Washington DC: IEEE Computer Society Press: 4–12.

Davenport, T. H. 1993. *Process Innovation: Reengineering Work through Information Technology.* Boston, MA: Harvard Business School Press.

Dobyns, L. and C. Crawford-Mason. 1991. *Quality or Else.* Boston, MA: Houghton-Mifflin.

Hammer, M. and J. Champy. 1993. *Reengineering the Corporation.* New York, NY: Harper Business.

Hanna, M. 1992. "Using documentation as a life-cycle tool." *Software Magazine,* (December): 41–46.

Hayley, K., J. Plewa, and M. Watts. 1993. "Reengineering Tops CIO Menu." *Datamation,* Vol. 38, No. 6 (April, 15): 73–74.

Jones, C. 1986. *Programming Productivity.* New York, NY: McGraw-Hill.

Martin, E. W., D. W. DeHayes, J. A. Hoffer, and W. C. Perkins, 1994. *Managing Information Technology: What Managers Need to Know.* New York, NY: Macmillan.

Moad, J. 1994. "After Reengineering: Taking Care of Business." *Datamation,* Vol. 40, No. 20 (October 15): 40–44.

Pressman, R. S. 1987. *Software Engineering: A Practitioner's Approach.* New York, NY: McGraw-Hill.

Pressman, R. S. 1992. *Software Engineering: A Practitioner's Approach.* 2d ed. New York, NY: McGraw-Hill.

Types of Information Systems

L E A R N I N G   O B J E C T I V E S

After studying this appendix, you should be able to:

■ Describe four types of information systems: transaction processing systems, management information systems, decision support systems, and expert systems.

■ Identify two additional types of information systems: scientific and office automation.

■ Contrast individual and group decision support systems.

INTRODUCTION

Traditionally, organizations have considered four types of information systems: transaction processing systems, management information systems, decision support systems, and expert systems. We will further break down decision support systems into those for individuals, groups, and executives. Then we will briefly outline two categories of information systems, scientific (or technical) computing and office automation systems, which are recognized in many organizations.

TRANSACTION PROCESSING SYSTEMS

The first type of information systems built were **transaction processing systems (TPS),** which are computer-based versions of manual processes used in organizations. Transaction processing systems automate the handling of **transactions**, which are individual simple events in the life of an organization. For example, in a sporting equipment store, a transaction occurs when a customer purchases a basketball. A record is made of each transaction that occurs. All of these records were originally kept on paper. In the early days of computerization, someone would later transfer the records to computer punched cards or to magnetic tape, so that the computer could then read and manipulate the data. Now, for medium- to large-size organizations, most transactions are captured in computer-readable form immediately at a point-of-sale terminal (see Figure A-1).

When a transaction processing system processes an organization's transactions, each transaction is available for recall later. More importantly to the organization,

Transaction processing systems (TPS): Computer-based versions of manual organization systems dedicated to handling the organization's transactions; e.g., payroll.

Transactions: Individual simple events in the life of an organization that contain data about organizational activity.

Figure A-1
Computer-based cash register.

the number and volume of transactions can be calculated for a given time period. For example, processing the number of sales per hour or per day at a sporting equipment store allows the store to more easily monitor inventory and thus more effectively manage inventory levels. Transactions also provide the official record of business activities, which drive other systems such as those which bill customers, pay vendors and employees, and reorder inventory or raw materials or stocked goods.

MANAGEMENT INFORMATION SYSTEMS

Management information systems (MIS): Computer-based systems designed to provide standard reports for managers about transaction data.

Management information systems (MIS) are designed to take the relatively raw data available through a TPS and convert them into a summarized and aggregated form for managers, usually in a report format (see Figure A-2). Several types of

Trend Analysis Report (Column Format)						
Projected Sales Per Month ($ 000)						
	Jan	Feb	Mar	Apr	May	Jun
Hardware	90.0	95.0	100.0	150.0	110.0	85.0
Software	35.0	40.0	45.0	70.0	40.0	35.0
Other	5.0	6.0	8.0	20.0	6.0	4.0
Total	130.0	141.0	153.0	240.0	156.0	124.0
Cost of Sales Per Month ($ 000)						
	Jan	Feb	Mar	Apr	May	Jun
Hardware	40	42.5	45	120	85	70
Software	15	18	22	30	18	15
Other	1	1.25	1.5	5	1.25	0.75
Total	56.0	61.8	68.5	155.0	104.3	85.8
Other Expenses Per Month ($ 000)						
	Jan	Feb	Mar	Apr	May	Jun
Marketing	7	6	5	5	6	8
Overhead	10	10	12	15	12	8
Total	17.0	16.0	17.0	20.0	18.0	16.0

reports can be produced. *Summary reports* present all activity over a given time period, geographic region, or other categorization in aggregate form. *Exception reports* only present information that is out of normal ranges. *On-demand reports* present anticipated summaries only when a manager wants or needs to check the status of activities. The precise contents of on-demand reports may change depending on the manager's immediate need (for example, only specified time periods, product lines, or geographical regions might be included). *Ad hoc reports* provide specific information as needed, the contents of which may change depending on the manager's needs. Ad hoc reports are unanticipated and may be one-time in nature.

DECISION SUPPORT SYSTEMS

Decision support systems (DSS) are designed to help organizational decision makers make decisions. DSS usually have three major components: a database, a model base, and a dialogue module (Sprague, 1980). Figure A-3 shows these components.

- The *database* contains data relevant to the decision to be made.

- The *model base* contains one or more models that can be used to analyze the decision situation.

- The *dialogue module* provides a way for the decision maker, usually a non-technical manager, to communicate with the DSS.

By running the data and possible decisions through one or more models, the decision maker can compare possible solutions to the problem at hand. The DSS allows

Decision support systems (DSS): Computer-based systems designed to help organization members make decisions; usually composed of a database, model base, and dialogue system.

Figure A-3
Components of a DSS
(Adapted from Sprague,
1980)

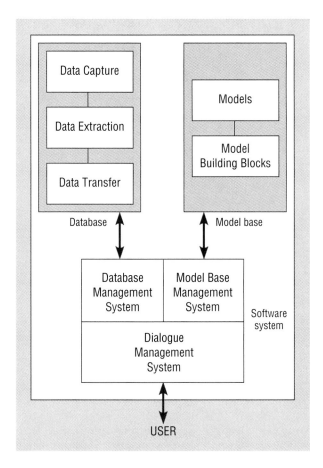

the manager to test or propose different solutions and see what the results may be before committing to any particular model.

The first decision support systems were designed to support individual decision makers. When computing technology was more primitive and more difficult for non-technical people to, an intermediary often used the DSS for the manager. The intermediary was usually a staff person who had the computer skills the manager lacked to work with the DSS. The manager would then use the output to help decide which course of action to take. Due to early technical limitations, each individual or specific DSS had to be designed and built one at a time. Now, many decision support systems run on microcomputers. The models are relatively easy to construct, change, and interpret using such software programs as electronic spreadsheets. Tools like spreadsheets and fourth generation languages (4GLs) are called **DSS generators** because they are general purpose tools that can be used to develop many specific DSS with relative ease.

DSS generators: General purpose computer-based tools used to develop specific decision support systems.

Group DSS

In the late 1970s, some companies began to develop DSS that would support groups of decision makers. All of these commercial systems failed. However, in the mid-1980s, group DSS (GDSS) were resurrected at several universities, including those in Arizona, Indiana, Michigan, and Minnesota (see Nunamaker, et al., 1991). Using

Figure A-4
A collaborative management
room at the University of
Arizona

a GDSS, a group of decision makers can use special software to contribute to a problem's solution. The most common form of a GDSS consists of a special-purpose room with a microcomputer for each group member and special presentation equipment (Figure A-4). The microcomputers are networked together so that group members can share information. Group members can also view aggregates of their work, such as the results of a group vote, on large screens. Researchers and packaged software houses, like Lotus Development Corp., are now working on making GDSS available and useful outside of these special-purpose rooms so that group members can join the process from their own offices on their own schedule.

Executive Support Systems

Another relatively new form of DSS is referred to as **executive support systems (ESS)** or **executive information systems (EIS).** Executive support systems are designed specifically for high-level executives who

1. may not have many computer skills, and

2. have very little time to devote to any given situation.

An ESS is relatively easy to manipulate and usually provides graphical presentations on several different pre-defined topics (see Figure A-5). Some executive support systems allow an executive to drill down into the data to a deeper, more specific level. For example, an executive who sees that sales have decreased for the month in the North American market may want to find out which segments of the market are doing best. The executive would then ask for the same information by segment and, seeing that the Western U.S. segment had the best performance, the executive may then want to see which sub-region had the best performance. Once the information is presented at this level, the executive would see that Southern California had done the best. The executive may then want to examine the information by city, and so on.

Executive support systems (ESS): Computer-based systems developed to support the information-intensive but limited-time decision making of executives (also referred to as *executive information systems*).

Figure A-5
Example screen for an executive information system (Source: Alter, 1992 / Courtesy Benjamin/Cummings Publishing Company, Inc.)

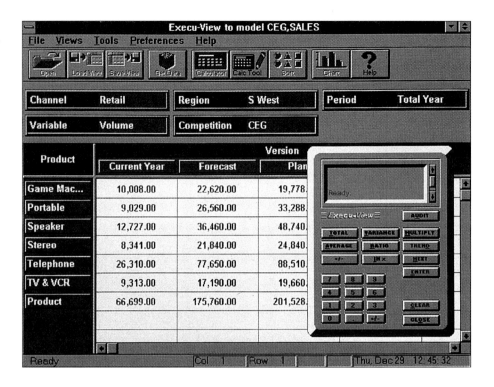

EXPERT SYSTEMS

Expert systems (ES): Computer-based systems designed to mimic the performance of human experts.

Different from any of the other classes of systems we have discussed so far, **expert systems (ES)** attempt to codify and manipulate knowledge rather than information. By knowledge, we mean understanding acquired through experience, deep and extensive learning. Expert systems are based on principles of artificial intelligence research. Artificial intelligence is the branch of computer science devoted to creating intelligence with machines. Typically users communicate with an ES through a dialogue during which the ES asks questions and the user supplies the answers. The answers are then used to determine which rules apply and the ES finishes with a recommendation based on its rules. One of the most difficult parts in building an ES is acquiring the knowledge of the expert in the particular problem domain.

Knowledge engineers: Computer professionals trained to elicit knowledge from domain experts in order to develop expert systems.

Specially trained people called **knowledge engineers** perform this knowledge acquisition. Knowledge engineers are similar to systems analysts; however, they are trained to use different techniques, as determining knowledge is considered more difficult than determining data.

SCIENTIFIC AND OFFICE INFORMATION SYSTEMS

Although we are concerned in this book primarily with the types of information systems used in the administration of organizations, for completeness we should also mention some other types of information systems. We will not describe how such systems are developed, however, as an understanding of this type of systems development requires skills beyond the scope of the book.

One broad classification of systems is based on *scientific* and *engineering computing*. These systems support engineers in the design of new products or improvement of older ones. Their computer support might require computer simulations or analytical models of physics or chemistry properties. *Computer-aided design (CAD) systems* allow engineers to create graphic simulations of the products they design. Engineers can then manipulate and observe these simulations, allowing the engineers to see what a product will look like and how the product will react without having to build the product first. Related to CAD is *computer-aided manufacturing (CAM) systems.* These systems help automate and control the manufacturing process in factories. Scientific computing allows scientists in many fields to simulate everything from molecular movements to global weather patterns, providing an understanding that would otherwise either not be possible or be cost-prohibitive.

Another class of systems consists of *office automation systems.* The term office automation promises more than the term delivers, as the term conjures up images of offices organized and run like automated factories. Office automation systems are usually quite basic and include such tools as word processing and accounting information systems. Integrated office systems that include electronic mail, calendaring features, and reminder files in addition to word processing are also available. *Electronic mail (e-mail)* allows office workers to send each other messages and files directly from their computers and is usually more convenient than trying to reach someone by telephone. *Calendaring* features allow office workers to coordinate their schedules, to reserve conference rooms, and to schedule meetings. *Reminder files* provide a means for conveniently reminding ourselves of meetings and other commitments. Office systems are rarely if ever developed in-house, but instead are purchased or leased from hardware or software producers.

APPENDIX REVIEW

KEY TERMS

Decision support systems
  (DSS)
DSS generators
Executive support systems
  (ESS)

Expert systems (ES)
Knowledge engineers
Management information
  systems (MIS)

Transaction processing
  systems (TPS)
Transactions

REVIEW QUESTIONS

1. Define each of the following terms:
   a. management information systems (MIS)
   b. exception report
   c. group DSS
2. Contrast on-demand and ad hoc reports.
3. List the three general components of a decision support system.
4. Contrast decision support systems and DSS generators.
5. Contrast decision support systems and expert systems.
6. Contrast decision support systems and group decision support systems.

PROBLEMS AND EXERCISES

1. Match the following terms to the appropriate definitions.

   _____ summary report
   _____ exception report
   _____ on-demand report
   _____ database
   _____ model base
   _____ decision support system
   _____ expert system

   a. system designed to mimic human experts
   b. set of models
   c. information out of normal range
   d. set of related data
   e. helps people make decisions
   f. information about all activity
   g. data when needed

2. Describe the potential links between transaction processing and management information systems. For example, consider an order processing system and a sales analysis system. How would these two systems be related to each other? List the pieces of data in this order processing system that might be used in the sales analysis system. Describe how these pieces of data would be used in the sales analysis system. Would these pieces of data take on new form in the sales analysis system or would they remain the same? Why or why not? What would happen if these two systems were designed independently of each other?

3. What unique features might a group DSS possess that a DSS designed to support only one person might not? What changes to an electronic spreadsheet package would have to be made if the spreadsheet files were made available to multiple users simultaneously? What security and data integrity issues would arise?

4. Discuss the importance of clearly defining a business transaction. Specifically, from the viewpoint of maintaining the integrity of information, what is the significance of the transaction concept?

5. For each of the following three group task types, describe a group task: (1) group members perform the task together simultaneously (2) group members work on the task apart from each other, yet in parallel (3) group members work on the task apart from each other and each performs his or her sub-task sequentially. For each task, what are each of the sub-tasks? How, when, and by whom are each performed? How could a group DSS be used to support each of these three group tasks? Would the group DSS help or hinder the group? Why?

6. Assume that you are the owner/manager of a small, start-up organization. Choose a product and/or service that you provide and a location for your firm. Describe the types of information that you will need to have to manage this organization effectively. What are the sources for this information? With what frequency and in what form will you need to have this information? At what level of detail will you want the information (for example, will you need to drill down from summaries and averages to specific data)? Does it make sense to have a microcomputer-based ESS to support your information needs for managing this company? Why or why not?

FIELD EXERCISES

1. Visit an office in your university or in another organization that uses an office automation system. What information processing functions does this system possess? Which ones do people use regularly? Which functions do you believe are underutilized and why?

2. Ask a friend or instructor to show you an expert system generator, the software used to build expert systems. How does one "program" an expert system? What are the components of an expert system generator?

3. Identify three DSS products. If you need to, speak directly with the vendors and/or users of these products. Are these tools for specific DSS applications or are they DSS generators? On what hardware platforms do these DSS products run? What is planned for these products? Is a microcomputer-based version offered or planned? Is a graphical, object-oriented version offered or planned? What other new features will come in future versions of the product? If no further versions or enhancements are planned, why not?

4. Identify three group DSS products. If you need to, speak directly with the vendors and/or users of these products. In what ways are these products similar? different? What advantages do these specific group DSS products seem to offer over other group DSS products? over other DSS tools? What is planned for these products? Can the software be run in a distributed fashion, with participants in different locations? If not, is this planned for? If this is neither offered or planned, why not?

5. Choose a work group in your university or another organization and determine which of their tasks can be better supported by a group DSS. Why? Which tasks cannot be better supported by a group DSS? Why not? What are the potential barriers to implementing a group DSS for this group?

REFERENCES

Alter, S. 1992. *Information Systems: A Management Perspective.* Redwood City, CA: Benjamin/Cummings Publishing.

Nunamaker, J. F., Jr., A. R. Dennis, J. S. Valacich, D. R. Vogel, and J. F. George. 1991. "Electronic Meeting Systems to Support Group Work." *Communications of the ACM* 34 (July): 40–61.

Sprague, R. H., Jr. 1980. "A Framework for the Development of Decision Support Systems." *MIS Quarterly* 4 (4): 1–26.

Rapid Application Development

After studying this appendix, you should be able to:

■ Define Rapid Application Development (RAD).

■ Contrast RAD with the systems development life cycle.

■ Explain the advantages and potential drawbacks to RAD as an exclusive systems development methodology.

INTRODUCTION

> *RAD [Rapid Application Development] has been demonstrated in many projects to be so superior to traditional development that it seems irresponsible to continue to develop systems the old way.*
>
> James Martin, 1991

James Martin is generally credited with inventing Rapid Application Development, or RAD, so it should be no surprise that he so enthusiastically supports the concept, as the above quotation shows. Not everyone shares his enthusiasm, however, as you will see. In this appendix, you will learn about RAD: what it is, the components it relies on to succeed, Martins's four-phase RAD life cycle, and the advantages and disadvantages of RAD. We end the appendix with an example of how the RAD approach can pay off handsomely if done correctly.

DEFINITIONS AND COMPONENTS OF RAD

Rapid Application Development (RAD) is a methodology for developing information systems which promises better and cheaper systems and more rapid deployment. These improvements are accomplished by having a small group of highly skilled systems developers and key end users work together jointly in real-time to develop systems. RAD grew out of the convergence of two trends: the increased speed and turbulence of doing business in the late 1980s and early 1990s and the ready availability of high-powered computer-based tools to support systems development. As the conditions of doing business in a changing, competitive global environment became more turbulent, management in many organizations began to question whether it made sense to wait two to three years to develop systems that would be obsolete

Rapid Application Development (RAD): Systems development methodology created to radically decrease the time needed to design and implement information systems. RAD relies on heavy user involvement, Joint Application Development sessions, prototyping, integrated CASE tools, and code generators.

upon completion. At the same time, CASE tools and prototyping software were diffusing throughout organizations, making it relatively easy for end users to see what their systems would look like before they were completed. Why not use these tools to more productively address the problems of developing systems in a rapidly changing business environment? And so RAD was born.

To succeed, RAD relies on bringing together several systems development components you have studied in this book. As you might have gathered from the definition, RAD depends on heavy *user involvement.* Much of that involvement takes place in the *prototyping* process, where end users and analysts work together to design and iteratively redesign interfaces, displays, and reports for new systems. The prototyping is conducted in sessions that resemble traditional *Joint Application Development (JAD)* sessions. In these frequent sessions, users perform detailed reviews of system prototypes and specifications. The primary difference is that in RAD, the prototype becomes the basis for the new system—the displays designed during prototyping become computer displays in the production system, not models for what the production system will do. This is accomplished through reliance on *integrated CASE tools* which include *code generators* for creating bug-free code from the designs end users and analysts create during prototyping. The code includes the interfaces as well as the application programs that use them. In many cases, the basic elements of the production system are being built even as users are talking about the system during development workshops. In many cases, end users can get hands-on experience with the developing system before the design workshops are over. To further help speed the process, the *reuse* of templates, components, or previous systems described in the CASE tool repository is strongly encouraged.

Tools alone are not enough to made RAD work, however (see Figure B-1). The other three pillars of RAD are people, who must be trained in the right skills; a coherent methodology which spells out the proper tasks to be done in the proper order; and the support and facilitation of management (Martin, 1991).

Martin suggests that no information systems organization should be converted to the RAD approach overnight. Instead, a small group of well-trained and dedicated professionals, called a RAD cell, should be created to demonstrate the viability of RAD through pilot projects. Over time, the cell can grow, gradually adding more people, skills, and projects, until RAD is the predominant approach of the information systems unit.

Figure B-1

The four necessary pillars for the RAD approach

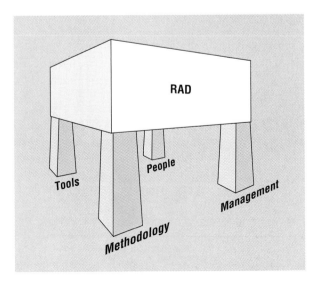

A FOUR-PHASE RAD LIFE CYCLE

Martin's RAD life cycle contains the basic phases of any life cycle: planning, analysis, design, and implementation, although he calls the phases Requirements Planning, User Design, Construction, and Cutover (see Figure B-2). Thus, in some organizations, RAD is treated as an adaptation of the SDLC which emphasizes continuous, iterative improvement rather than a strictly sequential step-by-step process. One way to view RAD is that several SDLC phases occur simultaneously; for example, the User Design RAD phase combines analysis with logical and physical design phases of the SDLC.

Requirements Planning incorporates elements of the traditional project identification and selection and analysis phases described in this text. During this phase, high-level end users determine system requirements, but the determination is done in the context of a discussion of business problems and business areas. Once specific systems have been identified for development, users and analysts conduct a Joint Requirements Planning workshop in order to reach agreement on system requirements.

During User Design, end users and information systems professionals participate in JAD workshops where those involved use integrated CASE tools to support the rapid prototyping of system design. Frequent client reviews of physical information systems are used to redefine requirements (see Figure B-3). In the traditional SDLC, we assume that clients can define their requirements early in the life cycle. With RAD, the user requirements can evolve throughout the User Design phase and the tools and highly skilled system builders can quickly modify the working system. Eventually, users sign off on the CASE design. Because User Design ends with agreement on a computer-based design, the gap between the end of design and the handing over of the new system to users might only take three months instead of the usual eighteen. As illustrated in Figure B-3, User Design is iterative or evolutionary. The high-level requirements from Requirements Planning are

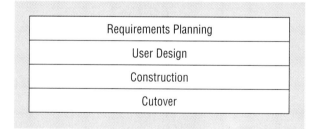

Figure B-2
Martin's four-phase RAD life cycle

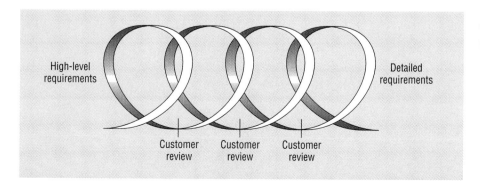

Figure B-3
The iterative nature of the RAD User Design phase

reviewed and, through successive iterations, more and more detailed versions of the system are reviewed and reworked. In contrast to the SDLC, concrete versions of the system, rather than abstract system specifications, often evolve. And in contrast to the use of prototyping during analysis or logical design, the final version of the system becomes the production system, not an approximation that still must evolve during subsequent life cycle phases.

In many instances, much of the User Design effort focuses on basic processes common to all components of a system. These components tend to be data-intensive, so the designs for basic components concentrate on user interfaces for adding, changing, deleting, and interrogating data. As these components can often be generated from a conceptual model of data (such as an entity-relationship diagram—explained in Chapter 11), initial JAD sessions for these components focus on the meaning and structure of data needed in the system. Later sessions confirm from prototypes that these common components are easy to use and capture the basic data handling functions. Other JAD sessions address more specialized functions which analyze or manipulate data in complex ways, such as to ask marketing research or financial analysis questions. For these more complex functions, process or decision logic models, rather than data models, are developed. These process models outline the logic of transforming data and the events that trigger certain data processing functions. Process models may be developed in structured English, state-transition diagrams, or other logic modeling notations (explained in Chapter 10).

In Construction, the same information systems professionals who created the design now generate production code using the CASE tools' code generator and manual coding as necessary. Much of the production code may have already been developed, but often each system component was developed separately. For example, a RAD process for developing a large manufacturing system might have separate teams of users and analysts working on the production scheduling, inventory control, and cost accounting components. In Construction, the systems builders assemble the complete system from the various components. Thus, new internal interface components that link separately developed components are written. Also, some highly complex components which could not be generated from CASE tools or code generators must be built. For example, part of the production scheduling system might include a mathematical workload balancing model for daily work scheduling. End users also participate in construction, validating new screens and other aspects of the design as the application system is being built. During Construction, the RAD teams also develop nonprogram components, such as user manuals and training materials. Construction can be combined with User Design into one phase for smaller systems.

Cutover means delivery of the new system to its end users. Because the RAD approach is so fast, planning for Cutover must begin early in the RAD process. Cutover involves many of the traditional activities of implementation, including testing, training users, dealing with organizational changes, and running the new and old systems in parallel; however, all of these activities occur on an accelerated basis. Project close-down occurs at the end of the Cutover phase, at which time a system developed by RAD is handled by the normal system maintenance processes.

According to Martin, RAD can produce a system in six months that would take 24 months to produce using the traditional systems development life cycle.

ADVANTAGES AND DISADVANTAGES OF RAD

The primary advantage of RAD seems obvious: information systems developed in as little as one-quarter the usual time. And shorter development cycles also mean cheaper systems, as fewer organizational resources need be devoted to develop any particular system. Martin (1991) points out that RAD also involves smaller development teams, which results in even more savings. Because of the intimate roles of users, RAD reduces project risks associated with misunderstanding user requirements and lack of customer commitment. System quality is also increased since much of the system is generated from code generators that have gone through extensive testing. Finally, because there is less time between the end of design and conversion, the new system is closer to current business needs and is therefore of higher quality than would be a similar system developed the traditional way.

Others point out, however, that although RAD works, it only works well for systems that *have to be* developed quickly (Gibson and Hughes, 1994). Such systems include those that are developed in response to quickly changing business conditions or new government regulations. Where RAD is the methodology of choice for all systems development, RAD's emphasis on what a system does and how it does it means that essential information about the business models behind the system is missing. In its highly accelerated analysis and design phases, RAD leaves little room for understanding the business area, what it does, what its functions are and who performs them, or what the people in the business area do in performing their jobs. The greater the reliance on RAD, the greater the risk that many systems will be out of alignment with the business, unless the clients involved in the RAD process are the most knowledgeable about future directions of the business area.

Another drawback to RAD also has to do with the very feature that makes it so attractive as a systems development methodology: its speed. Because the RAD process puts such an emphasis on speed, many important software engineering concepts, such as interface consistency, programming standards, module reuse, scalability, and systems administration, are overlooked (Bourne, 1994). For example, consider the following:

- *Consistency:* In their efforts to quickly paint screens, RAD analysts often ignore the need to be consistent both within an application and across a suite of related applications. Areas of concern include window size and color, consistent format masks, and using the same error message for the same offense.

- *Programming standards:* Documentation standards and data naming standards should be established early in RAD, or it may be difficult to implement them later.

- *Module reuse:* Many times during prototyping, analysts forget that similar display or report designs may have already been created. Often in RAD there is no mechanism in place that allows analysts to easily determine whether modules that can be reused are already in existence.

- *Scalability:* If the system designed during RAD is useful, its use will gradually spread beyond those initial users who helped build it. Developers should build such growth into their initial system. Scalability also applies to hardware; system scope; the number, type, and users of reports; growth in the software team developing and maintaining the system as it develops; user training as system use expands; and security.

- *Systems administration:* System administration is typically ignored altogether during RAD, as the emphasis is usually on the excitement of seeing a new application system develop before users' eyes. Important system administration issues include database maintenance and reorganization, backup and recovery, distribution or installation of application updates, and scheduling and implementing planned system downtime and restarts.

AN EXAMPLE OF SUCCESSFUL RAD USE

Despite the potential drawbacks of RAD, when it is used successfully, the results can be dramatic. A company in the Boston area, called Cambridge Technology, has made RAD the basis of its highly successful Rapid Solutions Workshops (Harrar, 1993). Using their RAD approach, Cambridge Technology was able to develop a decision support system for Hughes Space and Communications Company in about half the time the traditional systems development methodology would have demanded.

The Cambridge Technology approach involves several steps. First, consultants spend two to three days identifying which of a client's applications would be suitable to the Rapid Solutions Workshop process. A team from Cambridge Technology then spends two weeks at the client's site, gathering information and observing the client's information systems operations. Then a client team spends a week at Cambridge Technology's offices in Cambridge, Massachusetts. Usually five to seven of the slots on the eight- to ten-person client team are filled with users, so the process is user-driven. The client team is given a very real incentive to cooperate: at the end of their week, the team must give a presentation of their newly developed system to client management. Often the managers present are very high-level. If the design is approved by management, Cambridge Technology then works to turn the design into a production system.

For Hughes Space and Communications Company, Cambridge Technology was able to create a complex decision support system prototype in only three weeks. The system integrated data from four databases that were previously thought to be unintegratable. After getting approval on the design, Cambridge Technology went on to complete requirements definition and design in less than two months. Ten months later, Hughes launched the production system. Hughes estimated that developing the same system in-house would have taken them at least two years. Hughes liked the Rapid Solutions Workshop approach so much that they had Cambridge Technology develop three more systems for them and subsequently train their own information systems people to conduct the workshops back at Hughes' offices in El Segundo, California.

SUMMARY

Rapid Application Development (RAD) is an alternative to the traditional systems development life cycle. All of the phases of the traditional life cycle are included in RAD, but the phases are executed at an accelerated rate. RAD relies on heavy user involvement, Joint Application Development sessions, prototyping, integrated CASE tools, and code generators to design and implement systems quickly. The abbreviated RAD life cycle begins with Requirements Planning, followed by User Design, Construction, and Cutover.

The primary advantage of RAD is the quick development of systems, but quick development may also lead to cost savings and higher-quality systems. RAD does have drawbacks: with its emphasis on developing systems quickly, the detailed business models that underlie information systems are often neglected, leading to the risk that systems may be out of alignment with the overall business. Similarly, the speed of development may lead to analysts overlooking such systems engineering concepts as consistency, programming standards, module reuse, scalability, and systems administration. If applied successfully, however, RAD may result in dramatic savings and improved performance. Systems may be designed and implemented in one-quarter to one-half the time needed for the traditional life cycle approach.

A P P E N D I X   R E V I E W

K E Y   T E R M S

Rapid Application Development (RAD)

R E V I E W   Q U E S T I O N S

1. Define each of the following terms:
    a. prototyping
    b. JAD
    c. rapid application development
    d. User Design
    e. Requirements Planning
2. List and briefly define the four phases of RAD, as defined by James Martin.
3. Explain the advantages and disadvantages of RAD.
4. What trends in information systems encouraged the discovery of the RAD approach to systems development?
5. Explain the concept of scalability and its influence on systems development.

P R O B L E M S   A N D   E X E R C I S E S

1. Match the following terms to the appropriate definitions:

    _____ Rapid Application Development

    _____ prototyping

    _____ code generator

    a. tool that allows the automatic writing of programs from system specifications
    b. iterative process of revising a system through close work between an analyst and users
    c. systems development methodology that relies on heavy user involvement, JAD, prototyping, and CASE

2. Compare RAD and prototyping. How are these methodologies different from one another? In what ways are they similar?
3. How might RAD be used in conjunction with the structured systems development life cycle? Are RAD and the SDLC totally different approaches or could they complement each other?
4. What types of tools are necessary to do RAD? Is it possible to do RAD without fourth-generation languages and other tools?
5. One of the criticisms of RAD is that RAD may cause a system to be out of alignment with the direction of the business. Explain how this may occur and suggest what might be done to overcome this potential hazard of RAD.
6. The example of the use of RAD summarized in this appendix involves the development of a decision support system. Do you think RAD could be applied to other types of systems besides DSSs? Why or why not? Are the criticisms of RAD outlined in this chapter more or less important depending on the type of information system being developed?

FIELD EXERCISES

1. Electronic Data Systems (EDS) uses a form of RAD which they call SLC-RISE (Systems Life Cycle-Rapid Iterative Systems Engineering). Contact a systems analyst at EDS or at a client organization and investigate the impact of this methodology on systems development projects. Have all systems professionals in the client organization embraced RISE? What does the RISE methodology do to address some of the potential RAD hazards outlined in this appendix?

2. Contact an organization that has done a RAD-based systems development project. Investigate why they chose to conduct that project using RAD as opposed to some other methodology also used in their organization. Explain what factors that organization considers when deciding to conduct a project using a RAD or an SDLC methodology.

REFERENCES

Bourne, K. C. 1994. "Putting Rigor Back in RAD." *Database Programming & Design* 7(8) (Aug.): 25–30.

Gibson, M. L., and Hughes, C. T. 1994. *Systems Analysis and Design: A Comprehensive Methodology with CASE.* Danvers, MA: Boyd & Fraser Publishing Company.

Harrar, G. 1993. "Welcome to IS Boot Camp." *Forbes ASAP* 152 (10): 112–18.

Martin, J. 1991. *Rapid Application Development.* New York, NY: Macmillan Publishing Company.

Advanced Topics in Conceptual Data Modeling

After studying this appendix, you should be able to:

■ Represent time-dependent data on an ER diagram.

■ Represent multiple relationships between the same entities and distinguish between exclusive and nonexclusive relationships on an ER diagram.

■ Represent generalization (IS-A) relationships between supertype and subtype entities on an ER diagram.

■ Represent aggregations on an ER diagram.

INTRODUCTION

A strong theme of this book is that a thorough description of data is an essential activity in the requirements structuring step of the analysis phase of the SDLC. Such data structuring should include as rich a meaning of data as can be captured and represented. Various experts have identified numerous semantics and ways to represent these in the E-R notation. In this appendix, you will see four extensions to the basic E-R modeling capabilities explained in Chapter 11: modeling time-dependent data, multiple relationships between the same entity types, and the data abstractions of generalization and aggregation. These semantics are now commonly represented in E-R data models. Other semantics exist, and the interested reader is referred to Batini, Ceri, and Navathe (1992) and Flavin (1981) for detailed coverage.

MODELING TIME-DEPENDENT DATA

Data values vary over time (see the fifth and eighth questions in Table 11-1). For example, the unit price for each product may change as material and labor costs and market conditions change. If only the current price is required, then only that value needs to be represented in a data model. For accounting, billing, and other purposes, however, we are likely to need a history of the prices and the time period over which each was in effect. As Figure C-1a shows, we can conceptualize this requirement as a series of prices and the effective date for each price. This results in repeating data that include the attributes PRICE and EFFECTIVE DATE. In

Figure C-1

Example of time stamping

(a) PRODUCT entity with
repeating group

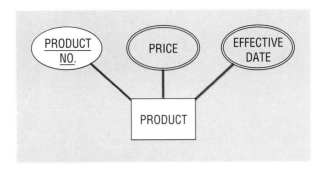

(b) PRODUCT
entity with PRICE HISTORY

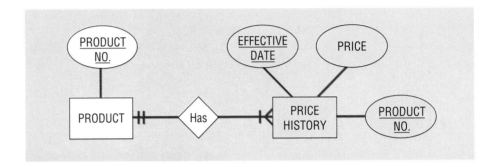

Figure C-1b, these repeating data have been replaced by a new (weak) entity named PRICE HISTORY, similar to how we modeled multivalued attributes (see section on Multivalued Attributes in Chapter 11). The relationship between PRODUCT and PRICE HISTORY is named Has.

In Figure C-1b, each value of the attribute PRICE is time stamped with its effective date. The use of simple time stamping (as in the above example) is often adequate for modeling time-dependent data. Time may introduce, however, more subtle complexities in data modeling. For example, Figure C-2a represents a portion of an E-R diagram for Pine Valley Furniture Company. Each product is assigned to one product line (a related group of products). Customer orders are processed throughout the year, and monthly summaries are reported by product line and by product within product line.

Suppose that in the middle of the year, due to a reorganization of the sales function, some products are reassigned to different product lines. The model shown in Figure C-2a is not designed to accommodate the reassignment of a product to a new product line. Thus, all sales reports will show cumulative sales for a product based on its *current* product line, rather than the one at the time of the sale. For example, a product may have total year-to-date sales of $50,000 and be associated with product line B, yet $40,000 of those sales may have occurred while the product was assigned to product line A. This fact will be lost using the model of Figure C-2a.

The simple design change shown in Figure C-2b will correctly recognize product reassignments. A new relationship (called Sales for product line) has been added between ORDER and PRODUCT LINE. As customer orders are processed, they are credited to both the correct product and product line as of the time of the sale using the Sales for product and Sales for product line relationships. When a product changes product lines, a change is made only in the Assigned relationship. In this way, orders implicitly change to the new product line through the Assigned relationship but stay associated to the product line at the time of sale through the Sales for product line relationship.

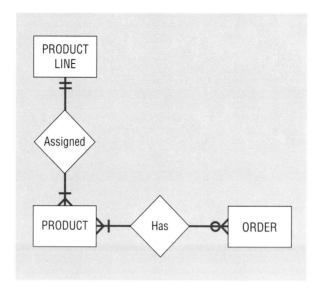

Figure C-2
Pine Valley Furniture product
database

(a) E-R diagram not
recognizing product
reassignment

(b) E-R diagram recognizing
product reassignment

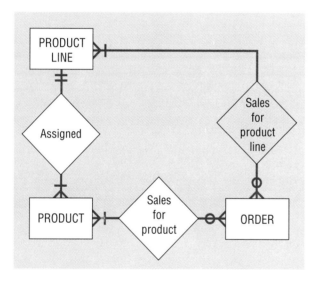

MULTIPLE RELATIONSHIPS BETWEEN ENTITIES

Suppose employees in a contract engineering firm are not all eligible to work on every project. Further assume that, as a new contract project is acquired, the chief project manager reviews the personnel files of all engineers to see who has the required skills, experience, and availability to be assigned to the project. As the project progresses, certain engineers are actually assigned to the project. This brief scenario describes a situation in which there are two relationships between the entity types of EMPLOYEE and PROJECT: Eligible for and Assigned to. Figure C-3 depicts these two many-to-many relationships between these entity types.

Such instances of two or more relationships between the same entities are common in organizations. Here are a few other example situations:

- Entity types VENDOR and PART—relationships of Potential domestic supplier and Potential international supplier

Figure C-3

Example of multiple relation-ships between a pair of entity types

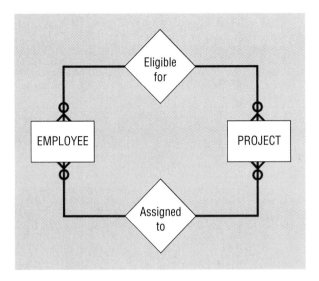

- Entity types EMPLOYEE and COURSE—relationships of Is taking, Is teaching, and Has taken equivalent of

- Entity types of MUSICAL ARTIST and SCORE—relationships of Composed, Arranged, Played, and Conducted

    In the example on page 837, it is possible that the same artist may have composed, arranged, played, and conducted the same score. However, in other cases when there are multiple relationships between the same pair of entity types, there may be restrictions on entity instances participating in more than one of the relationships at the same time. For example, consider the situation depicted in Figure C-4. This examples shows that a DOCTOR and a NURSE may be Married to each other or may Work for each other, but not both at the same time. The restriction of participating in only one of several relationships is called **exclusive relationships.** The notation for exclusive relationships, an arc crossing the lines for the exclusive relationships, was indicated in Figure 11-5.

Exclusive relationships: A set of relationships for which an entity instance can participate in only one of the relationships at a time.

Figure C-4

Example of exclusive relationships between a pair of entity types

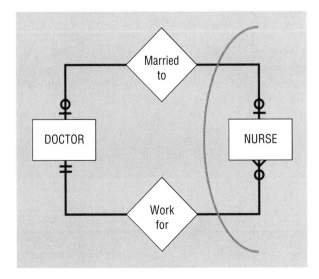

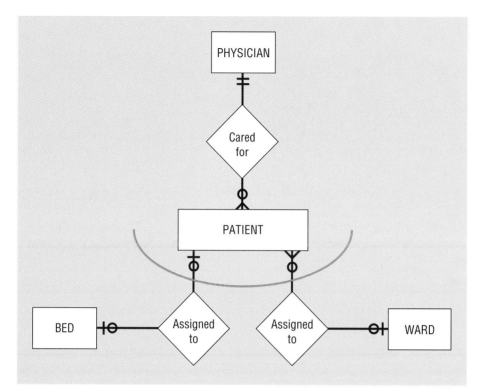

Figure C-5
Example of exclusive and
non-exclusive relationships
between entity types

An exclusive relationship can apply not only to multiple relationships between the same pair of entity types but also to multiple relationships between an entity type and several other entity types. Consider the situation depicted in Figure C-5. This example shows that a PATIENT can be Assigned to either a BED or a WARD, but not both, whereas a PATIENT simultaneously is Cared for by a PHYSICIAN.

GENERALIZATION

One of the unique aspects of human intelligence is its ability to classify objects and experiences and to generalize their properties. If we see a robin or an eagle, for example, we immediately classify each as a bird. Even if the bird is searching for worms in the grass, we know that it can fly since most birds can fly (we memorize the exceptions at an early age). Similarly, if we see a Porsche, we assume that it is both fast and expensive since these are properties of luxury sports cars.

Business entities are often best modeled using the concepts of generalization and categorization. **Generalization** is the concept that some things (entities) are subtypes of other, more general, things (see the sixth question in Table 11-1). For example, to an airline a business passenger is one subtype of the more general type called *passenger*.

Generalization: The concept that some things (entities) are subtypes of other, more general, things.

Subtypes and Supertypes

One of the major challenges in data modeling is to recognize and clearly represent entities that are almost the same; that is, entity types that share common properties but also have one or more distinct properties. For example, suppose that an organization has three basic types of employees: hourly employees, salaried employees, and contract consultants. Some of the important attributes for these types of employees are the following:

1. Hourly employees: EMPLOYEE NO., NAME, ADDRESS, DATE HIRED, HOURLY RATE

2. Salaried employees: EMPLOYEE NO., NAME, ADDRESS, DATE HIRED, ANNUAL SALARY, STOCK OPTION

3. Contract consultants: EMPLOYEE NO., NAME, ADDRESS, DATE HIRED, CONTRACT NUMBER, DAILY RATE

Notice that all of the employee types have several attributes in common (EMPLOYEE NO., NAME, ADDRESS, DATE HIRED). In addition, each type has one or more unique attributes that distinguish it from the other types (for example, HOURLY RATE is unique to hourly employees).

Supertype: A generic entity type that is subdivided into subtypes.

Figure C-6 shows a representation of the EMPLOYEE supertype with its subtypes, using E-R notation. A **supertype** is a generic entity type (such as EMPLOYEE) that is subdivided into subtypes. A **subtype** is a subset of a supertype (that is, the instances of entities in the subtype are a subset of the instances of its associated supertype). The members of a subtype have some attributes or relationships distinct from members of the other subsets. Entity subtypes behave in exactly the same way

Subtype: A subset of a supertype that shares common attributes or relationships distinct from other subsets.

Figure C-6
EMPLOYEE supertype with subtypes

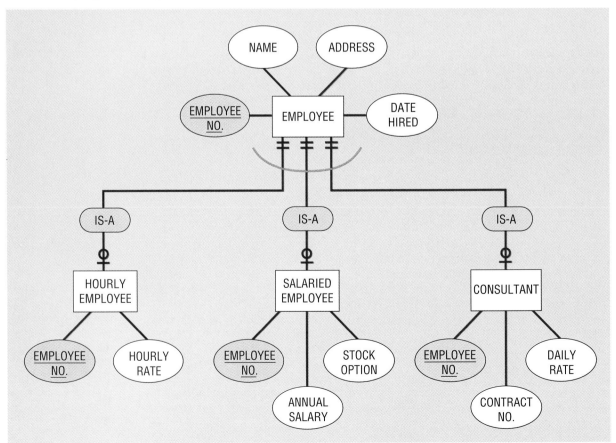

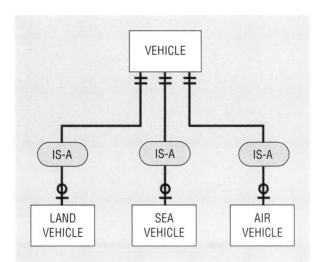

Figure C-7
Example of non-exclusive subtypes

as any entity type. For example, entity subtypes have attributes and may have relationships with other entity types.

In Figure C-6, common attributes for all employees are included with the EMPLOYEE entity type. The primary key for EMPLOYEE, as well as for each of the subtypes, is EMPLOYEE NO. Attributes that are peculiar to each subtype are included with that subtype only.

The relationship between each subtype and supertype is called an **IS-A relationship.** A rectangle with rounded ends is used to designate IS-A relationships. This relationship is read from the subtype to the supertype; for example, "HOURLY EMPLOYEE IS-A EMPLOYEE" (correct grammar is not enforced). As shown in Figure C-6, the cardinality of the relationship from a subtype to the supertype is mandatorily one. The cardinality is mandatorily one because an instance of a subtype is *always* an instance of the supertype (a salaried employee is always an employee). On the other hand, the relationship from the supertype to a subtype is optionally zero or one (an employee may or may not be a salaried employee). Since these cardinality relationships are always the same in IS-A relationships, you will omit the cardinality notation in such diagrams.

IS-A relationship: The relationship between each subtype and its supertype.

The IS-A relationships in Figure C-6 are exclusive, but this is not always true for generalizations. Consider the situation depicted in Figure C-7. Here VEHICLE has subtypes LAND, SEA, and AIR. Since some vehicles can travel in multiple mediums, the same VEHICLE instance may participate in several of these IS-A relationships. Consider an amphibious vehicle. This vehicle would have instances of the VEHICLE supertype and the LAND and SEA subtypes, but not the AIR subtype.

Inheritance

Inheritance is the property by which all attributes of a supertype become attributes of its subtypes. The term inheritance is also more generally used in object-oriented programming languages and covers not only the adoption of attributes by subtypes but also methods, or behaviors, of the supertype (see Appendix D). Thus in Figure C-6, the attributes NAME, ADDRESS, and DATE HIRED are inherited by all three employee subtypes (which is why these attributes are not explicitly attached to the subtypes). Except for the primary key, only attributes that are unique to a subtype are associated with that subtype.

AGGREGATION

Aggregation (aggregate entity): A collection of entities that together form a higher-order concept.

An **aggregation** (or aggregate entity) is a collection of entities that together form a higher-order concept. An instance of a WORK ORDER entity in Pine Valley Furniture, for example, would be the collection of related RAW MATERIAL, TOOL, WORK CENTER, and FACTORY WORKER entity instances needed to produce a certain piece of furniture. In addition to inheriting the attributes about component entities, an aggregate entity can have attributes not found in its component entities. For example, a work order might have a promised completion date as well as material descriptions, tool codes, and so on from the component entities. An aggregate may also have attributes which summarize characteristics of the component entities. For example, we might want to know the number of tools used on a work order. Finally, an aggregate may participate in other relationships. For example, a work order is likely for the purpose of building a particular product. This complete situation involving a work order is depicted in Figure C-8.

Note that the aggregate entity type, WORK ORDER, is shown as a gerund in Figure C-8. This is perfectly consistent with the application of gerunds as explained in Chapter 11. The computed attribute of WORK ORDER, NUMBER OF TOOLS, is shown in a dashed oval. The component entities are composed into a quaternary relationship, the WORK ORDER gerund. There is a separate binary relationship between WORK ORDER and PRODUCT, since a product is sometimes built via a work order, and a work order must describe the building of some product.

SUMMARY

The purpose of conceptual data modeling is to capture into a structured form as much meaning about data as you can understand. Any semantic, or meaning, of data would ideally be represented in a data model diagram, such as an E-R dia-

Figure C-8
Example of an aggregate entity type

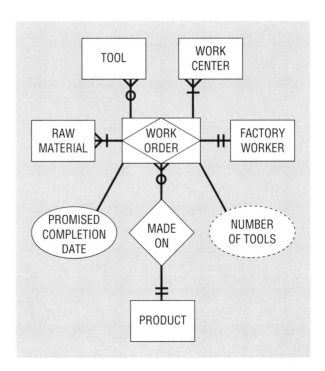

gram. A thorough coverage of all semantics that have been identified is beyond the scope of this book, but this appendix has extended the discussion from Chapter 11 to include four frequently represented, advanced data modeling concepts: time in data modeling, multiple relationships between the same entities, and the data abstractions of generalization and aggregation.

Time can enter a data model in a wide variety of ways, two of which were discussed. First, a time series or history of the same attribute(s) may need to be maintained for an entity type. Second, the original association between an entity when it was created and another entity as well as the current association may need to be kept.

There may exist multiple associations between the same pair of entity types, and these relationships may or may not be exclusive. If several relationships (between an entity and one or more other entities) are exclusive, then instances may participate in only one relationship at a time (for example, an employee could be taking a course or teaching a course, but not both at the same time). Alternatively, in a non-exclusive set of relationships, instances may participate in multiple relationships simultaneously. For example, an artist could play, conduct, and arrange the same musical score.

Generalization occurs when one or more entity types are special cases of another more inclusive entity type. For example, an employee supertype may have clerical, technical, and professional employee subtypes. A supertype-subtype relationship is called an IS-A relationship, and as with other types of relationships, IS-A relationships may be exclusive or non-exclusive.

An aggregate entity can be defined as the composition of several more detailed entities. For example, an airline reservation is composed of one or more passengers on an itinerary of flights, possibly with assigned seats on each flight. An aggregate entity can have its own attributes and participate in relationships besides those associated with its component entities.

A P P E N D I X   R E V I E W

K E Y   T E R M S

Aggregation (aggregate entity)
Exclusive relationships

Generalization
IS-A relationship

Subtype
Supertype

R E V I E W   Q U E S T I O N S

1. Define each of the following terms:
   a. gerund
   b. quaternary relationship
   c. subtype
   d. time stamp
2. Describe two general situations in which time plays a role in modeling data.

3. What is the difference between exclusive and non-exclusive relationships? Give an example of each.

4. What data modeling construct defined in Chapter 11 is used to represent an aggregate entity type?

5. Describe the property of inheritance as it applies to entity subtypes and supertypes.

PROBLEMS AND EXERCISES

1. Consider an aggregate entity of an airline reservation with component entities passenger, flight, and seat. Pick a few attributes for each component entity and for the aggregate and draw this situation with an E-R diagram.

2. Suppose employees in a company are all described by their employee ID, name, and phone number. An employee is either a clerical, technical, or professional employee. For clerical employees, we want to know their typing speed and which word processors they can use. For technical employees, we need to know their highest degree title. For professional employees, we need to know what other employees they supervise. Depict this situation with an E-R diagram.

3. In a club of computer information systems students, members can chair activities and participate in activities. A club member can also be assigned to evaluate club activities, but not those he or she chairs. Show this situation with an E-R diagram.

4. Employees are assigned to departments. We want to know to which department an employee was first assigned and to which department an employee is currently assigned. Show this situation with an E-R diagram using two separate relationships, one for first assignment and one for current assignment.

5. Employees are assigned to departments, and employees may be reassigned to different departments over their career. We must keep track of this history of assignments, and for each assignment we need to know the start and end date of the assignment. Depict this employment situation with an E-R diagram.

6. Refer to Problem and Exercise 9 in Chapter 11. Since both "customers" and "case workers" are employees of the company, redraw your answer to this exercise using one entity type for employees and subtypes for customers and case workers. Indicate on this diagram the business rule that a case worker may not be assigned to a purchase request if that case worker is also the customer of that request.

FIELD EXERCISES

1. Obtain access to a CASE tool and investigate how you could model IS-A relationships with this CASE tool. Is there a direct way to model IS-A relationships or must you use other constructs to simulate an IS-A relationship? Can you model whether IS-A relationships are exclusive or not?

2. Talk to a systems or data analyst in an organization in which you have contacts. Show this person Figure C-2 and explain the situation depicted in this figure. Ask them if there are any similar situations in their organization and how the data models for their systems deal with dynamic relationships. If their answer is that such dynamics are ignored, have them explain what the potential problems might be

with ignoring this issue. Write a summary of what they said and discuss how you would suggest they change the data model to handle the dynamic data entity relationships in their organization.

R E F E R E N C E S

Batini, C., S. Ceri, and S. B. Navathe. 1992. *Conceptual Database Design.* Redwood City, CA: Benjamin/Cummings Publishing.

Flavin, M. 1981. *Fundamental Concepts of Information Modeling.* New York, NY: Yourdon Press.

Object-Oriented Analysis and Design: An Overview

LEARNING OBJECTIVES

After studying this appendix, you should be able to:

- Contrast the object-oriented and structured approaches to systems analysis and design.

- Define the core concepts of object orientation: object, class, encapsulation, inheritance, polymorphism, and Gen-Spec, Whole-Part, and instance connections.

- Read and draw static and dynamic object-oriented models of system requirements.

- Outline the full set of system specifications associated with OOAD.

INTRODUCTION

> *We have no doubt that one could arrive at the same results using different methods; but it has also been our experience that the thinking process, the discovery process, and the communication between user and analyst are fundamentally different with OOAD than with structured analysis.*
>
> Coad and Yourdon, 1991a

Object-oriented analysis and design (OOAD) has been presented by most authors as an alternative approach for requirements determination, structuring, and subsequent system design distinct from the structured techniques and methods presented in the book chapters. OOAD is not strictly an alternative to a structured approach since, as we will see, there are some similarities between structured and OO techniques (data modeling is quite similar), and some OO techniques, like state-transition diagrams, are now readily accepted within the structured approach. OOAD is of significant current interest in the systems development field because of the growth of object-oriented and event-driven system implementation environments and the desire to drastically improve the reusablility of software. It is argued that the use of OO principles during systems analysis, design, and implementation should make the whole systems development process easier by making the thought processes and notations consistent between your study of the problem space (analysis) and the solution space (design and implementation). OOAD also promotes reusability of system components (such as data objects, user interface elements, and program modules) in analysis, design, and implementation.

Although different from structured methodologies, is OOAD better? Meyer (1989), Henderson-Sellers and Edwards (1990), and others claim that a purely structured approach to systems design, characterized by top-down decomposition of functions, is flawed as follows:

- It handles evolutionary changes poorly

- It is tied to the notion of a single function (the highest level DFD) and associated decompositions, which are unstable, rather than multiple business objects, which are more stable

- It unnaturally separates data from the processes that manipulate those data, and concentrates on process design rather than data design (for example, creating system structure from transformation of DFDs into function charts)

- It does not encourage reusability of components

On the other hand, even OOAD advocates recognize the value of functional decomposition for the design of individual code modules that are within object classes. Fichman and Kemerer (1992) conclude, after a thorough review of many different structured and object-oriented systems analysis and design methodologies, that OOAD is a radical change from the process-oriented methodologies (like data flow diagramming) but only an incremental change over data-oriented methodologies (like information engineering using the techniques of entity-relationship diagramming and planning matrices). Fichman and Kemerer state (p. 23) that "The main differences between OOAD and data-oriented conventional methodologies arise from the principle of encapsulation of data and behavior." Thus, rather than grouping functions together as steps in higher-level functions, in OOAD you model functions as associated with the data on which they operate. Even with this fundamental difference, you will see that the concepts and notation used for static object-oriented models are very similar to the concepts and notation for entity-relationship diagramming.

In this appendix, we present an overview of the techniques and methodologies of OOAD. In the next section you will review the core concepts on which OOAD is built: object, class, encapsulation of data and function, inheritance of properties, polymorphism, and various forms of connecting objects and object classes. A notation for representing these concepts and subsequent models is also introduced in this section. In the following sections you will see how static and dynamic models of system components and behavior are built using OOAD techniques. The final section summarizes the OOAD process as different from the structured SA&D process. You should note in this presentation that most organizations use a blended set of techniques and methods for systems analysis and design, so you are likely to encounter aspects of both structured and OO analysis and design in your professional work.

CORE CONCEPTS OF OOAD

Figure D-1 summarizes with examples the basic OOAD notation illustrated in this appendix. This OOAD notation as well as the methodology for OOAD outlined in this appendix are a variation of that developed by Coad and Yourdon (1991a), which is arguably the most widely used. We use the crow's foot notation to represent cardinality, consistent with the E-R notation from Chapter 11. Other popular OO notations by Booch (1994), Martin and Odell (1992), Rumbaugh et al. (1991), and Shlaer and Mellor (1992) exist. Monarchi and Puhr (1992) provide an extensive

Figure D-1

Basic object-oriented notation

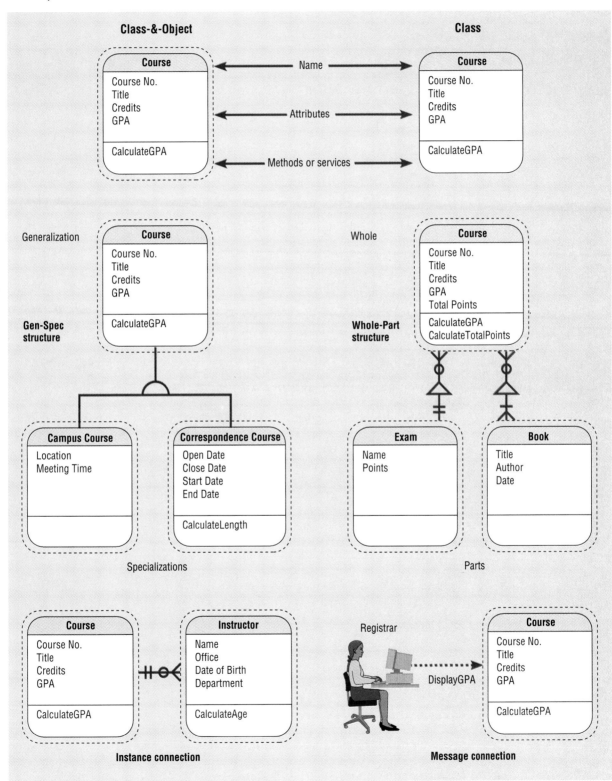

review of thirteen different OO notations and tool sets. All of these notations are built on several core concepts of OOAD: object, class, encapsulation, inheritance, and polymorphism.

Object

An *object* (or instance) is an abstraction of something from the problem domain. For example, an object is a particular course (for example, MIS 401), instructor, or student. An object is a "package" which has a unique name (Course), attributes (Course No., Title, etc.), and **methods** or services (CalculateGPA), which are the behaviors or functions associated with that object. In object orientation, functions (methods) cannot stand alone but rather must be stored with an object, whether the object be a business object, such as a university course or instructor, or a user interface object, such as a computer display form which shows attributes of a course to a system user.

Each object has an identity, or identifier value, which is separate from all of the object's attributes. In addition, an object may have constraints or business rules (as described in Chapter 11) which restrict values of the object's attributes or its behavior.

Since object orientation is applied consistently throughout the systems development process, each attribute is itself designated as an instance (object) of some kind of object, like an integer or character string, at the time of implementation. However, during analysis, each attribute is considered to be atomic—irreducible.

Method: A specific behavior that an object is responsible for performing. Also called *service*.

Class

A *class* is a description (a template) of objects that have the same attributes *and* behaviors. This description includes how to create new objects in the class. In Figure D-1, a class is represented by a rectangle with rounded corners and solid line edges. When we want to represent a class *and* its associated objects, we enclose the class rectangle within another rectangle with dashed (or shadowed) edges. As we will see, this subtle distinction is very important, since some structures link objects and other structures link classes. Most classes have multiple objects, but some classes have no objects or have a maximum of one object, as necessary to represent the semantics of the problem domain.

The classes which define most organizations are quite stable (for example, a university will always have courses, instructors, and exams). In contrast, the attributes and services of these classes will change (for example, a historical average class size attribute is necessary only when the university begins to include this in its bulletin or other documentation).

Encapsulation

Encapsulation: The property that the structure of attributes and methods of an object are hidden from the outside world and do not have to be understood to access its data values or invoke its methods.

Central to the discussion above is the notion that an object and its attributes and methods are inseparable. **Encapsulation** means that the structure of attributes and methods of an object are hidden from the outside world and do not have to be understood to access its data values or invoke its methods. A consequence of encapsulation is that an object cannot change any other object but must request that the other object change itself. Figure D-2 represents the encapsulation concept. In the example in Figure D-2, we see that a Course object includes several attributes (including Status) and methods (including Cancel). When the university registrar decides that a course should be cancelled, a message (along with any necessary pa-

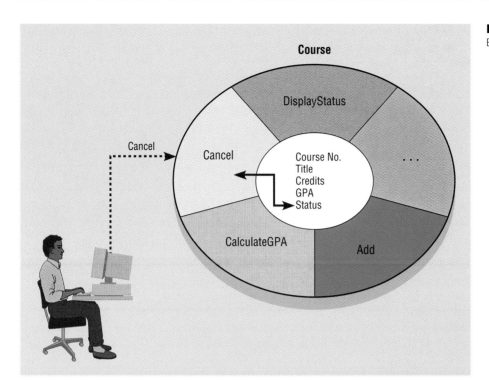

Figure D-2
Encapsulation concept

rameters) is sent to the associated course object, which invokes its Cancel method. The Cancel method modifies the attributes of the course as needed, in this case changing the value of the Status attribute, and returns a response to the message's sender. As we will elaborate later, such message connections combine event-response (see state-transition diagrams in Chapter 10) and data flow (see data flow diagrams in Chapter 9) perspectives.

Encapsulation enforces data and function integrity. Data are protected from contamination from outside agents, which are likely to eventually mishandle the data. Further, other objects are protected from having to understand the evolving internal structure of an object. With encapsulation, other objects only need to know what service to request from a given object. The real payoff from encapsulation is this protection of one object from changes in the internal structure of other objects.

Figure D-2 also emphasizes the philosophical similarities between object-orientation and client/server concepts (see Chapter 18), which is why these two trends in systems development are often associated. In a simplified view of systems development in an object-oriented environment, the division of responsibilities between different objects (the Registrar and the Course) can be easily extended to a user interface object (computer program) running on a local (client) personal computer (for example, a data entry screen) and a database object (for example, a Course) located on a server. In reality, more objects would be involved, in which many client-server relationships would exist.

Inheritance

The idea that an object is an instance of a class implies the notion of inheritance. *Inheritance* means that an object or object class takes on the same properties of the object class from which it is instantiated. An object must be an instance of some object class. For example, course MIS 412 is an instance of the class Course and hence has all the same attributes (but not values) and methods of any Course instance. Further, if we were to create specialty courses, such as Campus Course and Correspondence Course (see Figure D-1), these subclasses of the superclass Course would inherit the properties of their superclasses. If you have ever programmed in an object-oriented or object-based programming language, like C++, Power-Builder, or ObjectPAL for Paradox for Windows, you should be familiar with the idea of inheritance.

Polymorphism

Polymorphism: The ability to refer to properties (attributes and methods) of instances from more than one class by the same name.

The final fundamental concept of OOAD is polymorphism. Literally, polymorphism means the ability to take more than one form. In object-oriented terms, **polymorphism** is the ability to send the same message to instances of different object classes, resulting in different behaviors. For example, both Student and Instructor classes might have an attribute Name and both Course and Registration classes might have a Cancel method. In these cases, an object (say Advisor) could send the same message to either Student or Instructor requesting its Name. Polymorphism is highly related to encapsulation since it is the information-hiding property of encapsulation that makes polymorphism possible. Polymorphism has more implications for implementation than analysis and design, so we will not refer to this concept in the remainder of this appendix.

OO ANALYSIS—MODELING SYSTEM REQUIREMENTS

An object-oriented representation of system requirements covers five layers of specification:

1. *Class/object layer:* The classes and objects from the problem domain

2. *Structure layer:* The inter-relationships between classes and objects (representations on this layer are very similar to entity-relationship diagrams)

3. *Service layer:* The methods of each object and the messages or events that cause methods to be performed

4. *Attribute layer:* The attribute properties of classes

5. *Subject layer:* A partitioning of the system requirements (sets of classes and their connections) into a scope for different projects or systems development teams

Static models are used to represent the class/object, structure, attribute, and subject layers while dynamic models are used to represent the service layer.

Static Modeling

The Class/object layer is documented by the Class-&-Object and Class symbols of Figure D-1. There is no Object symbol since objects may not exist without an associated class. Most classes are represented by the Class-&-Object symbol, meaning that any number of instances of the class may exist. Occasionally, a class may exist without any objects; we will see an example of how this situation arises when we review the Gen-Spec structure below.

Static structures (Gen-Spec, Whole-Part, and instance connections—see Figure D-1) express the problem-domain complexity, pertinent to the system's responsibilities (Coad and Yourdon, 1991a). Although the notation is slightly different, you have probably noticed the strong similarities between these static structures in Figure D-1 and the E-R diagramming notation discussed in Chapter 11 and Appendix C. Specifically, we use the same crow's foot notation for cardinality. The convention with OO models is, however, to place the cardinality at the opposite ends of the relationship than we have done with E-R diagrams and to not name the connection. For example, the instance connection between Course and Instructor has no name and this connection says that a Course has exactly one Instructor and an Instructor can be associated with zero to many Courses. One difference between OO and E-R data models is that OO models use special connection symbols to differentiate each type of connection (a half circle for Gen-Spec, a triangle for Whole-Part, and only a line for instance connections). Due to the similarities with E-R notation, we will not extensively discuss these OO modeling notations.

One subtlety of OO static modeling is that the connection lines may touch either the Class or Class-&-Object rectangles. Specifically, Gen-Spec and some message connections map a Class with another Class. For example, in Figure D-1, the class Course is related to two specialization classes, Campus Course and Correspondence Course. The point is that classes have specializations, not objects. Whole-Part, instance connections, and most message connections map an Object to another Object. For example, in Figure D-1, a specific Course object will include zero or many particular Exam objects and zero or many particular Book objects. Finally, some message connections map an Object to a Class.

Gen-Spec Connections   A relationship between a generalized class and one or more specialized subclasses is called a **Gen-Spec connection** (also called an 'is a' connection); this is identical to the IS-A relationship discussed in Appendix C on advanced E-R modeling notations. A specialized subclass inherits all of the attributes and properties of its superclasses. For example, in Figure D-1 the Campus Course class has attributes Course No., Title, Credits, and GPA and method CalculateGPA. In addition, it has Location and Meeting Time attributes whereas only Correspondence Course objects have Open Date, Close Date, Start Date, and End Date attributes. The Class-&-Object symbol is always used for a class that has no subclasses (if there were no instance of such a class, there would be no reason for it to exist); either the Class-&-Object or Class symbol may be used for classes that have specializations.

> Gen-Spec connection: A linkage between a class and one or more specialized subclasses. Also called an 'is a' connection. A Gen-Spec connection is functionally equivalent to the concept of generalization used in conceptual data modeling.

Additional examples of Gen-Spec connections appear in Figure D-3. Figure D-3a shows the case, called exhaustive specializations, where there are no objects for the Patient generalization class since every patient object must be either an In-Patient or Out-Patient object. In contrast, the Gen-Spec connection in Figure D-1 is non-exhaustive, since a Class-&-Object symbol is used to represent Course. This Gen-Spec connection shows that it is possible for a Course object to be neither a Campus Course or a Correspondence Course and that such courses have the attributes and methods of only the Course class. Figure D-3b illustrates a situation for which only one specialized class exists; in this example, Residential Student objects have some additional properties other student objects do not have.

Figure D-3c shows that Gen-Spec structures can form a lattice, not just a hierarchy of classes, in which a specialized class inherits attributes of several more general classes. Lattice structures are necessary when subclasses are not mutually exclusive. In this example, a Student object may be an instance of any of six classes: (regular) Student, Student Athlete, Honors Student, Student Employee, Honors Athlete, or Honors Employee. A Student Employee has attributes Name, Address, Employee No., and Max Hours/Week (how many hours he or she is allowed to work per week). An Honors Employee has attributes Name, Address, Employee

Figure D-3

Gen-Spec examples

(a) Exhaustive specializations

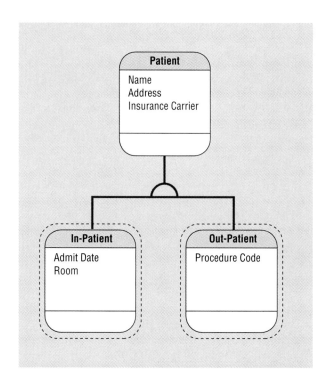

(b) One specialized class

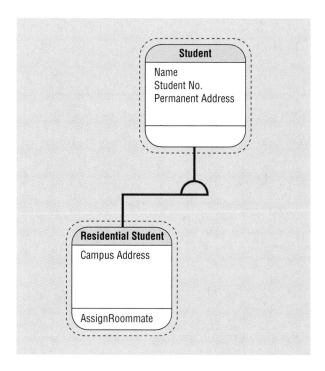

No., Max Hours/week, and Minimum GPA; such a student must maintain the Min-imum GPA to keep his or her job. Although all Student Employees have a Max Hours/week attribute, an Honors Employee will have this attribute definition overridden, which is why this attribute appears again within the more specialized Honors Employee class. An Honors Athlete has attributes Name, Address, Sport,

Figure D-3 *(continued)*

(c) Gen-Spec lattice

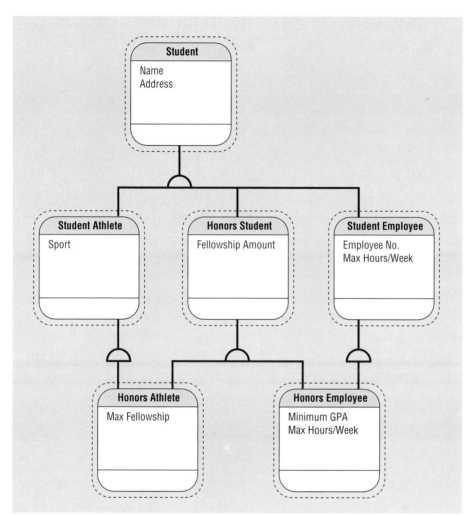

Fellowship Amount, and Max Fellowship (the fellowship for an Honors Athlete is restricted by NCAA regulations). This lattice also shows that there is no student athlete employee class, since again NCAA regulations do not permit student athletes to be paid for work at the university. This example also illustrates that a Gen-Spec connection does not imply that specializations are mutually exclusive. Some OO notations differentiate exclusive from non-exclusive specializations (see McFadden and Hoffer, 1994).

Whole-Part Connections   One of the most common connections between objects occurs when an object is composed of other objects (for example, an automobile is composed of an engine, a transmission, and possibly a radio). A **Whole-Part connection** (also called an assembly or 'has a' connection) represents the aggregation of objects to form a composite object. The ends of a Whole-Part connection are annotated to indicate the *cardinality* of objects involved in the connection. For example, in Figure D-1, a Course object contains zero to many Exam objects and zero to many Book objects. The purpose of a Whole-Part connection is the same as aggregation in conceptual data modeling (see Appendix C). A Whole-Part connection links objects, not classes.

Whole-Part connection: A linkage between an object and its components. Also called an assembly or 'has a' connection. Whole-Part connections are equivalent to aggregation relationships in conceptual data modeling.

Whole-Part connections might represent any of the following aggregate to components relationships:

- assembly-parts (for example, a product and its subassemblies and parts)

- container-contents (for example, a postal package and its contents)

- collection-members (for example, a student club and its members, or a travel itinerary and its segments)

Key to understanding Whole-Part connections is recognizing that the "whole" object is more than the collection of the "part" objects. For example, a postal package might have attributes for freight cost and the name of the person who wrapped the package, as well as a method to calculate total weight. In addition, the postal package has all the properties of its individual content items.

Whole-Part connections may have constraints. Besides minimum and maximum cardinality values (zero, one, many, or a specific value, just as in E-R diagramming), other constraints may exist. For example, a constraint may state that once an item component is placed in a specific package, it may not be removed from that package and placed in another package. Or, the total weight of a package may not be more than 75 pounds (the total weights for its content items). Constraints are usually provided in text in documentation (repository entries) which describe the elements of an OO static model.

As in E-R modeling, unary Whole-Part connections may exist. For example, Figure D-4 shows how a bill-of-materials structure would be modeled in OO notation. In Figure D-4a the nesting of items within items is shown by indicating that the Item No. and Quantity attributes repeat (the } symbol) and a unary Whole-Part

Figure D-4
Bill-of-materials structures

(a) Unary Whole-Part connections

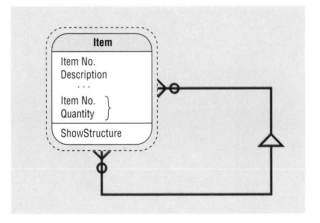

(b) Two Whole-Part connections

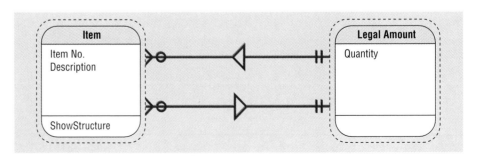

connection. In Figure D-4b the repeating data is contained in a separate object, the equivalent of a gerund from E-R diagramming, and two Whole-Part connections are included.

Instance Connections   The values of attributes represent an object's state. A state also includes which other objects that object is connected to. An **instance connection** extends the representation of an object's state by showing the relationship one object has with another object in order to fulfill its responsibilities. Note: Instance connections link objects, not classes. Instance connections are identical to relationships from conceptual data modeling. Crow's foot notation is used to depict cardinalities. For example, the Instance connection in Figure D-1 shows that a given Course object needs exactly one Instructor object, and an Instructor object needs zero to many Course objects. Most OO modeling notations directly cover unary and binary instance connections, but not higher-order connections. Figure D-5 shows how a ternary Instance connection between Vendor, Part, and Warehouse objects could be represented.

In OO modeling, you must explicitly distinguish Whole-Part from instance connections. Although similar (for example, in Figure D-1, a course has many exams and an instructor teaches many courses), a Whole-Part connection is a stronger semantic statement than an instance connection. As with E-R modeling, gerund-like objects are necessary when attributes or methods are properties of a many-to-many instance connection.

> **Instance connection**: The relationship one object has with another object in order to fulfill its responsibilities. An instance connection is identical to a relationship in conceptual data modeling.

Figure D-5
Ternary instance connection

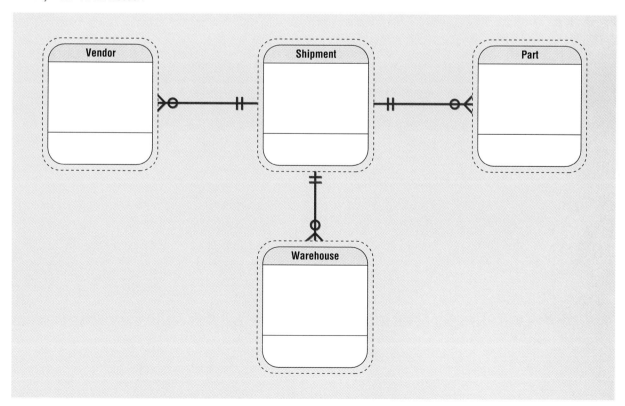

Dynamic Modeling

Whereas data flow diagrams, Structured English, decision tables, and other techniques support structured analysis methods, the dynamic nature of a system is represented by different constructs in OO analysis. At the heart of dynamic, or behavior, modeling is services. A service (or *method*) is a specific behavior (a function) that an object is reponsible for exhibiting. For example, a person class might have one service to calculate age based on a date of birth attribute and today's date and another service to delete an instance (object) of that class. Remember, due to the core OO principle of encapsulation, an object has to contain all functions that operate on attributes of that object since the attributes and services of an object are hidden from other objects. You represent services in the following ways in OO modeling:

- You list a service in the lowest section of an Object-&-Class symbol, identifying with which class that service is associated.

- You use Structured English or any specification language (e.g., pseudocode) to outline the detailed steps of a service. Structured English was discussed in Chapter 10, so we do not illustrate such specifications here.

- You draw *message connections* between objects to show which objects call upon the services of other objects. Message connections can be between data objects or between a user interface object (like a form in which data are displayed to the Registrar in Figure D-1) and a data object.

- You draw a state-transition diagram for an object to specify what services are performed for each object state and event associated with that object. We do not elaborate on this part of modeling the dynamics of a system since state-transitions diagrams are addressed in Chapter 10.

Services   There are three types of services: simple, complex, and monitoring/trigger. Simple services are implicit with all classes and are not shown in Object-&-Class symbols. Simple (also called occur or mandatory) services are the following:

- *Create (Add)*. Creates and initializes a new object in the class. Create is a method of a class, not of an object.

- *Connect*. Connects (and disconnects) an object in the class with another object for an instance or Whole-Part connection.

- *Access (Display)*. Show the attribute values of an object in the class.

- *Release (Delete)*. Disconnects an object in the class from all connections and deletes the object.

A complex (also called calculate) service mathematically or logically manipulates attributes of a class or object and may call upon services of other objects to display attributes or perform other calculations. For example, consider the partial OO model of Figure D-6. In this example, a Customer Order object has a service, CalculateValue, to calculate the total value of the order. This service would request each Line Item object connected to it via a Whole-Part connection to perform its CalculateExtendPrice service (to calculate price times quantity). The CalculateExtendPrice service of each Line Item object would request the simple Display service of the associated Item object to reveal the Price attribute. We will see shortly how to use message maps to indicate the sequence of these requests for services.

Monitoring/trigger services provide surveillance of an object. For example, the InventoryCheck service of the Item object in Figure D-6 monitors the Price at-

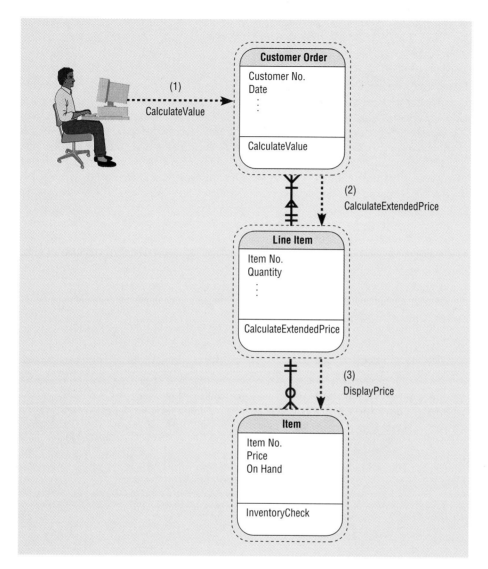

Figure D-6
Example complex services

tribute value and warns users when the stock level falls below a pre-determined limit. Other types of trigger services watch for external events (such as a button push on a GUI screen).

One analysis and design issue with services is where to place a service within Gen-Spec structures. Although performance issues can sometimes suggest otherwise, the general rule is to locate the service at the level at which it most generally applies. Consider again the Gen-Spec structure in Figure D-1. The CalculateGPA service is located in the Course superclass since this same service applies to all types of courses. By locating this service in Course, when we want to change the specification of CalculateGPA, we need to do this in only one place. In contrast, consider the method CalculateLength in the Correspondence Course class. This method computes how long a student has been taking a course by finding the difference between the End and Start Dates. This method applies only to Correspondence Courses, so it should not be placed in Course. As in the example of Figure D-3c which showed that attributes can be overridden, services can also be overridden in specializations. Thus, when a method is to be performed, a system built using object-oriented technology searches for the desired method starting at the lowest specialized class to which the method might apply, and moves up the lattice only when the desired method is not found.

Message Connections   Often an object does not possess all of the services it needs because these services are the responsibility of other objects. A **message connection** represents a dependency of an object on the services of another object or class. As depicted in Figure D-1, we can think of a message connection as a request to perform a service; in this case, a Registrar's user interface object requests a Course object to perform its DisplayGPA service. A message might include parameters if required by the requested service in the receiver object, and implicitly every message has a return value—some response to the sender object. This response may be an attribute value or an indicator of when or whether the service was performed.

Message connection: A dependency of an object on the services of another object or class.

You can also show the sequence of messages required. Look again at Figure D-6 and note that the message connections have been numbered to indicate the sequence in which messages are sent. Such a message map would be developed for each high-level function that generates a sequence of messages. Besides sequence numbers, it is also advisable to include the name of the requested service on each message connection, otherwise it is not clear which of the several services of the receiver object is requested.

Dynamics of User Interface Objects   A user interface object (UIO) is a computer display form or report which captures data to be sent (via messages) to data objects or reveals the properties of objects to system users. A user interface object could also be a menu display, which might call another menu UIO, and so on. A specific UIO is an object from a class of similar UIOs, just as data objects come from data classes. User interface objects are represented by both static and dynamic structures. For example, consider the vendor price quote UIO form in Figure D-7a. This form was produced by Paradox for Windows 5.0. We can define the contents of this

Figure D-7
Model of user interface object

(a) User interface form

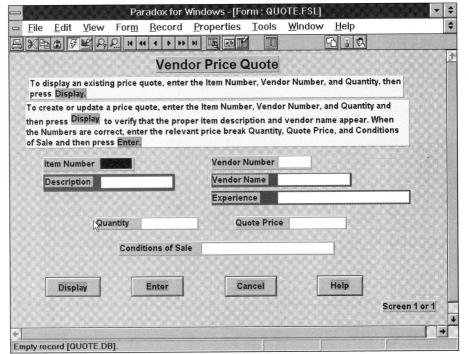

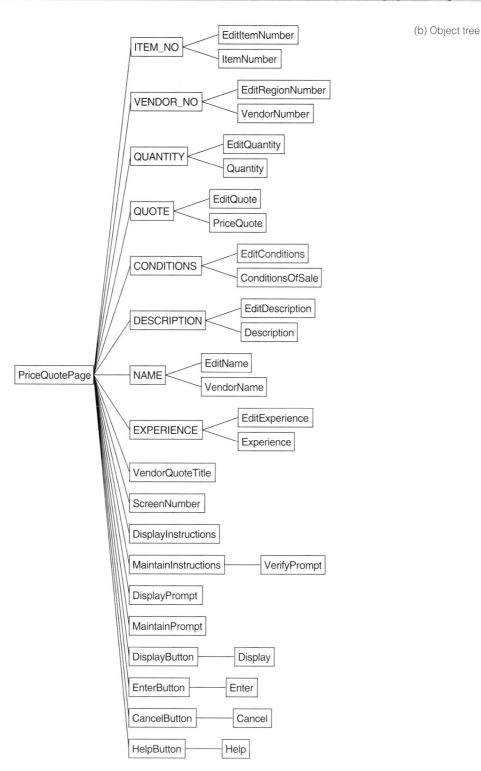

(b) Object tree

form by a static object model showing the data and display objects contained in the form (Figure D-7b); this is called the object tree in Paradox. The object tree shows every object and the hierarchy of objects contained within other objects. We can also use a data model to explain the static structure of data used in the form

Figure D-7 *(continued)*

(c) Data model

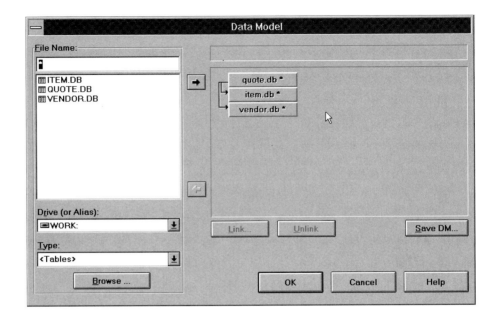

(Figure D-7c). Such a data model explains the Gen-Spec, Whole-Part, and instance connections between the data objects referenced on this UIO. Finally, we can also use a state-transition diagram to explain the behavior of this form object and each of the contained objects (this use of state-transition diagrams is discussed in Chapter 14). Thus, you can use the various object-oriented and structured documentation notations we have reviewed in this text to document both data and user interface objects.

The Complete Object-Oriented Specification

The result of an OO analysis of a problem domain is a set of object specifications and a structure chart showing all the classes and objects; Gen-Spec, Whole-Part, instance, and message connections; and subject areas for separate projects or work teams. The additional specifications include a thorough description of each Class-&-Object. A template for each Class-&-Object specification appears in Figure D-8. Most of the elements of this template have been discussed in this appendix. Exter-

Figure D-8
Typical Class-&-Object specification template

> **Class Specification Template** (repeat for each class)
>   Class-&-Object name
>   List of attributes with integrity rules (e.g., legal range of values)
>   External input
>   External output
>   State-transition diagram
>   Additional constraints (e.g., rules which govern simple methods, such as
>   whether objects can be disconnected and reconnected to other Whole-
>   Part objects to which it is connected)
>   Narrative explanation
>   List of services for each:
>   Expected input parameters
>   Response values
>   Structured English, flow chart, decision table, or similar descriptions of
>   processing steps and logic

nal input (output) is a list of data the object receives from (sends to) outside the model (such as a human user or physical device). Other specifications will be added to this template during design, such as a list of error codes, process time requirements, or computer memory limitations.

THE OBJECT-ORIENTED SA&D PROCESS

The prior sections of this appendix have concentrated on OO analysis. Since the concepts and principles of object orientation apply throughout the SA&D process, we do not have to introduce new concepts or techniques to discuss design and implementation activities within an OO approach to SA&D. Instead, this section addresses the linkages between analysis, design, and implementation within an OO development environment.

Subject Layer

You probably noticed that of the five layers of OO specifications listed earlier, we have not, yet, discussed the subject layer. A subject is a subset of an OO structure model to which you wish to direct attention. Thus, the subject layer is more for the convenience of system designers than for any modeling purpose. For large systems development projects or for a comprehensive object model of an organization, each subject will likely be the domain of a separate design project team. If there is only one project team, different subjects can be used to prioritize work.

A subject is shown on an object structure diagram by enclosing one or more Class-&-Objects within a rectangle, as illustrated in Figure D-9 for a simple view of a hospital. Each subject is identified by placing a unique number in the corners of the rectangle. Occasionally, each subject will be disconnected from each other subject but, as in the case of Figure D-9, this is not usually true. However, you should try to create subjects which are as independent as possible from each other, so that the separate project teams have minimal coordination. In Figure D-9, the two subjects are linked only by the many-to-many instance connection between Patient and Item classes. Often each subject corresponds to a class to which most of the other classes within the subject are attached. In Figure D-9, subject 1 is centered on the Patient class and subject 2 is centered on the Item class.

OO Design

Whereas OO analysis models the problem domain, OO design models the solution space. Because OO analysis and design both rely on the same modeling concepts, the transition from analysis to design is usually smooth (see Fichman and Kemerer, 1992, for an overview of OO design methods). During OO design, the internal structure of each object is specified. Where possible, methods, attributes, or whole objects are reused from current systems. For example, attributes are defined as instances of more primitive data type classes, such as a string or a floating point number. User interface objects are built from libraries of objects, such as forms, scroll bars, buttons, radio dials, and dialog boxes. Each class and UIO may be individually tested at this stage.

Analysis and design are very iterative. For example, as you refine the classes from analysis during design, you may identify the need for new Gen-Spec connections. As you write the methods for a class, you may discover that you use various IF statements, explaining how the object behaves under different conditions. What you may be identifying are new subclasses of the object, such that each IF clause applies to each subclass.

Figure D-9
Indicating subjects on an object structure diagram

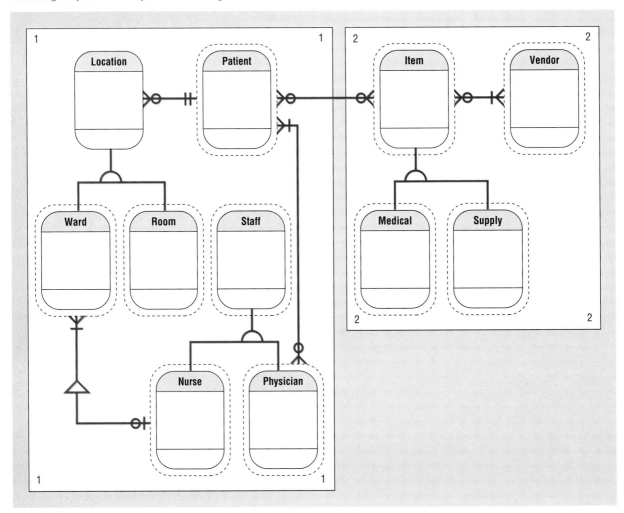

High-level object specifications from analysis allow designers and programmers to begin work on designing and implementing individual objects. This allows for considerable parallelism between analysis, design, and implementation. An approach by Meyer (1989) suggests that a cluster of highly related classes should be worked on together. First, the design specifications for each object in a cluster should be developed. Then, this cluster of classes is implemented and validated. Finally, the cluster is generalized so that other clusters can draw upon previously created objects. Once all clusters go through this cycle, the system is implemented by bringing together a library of classes, specializing a class as necessary to build a specific system.

SUMMARY

This appendix has given an overview of the techniques and systems development process for object-oriented analysis and design (OOAD). You have learned that OOAD is based on the core concepts of object, class, encapsulation, inheritance, and polymorphism. A Class-&-Object is described by a name and a set of attributes and methods. The attributes and methods or services of an object are hidden inside the object such that an object does not need to know the internal structure of any other object.

The requirements for a system developed following an OO approach are determined using any of the methods discussed in Chapter 8 and Appendix B. Once identified, requirements are structured into static and dynamic models of system problem domain behavior. OO models represent generalization-specialization (Gen-Spec), aggregation (Whole-Part), and instance connection between classes and objects. Gen-Spec connections indicate that a class is a special form of another class; specialized subclasses inherit all the properties of their superclasses, and may add or override properties. Whole-Part connections show that an object has one or more components, which together form the details of the object. Instance connections show additional relationships between objects and classes.

The dynamic behavior of a system is represented by the services of objects and the passing of messages between objects. Message maps and state-transition diagrams are used to indicate when services are performed. Related classes are grouped into subjects, and each subject may be worked on by a separate design team.

OOAD is characterized by iteration and reusable classes. As a class or cluster of classes are analyzed, designers begin developing implementation of that class or cluster, possibly by specializing existing classes. As the internal logic of a class is developed, new properties or specializations of the class may be identified; design often leads to refinement of the structural model.

Research on the real benefits of OOAD is limited. One laboratory study (Vessey and Conger, 1994) calls into question the claimed benefits of OOAD. In this study, process modeling (DFDs), data modeling (a variation of E-R diagrams), and object structure diagramming were compared using inexperienced (novice) analysts. The results showed that the subjects were able to use process modeling the best and used object modeling the least proficiently. In addition, subjects using process modeling improved more with repeated use of the technique than did subjects using object models. Since this is only one study using novice analysts, the results should be interpreted with caution. Further, since many analysts are trained first on structured methods, it would also be interesting to know about the proficiency of experienced analysts.

Fichman and Kemerer (1992) have also criticized OO modeling as being of limited value in macro redesign of business processes, since OO modeling concentrates on detailed processes (objects), not the big picture. Process modeling techniques seem to be more appropriate for re-engineering business processes (see Chapter 21 for a discussion of business process re-engineering).

For a more thorough survey of a variety of OOAD methods, see Taylor (1990) and Wilkie (1993).

APPENDIX   REVIEW

KEY TERMS

Encapsulation	Message connection	Polymorphism
Gen-Spec connection	Method	Whole-Part connection
Instance connection		

REVIEW QUESTIONS

1. Define each of the following terms:
   a. state-transition diagram
   b. subject
   c. user interface object
   d. overriding
   e. monitoring/trigger service

2. In what ways are structured methods claimed to be flawed?

3. Explain the importance of encapsulation in OOAD.

4. Explain the difference between two Gen-Spec connections, one which has a class as the superclass and the other which has a Class-&-Object as the superclass.

5. Define the five layers of an OO representation.

6. Which types of connections (Gen-Spec, Whole-Part, instance, and message) are between classes and which are between objects?

7. Define and contrast simple, complex, and monitoring/trigger services.

8. List four types of simple services.

9. What are some criticisms of the OO approach to systems analysis and design?

10. What OO models are used to represent a user interface object?

PROBLEMS AND EXERCISES

1. Match the following terms to the appropriate definitions.

   _____ Whole-Part connection   a. an "is-a" relationship
   _____ Gen-Spec connection   b. the hiding of properties inside an object
   _____ polymorphism   c. depicts the requests for services between
                                         objects and classes
   _____ encapsulation   d. the ability to refer to the same name in
                                         multiple contexts
   _____ message map   e. a "has-a" relationship

2. Explain why OO systems development is often iterative. What are the consequences of such an iterative development process?

3. Explain how reusability is an important consequence of OOAD. What is reusable?

4. Draw an object static model for each of the following situations. Use explicit notation for each type of structure you understand to exist.

   a. A company has a number of employees. The attributes of EMPLOYEE include Name, Address, Birth Date, and Date Hired. One method that is required of all employees is CalculateYearsOfService. The company also has several projects. Attributes of PROJECT include Code, Description, and Start Date. Each employee may be assigned to one or more projects, or may not be assigned to a project. A project must have at least one employee assigned, and may have several employees assigned. One method required of all projects is CalculateTotalCostToDate.

   b. In a vehicle-licensing application, there are three types of vehicles: passenger, truck, and trailer. Vehicle ID is an attribute of all vehicle types. Truck and trailer (but not passenger) vehicles have an attribute named Gross Capacity. The passenger and truck vehicle types require a method named PerformSmogCheck.

c. In a military operation, a mission consists of a number of flights. Each flight in turn consists of a number of shipments. There are two types of shipments: passenger and cargo item. Following are some attributes of each of these entities:
   MISSION: Codename, Description, Date
   FLIGHT: No., Origin, Destination
   SHIPMENT: No., Location
   PASSENGER: Name, Rank
   CARGO ITEM: Weight, Dimensions, Description

d. A college course may have one or more scheduled sections, or may not have a scheduled section. Attributes of COURSE include Course ID, Name, and Units. Attributes of SECTION include Section No. and Instructor. A method that is required of all courses is ChangeCourseDescription. A method that is required of all sections is one that will delete a section.

5. In a library application, there are three types of books: text, reference, and trade. Draw an object-oriented diagram to represent each of the following:

   a. Any book must be exactly one of the three types of books.
   b. Any book may be one or more of the three types of books.

6. Obtain a copy of a credit card statement, phone bill, or some other common business form. Prepare an object-oriented diagram to represent this form.

7. Consider Figure D-6 and an external input message of the Item object which is a request from a user for a list of customer numbers for customers who have bought a specified item (designated by Item No.) within the past week. Create the necessary methods and message connections to handle this request for service.

8. Review Figure D-9.

   a. What is the meaning of Location as only a class, not a Class-&-Object?
   b. Why is the connection between Nurse and Ward shown as a Whole-Part connection rather than an instance connection?
   c. Modify this diagram to show three subjects rather than two. Why did you choose the three subjects you have identified?

FIELD EXERCISES

1. Interview a systems analyst who has used OOAD techniques on a systems development project. What did this person find difficult about using these techniques? If this person had used structured methods prior to using OO methods, what difficulties did they have in the transition between these approaches? Has this person's organization adopted OOAD as a standard? Why or why not?

2. Find a book in the library about object-oriented programming (either the general principles of OO programming or a specific language). From studying this book, what is a class library? How is a class library used during system implementation? How does an organization manage a class library? Does this book discuss versioning of software modules? If so, explain what versioning is, and discuss the difficulties of versioning in an OO development environment.

3. Find in your university library the book by Wilkie (1993) listed in the References at the end of this chapter. If you cannot find this book, find any book which compares different OO notations. Develop a table in which the rows are different OO modeling constructs (like object, class, Gen-Spec connection, etc.) and each column is a different notation. Use this table to compare how at least two different OO modeling notations represent the same constructs (or, possibly, that a notation cannot model certain constructs).

REFERENCES

Booch, G. 1994. *Object-Oriented Design with Applications.* Redwood City, CA: Benjamin/Cummings.

Coad, P., and E. Yourdon. 1991a. *Object-Oriented Analysis.* 2d ed. Englewood Cliffs, NJ: Prentice-Hall.

Coad, P., and E. Yourdon. 1991b. *Object-Oriented Design.* Englewood Cliffs, NJ: Prentice-Hall.

Fichman, R. G., and C. F. Kemerer. 1992. "Object-Oriented and Conventional Analysis and Design Methodologies: Comparison and Critique." *Compute* (October): 22–39.

Henderson-Sellers, B., and J. M. Edwards. 1990. "The Object-Oriented Systems Life Cycle." *Communications of the ACM* 33 (September): 143–159.

Martin, J., and J. Odell. 1992. *Object-Oriented Analysis and Design.* Englewood Cliffs, NJ: Prentice-Hall.

McFadden, F. R., and J. A. Hoffer. 1994. *Modern Database Management.* 4th ed. Redwood City, CA: Benjamin/Cummings.

Meyer, B. 1989. From Structured Programming to Object-Oriented Design: The Road to Eiffel." *Structured Programming.* 1: 19–39.

Monarchi, D. E., and G. I Puhr. 1992. "A Research Typology for Object-Oriented Analysis and Design." *Communications of the ACM.* 35 (September): 35–47.

Rumbaugh, J., M. Blaha, W. Premerlani, F. Eddy, and W. Lorensen. 1991. *Object-Oriented Modeling and Design.* Englewood Cliffs, NJ: Prentice-Hall.

Shlaer, S., and S. J. Mellor. 1992. *Object Lifetimes: Modeling the World in States.* New York, NY: Yourdon Press.

Taylor, D. A. 1990. *Object-Oriented Technology: A Manager's Guide.* Reading, MA: Addison-Wesley.

Vessey, I., and S. A. Conger. 1994. "Requirements Specification: Learning Object, Process, and Data Methodologies." *Communications of the ACM* 37 (May): 102–13.

Wilkie, G. 1993. *Object-Oriented Software Engineering: A Professional Developer's Guide.* Reading, MA: Addison-Wesley.

Glossary of Terms

Note: Chapter or Appendix in which term is defined is listed in parentheses after the definition.

Acceptance testing The process whereby actual users test a completed information system, the end result of which is the users' acceptance of the system. (19) *See also* System testing.

Access method An operating system algorithm for storing and locating data in secondary memory. (16) *See also* File organization.

Action stubs That part of a decision table that lists the actions that result for a given set of conditions. (10)

Adaptive maintenance Changes made to a system to evolve its functionality to changing business needs or technologies. (21)

Afferent module A module of a structure chart related to input to the system. (17) *See also* Efferent module.

Affinity clustering The process of arranging planning matrix information so that clusters of information with some pre-determined level or type of affinity are placed next to each other on a matrix report. (6)

Aggregation (aggregate entity) A collection of entities that together form a higher-order concept. (C) *See also* Whole-Part connection.

Alias An alternative name given to an attribute. (15)

Alpha testing User testing of a completed information system using simulated data. (19) *See also* Beta testing, System testing.

Analysis The third phase of the SDLC in which the current system is studied and alternative replacement systems are proposed. (1)

Analysis tools CASE tools that enable automatic checking for incomplete, inconsistent, or incorrect specifications in diagrams, forms, and reports. (5)

Anomalies Errors or inconsistencies that may result when a user attempts to update a table that contains redundant data. There are three types of anomalies: insertion, deletion, and modification anomalies. (15) *See also* Normalization.

Application independence The separation of data and the definition of data from the applications that use these data. (1)

Application program interface (API) Software which allows a specific front-end program development platform to communicate with a particular back-end database engine, even when the front-end and back-end were not built to be compatible. (18)

Application software Computer software designed to support organizational functions or processes. (1)

Attribute A named property or characteristic of an entity that is of interest to the organization. (11)

Audit trail A list of changes to a data file which allows business transactions to be traced. Both the updating and use of data should be recorded in the audit trail, since the consequences of bad data should be discovered and corrected. (16)

Authorization rules Controls incorporated to restrict access to systems and data and also to restrict the actions that people may take once in the system. (14)

Backward recovery (rollback) An approach to rebuilding a file in which before images of changed records are restored to the file in reverse order until some earlier state is achieved. (16) *See also* Forward recovery (rollforward).

Balancing The conservation of inputs and outputs to a data flow diagram process when that process is decomposed to a lower level. (9)

Baseline modules Software modules that have been tested, documented, and approved to be included in the most recently created version of a system. (21)

Baseline Project Plan A major outcome and deliverable from the project initiation and planning phase which contains the best estimate of a project's scope, benefits, costs, risks, and resource requirements. (7)

Batch processing Information that is collected or generated at some pre-determined time interval and can be accessed via hard copy or on-line devices. (13)

Beta testing User testing of a completed information system using real data in the real user environment. (19) *See also* Alpha testing.

Binary relationship A relationship between instances of two entity types. This is the most common type of relationship encountered in data modeling. (11)

Biometric device An instrument that detects personal characteristics such as fingerprints, voice prints, retina prints, or signature dynamics. (14)

Blocking factor The number of physical records per page. (16)

Bottom-up planning A generic information systems planning methodology that identifies and defines IS development projects based upon solving operational business problems or taking advantage of some business opportunities. (6) *See also* Top-down planning, Corporate strategic planning.

Boundary The line that marks the inside and outside of a system and which sets off the system from its environment. (3)

Build routines Guidelines that list the instructions to construct an executable system from the baseline source code. (21)

Business case The justification for an information system, presented in terms of the tangible and intangible economic benefits and costs, and the technical and organizational feasibility of the proposed system. (2)

Business Process Re-engineering (BPR) The search for, and implementation of, radical change in business processes to achieve breakthrough improvements in products and services. (21) *See also* Disruptive technologies.

Business rules Specifications that preserve the integrity of a conceptual or logical data model. (11)

Calculated field A field which can be derived from other database fields. Also called computed or derived field. (16)

Candidate key An attribute (or combination of attributes) that uniquely identifies each instance of an entity type. (11) *See also* Primary key.

Cardinality The number of instances of entity B that can (or must) be associated with each instance of entity A. (11)

CASE *See* Computer-aided software engineering.

Central transform The area of a transform-centered information system where the most important derivation of new information takes place. (17) *See also* Transform analysis.

Client The (front-end) portion of the client/server database system that provides the user interface and data manipulation functions. (18)

Client/server architecture A LAN-based computing environment in which a central database server or engine performs all database commands sent to it from client workstations, and application programs on each client concentrate on user interface functions. (18)

Closed-ended questions Questions in interviews and on questionnaires that ask those responding to choose from among a set of pre-specified responses. (8) *See also* Open-ended questions.

Closed system A system that is cut off from its environment and does not interact with it. (3) *See also* Open system.

Code generators CASE tools that enable the automatic generation of program and database definition code directly from the design documents, diagrams, forms, and reports stored in the repository. (5)

Cohesion The extent to which a system or a subsystem performs a single function. (3)

Command language interaction A human-computer interaction method where users enter explicit statements into a system to invoke operations. (14)

Competitive strategy The method by which an organization attempts to achieve its mission and objectives. (6)

Component An irreducible part or aggregation of parts that make up a system, also called a subsystem. (3) See also *Inter-related components*.

Computer-aided software engineering (CASE) Software tools that provide automated support for some portion of the systems development process. (5)

Computing infrastructure All the resources and practices required to help people adequately use computer systems to do their primary work. (20) *See also* Support.

Conceptual data model A detailed model that captures the overall structure of organizational data while being independent of any database management system or other implementation considerations. (11) *See also* Entity-relationship data model, Logical data model.

Concurrency control A method for preventing loss of data integrity due to interference between users in a multi-user environment. (18)

Condition stubs That part of a decision table that lists the conditions relevant to the decision. (10)

Configuration management The process of assuring that only authorized changes are made to a system. (21)

Constraint A limit to what a system can accomplish. (3)

Context diagram An overview of an organizational system that shows the system boundary, external entities that interact with the system, and the major information flows between the entities and the system. (2) *See also* Data flow diagram.

Conversion *See* Installation.

Corporate strategic planning An ongoing process that defines the mission, objectives, and strategies of an organization. (6)

Corrective maintenance Changes made to a system to repair flaws in its design, coding, or implementation. (21)

Coupling The extent to which subsystems depend on each other. (3)

Critical path scheduling A scheduling technique where the order and duration of a sequence of activities directly affect the completion date of a project. (4)

Cross life cycle CASE CASE tools designed to support activities that occur *across* multiple phases of the systems development life cycle. (5) *See also* Lower CASE, Upper CASE.

Cross referencing A feature performed by a data dictionary that enables one description of a data item to be stored and accessed by all individuals so that a single definition for a data item is established and used. (5)

Data Raw facts about people, objects, and events in an organization. (1)

Data compression technique Pattern matching and other methods which replace repeating strings of characters with codes of shorter length. (16)

Data couple A diagrammatic representation of the data exchanged between two modules in a structure chart. (17) *See also* Flag.

Data dictionary The repository of all data definitions for all organizational applications. (5)

Data flow Data in motion, moving from one place in a system to another. (1)

Data flow diagram A picture of the movement of data between external entities and the processes and data stores within a system. (2)

Data-oriented approach An overall strategy of information systems development that focuses on the ideal organization of data rather than where and how data are used. (1) *See also* Process-oriented approach.

Data store Data at rest, which may take the form of many different physical representations. (9)

Data type A detailed coding scheme recognized by system software for representing organizational data. (16)

Database A shared collection of logically related data designed to meet the information needs of multiple users in an organization. (1)

Database engine The (back-end) portion of the client / server database system running on the server and providing database processing and shared access functions. (18)

Database management system (DBMS) Software that is used to create, maintain, and provide controlled access to user databases. (16)

Decision support systems (DSS) Computer-based systems designed to help organization members make decisions; usually composed of a database, model base, and dialogue system. (A) *See also* DSS Generators.

Decision table A matrix representation of the logic of a decision, which specifies the possible conditions for the decision and the resulting actions. (10) *See also* Action stubs, Condition stubs, Rules.

Decision tree A graphical representation of a decision situation in which decision points (nodes) are connected together by arcs (one for each alternative on a decision) and terminate in ovals (the action which is the result of all of the decisions made on the path that leads to that oval). (10)

Decomposition *See* Functional decomposition.

Default value A value a field will assume unless an explicit value is entered for that field. (16)

Degree The number of entity types that participate in a relationship. (11)

Design strategy A high-level statement about the approach to developing an information system. It includes statements on the system's functionality, hardware and system software platform, and method for acquisition. (12)

Desk checking A testing technique in which the program code is sequentially executed manually by the reviewer. (19)

DFD completeness The extent to which all necessary components of a data flow diagram have been included and fully described. (9)

DFD consistency The extent to which information contained on one level of a set of nested data flow diagrams is also included on other levels. (9)

Diagramming tools CASE tools that support the creation of graphical representations of various system elements such as process flow, data relationships, and program structures. (5)

Dialogue The sequence of interaction between a user and a system. (14)

Dialogue diagramming A formal method for designing and representing human-computer dialogues using box and line diagrams. (14)

Direct installation Changing over from the old information system to a new one by turning off the old system as the new one is turned on. (19)

Discount rate The rate of return used to compute the present value of future cash flows. (7)

Disruptive technologies Technologies that enable the breaking of long-held business rules that inhibit organizations from making radical business changes. (21) *See also* Business Process Re-engineering (BPR).

Distributed database A single logical database that is spread across computers in multiple locations which are connected by a data communications link. (18) *See also* Location transparency.

Documentation *See* External documentation, Internal documentation, System documentation, User documentation.

Documentation generators CASE tools that enable the easy production of both technical and user documentation in standard formats. (5)

Domain The set of all data types and values that an attribute can assume. (11)

Drop-down menu A menu positioning method that places the access point of the menu near the top line of the display; when accessed, menus open by dropping down onto the display. (14) *See also* Pop-up menu.

DSS generators General purpose computer-based tools used to develop specific decision support systems. (A)

Economic feasibility A process of identifying the financial benefits and costs associated with a development project. (7)

Efferent module A module of a structure chart related to output from the system. (17) *See also* Afferent module.

Electronic performance support system (EPSS) Component of a software package or application in which training and educational information is embedded. An EPSS can take several forms, including a tutorial, an expert system shell, and hypertext jumps to reference material. (20)

Encapsulation The property that the structure of attributes and methods of an object are hidden from the outside world and do not have to be understood to access its data values or invoke its methods. (D)

Encryption The coding (or scrambling) of data so that they cannot be read by humans. (16)

End-user development An approach to systems development in which users who are not computer experts satisfy their own computing needs through the use of high-level software and languages such as electronic spreadsheets and relational database management systems. (1) *See also* Help desk, Support.

End users Non-information system professionals in an organization who specify the business requirements for and use software applications. End users often request new or modified applications, test and approve applications, and may serve on project teams as business experts. (1)

Entity instance (instance) A single occurrence of an entity type. (11)

Entity-relationship data model (E-R model) A detailed, logical representation of the entities, associations, and data elements for an organization or business area. (11)

Entity-relationship diagram (E-R diagram) A graphical representation of an E-R model. (11)

Entity type A collection of entities that share common properties or characteristics. (11)

Environment Everything external to a system which interacts with the system. (3)

Exclusive relationships A set of relationships for which an entity instance can participate in only one of the relationships at a time. (C)

Executive information systems See *Executive support systems*.

Executive support systems Computer-based systems developed to support the information-intensive but limited-time decision making of executives (also referred to as *executive information systems*). (A)

Expert systems Computer-based systems designed to mimic the performance of human experts. (A)

External documentation System documentation that includes the outcome of such structured diagramming techniques as data flow and entity-relationship diagrams. (20) *See also* Internal documentation.

External information Information that is collected from or created for individuals and groups external to an organization. (13) *See also* Internal information.

Feasibility *See* Economic feasibility, Legal and contractual feasibility, Operational feasibility, Political feasibility, Schedule feasibility, Technical feasibility.

Field The smallest unit of named application data recognized by system software. (16)

File organization A technique for physically arranging the records of a file on secondary storage devices. (16) *See also* Access method, Hashed file organization, Indexed file organization, Sequential file organization.

File server A device that manages file operations and is shared by each client PC attached to a LAN. (18) *See also* Client/server architecture.

First normal form (1NF) A relation that contains no repeating data. (15) *See also* Normalization.

Flag A diagrammatic representation of a message passed between two modules. (17) *See also* Data couple.

Foreign key An attribute that appears as a nonkey attribute in one relation and as a primary key attribute (or part of a primary key) in another relation. (15)

Form A business document that contains some predefined data and may include some areas where additional data are to be filled in. An instance of a form is typically based on one database record. (13)

Form and report generators CASE tools that support the creation of system forms and reports in order to prototype how systems will "look and feel" to users. (5)

Form interaction A highly intuitive human-computer interaction method whereby data fields are formatted in a manner similar to paper-based forms. (14)

Formal system The official way a system works as described in organizational documentation. (8) *See also* Informal system.

Forward recovery (rollforward) An approach to rebuilding a file in which one starts with an earlier version of the file and either reruns prior transactions or replaces a record with its image after each transaction. (16) *See also* Backward recovery (rollback).

Functional decomposition An iterative process of breaking the description of a system down into finer and finer detail which creates a set of charts in which one process on a given chart is explained in greater detail on another chart. (2)

Functional dependency A particular relationship between two attributes. For any relation R, attribute B is functionally dependent on attribute A if, for every valid instance of A, that value of A uniquely determines the value of B. The functional dependence of B on A is represented as A $\rightarrow$ B. (15) *See also* Partial functional dependency, Transitive dependency.

Gantt chart A graphical representation of a project that shows each task activity as a horizontal bar whose length is proportional to its time for completion. (4)

Gen-Spec connection A linkage between a class and one or more specialized subclasses. Also called an "is a" connection. A Gen-Spec connection is functionally equivalent to the concept of generalization used in conceptual data modeling. (D)

Generalization The concept that some things (entities) are subtypes of other, more general, things. (C) *See also* IS-A relationship, Gen-Spec connection.

Gerund A many-to-many (or one-to-one) relationship that the data modeler chooses to model as an entity type with several associated one-to-many relationships with other entity types. (11)

Hashed file organization The address for each record is determined using a hashing algorithm. (16)

Hashing algorithm A routine that converts a primary key value into a relative record number (or relative file address). (16)

Help desk A single point of contact for all user inquiries and problems about a particular information system or all users in a particular department. (20) *See also* Computing infrastructure, Information center, Support.

Homonym A single name that is used for two or more different attributes (for example, the term *invoice* to refer to both a customer invoice and a supplier invoice). (15)

Horizontal partitioning Distributing the rows of a table into several separate tables. (18) *See also* Vertical partitioning.

I-CASE An automated systems development environment that provides numerous tools to create diagrams, forms, and reports; provides analysis, reporting, and code generation facilities; and seamlessly shares and integrates data across and between tools. (5)

Icon Graphical pictures that represent specific functions within a system. (14) *See also* Object-based interaction.

Implementation The sixth phase of the SDLC in which the information system is coded, tested, installed, and supported in the organization. (1)

Incremental commitment A strategy in systems analysis and design in which the project is reviewed after each phase and continuation of the project is rejustified in each of these reviews. (2)

Index A table or other data structure used to determine the location of rows in a file that satisfy some condition. (16)

Indexed file organization The records are either stored sequentially or non-sequentially and an index is created that allows software to locate individual records. (16)

Indifferent condition In a decision table, a condition whose value does not affect which actions are taken for two or more rules. (10)

Informal system The way a system actually works. (8) *See also* Formal system.

Information Data that have been processed and presented in a form suitable for human interpretation, often with the purpose of revealing trends or patterns. (1)

Information center An organizational unit whose mission is to support users in exploiting information technology. (20) *See also* Computing infrastructure, Help desk, Support.

Information repository Automated tools to manage and control access to organizational business information and application portfolios as components within a comprehensive repository. (5)

Information systems analysis and design The complex organizational process whereby computer-based information systems are developed and maintained. (1)

Information systems planning (ISP) An orderly means of assessing the information needs of an organization and defining the systems, databases, and technologies that will best satisfy those needs. (6) *See also* Corporate strategic planning, Top-down planning.

Inheritance The property that occurs when entity types or object classes are arranged in a hierarchy and each entity type or object class assumes the attributes and methods of its ancestors; that is, those higher up in the hierarchy. Inheritance allows new but related classes to be derived from existing classes. (1) *See also* Gen-Spec connection, Generalization, IS-A relationship.

Input Whatever a system takes from its environment in order to fulfill its purpose. (3)

Inspections A testing technique in which participants examine program code for predictable language-specific errors. (19)

Installation The organizational process of changing over from the current information system to a new one. (19) *See also* Direct installation, Parallel installation, Phased installation, Single location installation.

Instance connection The relationship one object has with another object in order to fulfill its responsibilities. An Instance connection is identical to a relationship in conceptual data modeling. (D)

Intangible benefit A benefit derived from the creation of an information system that cannot be easily measured in dollars or with certainty. (7) *See also* Tangible benefit.

Intangible cost A cost associated with an information system that cannot be easily measured in terms of dollars or with certainty. (7) *See also* Tangible cost.

Integrated CASE *See* I-CASE.

Integration testing The process of bringing together all of the modules that comprise a program for testing purposes. Modules are typically integrated in a top-down, incremental fashion. (19)

Interface In systems theory, the point of contact where a system meets its environment or where subsystems meet each other. (3) In human-computer interaction, a method by which users interact with information systems. (14)

Internal documentation System documentation that is part of the program source code or is generated at compile time. (20) *See also* External documentation.

Internal information Information that is collected, generated, or consumed within an organization. (13) *See also* External information.

Inter-related components Dependence of one subsystem on one or more subsystems. (3)

IS-A relationship The relationship between each subtype and its supertype. (C) *See also* Gen-Spec connection, Generalization, Inheritance.

JAD session leader The trained individual who plans and leads Joint Application Design sessions. (8)

Joint Application Design (JAD) A structured process in which users, managers, and analysts work together for several days in a series of intensive meetings to specify or review system requirements. (1)

Key business processes The structured, measured set of activities designed to produce a specific output for a particular customer or market. (21) *See also* Business Process Re-engineering (BPR).

Knowledge elicitation The process of obtaining knowledge requirements for an expert system. (8)

Knowledge engineers Computer professionals whose job it is to elicit knowledge from domain experts in order to develop expert systems. (A)

Legal and contractual feasibility The process of assessing potential legal and contractual ramifications due to the construction of a system. (7)

Level-n diagram A DFD that is the result of n nested decompositions of a series of sub-processes from a process on a level-0 diagram. (9)

Level-0 diagram   A data flow diagram that represents a system's major processes, data flows, and data stores at a high level of detail. (9)

Local area network (LAN)   The cabling, hardware, and software used to connect workstations, computers, and file servers located in a confined geographical area (typically within one building or campus). (18)

Location transparency   A design goal for a distributed database which says that a user (or user program) requesting data need not know at which site those data are located. (18)

Logical data model   A description of data using a notation that corresponds to an organization of data used by database management systems. (15) *See also* Relational database model.

Logical design   The fourth phase of the SDLC in which all functional features of the system chosen for development in analysis are described independently of any computer platform. (1)

Logical system description   Description of a system that focuses on the system's function and purpose without regard to how the system will be physically implemented. (3)

Lower CASE   CASE tools designed to support the implementation and maintenance phases of the systems development life cycle. (5) *See also* Upper CASE.

Maintainability   The ease with which software can be understood, corrected, adapted, and enhanced. (21)

Maintenance   The final phase of the SDLC in which an information system is systematically repaired and improved; or changes made to a system to fix or enhance its functionality. (1; 21) *See also* Adaptive maintenance, Corrective maintenance, Perfective maintenance, Preventive maintenance.

Management information systems (MIS)   Computer-based systems designed to provide standard reports for managers about transaction data. (A)

Mean time between failures (MTBF)   A measurement of error occurrences that can be tracked over time to indicate the quality of a system. (21)

Menu interaction   A human-computer interaction method where a list of system options is provided and a specific command is invoked by user selection of a menu option. (14) *See also* Drop-down menu, Pop-up menu.

Message connection   A dependency of an object on the services of another object or class. (D)

Method   A specific behavior that an object is responsible for performing. Also called *service*. (D)

Mission statement   A statement that makes it clear what business a company is in. (6)

Modularity   Dividing a system up into chunks or modules of a relatively uniform size. (3) *See also* Cohesion, Coupling.

Module   A self-contained component of a system, defined by function. (17)

Multivalued attribute   An attribute that can have more than one value for each entity instance. (11)

Natural language interaction   A human-computer interaction method where inputs to and outputs from a computer-based application are in a conventional speaking language such as English. (14)

Normal form   A state of a relation that can be determined by applying simple rules regarding dependencies to that relation. (15) *See also* Functional dependency.

Normalization   The process of converting complex data structures into simple, stable data structures. (15)

Null value   A special field value, distinct from 0, blank, or any other value, that indicates that the value for the field is missing or otherwise unknown. (16)

Object   A structure that encapsulates (or packages) attributes and methods that operate on those attributes. An object is an abstraction of a real-world thing in which data and processes are placed together to model the structure and behavior of the real-world object. (1)

Object-based interaction   A human-computer interaction method where symbols are used to represent commands or functions. (14) *See also* Icon.

Object class   A logical grouping of objects that have the same (or similar) attributes and behaviors (methods). (1)

Object-oriented analysis and design (OOAD)   Systems development methodologies and techniques based on objects rather than data or processes. (1)

Objective statements   A series of statements that express an organization's qualitative and quantitative goals for reaching a desired future position. (6)

One-time cost   A cost associated with project start-up and development, or system start-up. (7)

On-line processing   The collection and delivery of the most recent available information, typically through an on-line workstation. (13)

Open-ended questions   Questions in interviews and on questionnaires that have no pre-specified answers. (8) *See also* Closed-ended questions.

Open system   A system that interacts freely with its environment, taking input and returning output. (3) *See also* Closed system.

Operational feasibility   The process of assessing the degree to which a proposed system solves business problems or takes advantage of business opportunities. (7)

Output   Whatever a system returns to its environment in order to fulfill its purpose. (3)

Outsourcing   The practice of turning over responsibility of some to all of an organization's information systems applications and operations to an outside firm. (12)

Page   The amount of data read or written in one secondary memory (disk) input or output operation. For I/O with a magnetic tape, the equivalent term is *record block*. (16) *See also* Blocking factor.

Parallel installation   Running the old information system and the new one at the same time until management decides the old system can be turned off. (19)

Partial functional dependency   A dependency in which one or more nonkey attributes are functionally dependent on part, but not all, of the primary key. (15)

Participatory Design (PD)   A systems development approach that originated in Northern Europe in which users and the improvement in their work lives are the central focus. (1)

Perfective maintenance   Changes made to a system to add new features or to improve performance. (21)

PERT chart   A diagram that depicts project activities and their inter-relationships. PERT stands for Program Evaluation Review Technique. (4)

Phased installation   Changing over from the old information system to a new one incrementally, starting with one or a few functional components and then gradually extending the installation to cover the whole new system. (19)

Physical design   The fifth phase of the SDLC in which the logical specifications of the system from logical design are transformed into technology-specific details from which all programming and system construction can be accomplished. (1)

Physical file   A named set of contiguous records. (16)

Physical record   A group of fields stored in adjacent memory locations and retrieved together as a unit. (16)

Physical system description   Description of a system that focuses on how the system will be materially constructed. (3)

Picture (or template)   A pattern of codes that restricts the width and possible values for each position of a field. (16)

Pilot installation   *See* Single location installation.

Pointer   A field of data that can be used to locate a related field or record of data. (16)

Political feasibility   The process of evaluating how key stakeholders within the organization view the proposed system. (7)

Polymorphism   The ability to refer to properties (attributes and methods) of instance from more than one class by the same name. (D)

Pop-up menu   A menu positioning method that places a menu near the current cursor position. (14)

Present value   The current value of a future cash flow. (7)

Preventive maintenance   Changes made to a system to avoid possible future problems. (21)

Primary key   A candidate key that has been selected as the identifier for an entity type. Primary key values may not be null. (11) Also called an *identifier.*

Primitive DFD   The lowest level of decomposition for a data flow diagram. (9)

Process   The work or actions performed on data so that they are transformed, stored, or distributed. (9)

Processing logic   The steps by which data are transformed or moved and a description of the events that trigger these steps. (1)

Process-oriented approach   An overall strategy to information systems development that focuses on how and when data are moved through and changed by an information system. (1) *See also* Data-oriented approach.

Project   A planned undertaking of related activities to reach an objective that has a beginning and an end. (4)

Project close-down   The final phase of the project management process that focuses on bringing a project to an end. (4)

Project execution   The third phase of the project management process in which the plans created in the prior phases (project initiation and planning) are put into action. (4)

Project identification and selection   The first phase of the SDLC in which an organization's total information system needs are identified, analyzed, prioritized, and arranged. (1)

Project initiation   The first phase of the project management process in which activities are performed to assess the size, scope, and complexity of the project and to establish procedures to support later project activities. (4)

Project initiation and planning   The second phase of the SDLC in which a potential information systems project is explained and an argument for continuing or not continuing with the project is presented; a detailed plan is also developed for conducting the remaining phases of the SDLC for the proposed system. (1)

Project management   A controlled process of initiating, planning, executing, and closing down a project. (4)

Project manager   An individual with a diverse set of skills—management, leadership, technical, conflict management, and customer relationship—who is responsible for initiating, planning, executing, and closing down a project. (4)

Project planning   The second phase of the project management process which focuses on defining clear, discrete activities and the work needed to complete each activity within a single project. (4)

Project workbook   An on-line or hard copy repository for all project correspondence, inputs, outputs, deliverables, procedures, and standards that is used for performing project audits, orientation of new team members, communication with management and customers, scoping future projects, and performing post-project reviews. (4) *See also* Repository.

Prototyping   An iterative process of systems development in which requirements are converted to a working system which is continually revised through close work between an analyst and users. (1) *See also* Rapid Application Development (RAD).

Pseudocode   A method for representing the instructions in a module with language very similar to computer programming code. (17)

Purpose   The overall goal or function of a system. (3)

Rapid Application Development (RAD)   Systems development methodology created to radically decrease the time needed to design and implement information systems. RAD relies on heavy user involvement, Joint Application Development sessions, prototyping, integrated CASE tools, and code generators. (B)

Record partitioning   The process of splitting logical records into separate physical segments based on affinity of use. (16) *See also* Horizontal partitioning, Vertical partitioning.

Recurring cost   A cost resulting from the ongoing evolution and use of a system. (7)

Recursive foreign key   A foreign key in a relation that references the primary key values of that same relation. (15)

Re-engineering   Automated tools that read program source code as input, perform an analysis of the program's data and logic, and then automatically, or interactively with a systems analyst, alter an existing system in an effort to improve its quality or performance. (5) *See also* CASE.

Referential integrity   An integrity constraint specifying that the value (or existence) of an attribute in one relation depends on the value (or existence) of an attribute in the same or another relation. (15)

Relation   A named, two-dimensional table of data. Each relation consists of a set of named columns and an arbitrary number of unnamed rows. (15)

Relational database model   A data model that represents data in the form of tables or relations. (15)

Relationship   An association between the instances of one or more entity types that is of interest to the organization. (11) *See also* Instance connection.

Repeating group   A set of two or more multivalued attributes that are logically related. (11)

Report   A business document that contains only predefined data; that is, it is a passive document used solely for reading or viewing. A report typically contains data from many unrelated records or transactions. (13)

Repository   A centralized database that contains all diagrams, form and report definitions, data structure, data definitions, process flows and logic, and definitions of other organizational and system components; it provides a set of mechanisms and structures to achieve seamless data-to-tool and data-to-data integration. (5) *See also* I-CASE, Information repository, Project workbook.

Resource   Any person, group of people, piece of equipment, or material used in accomplishing an activity. (4)

Reusability   The ability to design software modules in a manner so that they can be used again and again in different systems without significant modification. (5)

Reverse engineering   Automated tools that read program source code as input and create graphical and textual representations of program design-level information such as program control structures, data structures, logical flow, and data flow. (5) *See also* CASE.

Rules   That part of a decision table that specifies which actions are to be followed for a given set of conditions. (10)

Scaffolding   Program modules and data written for debugging purposes but never intended to be part of the final product delivered to users or customers. (19)

Schedule feasibility   The process of assessing the degree to which the potential timeframe and completion dates for all major activities within a project meet organizational deadlines and constraints for affecting change. (7)

Scribe   The person who makes detailed notes of the happenings at a Joint Application Design session. (8)

Second normal form (2NF)   A relation is in second normal form if it is in first normal form and every nonkey attribute is fully functionally dependent on the primary key. Thus no nonkey attribute is functionally dependent on part (but not all) of the primary key. (15) *See also* Functional dependency, Partial functional dependency.

Secondary key   One or a combination of fields for which more than one record may have the same combination of values. (16)

Sequential file organization   The records in the file are stored in sequence according to a primary key value. (16)

Service   *See* Method.

Single location installation   Trying out a new information system at one site and using the experience to decide if and how the new system should be deployed throughout the organization. (19)

Slack time   The amount of time that an activity can be delayed without delaying the project. (4)

Smart card   A thin plastic card the size of a credit card with an embedded microprocessor and memory. (14)

Source/sink   The origin and/or destination of data, sometimes referred to as external entities. (9)

Stakeholder   A person who has an interest in an existing or new information system. A stakeholder is someone who is involved in the development of a system, in the use of a system, or someone who has authority over the parts of the organization affected by the system. (2)

State-transition diagram   A diagram that illustrates how processes are related to each other in time. State-transition diagrams illustrate the states a system component can have and the events that cause change from one state to another. (10)

State-transition table   A table that illustrates how processes are related to each other in time. State-transition tables illustrate the states a system component can have and the events that cause change from one state

to another. All possible state-event combinations are explored. (10)

Statement of Work (SOW)   Document prepared for the customer during project initiation and planning that describes what the project will deliver and outlines generally at a high level all work required to complete the project. (7)

Structure chart   Hierarchical diagram that shows how an information system is organized. (17)

Structured English   Modified form of the English language used to specify the logic of information system processes. Although there is no single standard, Structured English typically relies on action verbs and noun phrases and contains no adjectives or adverbs. (10)

Stub testing   A technique used in testing modules, especially where modules are written and tested in a top-down fashion, where a few lines of code or stubs are used to substitute for subordinate modules. (19)

Subtype   A subset of a supertype that shares common attributes or relationships distinct from other subsets. (C) *See also* Generalization.

Supertype   A generic entity type that is subdivided into subtypes (C). *See also* Generalization.

Support   Providing on-going educational and problem solving assistance to information system users. For in-house developed systems, support materials and jobs will have to be prepared or designed as part of the implementation process. (20) *See also* Computing infrastructure, Help desk, Information center.

Synonyms   Two different names that are used to refer to the same data item (for example, *car* and *automobile*). (15)

System   An inter-related set of components, with an identifiable boundary, working together for some purpose. (3)

System documentation   Detailed information about a system's design specifications, its internal workings, and its functionality. (20)

System librarian   A person responsible for controlling the checking-out and checking-in of baseline modules for a system when a system is being developed or maintained. (21)

System testing   The bringing together of all the programs that comprise a system for testing purposes. Programs are typically integrated in a top-down, incremental fashion. (19) *See also* Acceptance testing, Alpha testing, Beta testing, Integration testing, Stub testing.

Systems analyst   The organizational role most responsible for the analysis and design of information systems. (1)

Systems development life cycle (SDLC)   The traditional methodology used to develop, maintain, and replace information systems. (1)

Systems development methodology   A standard process followed in an organization to conduct all the steps necessary to analyze, design, implement, and maintain information systems. (2)

Tangible benefit   A benefit derived from the creation of an information system that can be measured in dollars and with certainty. (7) *See also* Intangible benefit.

Tangible cost   A cost associated with an information system that can be measured in terms of dollars and with certainty. (7) *See also* Intangible cost.

Technical communicator   Skilled individual who is trained to write system and user documentation about mechanical and information systems. (20)

Technical feasibility   A process of assessing the development organization's ability to construct a proposed system. (7)

Ternary relationship   A simultaneous relationship among instances of three entity types. (11)

Third normal form (3NF)   A relation is in third normal form if it is in second normal form and no transitive dependencies exist. (15)

Top-down planning   A generic information systems planning methodology that attempts to gain a broad understanding of the information system needs of the entire organization. (6) *See also* Bottom-up planning.

Total Quality Management (TQM)   A systematic approach to improving business processes through a philosophy of continuous improvement. (21)

Transaction analysis   The process of turning data flow diagrams of a transaction-centered system into structure charts. (17)

Transaction-centered system   An information system that has as its focus the dispatch of data to their appropriate locations for processing. (17)

Transaction processing systems (TPS)   Computer-based versions of manual organization systems dedicated to handling the organization's transactions; e.g., payroll. (A)

Transactions   Individual, simple events in the life of an organization that contain data about organizational activity. (A)

Transform analysis   The process of turning data flow diagrams of a transform-centered system into structure charts. (17)

Transform-centered system   An information system that has as its focus the derivation of new information from existing data. (17) *See also* Central transform.

Transitive dependency   A functional dependency between two (or more) nonkey attributes in a relation. (15) *See also* Third normal form.

Triggering operation (trigger)   An assertion or rule that governs the validity of data manipulation operations such as insert, update, and delete. (11)

Turnaround document   Information that is delivered to an external customer as an output that can be returned to provide new information as an input to an information system. (13)

Unary relationship (recursive relationship)   A relationship between the instances of one entity type. (11)

Upper CASE   CASE tools designed to support information planning and the project identification and selection, project initiation and planning, analysis, and design phases of the systems development life cycle. (5) *See also* Lower CASE.

Usability   An overall evaluation of how a system performs in supporting a particular user for a particular task. (13)

User documentation   Written or other visual information about an application system, how it works, and how to use it. (20)

Vertical partitioning   Distributing the columns of a table into several separate tables. (18) *See also* Horizontal partitioning.

View   A subset of the database that is presented to one or more users. (14)

Walkthrough   A peer group review of any product created during the systems development process. Also called structured walkthrough. (7)

Well-structured relation   A relation that contains a minimum amount of redundancy and allows users to insert, modify, and delete the rows in a table without errors or inconsistencies. (15) *See also* Normalization.

Whole-Part connection   A linkage between an object and its components. Also called an assembly or "has a" connection. Whole-Part connections are equivalent to aggregation relationships in conceptual data modeling. (D)

Work breakdown structure   The process of dividing the project into manageable tasks and logically ordering them to ensure a smooth evolution between tasks. (4)

Glossary of Acronyms

Note: Some acronyms are abbreviations for entries in the Glossary of Terms. For these and some other acronym entities, we list in parenthesis the chapter or appendix in which the associated term is defined. Other acronyms are generally used in the information systems field and are included here for your convenience.

ACM Association for Computing Machinery (3)

ADW Application Development Workbench (5)

API Application Program Interface (18)

ASM Association for Systems Management

ATM Automated Teller Machine (13)

AT&T American Telephone & Telegraph

BEC Broadway Entertainment Company (5)

BPP Baseline Project Plan (7)

BPR Business Process Re-engineering (21)

BSP Business Systems Planning (6)

CAD Computer-Aided Design (A)

CAM Computer-Aided Manufacturing (A)

CASE Computer-Aided Software Engineering (5)

CATS Customer Activity Tracking System (6)

CD Compact Disk (3)

CDP Certified Data Processing

CD-ROM Compact Disk-Read Only Memory

COBOL COmmon Business Oriented Language

COCOMO COnstruction COst MOdel

CRT Cathode Ray Tube

C/S Client/Server (18)

CUA Common User Access (14)

DB2 Data Base 2

DBMS Database Management System (16)

DCR Design Change Request (19)

DFD Data Flow Diagram (2)

DPMA Data Processing Managers Association (3)

DSS Decision Support System (A)

EDS Electronic Data Systems

EFT Electronic Funds Transfer (14)

EIS Executive Information System (A)

EPSS Electronic Performance Support System (20)

E-R Entity-Relationship (2)

ERD Entity-Relationship Diagram (11)

ES Expert System (A)

ESS Executive Support System (A)

ET Estimated Time (4)

EUC End User Computing (20)

FORTRAN FORmula TRANslator

GDSS Group Decision Support System (A)

GSS Group Support System (8)

GUI Graphical User Interface

IBM International Business Machines

I-CASE Integrated Computer-Aided Software Engineering (5)

IDSS Integrated Development Support System (19)

IE Information Engineering (6)

IEF Information Engineering Facility (5)

I/O Input/Output

IS Information System

ISA Information Systems Architecture

ISAM Indexed Sequential Access Method

ISP Information Systems Planning (6)

IT Information Technology

IU Indiana University

JAD Joint Application Design (1)

KLOC   Thousand Lines of Code (19)
LAN   Local Area Network (18)
MIS   Management Information System (A)
M:N   Many-to-Many
MRP   Material Requirements Planning (19)
MTBF   Mean Time Between Failures (21)
MTTR   Mean Time to Repair Defect (19)
NPV   Net Present Value (7)
OA   Office Automation
OO   Object-Oriented
OOAD   Object-Oriented Analysis and Design (1)
PC   Personal Computer
PD   Participatory Design (1)
PERT   Project Evaluation and Review Technique (4)
PIP   Project Initiation and Planning (7)
POS   Point-of-Sale
PTR   Problem Tracking Report (19)
PV   Present Value (7)
PVF   Pine Valley Furniture (1)
RAD   Rapid Application Development (B)
RAM   Random Access Memory (18)
R&D   Research and Development
RFP   Request for Proposal (12)
RFQ   Request for Quote (12)

ROI   Return on Investment (7)
ROM   Read Only Memory
SDLC   Systems Development Life Cycle (1)
SDM   Systems Development Methodology (2)
SNA   System Network Architecture
SOW   Statement of Work (7)
SPTS   Sales Promotion Tracking System (4)
SQL   Structured Query Language (2)
SSR   System Service Request (1)
STD   State-Transition Diagram (10)
TI   Texas Instruments
TPS   Transaction Processing System (A)
TQM   Total Quality Management (21)
TVM   Time Value of Money (7)
VAW   Visible Analyst Workbench (5)
VSAM   Virtual Sequential Access Method
WIP   Work in Process (19)
1:1   One-to-One
1:*M*   One-to-Many
1NF   First Normal Form (15)
2NF   Second Normal Form (15)
3NF   Third Normal Form (15)
4GL   Fourth-Generation Language (1)

Index

(*Italicized* page number indicates that term is defined on that page.)